Singularities in Generic Geometry

ADVANCED STUDIES
IN PURE MATHEMATICS 78

Chief Editor of the Series: Ken'ichi Ohshika (Osaka University)

Singularities in Generic Geometry

Edited by

Shyuichi Izumiya (Hokkaido University)
Goo Ishikawa (Hokkaido University)
Minoru Yamamoto (Hirosaki University)
Kentaro Saji (Kobe University)
Takahiro Yamamoto (Tokyo Gakugei University)
Masatomo Takahashi (Muroran Institute of Technology)

Mathematical Society of Japan

This book was typeset by $\mathcal{A}_\mathcal{M}\mathcal{S}$-TeX and $\mathcal{A}_\mathcal{M}\mathcal{S}$-LaTeX, the TeX macro systems of the American Mathematical Society, together with the style files aspm.sty *and* aspmfm.sty *for $\mathcal{A}_\mathcal{M}\mathcal{S}$-TeX written by Dr. Chiaki Tsukamoto and* aspmproc.sty *for $\mathcal{A}_\mathcal{M}\mathcal{S}$-LaTeX written by Dr. Akihiro Munemasa and* aspm.cls *for $\mathcal{A}_\mathcal{M}\mathcal{S}$-LaTeX provided by Livretech Co., Ltd.*

TeX is a trademark of the American Mathematical Society.

Edited by the Mathematical Society of Japan.

Published by the Mathematical Society of Japan.

Distributed by the Mathematical Society of Japan, the American Mathematical Society and the World Scientific Publishing Co., Ltd.
Distributed exclusively in North America by the American Mathematical Society.
Partially supported by Grant-in-Aid for Publication of Scientific Research Results, Japan Society for the Promotion of Science, Grant Number 17HP1002.

Advanced Studies in Pure Mathematics 78
ISBN 978-4-86497-055-6

PRINTED IN JAPAN
by Livretech Co., Ltd.

2010 Mathematics Subject Classification.
Primary 58K99.
Secondary 53A99.

Preface

The workshop "Singularities in Generic Geometry and Applications –Kobe-Kyoto 2015 (Valencia IV)–" was the fourth in a sequence of biennial workshops, the first being at Valencia, Spain in 2009, the second at Bedlewo, Poland, in 2011 and the third at Edinburgh, Scotland in 2013. The entire sequence is a tribute to the inspirational work of Carmen Romero Fuster (Valencia), and the fifth in the sequence is scheduled to take place in Mexico in 2017.

The fourth workshop was held at Kobe University in Kobe, Japan from 3 to 6 June 2015 and the Research Institute of Mathematical Science (RIMS) in Kyoto, Japan from 8 to 10 June 2015. The workshop had 81 mathematicians as participants, with 23 participants from abroad coming from nine different countries (Brazil, China, France, Italy, Mexico, Poland, Russia, Spain and the USA), and there were 33 talks and 16 poster presentations. The program focused on applications of singularity theory to differential geometry and differential topology. The scientific committee consisted of Professors A. Davydov (Moscow-Vladimir), T. Fukui (Saitama), P. Giblin (Liverpool), T. Nishimura (Yokohama), T. Ohmoto (Sapporo), M. C. Romero Fuster (Valencia), O. Saeki (Fukuoka), M. A. Soares Ruas (Saõ Carlos), F. Sánchez Bringas (Mexico city) and F. Tari (Saõ Carlos).

Eighteen of the speakers now present their works in written forms, as research articles, surveys or expositions in this volume. All papers here were strictly peer-reviewed and selected following the standards of mathematical research journals. We would like to express our gratitude to the authors for their contributions, and to the referees for contributing to the selection process.

We gratefully acknowledge financial support for the conference from Grant-in-Aid for Scientific Research (JSPS), Faculty of Science (Kobe University) and RIMS (Kyoto University). We also thank the staff members of the Mathematical Society of Japan involved with the preparation of this volume. Finally, it is our great pleasure to thank all the participants of the workshop for their support.

June 2017

Shyuichi Izumiya, Goo Ishikawa, Minoru Yamamoto, Kentaro Saji,
Takahiro Yamamoto, Masatomo Takahashi
Editors

All papers in this volume have been refereed and are in final form. No version of any of them will be submitted for publication elsewhere.

CONTENTS

I. Survey Articles

II. Research Articles

Advanced Studies in Pure Mathematics 78, 2018
Singularities in Generic Geometry
pp. 1–53

Spherical method for studying Wulff shapes and related topics

Huhe Han and Takashi Nishimura

Abstract.

This is a survey article on the spherical method for studying Wulff shapes and related topics. The spherical method, which seems less common, is a powerful tool to study Wulff shapes and their related topics. It is verified how powerful the spherical method is by various results which seem difficult to be obtained without using the method. In this survey, the spherical method is explained in detail, and results obtained by using the spherical method until April 2016 are explained as well.

Contents

Received November 20, 2015.
Revised February 18, 2017.
2010 *Mathematics Subject Classification.* Primary 52A55; Secondary 52A20, 57R45, 82D25.
Key words and phrases. Wulff shape, Convex integrand, Pedal, Spherical method, Spherical dual, Spherical pedal, Spherical polar set, Spherical Wulff shape, Spherical caustic, Spherical symmetry set.
This work was partially supported by JSPS KAKENHI Grant Numbers 26610035, 17K05245.

§1. Introduction

It is known that, in the last half of 19th century, J. W. Gibbs and P. Curie had already pointed out that the shape of a crystal at equilibrium is determined so that the total surface energy is minimized amongst all crystals having the same total volume of the given crystal (see for instance [61, 63]). At the beginning of 20th century, more precisely in 1901, in the epoch-making paper [73], G. Wulff gave an important formula, called *Wulff's theorem* later, for a crystalline (namely, the surface energy function for a crystal of polytope type) at equilibrium. This formula leads to a geometric method to construct the shape of crystal which satisfies Wulff's theorem, called *Wulff's construction*[1], from the

---

[1]For the surface energy of a general crystal, it has a long history to show that "Wulff's construction gives the unique solution for the minimizing problem amongst all crystals having the same volume as the given crystal". The following is a quotation from [20].

> A. Dinghas (1944, [14]) gave a formal proof. J. Taylor (1978, [69]) gave a precise proof for very general surface energies and a very general class of set for which the surface energy is defined by using geometric measure theory. B. Dacorogna and C. E. Pfister (1992, [13]) gave an analytic proof when $n = 1$. I. Fonseca (1991, [16]) and I. Fonseca and S. Müller (1991, [17]) gave a simpler proof for arbitrary dimensions.

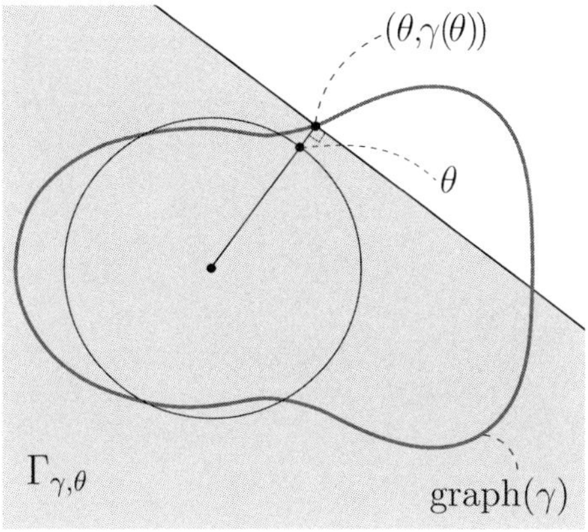

Fig. 1. The figure of $\Gamma_{\gamma,\theta}$.

known energy density $\gamma : S^n \to \mathbb{R}_+$. Nowadays, the shape of a crystal constructed by Wulff's construction is called the *Wulff shape*. Wulff shapes and their related topics are our main objects in this article. Our main tool to investigate them is the spherical method, which is a new and powerful tool. The main purpose of this article is to give a survey on "what the spherical method is" and "how powerful the spherical method is".

We start from explaining Wulff shapes. The most main character for studying Wulff shapes is a *continuous* function $\gamma : S^n \to \mathbb{R}_+$, where n, S^n and $\mathbb{R}_+$ are a positive integer, the unit sphere in $\mathbb{R}^{n+1}$ and the set consisting of positive real numbers respectively. Thus, first of all, we assume that a continuous function $\gamma : S^n \to \mathbb{R}_+$ is given. For the given γ and a point $\theta \in S^n$, the half space $\Gamma_{\gamma,\theta}$ is defined as follows, where $x \cdot \theta$ means the standard scalar product of two vectors x, θ of the vector space $\mathbb{R}^{n+1}$ (see Figure 1).

$$\Gamma_{\gamma,\theta} = \{x \in \mathbb{R}^{n+1} \mid x \cdot \theta \leq \gamma(\theta)\}.$$

Then, the *Wulff shape associated with* γ, denoted by $\mathcal{W}_\gamma$, is constructed as follows (see Figure 2).

$$\mathcal{W}_\gamma = \bigcap_{\theta \in S^n} \Gamma_{\gamma,\theta}.$$

By definition, it is clear that $\mathcal{W}_\gamma$ is compact, convex and the origin $\mathbf{0}$ of $\mathbb{R}^{n+1}$ is contained in $\mathcal{W}_\gamma$ as an interior point of $\mathcal{W}_\gamma$. The constructed

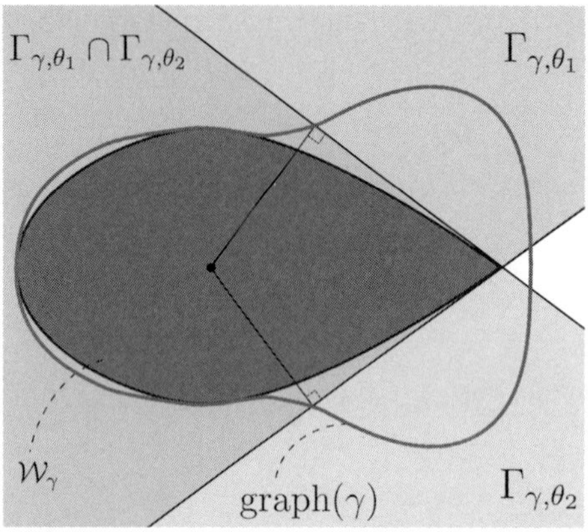

Fig. 2. The Wulff shape $\mathcal{W}_\gamma$ associated with γ.

shape $\mathcal{W}_\gamma$ is a geometric model of a crystal. Then, a natural question arises as follows:

Question 1. What does the given continuous function γ mean physically ?

Question 1 is answered as follows. Let $\partial\mathcal{W}_\gamma$ be the boundary of $\mathcal{W}_\gamma$. For any $\theta \in S^n$, the real number $\gamma(\theta)$ means the surface energy density of $\partial\mathcal{W}_\gamma$ at the direction $\theta \in S^n$, namely, at the point $\partial\mathcal{W}_\gamma \cap L_\theta$ where L_θ is the half line $\{r\theta \in \mathbb{R}^{n+1} \mid r \in \mathbb{R}_+\}$.

Let $C^0(S^n, \mathbb{R}_+)$ be the set consisting of continuous function $S^n \to \mathbb{R}_+$. The set $C^0(S^n, \mathbb{R}_+)$ is a metric space endowed with the uniformly convergent metric. Let $\mathcal{H}_{conv,\mathbf{0}}\left(\mathbb{R}^{n+1}\right)$ be the set consisting of compact and convex subsets of $\mathbb{R}^{n+1}$ containing the origin $\mathbf{0}$ as an interior point. Let $\mathcal{H}\left(\mathbb{R}^{n+1}\right)$ be the set consisting of non-empty compact sets of $\mathbb{R}^{n+1}$. It is known that $\mathcal{H}\left(\mathbb{R}^{n+1}\right)$ is a complete metric space by the Pompeiu-Hausdorff metric (see for instance [6, 15]). The set $\mathcal{H}_{conv,\mathbf{0}}\left(\mathbb{R}^{n+1}\right)$ is a subspace of $\mathcal{H}\left(\mathbb{R}^{n+1}\right)$. Notice that $\mathcal{H}_{conv,\mathbf{0}}\left(\mathbb{R}^{n+1}\right)$ is not complete. Then, by the Wulff construction, the following mapping defined by $\mathcal{W}(\gamma) = \mathcal{W}_\gamma$ is well-defined.

$$\mathcal{W} : C^0(S^n, \mathbb{R}_+) \to \mathcal{H}_{conv,\mathbf{0}}\left(\mathbb{R}^{n+1}\right).$$

Although it is almost clear that the well-defined mapping $\mathcal{W}$ is continuous under the above topologies, the following fundamental questions on $\mathcal{W}$ naturally arise.

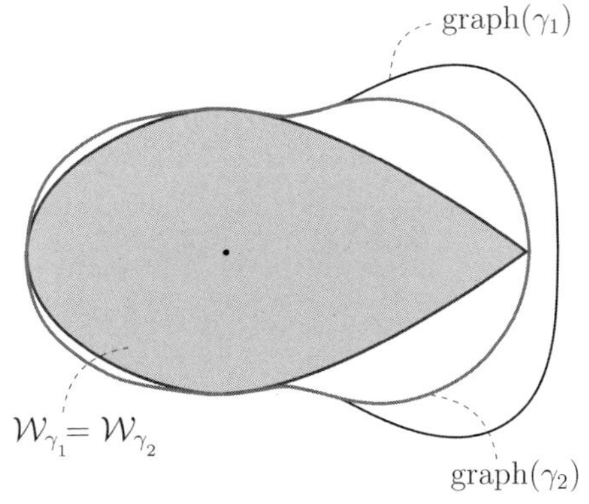

Fig. 3. A counterexample for Question 3.

Question 2. Is $\mathcal{W}$ surjective ?

Question 3. Is $\mathcal{W}$ injective ?

The affirmative answer to Question 2 can be found in [69]. Namely, for any $W \in \mathcal{H}_{conv,\mathbf{0}}\left(\mathbb{R}^{n+1}\right)$, it is known that $\mathcal{W}^{-1}(W) \neq \emptyset$. On the other hand, the answer to Question 3 is negative in general as follows. For any $\gamma \in C^0(S^n, \mathbb{R}_+)$, set

$$\mathrm{graph}(\gamma) = \{(\theta, \gamma(\theta)) \in \mathbb{R}^{n+1} - \{\mathbf{0}\} \mid \theta \in S^n\},$$

where $(\theta, \gamma(\theta))$ is the polar plot expression for a point of $\mathbb{R}^{n+1} - \{\mathbf{0}\}$. The mapping $\mathrm{inv} : \mathbb{R}^{n+1} - \{\mathbf{0}\} \to \mathbb{R}^{n+1} - \{\mathbf{0}\}$ defined as follows is called the *inversion* with respect to the origin of $\mathbb{R}^{n+1}$.

$$\mathrm{inv}(\theta, r) = \left(-\theta, \frac{1}{r}\right).$$

Let Γ_γ be the boundary of the convex hull of $\mathrm{inv}(\mathrm{graph}(\gamma))$. Then, the following is known.

Proposition 1 ([69, 23]). *Let γ_1, γ_2 be two elements of $C^0(S^n, \mathbb{R}_+)$ such that $\Gamma_{\gamma_1} = \Gamma_{\gamma_2}$. Then, the equality $\mathcal{W}_{\gamma_1} = \mathcal{W}_{\gamma_2}$ holds.*

By Proposition 1, counterexamples for Question 3 are easily constructed (see Figure 3).

As the next question, the following question naturally arises.

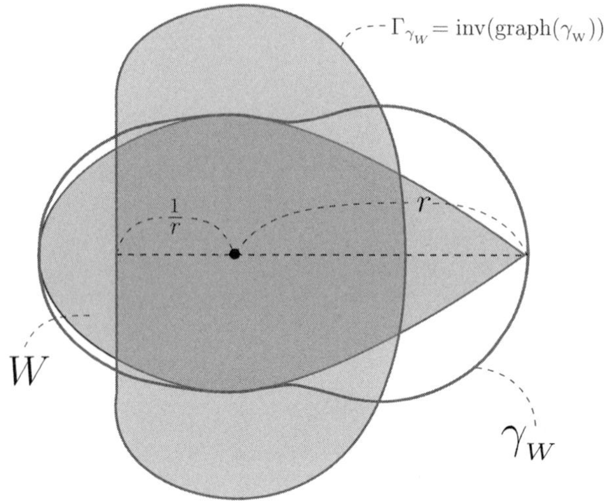

Fig. 4. The convex integrand γ_W of W.

Question 4. Let W be an element of $\mathcal{H}_{conv,\mathbf{0}}\left(\mathbb{R}^{n+1}\right)$, Then, characterize the best efficient element of $\mathcal{W}^{-1}(W)$ if it exists.

There is the well-known notion which may be considered as the ultimate answer to Question 4. Proposition 1 may be a clue to reach the following notion.

Definition 1. let W be an element of $\mathcal{H}_{conv,\mathbf{0}}\left(\mathbb{R}^{n+1}\right)$. Let $\gamma_W : S^n \to \mathbb{R}_+$ be the continuous function satisfying $\mathrm{inv}(\mathrm{graph}(\gamma_W)) = \Gamma_{\gamma_W}$. Then, the function γ_W is called the *convex integrand* of W (see Figure 4).

The notion of convex integrand was firstly introduced by J. Taylor in [69] and it plays a key role for studying Wulff shapes (for details on convex integrands, see for instance [51, 69]). For any Wulff shape W and any $\gamma \in \mathcal{W}^{-1}(W)$, since $\gamma(\theta)$ means the surface energy density of ∂W at the direction $\theta \in S^n$, integrating the convex integrand γ over S^n represents the surface energy of W. Hence, γ is called an *integrand* of W. And moreover, since $\mathrm{inv}(\mathrm{graph}(\gamma_W))$ is exactly the boundary of a convex set, γ_W is called the *convex* integrand of W.

Proposition 2. [69, 23] *Let W be an element of $\mathcal{H}_{conv,\mathbf{0}}\left(\mathbb{R}^{n+1}\right)$. Then, for any $\gamma \in \mathcal{W}^{-1}(W)$ and any $\theta \in S^n$, the following inequality* *holds.*

$$\gamma_W(\theta) \leq \gamma(\theta).$$

By Proposition 2, the convex integrand γ_W may be regarded as the best efficient continuous function in $\mathcal{W}^{-1}(W)$.

Another question which can naturally and easily arises is the following.

Question 5. Characterize the element $W \in \mathcal{H}_{conv,\mathbf{0}}\left(\mathbb{R}^{n+1}\right)$ such that $\mathcal{W}^{-1}(W)$ consists of only the convex integrand of W.

In Subsection 3.5, a partial answer to Question 5 is given with its proof.

In order to understand the deep relation between a Wulff shape W and its convex integrand γ_W well, we introduce the notion of dual convex integrand for a given convex integrand. Given a convex integrand $\gamma : S^n \to \mathbb{R}_+$, define the continuous function $\delta : S^n \to \mathbb{R}_+$ by the following equality, where $\partial\mathcal{W}_\gamma$ denotes the boundary of $\mathcal{W}_\gamma$.

$$\mathrm{inv}(\mathrm{graph}(\delta)) = \partial\mathcal{W}_\gamma.$$

The defined function $\delta : S^n \to \mathbb{R}_+$ is clearly a convex integrand, and it is called the *dual convex integrand* or *dual* of γ. The *dual Wulff shape of* $\mathcal{W}_\gamma$, denoted by $\mathcal{DW}_\gamma$, is the Wulff shape associated with δ.

$$\mathcal{DW}_\gamma = \mathcal{W}_\delta.$$

Concerning convex integrands and duals, the following question seems to be the most natural.

Question 6. (1) Is the dual convex integrand of δ γ ?
(2) Does the following involutive property hold ?

$$\mathcal{DDW}_\gamma = \mathcal{W}_\gamma.$$

The spherical method solves Question 6 affirmatively. See Proposition 8 in Subsection 2.5.

The next natural question on convex integrands and their duals is the following.

Question 7. *Characterize the convex integrand* γ *such that* $\gamma = \delta$ *in terms of the Wulff shape* $\mathcal{W}_\gamma$.

Question 7 is equivalent to say that "Characterize the Wulff shape $\mathcal{W}_\gamma$ satisfying $\mathcal{W}_\gamma = \mathcal{DW}_\gamma$ in terms of $\mathcal{W}_\gamma$, where $\gamma : S^n \to \mathbb{R}_+$ is a given convex integrand". Thus, Question 7 is the characterization question of a self-dual Wulff shape. A complete answer to this question with a rigorous proof and with many examples, which is summarized in Subsection 3.6, is obtained in [24].

The following Question 8, which is also natural, is a characterization question on differentiability of a convex integrand.

Question 8. Let k be a positive integer or ∞. Then, characterize a C^k convex integrand γ_W in terms of the Wulff shape W.

In the case $k = 1$, a partial answer to Question 8 can be found in [51]. In [23], a complete answer to this question with a rigorous proof is obtained, and it is summarized in Subsection 3.4.

In the case $1 < k < \infty$, as of May, 2016, it seems that Question 8 is still open. In Section 4, Question 8 is explained in detail with its history.

In the case $k = \infty$, independent partial answers to Question 8 can be found in [1, 51, 68]. In [8], a complete answer to this question, which is summarized in Subsection 3.8, is obtained.

In the case that a given convex integrand $\gamma : S^n \to \mathbb{R}_+$ is of class C^∞, the notion of stability is naturally defined. Let $C^\infty(S^n, \mathbb{R}_+)$ be the set consisting of C^∞ functions $S^n \to \mathbb{R}_+$. For the set $C^\infty(S^n, \mathbb{R}_+)$, there exists the natural topology, called *Whitney C^∞ topology*. For details on Whitney C^∞ topology see Subsection 3.7. Notice that $C^\infty(S^n, \mathbb{R}_+)$ may be regarded as a topological subspace of the topological space $C^0(S^n, \mathbb{R}_+)$. However, since any element of $C^\infty(S^n, \mathbb{R}_+)$ is differentiable as many times as we want and we would like to investigate the stability of C^∞ convex integrand, we prefer to regard $C^\infty(S^n, \mathbb{R}_+)$ as the topological space endowed with Whitney C^∞ topology. Let $S^\infty(S^n, \mathbb{R}_+)$ be the subspace of $C^\infty(S^n, \mathbb{R}_+)$ consisting of stable functions $S^n \to \mathbb{R}_+$. For details on stable functions also, see Subsection 3.7. Let $C^\infty_{\mathrm{conv}}(S^n, \mathbb{R}_+)$ be the subspace of $C^\infty(S^n, \mathbb{R}_+)$ consisting of C^∞ convex integrand. By Mather's profound works on stability of C^∞ mappings ([40, 41, 42, 43, 44, 45]), it is well-known that $S^\infty(S^n, \mathbb{R}_+)$ is dense in $C^\infty(S^n, \mathbb{R}_+)$. However, this well-known fact is useless to answer the following question.

Question 9. Is the following intersection dense in $C^\infty_{\mathrm{conv}}(S^n, \mathbb{R}_+)$?

$$S^\infty(S^n, \mathbb{R}_+) \cap C^\infty_{\mathrm{conv}}(S^n, \mathbb{R}_+).$$

Question 9 may be considered as the density question of stable functions in a restricted function space. Researches arisen from similar motivations as Question 9 can be found in [46, 12, 26, 27, 28, 29, 30]. Thus, from the viewpoint of Singularity Theory, Question 9 may be regarded as a natural question. A complete answer to Question 9, which is summarized in Subsection 3.7, can be found in [7].

The following Question 10, which is also natural, is the characterization question on stability of a C^∞ convex integrand.

Question 10. Characterize a stable convex integrand γ in terms of its dual convex integrand δ.

In [8], a complete answer to Question 10, which is summarized in Subsection 3.8, is given. The answer given in [8] has the form that a C^∞ convex integrand is stable if and only if its dual convex integrand δ is stable. By this answer to Question 10, the following question naturally arises.

Question 11. Let $\gamma : S^n \to \mathbb{R}_+$ be a stable convex integrand.

(1) Suppose that θ_0 is a non-degenerate critical point of γ. Then, is $-\theta_0$ a non-degenerate critical ponit of δ ?

(2) If the answer to (1) is affirmative, then are there some relations between the Morse index of γ at θ_0 and the Morse index of δ at $-\theta_0$?

A complete answer to Question 11, too, can be found in [8]. Summary of this result, too, is given in Subsection 3.8.

For the topics on Wulff shapes which are not treated in this survey article, refer to [20, 51, 61, 63, 69, 70].

This article is organized as follows. In Section 2, what is the spherical method is explained. In Section 3, our results obtained by using the spherical method until April of 2016 are gathered together. By Section 3, it is expected that the readers can understand how powerful the spherical method is. Finally, in Section 4, it is posed several questions which are related to the topics of this article and seem to be open.

§2. Spherical method

2.1. Pedals

In Physics, pedals seem to be one of common backgrounds to study crystals (see for instance, [61, 63]). From our viewpoint as well, the notion of pedal for a smooth hypersurface in $\mathbb{R}^{n+1}$ is very important. Therefore, firstly in Section 2, pedals are quickly explained. As for the references of pedals, the authors recommend [2, 4, 10] as excellent references for pedals from mathematical side.

Definition 2. Let $\Phi : S^n \to \mathbb{R}^{n+1} - \{\mathbf{0}\}$ be a C^1 embedding. Then, the *pedal* relative to the *pedal point* $\mathbf{0}$ for Φ, denoted by $ped_{\Phi,\mathbf{0}} : S^n \to \mathbb{R}^{n+1}$, is the mapping which maps $\theta \in S^n$ to the unique nearest point of

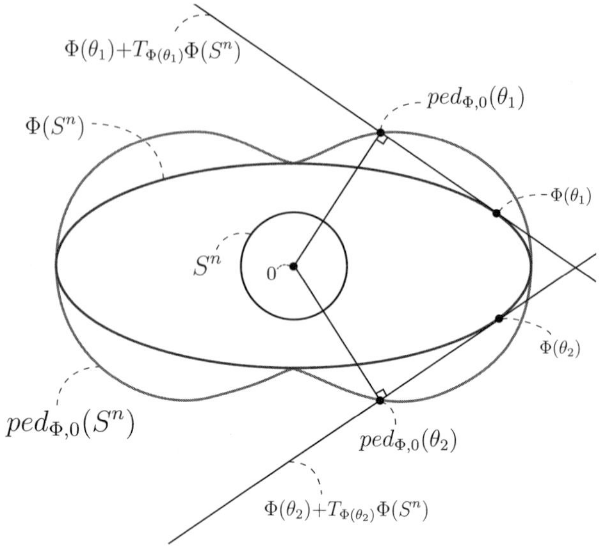

Fig. 5. The pedal relative to the pedal point $\mathbf{0}$ for Φ.

$\Phi(\theta) + T_{\Phi(\theta)}\Phi(S^n)$ from the origin $\mathbf{0}$, where $\Phi(\theta) + T_{\Phi(\theta)}\Phi(S^n)$ stands for the affine tangent hyperplane to $\Phi(S^n)$ at $\Phi(\theta)$ (see Figure 5).

Why is the notion of pedal important ? Let $\gamma : S^n \to \mathbb{R}_+$ be a convex integrand. Suppose that the boundary of a Wulff shape $\partial\mathcal{W}_\gamma$ is the image of a C^∞ embedding $\Phi : S^n \to \mathbb{R}^{n+1} - \{\mathbf{0}\}$. Then, by definition, the graph of the given convex integrand $\gamma : S^n \to \mathbb{R}^{n+1} - \{\mathbf{0}\}$ may be considered as the pedal relative to the pedal point $\mathbf{0}$ for Φ. By using the spherical method, it is proved that γ is of class C^∞ (see Subsection 3.8). Since graph(γ) does not contain the origin, from the information of $ped_{\Phi,\mathbf{0}} : S^n \to \mathbb{R}^{n+1}$, the family of affine tangent hyperplanes to $\Phi(S^n)$ can be uniquely restored. In other words, $ped_{\Phi,\mathbf{0}} : S^n \to \mathbb{R}^{n+1}$ is a method to store the family of affine tangent hyperplanes to $\Phi(S^n)$. In this sense, $ped_{\Phi,\mathbf{0}} : S^n \to \mathbb{R}^{n+1}$ itself may be considered as a sort of Legendre transform for the hypersurface $\Phi(S^n)$. Since $ped_{\Phi,\mathbf{0}}(\theta)$ can be expressed as $(\theta, \gamma(\theta))$ by using the polar plot expression and inv(graph(δ))$= \partial\mathcal{W}_\gamma$ where δ is the dual convex integrand of γ, it follows that γ may be regarded as the very Legendre transform of $\widehat{\delta}$ where $\hat{\delta} : S^n \to \mathbb{R}_+$ is defined by

$$\hat{\delta}(\theta) = \frac{1}{\delta(-\theta)} \quad (\forall\theta \in S^n).$$

And, in Physics, it is well-known Legendre transform is a very important notion for studying Wulff shapes (for instance, see pp.46–50 of [63]). Moreover, by using the spherical method, it can be proved out that $\mathcal{W}_\gamma$ is strictly convex if and only if the convex integrand γ is of class C^1 (see Theorem 8 in Subsection 3.4). Thus, in our situation, if both γ and δ are of class C^∞ , then both of $\mathcal{W}_\gamma$ and $\mathcal{W}_\delta$ are strictly convex. Therefore, we can expect that Wulff shapes with C^∞ boundary are very nice objects so that the Legendre transform works very well. By these observations, one can understand why the notion of pedal is one of common backgrounds in physics.

Observe that both of Wulff shapes and pedals are defined by using perpendicular properties. Therefore, the unit sphere S^{n+1} in $\mathbb{R}^{n+2}$ seems to be more suitable than $\mathbb{R}^{n+1}$ as the space where perpendicular properties are considered. This observation leads us to adopt the spherical method which deals with spherical counterparts of Wulff shapes, pedals etc. inside S^{n+1} rather than inside $\mathbb{R}^{n+1}$.

2.2. Spherical duals and spherical pedals

Given a C^∞ embedding $\Phi : S^n \to \mathbb{R}^{n+1}$, a C^∞ embedding $\widetilde{\Phi} : S^n \to S^{n+1}$ is firstly constructed, where S^{n+1} is the unit sphere in $\mathbb{R}^{n+2}$. The north pole of S^{n+1} is denoted by N, namely,

$$N = (0, \ldots, 0, 1) \in S^{n+1} \subset \mathbb{R}^{n+2}.$$

The northern hemisphere of S^{n+1} is denoted by $S^{n+1}_{N,+}$, namely,

$$S^{n+1}_{N,+} = \{P \in S^{n+1} \mid N \cdot P > 0\}$$

where $N \cdot P$ stands for the standard scalar product of $(n+2)$-dimensional two vectors $N, P \in \mathbb{R}^{n+2}$. The mapping

$$\mathbb{R}^{n+1} \ni x \mapsto (x, 1) \in \mathbb{R}^{n+1} \times \{1\} \subset \mathbb{R}^{n+2}$$

is denoted by Id, namely,

$$Id(x) = (x, 1).$$

The *central projection with respect to* N is the mapping $\alpha_N : S^{n+1}_{N,+} \to \mathbb{R}^{n+1} \times \{1\} \subset \mathbb{R}^{n+2}$ defined by

$$\alpha_N(P_1, \ldots, P_{n+1}, P_{n+2}) = \left(\frac{P_1}{P_{n+2}}, \ldots, \frac{P_{n+1}}{P_{n+2}}, 1\right),$$

where $P = (P_1, \ldots, P_{n+1}, P_{n+2})$ is a point of $S^{n+1}_{N,+}$ (see Figure 6). Then,

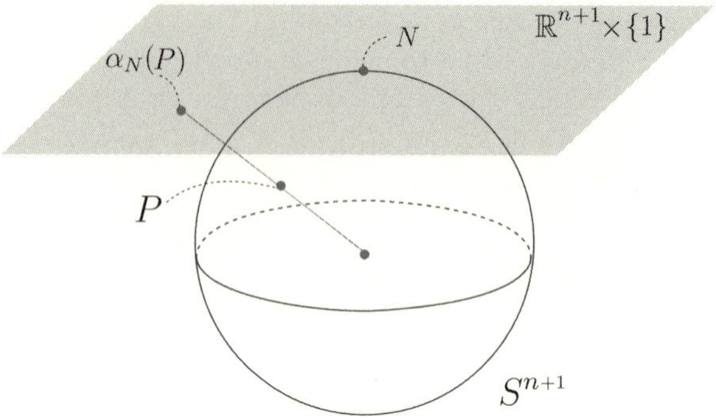

Fig. 6. The central projection with respect to N.

the mapping $\widetilde{\Phi} : S^n \to S^{n+1}_{N,+} \subset S^{n+1}$ derived from the given C^∞ embedding Φ is defined as follows.

$$\widetilde{\Phi} = \alpha_N^{-1} \circ Id \circ \Phi : S^n \to S^{n+1}_{N,+} \subset S^{n+1}.$$

Definition 3. Let $\Phi : S^n \to \mathbb{R}^{n+1}$ be a C^∞ embedding. Let $\widetilde{\Phi} : S^n \to S^{n+1}_{N,+}$ be the C^∞ embedding defined by $\widetilde{\Phi} = \alpha_N^{-1} \circ Id \circ \Phi$. Then, the mapping which maps $\theta \in S^n$ to the unique nearest point of $GH_{\widetilde{\Phi}(\theta)}\widetilde{\Phi}(S^n)$ from the north pole N is denoted by $s\text{-}ped_{\widetilde{\Phi},N} : S^n \to S^{n+1}$ and is called the *spherical pedal* relative to the *pedal point* N for $\widetilde{\Phi}$ (see Figure 7). Here, $GH_{\widetilde{\Phi}(\theta)}\widetilde{\Phi}(S^n)$ means the great hypersphere which is tangent to $\widetilde{\Phi}(S^n)$ at $\widetilde{\Phi}(\theta)$.

The spherical pedal relative to N for $\widetilde{\Phi}$ is known to be decomposed into two simple mappings. In order to explain this decomposition, the spherical dual $D\widetilde{\Phi} : S^n \to S^{n+1}$ of $\widetilde{\Phi}$ is necessary.

Definition 4. The mapping $D\widetilde{\Phi} : S^n \to S^{n+1}$, called the *spherical dual of* $\widetilde{\Phi}$, is defined as the mapping which maps a point $\theta \in S^n$ to the point in $S^{n+1}_{N,+}$ such that $D\widetilde{\Phi}(\theta)$ is perpendicular to any $P \in GH_{\widetilde{\Phi}(\theta)}\widetilde{\Phi}(S^n)$ (see Figure 8).

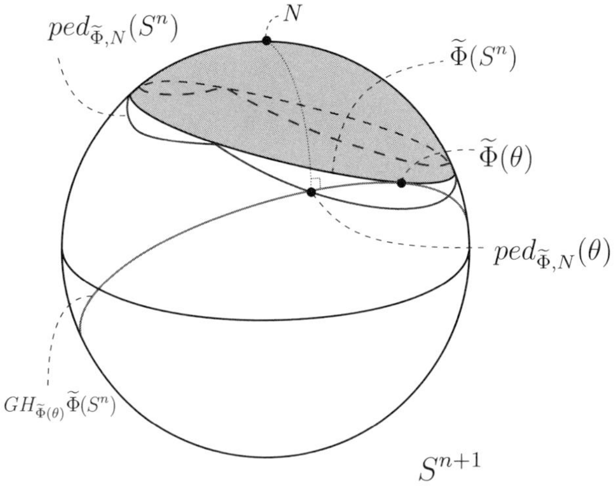

Fig. 7. The spherical pedal relative to N for $\widetilde{\Phi}$.

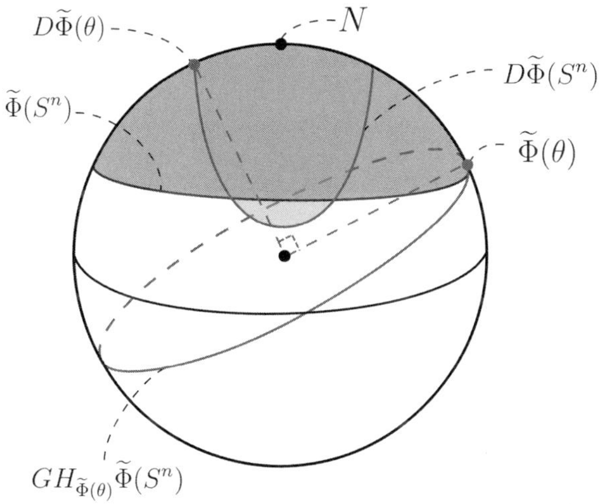

Fig. 8. The spherical dual of $\widetilde{\Phi}$.

It was V. I. Arnol'd who firstly defined the notion of spherical dual (for a spherical curve) (see [3]). Definition 4 is a natural generalization of his notion to spherical hypersurfaces.

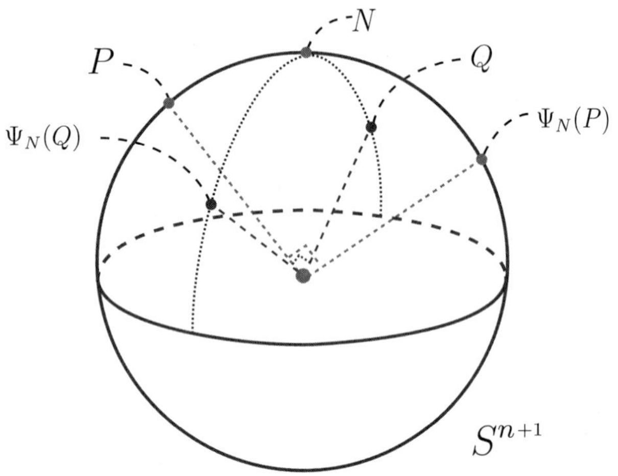

Fig. 9. The spherical blow-up relative to N.

Next, we consider the mapping $\Psi_N : S^{n+1} - \{\pm N\} \to S^{n+1}$ defined by

$$\Psi_N(P) = \frac{1}{\sqrt{1-(N\cdot P)^2}}(N - (N\cdot P)P)$$

and depicted in Figure 9. By using $D\widetilde{\Phi}$ and Ψ_N, $s\text{-}ped_{\widetilde{\Phi},N}$ can be decomposed as follows.

Proposition 3 ([54]). $s\text{-}ped_{\widetilde{\Phi},N} = \Psi_N \circ D\widetilde{\Phi}$.

Notice that Proposition 3 has been proved only for spherical pedal curves in [54]. However, by the same proof, even when n is greater than 1, $s\text{-}ped_{\widetilde{\Phi},N}$ can be decomposed as in Proposition 3. The mapping Ψ_N has the following nice properties.

(1) For any $P \in S^{n+1} - \{\pm N\}$, the equality $P \cdot \Psi_N(P) = 0$ holds,
(2) for any $P \in S^{n+1} - \{\pm N\}$, the property $\Psi_N(P) \in \mathbb{R}N + \mathbb{R}P$ holds,
(3) for any $P \in S^{n+1} - \{\pm N\}$, the property $N \cdot \Psi_N(P) > 0$ holds,
(4) the restriction $\Psi_N|_{S^{n+1}_{N,+}-\{N\}} : S^{n+1}_{N,+} - \{N\} \to S^{n+1}_{N,+} - \{N\}$ is a C^∞ diffeomorphism.

By these properties, it is reasonable to call the mapping Ψ_N the *spherical blow-up relative to* N. The mapping Ψ_N has been already used to study many topics related to perpendicularity. For instance, it was used for studying singularities of spherical pedal curves in [36, 53, 54, 55], for studying spherical pedal unfoldings in [56], for studying hedgehogs and no-silhouettes in [59], for studying (spherical) Wulff shapes in [60, 22, 23], and for studying the aperture of plane curves in [35]. A hyperbolic version of Ψ_N is also useful in hyperbolic situation (see [34]). The above properties clearly derives the following:

Lemma 2.1 ([35]). *The inversion* inv : $\mathbb{R}^{n+1} - \{\mathbf{0}\} \to \mathbb{R}^{n+1} - \{\mathbf{0}\}$ *can be decomposed as follows.*

$$\mathrm{inv} = Id^{-1} \circ \alpha_N \circ \Psi_N \circ \alpha_N^{-1} \circ Id.$$

In p. 50 of [2] and p. 91 of [4], the following assertion can be found.

Proposition 4 ([2, 4]). *Let* $\Phi : S^n \to \mathbb{R}^{n+1} - \{\mathbf{0}\}$ *be a* C^∞ *embedding. Then, the following equality holds.*

$$ped_{\Phi,\mathbf{0}} = \mathrm{inv} \circ Id^{-1} \circ \alpha_N \circ D\widetilde{\Phi}.$$

Here, $\widetilde{\Phi} = \alpha_N^{-1} \circ Id \circ \Phi$.

In the case that $\mathbf{0} \notin \Phi(S^n)$ and $\widetilde{\Phi} = \alpha_N^{-1} \circ Id \circ \Phi$, combining Lemma 2.1 and Proposition 4, one can easily obtain the proof of Proposition 3 as follows.

$$\begin{aligned} s\text{-}ped_{\widetilde{\Phi},N} &= \alpha_N^{-1} \circ Id \circ ped_{\Phi,\mathbf{0}} \\ &= \alpha_N^{-1} \circ Id \circ \mathrm{inv} \circ Id^{-1} \circ \alpha_N \circ D\widetilde{\Phi} \\ &= \alpha_N^{-1} \circ Id \circ Id^{-1} \circ \alpha_N \circ \Psi_N \circ \alpha_N^{-1} \circ Id \circ Id^{-1} \circ \alpha_N \circ D\widetilde{\Phi} \\ &= \Psi_N \circ D\widetilde{\Phi}. \end{aligned}$$

However, in the case that $\mathbf{0} \in \Phi(S^n)$, it is impossible to apply Proposition 4. Moreover, in such a case, it seems impossible to investigate $ped_{\Phi,\mathbf{0}}$ by the standard method. Nevertheless, via $\alpha_N^{-1} \circ Id$, we can investigate $ped_{\Phi,\mathbf{0}}$ by using the spherical method. This is one of merits of the spherical method, and this plays one of the most important roles in Subsection 3.2.

2.3. Spherical polar sets

The notion of spherical polar set seems to be less common though it seems that the notion of polar set in $\mathbb{R}^{n+1}$ is relatively common (for instance, see [47]). Since the notion of spherical polar set is one of the major ingredients for the spherical method, in this subsection, properties of spherical polar sets in S^{n+1} are quickly reviewed.

Given a point P of S^{n+1}, $H(P)$ is the following set:

$$H(P) = \{Q \in S^{n+1} \mid P \cdot Q \geq 0\}.$$

Namely, $H(P)$ is the hemisphere with boundary centered at P.

Definition 5. For any subset $\widetilde{X}$ of S^{n+1}, the set $\widetilde{X}^\circ$ defined by

$$\widetilde{X}^\circ = \bigcap_{P \in \widetilde{X}} H(P)$$

is called the *spherical polar set of* $\widetilde{X}$.

It is clearly seen that for any $\widetilde{X} \subset S^{n+1}$, $\widetilde{X}^\circ$ is closed.

Lemma 2.2 ([60]). *Given two subsets $\widetilde{X}, \widetilde{Y} \subset S^{n+1}$, we suppose that the inclusion $\widetilde{X} \subset \widetilde{Y}$ holds. Then, for the spherical polar sets $\widetilde{X}^\circ, \widetilde{Y}^\circ$, the inclusion $\widetilde{Y}^\circ \subset \widetilde{X}^\circ$ holds.*

Lemma 2.3 ([60]). *Let $\widetilde{X}$ be a subset of S^{n+1}. Then, the inclusion $\widetilde{X} \subset \widetilde{X}^{\circ\circ}$ is satisfied.*

Definition 6. For a subset $\widetilde{X} \subset S^{n+1}$, if there exists a point $P \in S^{n+1}$ such that $H(P) \cap \widetilde{X} = \emptyset$, then $\widetilde{X}$ is said to be *hemispherical.*

For a hemispherical two-point set $\{P, Q\}$ of S^{n+1}, PQ stands for the following arc:

$$PQ = \left\{ \frac{(1-t)P + tQ}{||(1-t)P + tQ||} \in S^{n+1} \;\middle|\; 0 \leq t \leq 1 \right\}.$$

Notice that since $\{P, Q\}$ is hemispherical in the above, it follows that $||(1-t)P + tQ|| \neq 0$ for any $t \in [0, 1]$.

Definition 7. (1) Let $\widetilde{X}$ be a hemispherical subset of S^{n+1}. Then, $\widetilde{X}$ is said to be *spherical convex* if $PQ \subset \widetilde{X}$ for any $P, Q \in \widetilde{X}$.

(2) Let $\widetilde{X}$ be a hemispherical subset of S^{n+1}. Then, $\widetilde{X}$ is said to be *strictly spherical convex* if $PQ - \{P, Q\}$ is a subset of $\mathrm{int}(\widetilde{X})$ for any $P, Q \in \widetilde{X}$, where $\mathrm{int}(\widetilde{X})$ is the set consisting of interior points of $\widetilde{X}$.

(3) Let $\widetilde{X}$ be a hemispherical subset of S^{n+1}. Then, $\widetilde{X}$ is called a *spherical convex body* if $\widetilde{X}$ is closed and spherical convex and it has an interior point.

In general, $\widetilde{X}^\circ$ is not necessarily spherical convex even if $\widetilde{X}$ is hemispherical (for instance if $\widetilde{X} = \{P\}$ then $\widetilde{X}^\circ = H(P)$ is not spherical convex). However, it is easily seen that if $\widetilde{X}$ is hemispherical and has an interior point, then $\widetilde{X}^\circ$ is spherical convex.

Lemma 2.4 ([60]). *Let* $\{\widetilde{X}_\lambda \subset S^{n+1}\}_{\lambda\in\Lambda}$ *be a family of spherical convex subsets. Then, the intersection*

$$\bigcap_{\lambda\in\Lambda} \widetilde{X}_\lambda$$

is spherical convex as well.

Definition 8. Let $\widetilde{X}$ be a hemispherical subset of S^{n+1}. Define the set s-conv$(\widetilde{X})$ as follows.

$$\text{s-conv}(\widetilde{X}) = \left\{ \frac{\sum_{i=1}^k t_iP_i}{||\sum_{i=1}^k t_iP_i||} \;\middle|\; P_i \in \widetilde{X},\ \sum_{i=1}^k t_i = 1,\ t_i \geq 0,\ k \in \mathbb{N} \right\}.$$

The set s-conv$(\widetilde{X})$ is called the *spherical convex hull* of $\widetilde{X}$.

It is clear that if $\widetilde{X}$ is spherical convex, then the equality s-conv$(\widetilde{X}) = \widetilde{X}$ holds. More generally, the following holds:

Lemma 2.5 ([60]). *Let* $\widetilde{X}$ *be a hemispherical subset of* S^{n+1}. *Then, the spherical convex hull of* $\widetilde{X}$ *is the smallest spherical convex set containing* $\widetilde{X}$, *that is to say, the intersection of all spherical convex sets containing* $\widetilde{X}$.

Definition 9. Let $\{P_1, \ldots, P_k\}$ be a hemispherical finite subset of S^{n+1} such that the spherical convex hull s-conv$(\{P_1, \ldots, P_k\})$ has an interior point. Then, s-conv$(\{P_1, \ldots, P_k\})$ is called the *spherical polytope generated by* $P_1, \ldots, P_k$.

Proposition 5 ([19, 60]). *Let* $\widetilde{X}$ *be a closed hemispherical subset of* S^{n+1}. *Then, the following equality holds:*

$$\text{s-conv}(\widetilde{X}) = (\text{s-conv}(\widetilde{X}))^{\circ\circ}.$$

Notice the following:

(1) For any closed hemispherical subset $\widetilde{X} \subset S^{n+1}$, s-conv$(\widetilde{X})$, too, is closed and hemispherical.
(2) For any subset $\widetilde{X} \subset S^{n+1}$, the inclusion $\widetilde{X} \subset \widetilde{X}^{\circ\circ}$ always holds by Lemma 2.3. On the other hand, even if $\widetilde{X}$ is closed and hemispherical, the inverse inclusion $\widetilde{X} \supset \widetilde{X}^{\circ\circ}$ is not satisfied in general.

Lemma 2.6 ([60]). *Let* $\widetilde{X} = \{P_1, \ldots, P_k\} \subset S^{n+1}$ *be a hemispherical finite subset of* S^{n+1}. *Then, the following equality holds:*

$$\left\{ \frac{\sum_{i=1}^k t_iP_i}{||\sum_{i=1}^k t_iP_i||} \;\middle|\; P_i \in \widetilde{X},\ \sum_{i=1}^k t_i = 1,\ t_i \geq 0 \right\}^\circ = H(P_1) \cap \cdots \cap H(P_k).$$

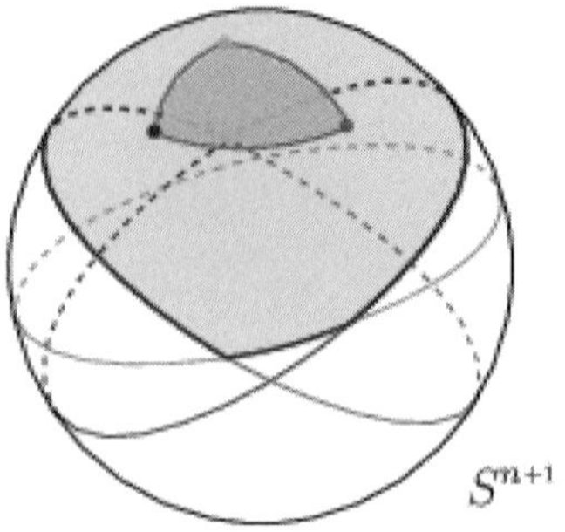

Fig. 10. Maehara's lemma.

Lemma 2.6 is depicted in Figure 10. The original form of Lemma 2.6 can be found in Maehara's book [39]. Thus, Lemma 2.6 is called *Maehara's lemma.*

2.4. Caustics and symmetry sets

When we want to study C^∞ convex integrands, especially the stability of them, we notice that spherical caustics and spherical symmetry sets are powerful tools. Thus, before explaining spherical caustics and spherical symmetry sets, in this subsection we quickly review the caustic and the symmetry set for a C^∞ embedding $S^n \to \mathbb{R}^{n+1} - \{\mathbf{0}\}$.

For any C^∞ embedding $\Phi : S^n \to \mathbb{R}^{n+1} - \{\mathbf{0}\}$, the following family of functions $F : \mathbb{R}^{n+1} \times S^n \to \mathbb{R}$ is considered.

$$F(v, \theta) = \frac{1}{2}||\Phi(\theta) - v||^2.$$

It is easily seen that F itself may be regarded as the mapping

$$F : \mathbb{R}^{n+1} \to C^\infty(S^n, \mathbb{R})$$

which maps $v \in \mathbb{R}^{n+1}$ to $f_v \in C^\infty(S^n, \mathbb{R})$, where f_v is the mapping defined by

$$f_v(\theta) = F(v, \theta).$$

Then, the *caustic* of Φ, denoted by $Caust(\Phi)$ (or $Caust(\Phi(S^n))$), is the set consisting of vectors v for which f_v has a degenerate critical point (see Figure 11).

$$Caust(\Phi) = \left\{v \in \mathbb{R}^{n+1} \mid f_v \text{ has a degenerate critical point}\right\}.$$

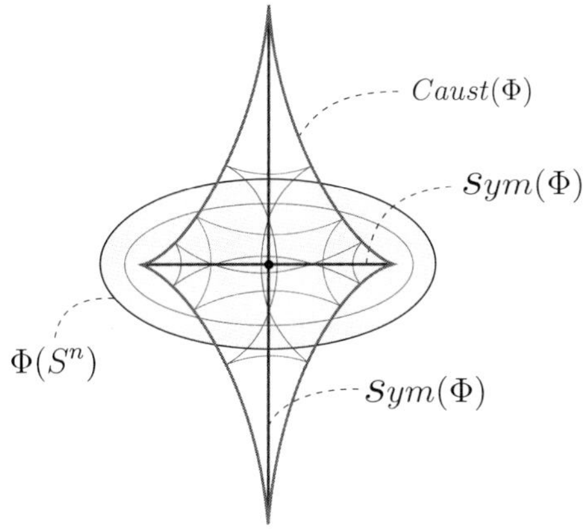

Fig. 11. The caustic and the symmetry set of Φ.

The caustic of Φ has been well-investigated in Singularity Theory. For details on caustics, for example see [2, 4, 5, 31, 32, 33].

Next, for any C^∞ embedding $\Phi : S^n \to \mathbb{R}^{n+1} - \{\mathbf{0}\}$, we review the symmetry set of Φ. The symmetry set of Φ is defined by using the same mapping f_v defined above. The set consisting of vectors v for which f_v has a multiple critical value is called the *symmetry set* of Φ and is denoted by $Sym(\Phi)$ (or $Sym(\Phi(S^n))$) (see Figure 11).

$$Sym(\Phi) = \left\{v \in \mathbb{R}^{n+1} \mid f_v \text{ has a multiple critical value}\right\}.$$

Although there are several literature on symmetry sets (for example [9, 10, 11]), it seems that, comparing with $Caust(\Phi)$, symmetry sets have been less-investigated.

Given a C^∞ embedding $\Phi : S^n \to \mathbb{R}^{n+1} - \{\mathbf{0}\}$, for any $t \in \mathbb{R}$, the C^∞ mapping $\Phi_t : S^n \to \mathbb{R}^{n+1}$ is defined as follows. Given a $\theta \in S^n$, $L_{\Phi(\theta)}$ is defined as the line passing through $\Phi(\theta)$ which is perpendicular to $\Phi(S^n)$ at $\Phi(\theta)$. For any non-zero real number t, there exist exactly two points $P_1(\theta), P_2(\theta) \in L_{\Phi(\theta)}$ satisfying

$$||P_1(\theta) - \Phi(\theta)||^2 = ||P_2(\theta) - \Phi(\theta)||^2 = t^2.$$

It is clear that, if t is sufficiently near zero, then exactly one of $P_1(\theta)$ and $P_2(\theta)$ must be inside the connected region of $\mathbb{R}^{n+1} - \Phi(S^n)$ containing $\mathbf{0}$. Thus, without loss of generality, we may assume that $P_1(\theta)$ is inside the connected region. Then, for any non-zero real number t, define the mapping $\Phi_t : S^n \to \mathbb{R}^{n+1}$ by

$$\Phi_t(\theta) = P_1(\theta) \text{ (resp., } \Phi_t(\theta) = P_2(\theta))$$

if t is positive (resp., t is negative). For $t = 0$, define

$$\Phi_0(\theta) = \Phi(\theta).$$

The mapping $\Phi_t : S^n \to \mathbb{R}^{n+1}$ is called the *wave front* of Φ. For any $t \in \mathbb{R}$, the wave front of Φ is clearly a C^∞ mapping. It is easily seen that the caustic of Φ and the symmetry set of Φ can be characterized by the wave fronts of Φ as follows (see Figure 11).

Proposition 6. (1)

$$Caust(\Phi) = \bigcup_{t \in \mathbb{R}} \Phi_t(S(\Phi_t)),$$

where $S(\Phi_t)$ is the set consisting of singular points of Φ_t.

(2)

$$Sym(\Phi) = \bigcup_{t \in \mathbb{R}} \{\Phi_t(\theta_1) = \Phi_t(\theta_2) \mid \theta_1, \theta_2 \in S^n, \theta_1 \neq \theta_2\}.$$

In Subsection 3.7, the union

$$Caust(\Phi) \bigcup Sym(\Phi)$$

will play an important role.

2.5. Spherical Wulff shapes, spherical caustics and spherical symmetry sets

Definition 10. For any Wulff shape $\mathcal{W}_\gamma$, the *spherical Wulff shape induced by* $\mathcal{W}_\gamma$, denoted by $\widetilde{\mathcal{W}_\gamma}$, is the image of $\mathcal{W}_\gamma$ by $\alpha_N^{-1} \circ Id : \mathbb{R}^{n+1} \to S^{n+1}_{N,+}$ (see Figure 13).

$$\widetilde{\mathcal{W}_\gamma} = \alpha_N^{-1} \circ Id\,(\mathcal{W}_\gamma)\,.$$

Any spherical Wulff shape $\widetilde{\mathcal{W}_\gamma}$ can be characterized by using the spherical blow-up Ψ_N and the spherical polar set operation as follows:

$$\widetilde{\mathcal{W}_\gamma} = \left(\Psi_N \circ \alpha_N^{-1} \circ Id\,(\text{graph}(\gamma))\right)^\circ.$$

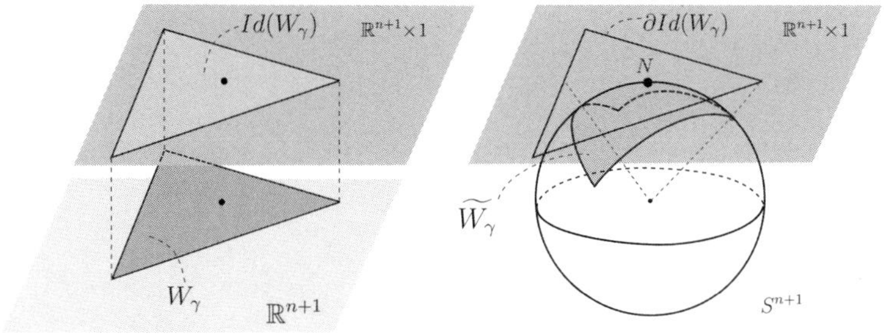

Fig. 12. The spherical Wulff shape associated with $\mathcal{W}_\gamma$.

Let $\widetilde{\mathcal{W}}_\gamma$ be a spherical Wulff shape. Then, the *spherical dual* of $\widetilde{\mathcal{W}}_\gamma$, denoted by $\mathcal{D}\widetilde{\mathcal{W}}_\gamma$, is the spherical polar set $\left(\widetilde{\mathcal{W}}_\gamma\right)^\circ$.

$$\mathcal{D}\widetilde{\mathcal{W}}_\gamma = \left(\widetilde{\mathcal{W}}_\gamma\right)^\circ .$$

It is clear that any spherical Wulff shape $\mathcal{W}_\gamma$ is closed, hemispherical and spherical convex. Hence, by Proposition 5 we have the following.

Proposition 7.

$$\mathcal{D}\mathcal{D}\widetilde{\mathcal{W}}_\gamma = \widetilde{\mathcal{W}}_\gamma .$$

By definition, it is clearly seen that $Id^{-1} \circ \alpha_N\left(\mathcal{D}\widetilde{\mathcal{W}}_\gamma\right) = \mathcal{D}\mathcal{W}_\gamma$. Hence, by Proposition 7 we have the assertion (1) of Proposition 8. The assertion (2) of Proposition 8 is a corollary of the assertion (1) of Proposition 8.

Proposition 8. (1)

$$\mathcal{D}\mathcal{D}\mathcal{W}_\gamma = \mathcal{W}_\gamma .$$

(2) *Suppose that $\gamma : S^n \to \mathbb{R}_+$ is a convex integrand. Then, the dual convex integrand of the dual convex integrand of γ is γ.*

Next, suppose that $\widetilde{\mathcal{W}}_\gamma$ be a spherical Wulff shape such that the boundary of it is the image of a C^∞ embedding $\widetilde{\Phi} : S^n \to S^{n+1}_{N,+}$. Then, we can define the *Spherical Caustic* and *Spherical Symmetry Set* for the embedding $\widetilde{\Phi}$ as follows. Let $d : S^{n+1} \times S^{n+1} - \triangle \to \mathbb{R}$ be the function such that $d\,(P_1, P_2)$ is the square of P_1P_2 where $\triangle = \{(P, -P) \in S^{n+1} \times S^{n+1} \mid P \in S^{n+1}\}$. We would like to consider the family of functions $\widetilde{F} : S^{n+1}_{N,+} \times S^n \to \mathbb{R}$ defined by

$$\widetilde{F}(v, \theta) = \frac{1}{2} d\left(\widetilde{\Phi}(\theta), v\right).$$

Then, $\widetilde{F}$ itself may be regarded as the mapping

$$\widetilde{F} : S^{n+1}_{N,+} \to C^\infty(S^n, \mathbb{R})$$

which maps $v \in S^{n+1}_{N,+}$ to $\widetilde{f}_v \in C^\infty(S^n, \mathbb{R}_+)$, where $\widetilde{f}_v$ is the mapping defined by

$$\widetilde{f}_v(\theta) = \widetilde{F}(v, \theta).$$

The *Spherical Caustic* of $\widetilde{\Phi}$, denoted by *Spherical-Caust*($\widetilde{\Phi}$) (or *Spherical-Caust*($\widetilde{\Phi}(S^n)$)), is the set consisting of vectors v for which $\widetilde{f}_v(\theta)$ has a degenerate critical point form.

$$\textit{Spherical-Caust}(\widetilde{\Phi}) = \left\{v \in S^{n+1}_{N,+} \mid \widetilde{f}_v \text{ has a degenerate critical point}\right\}.$$

The *Spherical Symmetry Set* of $\widetilde{\Phi}$, denoted by *Spherical-Sym*($\widetilde{\Phi}$) (or *Spherical-Sym*($\widetilde{\Phi}(S^n)$)), is the set consisting of vectors v for which $\widetilde{f}_v(\theta)$ has a multiple critical value.

$$\textit{Spherical-Sym}(\widetilde{\Phi}) = \left\{v \in S^{n+1}_{N,+} \mid \widetilde{f}_v \text{ has a multiple critical value}\right\}.$$

Given a C^∞ embedding $\widetilde{\Phi} : S^n \to S^{n+1}_{N,+} - \{N\}$, for any $t \in \mathbb{R}$ ($-\pi < t < \pi$), the C^∞ mapping $\widetilde{\Phi}_t : S^n \to S^{n+1}$ is defined as follows. Given a $\theta \in S^n$, $GC_{\widetilde{\Phi}(\theta)}$ is defined as the great circle passing through $\widetilde{\Phi}(\theta)$ which is perpendicular to $\widetilde{\Phi}(S^n)$ at $\widetilde{\Phi}(\theta)$. For any non-zero real number t ($-\pi < t < \pi$), there exist exactly two points $P_1(\theta), P_2(\theta) \in GC_{\widetilde{\Phi}(\theta)}$ satisfying

$$d(P_1(\theta), \widetilde{\Phi}(\theta)) = d(P_2(\theta), \widetilde{\Phi}(\theta)) = t^2.$$

It is clear that, if t is sufficiently near zero, then exactly one of $P_1(\theta)$ and $P_2(\theta)$ must be inside the connected region of $S^{n+1} - \widetilde{\Phi}(S^n)$ containing N. Thus, without loss of generality, we may assume that $P_1(\theta)$ is inside the connected region. Then, for any non-zero real number t ($-\pi < t < \pi$), define the mapping $\widetilde{\Phi}_t : S^n \to S^{n+1}$ by

$$\widetilde{\Phi}_t(\theta) = P_1(\theta) \text{ (resp., } \widetilde{\Phi}_t(\theta) = P_2(\theta))$$

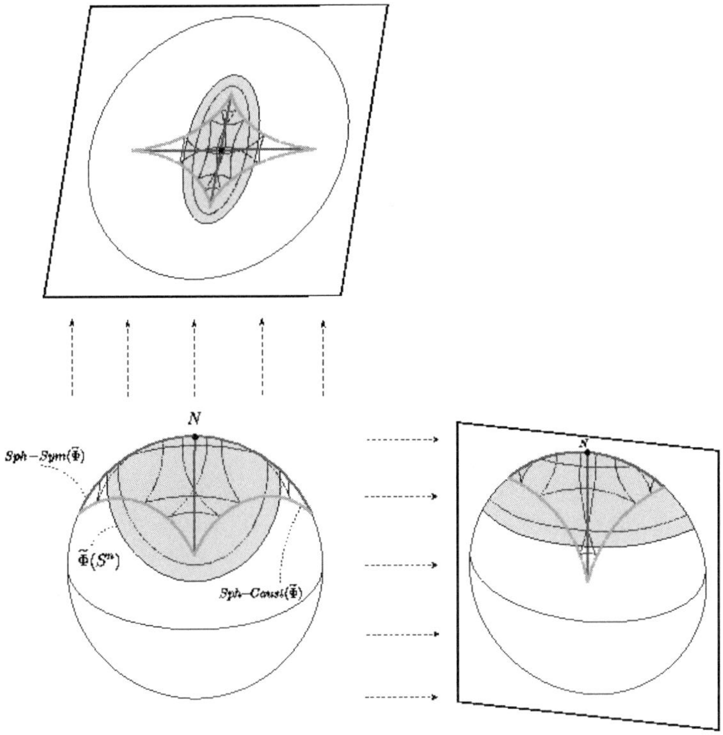

Fig. 13. The spherical caustic and the spherical symmetry set of $\widetilde{\Phi}$.

if t is positive (resp., t is negative). For $t = 0$, define

$$\widetilde{\Phi}_0(\theta) = \widetilde{\Phi}(\theta).$$

The mapping $\widetilde{\Phi}_t : S^n \to S^{n+1}$ is called the *spherical wave front* of $\widetilde{\Phi}$. For any $t \in \mathbb{R}$, the spherical wave front of $\widetilde{\Phi}$ is clearly a C^∞ mapping. Similarly as the case of C^∞ embedding $\Phi : S^n \to \mathbb{R}^{n+1} - \{\mathbf{0}\}$, it is easily seen that the spherical caustic of $\widetilde{\Phi}$ and the spherical symmetry set of $\widetilde{\Phi}$ can be characterized by the spherical wave fronts of $\widetilde{\Phi}$ as follows (see Figure 13).

Proposition 9. (1)

$$\textit{Spherical-Caust}(\widetilde{\Phi}) = \bigcup_{-\pi<t<\pi} \widetilde{\Phi}_t(S(\widetilde{\Phi}_t)),$$

where $S(\widetilde{\Phi}_t)$ is the set consisting singular points of $\widetilde{\Phi}_t$.

(2)

$$\textit{Spherical-Sym}(\widetilde{\Phi}) = \bigcup_{-\pi<t<\pi} \{\widetilde{\Phi}_t(\theta_1) = \widetilde{\Phi}_t(\theta_2) \mid \theta_1, \theta_2 \in S^n, \theta_1 \neq \theta_2\}.$$

In the case of $\widetilde{\Phi} : S^n \to S^{n+1}_{N,+} - \{N\}$, we have more. It is clear that

$$\widetilde{\Phi}_{\pi/2}(S^n) = \partial \mathcal{D}\widetilde{\mathcal{W}}_\gamma.$$

By this fact and proposition 7, considering inside the sphere S^{n+1} seems to derive a reasonable situation. Moreover, notice that, by [24], the self-dual Wulff shapes are strongly related to the angle $\pi/2$ (see Subsection 3.6). Thus, the angle $\pi/2$ may be considered as a significant number for studying Wulff shapes, although in $\mathbb{R}^{n+1}$ there are no such significant real numbers for studying Wulff shapes.

Definition 11. A *Legendrian* map-germ is a C^∞ map-germ $f : (\mathbb{R}^n, \mathbf{0}) \to (\mathbb{R}^{n+1}, \mathbf{0})$ such that there exists a germ of C^∞ vector field ν_f along f satisfying the following two:

(1)

$$\frac{\partial f}{\partial x_1}(x) \cdot \nu_f(x) = \cdots = \frac{\partial f}{\partial x_n}(x) \cdot \nu_f(x) = 0.$$

(2) The map-germ $L_f : (\mathbb{R}^n, \mathbf{0}) \to T_1\mathbb{R}^{n+1}$ defined as follows is non-singular, where $T_1\mathbb{R}^{n+1}$ means the unit tangent bundle of $\mathbb{R}^{n+1}$.

$$L_f(x) = (f(x), \nu_f(x)).$$

It is wll-known that the germ of spherical wave front $\widetilde{\Phi}_t : (S^n, \theta) \to S^{n+1}_{N,+} - \{N\}$ is Legendrian for any real number t ($-\pi < t < \pi$) and any $\theta \in S^n$. Some Legendrian map-germs are closely related with pedal hypersurface germs ([57, 58]). For more details on Legendrian map-germs, for example see [2, 4, 5, 31, 32, 33].

§3. Several results obtained by using the spherical method

3.1. Crystallines

Let $\gamma : S^n \to \mathbb{R}_+$ be a continuous function. Recall that, in Subsection 2.5, the spherical Wulff shape $\widetilde{\mathcal{W}_\gamma}$ associated with the Wulff shape $\mathcal{W}_\gamma$ was defined as follows.

$$\widetilde{\mathcal{W}_\gamma} = \alpha_N^{-1} \circ Id\,(\mathcal{W}_\gamma)\,.$$

Moreover, the spherical dual $\mathcal{D}\widetilde{\mathcal{W}_\gamma}$ of $\widetilde{\mathcal{W}_\gamma}$ was defined as follows.

$$\mathcal{D}\widetilde{\mathcal{W}_\gamma} = \left(\widetilde{\mathcal{W}_\gamma}\right)^\circ = \left(\alpha_N^{-1} \circ Id\,(\mathcal{W}_\gamma)\right)^\circ.$$

As explained in Subsection 2.5, the dual Wulff shape $\mathcal{DW}_\gamma$ is expressed in terms of $\mathcal{D}\widetilde{\mathcal{W}_\gamma}$ as follows.

$$\mathcal{DW}_\gamma = Id^{-1} \circ \alpha_N \left(\mathcal{D}\widetilde{\mathcal{W}_\gamma}\right) = Id^{-1} \circ \alpha_N \left(\left(\alpha_N^{-1} \circ Id\,(\mathcal{W}_\gamma)\right)^\circ\right).$$

A convex integrand γ is called a *crystalline* if the Wulff shape $\mathcal{W}_\gamma$ is a polytope. Notice that, inside S^{n+1}, we have Maehara's lemma (Lemma 2.6) for the spherical polar operation; and Maehara's lemma works very well for spherical polytopes. Thus, thanks to Maehara's lemma, we relatively easily have the following theorem.

Theorem 1 ([60]). *Let $\gamma : S^n \to \mathbb{R}_+$ be a continuous function. Then, γ is a crystalline if and only if its dual convex integrand δ is a crystalline.*

Since it is easy to see that the dual Wulff shape of $\mathcal{W}_\gamma$ is exactly the convex hull of $\frac{1}{\gamma}$ polar plot, it follows that the dual Wulff shape of $\mathcal{W}_\gamma$ may be regarded as a generalization of Frank-Meijering construction (for the Frank-Meijering construction, see [18, 48, 61]).

Although the proof of Theorem 1 is relatively easy, Theorem 1 is useful. For instance, the following result may be regarded as one of applications of Theorem 1.

Theorem 2 ([60]). *Let $\gamma : S^n \to \mathbb{R}_+$ be a function of class C^1. Then the Wulff shape $\mathcal{W}_\gamma$ is never a polytope.*

For the proof of Theorem 2, we use not only the spherical method and Theorem 1 but also the fact that the boundary of the convex hull of the graph of a C^1 function $S^n \to \mathbb{R}_+$ is a C^1 submanifold (for this fact, see [65, 74]).

3.2. Aperture of plane curves

Let $\mathbf{r} : S^1 \to \mathbb{R}^2$ be a C^∞ immersion. For the $\mathbf{r}$, we consider its inside region $\mathcal{NS}_\mathbf{r}$.

$$\mathcal{NS}_\mathbf{r} = \mathbb{R}^2 - \bigcup_{s\in S^1} \left(\mathbf{r}(s) + d\mathbf{r}_s(T_s(S^1))\right).$$

Since the perspective projection of $\mathbf{r}(S^1)$ from any point of the inside region is non-singular, $\mathcal{NS}_\mathbf{r}$ is called *no-silhouette of* $\mathbf{r}$.

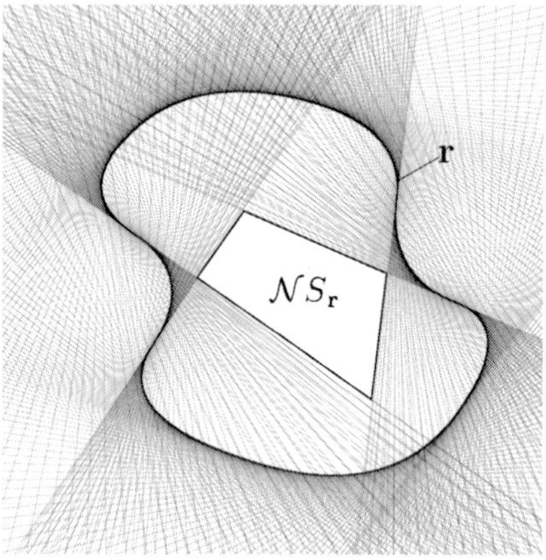

Fig. 14. Taken from [35]. The no-silhouette $\mathcal{NS}_\mathbf{r}$.

Next rotate all tangent lines about their tangent points simultaneously with the same angle, and consider the following inside region.

$$\mathcal{NS}_{\theta,\mathbf{r}} = \mathbb{R}^2 - \bigcup_{s\in S^1} \left(\mathbf{r}(s) + R_\theta\left(d\mathbf{r}_s(T_s(S^1))\right)\right),$$

where $R_\theta : \mathbb{R}^2 \to \mathbb{R}^2$ is the rotation defined by $R_\theta(x, y) = (x\cos\theta - y\sin\theta, x\sin\theta + y\cos\theta)$.
The *aperture of* S^1 is the animation

$$\left(\bigcup_{0\le\theta<\pi/2} (\mathcal{NS}_{\theta,\mathbf{r}} \times \{\theta\})\right) \bigcup (\{(0,0)\} \times \{\pi/2)\}) \subset \mathbb{R}^2 \times \mathbb{R}.$$

We are interested in how the topological closure of $\mathcal{NS}_{\theta,\mathbf{r}}$ is growing as θ increases for general $\mathbf{r} : S^1 \to \mathbb{R}^2$ such that $\mathcal{NS}_\mathbf{r} \neq \emptyset$.

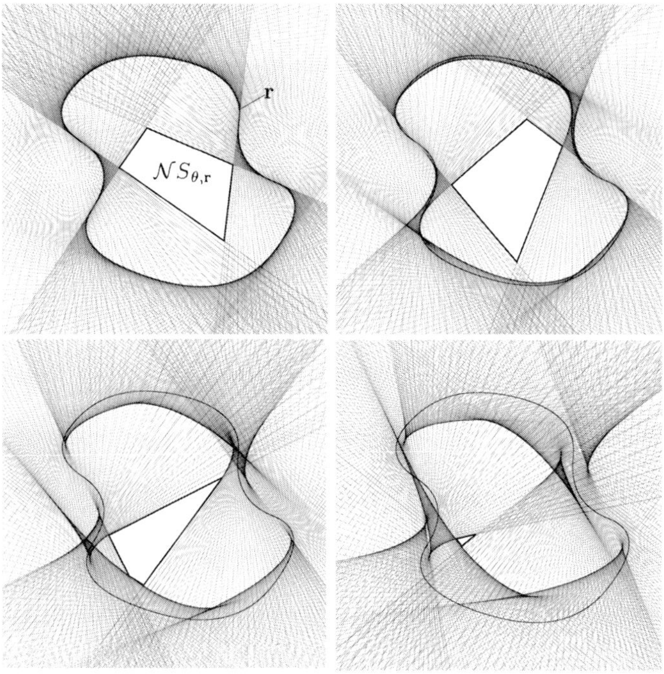

Fig. 15. Taken from [35]. Left top : $\theta = 0$, right top : $\theta = \pi/12$, left bottom : $\theta = \pi/6$, right bottom : $\theta = \pi/4$.

Lemma 3.1 ([35]). *For any C^∞ immersion $\mathbf{r} : S^1 \to \mathbb{R}^2$, $\mathcal{NS}_{\frac{\pi}{2},\mathbf{r}}$ is the empty set.*

Since our object is moving convex bodies (and its limit) $\overline{\mathcal{NS}_{\theta,\mathbf{r}}}$, it is better to treat the spherical polar operation as a transform of a space consisting of reasonable subsets of S^2. Let $\mathcal{H}(S^{n+1})$ be the set consisting of non-empty closed subsets of S^{n+1}. It is well-known that $\mathcal{H}(S^{n+1})$ is a complete metric space with respect to the Pompeiu-Hausdorff metric (see [6, 15]). Let $\mathcal{H}^\circ(S^{n+1})$ be the subspace of $\mathcal{H}(S^{n+1})$ consisting of non-empty closed subset $\widetilde{W}$ of S^{n+1} such that $\widetilde{W}^\circ \neq \emptyset$.

$$\mathcal{H}^\circ(S^{n+1}) = \left\{ \widetilde{W} \in \mathcal{H}(S^{n+1}) \mid \widetilde{W}^\circ \neq \emptyset \right\}.$$

Then, the *spherical polar transform*

$$\bigcirc : \mathcal{H}^\circ(S^{n+1}) \to \mathcal{H}^\circ(S^{n+1})$$

is defined as follows.

$$\bigcirc(\widetilde{W}) = \widetilde{W}^{\circ}.$$

By Lemma 2.3, it follows that $\widetilde{W}^{\circ}$ is contained in $\mathcal{H}^{\circ}(S^{n+1})$ for any $\widetilde{W} \in \mathcal{H}^{\circ}(S^{n+1})$. Thus, the spherical polar transform $\bigcirc$ is well-defined. The following proposition is proved in [35].

Proposition 10 ([35]). *In the case* $n = 1$, *the spherical polar transform* $\bigcirc : \mathcal{H}^{\circ}(S^2) \to \mathcal{H}^{\circ}(S^2)$ *is continuous.*

It is easily seen that Proposition 10 holds for any $n \in \mathbb{N}$. In the next subsection, a result obtained in [22] which is stronger than Proposition 10 (for general n) is surveyed. Namely, in Subsection 3.3, it shall be stated that the spherical polar transform $\bigcirc : \mathcal{H}^{\circ}(S^{n+1}) \to \mathcal{H}^{\circ}(S^{n+1})$ is Lipschitz.

By using Proposition 10, the following is obtained.

Proposition 11 ([35]). *With respect to the Pompeiu-Hausdorff metric, the topological closure of* $\mathcal{NS}_{\theta,\mathbf{r}}$ *varies continuously depending on* θ *while* $\mathcal{NS}_{\theta,\mathbf{r}}$ *is not empty.*

Therefore, by Lemma 3.1, the notion of aperture angle $\theta_{\mathbf{r}}$ $(0 < \theta_{\mathbf{r}} \leq \pi/2)$ is well-defined as follows.

Definition 12 ([35]). Let $\mathbf{r} : S^1 \to \mathbb{R}^2$ be a C^{∞} immersion with its no-silhouette $\mathcal{NS}_{\mathbf{r}} \neq \emptyset$. Then, $\theta_{\mathbf{r}}$ $(0 < \theta_{\mathbf{r}} \leq \frac{\pi}{2})$ is defined as the largest angle which satisfies $\mathcal{NS}_{\theta,\mathbf{r}} \neq \emptyset$ for any θ $(0 \leq \theta < \theta_{\mathbf{r}})$. The angle $\theta_{\mathbf{r}}$ is called the *aperture angle* of the given $\mathbf{r}$.

It is not difficult to see that $\overline{\mathcal{NS}_{\theta,\mathbf{r}}}$ is a Wulff shape for any θ $(0 \leq \theta < \theta_{\mathbf{r}})$. Thus, we may regard that $\{\overline{\mathcal{NS}_{\theta,\mathbf{r}}} \mid 0 \leq \theta \leq \theta_{\mathbf{r}}\}$ is a simple geometric model of crystal growth.
We are interested in how our simple geometric model of crystal growth melts as θ goes to $\theta_{\mathbf{r}}$ from 0.

Theorem 3 ([35]). *Let* $\mathbf{r} : S^1 \to \mathbb{R}^2$ *be a* C^{∞} *immersion with its no-silhouette* $\mathcal{NS}_{\mathbf{r}} \neq \emptyset$. *Then, for any* θ $(0 < \theta < \theta_{\mathbf{r}})$, $\overline{\mathcal{NS}_{\theta},\mathbf{r}}$ *is never a polygon even if the given* $\overline{\mathcal{NS}_{\mathbf{r}}}$ *is a polygon.*

By Theorem 3, among $\overline{\mathcal{NS}_{\frac{\pi}{12},\mathbf{r}}}$, $\overline{\mathcal{NS}_{\frac{\pi}{6},\mathbf{r}}}$ and $\overline{\mathcal{NS}_{\frac{\pi}{4},\mathbf{r}}}$ in Figure 15, there are no polygons; although $\overline{\mathcal{NS}_{0,\mathbf{r}}}$ is a polygon constructed by four tangent lines to $\mathbf{r}$ at four inflection points.

Theorem 4 ([35]). *Let* $\mathbf{r} : S^1 \to \mathbb{R}^2$ *be a* C^{∞} *immersion with its no-silhouette* $\mathcal{NS}_{\mathbf{r}} \neq \emptyset$. *Then, there exists the unique point* $P_{\mathbf{r}} \in \mathbb{R}^2$ *such* *that for any sequence* $\{\theta_i\}_{i=1,2,\ldots} \subset [0, \theta_{\mathbf{r}})$ *converging to the aperture angle* $\theta_{\mathbf{r}}$*, the following holds:*

$$\lim_{i\to\infty} d_H(\overline{\mathcal{NS}_{\theta_i},\mathbf{r}}, P_{\mathbf{r}}) = 0.$$

Here, $d_H : \mathcal{H}(\mathbb{R}^2) \times \mathcal{H}(\mathbb{R}^2) \to \mathbb{R}$ is the Pompeiu-Hausdorff metric. Theorem 4 justifies the following definition.

Definition 13 ([35]). Let $\mathbf{r} : S^1 \to \mathbb{R}^2$ be a C^∞ immersion with its no-silhouette $\mathcal{NS}_{\mathbf{r}} \neq \emptyset$. Then, the unique point $P_{\mathbf{r}} = \lim_{\theta\to\theta_{\mathbf{r}}} \overline{\mathcal{NS}_{\theta,\mathbf{r}}}$ is called the *aperture point* of $\mathbf{r}$.

The simplest example is a circle. The aperture point of it is its center. In this case, the aperture angle is $\pi/2$. In general, in the case of curves with no inflection points, the crystal growth is relatively simpler than curves with inflections as follows.

Theorem 5 ([35]). *Let* $\mathbf{r} : S^1 \to \mathbb{R}^2$ *be a* C^∞ *immersion with its no-silhouette* $\mathcal{NS}_{\mathbf{r}} \neq \emptyset$*. Suppose that* $\mathbf{r}$ *has no inflection points. Then, for any two* θ_1, θ_2 *satisfying* $0 \leq \theta_1 < \theta_2 < \theta_{\mathbf{r}}$*, the following inclusion holds:*

$$\mathcal{NS}_{\theta_1,\mathbf{r}} \supset \mathcal{NS}_{\theta_2,\mathbf{r}}.$$

Figure 15 shows that in general it is impossible to expect the same property as Theorem 5 for a curve with inflection points.

3.3. The spherical polar transform

In Subsection 3.1, the spherical polar set played an essential role for investigating a Wulff shape. In Subsection 3.2, moving convex bodies (and their limits) in $\mathbb{R}^2$ were investigated. One of essential tools in Subsection 3.2 was treating any spherical polar set as the image of a non-empty closed subset by the map called the *spherical polar transform.* In [22], the spherical polar transform is investigated more and more by using the spherical method explained in Section 2. In this subsection, we survey results on the spherical polar transform obtained in [22].

Recall that $\mathcal{H}^\circ(S^{n+1})$ is the set consisting of non-empty closed subsets $\widetilde{W}$ of S^{n+1} such that $\widetilde{W}^\circ \neq \emptyset$ and that the spherical polar transform $\bigcirc : \mathcal{H}^\circ(S^{n+1}) \to \mathcal{H}^\circ(S^{n+1})$ is defined by $\bigcirc(\widetilde{W}) = \widetilde{W}^\circ$.

Lemma 3.2. *For any* $n \in \mathbb{N}$*, the mapping* $\bigcirc : \mathcal{H}^\circ(S^{n+1}) \to \mathcal{H}^\circ(S^{n+1})$ *is continuous.*

As we already announced in Subsection 3.2, Lemma 3.2 can be improved as follows.

Definition 14. Let $(X, d_X), (Y, d_Y)$ be metric spaces. A mapping $f : X \to Y$ is said to be *Lipschitz* if there exists a positive real number $K \in \mathbb{R}_+$ such that the following holds for any $x_1, x_2 \in X$:

$$d_Y(f(x_1), f(x_2)) \leq K d_X(x_1, x_2).$$

The positive real number $K \in \mathbb{R}_+$ for a Lipschitz mapping is called the *Lipschitz coefficient* of f.

Proposition 12 ([22]). *Let n be a positive integer. Then, the spherical polar transform $\bigcirc : \mathcal{H}^\circ(S^{n+1}) \to \mathcal{H}^\circ(S^{n+1})$ is Lipschitz with respect to the Pompeiu-Hausdorff metric.*

Proposition 12 suggests that the spherical polar transform $\bigcirc$: $\mathcal{H}^\circ(S^{n+1}) \to \mathcal{H}^\circ(S^{n+1})$ might have many nice properties. However, it is easily seen that $\bigcirc$ is neither injective nor surjective. Hence, in order to obtain nice properties of the spherical polar transform, we naturally reach restrictions of $\bigcirc$ to reasonable subspaces of $\mathcal{H}^\circ(S^{n+1})$.

Definition 15. (1) A subset $\widetilde{W}$ of S^{n+1} is called a *spherical convex body* if $\widetilde{W}$ is hemispherical, closed and spherical convex and it has an interior point.

(2) For any point P of S^{n+1}, let $\mathcal{H}_{\text{Wulff}}(S^{n+1}, P)$ be the following set:

$$\mathcal{H}_{\text{Wulff}}(S^{n+1}, P) = \left\{\widetilde{W} \in \mathcal{H}(S^{n+1}) \;\middle|\; \widetilde{W} \cap H(-P) = \emptyset, P \in \text{int}(\widetilde{W}), \right.$$
$$\left. \widetilde{W} \text{ is a spherical convex body}\right\},$$

where $\text{int}(\widetilde{W})$ stands for the set consisting of interior points of $\widetilde{W}$. The topological closure of $\mathcal{H}_{\text{Wulff}}(S^{n+1}, P)$ is denoted by $\overline{\mathcal{H}_{\text{Wulff}}(S^{n+1}, P)}$.

(3) Let $\mathcal{H}_{\text{S-conv}}(S^{n+1})$ be the set consisting of spherical convex closed sets. The topological closure of $\mathcal{H}_{\text{S-conv}}(S^{n+1})$ is denoted by $\overline{\mathcal{H}_{\text{S-conv}}(S^{n+1})}$.

Recall that $N \in S^{n+1}$ stands for the north pole $= (0, \ldots, 0, 1) \in S^{n+1}$. Notice that in the case $P = N$, $\mathcal{H}_{\text{Wulff}}(S^{n+1}, N)$ is nothing but the set consisting of spherical Wulff shapes defined in Subsection 2.5. Thus, we have the following characterization.

$$\mathcal{H}_{\text{Wulff}}(S^{n+1}, N) = \left\{\widetilde{\mathcal{W}_\gamma} = \alpha_N^{-1} \circ Id(\mathcal{W}_\gamma) \mid \gamma \in C^0(S^n, \mathbb{R}_+)\right\}.$$

The next proposition explains the relation between $\overline{\mathcal{H}_{\text{Wulff}}(S^{n+1}, P)}$ and $\overline{\mathcal{H}_{\text{S-conv}}(S^{n+1})}$.

Proposition 13 ([22]).

$$\bigcup_{P \in S^{n+1}} \overline{\mathcal{H}_{\mathrm{Wulff}}(S^{n+1}, P)} = \overline{\mathcal{H}_{\mathrm{S\text{-}conv}}(S^{n+1})}.$$

Proposition 14 ([22]). (1) *Let P be a point of S^{n+1}. Then, the set $\overline{\mathcal{H}_{\mathrm{Wulff}}(S^{n+1}, P)}$ is contained in $\mathcal{H}^{\circ}(S^{n+1})$.*

(2) *For any point $P \in S^{n+1}$, the following holds:*

$$\bigcirc(\mathcal{H}_{\mathrm{Wulff}}(S^{n+1}, P)) = \mathcal{H}_{\mathrm{Wulff}}(S^{n+1}, P).$$

(3) *For any point $P \in S^{n+1}$, the following holds:*

$$\bigcirc(\overline{\mathcal{H}_{\mathrm{Wulff}}(S^{n+1}, P)}) = \overline{\mathcal{H}_{\mathrm{Wulff}}(S^{n+1}, P)}.$$

(4) *For any point $P \in S^{n+1}$, the following restriction of $\bigcirc$ is injective:*

$$\bigcirc|_{\overline{\mathcal{H}_{\mathrm{Wulff}}(S^{n+1}, P)}} : \overline{\mathcal{H}_{\mathrm{Wulff}}(S^{n+1}, P)} \to \overline{\mathcal{H}_{\mathrm{Wulff}}(S^{n+1}, P)}.$$

(5) $\overline{\mathcal{H}_{\mathrm{S\text{-}conv}}(S^{n+1})} \subset \mathcal{H}^{\circ}(S^{n+1})$.

(6) $\bigcirc(\mathcal{H}_{\mathrm{S\text{-}conv}}(S^{n+1})) \neq \mathcal{H}_{\mathrm{S\text{-}conv}}(S^{n+1})$.

(7) $\bigcirc(\overline{\mathcal{H}_{\mathrm{S\text{-}conv}}(S^{n+1})}) = \overline{\mathcal{H}_{\mathrm{S\text{-}conv}}(S^{n+1})}$.

(8) *The restriction of $\bigcirc$ to $\overline{\mathcal{H}_{\mathrm{S\text{-}conv}}(S^{n+1})}$ is injective.*

By Proposition 14, we have the following:

Lemma 3.3 ([22]). *Each of the following is well-defined bijective mapping.*

$$\begin{aligned}
\bigcirc|_{\mathcal{H}_{\mathrm{Wulff}}(S^{n+1},P)} : \mathcal{H}_{\mathrm{Wulff}}(S^{n+1}, P) &\to \mathcal{H}_{\mathrm{Wulff}}(S^{n+1}, P), \\
\bigcirc|_{\overline{\mathcal{H}_{\mathrm{Wulff}}(S^{n+1},P)}} : \overline{\mathcal{H}_{\mathrm{Wulff}}(S^{n+1}, P)} &\to \overline{\mathcal{H}_{\mathrm{Wulff}}(S^{n+1}, P)}, \\
\bigcirc|_{\overline{\mathcal{H}_{\mathrm{S\text{-}conv}}(S^{n+1})}} : \overline{\mathcal{H}_{\mathrm{S\text{-}conv}}(S^{n+1})} &\to \overline{\mathcal{H}_{\mathrm{S\text{-}conv}}(S^{n+1})}.
\end{aligned}$$

Definition 16. Let $(X, d_X), (Y, d_Y)$ be metric spaces.

(1) A mapping $f : X \to Y$ is said to be *bi-Lipschitz* if f is bijective and there exist positive real numbers $K, L \in \mathbb{R}_+$ such that the following hold for any $x_1, x_2 \in X$ and any $y_1, y_2 \in Y$:

$$\begin{aligned}
d_Y(f(x_1), f(x_2)) &\leq K d_X(x_1, x_2), \\
d_X(f^{-1}(y_1), f^{-1}(y_2)) &\leq L d_Y(y_1, y_2),
\end{aligned}$$

(2) A mapping $f : X \to Y$ is called an *isometry* if f is bijective and the following holds for any $x_1, x_2 \in X$:

$$d_Y(f(x_1), f(x_2)) = d_X(x_1, x_2).$$

Theorem 6 ([22]). *Let P be a point of S^{n+1}. Then, with respect to the Pompeiu-Hausdorff metric, the following two hold:*

(1) *The restriction of $\bigcirc$ to $\mathcal{H}_{\mathrm{Wulff}}(S^{n+1}, P)$*

$$\bigcirc|_{\mathcal{H}_{\mathrm{Wulff}}(S^{n+1},P)} : \mathcal{H}_{\mathrm{Wulff}}(S^{n+1}, P) \to \mathcal{H}_{\mathrm{Wulff}}(S^{n+1}, P)$$

is an isometry.

(2) *The restriction of $\bigcirc$ to $\overline{\mathcal{H}_{\mathrm{Wulff}}(S^{n+1}, P)}$*

$$\bigcirc|_{\overline{\mathcal{H}_{\mathrm{Wulff}}(S^{n+1},P)}} : \overline{\mathcal{H}_{\mathrm{Wulff}}(S^{n+1}, P)} \to \overline{\mathcal{H}_{\mathrm{Wulff}}(S^{n+1}, P)}$$

is an isometry.

For any positive real number r, let D_r be the set consisting of $x \in \mathbb{R}^{n+1}$ satisfying $||x|| \leq r$. Then, D_r is a Wulff shape for any $r \in \mathbb{R}$ $(r > 0)$ and it is well-known that the dual Wulff shape of D_r is $D_{\frac{1}{r}}$. Moreover, it is easily seen that $h(D_{r_1}, D_{r_2}) = |r_1 - r_2|$ holds for any $r_1, r_2 \in \mathbb{R}$ $(r_1, r_2 > 0)$, where h is the Pompeiu-Hausdorff metric. Thus, it is impossible to expect the Euclidean counterpart of the assertion (1) of Theorem 6. This shows an advantage of studying the spherical version of Wulff shapes. Moreover, the Euclidean counterpart of $\overline{\mathcal{H}_{\mathrm{Wulff}}(S^{n+1}, P)}$ is not well-defined. This, too, shows an advantage of studying the spherical version of Wulff shapes.

Theorem 7 ([22]). *With respect to the Pompeiu-Hausdorff metric, the restriction of the spherical polar transform*

$$\bigcirc|_{\overline{\mathcal{H}_{\text{S-conv}}(S^{n+1})}} : \overline{\mathcal{H}_{\text{S-conv}}(S^{n+1})} \to \overline{\mathcal{H}_{\text{S-conv}}(S^{n+1})}$$

is bi-Lipschitz but never an isometry.

3.4. Convex integrands of class C^1

Recall that the following mapping

$$\mathcal{W} : C^0(S^n, \mathbb{R}_+) \to \mathcal{H}_{conv,\mathbf{0}}\left(\mathbb{R}^{n+1}\right)$$

defined by $\mathcal{W}(\gamma) = \mathcal{W}_\gamma$ has been introduced and properties of $\mathcal{W}$ has been studied in Section 1. Recall that $\mathcal{W}$ is surjective. Thus, for any convex body $W \subset \mathbb{R}^{n+1}$ containing the origin of $\mathbb{R}^{n+1}$ as an interior point of W, it follows that

$$\mathcal{W}^{-1}(W) \neq \emptyset.$$

Recall moreover that there uniquely exists the most efficient C^0 function, called the convex integrand of W and denoted by γ_W. By using the spherical method explained in Section 2, we can obtain a characterization of a C^1 convex integrand as follows.

Theorem 8 ([23]). *Let $W \subset \mathbb{R}^{n+1}$ be a convex body containing the origin of $\mathbb{R}^{n+1}$ as an interior point of W. Then, W is strictly convex if and only if its convex integrand γ_W is of class C^1.*

Remarks should be stated on Theorem 8. The first remark concerns [52]. Thanks to [52], the authors learned that Frank Morgan has shown a result in [51] which is similar as Theorem 8. His result is the following Theorem 9.

Definition 17. (1) Let W be a strictly convex convex body in $\mathbb{R}^{n+1}$ containing the origin of $\mathbb{R}^{n+1}$ as its interior point. Then, W is said to be *uniformly convex* if the mapping $f : S^n \to S^n$ defined by $f(\mathbf{n}) = \theta$ is Lipschitz, where $\mathbf{n} \in S^n$ is the unit normal vector of ∂W at the intersection $\mathbb{R}_+\theta \cap \partial W$.

(2) A continuous function $\gamma : S^n \to \mathbb{R}_+$ is said to be *of class $C^{1,1}$* if it is of class C^1 and all first partial derivatives are Lipschitz.

Theorem 9 ([51]). *Let $W \subset \mathbb{R}^{n+1}$ be a convex body containing the origin of $\mathbb{R}^{n+1}$ as an interior point of W. Then, W is uniformly convex if and only if its convex integrand γ_W is of class $C^{1,1}$.*

As Figure 16 shows, the notion of "uniform convexity" (resp., "class $C^{1,1}$") is actually stronger than "strict convexity" (resp., "class C^1"). Moreover, "strict convexity" (resp., "class C^1") is much more common, natural and easy to treat than "uniform convexity" (resp., "class $C^{1,1}$"). Thus, Theorem 8 is a more applicable and more desirable result than Theorem 9.

The next remark is that Theorem 8 is much stronger than Theorem 2 given in Subsection 3.1. This is explained as follows. Suppose that a C^1 function $\gamma : S^1 \to \mathbb{R}_+$ is given. Then, since inv: $\mathbb{R}^{n+1} - \{\mathbf{0}\} \to \mathbb{R}^{n+1} - \{\mathbf{0}\}$ is a C^∞ diffeomorphism, it follows that the boundary of the convex hull of inv(graph(γ)) is the graph of a C^1 function $S^n \to \mathbb{R}_+$ (for instance, refer to [65, 74]). By definition, we have that the convex integrand $\gamma_{\mathcal{W}_\gamma}$ is of class C^1. Therefore, by Theorem 8, the Wulff shape $\mathcal{W}_\gamma$ must be strictly convex. Hence, $\mathcal{W}_\gamma$ is never a polytope. Theorem 8 asserts that

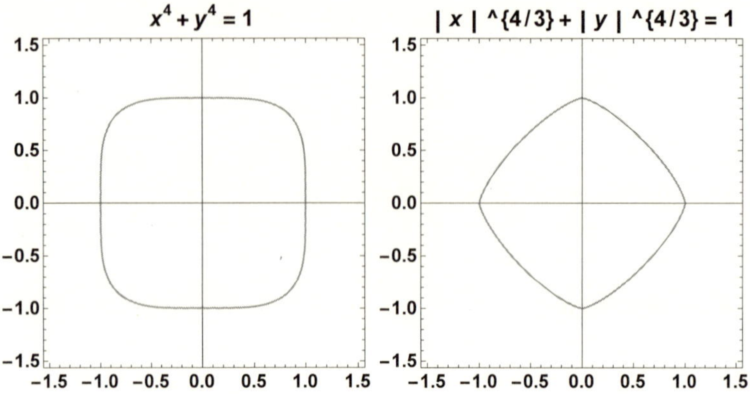

Fig. 16. These two examples are taken from [51]. The convex hull of left (resp., right) curve is called the unit L_4 (resp., $L_{4/3}$) ball. The unit L_4 ball is not uniformly convex but strictly convex and has $C^{1,1}$ boundary. On the other hand, the unit $L_{4/3}$ ball does not have $C^{1,1}$ boundary but does have C^1 boundary and is uniformly convex.

even the converse holds if γ is a convex integrand. Thus, Theorem 8 is much stronger than Theorem 2.

Moreover, Theorem 8 has many applications. For instance, we have the following as a direct application.

Corollary 1 ([23]). *A Wulff shape in $\mathbb{R}^{n+1}$ is strictly convex if and only if the boundary of its dual Wulff shape is C^1 diffeomorphic to S^n.*

In particular, we have the following:

Corollary 2 ([23]). *A Wulff shape in $\mathbb{R}^{n+1}$ is strictly convex and its boundary is C^1 diffeomorphic to S^n if and only if its dual Wulff shape is strictly convex and the boundary of it is C^1 diffeomorphic to S^n.*

It seems interesting to compare Corollary 2 and Theorem 1.

There is also an application of Theorem 8 from the viewpoint of pedal. Firstly, we recall the definition of pedal given in Subsection 2.1.

Definition 2. Let $\Phi : S^n \to \mathbb{R}^{n+1} - \{\mathbf{0}\}$ be a C^1 embedding. Then, the *pedal* relative to the *pedal point* $\mathbf{0}$ for Φ, denoted by $ped_{\Phi,\mathbf{0}} : S^n \to \mathbb{R}^{n+1}$, is the mapping which maps $\theta \in S^n$ to the unique nearest point of $\Phi(\theta) + T_{\Phi(\theta)}\Phi(S^n)$ from the origin $\mathbf{0}$, where $\Phi(\theta) + T_{\Phi(\theta)}\Phi(S^n)$ stands for the affine tangent hyperplane to $\Phi(S^n)$ at $\Phi(\theta)$.

Let $W \subset \mathbb{R}^{n+1}$ be a Wulff shape. Suppose that ∂W is C^1 diffeomorphic to S^n. Then, ∂W may be regarded as the image of a certain C^1 embedding $F : S^n \to \mathbb{R}^{n+1} - \{\mathbf{0}\}$, and γ_W is exactly the pedal of ∂W relative to the origin. Theorem 8 gives a sufficient condition for the pedal of ∂W relative to the origin to be smooth:

Corollary 3 ([23]). *Suppose that a Wulff shape W in $\mathbb{R}^{n+1}$ is strictly convex and its boundary is C^1 diffeomorphic to S^n. Then, $ped_{\Phi,\mathbf{0}} : S^n \to \mathbb{R}^{n+1}$ is of class C^1.*

3.5. Wulff shapes with C^1 boundary

The following result is an application of Theorem 8 as well. Thus, Theorem 10, too, may be regarded as a result of the spherical method.

Theorem 10. *Let $\gamma : S^n \to \mathbb{R}_+$ be a continuous function and let $\mathcal{W}_\gamma$ be the Wulff shape associated with γ. Suppose that the boundary of $\mathcal{W}_\gamma$ is a C^1 submanifold. Then, γ must be the convex integrand of $\mathcal{W}_\gamma$. In other words, the following holds.*

$$\mathcal{W}^{-1}(\mathcal{W}_\gamma) = \{\gamma\},$$

where $\mathcal{W} : C^0(S^n, \mathbb{R}_+) \to \mathcal{H}_{conv,\mathbf{0}}\left(\mathbb{R}^{n+1}\right)$. is the mapping defined by $\mathcal{W}(\gamma) = \mathcal{W}_\gamma$.

Theorem 10 asserts that the given continuous function γ is uniquely determined only bythe shape of crystal if the boundary of $\mathcal{W}_\gamma$ is of class C^1. In other words, Theorem 10 is equivalet to say that the following inclusion holds, where $CI(S^n, \mathbb{R}_+)$ is the subset of $C^0(S^n, \mathbb{R}_+)$ consisting of convex integrands and $\mathcal{H}^1_{conv,\mathbf{0}}(\mathbb{R}^{n+1})$ is the set of convex bodies W in $\mathbb{R}^{n+1}$ such that the origin of $\mathbb{R}^{n+1}$ is an interior point of W and the boundary of W is a C^1 submanifold.

$$CI(S^n, \mathbb{R}_+) \supset \mathcal{W}^{-1}(\mathcal{H}^1_{conv,\mathbf{0}}(\mathbb{R}^{n+1})).$$

The converse of Theorem 10 does not hold in general except for $n = 1$.

Proof. Define $\gamma_{\mathcal{W}_\gamma} : S^n \to \mathbb{R}_+$ so that the graph of $\gamma_{\mathcal{W}_\gamma}$ satisfies the following:

$$\Gamma_\gamma = \mathrm{inv}(\mathrm{graph}(\gamma_{\mathcal{W}_\gamma})).$$

Here, as defined in Section 1, Γ_γ is the boudary of the convex hull of $\mathrm{inv}(\mathrm{graph}(\gamma))$. Then, $\gamma_{\mathcal{W}_\gamma}$ is the convex integrand of $\mathcal{W}_\gamma$, and the Wulff shape associated with $\gamma_{\mathcal{W}_\gamma}$ is exactly $\mathcal{W}_\gamma$.

Suppose that $\gamma_{\mathcal{W}_\gamma} \neq \gamma$. Then, there exists at least one point $x \in \Gamma_\gamma$ such that $x \notin \mathrm{inv}(\mathrm{graph}(\gamma))$. Since Γ_γ is the boundary of the convex hull of inv(graph(γ)), this implies that there exist two distinct points x_1, x_2 in $\Gamma_\gamma \cap$ inv(graph(γ)) and a real number t $(0 < t < 1)$ such that $x = (1-t)x_1 + tx_2$.

On the other hand, since the boundary of $\mathcal{W}_\gamma$ is a C^1 submanifold, by Theorem 8, the convex hull of inv(graph(γ)) must be strictly convex. Hence, we have a contradiction. Q.E.D.

3.6. Self-dual Wulff shapes and spherical Wulff shapes of constant width $\pi/2$

Recall that a subset $\widetilde{W}$ of S^{n+1} is called a *spherical convex body* if $\widetilde{W}$ is closed and spherical convex and it has an interior point (see Definition 7 in Subsection 2.3).

Definition 18. Let $\widetilde{W} \subset S^{n+1}$ be a spherical convex body. Then, a hemisphere $H(P)$ is said to *support* $\widetilde{W}$ if $H(P)$ contains $\widetilde{W}$ and there exists a point Q of the boundary of $\widetilde{W}$ which is contained in the boundary of $H(P)$.

For a spherical convex body $\widetilde{W}$ and a hemisphere $H(P)$ supporting $\widetilde{W}$, following [37, 38], the width of $\widetilde{W}$ determined by $H(P)$ is defined as follows. For any two $P, Q \in S^{n+1}$ $(P \neq \pm Q)$, the intersection $H(P) \cap H(Q)$ is called a *lune* of S^{n+1}. The *thickness of the lune* $H(P) \cap H(Q)$, denoted by $\triangle(H(P) \cap H(Q))$, is the real number $\pi - |PQ|$, where $|PQ|$ stands for the length of the arc PQ. For a spherical convex body $\widetilde{W}$ and a hemisphere $H(P)$ supporting $\widetilde{W}$, the *width of* $\widetilde{W}$ *determined by* $H(P)$, denoted by $\mathrm{width}_{H(P)}\widetilde{W}$, is the minimum of the following set:

$$\left\{\triangle(H(P)\cap H(Q)) \;\middle|\; \widetilde{W} \subset H(P)\cap H(Q),\, H(Q) \text{ supports } \widetilde{W}\right\}.$$

For any $\rho \in \mathbb{R}_+$ less than π, a spherical convex body $\widetilde{W} \subset S^{n+1}$ is said to be *of constant width* ρ if $\mathrm{width}_{H(P)}\widetilde{W} = \rho$ for any $H(P)$ supporting $\widetilde{W}$.

Let $\mathcal{W}_\gamma \subset \mathbb{R}^{n+1}$ be a Wulff shape. In Subsection 2.5, the following set was called the spherical Wulff shape induced by $\mathcal{W}_\gamma$.

$$\widetilde{\mathcal{W}}_\gamma = \alpha_N^{-1} \circ Id\,(\mathcal{W}_\gamma)\,.$$

Here, $Id : \mathbb{R}^{n+1} \to \mathbb{R}^{n+1} \times \{1\} \subset \mathbb{R}^{n+2}$ is the mapping defined by $Id(x) = (x, 1)$, $N \in S^{n+1}$ is the north pole of S^{n+1} and $\alpha_N : S^{n+1} - H(-N) \to \mathbb{R}^{n+1} \times \{1\} \subset \mathbb{R}^{n+2}$ is the central projection defined as follows.

$$\alpha_N(P_1, \ldots, P_{n+1}, P_{n+2}) = \left(\frac{P_1}{P_{n+2}}, \ldots, \frac{P_{n+1}}{P_{n+2}}, 1\right)$$
$$(\forall (P_1, \ldots, P_{n+1}, P_{n+2}) \in S^{n+1} - H(-N)).$$

Then, for any $\mathcal{W}_\gamma$, it is clear that the spherical Wulff shape induced by $\mathcal{W}_\gamma$ is a spherical convex body.

Definition 19. A Wulff shape $\mathcal{W}_\gamma$ is said to be *self-dual* if the following equality holds.

$$\widetilde{\mathcal{W}_\gamma} = \mathcal{D}\widetilde{\mathcal{W}_\gamma} = \left(\widetilde{\mathcal{W}_\gamma}\right)^\circ.$$

Theorem 11 ([24]). *Let $\gamma : S^n \to \mathbb{R}_+$ be a continuous function. Then, the Wulff shape W_γ is self-dual if and only if the spherical Wulff shape induced by W_γ is of constant width $\pi/2$.*

The unit disc $D^{n+1} = \{x \in \mathbb{R}^{n+1} \mid ||x|| \leq 1\}$ of $\mathbb{R}^{n+1}$ is clearly self-dual. Let R be a rotation of $\mathbb{R}^{n+2}$ about an n dimensional linear subspace with a small angle. Then, since the property of constant width is an invariant property by R, by Theorem 11, $Id^{-1} \circ \alpha_N\left(R\left(\alpha_N^{-1} \circ Id(D^{n+1})\right)\right)$ is self-dual as well (see Figure 17). Moreover, let $\widetilde{\triangle}$ be a spherical triangle of constant width $\pi/2$ in S^2 containing N as an interior point. Then, by Theorem 11, not only $Id^{-1} \circ \alpha_N\left(\widetilde{\triangle}\right)$ itself, but also any $Id^{-1} \circ \alpha_N\left(R\left(\widetilde{\triangle}\right)\right)$ is self-dual (see Figure 18). For more consideration on simple, explicit examples, see [24].

On the other hand, any Reuleaux triangle in $\mathbb{R}^2$ containing the origin as an interior point (see Figure 19) is not a self-dual Wulff shape, although it is a Wulff shape of constant width in $\mathbb{R}^2$. This is because any Reuleaux triangle is strictly convex, and thus the boundary of it must be smooth by Theorem 8 if it is self-dual. However, there are three non-smooth points for any Reuleaux triangle in $\mathbb{R}^2$. By Theorem 11, its spherical convex body is not of constant width $\pi/2$.

3.7. Stability of C^∞ convex integrands

For the proof of the main theorem of this subsection (Theorem 13), the spherical method is not required. However, since the results in Subsection 3.8 may be regarded as the next step of Theorem 13 and they heavily depends on the spherical method, for the sake of readers' convenience, stability of C^∞ convex integrands is quickly reviewed in this subsection.

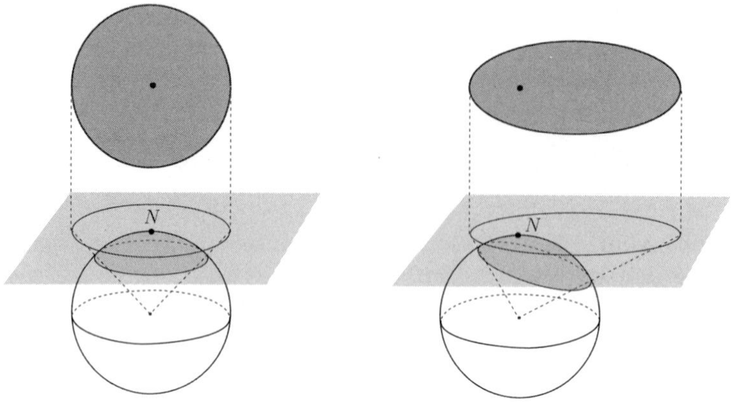

Fig. 17. Self-dual Wulff shapes include central projections of spherical caps of width $\pi/2$.

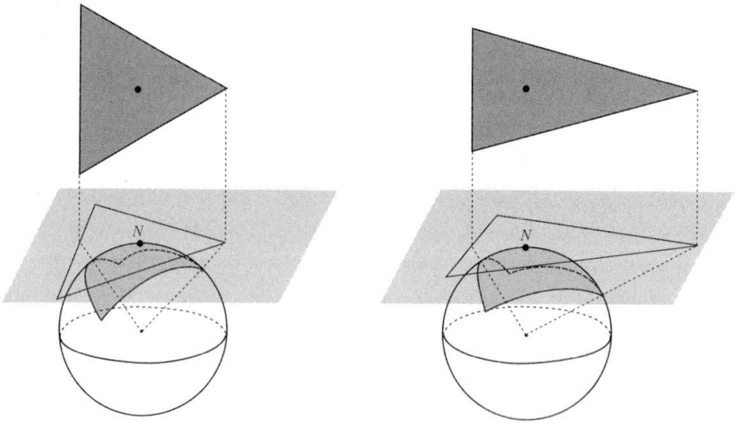

Fig. 18. Self-dual Wulff shapes include triangles which are central projections of constant-width spherical triangles of width $\pi/2$.

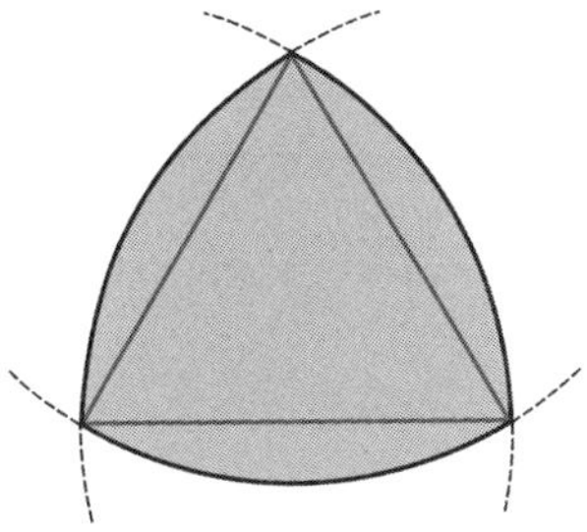

Fig. 19. Reuleaux triangle.

We start to review the definition of Whitney C^∞ topology. Let m, p be positive integers and let M (resp., P) be a C^∞ manifold of dimension m (resp., dimension p). For any positive integer r, following [21], we define the r jet bundle $J^r(M, P)$ as follows.

Definition 20 ([21]). Let x be a point of M. Let $f, g : M \to P$ be C^∞ mappings.

(1) f has *zero order contact with* g at x if there exists a point $y \in P$ such that $f(x) = g(x) = y$.

(2) Suppose that f has zero order contact with g at x. Then, f has *first order contact with* g at x if $(df)_x = (dg)_x$ as mappings of $T_xM \to T_yP$. This is written as $f \sim_0 g$ at x.

(3) f has *r-th order contact with* g at x if $(df) : TM \to TP$ has $(r-1)$-th order contact with g at every point of T_xM. This is written as $f \sim_r g$ at x.

(4) Let $J^0(M, P)_{x,y}$ denote the set consisting of equivalence classes under "$\sim_0$ at x" of C^∞ mappings $f : M \to P$.

(5) Let $J^r(M, P)_{x,y}$ denote the set consisting of equivalence classes under "$\sim_r$ at x" of C^∞ mappings $f : M \to P$ where $f(x) = y$.

(6) Let $J^r(M, P) = \bigcup_{(x,y)\in M\times P} J^r(M, P)_{x,y}$ (disjoint union). An element σ in $J^r(M, P)$ is called a *r-jet of mappings* (or just a *r-jet*) from M to P.

(7) Given a C^∞ mapping $f : M \to P$, the mapping $j^r : M \to J^r(M, P)$, called the *r-jet extension of* f, is defined by $j^r f(x) =$ equivalence class of f in $J^r(M, P)_{x,f(x)}$.

Definition 21 ([21]). (1) Denote by $C^\infty(M, P)$, the set consisting of C^∞ mappings from M to P.

(2) Fix a non-negative integer r. Let U be a subset of $J^r(M,P)$. Then, denote by $V(U)$ the set

$$\{f \in C^\infty(M,P) \mid j^r f(M) \subset U\}.$$

Notice that $V(U_1) \cap V(U_2) = V(U_1 \cap U_2)$.

(3) The family of sets $\{V(U)\}$ form a basis for a topology on $C^\infty(M,P)$, where U is an open subset of $J^r(M,P)$. This topology is called the *Whitney C^r topology.* Denote by W_r the set of open subsets of $C^\infty(M,P)$ in the Whitney C^r topology.

(4) The *Whitney C^∞ topology* on $C^\infty(M,P)$ is the topology whose basis is $W = \bigcup_{r=0}^\infty W_r$. This is a well-defined basis since $W_{r_1} \subset W_{r_2}$ whenever $r_1 \le r_2$.

Definition 22 ([21]). (1) Let f, g be elements of $C^\infty(M,P)$. Then f is said to be *$\mathcal{A}$-equivalent to g* if there exist C^∞ diffeomorphisms $h : M \to M$ and $H : P \to P$ such that the equality $f = H^{-1} \circ g \circ h$ holds.

(2) Let f be in $C^\infty(M,P)$. Then, f is said to be *stable* if there exists a neighborhood W_f of f in $C^\infty(M,P)$ such that each g in W_f is $\mathcal{A}$-equivalent to f.

(3) Denote the set consisting of stable mappings $f : M \to P$ by $S^\infty(M,P)$.

For more details on Jet bundles, the Whitney C^∞ topology and stable mappings, for instance refer to [5, 21]. By definition, it is clear that $S^\infty(M,P)$ is an open subset of $C^\infty(M,P)$.

Next, we review Mather's well-known answer to the structual stability problem posed by R. Thom in [71].

Definition 23. Let M, P be C^∞ manifolds.

(1) A C^∞ mapping $f : M \to P$ is said to be *proper* if $f^{-1}(C)$ is compact for any compact subset $C \subset P$.

(2) The subset of $C^\infty(M,P)$ consisting of proper mappings is denoted by $C^\infty_{\mathrm{pro}}(M,P)$.

Question 12. Is the intersection $S^\infty(M,P) \cap C^\infty_{\mathrm{pro}}(M,P)$ dense in $C^\infty_{\mathrm{pro}}(M,P)$?

Question 12 was completely solved by J. Mather ([45]). Mather's answer has the following surprizing form.

Theorem 12 ([45]). *Suppose that $C^\infty_{\mathrm{pro}}(M,P)$ is not empty. Then, $S^\infty(M,P) \cap C^\infty_{\mathrm{pro}}(M,P)$ is dense in $C^\infty_{\mathrm{pro}}(M,P)$ if and only if the dimension-pair (m,p) satisfies one of the following conditions.*

(1) $m < \frac{6}{7}p + \frac{8}{7}$ *and* $p - m \geq 4$.
(2) $m < \frac{6}{7}p + \frac{9}{7}$ *and* $3 \geq p - m \geq 0$.
(3) $p < 8$ *and* $p - m = -1$.
(4) $p < 6$ *and* $p - m = -2$.
(5) $p < 7$ *and* $p - m \leq -3$.

Notice that $M = S^n$ and $P = \mathbb{R}_+$ in our case. Thus, $(m, p) = (m, 1)$ in our case. In this case, one of the conditions (2)–(5) of Theorem 12 is satisfied. Moreover, since S^n is compact, any C^∞ function $\gamma : S^n \to \mathbb{R}_+$ is proper. Hence, we have the following.

Corollary 4 ([45]). *Let n be a positive integer. Then, the open subset $S^\infty(S^n, \mathbb{R}_+)$ is dense in $C^\infty(S^n, \mathbb{R}_+)$.*

For proper C^∞ mappings of special type, it is natural to ask the similar question, namely to ask "Are generic proper mappings of special type stable ? ". Such investigations, for instance, can be found in [46] for generic projections of submanifolds, in [12] for generic projections of stable mappings and in [26, 27, 28, 29, 30] for generic distance-squared mappings and their generalizations. Motivated by these researches, we posed Question 9 in Section 1. We have a complete answer to Question 9 as follows.

Theorem 13 ([7]). *The open subset $S^\infty(S^n, \mathbb{R}_+) \cap C^\infty_{\mathrm{conv}}(S^n, \mathbb{R}_+)$ is dense in $C^\infty_{\mathrm{conv}}(S^n, \mathbb{R}_+)$.*

In order to prove Theorem 13, it is better to replace an element of $S^\infty(S^n, \mathbb{R}_+)$ with an easily treatable function. Thus, we want to have a geometric characterization of an element of $S^\infty(S^n, \mathbb{R}_+)$. Mather's celebrated geometric characterization of a proper stable mapping can be found in [44]. Notice that any C^0 function $S^n \to \mathbb{R}_+$ is proper. Thus, it is easily seen that the following well-known geometric characterization of a stable function $S^n \to \mathbb{R}_+$ is derived from the geometric characterization given in [44].

Proposition 15 ([44]). *Let $\gamma : S^n \to \mathbb{R}_+$ be a C^∞ function. Then, γ is stable if and only if all critical points of γ are non-degenerate and $\gamma(\theta_1) \neq \gamma(\theta_2)$ is satisfied for any two distinct critical points $\theta_1, \theta_2 \in S^n$.*

Notice that a proper stable function seems to be usually called a *Morse function.* However, a Morse function defined (for instance) in [21] or [49] is a C^∞ function having only non-degenerate critical points, and thus it is a weaker notion than the notion of stable function in the sense of Mather. Therefore, in order to avoid unnecessary confusion, stable functions and Morse functions are distinguished in this survey article.

Definition 24 ([49]). Let $\gamma : S^n \to \mathbb{R}_+$ be a C^∞ function and let $\theta \in S^n$ be a non-degenerate critical point of γ. Then, there exists a coordinate neighbourhood (U, φ) of θ such that $\varphi(\theta) = \mathbf{0}$ and the following equality holds:

$$\gamma \circ \varphi^{-1}(x_1, \ldots, x_n) = \gamma(\theta) - x_1^2 - \cdots - x_i^2 + x_{i+1}^2 + \cdots + x_n^2.$$

The integer i $(0 \le i \le n)$ does not depend on the particular choice of the coordinate neighbourhood (U, φ) and it is called the *Morse index of* γ *at* θ.

3.8. Simultaneous stability of C^∞ convex integrands and their duals

Let $\gamma : S^n \to \mathbb{R}_+$ be a convex integrand. Recall that the dual convex integrand of γ is $\delta : S^n \to \mathbb{R}_+$ such that the following holds (see Section 1).

$$\mathrm{inv}(\mathrm{graph}(\delta)) = \partial \mathcal{W}_\gamma.$$

In this subsection, results on simultaneous stability of γ and δ are summarized.

Define the functions $\widehat{\gamma}, \widehat{\delta} : S^n \to \mathbb{R}_+$ by

$$\widehat{\gamma}(\theta) = \frac{1}{\gamma(-\theta)} \quad \text{and} \quad \widehat{\delta}(\theta) = \frac{1}{\delta(-\theta)} \qquad (\forall \theta \in S^n)$$

respectively. By Proposition 8 of Subsection 2.5, we have the following.

$$\partial \mathcal{D}\mathcal{W}_\gamma = \partial \mathcal{W}_\delta = \mathrm{inv}(\mathrm{graph}(\gamma)) = \mathrm{graph}(\hat{\gamma}).$$

Notice that for any convex integrand $\gamma : S^n \to \mathbb{R}_+$, the Wulff shape $\mathcal{W}_\gamma$ is a convex body such that the origin is contained in it's interior. In Convex Body Theory, there is the notion of *dual* for a convex body containing the origin as an interior point. Namely, in Convex Body Theory, the boundary of the *dual* of $\mathcal{W}_\gamma$ is the following set (see for example [67]).

$$\left\{ \left(\theta, \frac{1}{\gamma(\theta)} \right) \,\middle|\, \theta \in S^n \right\}.$$

However, the notion of dual in this sense seems to have less relations with the notion of pedal which is very important for this article and a common background in Physics (for instance, see [63]). On the other hand, the notion of dual Wulff shapes in our sense is closely related to the notion of pedal. Moreover, via the central projection, the pedal of C^∞ embedding $\Phi : S^n \to \mathbb{R}^{n+1} - \{\mathbf{0}\}$ defined by $\Phi_\delta(\theta) = \left(\theta, \hat{\delta}(\theta)\right)$

relative to the origin is characterized by using the spherical dual of the corresponding embedding $\widetilde{\Phi} : S^n \to S^{n+1}$; and the spherical dual is a well-known notion in Singularity Theory (for details, see Subsection 2.2). Since the notion of pedal is very important for our study, we adopt $\mathcal{DW}_\gamma$ as the notion of dual Wulff shape of $\mathcal{W}_\gamma$.

For the simultaneous stability of C^∞ convex integrands and their duals, we have the following:

Theorem 14 ([8]). *Let $\gamma : S^n \to \mathbb{R}_+$ be a stable convex integrand and let δ be the dual convex integrand of γ. Then, δ is stable.*

In order to prove Theorem 14, the following three assertions are needed.

Define the function $\widetilde{\gamma} : S^n \to \mathbb{R}_+$ by $\widetilde{\gamma}(\theta) = d(\widetilde{\Psi}_{\widetilde{\gamma}}(\theta), N)$.

Lemma 3.4 ([8]). *There exists a point θ of S^n which is a degenerate critical point of $\widetilde{\gamma}$ if and only if the north pole N is contained in the spherical caustic of $\widetilde{\Psi}_{\widetilde{\gamma}}$.*

For the proof of Lemma 3.4, we need the well-known fact that the singular set $S(\Phi)$ of the following mapping Φ is an $(n+1)$-dimensional C^∞ submanifold of $S^n \times S^{n+1}_{N,+}$ (see for instance [64, 66]).

$$\begin{aligned} \Phi : S^n \times S^{n+1}_{N,+} &\to \mathbb{R} \times S^{n+1}_{N,+} \\ (\theta, P) &\mapsto \left(d\left(\widetilde{\Psi}_{\widetilde{\gamma}}(\theta), P\right), P\right). \end{aligned}$$

Lemma 3.5 ([8]). *The origin $\mathbf{0}$ is a point of symmetry set of $\partial\mathcal{W}$ if and only if the north polar N is a point of spherical symmetry set of $\partial\widetilde{\mathcal{W}}$.*

Remark 3.1. *Notice that $\gamma(\theta) = \tan(\widetilde{\gamma}(\theta))$ for any $\theta \in S^n$. Since the function $\tan : (0, \pi/2) \to \mathbb{R}_+$ is a C^∞ diffeomorphism, it follows that θ is a non-degenerate critical point of γ if and only if θ is a non-degenerate critical point of $\widetilde{\gamma}$.*

Proposition 16 ([8]). *Let $\mathcal{W}_\gamma$ be the Wulff shape associated with γ and let $\widetilde{\mathcal{W}}_\gamma$ be the spherical Wulff shape associated with $\mathcal{W}_\gamma$. Then the following holds:*

(1) *Spherical-Caust* $\left(\partial\widetilde{\mathcal{W}_\gamma}\right) =$ *Spherical-Caust* $\left(\partial\widetilde{\mathcal{DW}_\gamma}\right)$.

(2) *Spherical-Sym* $\left(\partial\widetilde{\mathcal{W}_\gamma}\right) =$ *Spherical-Sym* $\left(\partial\widetilde{\mathcal{DW}_\gamma}\right)$.

Proposition 16 is one of merits of the spherical method because the corresponding Euclidean properties do not hold in general.

By Theorem 14, we have the following corollary.

Corollary 5 ([8]). *Let $\gamma : S^n \to \mathbb{R}_+$ be a convex integrand and let $\delta : S^n \to \mathbb{R}_+$ be the dual convex integrand of γ. Then, the following are equivalent.*

(1) *The convex integrand γ is stable.*
(2) *The convex integrand δ is stable.*
(3) *The function $\widehat{\gamma}$, whose graph is exactly $\partial\mathcal{W}_\delta = \partial\mathcal{DW}_\gamma$, is stable.*
(4) *The function $\widehat{\delta}$, whose graph is exactly $\partial\mathcal{W}_\gamma = \partial\mathcal{DW}_\delta$, is stable.*

Next, more detailed dual relationships for stable convex integrands γ, δ are explained.

Theorem 15 ([8]). *Let $\gamma : S^n \to \mathbb{R}_+$ be a stable convex integrand and let $\delta : S^n \to \mathbb{R}_+$ be the dual convex integrand of γ. Then, the following hold:*

(1) *A point $\theta_0 \in S^n$ is a non-degenerate critical point of γ if and only if $-\theta_0 \in S^n$ is a non-degenerate critical point of δ.*
(2) *Suppose that $\theta_0 \in S^n$ is a non-degenerate critical point of γ. Then, for any i $(0 \le i \le n)$, the Morse index of γ at θ_0 is i if and only if the Morse index of δ at $-\theta_0$ is $(n-i)$.*

For the proof of Theorem 15, not only the spherical method, but also Andrews formulas ([1]), which is explained quickly below, is needed.

Let $\gamma : S^n \to \mathbb{R}_+$ be a C^∞ convex integrand and let $\delta : S^n \to \mathbb{R}_+$ be the dual convex integrand of γ. Notice that $\partial\mathcal{W}_\gamma$ (resp., $\partial\mathcal{DW}_\gamma$) is the image of the embedding Φ_δ (resp., Φ_γ) defined by $\Phi_\delta(\theta) = \left(\theta, \widehat{\delta}(\theta)\right)$ (resp., $\Phi_\gamma(\theta) = (\theta, \widehat{\gamma}(\theta))$). Define the mapping $h_\delta : S^n \to S^n$ (resp., $h_\gamma : S^n \to S^n$) so that $\gamma(\theta)$ (resp., $\delta(\theta)$) is the perpendicular distance from the origin $\mathbf{0}$ to the affine tangent hyperplane to $\Phi_\delta(S^n)$ (resp., $\Phi_\gamma(S^n)$) at $\Phi_\delta(h_\delta(\theta))$ (resp., $\Phi_\gamma(h_\gamma(\theta))$). Then, the following holds.

Proposition 17 ([8]).

(1) *Both of h_δ and h_γ are C^∞ diffeomorphisms.*
(2) *Both of $\Phi_\delta(S^n)$ and $\Phi_\gamma(S^n)$ are strictly locally convex.*

Therefore, by Andrews [1], the following equalities hold for any $\theta \in S^n$ (see Figure 20 where T_1 (resp., T_2) denotes the affine tangent hyperplane to $\Phi_\delta(S^n)$ (resp., $\Phi_\gamma(S^n)$) at $\Phi_\delta(h_\delta(\theta))$ (resp., $\Phi_\gamma(h_\gamma(\theta))$)). These two equalities are called *Andrews's formulas.*

$$\begin{aligned}\Phi_\delta(h_\delta(\theta)) &= \gamma(\theta)\theta + \nabla\gamma(\theta),\\ \Phi_\gamma(h_\gamma(\theta)) &= \delta(\theta)\theta + \nabla\delta(\theta).\end{aligned}$$

Here, $\nabla\gamma(\theta)$ (resp., $\nabla\delta(\theta)$) stands for the gradient vector of γ (resp., δ) at $\theta \in S^n$ with respect to the standard metric on S^n.

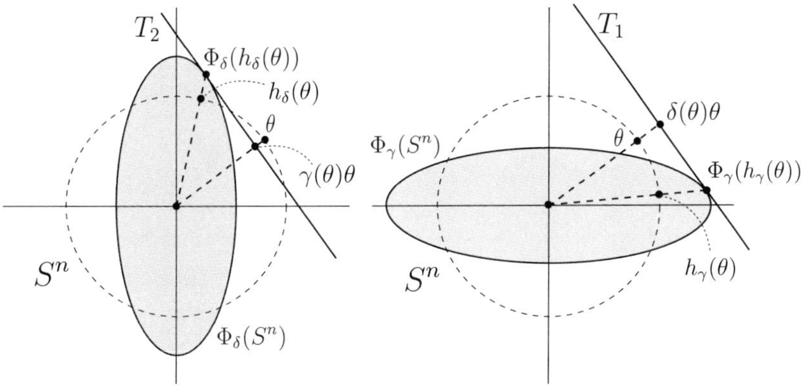

Fig. 20. Andrews's formulas

It is clear that, by Theorem 15, we have the following corollary.

Corollary 6 ([8]). *Let $\gamma : S^n \to \mathbb{R}_+$ be a stable convex integrand and let $\delta : S^n \to \mathbb{R}_+$ be the dual convex integrand of γ. Moreover, let θ_0 be a point of S^n and let i be an integer such that $0 \le i \le n$. Then, the following are equivalent.*

(1) *The point $\theta_0 \in S^n$ is a non-degenerate critical point of γ with Morse index i.*

(2) *The point $-\theta_0 \in S^n$ is a non-degenerate critical point of δ with Morse index $(n - i)$.*

(3) *The point $-\theta_0 \in S^n$ is a non-degenerate critical point of $\widehat{\gamma}$ with Morse index $(n - i)$.*

(4) *The point $\theta_0 \in S^n$ is a non-degenerate critical point of $\widehat{\delta}$ with Morse index i.*

§4. Questions

In this section, we pose questions related to the topics in Section 3. To the best of authors' knowledge, all questions posed in this section except for Questions 14, 15 seem to be open.

4.1. On aperture of higher dimensional hypersurfaces

Question 13. *Let M be a compact C^∞ manifold of dimension n. Then,* is it possible to obtain similar results as Theorems 3 –5 for a C^∞ immersion $\mathbf{r} : M \to \mathbb{R}^{n+1}$ with $n \geq 2$?

In our setting, at the initial stage(= before rotating all tangent spaces), the no-silhouette of $\mathbf{r}$ must exists, namely the set

$$\mathcal{NS}_{\mathbf{r}} = \mathbb{R}^{n+1} - \bigcup_{s \in M} \left(\mathbf{r}(s) + d\mathbf{r}_s \left(T_s M\right)\right)$$

must be non-empty.

Proposition 18 ([59])**.** *Let M be a closed C^∞ manifold of dimension n, and let $\mathbf{r} : M \to \mathbb{R}^{n+1}$ be a C^∞ immersion with its no-silhouette. Suppose that $n \geq 2$. Then, the immersion $\mathbf{r}$ must be an embedding and M must be C^∞ diffeomorphic to S^n.*

Firstly, we examine the case $n = 2$. By this proposition, our initial situation is as follows:
Given a C^∞ embedding $S^2 \to \mathbb{R}^3$ such that

$$\mathcal{NS}_{\mathbf{r}} = \mathbb{R}^3 - \bigcup_{s \in S^2} \left(\mathbf{r}(s) + d\mathbf{r}_s \left(T_s(S^2)\right)\right) \neq \emptyset.$$

Next, from the initial situation, we would like to rotate all tangent planes about their rotation axes simultaneously with the same angle. So, the following question arises:

Question 14. Is it possible to choose rotation axes depending on $s \in S^2$ smoothly ?

The answer to this question is "NO" by the following well-known result due to H. Poincaré.

Proposition 19 (H. Poincaré)**.** *There is no C^∞ non-zero vector field on S^2.*

As for Poincaré's result, for instance refer to [50]. Secondly, we consider the case $n \geq 3$. By Proposition 18, our initial situation is as follows: "Given a C^∞ embedding $S^n \to \mathbb{R}^{n+1}$ such that

$$\mathcal{NS}_{\mathbf{r}} = \mathbb{R}^{n+1} - \bigcup_{s \in S^n} \left(\mathbf{r}(s) + d\mathbf{r}_s \left(T_s(S^n)\right)\right) \neq \emptyset,$$

where $n \geq 3$."

Next, from the initial situation, we would like to rotate all tangent spaces (n-dimensional vector spaces $d\mathbf{r}_s \left(T_s(S^n)\right)$) about their rotation axes ($(n-1)$-dimensional vector spaces) simultaneously with the same angle. Then, we may ask the same question as in the case $n = 2$.

Question 15. Is it possible to choose rotation axes depending on $s \in S^n$ smoothly ?

Fortunately, by the celebrated result due to W. P. Thurston (Theorem 16), we already have the complete answer to Question 15.

Theorem 16 ([72]). *Let M be a compact C^∞ manifold of dimension n. Then, the following hold.*

(1) *The given manifold M has a C^∞ codimension one foliation if and only if $\chi(M) = 0$, where $\chi(M)$ denotes the Euler characteristic of M.*
(2) *Every $(n-1)$-plane field τ^{n-1} on the given M is homotopic to the tangent plane field of a C^∞ codimension one foliation.*

Corollary 7. *It is possible to choose rotation axes depending on $s \in S^n$ smoothly if and only if n is odd.*

Let $\mathbf{r} : S^{2k+1} \to \mathbb{R}^{2k+2}$ be a C^∞ embedding such that $\mathcal{NS}_{\mathbf{r}} \neq \emptyset$. Moreover, we let $\mathcal{F}$ be a C^∞ codimension one foliation of S^{2k+1}. Then, the following *rotated no-silhouette* is well-defined:

$$\mathcal{NS}_{\theta,\mathbf{r},\mathcal{F}} = \mathbb{R}^{2k+2} - \bigcup_{s \in S^{2k+1}} \left(\mathbf{r}(s) + R_{\theta,L_s} d\mathbf{r}_s \left(T_s(S^{2k+1})\right)\right),$$

where L_s is the leaf of $\mathcal{F}$ containing $s \in S^{2k+1}$ and

$$R_{\theta,L_x} : \mathbb{R}^{2k+2}/d\mathbf{r}_s(T_s(L_s)) \to \mathbb{R}^{2k+2}/d\mathbf{r}_s(T_s(L_s))$$

is the rotation defined by $R_{\theta,L_x}([x]) = [x]\begin{pmatrix} \cos\theta & \sin\theta \\ -\sin\theta & \cos\theta \end{pmatrix}$. Thus, we may ask

Question 16. How is $\mathcal{NS}_{\theta,\mathbf{r},\mathcal{W}}$ growing as θ increases ?

Some of results in the plane curve case works as well. For instance, we have the following:

Lemma 4.1. *Let $\mathbf{r} : S^{2k+1} \to \mathbb{R}^{2k+2}$ (resp., $\mathcal{F}$) be a C^∞ embedding (resp., C^∞ codimension one foliation of S^{2k+1}). Then, $\mathcal{NS}_{\frac{\pi}{2},\mathbf{r},\mathcal{F}}$ is the empty set.*

However, there are still many open questions. For instance,

Question 17. Is the aperture point well-defined in higher dimensional case ?

Question 18. Apart from no-silhouettes, investigate "rotated envelopes". Namely, for a given compact C^∞ manifold M of dimension n satisfying $\chi(M) = 0$, a given C^∞ codimension one foliation $\mathcal{F}$ of M, a given C^∞ immersion $\mathbf{r} : M \to \mathbb{R}^{n+1}$ and a given angle $\theta \in \mathbb{R}$, investigate the envelope of the family of n-planes in $\mathbb{R}^{n+1}$:

$$\{\mathbf{r}(s) + R_{\theta, L_s} d\mathbf{r}_s (T_s M)\}_{s \in M}$$

where L_s is the leaf of $\mathcal{F}$ containing $s \in M$.

4.2. On the space consisting of convex integrands and the space consisting of Wulff shapes

Recall again that the mapping

$$\mathcal{W} : C^0(S^n, \mathbb{R}_+) \to \mathcal{H}_{conv,\mathbf{0}}\left(\mathbb{R}^{n+1}\right)$$

defined by $\mathcal{W}(\gamma) = \mathcal{W}_\gamma$ has been introduced and properties of $\mathcal{W}$ has been studied in Section 1. In more details, it has been confirmed that the restriction of $\mathcal{W}$ to $CI(S^n, \mathbb{R}_+)$

$$\mathcal{W}|_{CI(S^n, \mathbb{R}_+)} : CI(S^n, \mathbb{R}_+) \to \mathcal{H}_{conv,\mathbf{0}}\left(\mathbb{R}^{n+1}\right)$$

is bijective, where $CI(S^n, \mathbb{R}_+)$ is the set consisting of convex integrands as defined in Subsection 3.5.

$$CI(S^n, \mathbb{R}_+) = \left\{\gamma \in C^0(S^n, \mathbb{R}_+) \mid \gamma : \text{Convex Integrand}\right\}$$

Moreover, it is not difficult to see that the restriction of $\mathcal{W}$ to $CI(S^n, \mathbb{R}_+)$ is continuous. We would like to study more detailed properties of the restriction of $\mathcal{W}$ to $CI(S^n, \mathbb{R}_+)$. Thus, we ask the following.

Question 19. Is the following mapping bi-Lipschitz ?

$$\mathcal{W}|_{CI(S^n, \mathbb{R}_+)} : CI(S^n, \mathbb{R}_+) \to \mathcal{H}_{conv,\mathbf{0}}\left(\mathbb{R}^{n+1}\right).$$

Question 19 asks whether or not the two spaces $CI(S^n, \mathbb{R}_+)$ and $\mathcal{H}_{conv,\mathbf{0}}\left(\mathbb{R}^{n+1}\right)$ cannot be distinguished in the sense of Lipschitz geometry. Notice that $CI(S^n, \mathbb{R}_+)$ is the space consisting of convex integrands $S^n \to \mathbb{R}_+$, whereas the space $\mathcal{H}_{conv,\mathbf{0}}\left(\mathbb{R}^{n+1}\right)$ is a space consisting of some compact sets in $\mathbb{R}^{n+1}$. Therefore, the metrics of these two spaces are completely different. Roughly speaking, the metric of $CI(S^n, \mathbb{R}_+)$ measures the difference of two functions $\gamma_1, \gamma_2 : S^n \to \mathbb{R}_+$ only by the difference from radial directions. On the other hand, the metric of $\mathcal{H}_{conv,\mathbf{0}}\left(\mathbb{R}^{n+1}\right)$ measures the difference of two convex bodies $W_1, W_2 \subset \mathbb{R}^{n+1}$ containing the origin as an interior point by the difference from all directions. Therefore, these two metrics are quite different. Nevertheless, to our surprise, there seems to be several signs which indicates that these two spaces might be bi-Lipschitz.[2]

4.3. On convex integrands of class C^k $(2 \leq k < \infty)$

Question 20. Let $\gamma : S^n \to \mathbb{R}_+$ be a convex integrand and let $\delta : S^n \to \mathbb{R}_+$ be the dual convex integrand of γ. Moreover, let k be a positive integer greater than or equal to 2. Then, are the following equivalent ?

(1) The convex integrand γ is of class C^k.
(2) The convex integrand δ is of class C^k.
(3) The function $\widehat{\gamma}$, whose graph is exactly $\partial\mathcal{W}_\delta = \partial\mathcal{D}\mathcal{W}_\gamma$, is of class C^k.
(4) The function $\widehat{\delta}$, whose graph is exactly $\partial\mathcal{W}_\gamma = \partial\mathcal{D}\mathcal{W}_\delta$, is of class C^k.

Notice that, as Figure 4 in Section 1 shows, it is impossible to replace C^k class $(2 \leq k < \infty)$ with C^1 class.

The background of Question 20 is as follows. Historically, F. Morgan in [51] seems to be the first mathematician who treated such a kind of question although his result requires a mild assumption and the conclusion of his result is weak. Few years later after [51], H. M. Soner gave the same conclusion as Question 20 in [68] under a mild assumption on γ (see also [20] where Soner's lemma (Lemma 1.7.2 in [20]) plays an important role for studying Wulff shapes by the level-set method). As for the C^∞ class, under some mild assumptions on γ, not only F. Morgan ([51]) and H. M. Soner ([68]) but also B. Andrews ([1]) obtains the same conclusionas Question 20.

Moreover, if we succeed to solve Question 20 affirmatively, then we can simultaneously solve affirmative following question completely.

Question 21. Obtain the lowest k so that γ is of class C^k if and only if $\partial\mathcal{W}_\gamma$ is a C^k submanifold.[3]

---

[2]**Note 1 added in revision.** After a few months from the first submission, Question 19 was solved affirmatively in [25]. It is proved in [25] that the restriction of $\mathcal{W}$ to $CI(S^n, \mathbb{R}_+)$ is an isometry. Thus, Question 19 has been solved in the strongest form.

[3]**Note 2 added in revision.** The authors learned from the referee that Andrews's formulas given in Subsection 3.8 works well if $\gamma : S^n \to \mathbb{R}_+$ is convex and of class C^k $(k \geq 2)$ (cf. [62]). Therefore, in this case, $\partial\mathcal{W}_\gamma$ is of class C^{k-1}.

§ Acknowledgements

The authors would like to express their sincere gratitude to the anonymous referee for careful reading of this article and making invaluable suggestions.

References

[1] B. Andrews, *Harnack inequalities for evolving hypersurfaces*, Math. Z., **217** (1994), no. 2, 179–197.
[2] V.I. Arnol'd, *Singularities of caustics and wave fronts*, Kluwer Academic Publishers Group, Dordrecht, 1990.
[3] V. I. Arnol'd, *The geometry of spherical curves and the algebra of quaternions*, Russian Math. Surveys, **50** (1995), 1–68.
[4] V. I. Arnol'd, V. V. Goryunov, O. V. Lyashko and V. A. Vasil'ev, *Dynamical Systems VIII*, Encyclopaedia of Mathematical Sciences, **39**, Springer-Verlag, Berlin Heidelberg New York, 1989.
[5] V.I. Arnol'd, S.M. Gusein-Zade and A.N. Varchenko, *Singularities of Differentiable Maps I*, Monographs in Mathematics, **82**, Birkhäuser, Boston Basel Stuttgart, 1985.
[6] M. Barnsley, *Fractals Everywhere 2nd edition*, Morgan Kaufmann Pub., San Fransisco, 1993.
[7] E. B. Batista, H. Han and T. Nishimura, *Stability of C^∞ convex integrands*, Kyushu J. Math., **71** (2017), 187–196.
[8] E. B. Batista, H. Han and T. Nishimura, *Simultaneous smoothness and simultaneous stability of a C^∞ strictly convex integrand and its dual*, preprint (available from arXiv:1707.02359 [math.GT]).
[9] J.W. Bruce and P.J. Giblin, *Growth, motion and 1-parameter families of symmetry sets*, Proc. Roy. Soc. Edinburgh Sect. A, **104** (1986), 179–204.
[10] J.W. Bruce and P.J. Giblin, *Curves and Singularities (second edition)*, Cambridge University Press, Cambridge, 1992.
[11] J.W. Bruce, P.J. Giblin and C.G. Gibson, *Symmetry sets*, Proc. Roy. Soc. Edinburgh Sect. A, **101**(12) (1985), 163–186.
[12] J.W. Bruce and N.P. Kirk, *Generic projections of stable mappings*, Bull. London Math. Soc., **32** (2000), 718–728.
[13] B. Dacorogna and C.E. Pfister, *Wulff theorem and best constant in Sobolev inequality*, J. Math. Pures Appl., **71**(1992), 97–118.
[14] A. Dinghas, *Über einen geometrischen Satz von Wulff für die Gleichgewichts form von Kristallen*, Z. Kristallogr., Mineral. Petrogr., **105** (1944), 304–314.
[15] K. Falconer, *Fractal Geometry–Mathematical Foundations and applications 2nd edition*, John Wiley & Sons Ltd., Chichester, West Sussex, 2003.
[16] I. Fonseca, *The Wulff theorem revisited*, Proc. Roy. Soc. London Ser. A, **432** (1991), 125–145.

[17] I. Fonseca and S. Müller, *A uniqueness proof for the Wulff theorem*, Proc. Roy. Soc. Edinburgh Sect. A, **119** (1991), 125–136.

[18] F. C. Frank, *Metal Surfaces*, ASM, Cleveland, OH, 1963.

[19] S. Gao, D. Hug and R. Schneider, *Intrinsic volumes and polar sets in spherical space*, Math. Notae, **41** (2001/02), 159–176(2003).

[20] Y. Giga, *Surface Evolution Equations*, Monographs of Mathematics, **99**, Springer, 2006.

[21] M. Golubitsky and V. Guillemin, *Stable mappings and their singularities*, Graduate Texts in Mathematics **14**, Springer, New York, 1973.

[22] H. Han and T. Nishimura, *The spherical dual transform is an isometry for spherical Wulff shapes*, to appear in Studia Math. (A preprint version is available from arXiv:1504.02845 [math.MG]).

[23] H. Han and T. Nishimura, *Strictly convex Wulff shapes and C^1 convex integrands*, Proc. Amer. Math. Soc., **145** (2017), 3997–4008.

[24] H. Han and T. Nishimura, *Self-dual Wulff shapes and spherical convex bodies of constant width $\pi/2$*, J. Math. Soc. Japan, **69** (2017), 1475–1484.

[25] H. Han and T. Nishimura, *The Wulff construction for convex integrands*, preprint (available from arXiv:1607.02885 [math.MG]).

[26] S. Ichiki and T. Nishimura, *Distance-squared mappings*, Topology Appl., **160** (2013), 1005–1016.

[27] S. Ichiki and T. Nishimura, *Recognizable classification of Lorentzian distance-squared mappings*, J. Geom. Phys., **81** (2014), 62–71.

[28] S. Ichiki and T. Nishimura, *Around distance-squared mappings*, RIMS Kôkyûroku, **1948** (2015), 28–37.

[29] S. Ichiki and T. Nishimura, *Generalized distance-squared mappings of $\mathbb{R}^{n+1}$ into $\mathbb{R}^{2n+1}$*, Real and complex singularities, 121–132, Contemp. Math., **675**, Amer. Math. Soc., Providence, RI, 2016.

[30] S. Ichiki, T. Nishimura, R. Oset Sinha and M.A.S. Ruas, *Generalized distance-squared mappings of the plane into the plane*, Adv. Geom., **16** (2016), 189–198.

[31] S. Izumiya, *Differential geometry from the viewpoint of Lagrangian or Legendrian singularity theory*, in Singularity Theory, Proceedings of the 2005 Marseille Singularity School and Conference, ed., D. Chéniót et al., World Scientic (2007), 241–275.

[32] S. Izumiya and G. Ishikawa, *Applied Singularity Theory*, Kyoritsu Syuppan Co., LTD, 1998 (in Japanese).

[33] S. Izumiya and M. Takahashi, *Caustics and wave front propagations: Applications to differential geometry*, Banach Center Publ., **82** (2008), 125–142.

[34] S. Izumiya and F. Tari, *Projections of hypersurfaces in the hyperbolic space to hyperhorospheres and hyperplanes*, Rev. Mat. Iberoam., **24** (2008), 895–920.

[35] D. Kagatsume and T. Nishimura, *Aperture of plane curves*, J. Singul., **12** (2015), 80–91.

[36] K. Kitagawa and T. Nishimura, *Classification of singularities of pedal curves in* S^2, The Natural Sciences, Journal of the Faculty of Education and Human Sciences, Yokohama National University, **10** (2008), 39–55.

[37] M. Lassak, *Width of spherical convex bodies*, Aequationes Math., **89** (2015), 555–567.

[38] M. Lassak, *Reduced spherical polygons*, Colloq. Math., **138** (2015), 205–216.

[39] H. Maehara, *Geometry of Circles and Spheres*, Asakura Publishing, 1998 (in Japanese).

[40] J.N. Mather, *Stability of* C^∞ *mappings*, I. *The division theorem*, Ann. of Math., **87** (1968), 89–104.

[41] J.N. Mather, *Stability of* C^∞ *mappings*, II. *Infinitesimal stability implies stability*, Ann. of Math., **89** (1969), 254–291.

[42] J.N. Mather, *Stability of* C^∞ *mappings*, III. *Finitely determined map-germs*, Publ. Math. Inst. Hautes Études Sci., **35** (1969), 127–156.

[43] J.N. Mather, *Stability of* C^∞ *mappings*, IV. *Classification of stable map-germs by* $\mathbb{R}$*-algebras*, Publ. Math. Inst. Hautes Études Sci., **37** (1970), 223–248.

[44] J.N. Mather, *Stability of* C^∞*-mappings* V. *Transversality*, Adv. in Math., **4** (1970), 301–336.

[45] J.N. Mather, *Stability of* C^∞*-mappings* VI. *The nice dimensions*, Lecture Notes in Math., **192** (1971), 207–253.

[46] J.N. Mather, *Generic projections*, Ann. of Math., **98** (1973), 226–245.

[47] J. Matousek, *Lectures on Discrete Geometry*, Springer, 2002.

[48] J. L. Meijering, *Usefulness of a* $1/\gamma$ *plot in the theory of thermal etching*, Acta Metallurgica, **11** (1963), 847–849.

[49] J. W. Milnor, *Morse Theory*, Annals of Mathematics Studies **51**, Princeton University Press, Princeton, New Jersey, 1963.

[50] J. W. Milnor, *Topology from the differentiable viewpoint. Revised version. Princeton Landmarks in Mathematics*, Princeton University Press, Princeton, New Jersey, 1997.

[51] F. Morgan, *The cone over the Clifford torus in* $\mathbb{R}^4$ *is* F*-minimizing*, Math. Ann., **289** (1991), 341–354.

[52] F. Morgan, *Private communications*, 2015.

[53] T. Nishimura, *Singularities of tangent pedal curves in* S^3, The Natural Sciences, Journal of the Faculty of Education and Human Sciences, Yokohama National University, **9** (2007), 43–50.

[54] T. Nishimura, *Normal forms for singularities of pedal curves produced by non-singular dual curve germs in* S^n, Geom Dedicata, **133** (2008), 59–66.

[55] T. Nishimura, *Singularities of pedal curves produced by singular dual curve germs in* S^n, Demonstratio Math., **43** (2010), 447–459.

[56] T. Nishimura, *Singularities of one-parameter pedal unfoldings of spherical pedal curves*, J. Singul., **2** (2010), 160–169.

[57] T. Nishimura, *Whitney umbrellas and swallowtails*, Pacific J. Math., **252** (2011), 459–471.

[58] T. Nishimura, *Wave front evolution and pedal evolution*, RIMS Kôkyûroku Bessatsu, **B38** (2013), 15–30.

[59] T. Nishimura and Y. Sakemi, *View from inside*, Hokkaido Math. J., **40** (2011), 361–373.
[60] T. Nishimura and Y. Sakemi, *Topological aspect of Wulff shapes*, J. Math. Soc. Japan, **66** (2014), 89–109.
[61] A. Ookawa, *Crystal growth. Reprinted Edition*, Kyoritsu Shuppan Co., Ltd., 2001 (in Japanese).
[62] B. Palmer, *Stability of the Wulff shape*, Proc. Amer. Math Soc., **126** (1998), 3661–3667.
[63] A. Pimpinelli and J. Villain, *Physics of Crystal Growth*, Monographs and Texts in Statistical Physics, Cambridge University Press, Cambridge New York, 1998.
[64] I. R. Porteous, *The normal singularities of a submanifold*, J. Diff. Geom., **5** (1971), 543–564.
[65] S. A. Robertson and M. C. Romero-Fuster, *The convex hull of a hypersurface*, Proc. London Math. Soc., **50** (1985), 370–384.
[66] M. C. Romero-Fuster and M. A. S. Ruas, *Some stability questions concerning caustics for different propagation laws*, Portugal. Math., **51** (1994), 595–605.
[67] R. Schneider, *Convex Bodies: The Brunn-Minkowski Theory* 2nd Edition, Encyclopedia of Mathematics and its Applications **44**, Cambridge University Press, Cambridge, 2013.
[68] H. M. Soner, *Motion of a set by the curvature of its boundary*, J. Differential Equations, **101** (1993), 313–372.
[69] J. E. Taylor, *Crystalline variational problems*, Bull. Amer. Math. Soc., **84** (1978), 568–588.
[70] J. E. Taylor, J. W. Cahn and C. A. Handwerker, *Geometric models of crystal growth*, Acta Metallurgica et Materialia, **40** (1992), 1443–1474.
[71] R. Thom and H. Levine, *Singularities of differentiable mappings*, Springer Lecture Notes in Mathematics, **192**, Springer-Verlag, Berlin, 1–89 (1971).
[72] W. P. Thurston, *Existence of codimension-one foliations*, Ann. of Math., **104** (1976), 249–268.
[73] G. Wulff, *Zur frage der geschwindindigkeit des wachstrums und der auflösung der krystallflachen*, Z. Kristallographine und Mineralogie, **34** (1901), 449–530.
[74] V. M. Zakalyukin, *Singularities of convex hulls of smooth maniifolds*, Functional Anal. Appl., **11** (1977), 225–227(1978).

College of Sciences,
Northwest Agriculture and Forestry University,
712100 Yangling, China
E-mail address: han-huhe@nwafu.edu.cn

Research Institute of Environment and Information Sciences,
Yokohama National University,
Yokohama 240-8501, Japan
E-mail address: nishimura-takashi-yx@ynu.jp

Advanced Studies in Pure Mathematics 78, 2018
Singularities in Generic Geometry
pp. 55–106

Singularities of frontals

Goo Ishikawa

§1. Introduction

In this survey article we introduce the notion of frontals, which provides a class of generalised submanifolds with singularities but with well-defined tangent spaces. We present a review of basic theory and known studies on frontals in several geometric problems from singularity theory viewpoints. In particular, in this paper, we try to give some of detailed proofs and related ideas, which were omitted in the original papers, to the basic and important results related to frontals.

We start with one of theoretical motivations for our notion "frontal". Let M be a C^∞ manifold of dimension m, which is regarded as an ambient space. Suppose $n \leq m$ and let $f : N \to M$ be an *immersion* of an n-dimensional C^∞ manifold N, which is regarded as a parameter space, to M. Then for each point $t \in N$, we have the n-plane $f_*(T_tN)$, the image of the differential map $f_* : T_tN \to T_{f(t)}M$ at t in the tangent space $T_{f(t)}M$. Thus we have a field of tangential n-planes $\{f_*(T_tN)\}_{t\in N}$ along the immersion f. Moreover if M is endowed with a Riemannian metric, then we have also a field of tangential $(m-n)$-planes $f_*(T_tN)^\perp$ along f. Taking those vector bundles we can develop differential topology, theory of characteristic classes and so on of immersed submanifolds. Besides, taking local adapted frames for immersions, we can develop differential geometry of immersed submanifolds in terms of frames. Then a natural and challenging problem arises to us on the possibility to find a natural class of singular mappings enjoying the same properties as immersed submanifolds and to develop generalised topological and geometric theories on them.

Received May 31, 2016.
Revised October 24, 2016.
This work was supported by JSPS KAKENHI No.15H03615 and No.15K13431.

In this paper we introduce such a class of generalised submanifolds in terms of Grassmannians: Let $\mathrm{Gr}(n, TM)$ denote the Grassmannian of tangential n-planes in the tangent bundle TM over an m-dimensional C^∞ manifold M with the canonical projection $\pi : \mathrm{Gr}(n, TM) \to M$ (see §3). Let N be a C^∞ manifold of dimension n with $0 \leq n \leq m$ and take a point $a \in N$. Then a C^∞ map-germ $f : (N, a) \to M$ is called a *frontal map-germ* or a *frontal* in short if there exists a "Legendre" lifting of f, that is, there exist an open neighbourhood U of a and a C^∞ lift $\widetilde{f} : U \to \mathrm{Gr}(n, TM)$ of f, $\pi \circ \widetilde{f} = f|_U$, such that the image of differential $f_*(T_t N)$ is contained in $\widetilde{f}(t)$, for any $t \in U$. Note that $\widetilde{f}(t)$ is an n-plane in $T_{f(t)}M$. Moreover a C^∞ mapping $f : N \to M$ is called a *frontal mapping* or a *frontal* in short if, the germ $f : (N, a) \to M$ at any point $a \in N$ is a frontal. See §4 for details. The formulation using Grassmannians is very natural and satisfactory from the viewpoint of differential systems and their geometric solutions as well. See for instance [101][52][53].

Note that, if $\dim(N) = 1$, then any frontal $f : N \to M$ has a global Legendre lift $\widetilde{f} : N \to \mathrm{Gr}(1, TM)$ (Lemma 12.3). However, if $\dim(N) = 2$, then a frontal $f : N \to M$ not necessarily has a global Legendre lift (Example 2.5). This fact seems to be found first in the present paper. Also note that any mapping $f : N \to M$ is a frontal if $\dim(N) = \dim(M)$ (Remark 4.2). Any constant mapping $f : N \to M$ is a frontal.

The notion of "frontals" was introduced already in many papers, e.g. [38][105][95][85][10][11], in the case of hypersurfaces as a natural generalisation of wave-fronts. See §2.

We are going to give a survey on local classification of singularities appearing in frontals in various geometric contexts. Basically we mean by the "singularities" of frontals, as usual, the equivalence classes of germs of frontals under the following equivalence relation:

Definition 1.1. Two map-germs $f : (N, a) \to (M, f(a))$ and $g : (N', a') \to (M', f'(a'))$ are *right-left equivalent* or *$\mathcal{A}$-equivalent* or *diffeomorphic*, if there exist diffeomorphism-germs $\varphi : (N, a) \to (N', a')$ and $\Phi : (M, f(a)) \to (M', f'(a'))$ such that the following diagram commutes:

$$\begin{array}{ccc} (N, a) & \xrightarrow{f} & (M, f(a)) \\ \varphi \downarrow & & \downarrow \Phi \\ (N', a') & \xrightarrow{g} & (M', f'(a')). \end{array}$$

As the typical singularities of frontals, we introduce cuspidal edges, swallowtails, folded umbrellas, open swallowtails, open folded umbrellas and so on.

The *cuspidal edge* is defined as the equivalence class of the map-germ $(\mathbf{R}^2, 0) \to (\mathbf{R}^m, 0)$, $m \geq 3$,

$$(t, s) \mapsto (t + s,\ t^2 + 2st,\ t^3 + 3st^2,\ 0,\ \dots,\ 0),$$

which is diffeomorphic to $(u, w) \mapsto (u, w^2, w^3, 0, \dots, 0)$. The cuspidal edge singularities are originally defined only in the three dimensional space. Here we are generalising the notion of the cuspidal edge in higher dimensional ambient space. It will be often emphasised it by writing "embedded" cuspidal edge.

The *folded umbrella* (or the *cuspidal cross cap*) is defined as the equivalence class of the map-germ $(\mathbf{R}^2, 0) \to (\mathbf{R}^3, 0)$,

$$(t, s) \mapsto (t + s,\ t^2 + 2st,\ t^4 + 4st^3),$$

which is diffeomorphic to $(u, t) \mapsto (u,\ t^2 + ut,\ t^4 + \frac{2}{3}ut^3)$.

The *open folded umbrella* is defined as the equivalence class of the map-germ $(\mathbf{R}^2, 0) \to (\mathbf{R}^m, 0)$, $m \geq 4$,

$$(t, s) \mapsto (t + s,\ t^2 + 2st,\ t^4 + 4st^3,\ t^5 + 5st^4,\ 0,\ \dots,\ 0),$$

which is diffeomorphic to $(u, t) \mapsto (u,\ t^2+ut,\ t^4+\frac{2}{3}ut^3,\ t^5+\frac{5}{8}ut^4,\ 0,\ \dots,\ 0)$. The open folded umbrella appeared for instance as a frontal-symplectic singularity in the paper [48].

The *swallowtail* is defined as the equivalence class of the map-germ $(\mathbf{R}^2, 0) \to (\mathbf{R}^3, 0)$,

$$(t, s) \mapsto (t^2 + s,\ t^3 + \tfrac{3}{2}st,\ t^4 + 2st^2),$$

which is diffeomorphic to $(u, t) \mapsto (u,\ t^3 + ut,\ t^4 + \frac{2}{3}ut^2)$.

The *open swallowtail* is defined as the equivalence class of the map-germ $(\mathbf{R}^2, 0) \to (\mathbf{R}^m, 0)$, $m \geq 4$,

$$(t, s) \mapsto (t^2 + s,\ t^3 + \tfrac{3}{2}st,\ t^4 + 2st^2,\ t^5 + \tfrac{5}{2}st^3,\ 0,\ \dots,\ 0),$$

which is diffeomorphic to $(u, t) \mapsto (u,\ t^3+ut, t^4+\frac{2}{3}ut^2,\ t^5+\frac{5}{9}ut^3,\ 0,\ \dots,\ 0)$. The open swallowtail singularity was introduced by Arnol'd (see [6]) as a singularity of Lagrangian varieties in symplectic geometry. Here we abstract its diffeomorphism class as the singularity of parametrised surfaces (see [26][43]).

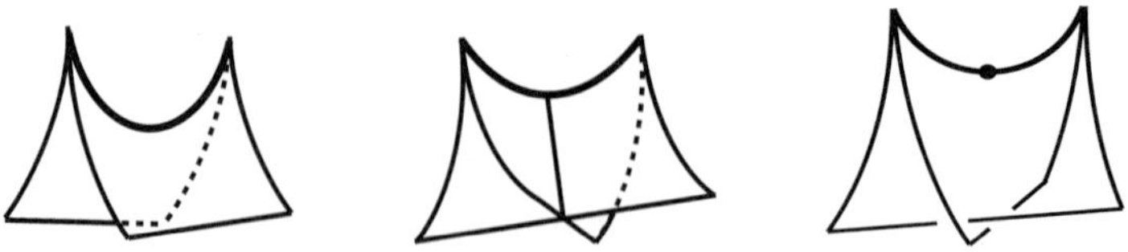

cuspidal edge folded umbrella open folded umbrella

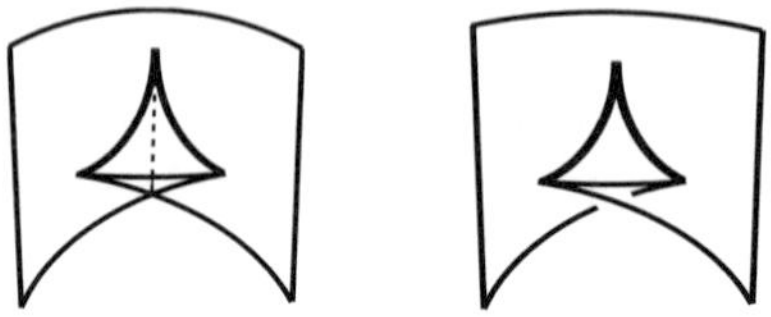

swallowtail open swallowtail

In Part I, we provide basic studies for an intrinsic understanding of frontals as parametrised singular submanifolds with well-defined tangent spaces.

We give the exact definition of frontals in §2 in the case of hypersurfaces and, after the description of Grassmannian bundles and canonical (or generalised contact) distributions in §3, we give the general definition in §4. In §5, we have introduced the density function as a main notion for the theory of frontals.

A frontal $f : N \to M$ is called a *proper frontal* in the present paper if the singular (non-immersive) locus $S(f)$ is nowhere dense in N (§6). In [44][45][46][47], "frontal" maps were defined as proper frontals, namely, the density of regular locus was assumed. Note that proper frontals are not generic in the space of all frontals for C^∞-topology in general (Remark 6.4). In §7, we introduce the tangent bundles to frontals.

Viewed from our generalisation, the notion of frontals turns to be closely related to the notion of *openings*. Though the notion of openings of mappings seems to be noticed naively in many previous contributions, it is introduced in the author's recent papers [44][45]. An opening separates the self-intersections of the original map-germ, preserving its singularities. For example, the swallowtail is an opening of the *Whitney's cusp* map-germ $(\mathbf{R}^2, 0) \to (\mathbf{R}^2, 0)$ defined by $(t, s) \mapsto (t^2 + s,\ t^3 + \frac{3}{2}st)$ which is diffeomorphic to $(u, t) \mapsto (u,\ t^3 + ut)$ and the open swallowtail is a versal opening of them. Openings of map-germs appear as typical singularities in several problems of geometry and its applications. Note that the process of unfoldings of map-germs $(\mathbf{R}^n, 0) \to (\mathbf{R}^m, 0)$ preserves the "relative dimension" $m - n$. On the other hand, the process of openings preserves n but changes m, and it gives bridges between map-germs of different relative dimensions. We recall also the related notions, "Jacobi modules" $\mathcal{J}_f$ and "ramification modules" $\mathcal{R}_f$. They play important role to analysis and classification of singularities of mappings f, in particular, the study on symplectic singularities, contact singularities and singularities of tangent surfaces ([29][30][31][34][35][37][39][40][49]). Moreover those notions seem to be related to recognition problem of singularities (see Definition 8.4). Note that we used the notation, for the ramification module of f, '$\mathcal{D}_f$' instead of $\mathcal{R}_f$ in [29], relating Mather's $\mathcal{C}$-equivalence, and we denoted it by 'H_f' in [30][31][33][34], because it can be regarded as a cohomological invariant. Note that the notion of openings, Jacobi modules and ramification modules for *multi-germs* is naturally introduced in the paper [44]. We give a review on the theory of opening related to frontals in §8 and §9. Moreover in §10 we give ideas of "subfrontals" and "superfrontals" related to frontals.

In [97], it is introduced the related notion of "coherent tangent bundles" as generalised Riemannian manifolds. Moreover Saji, Umehara, Yamada are developing the intrinsic studies of frontals in terms of singular metrics introduced by Kossowski [72]. We intend to give an abstract differential-topological feature of frontals, which is invariant under diffeomorphisms, by proving another way to study intrinsically frontals in terms of the theory of C^∞-rings (§11).

In part II, we give a survey of several results on frontals as an application of the basic theory presented in part I.

In §12, we treat frontal curves and give basic results on them. Let $\gamma : N \to \mathbf{R}^2$ be a planar frontal curve with $\dim(N) = 1$. By Lemma 12.3, there exists a global Legendre lifting $\widetilde{\gamma} : I \to P(T^*\mathbf{R}^2)$. Thus it is possible to perform differential geometry of planar frontals, as a generalised differential geometry of planar immersions, in terms of Legendre curves covering frontals. In fact geometric studies on planar frontals, evolutes and involutes, are given in a series of papers [21][22][23][24]. As a related topics to planar frontals, we gave a review the "Goursat Monster tower" found by Zhitomirskii, Montgomery, Mormul and others (cf. [80][81][16]) and the "Legendre-Goursat duality" related to it in [46].

The study on singularities and bifurcations of wavefronts based on Legendre singularity theory are established by Arnold-Zakalyukin's theory ([5][7][103][104]). The application of singularity theory to differential geometry has been developed by many authors (see for instance [90][91][14][63]). The geometric study of submanifolds in hyperbolic space H^{n+1} based on singularity theory was initiated by Izumiya et al.([66][67][68]). The Legendre duality developed in [62][18] enables us to unify the theory of framed curves in any space form as describes in [42].

We recall Legendre duality (see [13][82][58][52][18]) in the framework of moving frames and flags and discuss its generalisation and relation with the theory of frontals in §13, §14 and §15.

Let $\gamma : I \to \mathbf{R}^3$ be a space frontal curve. Then the tangent surface (tangent developable) $\text{Tan}(\gamma)$ is defined as the surface ruled by tangent lines to γ. Then the tangent surface has zero Gaussian curvature, therefore it is flat with respect to Euclidean metric of $\mathbf{R}^3$ at least off the singular locus. Thus the tangent surfaces serve main parts of "flat frontals" (§16). Flat fronts or flat frontals are studied also in [83][84].

The notion of tangent surfaces ruled by "tangent lines" to directed curves is naturally generalised in various ways: For a curve in a projective space, we regard tangent projective lines as "tangent lines". The classification is generalised to A_n-geometry (§17). For a curve in a Riemannian manifold, we regard tangent geodesics as "tangent lines. In fact, tangent surfaces are defined for proper frontal curves (directed curves) in a manifold with an affine connection (§18). After discussing useful criteria of singularities in §19, we define null tangent surface to a null curve of a semi (pseudo)-Riemannian manifold, regarding null geodesics as "tangent lines" (see §20). In particular we pick up several results related to D_n-geometry ([57]). For a horizontal curve of a sub-Riemannian manifold, we regard "tangent lines" by abnormal geodesics (see §21). In particular the classification result of singularities of tangent surfaces to generic integral curves to Cartan distribution with G_2-symmetry is introduced.

Speaking of G_2, we note that the work on frontals may be related to the rolling ball problem [1][9][8][79]. We will treat "rolling frontals" as a generalisation of rolling bodies [4] in a forthcoming paper.

In the last section (§22), as an appendix, we show the Malgrange's preparation theorem on differentiable algebras ([74]) from the ordinary Malgrange-Mather's preparation theorem (see for example [15]), relating to the theory of C^∞-rings which we have utilised in this paper.

The author hopes very much that this survey paper helps to raise wider reader's interest to the mathematics on frontals.

The author would like to thank the anonymous referee for the valuable comment to improve the paper.

In this paper a manifold or a mapping is supposed to be of class C^∞ unless otherwise stated. The symbol $\subseteq$ of inclusion is often used, which has the same meaning as $\subset$ just to stress that the equality may occur.

Part I. Basic Theory

§2. The case of hypersurfaces

Let M be a manifold of dimension m. Let $P(T^*M)$ denote the projective cotangent bundle of M, which consists of non-zero cotangent vectors somewhere on M considered up to a non-zero scalar multiplication. Note that $P(T^*M)$ is naturally identified with the Grassmannian bundle $\mathrm{Gr}(m-1, TM)$ (see §3) by sending each class $(x, [\alpha]) \in P(T^*M)$ of a non-zero covector $\alpha \in T^*M$ to its kernel $\mathrm{Ker}(\alpha) \in \mathrm{Gr}(m-1, T_xM)$. Note that $\alpha \in T_x^*M$, that $\alpha : T_xM \to \mathbf{R}$ is a non-zero linear map, and that $\mathrm{Ker}(\alpha) \subset T_xM$ is an $(m-1)$-plane. Then the $(2m-1)$-dimensional manifold $P(T^*M)$ has a canonical contact structure $D \subset TP(T^*M)$. In fact it is defined by $D = \bigcup_{(x,[\alpha])} D_{(x,[\alpha])}$, and $D_{(x,[\alpha])} = \pi_*^{-1}(\mathrm{Ker}(\alpha))$, where $\pi : P(T^*M) \to M$ is the canonical projection.

We recall the coordinate description of the contact structure, which will be needed for the detailed computation on singularities.

Let $(x^1, x^2, \dots, x^m)$ be a local coordinate system on an open subset U of M. Let

$$(x^1, x^2, \dots, x^m, p_1, p_2, \dots, p_m)$$

be the associated system of coordinates on T^*U such that any element $\alpha \in T^*U$ is expressed as

$$\alpha = p_1 dx^1 + p_2 dx^2 + \cdots + p_m dx^m,$$

by its coordinates. Set $V_i = \{p_i \neq 0\} \subset T^*U, 1 \leq i \leq m$. Then we have a local system of coordinates of $P(T^*M)$ associated to V_i,

$$x^1, x^2, \dots, x^m, -p_1/p_i, \dots, -p_{i-1}/p_i, -p_{i+1}/p_i, \dots, -p_m/p_i.$$

To avoid non-essential complexity, we will discuss just for $i = m$ in what follows. Then set $a_i = -p_i/p_m, 1 \leq i \leq m-1$. Then

$$x^1, x^2, \dots, x^m, a_1, a_2, \dots, a_{m-1}$$

give a local system of coordinates of $P(T^*M)$ and the contact structure $D \subset TP(T^*M)$ is given locally by

$$dx^m - (a_1 dx^1 + a_2 dx^2 + \cdots + a_{m-1} dx^{m-1}) = 0.$$

Let N be a submanifold of dimension n with $n < m$. Then the submanifold N induces the *projective conormal bundle*

$$\widetilde{N} = P(T_N^*M) = \{(x, [\alpha]) \in P(T^*M) \mid \alpha|_{T_xN} = 0\},$$

which satisfies that $T\widetilde{N} \subset D$ and $\dim(\widetilde{N}) = m - 1$, in other words, a Legendre submanifold in the contact manifold $P(T^*M)$.

In particular, suppose $n = m - 1$, that is, N is a hypersurface of M. Then $\pi|_{\widetilde{N}} : \widetilde{N} \to N$ is a diffeomorphism. Its inverse $N \to \widetilde{N}$ is given by $x \mapsto (x, T_xN)$.

Let $f : N \to M$ be an immersion of an $(m-1)$-dimensional manifold N to an m-dimensional manifold M. Then we have an immersion $\widetilde{f} : N \to P(T^*M)$ defined by $\widetilde{f}(t) = (f(t), f_*(T_tN))$. Then $\widetilde{f}$ is a lift of f and $\widetilde{f}$ is D-integral, i.e. $\widetilde{f}_*(T_tN) \subset D_{\widetilde{f}(t)}$ for any $t \in N$. In other words, $\widetilde{f}$ is a Legendre immersion.

Remark 2.1. Set $\widetilde{f}(t) = (f(t), [\alpha(t)]) \in P(T^*_{f(t)}M)$. Then the condition $\widetilde{f}_*(T_tN) \subset D_{\widetilde{f}(t)}$ is equivalent to that $\alpha(t)|_{f_*(T_tN)} = 0$.

Definition 2.2. Let N be a manifold of dimension $m - 1$ and M a manifold of dimension m. A map-germ $f : (N, a) \to M$ is called a *wave-front* or a *front* in short if there exists a germ of Legendre immersion $\widetilde{f} : (N, a) \to P(T^*M)$ with $\pi \circ \widetilde{f} = f$.

A mapping $f : N \to M$ is called a *wave-front* or a *front* in short if, for any point $a \in N$, the germ of f at a is a front.

A map-germ $f : (N, a) \to M$ is a front if and only if there exists a representative of f, which is a front.

Remark 2.3. In the original and naive context, the image $f(N)$ was called a wave-front rather than the parametrisation f itself.

Definition 2.4. Let N be a manifold of dimension $m - 1$ and M a manifold of dimension m. Let $a \in N$. A map-germ $f : (N, a) \to M$ is called a *frontal map-germ* or a *frontal* in short if there exist a germ of Legendre lifting $\widetilde{f} : (N, a) \to P(T^*M)$ of f, that is, there exist an open neighbourhood U of a, a representative $f : U \to M$ of f and a Legendre lifting $\widetilde{f} : U \to P(T^*M)$ of $f|_U$, i.e. $\widetilde{f}_*(T_tN) \subset D_{\widetilde{f}(t)}$ for any $t \in U$ and $\pi \circ \widetilde{f} = f|_U$. Here we do not assume that $\widetilde{f}$ is an immersion.

A mapping $f : N \to M$ is called a *frontal mapping* or a *frontal* in short if, for any $a \in N$, the germ of f at a is a frontal.

A map-germ $f : (N, a) \to M$ is a frontal if and only if there exists a representative of f, which is a frontal.

In Definition 2.4 we have defined the notion of frontals by the *local* existence of its Legendre liftings. A frontal $f : N \to M$ not necessarily has its *global* Legendre lifting $\widetilde{f} : N \to P(T^*M)$.

Example 2.5. Define a C^∞ function $\varphi : \mathbf{R} \to \mathbf{R}$ by $\varphi(t) = e^{-1/t^2}(t > 0), \varphi(t) = 0(t \leq 0)$. Then define $h : \mathbf{R}^2 \to \mathbf{R}^3$ by $h(t_1, t_2) = (t_1, t_2^2, t_2^3 + \varphi(t_1)t_2)$, which we will call a *half cuspidal edge*.

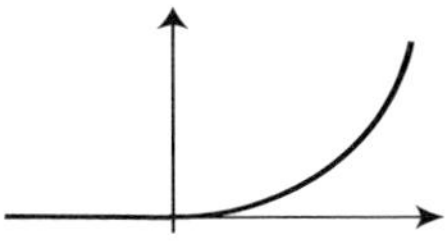

The graph of $\varphi(t)$.

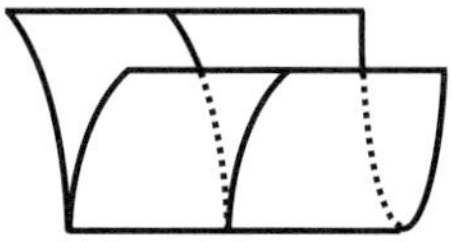

The image of the half cuspidal edge.

The mapping h is not frontal. In fact the local existence of Legendre lift for h does not hold at the origin $(t_1, t_2) = (0, 0)$. Moreover h is a front on $\mathbf{R}^2 \setminus \{(0,0)\}$ with cuspidal edge along $\{(t_1, 0) \mid t_1 < 0\}$ and the Legendre lifting $\widetilde{h} : \mathbf{R}^2 \setminus \{(0,0)\} \to P(T^*\mathbf{R}^3) \cong \mathbf{R}^3 \times \mathbf{R}P^2$ is not homotopically trivial. In fact $\widetilde{h}$ restricted to a loop around the origin of $\mathbf{R}^2$ generates the fundamental group $\pi_1(P(T^*\mathbf{R}^3)) \cong \pi_1(\mathbf{R}P^2) \cong \mathbf{Z}/2\mathbf{Z}$.

Define $k : \mathbf{R}^2 \to \mathbf{R}^2$ by $k(t_1, t_2) = \varphi(t_1^2 + t_2^2 - 1)(t_1, t_2)$. Then k is a C^∞ mapping which collapses the unit disc to the origin and maps the outside of the unit disc to $\mathbf{R}^2 \setminus \{(0,0)\}$ diffeomorphically. Set $f = h \circ k : \mathbf{R}^2 \to \mathbf{R}^3$. Then
(1) f is a frontal.
(2) There does not exist a global Legendre lifting $\widetilde{f} : \mathbf{R}^2 \to P(T^*\mathbf{R}^3)$ of f.
To see (1), let $t = (t_1, t_2)$ satisfy $t_1^2 + t_2^2 < 1$. Then the germ of f at t is constant and therefore it is a frontal. Let t satisfy $t_1^2 + t_2^2 > 1$. Then the germ of f at t is right equivalent to h at $k(t) \in \mathbf{R}^2 \setminus \{(0,0)\}$, which is a frontal. Let t satisfy $t_1^2 + t_2^2 = 1$. Then any local extension of $\widetilde{h} \circ k$ to $(\mathbf{R}^2, t)$ turns to be a Legendre lift of the germ of f at t. Therefore f is a frontal. Thus we have (1). To see (2), it is sufficient to observe that $\widetilde{h} \circ k : \mathbf{R}^2 \setminus D^2 \to P(T^*\mathbf{R}^3)$, which is the unique Legendre lift of f restricted to $\mathbf{R}^2 \setminus D^2$, is never extended continuously to $\mathbf{R}^2$.

§3. Grassmannian bundle and generalised contact distribution

Let N be a manifold of dimension n and M a manifold of dimension m with $n \leq m$. Note that in the previous section we assumed $m = n+1$. However in the next section we treat the general case under the weaker condition $n \leq m$.

To treat the general cases, we recall the Grassmannian bundle associated to the tangent bundle TM of M. For each $x \in M$, $\mathrm{Gr}(n, T_xM)$ denotes the Grassmannian manifold consisting of n-dimensional subspaces of T_xM. Then let $\mathrm{Gr}(n, TM) = \bigcup_{x \in M} \mathrm{Gr}(n, T_xM)$. Note that

$\mathrm{Gr}(n, TM)$ is a bundle over M with fibres $\mathrm{Gr}(n, T_xM)$ and that the dimension $\dim(\mathrm{Gr}(n, T_xM)) = n(m-n)$. Note also that $\mathrm{Gr}(n, T_xM)$ is identified with $\mathrm{Gr}(m-n, T_x^*M)$ and therefore that, when $m = n+1$, $\mathrm{Gr}(n, TM)$ is identified with $P(T^*M)$. Let $\pi : \mathrm{Gr}(n, TM) \to M$ be the canonical projection, $\pi(x, V) = x$ for any $(x, V) \in \mathrm{Gr}(n, TM)$ with $V \in \mathrm{Gr}(n, T_xM), x \in M$. If $n = m$, then $\pi : \mathrm{Gr}(m, TM) \to M$ is a diffeomorphism.

Lemma 3.1. *Let $\Phi : M \to M'$ be a diffeomorphism. Let n be an integer with $0 \le n \le m = \dim(M)$. Let $\Phi_\sharp : \mathrm{Gr}(n, TM) \to \mathrm{Gr}(n, TM')$ denote the diffeomorphism induced by the differential map Φ_* which is regarded as the bundle isomorphism $\Phi_* : TM \to TM'$ covering Φ. Then we have $\pi \circ \Phi_\sharp = \Phi \circ \pi : \mathrm{Gr}(n, TM) \to M'$. Here π means the canonical projection $\mathrm{Gr}(n, TM') \to M'$ as well as $\mathrm{Gr}(n, TM) \to M$.*

Proof: Let $(x, V) \in \mathrm{Gr}(n, TM)$. Then $\Phi_\sharp(x, V) = (\Phi(x), \Phi_*(V))$. Therefore $(\pi \circ \Phi_\sharp)(x, V) = \pi(\Phi(x), \Phi_*(V)) = \Phi(x) = (\Phi \circ \pi)(x, V)$. □

We recall the coordinate description of Grassmannians. Let $(x_0, V_0) \in \mathrm{Gr}(n, TM)$. Here $x_0 \in M$ and $V_0 \in \mathrm{Gr}(n, T_{x_0}M)$ so that $V_0 \subset T_{x_0}M$ is a fixed n-plane. The suffix 0 is used to indicate that (x_0, V_0) becomes the centre of the local coordinate system we are going to provide. Let us take a local coordinate system $(x^1, \dots, x^n, x^{n+1}, \dots, x^m)$ on a coordinate neighbourhood $U \subset M$ with the centre at x_0 such that $\partial/\partial x^1, \dots, \partial/\partial x^n$ generate V_0 at x_0. Let $\pi' : U \to \mathbf{R}^n$ denote the coordinate projection defined by $(x^1, \dots, x^n, \dots, x^m) \mapsto (x^1, \dots, x^n)$. Let $\Omega \subset \pi^{-1}(U)$ be the set of (x, V) with $x \in U, V \in \mathrm{Gr}(n, T_xM)$ such that V is mapped isomorphically by $\pi'_* : TU \to T\mathbf{R}^n$ to $\pi'_*(V)$. Then, for any (x, V), there exist unique real numbers $a_j^k, (1 \le j \le n, n+1 \le k \le m)$ such that the n-plane V has the basis $h_1, h_2, \dots, h_n$ of the form

$$\begin{cases} h_1 = \dfrac{\partial}{\partial x^1}(x) + a_1^{n+1}\dfrac{\partial}{\partial x^{n+1}}(x) + \cdots + a_1^m \dfrac{\partial}{\partial x^m}(x), \\ h_2 = \dfrac{\partial}{\partial x^2}(x) + a_2^{n+1}\dfrac{\partial}{\partial x^{n+1}}(x) + \cdots + a_2^m \dfrac{\partial}{\partial x^m}(x), \\ \vdots \qquad \ddots \\ h_n = \dfrac{\partial}{\partial x^n}(x) + a_n^{n+1}\dfrac{\partial}{\partial x^{n+1}}(x) + \cdots + a_n^m \dfrac{\partial}{\partial x^m}(x). \end{cases}$$

Thus we have a system of coordinates $(x^1, \dots, x^m, a_j^k, (1 \le j \le n, n+1 \le k \le m))$ on Ω of $\mathrm{Gr}(n, TM)$ with the centre at (x_0, V_0).

We call the coordinate systems constructed as above *Grassmannian coordinates.*

The *canonical distribution* $D \subset T(\mathrm{Gr}(n, TM))$ on the Grassmann bundle $\mathrm{Gr}(n, TM)$ is defined by $D = \bigcup_{(x,V)} D_{(x,V)}$ where (x, V) runs over $\mathrm{Gr}(n, TM)$, V being an n-plane of T_xM, $x \in M$, and, for $v \in T_{(x,V)}(\mathrm{Gr}(n, TM))$,

$$v \in D_{(x,V)} \Longleftrightarrow \pi_*(v) \in V(\subset T_xM).$$

We call the canonical distribution D on $\mathrm{Gr}(n, TM)$ also the *canonical differential system* and also the *contact distribution*, in a generalised and wider sense. If $n = m-1$, then D is the contact distribution in the strict sense. Note that, if $n = m$, then $D = T(\mathrm{Gr}(n, TM)) \cong TM$.

Definition 3.2. A mapping $F : N \to \mathrm{Gr}(n, TM)$ is called an *integral mapping* of the contact distribution $D \subset T\mathrm{Gr}(n, TM)$ or a *D-integral mapping* if $F_*(TN) \subset D$. If $\dim(N) = n$, then we call an integral mapping $f : N^n \to \mathrm{Gr}(n, TM)$ of the contact structure $D \subset T\mathrm{Gr}(n, TM)$ a *Legendre mapping* in a generalised and wider sense.

Lemma 3.3. *A mapping $F : N^n \to \mathrm{Gr}(n, TM)$ is a Legendre mapping if and only if, $(\pi \circ F)_*(T_tN) \subseteq F(t), (t \in N)$. If F is Legendre and $\pi \circ F$ is an immersion at $t \in N$, then $F(t) = (\pi \circ F)_*(T_tN)$.*

Proof: By definition, F is Legendre if and only if, for any $t \in N$, $F_*(T_tN) \subset D_{F(t)}$. Since $D_{F(t)} = \pi_*^{-1}(F(t))$, regarding $F(t)$ as an n-plane in $T_{(\pi\circ F)(t)}M$, the condition is equivalent to that $\pi_*(F_*(T_tN)) \subseteq F(t)$, that is, $(\pi \circ F)_*(T_tN) \subseteq F(t)$, for any $t \in N$. Moreover if $\pi \circ F$ is an immersion at $t \in N$, then $\dim((\pi \circ F)_*(T_tN)) = n$. Therefore we have $(\pi \circ F)_*(T_tN) = F(t)$. □

The following result shows one of fundamental properties of the canonical differential systems (the generalised contact distributions).

Proposition 3.4. *Let $\Phi : M \to M'$ be a diffeomorphism. Let $0 \leq n \leq m = \dim(M)$. Let D denote the contact distribution of $\mathrm{Gr}(n, TM')$ as well as that of $\mathrm{Gr}(n, TM)$. Then, for any $(x, V) \in \mathrm{Gr}(n, TM)$, we have*

$$(\Phi_\sharp)_*(D_{(x,V)}) = D_{(\Phi(x),\Phi_*(V))}.$$

In particular we have $(\Phi_\sharp)_*(D) = D \subset T(\mathrm{Gr}(n, TM'))$ (see Lemma 3.1).

Proof: Let $v \in D_{(x,V)}$. Then $\pi_*v \in V$. Then we have, by Lemma 3.1,

$$\pi_*((\Phi_\sharp)_*(v)) = (\pi \circ \Phi_\sharp)_*(v) = (\Phi \circ \pi)_*(v) = \Phi_*(\pi_*v) \in \Phi_*(V).$$

Therefore we have $(\Phi_\sharp)_*(D_{(x,V)}) \subseteq D_{(\Phi(x),\Phi_*(V))}$. The converse inclusion is obtained by considering Φ^{-1}, or, by counting the dimension of the vector spaces. □

We conclude this section by the coordinate description of the contact distribution: Take the Grassmannian coordinates $(x^1, \ldots, x^m, a^k_j, (1 \leq j \leq n, n+1 \leq k \leq m))$ of $\mathrm{Gr}(n, TM)$ on an open set $\Omega \subset \mathrm{Gr}(n, TM)$. Set

$$\theta^k := dx^k - \sum_{j=1}^{n} a^k_j dx^j, \ (n+1 \leq k \leq m).$$

Lemma 3.5. *Let* $0 \leq n \leq \dim(M)$. *The local description of the contact distribution* D *of* $\mathrm{Gr}(n, TM)$ *is given by*

$$D|_{T\Omega} = \{v \in T\Omega \mid \theta^{n+1}(v) = 0, \ldots, \theta^m(v) = 0\}.$$

Proof: Let $(x, V) \in \Omega$ and $v = \sum_{i=1}^m b^i \partial/\partial x^i + \sum_{j,k} c^k_j \partial/\partial a^k_j \in T_{(x,V)}\Omega$. Then $v \in D_{(x,V)}$ if and only if $\pi_*(v) \in V$. Now $V = \langle h_1, \ldots, h_n \rangle_{\mathbf{R}}$ in terms of the above basis (described after Lemma 3.1). Then the condition is equivalent to that $\sum_{i=1}^m b^i \partial/\partial x^i = \sum_{j=1}^n \lambda^j h_j$ for some $\lambda^1, \ldots, \lambda^n \in \mathbf{R}$, which is equivalent to that $b^j = \lambda^j, 1 \leq j \leq n$ and $b^k = \sum_{j=1}^n b^j a^k_j, n+1 \leq k \leq m$, and thus equivalent to that $\theta^k(v) = 0$, $n+1 \leq k \leq m$. □

§4. Generalised frontals

We give the exact definition of our main notion in this paper:

Definition 4.1. Let N be an n-dimensional manifold and M an m-dimensional manifold with $n \leq m$. A map-germ $f : (N, a) \to M$ is called a *frontal map-germ* or a *frontal* in a generalised sense, if there exists a germ of Legendre lift $\widetilde{f} : (N, a) \to \mathrm{Gr}(n, TM)$ of f, that is, if there exists an open neighbourhood U of a and a D-integral lift $\widetilde{f} : U \to \mathrm{Gr}(n, TM)$ of f for the canonical distribution $D \subset T\mathrm{Gr}(n, TM)$ and for the canonical projection $\pi : \mathrm{Gr}(n, TM) \to M$, which satisfies that $f_*(T_tN) \subseteq \widetilde{f}(t)$ for any $t \in U$ and $\pi \circ \widetilde{f} = f$.

We call a mapping $f : N^n \to M^m$ a *frontal mapping* or a *frontal* in a generalised sense, if, for any point $a \in N$, the germ of f at a is a frontal.

Remark 4.2. Note that, in the equi-dimensional case $n = m$, any mapping $f : N \to M$ is a frontal. In fact the mapping $\widetilde{f} : N \to \mathrm{Gr}(m, TM)$ defined by $\widetilde{f}(t) := T_{f(t)}M$ is a Legendre lift of f.

Proposition 4.3. *Let $f : (N, a) \to (M, f(a))$ and $g : (N', a') \to (M', f(a'))$ be map-germs. If f is a frontal and g is right-left equivalent to f, then g is a frontal.*

Proof: Suppose $g \circ \varphi = \Phi \circ f$ for some diffeomorphism-germs $\varphi : (N, a) \to (N', a')$ and $\Phi : (M, f(a)) \to (M', f(a'))$. Let $\widetilde{f} : (N, a) \to \mathrm{Gr}(n, TM)$ be a Legendre lift of f. Set $\widetilde{g} := \Phi_\sharp \circ \widetilde{f} \circ \varphi^{-1} : (N', a') \to \mathrm{Gr}(n, TM')$. For $t' \in (N, a')$, we have, by Proposition 3.4,

$$\widetilde{g}_*(T_{t'}N') = (\Phi_\sharp)_*(\widetilde{f}_*(\varphi_*^{-1}(T_{t'}N))) = (\Phi_\sharp)_*(\widetilde{f}_*(T_{\varphi^{-1}(t)}N)) \subset (\Phi_\sharp)_*D = D.$$

Therefore $\widetilde{g}$ is Legendre. Moreover, by Lemma 3.1, we have

$$\pi \circ \widetilde{g} = \pi \circ \Phi_\sharp \circ \widetilde{f} \circ \varphi^{-1} = \Phi \circ \pi \circ \widetilde{f} \circ \varphi^{-1} = \Phi \circ f \circ \varphi^{-1} = g.$$

Therefore $\widetilde{g}$ is a Legendre lifting of g, and hence g is a frontal. □

Definition 4.4. A map-germ $f : (N, a) \to M$ is called a *front* in the generalised sense if there exists a Legendre lift $\widetilde{f} : (N, a) \to \mathrm{Gr}(n, TM)$ of f such that $\widetilde{f}$ is an immersion-germ. A mapping $f : N^n \to M^m$ is called a *front* in the generalised sense if, for any $a \in N$, the germ of f at a is a front.

A map-germ $f : (N, a) \to M$ is a front in the generalised sense if and only if there exists a representative of f which is a front. The condition that $f : N \to M$ is a front in the generalised sense is equivalent to the local existence, at each point of N, of an immersive lift $\widetilde{f} : U \to \mathrm{Gr}(n, TM)$ of f satisfying $f_*(T_tN) \subset \widetilde{f}(t), (t \in U)$.

§5. Density function

The notion of density functions is a key to understand the geometry of frontals, which was introduced in [71][25][95] first. We introduce its generalisation (see also [59][60]):

Proposition 5.1. *Let $f : (N, a) \to M$ be a map-germ with* $\dim(N) = n \leq m = \dim(M)$. *Then the following conditions are equivalent:*
(1) *f is a frontal map-germ.*
(2) *There exists a frame $h_1, h_2, \dots, h_n : (N, a) \to TM$ along f and a function-germ $\sigma : (N, a) \to \mathbf{R}$ such that*

$$\left(\frac{\partial f}{\partial t_1} \wedge \frac{\partial f}{\partial t_2} \wedge \cdots \wedge \frac{\partial f}{\partial t_n}\right)(t) = \sigma(t)(h_1 \wedge h_2 \wedge \cdots \wedge h_n)(t),$$

as germs of n-vector fields $(N, a) \to \wedge^n TM$ over f. Here $t_1, t_2, \dots, t_n$ are coordinates on (N, a).

The function $\sigma : (N, a) \to \mathbf{R}$ in Proposition 5.1 is called a *signed area density function* or briefly an *s-function* of the frontal f associated with the frame. Note that the function σ is essentially the same thing with the function λ introduced in [71][25] in the case $\dim(M) = 3$.

Two function-germs $\sigma, \widetilde{\sigma} : (N, a) \to \mathbf{R}$ are called *$\mathcal{K}$-equivalent* if there exists a diffeomorphism-germ $T : (N, a) \to (N, a)$ and a non-vanishing function-germ $c : (N, a) \to \mathbf{R}$, $c(a) \neq 0$, such that $\widetilde{\sigma}(T(t)) = c(t)\sigma(t), (t \in (N, a))$ (see [75]).

Lemma 5.2. *The $\mathcal{K}$-equivalence class of a signed area density function σ is independent of the choice of the frame $h_1, h_2, \ldots, h_n$ and of the coordinates $t_1, t_2, \ldots, t_n$ on $(\mathbf{R}^n, a)$ and depend only on the frontal f.*

Proof: Let us take another frame $k_1, \ldots, k_n$. Then there exists $A = (a_{ij}) : (\mathbf{R}^n, a) \to \mathrm{GL}(n, \mathbf{R})$ such that $(h_1, \ldots, h_n) = (k_1, \ldots, k_n)A$. Then $h_1 \wedge h_2 \wedge \cdots \wedge h_n = (\det A)(k_1 \wedge k_2 \wedge \cdots \wedge k_n)$. Therefore σ is transformed to $(\det A)\sigma$. Let us take another coordinates $T_1, T_2, \ldots, T_n$ on $(\mathbf{R}^n, a)$. Then

$$\left(\frac{\partial f}{\partial T_1} \wedge \frac{\partial f}{\partial T_2} \wedge \cdots \wedge \frac{\partial f}{\partial T_n}\right)(T(t)) = J(t)\left(\frac{\partial f}{\partial t_1} \wedge \frac{\partial f}{\partial t_2} \wedge \cdots \wedge \frac{\partial f}{\partial t_n}\right)(t),$$

where $J(t)$ is the Jacobian function $\partial(T_1, \ldots, T_n)/\partial(t_1, \ldots, t_n)$ at t. Therefore $\sigma(t)$ is transformed to the function $J(t)\sigma(T(t))$. Thus we have the required result. □

We call the signed density function of a frontal, considered up to $\mathcal{K}$-equivalence, a *density function* of the frontal. The singular locus (non-immersive locus) $S(f)$ of f coincides with the zero locus $\{\sigma = 0\}$ of the density function σ.

§6. Proper frontals

Frontals can be collapsing in general. For example, any constant mapping $f : N \to M$ is a frontal. In fact any lifting $F : N \to \mathrm{Gr}(n, TM)$ of f is Legendre in that case. See also Example 2.5.

Definition 6.1. A frontal $f : N \to M$ is called a *proper frontal* if the regular locus

$$R(f) := \{t \in N \mid f_* : T_tN \to T_{f(t)}M \text{ is injective.}\}$$

of f is dense in N. A germ of frontal $f : (N, a) \to M$ is called a *germ of proper frontal* if there exists a representative of f which is a proper frontal.

Note that $R(f)$ is an open subset of N in general. Then the condition that f is a proper frontal requires that $R(f)$ is open and dense.

The fundamental property of proper frontals is the following:

Proposition 6.2. *Let $f : N \to M$ be a proper frontal. Then there exists the unique global Legendre (i.e. D-integral) lift $\widetilde{f} : N \to$ $\mathrm{Gr}(n, TM)$ of f, for the canonical projection $\pi : \mathrm{Gr}(n, TM) \to M$, $\pi \circ \widetilde{f} = f$. Here D is the contact distribution on* $\mathrm{Gr}(n, TM)$, $n = \dim(N)$, *introduced in §4.*

Proof: Consider the mapping $F : R(f) \to \mathrm{Gr}(n, TM)$ defined by $F(t) = f_*(T_tN) \in \mathrm{Gr}(n, T_{f(t)}M) \subset \mathrm{Gr}(n, TM)$. Then F is a D-integral mapping and $\pi \circ F = f|_{R(f)}$. By Lemma 3.3, F is a unique Legendre lifting of $f|_{R(f)}$. Since f is a frontal, for any $a \in N$, there exists an open neighbourhood U of a and a D-integral lift $\widetilde{f} : U \to \mathrm{Gr}(n, TM)$ of f. Then by the uniqueness of F, we have $\widetilde{f} = F$ on $U \cap R(f)$. Since f is a proper frontal, $R(f)$ is dense in N, and therefore $U \cap R(f)$ is dense in U. Thus the Legendre lift F of f is uniquely extended to $U \cup R(f)$. Since a is arbitrary, we have the unique Legendre lift $\widetilde{f} : N \to \mathrm{Gr}(n, TM)$ of f. □

Proposition 6.3. *If $f : N \to M$ is a frontal and it has a unique Legendre lift $\widetilde{f} : N \to \mathrm{Gr}(n, TM)$, then f is a proper frontal.*

Proof: Suppose the regular locus $R(f)$ of f is not dense in N. Then there exists a non-void open subset $U \subset N$ such that the maximal rank of $f|_U$ is $\ell < n$. Then there exists a non-void open subset $V \subset U$ such that $f|_V$ is of constant rank ℓ. Then there exists a non-void open subset $W \subset V$ and an open subset $\Omega \subset M$ such that $f|_W : W \to \Omega$ is right-left equivalent to $h : \mathbf{R}^n \to \mathbf{R}^m$ which is defined by $h(s_1, \dots, s_\ell, s_{\ell+1}, \dots, s_n) = (s_1, \dots, s_\ell, 0, \dots, 0)$ ("Rank theorem", see [15]). Let $\widetilde{f} : N \to \mathrm{Gr}(n, TM)$ be a Legendre lift and $\widetilde{h} : \mathbf{R}^n \to \mathrm{Gr}(n, T\mathbf{R}^m)$ be the induced lift of h by $\widetilde{f}|_W$ (cf. Proposition 4.3). Then $T_{h(s)}(\mathbf{R}^\ell \times \{0\}) \subset \widetilde{h}(s), (s \in \mathbf{R}^n)$. Then there exists a non-trivial perturbation of $\widetilde{h}$ therefore of $\widetilde{f}$ with compact support. □

Remark 6.4. *Proper frontals are not generic* in C^∞-topology in general. In fact the frontal mapping $f : \mathbf{R}^2 \to \mathbf{R}^3$ constructed in Example 2.5 can not be approximated by any proper frontal.

Now we introduce the notion of non-degenerate frontals which was originated in [71].

Definition 6.5. We say that a frontal $f : (N, a) \to M$ has a *non-degenerate* singular point at a if the density function σ of f satisfies that

$\sigma(a) = 0$ and $d\sigma(a) \neq 0$. Note that the condition is invariant under the $\mathcal{K}$-equivalence of σ (see Proposition 5.2).

To study the property of non-degenerate singular points of frontals, we recall the following result.

Lemma 6.6. *Let N be a manifold of dimension n. Let $g : (N, a) \to (N, g(a))$ be a map-germ. Let J_g denote the Jacobi matrix of g and $\det(J_g) : (N, a) \to \mathbf{R}$ the Jacobian determinant of g. Suppose $(\det J_g)(a) = 0$. Then $(d\det(J_g))(a) = 0$ if $\mathrm{rank}(J_g)(a) \leq n - 2$, that is, if g is of corank ≥ 2 at a.*

Proof: It is easy to see, as a fundamental fact in the linear algebra, for the determinant function $\det : M(n, n; \mathbf{R})$ on the space of $n \times n$-matrices, and for any $A \in M(n, n; \mathbf{R})$ with $\det(A) = 0$, $(d\det)(A) = 0$ if and only if $\mathrm{rank}(A) \leq n - 2$. Then we have, if $\mathrm{rank}(J_g) \leq n - 2$, then $(d\det(J_g))(p) = (J_g)^*(d\det)(p) = 0$. □

Lemma 6.7. *If a frontal $f : (N, a) \to M$ has a non-degenerate singular point at a, then f is of corank 1 such that the singular locus $S(f) \subset (N, a)$ is a regular hypersurface.*

Proof: Let us take a representative $f : U \to M$ of f, using the same symbol, satisfying that $d\sigma(t) \neq 0$ for any $t \in U$. Then $S(f) = \{t \in U \mid \sigma(t) = 0\}$ is a regular hypersurface of U. In particular $S(f)$ is nowhere dense in U. Therefore f is a proper frontal. Let $\widetilde{f} : U \to \mathrm{Gr}(n, TM)$ be the unique Legendre lifting of f. Set $V = \widetilde{f}(a) \subset T_{f(p)}M$. Take a local coordinate system $(x^1, \dots, x^n, x^{n+1}, \dots, x^m)$ around $f(a)$ of M such that $V = \langle (\partial/\partial x^1)(f(a)), \dots, (\partial/\partial x^n)(f(a)) \rangle_{\mathbf{R}}$ Define $g : U \to \mathbf{R}^n$ by $g = (x^1 \circ f, \dots, x^n \circ f)$, deleting U if necessary. Then the rank of g_* at a is equal to the rank of f_* at a. Moreover the signed area density function of g is $\mathcal{K}$-equivalent to that of f. Note that the signed area density function of g is $\mathcal{K}$-equivalent to the Jacobian determinant of g. Suppose the rank of g_* at p is less than $n - 1$. Then by Lemma 6.6 we see that $(d\sigma)(a) = d(\det(J_g))(a) = 0$. This leads a contradiction to the assumption of non-degeneracy. Therefore we have that the rank of g_* is equal to $n - 1$. Thus we have the required result. □

§7. Tangent bundles and complementary bundles

Let $f : N \to M$ be a proper frontal. Let $\dim(N) = n$ and $\widetilde{f} : N \to \mathrm{Gr}(n, TM)$ be the unique Legendre lifting of f (Proposition 6.2). Then we have a subbundle T_f of the pull-back bundle f^*TM over N defined by

$$T_f := \{(t, v) \in N \times TM \mid v \in \widetilde{f}(t)\} \subset f^*TM.$$

We call T_f the *tangent bundle* to the proper frontal f. Moreover we call the quotient bundle $Q_f := f^*TM/T_f$ the *complementary bundle.*

Definition 7.1. A proper frontal $f : N \to M$ is called *oriented* (resp. *co-oriented*) if the bundle T_f (resp. Q_f) is oriented. f is called *orientable* (resp. *co-orientable*) if T_f (resp. Q_f) is orientable.

Example 7.2. The proper front $f : S^1(\subset \mathbf{C}) \to \mathbf{R}^2(= \mathbf{C})$ defined by $z \mapsto 2z - \overline{z}^2$, for $z \in \mathbf{C}, |z| = 1$ ("cardioid") is not orientable nor co-orientable. The half cuspidal edge $h : \mathbf{R}^2 \setminus \{(0,0)\} \to \mathbf{R}^3$ (see Example 2.5) restricted to $\mathbf{R}^2 \setminus \{(0,0)\}$ is a proper front which is not orientable nor co-orientable. The mapping $\mathbf{R}^2 \to \mathbf{R}^3$ defined by the normal form of the cuspidal edge (resp. folded umbrella) is a proper front (resp. frontal) which is orientable and co-orientable.

Let $f : N \to M$ be a proper frontal. Then the bundle homomorphism $\varphi_f : TN \to T_f, \varphi(t,v) = (t, f_*(v))$ is induced. Then we have

$$\begin{aligned} R(f) &= \{t \in N \mid f_* : T_tN \to T_{f(t)}M \text{ is injective}\} \\ &= \{t \in N \mid \varphi_f \text{ is injective at } t\}. \end{aligned}$$

The notion of frontals will play important role in differential geometry. Therefore the following observations are important. First we treat the case of hypersurfaces ($m = n + 1$).

Lemma 7.3. *If M is endowed with a Riemannian metric, then $f : N^n \to M^{n+1}$ is a frontal if and only if, for any $a \in N$, there exists an open neighbourhood U of a and a unit vector field ν along f such that $\nu(t)$ is normal to the subspace $f_*(T_tN)$ for any $t \in U$.*

Proof: Let f be frontal. Let $\widetilde{f}(t) = (f(t), [\alpha(t)])$ be a Legendre lifting of f. It defines the local integral tangential hyperplane field $\mathrm{Ker}(\alpha(t))$ along f. Then we associate the normal line field $\mathrm{Ker}(\alpha(t))^{\perp}$ with $\widetilde{f}(t)$ and take a local unit frame $\nu(t)$ of $\mathrm{Ker}(\alpha(t))^{\perp}$. Conversely let $\nu(t)$ be a local unit normal field along f with $f_*(T_tN)$. Regarding the metric, we associate a non-zero cotangent vector field $\alpha(t)$ with $\nu(t)$ so that $\mathrm{Ker}(\alpha(t)) = \nu(t)^{\perp}$. Then the tangential hyperplane field $\nu(t)^{\perp}$ satisfies the condition $f_*(T_tN) \subseteq \nu(t)^{\perp}$. The condition is equivalent to that $\widetilde{f}(t) = (f(t), [\alpha(t)])$ is a Legendre map. □

Lemma 7.4. *If M is endowed with a Riemannian metric, then $f : N^n \to M^{n+1}$ is a front if and only if locally there exists a normal unit vector field ν along f such that (f, ν) is an immersion to the unit tangent bundle T_1M.*

Proof: Regarding each unit vector $\nu \in T_xM$ as an element of T_x^*M by $v \mapsto \nu \cdot v$, we have the natural double covering $T_1M \to P(T^*M)$. Therefore we have required result by Lemma 7.3. □

In generalised cases, we have:

Lemma 7.5. *If M is endowed with a Riemannian metric, then $f : N^n \to M^m$ is a frontal if and only if, for any $a \in N$, there exists an open neighbourhood U of a and a system of orthonormal vector fields $\nu_1, \dots, \nu_{m-n}$ over U along f such that $\nu_i(t)$ is normal to the subspace $f_*(T_tN)$ for any $t \in U$, $i = 1, \dots, m-n$.*

Proof: Suppose f is a frontal. For any a, let $\widetilde{f} : U \to \mathrm{Gr}(n, TM)$ be a Legendre local lifting of $f|_U$. Deleting U if necessary, take an orthonormal frame $h_1, \dots, h_n, \nu_1, \dots, \nu_{m-n}$ on U such that $h_1(t), \dots, h_n(t)$ form a basis of $\widetilde{f}(t) \subset T_{f(t)}M$ for any $t \in U$. Then $\nu_1, \dots, \nu_{m-n}$ satisfy the required condition. Conversely we may set $\widetilde{f}(t) = \langle \nu_1(t), \dots, \nu_{m-n}(t)\rangle^{\perp}$. Then $\pi \circ \widetilde{f} = f$ and $(\pi \circ \widetilde{f})_*(T_tN) = f_*(T_tN) \subset \widetilde{f}(t)$, hence $\widetilde{f}$ is Legendre by Lemma 3.3. □

The following is clear:

Lemma 7.6. *If M is a Riemannian manifold, then the condition that $f : N^n \to M^m$ is a front is equivalent to the local existence of an orthonormal unit frame $\nu_1, \dots, \nu_n$ along f such that $t \mapsto (f(t), \langle \nu_1(t), \dots, \nu_n(t)\rangle^{\perp})$ is an immersion to $\mathrm{Gr}(n, TM)$.*

Let $f : N \to M$ be a proper frontal. If M is endowed with a Riemannian metric, then we define the *normal bundle* to f by

$$N_f := \{(t, w) \in N \times TM \mid w \in \widetilde{f}(t)^{\perp}\} \subset f^*TM,$$

which is isomorphic to the complementary bundle Q_f (see §7). Note that both bundles T_f and N_f have induced Riemannian bundle structures from TM.

§8. Openings and frontals

In this section, we review the known results on "geometric" openings.

We denote by $\mathcal{E}_{N,a}$ the $\mathbf{R}$-algebra of C^∞ function-germs on (N, a) with the maximal ideal $\mathfrak{m}_{N,a}$. If $(N, a) = (\mathbf{R}^n, 0)$ is the origin, then we use $\mathcal{E}_n, \mathfrak{m}_n$ instead of $\mathcal{E}_{N,a}, \mathfrak{m}_{N,a}$ respectively.

Definition 8.1. ([30][35]) Let $f : (N, a) \to (M, b)$ be a C^∞ map-germ with $\dim(N) = n \leq m = \dim(M)$. We define the *Jacobi module* of f:

$$\mathcal{J}_f := \mathcal{E}_{N,a}\, d(f^*\Omega_{M,b}) = \left\{ \sum_{j=1}^{m} a_j \; df^j \mid a_j \in \mathcal{E}_{N,a}, 1 \leq j \leq m \right\}$$

in the space $\Omega^1_{N,a}$ of 1-form germs on (N, a). Here $f^j = x^j \circ f$, for a system of coordinates $(x^1, \ldots, x^m)$ of (M, b). Further we define the *ramification module* $\mathcal{R}_f$ by

$$\mathcal{R}_f := \{h \in \mathcal{E}_{N,a} \mid dh \in \mathcal{J}_f\}.$$

Example 8.2. Let μ be a positive integer and $g : (\mathbf{R}, 0) \to (\mathbf{R}, 0)$ a map-germ defined by $g(t) = t^\mu$. Then $\mathcal{J}_g = \mathfrak{m}_1^{\mu-1} dt$ and $\mathcal{R}_g = \mathbf{R} + \mathfrak{m}_1^\mu$. Here $\mathfrak{m}_1^\mu = t^\mu \mathcal{E}_1 = \{h \in \mathcal{E}_1 \mid \frac{d^k h}{dt^k}(0) = 0, (0 \leq k \leq \mu)\}$. In fact, since $dg = \mu t^{\mu-1} dt$, we gave $\mathcal{J}_g = \mathfrak{m}_1^{\mu-1} dt$. Moreover, for a $k \in \mathcal{E}_1$, we have that $k \in \mathcal{R}_f$ if and only if $\frac{dk}{dt} \in \mathfrak{m}_1^{\mu-1}$ if and only if $k \in \mathbf{R} + \mathfrak{m}_1^\mu$.

Note that $\mathcal{J}_f$ is just the first order component of the graded differential ideal $\mathcal{J}_f^\bullet$ in $\Omega^\bullet_{N,a}$ generated by $df^1, \ldots, df^m$. Then the singular locus is given by $S(f) = \{x \in (N, a) \mid \operatorname{rank} \mathcal{J}_f(x) < n\}$. Also we consider the *kernel field* $\operatorname{Ker}(f_* : TN \to TM)$, of f near a. Then we see that, for another map-germ $f' : (N, a) \to (M', b')$ with $\mathcal{J}_{f'} = \mathcal{J}_f$, $n \leq m'$, we have $\mathcal{S}_{f'} = \mathcal{S}_f$ and $\operatorname{Ker}(f'_*) = \operatorname{Ker}(f_*)$. Note that related notion was introduced in [78].

Lemma 8.3. *Let $f : (N, a) \to (M, b)$ be a map-germ. Then we have:*
(1) $f^*\mathcal{E}_{M,b} \subset \mathcal{R}_f \subset \mathcal{E}_{N,a}$ and $\mathcal{R}_f$ is an $\mathcal{E}_{M,b}$-module via f^*.
(2) For another map-germ $f' : (N, a) \to (M', b')$, $\mathcal{J}_{f'} = \mathcal{J}_f$ if and only if $\mathcal{R}_{f'} = \mathcal{R}_f$.
(3) If $\tau : (M, b) \to (M', b')$ is a diffeomorphism-germ, then $\mathcal{R}_{\tau \circ f} = \mathcal{R}_f$. If $\sigma : (N', a') \to (N, a)$ is a diffeomorphism-germ, then $\mathcal{R}_{f \circ \sigma} = \sigma^*(\mathcal{R}_f)$.

Proof: (1) follows from that, if $h \in \mathcal{R}_f$ and $dh = \sum_{j=1}^m p_j df_j$, then we have

$$d\{(k \circ f)h\} = \sum_{j=1}^{m} \{(k \circ f)p_j + h\,(\partial k/\partial y_j)\}\, df_j.$$

(2) It is clear that $\mathcal{J}_{f'} = \mathcal{J}_f$ implies $\mathcal{R}_{f'} = \mathcal{R}_f$. Conversely suppose $\mathcal{R}_{f'} = \mathcal{R}_f$. Then any component f'_j of f' belongs to $\mathcal{R}_{f'} = \mathcal{R}_f$, hence $df_j \in \mathcal{J}_f$. Therefore $\mathcal{J}_{f'} \subset \mathcal{J}_f$. By the symmetry we have $\mathcal{J}_{f'} = \mathcal{J}_f$.

(3) follows from that $\mathcal{J}_{\tau \circ f} = \mathcal{J}_f$ and $\mathcal{J}_{f \circ \sigma} = \sigma^*(\mathcal{J}_f)$. □

Definition 8.4. Let $f : (N, a) \to M$ and $g : (N', a') \to M'$ be map-germs. Then f and g are called *$\mathcal{J}$-equivalent* if there exists a diffeomorphism-germ $\sigma : (N, a) \to (N', a')$ such that $\mathcal{J}_{g\circ\sigma} = \mathcal{J}_f$. Here $\mathcal{J}_f = \mathcal{E}_{N,t_0} f^*\Omega^1_{M,f(t_0)}$ (see Definition 8.1). Note that $\dim(M)$ and $\dim(M')$ can be different.

Definition 8.5. Let $f : (\mathbf{R}^n, a) \to (\mathbf{R}^m, b)$ be a map-germ. Let $h_1, \dots, h_r \in \mathcal{R}_f$. Then the map-germ $F : (\mathbf{R}^n, a) \to \mathbf{R}^m \times \mathbf{R}^r = \mathbf{R}^{m+r}$ defined by

$$F = (f_1, \dots, f_m, h_1, \dots, h_r)$$

is called an *opening* of f, while f is called a *closing* of F.

Proposition 8.6. *Let M, N be a manifold of dimension m, n respectively. A map-germ $f : (N, a) \to M$ is a frontal if and only if f is right-left equivalent to an opening of a map-germ $g : (\mathbf{R}^n, 0) \to (\mathbf{R}^n, 0)$.*

Proof: Suppose $f : (N, a) \to M$ be a frontal map-germ. Then, since $\widetilde{f}(a)$ is an n-dimensional vector subspace of $T_{f(a)}M$, there exists of a system of local coordinates of $(M, f(a))$

$$y_1, \dots, y_n, z_1, \dots, z_k, (k = m - n),$$

such that, for $g = (y_1 \circ f, \dots, y_n \circ f)$, the component $z_j \circ f, (1 \leq j \leq k)$ belongs to the $\mathcal{R}_g$ and therefore f is right-left equivalent to an opening of g. Conversely, suppose f is right-left equivalent to an opening $G : (\mathbf{R}^n, a) \to \mathbf{R}^{n+k} = \mathbf{R}^m$ of a germ $g = (g_1, \dots, g_n) : (\mathbf{R}^n, a) \to (\mathbf{R}^n, g(a))$. Set $G = (g_1, \dots, g_\ell, h_1, \dots, h_k)$. Then, since $h_j \in \mathcal{R}_g, 1 \leq j \leq n$ $dh_j = \sum_{i=1}^n a^i_j dg_i$, for some function-germs $a^i_j : (\mathbf{R}^n, a) \to \mathbf{R}$. Define $\widetilde{G} : (\mathbf{R}^n, a) \to \mathrm{Gr}(n, T\mathbf{R}^m)$, in terms of Grassmannian coordinates,

$$\widetilde{G}(t) = \left(g_1(t), \dots, g_\ell(t), h_1(t), \dots, h_k(t), a^i_j(t)\right), \quad (t \in (\mathbf{R}^n, a)).$$

Then, by Lemma 3.5, $\widetilde{G}$ is a D-integral lift of g. Therefore G is an ℓ-frontal, and so is f. □

§9. Versal openings

Definition 9.1. An opening $F = (f, h_1, \dots, h_r)$ of f is called a *versal opening* (resp. a *mini-versal opening*) of $f : (\mathbf{R}^n, a) \to (\mathbf{R}^m, b)$, if $1, h_1, \dots, h_r$ form a (minimal) system of generators of $\mathcal{R}_f$ as an $\mathcal{E}_{\mathbf{R}^m,b}$-module via $f^* : \mathcal{E}_{\mathbf{R}^m,b} \to \mathcal{E}_{\mathbf{R}^n,a}$.

A C^∞ map-germ $f : (\mathbf{R}^n, a) \to (\mathbf{R}^m, b)$ is called *analytic* if f is right-left equivalent to a real analytic map-germ ([35]). Moreover f is called a *finite map-germ* if $\mathcal{E}_{\mathbf{R}^n,a}$ is a finite $f^*(\mathcal{E}_{\mathbf{R}^m,b})$-module. Then f is finite if and only if $\dim_{\mathbf{R}}(\mathcal{E}_{\mathbf{R}^n,a}/\langle f_1, \dots, f_m\rangle_{\mathcal{E}_{\mathbf{R}^n,a}} < \infty$. If f is analytic, then f is finite if and only if its complexification has isolated zero set ([100]). By $\mathfrak{m}_{\mathbf{R}^m,a}$, we denote the maximal ideal of $\mathcal{E}_{\mathbf{R}^m,a}$ which consists of function-germs vanishing at 0. By the projection $\pi_m : \mathbf{R}^{m+r} = \mathbf{R}^m \times \mathbf{R}^r \to \mathbf{R}^m$ we regard $\mathbf{R}^{m+r}$ as an affine bundle over $\mathbf{R}^m$. If f is finite and analytic, then, in the analytic category, $\mathcal{R}^\omega f$ is a finite $\mathcal{O}_{\mathbf{R}^m,b}$-module.

We summarise the known results on the existence of versal openings:

Theorem 9.2. *Let $f : (\mathbf{R}^n, a) \to (\mathbf{R}^m, b)$ be a map-germ. Suppose that* (I) *f is finite and of corank at most one, or* (II) *f is finite analytic. Then we have*
(1) *The ramification module $\mathcal{R}_f$ of f is a finitely generated $f^*(\mathcal{E}_{\mathbf{R}^m,b})$-module. In particular f has a versal opening.*
(2) *$1, h_1, \dots, h_r \in \mathcal{R}_f$ form a system of generators of $\mathcal{R}_f$ as a $f^*(\mathcal{E}_{\mathbf{R}^m,b})$-module if and only if the residue classes $1, \overline{h}_1, \dots, \overline{h}_r$ form an $\mathbf{R}$-basis of the vector space $V = \mathcal{R}_f/f^*(\mathfrak{m}_{\mathbf{R}^m,b})\mathcal{R}_f$. In particular there exists a versal opening $F = (f, h_1, \dots, h_r)$ of f. If $r = \dim_{\mathbf{R}} V - 1$, then F is called a mini-versal opening of f.*
(3) *For any versal opening $F : (\mathbf{R}^n, a) \to \mathbf{R}^{m+r}$ of f and for any opening $G : (\mathbf{R}^n, a) \to \mathbf{R}^{m+s}$ of f, there exists an affine bundle map $\Psi : (\mathbf{R}^{m+r}, F(a)) \to (\mathbf{R}^{m+s}, G(a))$ such that $G = \Psi \circ F$.*
(4) *For any two mini-versal openings $F, F' : (\mathbf{R}^n, a) \to \mathbf{R}^{m+r}$ of f, there exists an affine bundle isomorphism $\Phi : (\mathbf{R}^{m+r}, F(0)) \to (\mathbf{R}^{m+r}, F'(0))$ such that $F' = \Phi \circ F$.*

Proof of Theorem 9.2: Case (I): (1) is proved as Lemma 2.1 of [35]. Then $f^* : \mathcal{E}_{\mathbf{R}^m,b} \to \mathcal{R}_f$ is a homomorphism of differentiable algebras in the sense of Malgrange [74]. Therefore, by Malgrange's preparation theorem ([74] Corollary 4.4) we have (2). Case (II): Let $1, h_1, \dots, h_r$ generate $\mathcal{R}^\omega f$ over $\mathcal{O}_{\mathbf{R}^m,b}$ via f^*. Then $1, h_1, \dots, h_r$ generate $\mathcal{R}_f$ over $\mathcal{E}_{\mathbf{R}^m,b}$ via f^* by Proposition 5.2 of [45]. Therefore we have (1) and (2). The assertion (3) is clear from the definitions. (4) follows from (2). □

We do not repeat the proofs of Lemma 2.1 in [35] nor Proposition 5.2 in [45]. However we give an exposition, relating the theory of C^∞-rings, on Malgrange's preparation theorem in §22 of this paper.

§10. Subfrontals and superfrontals

Relating the theory of frontals with that of openings, we are led to the following generalisations of frontals naturally.

Definition 10.1. Let M be a manifold and ℓ an integer with $0 \leq \ell \leq \dim(M)$. A map-germ $f : (N, a) \to M$ is called an *ℓ-frontal* if there exists a D-integral lift $\widetilde{f} : (N, a) \to \mathrm{Gr}(\ell, TM)$ of f. Here we do not assume that $\ell = \dim(N)$ and D is the contact distribution on $\mathrm{Gr}(\ell, TM)$. The condition on the C^∞ mapping $\widetilde{f}$ is that $f_*(T_t N) \subset \widetilde{f}(t) \in \mathrm{Gr}(\ell, T_{f(t)}M)$ for any $t \in (N, a)$. If $0 < \ell < \dim(N)$, then an ℓ-frontal is called a *subfrontal.* If $\dim(N) < \ell < \dim(M)$, then an an ℓ-frontal is called a *superfrontal.*

Proposition 10.2. *Let $f : (N, a) \to M$ be an ℓ-frontal and $f' : (N', a') \to M'$ be right-left equivalent to f. Then also f' is an ℓ-frontal.*

Proof: The proof is performed similarly to Proposition 4.3 using Proposition 3.4. □

Proposition 10.3. *Let M, N be a manifold of dimension m, n respectively and ℓ an integer with $0 \leq \ell \leq m$. A map-germ $f : (N, a) \to M$ is an ℓ-frontal if and only if f is right-left equivalent to an opening of a map-germ $g : (\mathbf{R}^n, 0) \to (\mathbf{R}^\ell, 0)$.*

Proof: Suppose $f : (N, a) \to M$ be an ℓ-frontal map-germ. Then, since $\widetilde{f}(a)$ is an ℓ-dimensional vector subspace of $T_{f(a)}M$, there exists of a system of local coordinates of $(M, f(a))$

$$y_1, \dots, y_\ell, z_1, \dots, z_k, (k = m - \ell),$$

such that, for $g = (y_1 \circ f, \dots, y_\ell \circ f)$, the component $z_j \circ f, (1 \leq j \leq k)$ belongs to the $\mathcal{R}_g$ and therefore f is right-left equivalent to an opening of g. Conversely, suppose f is right-left equivalent to an opening $G : (\mathbf{R}^n, a) \to \mathbf{R}^{\ell+k} = \mathbf{R}^m$ of a germ $g = (g_1, \dots, g_\ell) : (\mathbf{R}^n, a) \to (\mathbf{R}^\ell, g(a))$. Set $G = (g_1, \dots, g_\ell, h_1, \dots, h_k)$. Then, since $h_j \in \mathcal{R}_g, 1 \leq j \leq \ell$, $dh_j = \sum_{i=1}^{\ell} a_j^i dg_i$, for some function-germs $a_j^i : (\mathbf{R}^n, a) \to \mathbf{R}$. Define $\widetilde{G} : (\mathbf{R}^n, a) \to \mathrm{Gr}(\ell, T\mathbf{R}^m)$, in terms of Grassmannian coordinates,

$$\widetilde{G}(t) = \left(g_1(t), \dots, g_\ell(t), h_1(t), \dots, h_k(t), a_j^i(t)\right), \quad (t \in (\mathbf{R}^n, a)).$$

Then, by Lemma 3.5, $\widetilde{G}$ is a D-integral lift of g. Therefore G is an ℓ-frontal, and so is f. □

§11. Algebraic openings

In this section we will utilise the notion of sheaves which describes locally-defined objects ([12]) to introduce an algebraic notion which is related to frontals.

Let N be a manifold. Let $\mathcal{E}_N$ denote the sheaf of C^∞ function-germs on N. For any open subset $U \subset N$, $\mathcal{E}_N(U) = C^\infty(U)$, the C^∞-ring of all real-valued C^∞ functions on U. Note that $\mathcal{E}_N$ has the natural structure of C^∞-ring sheaf. See §22 for the notion of C^∞-rings.

Definition 11.1. Let $\mathcal{F}$ be a sub C^∞-ring sheaves of $\mathcal{E}_N$. The *versal opening* $\widetilde{\mathcal{F}}$ of $\mathcal{F}$ is defined as follows: For any open subset $U \subset N$, $\widetilde{\mathcal{F}}(U)$ is the set of $h \in \mathcal{E}_N(U)$ satisfying that, for any $p \in U$, there exists $g_1, \ldots, g_r \in \mathcal{F}_p$ and $a_1, \ldots, a_r \in \mathcal{E}_{N,p}$ such that

$$dh = \sum_{i=1}^{r} a_i dg_i$$

in $\Omega^1_{N,p}$. Here Ω^1_N means the sheaf of C^∞ 1-form-germs on N, $\mathcal{E}_{N,p}$ (resp. $\Omega^1_{N,p}$) the stalk of $\mathcal{E}_N$ (resp. of Ω^1_N) at p, i.e. the set of germs at p, and $d : \mathcal{E}_{N,p} \to \Omega^1_{N,p}$ the exterior differential.

Let $\mathcal{F}, \mathcal{G}$ be a sub C^∞-ring sheaves of $\mathcal{E}_N$. Then $\mathcal{G}$ is called an *opening* of $\mathcal{F}$ if $\mathcal{F} \subseteq \mathcal{G} \subseteq \widetilde{\mathcal{F}}$.

We have that following basic properties of algebraic openings.

Proposition 11.2. *Let N be a C^∞ manifold and let $\mathcal{E}_N$ denote the sheaf of C^∞ function-germs on N. Let $\mathcal{F}$ be a sub C^∞-ring sheaf of $\mathcal{E}_N$. Then we have*
(1) *$\widetilde{\mathcal{F}}$ is a sub C^∞-ring sheaf of $\mathcal{E}_N$.* (2) *$\mathcal{F} \subset \widetilde{\mathcal{F}}$.* (3) *$\widetilde{\widetilde{\mathcal{F}}} = \widetilde{\mathcal{F}}$.*

Proof: (1) Let $p \in N$. Let $h_1, \ldots, h_r \in \widetilde{\mathcal{F}}_p$ and $f \in C^\infty(\mathbf{R}^r)$. Let $dh_i = \sum_{j=1}^{s_i} a_{ij} dg_{ij}$ for some $a_{ij} \in \mathcal{E}_{N,p}$, $g_{ij} \in \mathcal{F}_p$. Then

$$d(f(h_1, \ldots, h_r)) = \sum_{i=1}^{r} \frac{\partial f}{\partial x_i}(h_1, \ldots, h_r)\, dh_i = \sum_{i=1}^{r} \sum_{j=1}^{s_i} \left(\frac{\partial f}{\partial x_i}(h_1, \ldots, h_r) a_{ij} \right) dg_{ij}.$$

Therefore $f(h_1, \ldots, h_r) \in \widetilde{\mathcal{F}}_p$. (2) Let $p \in N$ and $g \in \mathcal{F}_p$. Then we have $dg = 1 \cdot dg$, and therefore $g \in \widetilde{\mathcal{F}}_p$. (3) Let $p \in N$ and $h \in \widetilde{\widetilde{\mathcal{F}}}_p$. Then $dh = \sum_{i=1}^{r} a_i dh_i$ for some $a_i \in \mathcal{E}_{N,p}, h_i \in \widetilde{\mathcal{F}}_p$. Since $h_i \in \widetilde{\mathcal{F}}_p$ for each i, $dh_i = \sum_{j=1}^{s_i} b_{ij} dg_{ij}$ for some $b_{ij} \in \mathcal{E}_{N,p}$ and $g_{ij} \in \mathcal{F}_p$. Then we have $dh = \sum_{i=1}^{r} \sum_{j=1}^{s_i} (a_i b_{ij}) dg_{ij}$, therefore $h \in \widetilde{\mathcal{F}}_p$. □

We call $\mathcal{F}$ *full* if $\widetilde{\mathcal{F}} = \mathcal{F}$. Then Proposition 11.2 shows that $\widetilde{\mathcal{F}}$ is the minimal full sheaf containing $\mathcal{F}$.

Let $\varphi : N' \to N$ be a C^∞ mapping and $\mathcal{F}$ a subsheaf of $\mathcal{E}_N$ on N. We define a subsheaf $\varphi^*\mathcal{F}$ of $\mathcal{E}_{N'}$ on N' by $(\varphi^*\mathcal{F})_q = \varphi^*(\mathcal{F}_{\varphi(q)})$, where $\varphi^* : \mathcal{E}_{N,\varphi(q)} \to \mathcal{E}_{N',q}$ is defined by $\varphi^*(h) = h \circ \varphi, (h \in \mathcal{E}_{N,\varphi(q)})$. If $\varphi = \Phi$ is a diffeomorphism, then $(\Phi^*\mathcal{F})(U') = \Phi^*(\mathcal{F}(\Phi(U')))$ for any open $U' \subset N'$.

Then we have the naturality of versal openings:

Proposition 11.3. *Let $\mathcal{F}$ be a sub C^∞-ring sheaves of $\mathcal{E}_N$. For any diffeomorphism $\Phi : N' \to N$ from another manifold N', we have $\widetilde{\Phi^*\mathcal{F}} = \Phi^*\widetilde{\mathcal{F}}$.*

Proof: Let $q \in N'$ and $h \in \widetilde{\Phi^*\mathcal{F}}_q$. Then $dh = \sum_{i=1}^r a_i d(\Phi^* g_i)$ for some $a_i \in \mathcal{E}_{N',q}$ and $g_i \in \mathcal{F}_{\Phi(q)}$. Then $d(\Phi^{-1*}h) = \Phi^{-1*}(\sum_{i=1}^r a_i d(\Phi^* g_i)) = \sum_{i=1}^r (\Phi^{-1*}a_i) dg_i$. Since $\Phi^{-1*}a_i \in \mathcal{E}_{N,\Phi(q)}$, we see $\Phi^{-1*}h \in \widetilde{\mathcal{F}}_{\Phi(q)}$, therefore $h \in \Phi^*((\widetilde{\mathcal{F}})_{\Phi(q)}) = (\Phi^*\widetilde{\mathcal{F}})_q$. Thus we have $\widetilde{\Phi^*\mathcal{F}}_q \subseteq (\Phi^*\widetilde{\mathcal{F}})_q$. Applying the same argument to Φ^{-1} and $\Phi^*\mathcal{F}$, then we have $\widetilde{\Phi^*\mathcal{F}}_q \supseteq (\Phi^*\widetilde{\mathcal{F}})_q$. Therefore we have the required equality. □

Definition 11.4. Let N be a manifold. Let $\mathcal{F}$ be a sub C^∞-ring sheaf of $\mathcal{E}_N$. A mapping $f : N \to M$ is called a *realisation* of $\mathcal{F}$ if $\mathcal{F} = f^*\mathcal{E}_M$.

The following is clear:

Proposition 11.5. *Let $f : (\mathbf{R}^n, a) \to (\mathbf{R}^m, b)$ be a map-germ. Let $\mathcal{F} = f^*\mathcal{E}_{\mathbf{R}^m,b}$ be the germ of subsheaf of $\mathcal{E}_{\mathbf{R}^n,a}$. Let $F : (\mathbf{R}^n, a) \to \mathbf{R}^{m+r}$ be an opening of f. Then F is a versal opening of f if and only if F is a realisation of the algebraic opening $\widetilde{\mathcal{F}}$ of $\mathcal{F}$.*

Definition 11.6. A mapping $f : N \to M$ is called *locally injective* if for any $a \in N$, there exists an open neighbourhood U of a in N such that $f|_U : U \to M$ is injective.

Proposition 11.7. *Let $f : N \to M$ be a finite mapping and $F : N \to M'$ a realisation of the versal opening $\widetilde{f^*\mathcal{E}_M}$ of $f^*\mathcal{E}_M$. Then F is locally injective.*

Proof: Let $a \in N$. Then $F^*\mathcal{E}_{M',F(a)} = \widetilde{f^*\mathcal{E}_M}_a = \mathcal{R}_{f,a}$. Then the germ of F at a is a versal opening of f. Therefore by Proposition 2.16 of [45] we have the result. □

Definition 11.8. Let $\mathcal{F}$ be a sub C^∞-ring of $\mathcal{E}_N$. We call $\mathcal{F}$ *locally injective* if for any $a \in N$, there exist $h_1, \dots, h_r \in \mathcal{F}_a$ such that $(h_1, \dots, h_r) : (N, a) \to \mathbf{R}^r$ has an injective representative.

Proposition 11.9. *If $f : N \to M$ is a realisation of a locally injective sub C^∞-ring $\mathcal{F}$ of $\mathcal{E}_N$, then f is locally injective.*

Proof: Let $a \in N$. There exist $h_1, \dots, h_r \in \mathcal{F}_a$ such that $(h_1, \dots, h_r) : (N, a) \to \mathbf{R}^r$ has an injective representative. There exists a $g_i \in \mathcal{E}_{M,f(a)}$ such that $h_i = g_i \circ f$ for each $i, 1 \leq i \leq r$. After taking representatives of germs we have $h_i = g_i \circ f : U \to \mathbf{R}, (1 \leq i \leq r)$ on an open neighbourhood of a. Deleting U if necessary, $(h_1, \dots, h_r) = (g_1, \dots, g_r) \circ f : U \to \mathbf{R}^r$ is injective. Therefore $f|_U$ is injective. □

Part II. Advanced studies and applications

§12. Frontal curves

Let us give several observations on frontal map-germs and frontal maps $N \to M$ with $\dim(N) = 1$.

Let $f : (N, a) \to M$ be a map-germ with $\dim(N) = 1$. We consider the classification problem of germs up to the right-left equivalence. To simplify this, let $(N, a) = (\mathbf{R}, 0)$ and $(M, f(a)) = (\mathbf{R}^m, 0)$. Let t be the coordinate of $(\mathbf{R}, 0)$ and $x^1, \dots, x^m$ of $(\mathbf{R}^m, 0)$. We define the *order* of f at 0 by

$$\operatorname{ord}(f) := \inf \left\{ k \in \mathbf{N} \;\middle|\; \frac{d^k f}{dt^k}(0) \neq 0 \right\}$$

If the Taylor infinite series of f is 0, then we set $\operatorname{ord}(f) = \infty$. It is easy to see that $\operatorname{ord}(f)$ is invariant under right-left equivalence.

Lemma 12.1. *If $\operatorname{ord}(f) < \infty$, then f is a frontal. Moreover f is right-left equivalent to an opening of the map-germ $g : (\mathbf{R}, 0) \to (\mathbf{R}, 0)$ defined by $t \mapsto t^\mu$, where $\mu = \operatorname{ord}(f)$.*

Proof: For a diffeomorphism-germ $\sigma : (\mathbf{R}, 0) \to (\mathbf{R}, 0)$ and a linear transformation $\Phi : (\mathbf{R}^m, 0) \to (\mathbf{R}^m, 0)$, $\Phi \circ f \circ \sigma$ is of form:

$$\Phi \circ f \circ \sigma = (t^\mu,\ h_2(t),\ \dots\ ,\ h_m(t)),$$

with $h_i \in \mathfrak{m}_1^{\mu+1}, 2 \leq i \leq m$. Set $g(t) = t^\mu$. Then $\mathcal{R}_g = \mathbf{R} + \mathfrak{m}_1^\mu$ and $h_i \in \mathcal{R}_g, 2 \leq i \leq m$ (see Example 8.2). Therefore $\Phi \circ f \circ \sigma$ is an opening of g. □

Since a constant map-germ is a frontal and a non-constant analytic curve-germ has a finite order, by Lemma 12.1 we have

Corollary 12.2. *If $f : (\mathbf{R}, a) \to \mathbf{R}^m$ is an analytic map-germ, then f is a frontal.*

As for a global result, we have:

Lemma 12.3. *Let* $\dim(N) = 1$ *and* $f : N \to M$ *a frontal. Then there exists a global Legendre lift* $\widetilde{f} : N \to M$.

Proof: Let $R(f)$ denote the immersion locus of f and set $S := N \setminus \overline{R(f)}$. We have the Legendre lift $F : R(f) \to \mathrm{Gr}(1, TM)$ of $f|_{R(f)}$ which is defined by $F(t) = f_*(T_t N)$. The mapping F is extended to $\overline{R(f)}$ continuously. Since f is a frontal. F is extended to a C^∞ Legendre lift of f on an open neighbourhood of $\overline{R(f)}$. Now take any connected component J of the open set S. Then J is diffeomorphic to S^1 or an open interval. In the case that J is diffeomorphic to S^1, then $f|_J$ is of constant rank 0 and it is a constant mapping. Let us consider the case that $J \subset N$ is diffeomorphic to an open interval. Take the closure $I = \overline{J}$ in N, which is diffeomorphic to an interval, $[0, 1], (0, 1]$ or $(0, 1)$. In the case I is diffeomorphic to $[0, 1]$, consider the boundary points of I, which belong to $R(f)$ necessarily. Since the fibre of $\pi : \mathrm{Gr}(1, TM) \to M$ is diffeomorphic to the projective space $\mathrm{Gr}(1, \mathbf{R}^m) = \mathbf{R}P^{m-1}$ which is connected, we can extend the given Legendre lift $F : R(f) \to \mathrm{Gr}(1, TM)$ to a Legendre lift on an open set containing $\overline{R(f)} \cup I$. The extension is performed independently for each connected component of S. Thus we have a global Legendre lift $\widetilde{f} : N \to M$. □

Remark 12.4. If $\dim(N) = 2$, then a frontal $f : N \to M$ does not necessarily have a global Legendre lifting. See Example 2.5.

Next we study the genericity problem of frontal curves. To simplify the story we treat frontals $f : \mathbf{R} \to \mathbf{R}^m$. Let $\widetilde{f} : \mathbf{R} \to \mathrm{Gr}(1, T\mathbf{R}^m) = P(T\mathbf{R}^m) = \mathbf{R}^m \times \mathbf{R}P^{m-1}$ be an integral lifting of f (see Lemma 12.3). Then, turning upside-down the view point, we start from an integral map $F : \mathbf{R} \to \mathrm{Gr}(1, T\mathbf{R}^m)$. Let F be, in terms of Grassmannian coordinates $x^1, \dots, x^m, a^2, \dots, a^m$,

$$F(t) = (x^1(t), \dots, x^m(t), a^2(t), \dots, a^m(t)),$$

which satisfies $F^*\theta^2 = 0, \dots, F^*\theta^m = 0$, namely that

$$dx^2 - a^2 dx^1 = 0, \dots, dx^m - a^m dx^1 = 0.$$

The condition is equivalent to that

$$\frac{dx^2}{dt}(t) = a^2(t)\frac{dx^1}{dt}(t), \ \dots, \ \frac{dx^m}{dt}(t) = a^m(t)\frac{dx^1}{dt}(t).$$

Therefore, if functions $x^1(t), a^2(t), \ldots, a^m(t)$ and values $x^2(0), \ldots, x^m(0)$ are arbitrarily given, then the integral mapping F is uniquely determined. Thus we can apply ordinary transversality theorem to discuss the genericity of frontal curves through Legendre curves.

Remark 12.5. In general we can apply a transversality argument to Legendre mappings of corank ≤ 1 and obtain the classification of generic singularities (see [33][40]). However the similar argument does not work for Legendre mappings having singularities of corank ≥ 2 (see Example 2.5, Remark 6.4).

§13. Frames and flags

As refinements of the notion of frontal curves, we consider framed curves or "flagged" curves. Flagged curves and framed curves in a space-form play important roles in topology, geometry and singularity theory. For example, as it is well-known, the self-linking number in 3-space is defined via framing ([88]). The fundamental theory of curves is formulated via osculation framing. Surface boundaries have adapted framings, etc. Two kinds of frames, adapted frames and osculating frames, are considered in [43] from the viewpoint of duality. We classify the singularities of envelopes associated to framed curves. The singularities of envelopes in E^3 were studied in [41] to apply to the flat extension problem of a surface with boundary. The problem on extensions by tangentially degenerate surfaces motivates to study the envelopes associated to framings on curves in a space form.

In this article already we have used Grassmannians to introduce the frontals. Then we are naturally led to the following definitions.

Let M be a manifold of dimension m and $\ell_1, \ldots, \ell_r$ integers with $0 \leq \ell_1 < \cdots < \ell_r \leq m$. Define the *flag bundle* $\mathrm{Fl}(\ell_1, \ldots, \ell_r; TM)$ over M of type $(\ell_1, \ldots, \ell_r)$ as the totality of flags $V_{\ell_1} \subset \cdots \subset V_{\ell_r} \subset T_xM$ with $\dim(V_{\ell_i}) = \ell_i, (1 \leq i \leq r)$, x running over M. Then $\pi : \mathrm{Fl}(\ell_1, \ldots, \ell_r; TM) \to M$ is a fibration with fibres of dimension

$$\ell_1(m - \ell_1) + (\ell_2 - \ell_1)(m - \ell_2) + \cdots + (\ell_r - \ell_{r-1})(m - \ell_r)$$

Moreover $\mathrm{Fl}(n; TM) = \mathrm{Gr}(n, TM)$.

Set $\mathcal{F} = \mathrm{Fl}(\ell_1, \ldots, \ell_r; TM)$. Suppose M is endowed with an affine connection ∇. Let $\gamma(t) = (x(t), V_{\ell_1}(t), \ldots, V_{\ell_r}(t))$ be a curve on $\mathcal{F}$. Let vectors $v_1(t), \ldots v_{\ell_r}(t) \in T_{x(t)}M$ satisfy that $V_{\ell_j}(t) = \langle v_1(t), \ldots, v_{\ell_j}(t) \rangle_{\mathbf{R}}$ for each t and $1 \leq j \leq r$. Then consider the condition

$$x'(t) \in V_{\ell_1}(t),\ \nabla v_1(t), \ldots, \nabla v_{\ell_j}(t) \in V_{\ell_{j+1}}(t), (1 \leq j < r),$$

at t. Here, for a vector field $v(t)$ along a curve $x(t)$ in M, we define $\nabla v(t) := \nabla_{x'(t)} v(t)$, the covariant derivative of $v(t)$ by the velocity vector $x'(t)$. By this condition we define the distribution $D \subset T\mathcal{F}$, which depends on the given affine connection.

If M is a projective space, then the above construction is more clarified ([43]). Let V be a real vector space of dimension $m+1$ and $n_1, \dots, n_s$ integers satisfying $0 < n_1 < \cdots < n_s \leq m+1$. Define the *flag manifold* $\mathrm{Fl}(n_1, \dots, n_s, V)$ of type $(n_1, \dots, n_s)$ by the totality of flags $V_{n_1} \subset \cdots \subset V_{n_s} \subset V$ of linear subspaces with $\dim(V_{n_i}) = n_i, (1 \leq i \leq s)$. Set $\mathcal{F} = \mathrm{Fl}(n_1, \dots, n_s, V)$. Then the *canonical distribution* $D \subset T\mathcal{F}$ is defined as follows: Denote by $\pi_i : \mathcal{F} \to \mathrm{Gr}(n_i, V)$ the canonical projection to the i-th member of the flag. Then, for $v \in T_{\mathbf{V}}\mathcal{F}, \mathbf{V} \in \mathcal{F}$,

$$v \in D_{\mathbf{V}} \Longleftrightarrow \pi_{i*}(v) \in T\mathrm{Gr}(n_i, V_{n_{i+1}}) (\subset T\mathrm{Gr}(n_i, V)), (1 \leq i \leq s-1).$$

Then D is a subbundle of $T\mathcal{F}$ with

$$\mathrm{rank}(D) = n_1(n_2-n_1)+(n_2-n_1)(n_3-n_2)+\cdots+(n_s-n_{s-1})(m+1-n_s).$$

Note that the flag bundle $\mathrm{Fl}(\ell_1, \dots, \ell_r; TP(V))$ is naturally identified with the flag manifold $\mathrm{Fl}(1, \ell_1+1, \dots, \ell_r+1, V)$. Therefore the canonical differential system on $\mathrm{Fl}(\ell_1, \dots, \ell_s; TP(V))$ is induced. Then the canonical distribution D on the Grassmannian bundle $\mathrm{Gr}(n, T(P(V))) = \mathrm{Fl}(1, n+1, V)$ introduced in §3 coincides with that introduced here.

Definition 13.1. Let V be a real vector space of dimension $m+1$. A curve-germ $f : (N, a) \to P(V), \dim(N) = 1$ is called a *flagged curve* if there exists a D-integral lift $\widetilde{f} : (N, a) \to \mathrm{Fl}(1, 2, \dots, m, V)$ of f with respect to the projection $\pi_1 : \mathrm{Fl}(1, 2, \dots, m, V) = \mathrm{Gr}(1, V) = P(V)$.

Let $\gamma : N \to \mathbf{R}P^m$ be a curve and $t_0 \in N$. Take a system of projective local coordinates $(x_1, x_2, \dots, x_m)$ of $\mathbf{R}P^m$ with the centre at $\gamma(t_0)$ and the local affine representation $(\mathbf{R}, t_0) \to (\mathbf{R}^m, 0)$,

$$\gamma(t) = {}^T(x_1(t), x_2(t), \dots, x_m(t))$$

of γ. Consider the $(m \times k)$-matrix

$$W_k(t_0) := \left(\gamma'(t_0), \gamma''(t_0), \cdots, \gamma^{(k)}(t_0)\right)$$

for any integer $k \geq 1$ and $k = \infty$. Note that the rank of $W_k(t_0)$ is independent of the choice on representations for γ.

Definition 13.2. We call γ *of finite type* at $t = t_0 \in N$ if the $(m \times \infty)$-matrix

$$W_\infty(t_0) = \Big(\gamma'(t_0),\ \gamma''(t_0),\ \cdots,\ \gamma^{(k)}(t_0),\ \cdots\cdots\Big)$$

is of rank m. Define, for $1 \leq i \leq m$, $a_i := \min\{k \mid \operatorname{rank} W_k(t_0) = i\}$. Then we have a sequence of natural numbers $1 \leq a_1 < a_2 < \cdots < a_m$, and we call γ *of type* $(a_1, a_2, \ldots, a_m)$ at $t = t_0 \in N$.

If $(a_1, a_2, \ldots, a_m) = (1, 2, \ldots, m)$, then $t = t_0$ is called an *ordinary point* of γ.

Let $f : N \to \mathbf{R}P^m$ be of finite type at $t_0 \in N$. Then the *osculating flag* to f at t_0 is defined by

$$O_1(t_0) \subset O_2(t_0) \subset \cdots \subset O_k(t_0) \subset \cdots \subset O_m(t_0) = T_{f(t_0)}\mathbf{R}P^m,$$

where O_r is the linear subspace of $T_{f(t_0)}\mathbf{R}P^m$ generated by

$$\gamma'(t_0),\ \gamma''(t_0),\ \cdots,\ \gamma^{(k)}(t_0).$$

The corresponding projective subspace through $f(t_0)$ to $O_k(t_0)$ is also regarded ([34]). Then there exists unique integral lift

$$\widetilde{f} : N \to \mathrm{Fl}(1, 2, \ldots, k, \ldots, m, V)$$

of f.

The classification results on singularities which are related to flagged curves are given in [41][42][43].

§14. Legendre duality

The Legendre duality is a natural geometric framework where the frontals play fundamental roles. In this section we review several studies of frontals in specified (semi-)Riemannian manifolds from [42][53].

Let $\mathbf{R}^{n,m}$ denote the metric vector space of signature (n, m), n minus and m plus ([27][87]).

We write $\mathbf{R}^{0,n}$ as $\mathbf{R}^n$ simply. Recall the space-models, the *sphere* and the *hyperbolic space*,

$$S^{n+1} = \{x \in \mathbf{R}^{n+2} \mid x{\cdot}x = 1\},\ H^{n+1} = \{x \in \mathbf{R}^{1,n+1} \mid x{\cdot}x = -1,\ x^0 > 0\},$$

where $\mathbf{R}^{1,n+1} = \mathbf{R}^{n+2}_1 = \{(x^0, x^1, \ldots, x^{n+1})\}$ is the Minkowski space of index $(1, n+1)$ (See for instance [66][27]). The inner product in $\mathbf{R}^{1,n+1}$ is defined by $x \cdot y = -x^0y^0 + \sum_{i=1}^{n+1} x^i y^i$. Moreover we identify Euclidean space E^{n+1} with $\{x \in \mathbf{R}^{n+2} \mid x^0 = 1\} \subset \mathbf{R}^{n+2}$ if necessary.

Let X denote S^{n+1}, H^{n+1} or E^{n+1}. Set $Z = \widetilde{\mathrm{Gr}}(n, TX)$, the oriented Grassmannian bundle over X. Then Z is a double covering of $\mathrm{Gr}(n, TX)$. The space Z is identified with T_1X, the unit tangent bundle to X. In fact,

$$\begin{aligned} T_1S^{n+1} &= \{(x,y) \in S^{n+1} \times S^{n+1} \mid x \cdot y = 0\}, \\ T_1H^{n+1} &= \{(x,y) \in H^{n+1} \times S^{1,n} \mid x \cdot y = 0\}, \end{aligned}$$

where $S^{1,n} = \{x \in \mathbf{R}^{1,n+1} \mid x \cdot x = 1\}$ is the *de Sitter space.* Note that $Z = T_1H^{n+1}$ is identified with $T_{-1}S^{1,n} = \{(y,v) \mid y \in S^{1,n}, v \in T_yS^{1,n}, v \cdot v = -1\}$. Moreover $T_1E^{n+1} = E^{n+1} \times S^n$. We set $Y = S^{n+1}, S^{1,n}, \mathbf{R} \times S^n$ corresponding to $S^{n+1}, H^{n+1}, E^{n+1}$ respectively. Define $\pi_1 : Z \to X$ by the projection to the first component in three cases. Define $\pi_2 : Z \to Y$ by the projection to the second component in the cases $(X,Y) = (S^{n+1}, S^{n+1}), (X,Y) = (H^{n+1}, S^{1,n})$. In the case $(X,Y) = (E^{n+1}, \mathbf{R} \times S^n)$, we define $\pi_2 : Z = E^{n+1} \times S^n \to \mathbf{R} \times S^n$ by $\pi_2(x,y) = (-x \cdot y,\, y)$. The space has the canonical contact structure and all fibres of π_1 and π_2 are Legendre submanifolds. Therefore π_1 and π_2 are Legendre fibrations. Then we have the double Legendre fibration in each case:

$$X \xleftarrow{\pi_1} Z \xrightarrow{\pi_2} Y.$$

As the model of duality, we do have the projective duality ([98][58]): We set

$$Z = \mathcal{I}_{n+2} := \{([x],[y]) \in P^{n+1} \times P^{n+1*} \mid x \cdot y = 0\}.$$

Here P^{n+1*} is the dual projective space and $\cdot$ means the natural paring. The contact structure on $\mathcal{I}_{n+2}$ is defined by $dx \cdot y = x \cdot dy = 0$ ([58]). The projections $\pi_1 : \mathcal{I}_{n+2} \to X = P^{n+1}, \pi_2 : \mathcal{I}_{n+2} \to Y = P^{n+1*}$ are both Legendre fibrations.

The following fact is basic to unify our treatment:

Proposition 14.1. ([52][53]) *All Legendre double fibrations $X \longleftarrow Z \longrightarrow Y$ constructed above are locally isomorphic to each other. In particular each of them is locally isomorphic to the double fibration of the projective duality $P^{n+1} \longleftarrow \mathcal{I}_{n+2} \longrightarrow P^{n+1*}$.*

Let $f : N^n \to X$ be a *co-oriented* proper frontal (see Definition 7.1). Then there arises naturally the Legendre lift $\widetilde{f} : N \to T_1X = Z$ for $\pi_1 : Z \to X$ by attaching the unit normal vector field along f. The *Legendre dual* of f is defined by $f^\vee := \pi_2 \circ \widetilde{f} : N \to Y$. Then $f^\vee$ is a frontal. If $f^\vee$ is a proper frontal. Then we have the equality $f^{\vee\vee} = f$.

Let $\gamma : I \to X$ be a C^∞ immersion from an interval or a circle I. In general, we mean by a *framing* of the immersed curve γ, an oriented orthonormal frame $(e_1, e_2, \ldots, e_{n+1})$ along γ. An immersion γ is called *framed* if a framing is given.

Remark 14.2. Note that in [62][18], more general framings are considered to treat also light cone in Minkowski space.

If $X = S^{n+1}$, then we set $e_0(t) = \gamma(t) \in S^{n+1}$, and we have the moving frame $\widetilde{\gamma} = (e_0, e_1, \ldots, e_{n+1}) : I \to G = SO(n+2) \subset \mathrm{GL}_+(n+2, \mathbf{R})$.

If $X = H^{n+1}$, then we set $e_0(t) = \gamma(t) \in H^{n+1}$, and we have the moving frame $\widetilde{\gamma} = (e_0, e_1, \ldots, e_{n+1}) : I \to G = SO(1, n+1) \subset \mathrm{GL}_+(n+2, \mathbf{R})$.

In any of three cases, the frame manifold G is identified with an open subset of the oriented flag manifold $\widetilde{\mathcal{F}}_{n+2}$ consisting of oriented complete flags

$$V_1 \subset V_2 \subset \cdots \subset V_{n+1} \subset \mathbf{R}^{n+2}$$

in $\mathbf{R}^{n+2}$. For each $g = (e_0, e_1, \ldots, e_{n+1}) \in \mathrm{GL}_+(n+2, \mathbf{R})$, we set the oriented subspace

$$V_i = \langle e_0, e_1, \ldots, e_{i-1} \rangle_{\mathbf{R}} \subset \mathbf{R}^{n+2}, \ (1 \leq i \leq n+1).$$

This induces an open embedding $G \to \widetilde{\mathcal{F}}_{n+2} = \mathrm{Fl}(1, 2, \ldots, n+1)$. Thus, for a framed curve $\gamma : I \to X$ in $X = E^{n+1}, S^{n+1}, H^{n+1}$, with the frame $(e_1, \ldots, e_{n+1})$, we have the flagged curve $\widetilde{\gamma}$ by setting

$$V_i(t) = \langle e_0(t), e_1(t), \ldots, e_{i-1}(t) \rangle_{\mathbf{R}} \subset \mathbf{R}^{n+2}, \ (1 \leq i \leq n+1).$$

Then $\widetilde{\gamma}$ is a lifting of γ for the projection $\pi_1 : \widetilde{\mathrm{Fl}}(1, 2, \ldots, n+1, \mathbf{R}^{n+2}) \to \widetilde{\mathrm{Gr}}(1, \mathbf{R}^{n+2})$ to Grassmannian of oriented lines in $\mathbf{R}^{n+2}$. Note that there is the natural open embedding $X \subset \widetilde{\mathrm{Gr}}(1, \mathbf{R}^{n+2})$ in each of three cases.

The projective duality plays an essential role, for instance, to formulate the famous Plücker-Klein's formula, to analyse generic projective hypersurface (Bruce, Platonova, Landis [7]), tangent surfaces and Monge-Ampère equations ([52]).

Let $f : N^n \longrightarrow \mathbf{R}P^{n+1}$ be a frontal. Then we have the Legendre lifting $\widetilde{f} : N \longrightarrow Gr(n, T\mathbf{R}P^{n+1}) = PT^*\mathbf{R}P^{n+1}$. Then we get the *projective dual* $f^\vee : N \longrightarrow \mathbf{R}P^{n+1*}$ of f by the composition of $\widetilde{f}$ with the projection $\pi^* : PT^*\mathbf{R}P^{n+1*} \longrightarrow \mathbf{R}P^{n+1*}$. If f is sufficiently generic, then $f^\vee$ is also frontal, and we get the presumable equality $f^{\vee\vee} = f$.

Viewed from Legendre duality, we consider the class of tangentially degenerate frontals.

Definition 14.3. Let $f : N \to \mathbf{R}P^{n+1}(S^{n+1}, H^{n+1}, E^{n+1})$ be a proper frontal. Then f is called *tangentially degenerate* if the regular locus $R(f^\vee) = \{t \in N \mid f^\vee \text{ is an immersion at } t\}$ of the dual $f^\vee$ of f is not dense in N.

See the basic text [3] for the tangentially degenerate submanifolds.

§15. Grassmannian frontals

With the notion of frontals, we are naturally led to the following generalization of the projective duality.

Let $f : N^n \longrightarrow \mathbf{R}P^m$ be a frontal of codimension $r = m - n$. Then, consider the Legendre lifting of f :

$$\begin{aligned}\widetilde{f} : N \longrightarrow \mathrm{Gr}(n, T\mathbf{R}P^m) &\hookrightarrow \mathrm{Gr}(1, \mathbf{R}^{m+1}) \times \mathrm{Gr}(n+1, \mathbf{R}^{m+1}) \\ &\cong \mathrm{Gr}(1, \mathbf{R}^{m+1}) \times \mathrm{Gr}(r, \mathbf{R}^{m+1*}).\end{aligned}$$

The Grassmannian bundle $\mathrm{Gr}(n, T\mathbf{R}P^m)$ is identified with

$$\mathcal{I} = \{(p, q) \in \mathrm{Gr}(1, \mathbf{R}^{m+1}) \times \mathrm{Gr}(r, \mathbf{R}^{m+1*}) \mid p \subseteq q^\vee\}.$$

Here, for $q \in \mathrm{Gr}(r, \mathbf{R}^{m+1*})$, we set $q^\vee := \{v \in \mathbf{R}^{m+1} \mid \alpha(v) = 0 (\alpha \in q)\}$.

Therefore we are naturally led to define the *Grassmannian dual* $f^\vee : N \longrightarrow \mathrm{Gr}(r, \mathbf{R}^{m+1*})$ of $f : N \longrightarrow \mathbf{R}P^m$ by $\widetilde{f}$ composed with the projection to the second component, $(p, q) \mapsto q$.

Definition 15.1. A proper (co-oriented) frontal $f : N^n \to \mathbf{R}P^m$ $(S^{n+1}, H^{n+1}, E^{n+1})$ is called *tangentially degenerate* if the regular locus $R(f^\vee) = \{t \in N \mid f^\vee \text{ is an immersion at } t\}$ of the Grassmannian dual $f^\vee$ of f is not dense in N.

Returning to the general case, we remark that the equality "$f^{\vee\vee} = f$" does not have any meaning, even if $f^\vee$ is a proper frontal in the sense of Definition 4.1. Therefore, for a mapping into a Grassmannian, it is natural to specialise the definition of frontals as follows:

Let $f : N \longrightarrow \mathrm{Gr}(r, \mathbf{R}^{m+1})$ be a C^∞ mapping with $n + r \leq m + 1$. Set $s = m + 1 - n - r$. Then f is called *Grassmannian frontal* if there exists the *unique* integral lift $\widetilde{f} : M \longrightarrow (\mathcal{I}, \mathcal{D})$ of f with respect to a fibration $\pi_1 : \mathcal{I} \longrightarrow \mathrm{Gr}(r, \mathbf{R}^{m+1})$ and a distribution $\mathcal{D}$ on $\mathcal{I}$ defined as follows: First set

$$\mathcal{I} := \{(p, q) \in \mathrm{Gr}(r, \mathbf{R}^{m+1}) \times \mathrm{Gr}(s, \mathbf{R}^{m+1*}) \mid p \subseteq q^\vee\},$$

and consider the projection $\pi_1 : I \to \mathrm{Gr}(r, \mathbf{R}^{n+2})$ (resp. $\pi_2 : I \to \mathrm{Gr}(s, \mathbf{R}^{m+1*})$). Moreover set

$$\mathcal{P} := \{(p, q, p') \in \mathrm{Gr}(r, \mathbf{R}^{m+1}) \times \mathrm{Gr}(s, \mathbf{R}^{m+1*}) \times \mathrm{Gr}(r, \mathbf{R}^{m+1}) \mid p \subseteq q^\vee, p' \subseteq q^\vee\},$$

and consider the projection $\rho : \mathcal{P} \to \mathcal{I}$ to the first and second factors (resp. $\varphi : \mathcal{P} \to \mathrm{Gr}(r, \mathbf{R}^{m+1})$ to the third factor). Then we get the double fibration (ρ, φ):

$$\mathcal{I} \xleftarrow{\rho} \mathcal{P} \xrightarrow{\varphi} \mathrm{Gr}(r, \mathbf{R}^{m+1}).$$

For each $c = (p, q) \in \mathcal{I}$, we consider $\rho^{-1}(c)$. Then we consider its projection

$$\varphi(\rho^{-1}(c)) = \{p' \in \mathrm{Gr}(r, \mathbf{R}^{m+1}) \mid p' \subseteq q^\vee\}$$

by φ, which is regarded as $\mathrm{Gr}(r, \mathbf{R}^{r+n})$. Note that $\dim q^\vee = r + n$, $p \in \varphi(\rho^{-1}(c))$ and that $\varphi(\rho^{-1}(c)) \subset \mathrm{Gr}(r, \mathbf{R}^{m+1})$ is a submanifold of codimension $r(m + 1 - r) - rn = rs$.

Define the *tautological subbundle* $\mathcal{D} \subset T\mathcal{I}$ of codimension rs, for each $c = (p, q) \in \mathcal{I}$, by

$$\mathcal{D}_c = \pi_*^{-1}(T_p(\varphi(\rho^{-1}(c)))) \subset T_c\mathcal{I}.$$

Note that, if $r \neq 1$, or, $r \neq n + 1$, then the "system of tangential linear subspaces" $\{\varphi(\rho^{-1}(c)) \mid c \in I\}$ in the Grassmannian $\mathrm{Gr}(r, \mathbf{R}^{m+1})$ defined by $\mathcal{D}$ does not represent general tangential linear subspaces of the Grassmannian.

If we take local Grassmannian coordinates $(a_{ij})_{1 \le i \le r, 1 \le j \le n+s}$ of $\mathrm{Gr}(r, \mathbf{R}^{m+1})$ and $(b_{k\ell})_{1 \le k \le n+r, 1 \le \ell \le s}$ of $\mathrm{Gr}(s, \mathbf{R}^{m+1*})$, then $\mathcal{I}$ is defined by the system of equations

$$b_{ij} + a_{i1} b_{r+1\,j} + \cdots + a_{in} b_{r+n\,j} + a_{i\,n+j} = 0, \ 1 \le i \le r, 1 \le j \le s,$$

and $\mathcal{D}$ is defined by the system of 1-forms

$$b_{r+1\,j} da_{i1} + \cdots + b_{r+n\,j} da_{in} + da_{i\,n+j} = 0, \ 1 \le i \le r, 1 \le j \le s.$$

The integral lifting $\widetilde{f}$ is called the *Legendre lifting* of f in the generalised sense. The relation to the original definition of frontals is as follows:

Lemma 15.2. *Let $F : (\mathbf{R}^n, 0) \longrightarrow (\mathcal{I}, (p_0, q_0))$ be an integral map-germ to the distribution $\mathcal{D} \subset T\mathcal{I}$. Then $f = \pi_1 \circ F : (\mathbf{R}^n, 0) \longrightarrow (\mathrm{Gr}(r, \mathbf{R}^{m+1}), p_0)$ is Grassmannian frontal if and only if $\kappa \circ f$ is proper, i.e. $S(\kappa \circ f) \subset (\mathbf{R}^n, 0)$ is nowhere dense, for some projection*

$$\kappa : (\mathrm{Gr}(r, \mathbf{R}^{m+1}), p_0) \hookrightarrow (Hom(\mathbf{R}^r, \mathbf{R}^{n+s}), 0) \xrightarrow{i^*} (Hom(\mathbf{R}, \mathbf{R}^{n+s}), 0) \hookrightarrow \mathbf{R}P^{n+s-1},$$

induced from a linear inclusion $i : \mathbf{R} \hookrightarrow \mathbf{R}^r$.

Now, from the duality, we have another distribution $\mathcal{D}' \subset T\mathcal{I}$ from the projection $\pi' : \mathcal{I} \longrightarrow \mathrm{Gr}(s, \mathbf{R}^{m+1*})$ to the second factor, setting

$$\mathcal{P}' = \{(q', p, q) \in \mathrm{Gr}(s, \mathbf{R}^{m+1*}) \times \mathrm{Gr}(r, \mathbf{R}^{n+2}) \times \mathrm{Gr}(s, \mathbf{R}^{m+1*}) \mid q \subseteq p^{\vee},\ q' \subseteq p^{\vee}\}.$$

Then the fundamental result is the following:

Proposition 15.3. *Two distributions $\mathcal{D}$ and $\mathcal{D}'$ on the incidental manifold $\mathcal{I}$ coincide.*

We conclude this section by the following observation:

Proposition 15.4. *Let $F : N^n \to I \subset Gr(r, \mathbf{R}^{m+1}) \times Gr(s, \mathbf{R}^{m+1*})$ be an integral mapping to the distribution $\mathcal{D}$ with $n + r + s = m + 1$. Suppose $\pi \circ F =: f$ and $\pi' \circ F =: f^{\vee}$ are Grassmannian frontals respectively. Then we have $f^{\vee\vee} = f$.*

§16. Tangent varieties

Given a curve in Euclidean 3-space $\mathbf{E}^3 = \mathbf{R}^3$, the embedded tangent lines to the curve draw a surface in $\mathbf{R}^3$, which is called the *tangent surface* (or *tangent developable*) to the curve ([17][35]). It is known that the tangent surfaces (tangent developables) are developable surfaces. Developable surfaces which are locally isometric to the plane keep on interesting many mathematicians, for instance, Monge (1764), Euler (1772), Cayley (1845), Lebesgue (1899). See [73] for details. Therefore the tangent surfaces are regarded as generalised solutions (with singularities) of the Monge-Ampère equation

$$\frac{\partial^2 z}{\partial x^2}\frac{\partial^2 z}{\partial y^2} - \left(\frac{\partial^2 z}{\partial x \partial y}\right)^2 = 0$$

on spacial surfaces $z = z(x, y)$. Tangent surfaces are flat in E^3. However they are not flat but "extrinsically flat" or tangentially degenerate in S^3 and H^3 (cf. [3][71]). See also §14. The notion of *types* (a_1, a_2, a_3) for a curve-germ is introduced (Definition 13.2). Then the cuspidal edge, (resp. the swallowtail, the cuspidal beaks (Mond surface), the cuspidal butterfly) is obtained as the tangent developable of a curve of type $(1, 2, 3)$ (resp. $(2, 3, 4)$, $(1, 3, 4)$, $(3, 4, 5)$).

This property is related to "projective duality": *The projective dual of a tangent surface collapse to a curve (the dual curve)*. See [36].

Let $\gamma : \mathbf{R} \to \mathbf{R}^3$ be an immersed curve. Then the tangent surface has the natural parametrization

$$\mathrm{Tan}(\gamma) : \mathbf{R}^2 \to \mathbf{R}^3, \quad \mathrm{Tan}(\gamma)(t, s) := \gamma(t) + s\gamma'(t).$$

The tangent surface necessarily has singularities at least along γ, "the edge of regression".

It is known that the tangent surface to a generic curve $\gamma : \mathbf{R} \to \mathbf{R}^3$ in $\mathbf{R}^3$ has singularities only along γ and is locally diffeomorphic to the cuspidal edge or to the folded umbrella (also called, the cuspidal cross cap), as is found by Cayley and Cleave (1980). Cuspidal edge singularities appear along ordinary points where $\gamma', \gamma'', \gamma'''$ are linearly independent, while the folded umbrellas appear at isolated points of zero torsion where $\gamma', \gamma'', \gamma'''$ are linearly dependent but $\gamma', \gamma'', \gamma''''$ are linearly independent.

In a higher dimensional space $\mathbf{R}^m, m \geq 4$, for an immersed curve $\gamma : \mathbf{R} \to \mathbf{R}^m$, we define the tangent surface $\mathrm{Tan}(\gamma) : \mathbf{R}^2 \to \mathbf{R}^m$ by $\mathrm{Tan}(\gamma)(t, s) := \gamma(t) + s\gamma'(t)$. Then we have generically that $\gamma', \gamma'', \gamma'''$ are linearly independent and $\mathrm{Tan}(\gamma)$ is locally diffeomorphic to the (embedded) cuspidal edge in $\mathbf{R}^m$. Now we give the general definition:

Definition 16.1. Let N be an n-dimensional manifold. Let $f : N^n \to \mathbf{R}^m$ be a proper frontal. Let $\widetilde{f} : N \to \mathrm{Gr}(n, T\mathbf{R}^m)$ be the Legendre lift of f. Then the *tangent mapping* $\mathrm{Tan}(f) : T_f \to \mathbf{R}^m$ of f is defined by, for $t \in N$ and $v \in \widetilde{f}(t) \subset T_{f(t)}\mathbf{R}^m$,

$$\mathrm{Tan}(f)(t, v) := f(t) + v, \quad (t, v) \in T_f,$$

using the affine structure of $\mathbf{R}^m$. Then we define the *tangent variety* of f as the parametrised variety which is defined by the right equivalence class of $\mathrm{Tan}(f)$. If $(t_1, \dots, t_n)$ is a system of local coordinates of N, and $(t_1, \dots, t_n, s_1, \dots, s_n)$ the induced system of local coordinates of T_f induced by a system of local frame $v_1(t), \dots, v_n(t)$ of $\widetilde{f}$, then $\mathrm{Tan}(f)$ is given by

$$\mathrm{Tan}(f)(t, s) = f(t) + \sum_{j=1}^{n} s_j v_j(t).$$

Also note that we can define similarly the tangent varieties of mappings to a projective space. Tangent varieties appear in various geometric problems and applications naturally ([3][14][54][55][56][64][69][65][90][73][102]). See [36][43], for the geometric exposition on the local classification problem of tangent varieties. In particular it is proved in [43][56] the following:

Proposition 16.2. *Let* $\gamma : (N, t_0) \to \mathbf{R}P^m$ *be a curve-germ of finite type (Definition 13.2). Then* $\mathrm{Tan}(\gamma) : (N \times \mathbf{R}, (t_0, 0)) \to \mathbf{R}P^m$ *is a proper frontal.*

A proper frontal $f : N \to M$ is called a *directed curve* if $\dim(N) = 1$ ([59][60][61]). A directed curve γ is called *orientable* if there exists a frame $u : N \to TM$, $u(t) \neq 0$, along γ such that $\gamma'(t) \in \langle\, u(t)\, \rangle_{\mathbf{R}}$, $t \in \mathbf{R}$, which projects to the unique Legendre lift $\widetilde{\gamma} : N \to P(TM) = \mathrm{Gr}(1, TM)$ of γ satisfying $\gamma'(t) \in \widetilde{\gamma}(t)$, $(t \in \mathbf{R})$.

Let $\gamma : N \to M$ be a directed curve and $\widetilde{\gamma} : N \to P(TM)$ the unique D-integral lift of f. Recall that the *tangent bundle* to f is defined by $T_\gamma := \{(t, v) \in N \times TM \mid v \in \widetilde{\gamma}(t)\}$, which is a line bundle over N (see §7). Let M be a manifold with an affine connection. We define the *tangent mapping* $\mathrm{Tan}(\gamma) : T_\gamma \to M$ by $(t, v) \to \exp(v)$, using the exponential map (see §18).

Remark 16.3. By Lemma 12.3, there exists a global Legendre lift $\widetilde{\gamma} : N \to P(TM)$ of f. Then the orientability condition means that the line bundle $T_{\widetilde{f}}$ over N is orientable.

Let $M = \mathbf{R}^m$ and $\gamma : N \to \mathbf{R}^m$ be a directed and orientable curve. Then the tangent surface $\mathrm{Tan}(\gamma) : N \times \mathbf{R} \to \mathbf{R}^m$ of a directed curve γ is defined by

$$\mathrm{Tan}(\gamma)(t, s) := \gamma(t) + s\, u(t)$$

The right equivalence class of $\mathrm{Tan}(\gamma)$ is independent of the choice of frame u.

The singularities of the tangent surface $\mathrm{Tan}(\gamma)$ for a generic directed curve $\gamma : \mathbf{R} \to \mathbf{R}^m$ on a neighbourhood of the curve are only the cuspidal edge, the folded umbrella, and swallowtail if $m = 3$, and the embedded cuspidal edge and the open swallowtail if $m \geq 4$. See [20][59]. Several degenerate cases are studied in [76][77][32][34][35][36].

§17. Grassmannian geometry

We will give a series of classification results of singularities of tangent surfaces in A_n-geometry, i.e. the geometry associated to the group $\mathrm{PGL}(n + 1, \mathbf{R})$ (see [57]).

Let $V = \mathbf{R}^{m+1}$ be the vector space of dimension $m + 1$ and consider a flag in V of the following type (a complete flag):

$$V_1 \subset V_2 \subset V_3 \subset \cdots \subset V_m \subset V, \quad \dim(V_i) = i.$$

The set of such flags form a manifold $\mathrm{Fl}(1, 2, 3, \dots, m)$ of dimension $\frac{n(n+1)}{2}$.

A curve $\gamma : \mathbf{R} \to P(V) = P(V^{m+1})$ induces a D-integral curve $\Gamma : \mathbf{R} \to \mathrm{Fl}(1, 2, 3, \dots, m)$ for the canonical distribution D on the flag manifold $T\mathrm{Fl}(1, 2, 3, \dots, m)$, if we regard its osculating planes: the curve itself is given by $V_1(t)$, the tangent line is given by $V_2(t)$, the osculating plane is given by $V_3(t)$ and so on.

Let $m = 2$. Let $V_1(t) \subset V_2(t) \subset V = \mathbf{R}^3$ be an admissible curve. For each a, planes V_2 satisfying $V_1(a) \subset V_2 \subset V$ form the tangent line to the curve $\{V_1(t)\}$ at $t = a$ in $P(V) = P^2$. Similarly lines V_1 satisfying $V_1 \subset V_2(a)$ form the tangent line to the dual curve $\{V_2(t)\}$ at $t = a$ in $\mathrm{Gr}(2, V) = P(V^*) = P^{2*}$, the dual projective plane. For a generic admissible curve, we have the duality on "tangent maps":

Let $m = 3$. Let $\Gamma : \mathbf{R} \to \mathrm{Fl}(1, 2, 3)$ be a D-integral curve. Set $\Gamma(t) = (V_1(t), V_2(t), V_3(t))$, $V_1(t) \subset V_2(t) \subset V_3(t) \subset V = \mathbf{R}^4$. Then Γ induces the curve $\pi_1 \circ \Gamma$ in $P^3 = P(\mathbf{R}^4)$, the curve $\pi_2 \circ \Gamma$ in $\mathrm{Gr}(2, \mathbf{R}^4)$ and the curve $\pi_3 \circ \Gamma$ in $P^{3*} = \mathrm{Gr}(3, \mathbf{R}^4)$. Then we have the following duality on their tangent surfaces in A_3-geometry:

③	④	③
Cuspidal Edge	Cuspidal Edge	Cuspidal Edge
Swallow Tail	Cuspidal Edge	Folded Umbrella
Mond Surface	Open Swallowtail	Mond Surface
Folded Umbrella	Cuspidal Edge	Swallow Tail

In general for a generic D-integral curve $\Gamma : \mathbf{R} \to \mathrm{Fl}(1, 2, 3, \ldots, m)$,

$$V_1(t) \subset V_2(t) \subset V_3(t) \subset \cdots \subset V_m(t) \subset V = \mathbf{R}^{m+1},$$

we have the classification of singularities of tangent surfaces:

Theorem 17.1. $(A_n, n \geq 4)$ *The classification list consists of* $n+1$ *cases for curves in Grassmannians*:

P^n	$\mathrm{Gr}(2,V)$	$\mathrm{Gr}(3,V)$	$\mathrm{Gr}(4,V)$	$\cdots$	$\mathrm{Gr}(n,V)$
CE	CE	CE	CE	$\cdots$	CE
OSW	CE	CE	CE	$\cdots$	CE
OM	OSW	CE	CE	$\cdots$	OFU
OFU	CE	OSW	CE	$\cdots$	OM
CE	CE	CE	OSW	$\cdots$	CE
$\vdots$	$\vdots$	$\vdots$	$\vdots$	$\ddots$	$\vdots$
CE	CE	CE	CE	$\cdots$	OSW

The *cuspidal edge* (resp. *open swallowtail*, *open Mond surface*, *open folded umbrella*) is defined as a diffeomorphism equivalence class of the tangent surface-germ to a curve of type $(1, 2, 3, \cdots)$ (resp. $(2, 3, 4, 5, \cdots)$, $(1, 3, 4, 5, \cdots)$, $(1, 2, 4, 5, \cdots)$) in an affine space.

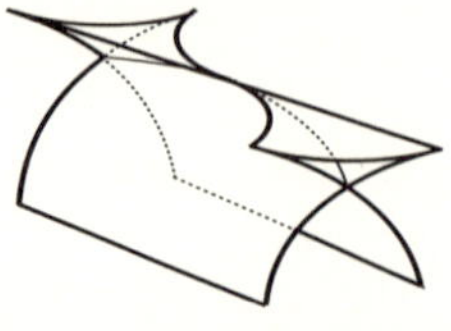

Mond surface (cuspidal-beaks)

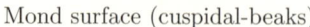

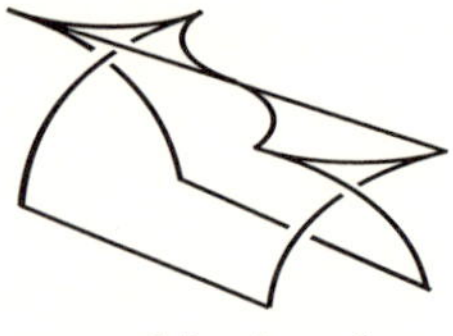

open Mond surface

§18. Affine connection and tangent surface

Now let us consider the case of directed curves in a Riemannian manifold, or more generally, the case of directed curves in a manifold with any affine connection, which is not necessarily projectively flat. For any directed curve, we have the well-defined tangent geodesic to each point of the curve. If we regard it as the "tangent line", then we have the well-defined tangent surface for the directed curve.

It is proved in [59], for any affine connection on a manifold of dimension $m \geq 3$, the singularities of the tangent surface to a generic directed curve on a neighbourhood of the curve are only the *cuspidal edge*, the *folded umbrella*, and *swallowtail* if $m = 3$, and the *embedded cuspidal edge* and the *open swallowtail* if $m \geq 4$. Moreover we have:

Theorem 18.1. ([59]) *Let ∇ be any torsion-free affine connection on a manifold M. Let $\gamma : \mathbf{R} \to M$ be a C^∞ curve.*

(1) *Let* $\dim(M) = 3$. *If $(\nabla\gamma)(a)$, $(\nabla^2\gamma)(a)$, $(\nabla^3\gamma)(a)$ are linearly independent at $t = a \in \mathbf{R}$, then the tangent surface* $\mathrm{Tan}(\gamma)$ *is locally diffeomorphic to the cuspidal edge at $(a, 0) \in \mathbf{R}^2$. If $(\nabla\gamma)(a)$, $(\nabla^2\gamma)(a)$, $(\nabla^3\gamma)(a)$ are linearly dependent, and $(\nabla\gamma)(a)$, $(\nabla^2\gamma)(a)$, $(\nabla^4\gamma)(a)$ are linearly independent, then the tangent surface* $\mathrm{Tan}(\gamma)$ *is locally diffeomorphic to the folded umbrella at $(a, 0) \in \mathbf{R}^2$. If $(\nabla\gamma)(a) = 0$ and $(\nabla^2\gamma)(a)$, $(\nabla^3\gamma)(a)$, $(\nabla^4\gamma)(a)$ are linearly independent, then the tangent surface* $\mathrm{Tan}(\gamma)$ *is locally diffeomorphic to the swallowtail at $(a, 0) \in \mathbf{R}^2$.*

(2) *Let* $\dim(M) \geq 4$. *If $(\nabla\gamma)(a), (\nabla^2\gamma)(a), (\nabla^3\gamma)(a)$ are linearly independent at $t = a \in \mathbf{R}$, then the tangent surface* $\mathrm{Tan}(\gamma)$ *is locally diffeomorphic to the embedded cuspidal edge at $(a, 0) \in \mathbf{R}^2$. If $(\nabla\gamma)(a) = 0$ and $(\nabla^2\gamma)(a), (\nabla^3\gamma)(a), (\nabla^4\gamma)(a), (\nabla^5\gamma)(a)$ are linearly independent at $t = a \in \mathbf{R}$, then the tangent surface* $\mathrm{Tan}(\gamma)$ *is locally diffeomorphic to the open swallowtail at $(a, 0) \in \mathbf{R}^2$.*

For the proof of Theorem 18.1, we apply the characterisation theorems found in [71][25][43].

In [59][60], singularities of tangent surfaces of torsionless curves are studied. In that case, so called fold singularities and $(2, 5)$-cuspidal edges appear. See also [28].

§19. Characterisation of frontal singularities

When we treat singularities in a general ambient space as in the previous section, we need the intrinsic characterisations of singularities. Note that the characterization of swallowtails was applied to hyperbolic geometry in [71] and to Euclidean and affine geometries in [52]. The characterization of folded umbrellas is applied to Lorenz-Minkowski geometry in [25]. In Theorem 18.1, we apply to non-flat projective geometry the characterisations and their some generalization via the notion of openings introduced in §8.

Let $f : (\mathbf{R}^2, p) \to M^3$ be a frontal with a non-degenerate singular point at p (see Lemma 6.7) and $\widetilde{f} : (\mathbf{R}^2, p) \to \mathrm{Gr}(2, TM)$ the integral lifting of f. Let $V_1, V_2 : (\mathbf{R}^2, p) \to TM$ be an associated frame with $\widetilde{f}$. Let $L : (\mathbf{R}^2, p) \to T^*M \setminus \zeta$ be an annihilator of $\widetilde{f}$. The condition is that $\langle L, V_1\rangle = 0, \langle L, V_2\rangle = 0$. Here ζ means the zero section. Let $c : (\mathbf{R}, t_0) \to (\mathbf{R}^2, p)$ be a parametrization of the singular locus $S(f)$, $p = c(t_0)$, and $\eta : (\mathbf{R}^2, p) \to T\mathbf{R}^2$ be a vector field which restricts to the kernel field of f_* on $S(f)$. Suppose that $V_2(p) \notin f_*(T_p\mathbf{R}^2)$. Then, for any affine connection ∇ on M, we define

$$\psi(t) := \langle L(c(t)), (\nabla^f_\eta V_2)(c(t))\rangle.$$

Note that the vector field $(\nabla^f_\eta V_2)(c(t))$ is independent of the extension η and the choice of affine connection ∇, since $\eta|_{S(f)}$ is a kernel field of f_*. We call the function $\psi(t)$ the *characteristic function* of f.

Then the following characterisations of cuspidal edges and folded umbrellas are given in [71][25]:

Theorem 19.1. (Theorem 1.4 of [25]). *Let $f : (\mathbf{R}^2, p) \to M^3$ be a germ of frontal with a non-degenerate singular point at p. Let $c : (\mathbf{R}, t_0) \to (\mathbf{R}^2, p)$ be a parametrization of the singular locus of f. Suppose $f_*c'(t_0) \neq 0$. Then, for the characteristic function ψ,*
(1) f is diffeomorphic to the cuspidal edge if and only if $\psi(t_0) \neq 0$.
(2) f is diffeomorphic to the folded umbrella if and only if $\psi(t_0) = 0$, $\psi'(t_0) \neq 0$.

We can summarise several known results as those on openings of the fold:

Theorem 19.2. ([59][60]) *Let $f : (\mathbf{R}^2, p) \to M^m, m \geq 2$ be a germ of frontal with a non-degenerate singular point at p, $\widetilde{f} : (\mathbf{R}^2, p) \to \mathrm{Gr}(2, TM)$ the integral lifting of f and $V_1, V_2 : (\mathbf{R}^2, p) \to TM$ an associated frame with $\widetilde{f}$. Let $c : (\mathbf{R}, t_0) \to (\mathbf{R}^2, p)$ be a parametrization of the singular locus of f. Suppose $f_*c'(t_0) \neq 0$. Then f is diffeomorphic to*

an opening of the fold, namely to the germ $(u, w) \mapsto (u, \frac{1}{2}w^2)$. *Moreover we have*:
(0) *Let* $m = 2$. *Then* f *is diffeomorphic to the fold.*
(1) *Let* $m \geq 3$. *Then* f *is diffeomorphic to the cuspidal edge if and only if* $\psi(t_0) \neq 0$.
(2) *Let* $m = 3$. *Then* f *is diffeomorphic to the folded umbrella if and only if* $\psi(t_0) = 0, \psi'(t_0) \neq 0$.

Based on results in [71] and [43], we summarise the characterization results on openings of the Whitney's cusp map-germ:

Theorem 19.3. ([59][61]) *Let* $f : (\mathbf{R}^2, p) \to M^m, m \geq 2$ *be a germ of frontal with a non-degenerate singular point at* p, $V_1, V_2 : (\mathbf{R}^2, p) \to TM$ *an associated frame with* $\widetilde{f}$ *with* $V_2(p) \not\in f_*(T_p\mathbf{R}^2)$, *and* $\eta : (\mathbf{R}^2, p) \to T\mathbf{R}^2$ *an extension of a kernel field along of* f_*. *Let* $c : (\mathbf{R}, t_0) \to (\mathbf{R}^2, p)$ *be a parametrization of the singular locus of* f. *Set* $\gamma = f \circ c : (\mathbf{R}, t_0) \to M$. *Suppose* $(\nabla\gamma)(t_0) = 0$ *and* $(\nabla^2\gamma)(t_0) \neq 0$. *Then* f *is diffeomorphic to an opening of Whitney's cusp, namely to the germ* $(u, t) \mapsto (u, t^3 + ut)$. *Moreover we have*

(0) *Let* $m = 2$. *Then* f *is diffeomorphic to Whitney's cusp.*

(1) *Let* $m = 3$. *Then* f *is diffeomorphic to the swallowtail if and only if*

$$V_1(c(t_0)),\ V_2(c(t_0)),\ (\nabla^f_\eta V_2)(c(t_0))$$

are linearly independent in $T_{f(p)}M$.

(2) *Let* $m \geq 4$. *Then* f *is diffeomorphic to the open swallowtail if and only if*

$$(V_1 \circ c)(t_0),\ (V_2 \circ c)(t_0),\ ((\nabla^f_\eta V_2) \circ c)(t_0),\ (\nabla^\gamma_{\partial/\partial t}((\nabla^f_\eta V_2) \circ c))(t_0)$$

are linearly independent in $T_{f(p)}M$.

Note that the conditions appeared in Theorem 19.1 are invariant under diffeomorphism equivalence introduced in Introduction. In fact the conditions are invariant under a weaker equivalence relation. In Definition 8.4, we have introduce the notion of $\mathcal{J}$-equivalence of map-germs.

Corollary 19.4. *Let* $f : (\mathbf{R}^2, 0) \to (\mathbf{R}^m, 0)$ *be a frontal. Then* f *is* $\mathcal{J}$*-equivalent to Whitney's cusp if and only if* f *is diffeomorphic to an opening of Whitney's cusp. Moreover. if* $m = 2$, *then* f *is diffeomorphic to Whitney's cusp. If* $m = 3$ *and* f *is a front, then* f *is diffeomorphic to swallowtail.*

The known criteria of singularities (see for instance [93][94][70]) seem to be closely related with frontals, openings and $\mathcal{J}$-equivalence. The detailed relations are still open to be studied.

§20. Null frontals

Let (M, g) be a semi-Riemannian manifold with an indefinite metric g. Denote by $\mathcal{C} \subset TM$ the null cone field associated with the indefinite metric g, i.e. $\mathcal{C}$ is the set of null vectors:

$$\mathcal{C} = \bigcup_{x \in M} \mathcal{C}_x, \quad \mathcal{C}_x = \{u \in T_x M \mid g_x(u, u) = 0\}.$$

Let $\pi : \mathcal{C} \to M$ be the canonical projection.

Definition 20.1. A mapping $f : N \to M$ is called *totally null* (resp. *null*) if the induced metric f^*g is identically zero (resp. f^*g is degenerate everywhere). The condition that f is totally null is equivalent to that $f_*(T_t N) \subset \mathcal{C}_{f(t)}$ (resp. $f_*(T_t N)$ is tangent to $\mathcal{C}_{f(t)}$), for any $t \in N$.

Definition 20.2. A curve-germ $\gamma : (\mathbf{R}, a) \to M$ is called *null* if $\gamma'(t) \in \mathcal{C} \quad (t \in (\mathbf{R}, a))$. Moreover $\gamma : (\mathbf{R}, a) \to M$ is called *null-directed* if there exists a lift $u : (\mathbf{R}, a) \to \mathcal{C}$ such that $\pi \circ u = \gamma, u(t) \neq 0, \gamma'(t) \in \langle u(t) \rangle_{\mathbf{R}}$, $t \in (\mathbf{R}, a)$.

A map-germ is null (resp. null-directed) if and only if it is totally null (resp. totally null frontal).

Definition 20.3. Let $\gamma : (\mathbf{R}, a) \to M$ be null-directed. Define the *null tangent surface* $\mathrm{Tan}(\gamma) : (\mathbf{R}^2, a \times \mathbf{R}) \to M$ of γ as the ruled surface by null geodesics through points $\gamma(t)$ with the directions $u(t)$.

The right equivalence class of $\mathrm{Tan}(\gamma)$ is independent of the choice of the lift u.

We have the following classification results. For the details see [56][57]: The singularities of tangent surface $\mathrm{Tan}(\gamma)$ for a generic null directed curve $\gamma : \mathbf{R} \to \mathbf{R}^{2,2}$ are cuspidal edges and open swallowtails. The singularities of tangent surface $\mathrm{Tan}(\gamma)$ for a generic null directed curve $\gamma : \mathbf{R} \to \mathbf{R}^{2,3}$ are cuspidal edges, open swallowtails, open Mond surfaces and unfurled folded umbrellas. The singularities of tangent surface $\mathrm{Tan}(\gamma)$ for a generic null directed curve $\gamma : \mathbf{R} \to \mathbf{R}^{3,3}$ (the projection of a generic "Engel integral" curve) are embedded cuspidal edges, open swallowtails and open Mond surfaces. See [43] for the normal forms and pictures of the singularities.

In general the tangent surface to a null curve is a ruled surface by null lines, which is not necessarily a totally null surface, but a null surface, which we call the *null tangent surface*.

Let X be a 3-dimensional Lorentzian manifold (with signature $(1, 2)$). A smooth map-germ $\varphi : (\mathbf{R}^2, 0) \to X$ is called a *null frontal surface* or a *null frontal* in short if there exists a smooth lift $\widetilde{\varphi} : (\mathbf{R}^2, 0) \to PT^*X = \mathrm{Gr}(2, TX)$ of φ such that $\widetilde{\varphi}(t)$ is a lightlike plane in $T_{\varphi(t)}X$ and $\varphi_*(T_t\mathbf{R}^2) \subset \widetilde{\varphi}(t)$, for any $t \in (\mathbf{R}^2, 0)$. The notion of null frontals is a natural generalization of null immersions to singular surfaces. We have presented several classification results of singularities which arise in null frontals up to local diffeomorphisms and up to $O(2,3)$-conformal transformations in the conformally flat case (cf. [54]). The classification is achieved by using the fact that null frontals are obtained as tangent surfaces to null curves in X, as well as "associated varieties" to Legendre curves in the space Y of null geodesics on X (cf. [55][56][57]). A related result is obtained in [19].

§21. Abnormal frontals

Let M be a 5-dimensional manifold and $\mathcal{D} \subset TM$ a distribution of rank 2. Then $\mathcal{D}$ is called a *Cartan distribution* if it has growth $(2, 3, 5)$, namely, if $\mathrm{rank}(\mathcal{D}^{(2)}) = 3$ and $\mathrm{rank}(\mathcal{D}^{(3)}) = 5$, where, we define in terms of Lie bracket, $\mathcal{D}^{(2)} = \mathcal{D} + [\mathcal{D}, \mathcal{D}]$ and $\mathcal{D}^{(3)} = \mathcal{D}^2 + [\mathcal{D}, \mathcal{D}^2]$. It is known that, for any point x of M and for any direction $\ell \subset \mathcal{D}_x$, there exists an abnormal geodesic, which is unique up to parametrisations, through x with the given direction ℓ (see [50][51]).

Then, for a given $\mathcal{D}$-directed curve γ, we define *abnormal tangent surface* of γ, which is ruled by abnormal geodesics through points $\gamma(t)$ with the directions $u(t)$.

On $\mathbf{R}^5$ with coordinates $(\lambda, \nu, \mu, \tau, \sigma)$, define the distribution $\mathcal{D} \subset T\mathbf{R}^5$ generated by the pair of vector fields

$$\begin{aligned} \eta_1 &= \frac{\partial}{\partial \lambda} + \nu \frac{\partial}{\partial \mu} - (\lambda\nu - \mu)\frac{\partial}{\partial \tau} + \nu^2 \frac{\partial}{\partial \sigma}, \\ \eta_2 &= \frac{\partial}{\partial \nu} - \lambda \frac{\partial}{\partial \mu} + \lambda^2 \frac{\partial}{\partial \tau} - (\lambda\nu + \mu)\frac{\partial}{\partial \sigma}. \end{aligned}$$

Then $\mathcal{D} \subset T\mathbf{R}^5$ is a Cartan distribution and it has maximal symmetry of dimension 14, maximal among all Cartan distributions, which is of type G_2, one of simple Lie algebras.

For a generic G_2-Cartan directed curve $\gamma : \mathbf{R} \to \mathbf{R}^5$, the tangent surfaces at any point $a \in \mathbf{R}$ is classified, up to local diffeomorphisms, into *embedded cuspidal edge*, *open Mond surface*, and *generic open folded pleat* (see [55] for details). The classification of singularities in abnormal tangent surfaces to generic Cartan directed curves for general Cartan distributions seems to be un-known yet.

§22. Appendix: Malgrange preparation theorem on differentiable algebras

We show the Malgrange's preparation theorem on differentiable algebras [74] from the ordinary Malgrange-Mather's preparation theorem (see for example [15]), relating to the theory of C^∞-rings which we have utilised in this paper.

An $\mathbf{R}$-algebra A is called *local* if it has a unique maximal ideal $\mathfrak{m}_A$.

Example 22.1. Let $\mathcal{E}_n$ denote the $\mathbf{R}$-algebra of C^∞-functions-germs $(\mathbf{R}^n, 0) \to \mathbf{R}$. Then $\mathcal{E}_n$ is a local $\mathbf{R}$-algebra with the unique maximal ideal $\mathfrak{m}_n = \{h \in \mathcal{E}_n \mid h(0) = 0\}$.

Definition 22.2. ([74]) A local $\mathbf{R}$-algebra A is called a *differentiable algebra* if a surjective $\mathbf{R}$-algebra homomorphism, mapping 1 to 1, $\pi : \mathcal{E}_m \to A$, for some $m \in \mathbf{N}$ is endowed.

A differentiable algebra A has the unique maximal ideal $\mathfrak{m}_A = \pi(\mathfrak{m}_m)$.

Let A and B be differentiable algebras with the surjective homomorphisms $\pi : \mathcal{E}_m \to A$ and $\psi : \mathcal{E}_n \to B$ respectively. An $\mathbf{R}$-algebra homomorphism $u : A \to B$ is called a *morphism* of differentiable algebras if there exists a C^∞ map-germ $g : (\mathbf{R}^n, 0) \to (\mathbf{R}^m, 0)$ such that the diagram

$$\begin{array}{ccc} \mathcal{E}_m & \overset{g^*}{\to} & \mathcal{E}_n \\ \pi \downarrow & & \downarrow \psi \\ A & \overset{u}{\to} & B \end{array}$$

commutes.

A morphism $u : A \to B$ of differentiable algebras is called *finite* (resp. *quasi-finite*) if B is a finite A-module via u (resp. $B/\mathfrak{m}_A B$ is a finite dimensional $\mathbf{R}$-vector space).

If u is finite, then it is quasi-finite. Then we have:

Theorem 22.3. (*Malgrange preparation theorem on differentiable algebras.* Theorem 4.1 in [74] p.73) *Let $u : A \to B$ be a morphism of differentiable algebras. Then u is finite if and only if it is quasi-finite. Moreover $b_1, \dots, b_r \in B$ generate B over A via u if and only if $\bar{b}_1, \dots, \bar{b}_r \in B/\mathfrak{m}_A B$ generate $B/\mathfrak{m}_A B$ over* $\mathbf{R}$.

Theorem 22.4. (*Malgrange-Mather's preparation theorem*: Theorem 6.5, Corollary 6.6 in [15]) *Let $f : (\mathbf{R}^n, 0) \to (\mathbf{R}^m, 0)$ be a C^∞ map-germ with the induced homomorphism $f^* : \mathcal{E}_m \to \mathcal{E}_n$. Let C be a finite $\mathcal{E}_n$-module. Then C is a finite $\mathcal{E}_m$-module via f^* if and only if*

$C/\mathfrak{m}_m C$ is a finite dimensional $\mathbf{R}$-vector space. Moreover $c_1, \dots, c_r \in C$ generate C over $\mathcal{E}_m$ via f^* if and only if $\bar{c}_1, \dots, \bar{c}_r \in C/\mathfrak{m}_m C$ generate $C/\mathfrak{m}_m C$ over $\mathbf{R}$.

Proof of the statement that Theorem 22.4 implies Theorem 22.3:
Let u be quasi-finite. Suppose $b_1, \dots, b_r \in B$ and $\bar{b}_1, \dots, \bar{b}_r \in B/\mathfrak{m}_A B$ generate $B/\mathfrak{m}_A B$ over $\mathbf{R}$. Let $g : (\mathbf{R}^n, 0) \to (\mathbf{R}^m, 0)$ and $g^* : \mathcal{E}_m \to \mathcal{E}_n$ cover $u : A \to B$. Note B is a finite $\mathcal{E}_n$-module via ψ. In fact $1 \in B$ generates B over $\mathcal{E}_n$ via the surjection ψ. Also note that $\mathfrak{m}_m B = \pi(\mathfrak{m}_m)B \subseteq \mathfrak{m}_A B$. Then $\bar{b}_1, \dots, \bar{b}_r \in B/\mathfrak{m}_m B$ generate $B/\mathfrak{m}_m B$ over $\mathbf{R}$ via $u \circ \pi = \psi \circ g^*$. Therefore, by Theorem 22.4, $b_1, \dots, b_r \in B$ generate B over A. Thus u is finite. This implies also the remaining statement naturally. □

Definition 22.5. A commutative ring A is called a *C^∞-ring* if the following conditions are satisfied:
(1) A contains the field $\mathbf{R}$ of real numbers.
(2) For any positive integer r, for any $a_1, \dots, a_r \in A$, and for any C^∞ function $f \in C^\infty(\mathbf{R}^r)$, an element $f(a_1, \dots, a_r) \in A$ is assigned, such that the equality

$$(g(f_1, \dots, f_s))(a_1, \dots, a_r) = g(f_1(a_1, \dots, a_r), \dots, f_s(a_1, \dots, a_r))$$

holds for any $g \in C^\infty(\mathbf{R}^s), f_1, \dots, f_s \in C^\infty(\mathbf{R}^n)$.
(3) The operations on A by C^∞ functions are compatible with the structure of $\mathbf{R}$-algebra on A, i.e. if f is a polynomial, $f = P(x_1, \dots, x_r) \in \mathbf{R}[x_1, \dots, x_r] \subset C^\infty(\mathbf{R}^r)$, then $f(a_1, \dots, a_r)$ is equal to the element $P(a_1, \dots, a_r)$ obtained just by substitutions (see [29]).

Note that by the condition (1), a C^∞-ring is naturally an $\mathbf{R}$-algebra. A C^∞-ring A is called a local C^∞-ring if A is a local $\mathbf{R}$-algebra. Let $\mathfrak{m}_A$ denote the unique maximal ideal of a local C^∞-ring. Let A be a C^∞-ring. We say that $a_1, \dots, a_n \in A$ generate A as the C^∞-ring if for any $a \in A$, there exists $f \in C^\infty(\mathbf{R}^n)$ such that $a = f(a_1, \dots, a_n)$. A is called a *finitely generated* C^∞-ring if there exists a finite number of elements generating A as the C^∞-ring. Let $\pi : A \to A/\mathfrak{m}_A$ denote the natural projection and $i : \mathbf{R} \to A$ the inclusion.

Lemma 22.6. *Let A be a differentiable algebra with a surjective $\mathbf{R}$-algebra homomorphism $\pi : \mathcal{E}_m \to A$. Then we have:*
(1) A has the induced structure of a local C^∞-ring.
(2) A is generated by $\pi(x_1), \dots, \pi(x_m)$ as the C^∞-ring. Here $(x_1, \dots, x_m)$ is a system of coordinates of $(\mathbf{R}^m, 0)$ with the centre at 0.
(3) $\pi \circ i : \mathbf{R} \to A/\mathfrak{m}_A$ is a bijection.

Proof: (1) For any positive integer r, for any $a_1, \ldots, a_r \in A$, and for any C^∞ function $f \in C^\infty(\mathbf{R}^r)$, we take a system of lifts $\widetilde{a}_1, \ldots, \widetilde{a}_r \in \mathcal{E}_m$ for π and define $f(a_1, \ldots, a_r) := \pi(f(\widetilde{a}_1, \ldots, \widetilde{a}_r))$. If we take another system of lifts $\widehat{a}_1, \ldots, \widehat{a}_r \in \mathcal{E}_m$ for π, we have

$$f(\widetilde{a}_1, \ldots, \widetilde{a}_r) - f(\widehat{a}_1, \ldots, \widehat{a}_r) = \sum_{i=1}^{r} g_i(\widetilde{a}, \widehat{a})(\widetilde{a}_i - \widehat{a}_i) \in \mathrm{Ker}(\pi),$$

for some C^∞ functions $g_i(\widetilde{x}_1, \ldots, \widetilde{x}_r; \widehat{x}_1, \ldots, \widehat{x}_r) \in C^\infty(\mathbf{R}^{2r}), 1 \leq i \leq r$. Thus $\pi(f(\widehat{a}_1, \ldots, \widehat{a}_r)) = \pi(f(\widetilde{a}_1, \ldots, \widetilde{a}_r))$. Moreover, take any $a \in A$. Then there exists $h \in \mathcal{E}_m$ such that $a = \pi(h)$. (2) Take an $H \in C^\infty(\mathbf{R}^m)$ having h as the germ at 0. Then $H(\pi(x_1), \ldots, \pi(x_m)) = \pi(H(x_1, \ldots, x_m)) = \pi(h) = a$. (3) $\mathbf{R} \cong \mathcal{E}_m/\mathfrak{m}_m \cong A/\mathfrak{m}_A$. □

Proposition 22.7. *Let $(A, \mathfrak{m}_A)$ be a local C^∞-ring. Then the following conditions are equivalent:*
(1) *A is finitely generated as C^∞-ring and the natural map $\pi \circ i : \mathbf{R} \to A/\mathfrak{m}_A$ is bijective.*
(2) *A is a differentiable algebra in the sense of Malgrange.*

Proof: (1) $\Rightarrow$ (2): Let $a_1, \ldots, a_m$ be a system of generators of A as C^∞-ring. Define $\Pi : C^\infty(\mathbf{R}^m) \to A$ by $\Pi(f) = f(a_1, \ldots, a_m)$. Then Π is surjective. Set $I = \Pi^{-1}(\mathfrak{m}_A)$ which is a maximal ideal of $C^\infty(\mathbf{R}^m)$ with $C^\infty(\mathbf{R}^m)/I \cong \mathbf{R}$. Then there exists a point $p \in \mathbf{R}^m$ such that $I = \{f \in C^\infty(\mathbf{R}^m) \mid f(p) = 0\}$ (see Proposition 2.1 [2] for instance). Moreover $\mathrm{Ker}(\Pi) \subset I$. Set $J = \{f \in C^\infty(\mathbf{R}^m) \mid \text{the germ of } f \text{ at } p \text{ is zero}\}$. We show that $J \subseteq \mathrm{Ker}(\Pi)$. Let $h \in J$. Then there exists $k \in C^\infty(\mathbf{R}^m)$ such that $k(p) \neq 0$ and $hk = 0$. Then

$$0 = (hk)(a_1, \ldots, a_m) = h(a_1, \ldots, a_m)k(a_1, \ldots, a_m).$$

On the other hand $k(a_1, \ldots, a_m) \notin \mathfrak{m}_A$. Hence $k(a_1, \ldots, a_m)$ is invertible. Then $\Pi(h) = h(a_1, \ldots, a_m) = 0$ and thus $h \in \mathrm{Ker}(\Pi)$. Now $\Pi : C^\infty(\mathbf{R}^n) \to A$ induces a surjective homomorphism $\pi' : \mathcal{E}_{\mathbf{R}^m, p} \to A$. Define $\pi : \mathcal{E}_m = \mathcal{E}_{\mathbf{R}^m, 0} \to A$ by $\pi(f) = \pi'(\widetilde{f})$, where $\widetilde{f}(x) = f(x - p)$. The implication (2) $\Rightarrow$ (1) follows by Lemma 22.6. □

Example 22.8. Let $I = \{h \in C^\infty(\mathbf{R}) \mid \exists n_0 \in \mathbf{N}, h(n) = 0 (n \in \mathbf{N}, n \geq n_0)\}$. Then I is a maximal ideal of $C^\infty(\mathbf{R})$. Let $A = C^\infty(\mathbf{R})_I$ be the localisation (a localisation at infinity). Then A is an $\mathbf{R}$-algebra with the unique maximal ideal $\mathfrak{m}_A$. However A is not a differentiable algebra in the sense of Malgrange. In fact, the quotient field $A/\mathfrak{m}_A \cong C^\infty(\mathbf{R})/I$ is a Robinson's *hyper-real number field* [92].

We call an $\mathbf{R}$-algebra homomorphism u a *C^∞-ring homomorphism* if

$$u(f(a_1, \dots, a_r)) = f(u(a_1), \dots, u(a_r)),$$

for any $r \geq 1$, for any $a_1, \dots, a_r \in A$ and for any $f \in C^\infty(\mathbf{R}^r)$.

Lemma 22.9. *Let $\varphi : \mathcal{E}_m \to \mathcal{E}_n$ be an $\mathbf{R}$-algebra homomorphism. Then the following conditions are equivalent*:
(1) *There exists a C^∞ map-germ $g : (\mathbf{R}^n, 0) \to (\mathbf{R}^m, 0)$ such that $\varphi = g^*$.*
(2) *φ is a C^∞-ring homomorphism.*

Proof: (1) $\Rightarrow$ (2): Let $a_1, \dots, a_r \in \mathcal{E}_m$ and $h \in C^\infty(\mathbf{R}^r)$. Then

$$\begin{aligned} h(\varphi(a_1), \dots, \varphi(a_r)) &= h(g^*a_1, \dots, g^*a_r) = h \circ (a_1, \dots, a_r) \circ g \\ &= g^*(h(a_1, \dots, a_r)) = \varphi(h(a_1, \dots, a_r)). \end{aligned}$$

(2) $\Rightarrow$ (1): Let $x_1, \dots, x_m$ be coordinates of $(\mathbf{R}^m, 0)$. Then we have $\varphi(x_1), \dots, \varphi(x_m) \in \mathfrak{m}_n$. Take representatives $\widetilde{g}_i : U \to \mathbf{R}$ of $\varphi(y_i)$ over a common open neighbourhood of 0 in $\mathbf{R}^n$, $(1 \leq i \leq m)$. We set $\widetilde{g} = (\widetilde{g}_1, \dots, \widetilde{g}_m) : U \to \mathbf{R}^m$. Then $\widetilde{g}(0) = 0$. Take the germ $g : (\mathbf{R}^n, 0) \to (\mathbf{R}^m, 0)$ of $\widetilde{g}$ at 0. Let $h \in \mathcal{E}_m$. Take a representative $\widetilde{h} \in C^\infty(\mathbf{R}^m)$. Then we have $\varphi(h) = \varphi(\widetilde{h}(x_1, \dots, x_m)) = \widetilde{h}(\varphi(x_1), \dots, \varphi(x_m)) = h \circ g = g^*(h)$. Therefore $\varphi = g^*$. □

Lemma 22.10. *Let $u : A \to B$ be an $\mathbf{R}$-algebra homomorphism of differentiable algebras. Then the following conditions are equivalent*:
(1) *u is a morphism of differentiable algebras.*
(2) *u is a C^∞-ring homomorphism.*

Proof: (1) $\Rightarrow$ (2): Let $a_1, \dots, a_r \in A$ and $f \in C^\infty(\mathbf{R}^r)$. Take $\widetilde{a}_i \in \mathcal{E}_m$ with $\pi(\widetilde{a}_i) = a_i$. Then $\psi(g^*\widetilde{a}_i) = u(a_i)$. Then $u(f(a_1, \dots, a_r) = u(\pi(f(\widetilde{a}_1, \dots, \widetilde{a}_r))) = \psi(f \circ (\widetilde{a}_1, \dots, \widetilde{a}_r) \circ g) = \psi(f(g^*\widetilde{a}_1, \dots, g^*\widetilde{a}_r)) = f(u(a_1), \dots, u(a_r))$.
(2) $\Rightarrow$ (1): Take $g_i \in \mathcal{E}_n$ with $u(\pi(x_i)) = \psi(g_i)$. Since $\psi(g_i) \in \mathfrak{m}_B$, we have $g_i \in \mathfrak{m}_m$. Set $g = (g_1, \dots, g_m) : (\mathbf{R}^n, 0) \to (\mathbf{R}^m, 0)$. Let $h \in \mathcal{E}_m$ and take a representative $H \in C^\infty(\mathbf{R}^m)$ of the germ h. Then we have

$$\begin{aligned} u(\pi(h)) &= u(H(\pi(x_1), \dots, \pi(x_m))) = H(u(\pi(x_1)), \dots, u(\pi(x_m))) \\ &= H(\psi(g_1), \dots, \psi(g_m)) = \psi(H(g_1, \dots, g_m) = \psi(g^*(h)). \end{aligned}$$

□

References

[1] A.A. Agrachev, *Rolling balls and octonions*, Proc. Steklov Inst. of Math, **258** (2007), 13–22.

[2] A.A. Agrachev, Y.L. Sachkov, *Control Theory from the Geometric Viewpoint*, Springer-Verlag, Berlin Heidelberg (2004).

[3] M.A. Akivis, V.V. Goldberg, *Differential geometry of varieties with degenerate Gauss maps*, CMS Books in Mathematics, **18**, Springer-Verlag, New York, (2004).

[4] D. An, P. Nurowski, *Twistor space for rolling bodies*, Commun. Math. Phys. **326** (2014), 393–414.

[5] V.I. Arnold, *Wave front evolution and equivariant Morse lemma*, Comm. Pure and Appl. Math., **29** (1976) 557–582.

[6] V.I. Arnol'd, *Lagrangian manifold singularities, asymptotic rays and the open swallowtail*, Funct. Anal. Appl., **15** (1981). 235–246.

[7] V.I. Arnold, *Singularities of Caustics and Wave Fronts*, Mathematics and its applications (Soviet series), **62**, Kluwer Academic Publishers., Dordrecht, (1990).

[8] J. C. Baez, J. Huerta, *G_2 and the rolling ball*, Trans. Amer. Math. Soc. **366** (2014), 5257–5293.

[9] G. Bor, R. Montgomery, *G_2 and the rolling distributions*, Enseign. Math. **55** (2009), 157–196.

[10] D. Brander, *Pseudospherical frontals and their singularities*, arXiv:1502.04876 [math.DG].

[11] D. Brander, *Pseudospherical surfaces with singularities*, arXiv:1502.04876v3 [math.DG].

[12] G.E. Bredon, *Sheaf theory*, Graduate Texts in Math., **170** (2nd ed.), Springer-Verlag (1997).

[13] J.W. Bruce, *Envelopes, duality and contact structures*, Proc. Symp. Pure Math., **40–1** (1983), 195–202.

[14] J.W. Bruce, P.J. Giblin, *Curves and Singularities*, Cambridge Univ. Press, (1984).

[15] Th. Bröcker, *Differentiable Germs and Catastrophes*, London Math. Soc. Lecture Note Series **17**, Cambridge Univ. Press (1975).

[16] A.L. Castro, R. Montgomery, *Spatial curve singularities and the monster/semple tower*, Israel Journal of Math., **192** (2012), 381–427.

[17] A. Cayley, Mémoire sur les coubes à double courbure et les surfaces développables, Journal de Mathematique Pure et Appliquees (Liouville), **10** (1845), 245–250 = The Collected Mathematical Papers vol. I, pp. 207–211.

[18] L. Chen, S. Izumiya, *A mandala of Legendrian dualities for pseudo-spheres in semi-Euclidean space*, Proc. Japan Acad., **85** Ser.A, (2009), 49–54.

[19] S. Chino, S. Izumiya, *Lightlike developables in Minkowski 3-space*, Demonstratio Mathematica **43–2** (2010), 387–399.

[20] J.P. Cleave, *The form of the tangent-developable at points of zero torsion on space curves*, Math. Proc. Cambridge Philos. Soc. **88–3** (1980), 403–407.

[21] T. Fukunaga, M. Takahashi, *Existence and uniqueness for Legendre curves*, Journal of Geometry, **104** (2013), 297–307.

[22] T. Fukunaga, M. Takahashi, *Evolutes of fronts in the Euclidean plane*, Journal of Singularities, **10** (2014), 92–107.

[23] T. Fukunaga, M. Takahashi, *Involutes of fronts in the Euclidean plane*, Beiträge zur Algebra und Geometrie, **57–3** (2016), 637–653, DOI: 10.1007/s13366-015-0275-1.

[24] T. Fukunaga, M. Takahashi, *Evolutes and involutes of frontals in the Euclidean plane*, Demonstratio Mathematica, **48** (2015), 147–166.

[25] S. Fujimori, K. Saji, M. Umehara, K. Yamada, *Singularities of maximal surfaces*, Math. Z. **259** (2008), 827–848.

[26] A.B. Givental, *Lagrangian imbeddings of surfaces and the open Whitney umbrella.* Funk. Anal. i Prilozhen, **20–3** (1986), 35–41.

[27] F. R. Harvey, *Spinors and Calibration*, Academic Press (1990).

[28] A. Honda, M. Koiso, K. Saji, *Fold singularities on spacelike CMC surfaces in Lorentz-Minkowski space*, arXiv:1509.03050 [math.DG], Hokkaido Mathematical Journal **47–2** (2018), 245–267.

[29] G. Ishikawa, *Families of functions dominated by distributions of C-classes of mappings*, Ann. Inst. Fourier **33–2** (1983), 199–217.

[30] G. Ishikawa, *Parametrization of a singular Lagrangian variety*, Trans. Amer. Math. Soc., **331–2** (1992), 787–798.

[31] G.Ishikawa, *The local model of an isotropic map-germ arising from one dimensional symplectic reduction*, Math. Proc. Camb. Philo. Soc., **111–1** (1992), 103–112.

[32] G. Ishikawa, *Determinacy of the envelope of the osculating hyperplanes to a curve*, Bull. London Math. Soc., **25**(1993), 603–610.

[33] G. Ishikawa, *Parametrized Legendre and Lagrange varieties*, Kodai Math. J., **17–3** (1994, October), pp.442–451.

[34] G. Ishikawa, *Developable of a curve and determinacy relative to osculation-type*, Quart. J. Math. Oxford, **46** (1995) 437–451.

[35] G. Ishikawa, *Symplectic and Lagrange stabilities of open Whitney umbrellas*, Invent. math., **126–2** (1996), 215–234.

[36] G. Ishikawa, *Singularities of developable surfaces*, London Math. Soc. Lect. Notes Series, **263** (1999), 403–418.

[37] G. Ishikawa, *Topological classification of the tangent developables of space curves*, J. London Math. Soc., **62–2** (2000), 583–598.

[38] G. Ishikawa, *Several questions on singularities: theories and applications*, RIMS Kōkyūroku, **1122** (2000), 35–48.

[39] G. Ishikawa, *Classifying singular Legendre curves by contactomorphisms*, J. of Geom. Physics, **52–2** (2004), 113-126.

[40] G. Ishikawa, *Infinitesimal deformations and stability of singular Legendre submanifolds*, Asian J. Math., **9–1** (2005), 133–166.

[41] G. Ishikawa, *Singularities of flat extensions from generic surfaces with boundaries*, Differential Geometry and its Applications **28** (2010), 341–354.

[42] G. Ishikawa, *Generic bifurcations of framed curves in a space form and their envelopes*, Topology and its Appl., **159** (2012), 492–500.

[43] G. Ishikawa, *Singularities of tangent varieties to curves and surfaces*, Journal of Singularities, **6** (2012), 54–83.

[44] G. Ishikawa, *Tangent varieties and openings of map-germs*, RIMS Kōkyūroku Bessatsu, **B38** (2013), 119–137.

[45] G. Ishikawa, *Openings of differentiable map-germs and unfoldings*, Topics on Real and Complex Singularities, Proceedings of the 4th Japanese-Australian Workshop (JARCS4), Kobe 2011, World Scientific (2014), 87–113.

[46] G. Ishikawa, *Classification problems on singularities of mappings and their applications*, Sugaku Exposition, **28–2** (2015), 189–214.

[47] G. Ishikawa, *Singularities of Curves and Surfaces in Various Geometric Problems*, CAS Lecture Notes 10, Exact Sciences, Warsaw University of Technology (2015).

[48] G. Ishikawa, S. Janeczko, *Symplectic bifurcations of plane curves and isotropic liftings,* Quart. J. Math., **54** (2003), 1–30.

[49] G. Ishikawa, S. Janeczko, *Bifurcations in symplectic space*, Banach Center Publ., **82** (2008), 111-124.

[50] G. Ishikawa, Y. Kitagawa, W. Yukuno, *Duality of singular paths for (2,3,5)-distributions*, Journal of Dynamical and Control Systems, **21** (2015), 155–171.

[51] G. Ishikawa, Y. Kitagawa, W. Yukuno, *Duality on geodesics of Cartan distributions and sub-Riemannian pseudo-product structures*, Demonstratio Mathematica. **48–2** (2015), 193–216.

[52] G. Ishikawa, Y. Machida, *Singularities of improper affine spheres and surfaces of constant Gaussian curvature*, Intern. J. of Math., **17–3** (2006), 269–293.

[53] G. Ishikawa, Y. Machida, *Monge-Ampère systems with Lagrangian pairs*, Symmetry, Integrability and Geometry: Methods and Applications (SIGMA), 11 (2015), 081, 32 pages.

[54] G. Ishikawa, Y. Machida, M. Takahashi, *Asymmetry in singularities of tangent surfaces in contact-cone Legendre-null duality*, Journal of Singularities, **3** (2011), 126–143.

[55] G. Ishikawa, Y. Machida, M. Takahashi, *Singularities of tangent surfaces in Cartan's split G_2-geometry*, Asian J. of Math., **20–2** (2016), 353–382.

[56] G. Ishikawa, Y. Machida, M. Takahashi, *Geometry of D_4 conformal triality and singularities of tangent surfaces*, J. of Singularities, **12** (2015), 27–52.

[57] G. Ishikawa, Y. Machida, M. Takahashi, *D_n-geometry and singularities of tangent surfaces*, Hokkaido University Preprint Series in Mathematics #1058, (2014). RIMS Kōkyūroku Bessatsu **B55** (2016), 67–87.

[58] G. Ishikawa, T. Morimoto, *Solution surfaces of Monge-Ampère equations*, Diff. Geom. Appl., **14–2** (2001), 113–124.

[59] G. Ishikawa, T. Yamashita, *Affine connections and singularities of tangent surfaces to space curves*, arXiv:1501.07341 [math.DG].

[60] G. Ishikawa, T. Yamashita, *Singularities of tangent surfaces to generic space curves*, arXiv:1602.02458 [math.DG], Journal of Geometry **108** (2017), 301–318.

[61] G. Ishikawa, T. Yamashita, *Singularities of tangent surfaces to directed curves*, Topology and its Appl. **234** (2018), 198–208.

[62] S. Izumiya, *Legendrian dualities and spacelike hypersurfaces in the lightcone*, Moscow Mathematical Journal **9** (2009), 325–357.

[63] S. Izumiya, Maria del Carmen Romero Fuster, Maria Aparecida Soares Ruas, F. Tari, *Differential Geometry from a Singularity Theory Viewpoint*, World Scientific Publishing Co. (2015).

[64] S. Izumiya, H. Katsumi, T. Yamasaki, *The rectifying developable and the spherical Darboux image of a space curve*, Caustics '98, Banach Center Publ., **50** (1999), 137–149.

[65] S. Izumiya, T. Nagai, K. Saji, *Great circular surfaces in the three-sphere*, Diff. Geom. and its Appl., **29–3** (2011), 409–425.

[66] S. Izumiya, D. Pei, T. Sano, *Singularities of hyperbolic Gauss maps*, Proc. London Math. Soc., **86** (2003), 485–512.

[67] S. Izumiya, D. Pei, T. Sano, *Horospherical surfaces of curves in hyperbolic space*, Publ. Math. Debrecen, **64** (2004), 1–13.

[68] S. Izumiya, D. Pei, M. Takahashi, *Singularities of evolutes of hypersurfaces in hyperbolic space*, Proc. Edinburgh Math. Soc., **47** (2004), 131–153.

[69] S. Izumiya, K. Saji, *The mandala of Legendrian dualities for pseudo-spheres in Lorenz-Minkowski space and "flat" spacelike surfaces*. J. of Singularities **2** (2010), 92–127.

[70] Y. Kabata, *Recognition of plane-to-plane map-germs,* Topology and its Applications, **202-1** (2016), 216–238.

[71] M. Kokubu, W. Rossman, K. Saji, M. Umehara, K. Yamada, *Singularities of flat fronts in hyperbolic space,* Pacific J. Math., **221–2** (2005), 303–351.

[72] M. Kossowski, *Realizing a singular first fundamental form as a nonimmersed surface in Euclidean* 3*-space*, J. Geom. **81** (2004), 101–113.

[73] S. Lawrence, *Developable surfaces, their history and application*, Nexus Network J., **13–3** (2011), 701–714.

[74] B. Malgrange, *Ideals of Differentiable Functions*, Oxford Univ. Press (1966).

[75] J.N. Mather, *Stability of C^∞ mappings III: Finitely determined map-germs*, Publ. Math. I.H.E.S., **35** (1968), 279–308.

[76] D. Mond, *On the tangent developable of a space curve*, Math. Proc. Cambridge Philos. Soc. **91–3** (1982), 351–355.

[77] D. Mond, *Singularities of the tangent developable surface of a space curve*, Quart. J. Math. Oxford Ser. (2) **40** (1989), 79–91.

[78] D. Mond, *Deformations which preserve the non-immersive locus of a map-germ*, Math. Scand., **66** (1990), 21–32.

[79] R. Montgomery, *A tour of subriemannian geometries, their geodesics and applications*, Mathematical Surveys and Monographs, **91**, Amer. Math. Soc., Providence, RI, (2002).

[80] R. Montgomery, M. Zhitomirskii, *Geometric approach to Goursat flags*, Ann. de L'institut Henri Poincaré, Analyse non Linéaire, **18–4** (2001), 459–493.

[81] R. Montgomery, M. Zhitomirskii, *Points and Curves in the Monster Tower*, Memoirs of Amer. Math. Soc., **956** (2010).

[82] T. Morimoto, *Monge-Ampère equations viewed from contact geometry*, Banach Center Publ., **39** (1998), pp. 105–121.

[83] S. Murata and M. Umehara, *Flat surfaces with singularities in Euclidean 3-space*, J. Differential Geom., **82–2** (2009), 279–316.

[84] K. Naokawa, *Singularities of the asymptotic completion of developable Möbius strips*, Osaka J. of Math., **50–2** (2013), 425–437.

[85] J.J. Nuño Ballesteros, *Unfolding plane curves with cusps and nodes*, Proc. Roy. Soc. Edinburgh Sect. A. **145–1** (2015), 161–174.

[86] J.J. Nuño Ballesteros, O. Saeki, *Singular surfaces in* 3-*manifolds, the tangent developable of a space curve and the dual of an immersed surface in* 3-*space,* Real and complex singularities (São Carlos, 1994), 49–64, Pitman Res. Notes Math. Ser., **333**, Longman, Harlow, (1995).

[87] B. O'Neill, *Semi-Riemannian Geometry With Applications to Relativity*, Pure and Applied Mathematics **103**, Academic Press (1983).

[88] W.F. Pohl, *The self-linking number of a closed space curve*, Journal of Mathematics and Mechanics, **17–10** (1968), 975–985.

[89] I.R. Porteous, *The normal singularities of a submanifold*, J. of Diff. Geom., **5** (1971), 543–564 .

[90] I.R. Porteous, *Geometric Differentiation, for the Intelligence of Curves and Surfaces*, Cambridge Univ. Press, Cambridge, (1994).

[91] I.R. Porteous, *Clifford Algebras and the Classical Groups*, Cambridge Studies in Adv. Math., **50**, Cambridge Univ. Press (1995).

[92] A. Robinson, *Non-standard Analysis*, Princeton University Press (1996).

[93] K. Saji, *Criteria for singularities of smooth maps from the plane into the plane and their applications*, Hiroshima Math. J. **40** (2010), 229–239.

[94] K. Saji, *Criteria for D_4 singularities of wave fronts*, Tohoku Math. J., **63–1** (2011), 137–147.

[95] K. Saji, M. Umehara, K. Yamada, *The geometry of fronts*, Annals of Math., **169–2** (2009), 491–529.

[96] K. Saji, M. Umehara, K. Yamada, *A_k singularities of wave fronts*, Math. Proc. Camb. Philos. Soc. **146–3** (2009), 731–746.

[97] K. Saji, M. Umehara, K. Yamada, *Coherent tangent bundles and Gauss-Bonnet formulas for wave fronts,* J. Geom.Anal., **22** (2012), 383–409.

[98] O.P. Shcherbak, *Projective dual space curves and Legendre singularities*, Trudy Tbiliss Univ., 232–233 (1982), 280–336.

[99] R. Thom, *Sur la théorie des enveloppes*, Journ. de Math., **41–2** (1962), 177–192.

[100] C.T.C. Wall, *Finite determinacy of smooth map-germs*, Bull. London Math. Soc., **13** (1981), 481–539.

[101] K. Yamaguchi, *Geometry of linear differential systems towards contact geometry of second order*, Symmetries and Overdetermined Systems of Partial Differential Equations, Vol. **144** of the series The IMA Volumes in Mathematics and its Applications (2008), pp 151–203.

[102] F.L. Zak, *Tangents and Secants of Algebraic Varieties*, Transl. of Math. Monographs **127**, Amer. Math. Soc., (1993).

[103] V. M. Zakalyukin, *Lagrangian and Legendrian singularities*, Funct. Anal. Appl., **10** (1976), 23-31.

[104] V. M. Zakalyukin, *Reconstructions of fronts and caustics depending on a parameter and versality of mappings*, J. Soviet Math., **27** (1983), 2713-2735.

[105] V.M. Zakalyukin, A.N. Kurbatskii, *Envelope singularities of families of planes in control theory*, Proc. Steklov Inst. Math., **262–1** (2008), 66–79.

Goo Ishikawa,
Department of Mathematics, Hokkaido University,
Sapporo 060-0810, Japan.
e-mail : ishikawa@math.sci.hokudai.ac.jp

Advanced Studies in Pure Mathematics 78, 2018
Singularities in Generic Geometry
pp. 107–161

The theory of graph-like Legendrian unfoldings : Equivalence relations

Shyuichi Izumiya

Abstract.

This is a half survey article on the recent development of the theory of graph-like Legendrian unfoldings which is the sequel to the previous surveys. The notion of big Legendrian submanifolds was introduced by Zakalyukin for describing the wave front propagations. Graph-like Legendrian unfoldings belong to a special class of big Legendrian submanifolds. In particular, natural equivalence relations among graph-like Legendrian unfoldings are introduced and geometric properties of these equivalence relations are investigated. Although this is a survey article, some new original results and proofs for some implicitly known results are given.

§1. Introduction

The notion of graph-like Legendrian unfoldings was introduced in [14]. It belongs to a special class of the big Legendrian submanifolds which Zakalyukin introduced in [38, 39]. There have been some developments on this theory during past two decades [14, 15, 19, 20, 21]. This is a sequel to a survey article on the theory of graph-like Legendrian unfoldings and its applications [24]. The results in the first half part of this paper have been already presented, implicitly or explicitly, in the above articles. The later half of this paper focuses on natural equivalence relations among big Legendrian submanifolds and graph-like Legendrian unfoldings as a special case. Some of the results here explain how the theory of graph-like Legendrian unfoldings is useful for applying to many situations related to the theory of Lagrangian singularities (caustics).

Received May 21, 2016.
Revised October 21, 2016.
2010 *Mathematics Subject Classification.* Primary 58K05; Secondary 57R45, 58K25.
Key words and phrases. Wave front propagations, Big wave fronts, graph-like Legendrian unfoldings, Caustics.

The caustic is described as the set of critical values of the projection of a Lagrangian submanifold from the phase space onto the configuration space. Moreover, it has been known that caustics equivalence (i.e., diffeomorphic caustics) does not imply Lagrangian equivalence. This is one of the main differences from the theory of Legendrian singularities. In the theory of Legendrian singularities, wave fronts equivalence (i.e., diffeomorphic wave fronts) implies Legendrian equivalence generically.

On the other hand, in the real world, the caustics given by refracted rays are visible. However, the wave front propagations are not visible. We give a picture of the caustic generated by the rays through a wine glass (cf. Fig.1).

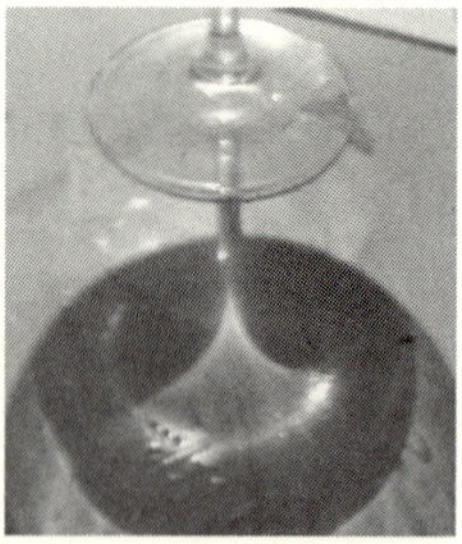

Fig.1: The caustic generated by the ray through a wine glass.
The picture was taken at A-TABLE in Sapporo.

We can observe the caustic but cannot observe the wave front propagation of the rays. However, if we draw the pictures of the parallels (cf. Fig.2) and the normal lines of a parabola (cf. Fig.3) respectively, we can observe the caustic (i.e. the evolute), the wave front propagation (i.e. the parallels) and the family of the rays (i.e. the lines) respectively. Therefore, we can say that there are hidden structures (i.e., wave front propagations and the family of the rays) on the picture of caustics in the real world.

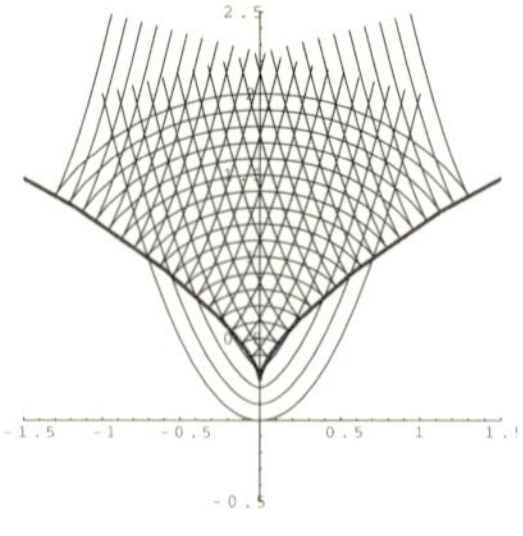

Fig.2: The parallels and the evolute of a parabola.

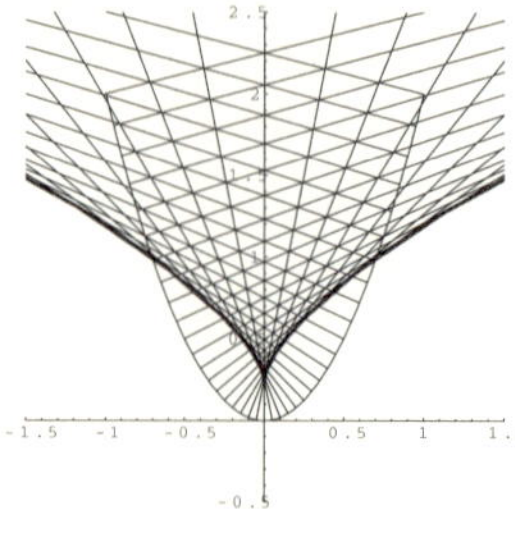

Fig.3: The normal lines and the evolute of a parabola.

In fact, caustics are a subject of classical physics (i.e. optics). The corresponding Lagrangian submanifold is, however, deeply related to the *semi-classical approximation* of quantum mechanics (cf. [9, 12, 29]). Moreover, it was believed around 1989 that the correct framework to describe the parallels of a curve is the theory of big wave fronts [2]. But it was pointed out that A_1 and A_2 bifurcations do not occur as the parallels of curves [3, 7]. Therefore, the framework of the theory of big wave fronts is too wide for describing the parallels of curves. The theory of the graph-like Legendrian unfoldings was introduced to construct the correct framework for the parallels of a curve in [14]. In order to understand such geometric properties, we introduce natural five equivalence relations among big Legendrian submanifolds. In particular, $S.P^+$-Legendrian equivalence is a key notion to understand the relations between the wave front propagation and the caustics, which was introduced in [15, 40] independently for different purposes. One of the main results in the theory of graph-like Legendrian unfoldings is Theorem 6.1 which reveals the relation between caustics and wave front propagations by using $S.P^+$-Legendrian equivalence. In Example 6.11 we give examples of Lagrangian submanifolds with diffeomorphic caustics but are not Lagrangian equivalent. Those examples are famous examples. However, we clarify the reason why these have diffeomorphic caustics but are not Lagrangian equivalent geometrically. Moreover, those examples explain the situation that even if the phase portraits of wave front propagations are different but those are Lagrangian equivalent (cf. Fig.7 and Fig.10). Here, the phase portrait means that the picture of the arrangement of both the caustics and the family of momentary fronts.

We give two examples of applications of the theory of wave front propagations in §8. One of the examples is a brief explanation of the results on the stability of caustics formulated in the framework of Hamiltonian systems by Jänich [27] and Wassermann [35]. They adopted the notion of universal unfoldings with respect to $\mathcal{A}$-equivalence (i.e. right-left equivalence). However, it is known that the Lagrangian stability of the caustic is equivalent to the universality of unfoldings with respect to $\mathcal{R}^+$-equivalence (i.e. right equivalence) [1, 9, 12, 37]. Therefore, their stability of caustics is not corresponding to the Lagrangian stability. In this paper we show that their stability for caustics is equivalent to the s-P-Legendrian stability for the corresponding graph-like Legendrian unfolding (cf. Theorems 7.9 and 8.11).

Another example is a survey on the caustics of world hyper-sheets in the Lorentz-Minkowski space-time [22, 23]. A *world hyper-sheet* in the Lorentz-Minkowski space-time is a timelike hypersurface formed by a one-parameter family of spacelike submanifolds of codimension two in the ambient space. Each spacelike submanifold in the world hyper-sheet is called a *momentary space*. We consider the family of lightlike hypersurfaces along monetary spaces in the world hyper-sheet. In [4, 5] Bousso and Randall considered that the locus of the singularities (the lightlike focal sets) of lightlike hypersurfaces along momentary spaces form a caustic in the Lorentz-Minkowski space-time. This construction is originally from the theoretical physics (the string theory, the brane world scenario, the cosmology, and so on). We call it a *BR-caustic* of the world hyper-sheet. We have no notion of the time constant in the relativity theory. Hence everything that is moving depends on the time. Therefore, we have to consider world hyper-sheets in the relativity theory. Even if we consider a fixed light source (i.e. a shining surface) in the Euclidean 3-space, it must be a world hyper-sheet in the 4-dimensional Lorentz-Minkowski space-time. So the caustic in the Euclidean 3-space is a slice of the BR-caustic of the world hyper-sheet with a spacelike hyperplane. Since the parameter of a world hyper-sheet is intrinsically given, we really need the theory of graph-like Legendrian unfoldings for the study of BR-caustics.

§2. Lagrangian and Legendrian singularities

We give a brief review of the local theory of Lagrangian and Legendrian singularities. We have already written surveys on these theory in several articles [17, 24, 26]. However, it is better to explain the basic results in those theories here again.

Firstly, we consider the cotangent bundle $\pi : T^*\mathbb{R}^n \to \mathbb{R}^n$ over $\mathbb{R}^n$. Let $(x,p) = (x_1, \dots, x_n, p_1, \dots, p_n)$ be the canonical coordinates on $T^*\mathbb{R}^n$. Then the canonical symplectic structure on $T^*\mathbb{R}^n$ is given by the *canonical two form* $\omega = \sum_{i=1}^n dp_i \wedge dx_i$. Let $i : L \subset T^*\mathbb{R}^n$ be a submanifold. We say that i is a *Lagrangian submanifold* if $\dim L = n$ and $i^*\omega = 0$. In this case, the set of critical values of $\pi \circ i$ is called the *caustic* of $i : L \subset T^*\mathbb{R}^n$, which is denoted by C_L. One of the main results in the theory of Lagrangian singularities is the description of Lagrangian submanifold germs by using families of function germs. Let $F : (\mathbb{R}^k \times \mathbb{R}^n, 0) \to (\mathbb{R}, 0)$ be an n-parameter unfolding of a function germ $f = F|_{\mathbb{R}^k \times \{0\}} : (\mathbb{R}^k, 0) \to (\mathbb{R}, 0)$. We say that F is a *Morse family of functions* if the map germ

$$\Delta F = \left(\frac{\partial F}{\partial q_1}, \dots, \frac{\partial F}{\partial q_k} \right) : (\mathbb{R}^k \times \mathbb{R}^n, 0) \to (\mathbb{R}^k, 0)$$

is non-singular, where $(q, x) = (q_1, \dots, q_k, x_1, \dots, x_n) \in (\mathbb{R}^k \times \mathbb{R}^n, 0)$. In this case, we have a smooth n-dimensional submanifold germ $C(F) =$

$(\Delta F)^{-1}(0) \subset (\mathbb{R}^k \times \mathbb{R}^n, 0)$ and a map germ $L(F) : (C(F), 0) \to T^*\mathbb{R}^n$ defined by

$$L(F)(q,x) = \left(x, \frac{\partial F}{\partial x_1}(q,x), \ldots, \frac{\partial F}{\partial x_n}(q,x)\right).$$

We can show that $L(F)(C(F))$ is a Lagrangian submanifold germ. It is known ([1], page 300) that all Lagrangian submanifold germs in $T^*\mathbb{R}^n$ are constructed by the above method. A Morse family of functions $F : (\mathbb{R}^k \times \mathbb{R}^n, 0) \to (\mathbb{R}, 0)$ is said to be a *generating family* of $L(F)(C(F))$.

We now define a natural equivalence relation among Lagrangian submanifold germs. Let $i : (L, p) \subset (T^*\mathbb{R}^n, p)$ and $i' : (L', p') \subset (T^*\mathbb{R}^n, p')$ be Lagrangian submanifold germs. Then we say that i and i' are *Lagrangian equivalent* if there exist a symplectic diffeomorphism germ $\hat{\tau} : (T^*\mathbb{R}^n, p) \to (T^*\mathbb{R}^n, p')$ and a diffeomorphism germ $\tau : (\mathbb{R}^n, \pi(p)) \to (\mathbb{R}^n, \pi(p'))$ such that $(\hat{\tau}(L), p') = (L', p')$ as set germs and $\pi \circ \hat{\tau} = \tau \circ \pi$, where $\pi : (T^*\mathbb{R}^n, p) \to (\mathbb{R}^n, \pi(p))$ is the canonical projection. Here $\hat{\tau}$ is said to be a *symplectic diffeomorphism germ* if it is a diffeomorphism germ such that $\hat{\tau}^*\omega = \omega$. Then the caustic C_L is diffeomorphic to the caustic $C_{L'}$ by the diffeomorphism germ τ. We say that L and L' are *caustics equivalent* if there is a diffeomorphism germ $\tau : (\mathbb{R}^n, \pi(p)) \to (\mathbb{R}^n, \pi(p'))$ such that $(\tau(C_L), \pi(p')) = (C_{L'}, \pi(p'))$ as set germs. It is known that caustic equivalence does not imply Lagrangian equivalence even generically (cf. Example 6.11).

We can interpret the Lagrangian equivalence by using the notion of generating families. Let $F, G : (\mathbb{R}^k \times \mathbb{R}^n, 0) \to (\mathbb{R}, 0)$ be function germs. We say that F and G are *P-$\mathcal{R}^+$-equivalent* if there exist a diffeomorphism germ $\Phi : (\mathbb{R}^k \times \mathbb{R}^n, 0) \to (\mathbb{R}^k \times \mathbb{R}^n, 0)$ of the form $\Phi(q,x) = (\phi_1(q,x), \phi_2(x))$ and a function germ $h : (\mathbb{R}^n, 0) \to (\mathbb{R}, 0)$ such that $G(q,x) = F(\Phi(q,x)) + h(x)$. For any $F_1 : (\mathbb{R}^k \times \mathbb{R}^n, 0) \to (\mathbb{R}, 0)$ and $F_2 : (\mathbb{R}^{k'} \times \mathbb{R}^n, 0) \to (\mathbb{R}, 0)$, F_1 and F_2 are said to be *stably P-$\mathcal{R}^+$-equivalent* if they become P-$\mathcal{R}^+$-equivalent after the addition to the arguments q_i of new arguments q_i' and to the functions F_i of non-degenerate quadratic forms Q_i in the new arguments, i.e., $F_1 + Q_1$ and $F_2 + Q_2$ are P-$\mathcal{R}^+$-equivalent.

Let $F : (\mathbb{R}^k \times \mathbb{R}^n, 0) \to (\mathbb{R}, 0)$ be a Morse family of functions and $\mathcal{E}_k$ the local ring of function germs of $q = (q_1, \ldots, q_k)$ variables at the origin with the unique maximal ideal $\mathfrak{M}_k = \{h \in \mathcal{E}_k \mid h(0) = 0\}$. We say that F is an *infinitesimally $\mathcal{R}^+$-versal unfolding* of $f = F|_{\mathbb{R}^k \times \{0\}}$ (cf. [6]) if

$$\mathcal{E}_k = J_f + \left\langle \frac{\partial F}{\partial x_1}|_{\mathbb{R}^k \times \{0\}}, \ldots, \frac{\partial F}{\partial x_n}|_{\mathbb{R}^k \times \{0\}} \right\rangle_{\mathbb{R}} + \langle 1 \rangle_{\mathbb{R}},$$

where $f = F|_{\mathbb{R}^k \times \{0\}}$ and

$$J_f = \left\langle \frac{\partial f}{\partial q_1}(q), \dots, \frac{\partial f}{\partial q_k}(q) \right\rangle_{\mathcal{E}_k}.$$

Remark 2.1. There is a definition of *Lagrangian stability* (cf. [1, §21.1]) of a Lagrangian submanifold germ. In this paper we do not need the original definition of Lagrangian stability, so that we omit to give the definition.

Then we have the following fundamental theorem of the theory of Lagrangian singularities (cf. [1]):

Theorem 2.2. *Let $F : (\mathbb{R}^k \times \mathbb{R}^n, 0) \to (\mathbb{R}, 0)$ and $G : (\mathbb{R}^{k'} \times \mathbb{R}^n, 0) \to (\mathbb{R}, 0)$ be Morse families of functions. Then we have the following*:
(1) *$L(F)(C(F))$ and $L(G)(C(G))$ are Lagrangian equivalent if and only if F and G are stably P-$\mathcal{R}^+$-equivalent.*
(2) *$L(F)(C(F))$ is Lagrangian stable if and only if F is an infinitesimally $\mathcal{R}^+$-versal unfolding of $f = F|_{\mathbb{R}^k \times \{0\}}$.*

On the other hand, we give a brief review on the theory of Legendrian singularities. Let $\overline{\pi} : PT^*(\mathbb{R}^m) \to \mathbb{R}^m$ be the projective cotangent bundle over $\mathbb{R}^m$. This fibration can be considered as a Legendrian fibration with the canonical contact structure K on $PT^*(\mathbb{R}^m)$. We now review geometric properties of this space. Consider the tangent bundle $\tau : TPT^*(\mathbb{R}^m) \to PT^*(\mathbb{R}^m)$ and the differential map $d\overline{\pi} : TPT^*(\mathbb{R}^m) \to T\mathbb{R}^m$ of $\overline{\pi}$. For any $X \in TPT^*(\mathbb{R}^m)$, there exists an element $\alpha \in T^*(\mathbb{R}^m)$ such that $\tau(X) = [\alpha]$. For an element $V \in T_x(\mathbb{R}^m)$, the property $\alpha(V) = \mathbf{0}$ does not depend on the choice of representative of the class $[\alpha]$. Thus we can define the canonical contact structure on $PT^*(\mathbb{R}^m)$ by $K = \{X \in TPT^*(\mathbb{R}^m) | \tau(X)(d\overline{\pi}(X)) = 0\}$. We have the trivialization $PT^*(\mathbb{R}^m) \cong \mathbb{R}^m \times P(\mathbb{R}^{m*})$ and we call $(x, [\xi])$ *homogeneous coordinates*, where $x = (x_1, \dots, x_m) \in \mathbb{R}^m$ and $[\xi] = [\xi_1 : \cdots : \xi_m]$ are homogeneous coordinates of the dual projective space $P(\mathbb{R}^{m*})$. It is easy to show that $X \in K_{(x,[\xi])}$ if and only if $\sum_{i=1}^m \mu_i \xi_i = 0$, where $d\overline{\pi}(X) = \sum_{i=1}^n \mu_i \frac{\partial}{\partial x_i}$. Let $\Phi : (\mathbb{R}^m, 0) \to (\mathbb{R}^m, 0)$ be a diffeomorphism germ. Then we have a unique contact diffeomorphism germ $\widehat{\Phi} : PT^*\mathbb{R}^m \to PT^*\mathbb{R}^m$ defined by $\widehat{\Phi}(x, [\xi]) = (\Phi(x), [\xi \circ d_{\Phi(x)}(\Phi^{-1})])$. We call $\widehat{\Phi}$ the *contact lift* of Φ.

A submanifold $i : \mathscr{L} \subset PT^*(\mathbb{R}^m)$ is said to be a *Legendrian submanifold* if $\dim \mathscr{L} = m - 1$ and $di_p(T_p\mathscr{L}) \subset K_{i(p)}$ for any $p \in \mathscr{L}$. We also call $\overline{\pi} \circ i = \overline{\pi}|_{\mathscr{L}} : \mathscr{L} \to \mathbb{R}^m$ a *Legendrian map* and $W(\mathscr{L}) = \overline{\pi}(\mathscr{L})$ a *wave front* of $i : \mathscr{L} \subset PT^*(\mathbb{R}^m)$. We say that a point $p \in \mathscr{L}$ is a

Legendrian singular point if $\operatorname{rank} d(\overline{\pi}|_{\mathscr{L}})_p < m-1$. In this case $\overline{\pi}(p)$ is the singular point of $W(\mathscr{L})$.

The main tool of the theory of Legendrian singularities is also the notion of generating families. Let $F : (\mathbb{R}^k \times \mathbb{R}^m, 0) \to (\mathbb{R}, 0)$ be a function germ. We say that F is a *Morse family of hypersurfaces* if the map germ

$$\Delta^* F = \left(F, \frac{\partial F}{\partial q_1}, \dots, \frac{\partial F}{\partial q_k} \right) : (\mathbb{R}^k \times \mathbb{R}^m, 0) \to (\mathbb{R} \times \mathbb{R}^k, 0)$$

is non-singular, where $(q, x) = (q_1, \dots, q_k, x_1, \dots, x_m) \in (\mathbb{R}^k \times \mathbb{R}^m, 0)$. In this case we have a smooth $(m-1)$-dimensional submanifold germ

$$\Sigma_*(F) = \left\{ (q, x) \in (\mathbb{R}^k \times \mathbb{R}^m, 0) \;\middle|\; F(q, x) = \frac{\partial F}{\partial q_i}(q, x) = 0, i = 1, \dots, k \right\}$$

and a map germ $\mathscr{L}_F : (\Sigma_*(F), 0) \to PT^*\mathbb{R}^m$ defined by

$$\mathscr{L}_F(q, x) = \left(x, \left[\frac{\partial F}{\partial x_1}(q, x) : \dots : \frac{\partial F}{\partial x_m}(q, x) \right] \right).$$

We can show that $\mathscr{L}_F(\Sigma_*(F)) \subset PT^*(\mathbb{R}^m)$ is a Legendrian submanifold germ. It is known ([1, page 320]) that all Legendrian submanifold germs in $PT^*(\mathbb{R}^m)$ are constructed by the above method. We call F a *generating family* of $\mathscr{L}_F(\Sigma_*(F))$. Therefore the wave front is given by

$$W(\mathscr{L}_F(\Sigma_*(F)) = \left\{ x \in \mathbb{R}^m \; | \exists q \in \mathbb{R}^k \text{ s.t } (q, x) \in \Sigma_*(F) \right\}.$$

Since the Legendrian submanifold germ $i : (\mathscr{L}, p) \subset (PT^*\mathbb{R}^n, p)$ is uniquely determined on the regular part of the wave front $W(\mathscr{L})$, we have the following simple but significant property of Legendrian submanifold germs [39].

Proposition 2.3 (Zakalyukin). *Let $i : (\mathscr{L}, p) \subset (PT^*\mathbb{R}^m, p)$ and $i' : (\mathscr{L}', p') \subset (PT^*\mathbb{R}^m, p')$ be Legendrian submanifold germs such that $\overline{\pi} \circ i, \overline{\pi} \circ i'$ are proper map germs and the sets singularities of these map germs are nowhere dense respectively. Then $(\mathscr{L}, p) = (\mathscr{L}', p')$ if and only if $(W(\mathscr{L}), \overline{\pi}(p)) = (W(\mathscr{L}'), \overline{\pi}(p'))$.*

In order to understand the ambiguity of generating families for a fixed Legendrian submanifold germ we introduce the following equivalence relation among Morse families of hypersurfaces. For function germs $F, G : (\mathbb{R}^k \times \mathbb{R}^m, 0) \to (\mathbb{R}, 0)$, we say that F and G are *strictly parametrized $\mathcal{K}$-equivalent* (briefly, *S.P-$\mathcal{K}$-equivalent*) if there exists a diffeomorphism germ $\Psi : (\mathbb{R}^k \times \mathbb{R}^m, 0) \to (\mathbb{R}^k \times \mathbb{R}^m, 0)$ of the form $\Psi(q, x) = (\psi_1(q, x), x)$ for $(q, x) \in (\mathbb{R}^k \times \mathbb{R}^m, 0)$ such that $\Psi^*(\langle F \rangle_{\mathcal{E}_{k+m}}) = \langle G \rangle_{\mathcal{E}_{k+m}}$.

Here $\Psi^* : \mathcal{E}_{k+m} \to \mathcal{E}_{k+m}$ is the pull back $\mathbb{R}$-algebra isomorphism defined by $\Psi^*(h) = h \circ \Psi$. The definition of *stably S.P-$\mathcal{K}$-equivalence* among Morse families of hypersurfaces is similar to the definition of stably P-$\mathcal{R}^+$-equivalence among Morse families of functions. The following is the key lemma of the theory of Legendrian singularities (cf. [1, 10, 37]).

Lemma 2.4 (Zakalyukin). *Let $F : (\mathbb{R}^k \times \mathbb{R}^m, 0) \to (\mathbb{R}, 0)$ and $G : (\mathbb{R}^{k'} \times \mathbb{R}^m, 0) \to (\mathbb{R}, 0)$ be Morse families of hypersurfaces. Then $(\mathscr{L}_F(\Sigma_*(F)), p) = (\mathscr{L}_G(\Sigma_*(G)), p)$ if and only if F and G are stably S.P-$\mathcal{K}$-equivalent.*

Let $F : (\mathbb{R}^k \times \mathbb{R}^m, 0) \to (\mathbb{R}, 0)$ be a Morse family of hypersurfaces and $\Phi : (\mathbb{R}^m, 0) \to (\mathbb{R}^m, 0)$ a diffeomorphism germ. We define $\Phi^* F : (\mathbb{R}^k \times \mathbb{R}^m, 0) \to (\mathbb{R}, 0)$ by $\Phi^* F(q, x) = F(q, \Phi(x))$. Then we have $(1_{\mathbb{R}^q} \times \Phi)(\Sigma_*(\Phi^* F)) = \Sigma_*(F)$ and

$$\mathscr{L}_{\Phi^* F}(\Sigma_*(\Phi^* F)) = \left\{ \left(x, \left[\left(\frac{\partial F}{\partial x}(q, \Phi(x)) \right) \circ d\Phi_x \right] \right) \,\middle|\, (q, \Phi(x)) \in \Sigma_*(F) \right\},$$

so that $\widehat{\Phi}(\mathscr{L}_{\Phi^* F}(\Sigma_*(\Phi^* F))) = \mathscr{L}_F(\Sigma_*(F))$ as set germs.

Proposition 2.5. *Let $F : (\mathbb{R}^k \times \mathbb{R}^m, 0) \to (\mathbb{R}, 0)$ and $G : (\mathbb{R}^{k'} \times \mathbb{R}^m, 0) \to (\mathbb{R}, 0)$ be Morse families of hypersurfaces. For a diffeomorphism germ $\Phi : (\mathbb{R}^m, 0) \to (\mathbb{R}^m, 0)$, $\widehat{\Phi}(\mathscr{L}_G(\Sigma_*(G))) = \mathscr{L}_F(\Sigma_*(F))$ if and only if $\Phi^* F$ and G are stably S.P-$\mathcal{K}$-equivalent.*

Proof. Since $\widehat{\Phi}(\mathscr{L}_{\Phi^* F}(\Sigma_*(\Phi^* F))) = \mathscr{L}_F(\Sigma_*(F))$, we have

$$\mathscr{L}_{\Phi^* F}(\Sigma_*(\Phi^* F)) = \mathscr{L}_G(\Sigma_*(G)).$$

By Lemma 2.4, the assertion holds. □

We say that $\mathscr{L}_F(\Sigma_*(F))$ and $\mathscr{L}_G(\Sigma_*(G))$ are *Legendrian equivalent* if there exists a diffeomorphism germ $\Phi : (\mathbb{R}^m, 0) \to (\mathbb{R}^m, 0)$ such that the condition in the above proposition holds. By Proposition 2.3, with the generic condition on F and G, $\Phi(W(\mathscr{L}_G(\Sigma_*(G)))) = W(\mathscr{L}_F(\Sigma_*(F)))$ if and only if $\widehat{\Phi}(\mathscr{L}_G(\Sigma_*(G))) = \mathscr{L}_F(\Sigma_*(F))$ for a diffeomorphism germ $\Phi : (\mathbb{R}^m, 0) \to (\mathbb{R}^m, 0)$.

For function germs $F, G : (\mathbb{R}^k \times \mathbb{R}^m, 0) \to (\mathbb{R}, 0)$, we say that F and G are *parametrized $\mathcal{K}$-equivalent* (briefly, *P-$\mathcal{K}$-equivalent*) if there exists a diffeomorphism germ $\Psi : (\mathbb{R}^k \times \mathbb{R}^m, 0) \to (\mathbb{R}^k \times \mathbb{R}^m, 0)$ of the form $\Psi(q, x) = (\psi_1(q, x), \psi_2(x))$ for $(q, x) \in (\mathbb{R}^k \times \mathbb{R}^m, 0)$ such that $\Psi^*(\langle F \rangle_{\mathcal{E}_{k+m}}) = \langle G \rangle_{\mathcal{E}_{k+m}}$. We also say that F is an *infinitesimally $\mathcal{K}$-versal unfolding* of $f = F|_{\mathbb{R}^k \times \{0\}}$ if

$$\mathcal{E}_k = T_e(\mathcal{K})(f) + \left\langle \frac{\partial F}{\partial x_1}|_{\mathbb{R}^k \times \{0\}}, \dots, \frac{\partial F}{\partial x_n}|_{\mathbb{R}^k \times \{0\}} \right\rangle_{\mathbb{R}},$$

where $f = F|_{\mathbb{R}^k \times \{0\}}$ and

$$T_e(\mathcal{K})(f) = J_f + \langle f \rangle_{\mathcal{E}_k} = \left\langle \frac{\partial f}{\partial q_1}(q), \ldots, \frac{\partial f}{\partial q_k}(q), f(q) \right\rangle_{\mathcal{E}_k}.$$

Remark 2.6. There is a definition of *Legendrian stability* (cf. [1, §21.1]) of a Legendrian submanifold germ. In this paper we do not need the original definition of Legendrian stability, so that we omit to give the definition.

Then we have the following fundamental theorem of the theory of Legendrian singularities (cf. [1, 37]):

Theorem 2.7. *Let* $F : (\mathbb{R}^k \times \mathbb{R}^m, 0) \to (\mathbb{R}, 0)$, $G : (\mathbb{R}^{k'} \times \mathbb{R}^m, 0) \to (\mathbb{R}, 0)$ *be Morse families of hypersurfaces. Then we have the following*:
(1) $\mathscr{L}_F(\Sigma_*(F))$ *and* $\mathscr{L}_G(\Sigma_*(G))$ *are Legendrian equivalent if and only if* F *and* G *are stably* P-$\mathcal{K}$*-equivalent.*
(2) $\mathscr{L}_F(\Sigma_*(F))$ *is Legendrian stable if and only if* F *is an infinitesimally* $\mathcal{K}$*-versal unfolding of* $f = F|_{\mathbb{R}^k \times \{0\}}$.

We have the following classification theorem as a corollary of Proposition 2.3 and Theorem 2.7 (cf. [16, Proposition A.4]).

Theorem 2.8. *Let* $F : (\mathbb{R}^k \times \mathbb{R}^m, 0) \to (\mathbb{R}, 0)$, $G : (\mathbb{R}^{k'} \times \mathbb{R}^m, 0) \to (\mathbb{R}, 0)$ *be Morse families of hypersurfaces. Suppose that* $\mathscr{L}_F(\Sigma_*(F))$ *and* $\mathscr{L}_G(\Sigma_*(G))$ *are Legendrian stable. Then the following conditions are equivalent*:
(1) $\mathscr{L}_F(\Sigma_*(F))$ *and* $\mathscr{L}_G(\Sigma_*(G))$ *are Legendrian equivalent,*
(2) $f = F|_{\mathbb{R}^k \times \{0\}}$ *and* $g = G|_{\mathbb{R}^{k'} \times \{0\}}$ *are stably* $\mathcal{K}$*-equivalent,*
(3) $W(\mathscr{L}_F(\Sigma_*(F)))$ *and* $W(\mathscr{L}_G(\Sigma_*(G)))$ *are diffeomorphic as set germs.*

Remark 2.9. We say that $f, g : (\mathbb{R}^k, 0) \to (\mathbb{R}, 0)$ are $\mathcal{K}$*-equivalent* if there exists a diffeomorphism germ $\phi : (\mathbb{R}^k, 0) \to (\mathbb{R}^k, 0)$ such that $\phi^*(\langle f \rangle_{\mathcal{E}_k}) = \langle g \rangle_{\mathcal{E}_k}$. We also say that f is r*-determined relative to* $\mathcal{K}$ if f and g are $\mathcal{K}$-equivalent for any $g \in \mathfrak{M}_k$ with $f - g \in \mathfrak{M}_k^{r+1}$. Suppose that f and g are r-determined relative to $\mathcal{K}$. Then it is known that f and g are $\mathcal{K}$-equivalent if and only if $Q_r(f)$ and $Q_r(g)$ are isomorphic as $\mathbb{R}$-algebras (cf. [30]), where $Q_r(f) = \mathcal{E}_k / (\langle g \rangle_{\mathcal{E}_k} + \mathfrak{M}_k^{r+1})$. Moreover, it is known that $f = F|_{\mathbb{R}^k \times \{0\}}$ is $m+1$-determined if $F : (\mathbb{R}^k \times \mathbb{R}^m, 0) \to (\mathbb{R}, 0)$ is an infinitesimally $\mathcal{K}$-versal unfolding of f (cf. [28]). Therefore, condition (2) in the above theorem can be replace by the following condition:
(2′) $Q_{m+2}(f)$ and $Q_{m+2}(g)$ are isomorphism as $\mathbb{R}$-algebras.

§3. Theory of the wave front propagations

In this section we give a brief survey of the theory of wave front propagations (for details, see [1, 14, 39, 40], etc). We consider one parameter families of wave fronts and their bifurcations. The principal idea is that a one parameter family of wave fronts is considered to be a wave front whose dimension is one dimension higher than each member of the family. This is called a *big wave front*. We consider the case when $m = n+1$ and distinguish space and time coordinates, so that we denote that $\mathbb{R}^{n+1} = \mathbb{R}^n \times \mathbb{R}$ and coordinates are denoted by $(x,t) = (x_1, \dots, x_n, t) \in \mathbb{R}^n \times \mathbb{R}$. Then we consider the projective cotangent bundle $\overline{\pi} : PT^*(\mathbb{R}^n \times \mathbb{R}) \to \mathbb{R}^n \times \mathbb{R}$. Because of the trivialization $PT^*(\mathbb{R}^n \times \mathbb{R}) \cong (\mathbb{R}^n \times \mathbb{R}) \times P((\mathbb{R}^n \times \mathbb{R})^*)$, we have homogeneous coordinates $((x_1, \dots, x_n, t), [\xi_1 : \cdots : \xi_n : \tau])$. We remark that $PT^*(\mathbb{R}^n \times \mathbb{R})$ is a fiber-wise compactification of the 1-jet space as follows: We consider an affine open subset $U_\tau = \{((x,t),[\xi : \tau]) | \tau \neq 0\}$ of $PT^*(\mathbb{R}^n \times \mathbb{R})$. For any $((x,t),[\xi : \tau]) \in U_\tau$, we have

$$((x_1, \dots, x_n, t), [\xi_1 : \cdots : \xi_n : \tau]) \\ = ((x_1, \dots, x_n, t), [-(\xi_1/\tau) : \cdots : -(\xi_n/\tau) : -1]),$$

so that we may adopt the corresponding *affine coordinates*

$$((x_1, \dots, x_n, t), (p_1, \dots, p_n)),$$

where $p_i = -\xi_i/\tau$. On U_τ we can easily show that $\theta^{-1}(0) = K|_{U_\tau}$, where $\theta = dt - \sum_{i=1}^n p_i dx_i$. This means that U_τ can be identified with the 1-jet space which is denoted by $J^1_{GA}(\mathbb{R}^n, \mathbb{R}) \subset PT^*(\mathbb{R}^n \times \mathbb{R})$. We call the above coordinates *a system of graph-like affine coordinates.* Throughout this paper, we use this identification.

For a Legendrian submanifold $i : \mathscr{L} \subset PT^*(\mathbb{R}^n \times \mathbb{R})$, the corresponding wave front $\overline{\pi} \circ i(\mathscr{L}) = W(\mathscr{L})$ is called a *big wave front*. We call

$$W_t(\mathscr{L}) = \pi_1(\pi_2^{-1}(t) \cap W(\mathscr{L})) \quad (t \in \mathbb{R})$$

a *momentary front* (or, a *small front*) for each $t \in (\mathbb{R}, 0)$, where $\pi_1 : \mathbb{R}^n \times \mathbb{R} \to \mathbb{R}^n$ and $\pi_2 : \mathbb{R}^n \times \mathbb{R} \to \mathbb{R}$ are the canonical projections defined by $\pi_1(x,t) = x$ and $\pi_2(x,t) = t$ respectively. In this sense, we call $\mathscr{L}$ a *big Legendrian submanifold.* We say that a point $p \in \mathscr{L}$ is a *space-singular point* if $\operatorname{rank} d(\pi_1 \circ \overline{\pi}|_{\mathscr{L}})_p < n$ and a *time-singular point* if $\operatorname{rank} d(\pi_2 \circ \overline{\pi}|_{\mathscr{L}})_p = 0$, respectively. By definition, if $p \in \mathscr{L}$ is a Legendrian singular point, then it is a space-singular point of $\mathscr{L}$. Even if we have no Legendrian singular points, we have space-singular points. In this case we have the following lemma.

Lemma 3.1 ([24]). *Let $i : \mathscr{L} \subset PT^*(\mathbb{R}^n \times \mathbb{R})$ be a big Legendrian submanifold without Legendrian singular points. If $p \in \mathscr{L}$ is a space-singular point of $\mathscr{L}$, then p is not a time-singular point of $\mathscr{L}$.*

The *discriminant of the family* $\{W_t(\mathscr{L})\}_{t\in(\mathbb{R},0)}$ is defined as the image of singular points of $\pi_1|_{W(\mathscr{L})}$. In the general case, the discriminant consists of three components: *the caustic* $C_{\mathscr{L}} = \pi_1(\Sigma(W(\mathscr{L}))$, where $\Sigma(W(\mathscr{L}))$ is the set of singular points of $W(\mathscr{L})$ (i.e, the critical value set of the Legendrian mapping $\overline{\pi}|_{\mathscr{L}}$), *the Maxwell stratified set* $M_{\mathscr{L}}$, the projection of the closure of the self intersection set of $W(\mathscr{L})$; and also of the critical value set $\Delta_{\mathscr{L}}$ of $\pi_1|_{W(\mathscr{L})\setminus\Sigma(W(\mathscr{L}))}$. In [19, 40], it has been stated that $\Delta_{\mathscr{L}}$ is the *envelope of the family of momentary fronts*. However, we remark that $\Delta_{\mathscr{L}}$ is not necessarily the envelope of the family of the projection of smooth momentary fronts $\overline{\pi}(W_t(\mathscr{L}))$. It can be happened that $\pi_2^{-1}(t)\cap W(\mathscr{L})$ is non-singular but $\pi_1|_{\pi_2^{-1}(t)\cap W(\mathscr{L})}$ has singularities, so that $\Delta_{\mathscr{L}}$ is the set of critical values of the family of mappings $\pi_1|_{\pi_2^{-1}(t)\cap W(\mathscr{L})}$ for smooth $\pi_2^{-1}(t)\cap W(\mathscr{L})$ (cf. [24, §5]).

For any Legendrian submanifold germ $i : (\mathscr{L}, p_0) \subset (PT^*(\mathbb{R}^n \times \mathbb{R}), p_0)$, there exists a generating family. Let $\mathcal{F} : (\mathbb{R}^k \times (\mathbb{R}^n \times \mathbb{R}), 0) \to (\mathbb{R}, 0)$ be a Morse family of hypersurfaces. In this case, we call $\mathcal{F}$ a *big Morse family of hypersurfaces*. Then $\Sigma_*(\mathcal{F}) = \Delta^*(\mathcal{F})^{-1}(0)$ is a smooth n-dimensional submanifold germ. By the previous arguments, we have a big Legendrian submanifold $\mathscr{L}_{\mathcal{F}}(\Sigma_*(\mathcal{F}))$ where

$$\mathscr{L}_{\mathcal{F}}(q,x,t) = \left(x, t, \left[\frac{\partial \mathcal{F}}{\partial x}(q,x,t) : \frac{\partial \mathcal{F}}{\partial t}(q,x,t)\right]\right),$$

and

$$\left[\frac{\partial \mathcal{F}}{\partial x}(q,x,t) : \frac{\partial \mathcal{F}}{\partial t}(q,x,t)\right] = \left[\frac{\partial \mathcal{F}}{\partial x_1}(q,x,t) : \cdots : \frac{\partial \mathcal{F}}{\partial x_n}(q,x,t) : \frac{\partial \mathcal{F}}{\partial t}(q,x,t)\right].$$

§4. Equivalence relations

We now consider five equivalence relations among big Legendrian submanifolds. Let $i : (\mathscr{L}, p_0) \subset (PT^*(\mathbb{R}^n \times \mathbb{R}), p_0)$ and $i' : (\mathscr{L}', p_0') \subset (PT^*(\mathbb{R}^n \times \mathbb{R}), p_0')$ be big Legendrian submanifold germs. Then we respectively say that $i : (\mathscr{L}, p_0) \subset (PT^*(\mathbb{R}^n \times \mathbb{R}), p_0)$ and $i' : (\mathscr{L}', p_0') \subset (PT^*(\mathbb{R}^n \times \mathbb{R}), p_0')$ are

(1) *strictly parametrized Legendrian equivalent* (or, briefly *S.P-Legendrian equivalent*) if there exists a diffeomorphism germ

$\Phi : (\mathbb{R}^n \times \mathbb{R}, \overline{\pi}(p_0)) \to (\mathbb{R}^n \times \mathbb{R}, \overline{\pi}(p'_0))$ of the form $\Phi(x,t) = (\phi_1(x), t)$ such that $\widehat{\Phi}(\mathscr{L}) = \mathscr{L}'$ as set germs,

(2) *space-time Legendrian equivalent* (briefly, *(s,t)-Legendrian equivalent*) if there exists a diffeomorphism germ $\Phi : (\mathbb{R}^n \times \mathbb{R}, \overline{\pi}(p_0)) \to (\mathbb{R}^n \times \mathbb{R}, \overline{\pi}(p'_0))$ of the form $\Phi(x,t) = (\phi_1(x), \phi_2(t))$ such that $\widehat{\Phi}(\mathscr{L}) = \mathscr{L}'$ as set germs,

(3) i and i' are *strictly parametrized$^+$ Legendrian equivalent* (briefly, *$S.P^+$-Legendrian equivalent*) if there exists a diffeomorphism germ $\Phi : (\mathbb{R}^n \times \mathbb{R}, \overline{\pi}(p_0)) \to (\mathbb{R}^n \times \mathbb{R}, \overline{\pi}(p'_0))$ of the form $\Phi(x,t) = (\phi_1(x), t + \alpha(x))$ such that $\widehat{\Phi}(\mathscr{L}) = \mathscr{L}'$ as set germs,

(4) i and i' are *time parametrized Legendrian equivalent* (briefly, *t-P-Legendrian equivalent*) if there exists a diffeomorphism germ $\Phi : (\mathbb{R}^n \times \mathbb{R}, \overline{\pi}(p_0)) \to (\mathbb{R}^n \times \mathbb{R}, \overline{\pi}(p'_0))$ of the form $\Phi(x,t) = (\phi_1(x,t), \phi_2(t))$ such that $\widehat{\Phi}(\mathscr{L}) = \mathscr{L}'$ as set germs,

(5) i and i' are *space parametrized Legendrian equivalent* (briefly, *s-P-Legendrian equivalent*) if there exists a diffeomorphism germ $\Phi : (\mathbb{R}^n \times \mathbb{R}, \overline{\pi}(p_0)) \to (\mathbb{R}^n \times \mathbb{R}, \overline{\pi}(p'_0))$ of the form $\Phi(x,t) = (\phi_1(x), \phi_2(x,t))$ such that $\widehat{\Phi}(\mathscr{L}) = \mathscr{L}'$ as set germs,

where $\widehat{\Phi} : (PT^*(\mathbb{R}^n \times \mathbb{R}), p_0) \to (PT^*(\mathbb{R}^n \times \mathbb{R}), p'_0)$ is the unique contact lift of Φ. We remark that (s,t)-Legendrian equivalence looks a natural equivalence relation among big Legendrian submanifold germs. It induces, however, the isomorphisms among divergent diagrams $\mathbb{R} \leftarrow \mathbb{R}^n \times \mathbb{R} \to \mathbb{R}^n$ on the base space which is not a geometric subgroup of $\mathcal{A}$ or $\mathcal{K}$ in the sense of Damon [8]. Although $S.P$-Legendrian equivalence relation gets rid of the difficulty for the (s,t)-Legendrian equivalence relation, there appear function moduli for generic classifications in very low dimensions (cf. [11], [24, §5]). In order to avoid the function moduli, we define the $S.P^+$-Legendrian equivalence among big Legendrian submanifolds, which has been independently introduced in [15, 40] for different purposes. If we have a generic classification of big Legendrian submanifold germs by $S.P^+$-Legendrian equivalence, then we have a classification by the $S.P$-Legendrian equivalence modulo function moduli. See [15, 40] for details. This equivalence relation plays an important role in the theory of graph-like Legendrian unfoldings. By definition, (1) implies (2) and (2) implies (4). Moreover, (1) implies (3) and (3) implies (5). The weakest equivalence (5) preserves the diffeomorphism type of $C_{\mathscr{L}} \cup M_{\mathscr{L}} \cup \Delta_{\mathscr{L}}$. Moreover (4) preserve the bifurcations of monetary fronts, which was deeply investigated and given a generic classification by Zakalyukin [39]. We used s-P-Legendrian equivalence among big Legendrian submanifolds for applying to the geometry of world sheets in in Lorentz-Minkowski space [22]. We can also define the notion of stability of Legendrian submanifold germs with respect to all the above equivalence relations which are analogous to the stability of Legendrian submanifold germs with respect to Legendrian equivalence (cf. [1, Part III]).

On the other hand, the assumption in Proposition 2.3 is a generic condition for i, i'. Here, we denote that $\mathcal{G} = S.P$, (s,t), $S.P^+$, t-P or s-P. Concerning the discriminant and the bifurcation of momentary fronts, we define the following equivalence relation among big wave front germs. Let $i : (\mathscr{L}, p_0) \subset (PT^*(\mathbb{R}^n \times \mathbb{R}), p_0)$ and $i' : (\mathscr{L}', p_0') \subset (PT^*(\mathbb{R}^n \times \mathbb{R}), p_0')$ be big Legendrian submanifold germs. We say that $W(\mathscr{L})$ and $W(\mathscr{L}')$ are *$\mathcal{G}$-diffeomorphic* if there exists a diffeomorphism germ $\Phi : (\mathbb{R}^n \times \mathbb{R}, \overline{\pi}(p_0)) \to (\mathbb{R}^n \times \mathbb{R}, \overline{\pi}(p_0'))$ of the form corresponding to the diffeomorphism germ in the list of $\mathcal{G}$-Legendrian equivalence such that $\Phi(W(\mathscr{L})) = W(\mathscr{L}')$ as set germs. We also call Φ a *$\mathcal{G}$-diffeomorphism germ*. We remark that a $\mathcal{G}$-diffeomorphism among big wave front germs preserves the diffeomorphism types of $C_{\mathscr{L}} \cup M_{\mathscr{L}} \cup \Delta_{\mathscr{L}}$. By Proposition 2.3, we have the following proposition.

Proposition 4.1. *Let $i : (\mathscr{L}, p_0) \subset (PT^*(\mathbb{R}^n \times \mathbb{R}), p_0)$ and $i' : (\mathscr{L}', p_0') \subset (PT^*(\mathbb{R}^n \times \mathbb{R}), p_0')$ be big Legendrian submanifold germs such that $\overline{\pi} \circ i, \overline{\pi} \circ i'$ are proper map germs and the sets of critical points of these map germs are nowhere dense respectively. Then i and i' are $\mathcal{G}$-Legendrian equivalent if and only if $(W(\mathscr{L}), \overline{\pi}(p_0))$ and $(W(\mathscr{L}'), \overline{\pi}(p_0'))$ are $\mathcal{G}$-diffeomorphic.*

Proof. By definition, if i and i' are $\mathcal{G}$-Legendrian equivalent, then $(W(\mathscr{L}), \overline{\pi}(p_0))$ and $(W(\mathscr{L}'), \overline{\pi}(p_0'))$ are $\mathcal{G}$-diffeomorphic. For the converse, suppose that there exists a $\mathcal{G}$-diffeomorphism germ $\Phi : (\mathbb{R}^n \times \mathbb{R}, \overline{\pi}(p_0)) \to (\mathbb{R}^n \times \mathbb{R}, \overline{\pi}(p_0'))$ such that $\Phi(W(\mathscr{L})) = W(\mathscr{L}')$ as set germs. Then $\widehat{\Phi}(\mathscr{L})$ is a big Legendrian submanifold such that $W(\widehat{\Phi}(\mathscr{L})) = \Phi(W(\mathscr{L})) = W(\mathscr{L}')$ as set germs. By Proposition 2.3, we have $\widehat{\Phi}(\mathscr{L}) = \mathscr{L}'$. This completes the proof. □

We explain s-P-Legendrian equivalence and $S.P^+$-Legendrian equivalence can be investigated by using the notion of generating families of Legendrian submanifold germs. For t-P-Legendrian equivalence, [39] is a good survey, so that we omit the detail here.

Let $\overline{f}, \overline{g} : (\mathbb{R}^k \times \mathbb{R}, 0) \to (\mathbb{R}, 0)$ be function germs. We say that $\overline{f}$ and $\overline{g}$ are *P-$\mathcal{K}$-equivalent* if there exists a diffeomorphism germ $\Phi : (\mathbb{R}^k \times \mathbb{R}, 0) \to (\mathbb{R}^k \times \mathbb{R}, 0)$ of the form $\Phi(q,t) = (\phi_1(q,t), \phi_2(t))$ such that $\langle \overline{f} \circ \Phi \rangle_{\mathcal{E}_{k+1}} = \langle \overline{g} \rangle_{\mathcal{E}_{k+1}}$. We also say that $\overline{f}$ and $\overline{g}$ are *$S.P$-$\mathcal{K}$-equivalent* if these are P-$\mathcal{K}$-equivalent by the diffeomorphism Φ of the form $\Phi(q,t) = (\phi_1(q,t), t)$. Let $\mathcal{F}, \mathcal{G} : (\mathbb{R}^k \times (\mathbb{R}^n \times \mathbb{R}), 0) \to (\mathbb{R}, 0)$

be function germs. We say that $\mathcal{F}$ and $\mathcal{G}$ are *space-P-$\mathcal{K}$-equivalent* (or, briefly, *s-P-$\mathcal{K}$-equivalent*) if there exists a diffeomorphism germ $\Psi : (\mathbb{R}^k \times (\mathbb{R}^n \times \mathbb{R}), 0) \to (\mathbb{R}^k \times (\mathbb{R}^n \times \mathbb{R}), 0)$ of the form $\Psi(q, x, t) = (\phi(q, x, t), \phi_1(x), \phi_2(x, t)))$ such that $\langle \mathcal{F} \circ \Psi \rangle_{\mathcal{E}_{k+n+1}} = \langle \mathcal{G} \rangle_{\mathcal{E}_{k+n+1}}$. We also say that $\mathcal{F}$ and $\mathcal{G}$ are *space-S.P$^+$-$\mathcal{K}$-equivalent* (or, briefly, *s-S.P$^+$-$\mathcal{K}$-equivalent*) if these are s-P-$\mathcal{K}$-equivalent by the diffeomorphism germ Ψ of the form $\Psi(q, x, t) = (\phi(q, x, t), \phi_1(x), t + \alpha(x))$. The notion of P-$\mathcal{K}$-versal unfoldings and S.P$^+$-$\mathcal{K}$-versal unfoldings play important roles for our purpose. We define the extended tangent spaces of $\overline{f} : (\mathbb{R}^k \times \mathbb{R}, 0) \to (\mathbb{R}, 0)$ relative to P-$\mathcal{K}$ by

$$T_e(P\text{-}\mathcal{K})(\overline{f}) = \left\langle \frac{\partial \overline{f}}{\partial q_1}, \dots, \frac{\partial \overline{f}}{\partial q_k}, \overline{f} \right\rangle_{\mathcal{E}_{k+1}} + \left\langle \frac{\partial \overline{f}}{\partial t} \right\rangle_{\mathcal{E}_1}$$

and the extended tangent spaces of $\overline{f}$ relative to S.P$^+$-$\mathcal{K}$ by

$$T_e(S.P^+\text{-}\mathcal{K})(\overline{f}) = \left\langle \frac{\partial \overline{f}}{\partial q_1}, \dots, \frac{\partial \overline{f}}{\partial q_k}, \overline{f} \right\rangle_{\mathcal{E}_{k+1}} + \left\langle \frac{\partial \overline{f}}{\partial t} \right\rangle_{\mathbb{R}},$$

respectively. Then we say that $\mathcal{F}$ an *infinitesimally P-$\mathcal{K}$-versal* unfolding of $\overline{f} = \mathcal{F}|_{\mathbb{R}^k \times \{0\} \times \mathbb{R}}$ if it satisfies

$$\mathcal{E}_{k+1} = T_e(P\text{-}\mathcal{K})(\overline{f}) + \left\langle \frac{\partial \mathcal{F}}{\partial x_1}|_{\mathbb{R}^k \times \{0\} \times \mathbb{R}}, \dots, \frac{\partial \mathcal{F}}{\partial x_n}|_{\mathbb{R}^k \times \{0\} \times \mathbb{R}} \right\rangle_{\mathbb{R}}$$

and it is an *infinitesimally S.P$^+$-$\mathcal{K}$-versal* unfolding of $\overline{f} = \mathcal{F}|_{\mathbb{R}^k \times \{0\} \times \mathbb{R}}$ if it satisfies

$$\mathcal{E}_{k+1} = T_e(S.P^+\text{-}\mathcal{K})(\overline{f}) + \left\langle \frac{\partial \mathcal{F}}{\partial x_1}|_{\mathbb{R}^k \times \{0\} \times \mathbb{R}}, \dots, \frac{\partial \mathcal{F}}{\partial x_n}|_{\mathbb{R}^k \times \{0\} \times \mathbb{R}} \right\rangle_{\mathbb{R}},$$

respectively. We can show the following theorem analogous to those in [15, 39, 40]. We only remark here that the proof is analogous to the proof of [1, Theorem in §21.4].

Theorem 4.2 ([15, 24, 40]). *Let $\mathcal{F} : (\mathbb{R}^k \times (\mathbb{R}^n \times \mathbb{R}), 0) \to (\mathbb{R}, 0)$ and $\mathcal{G} : (\mathbb{R}^{k'} \times (\mathbb{R}^n \times \mathbb{R}), 0) \to (\mathbb{R}, 0)$ be big Morse families of hypersurfaces. Then*
(1) $\mathscr{L}_{\mathcal{F}}(\Sigma_*(\mathcal{F}))$ and $\mathscr{L}_{\mathcal{G}}(\Sigma_*(\mathcal{G}))$ are s-P-Legendrian equivalent if and only if $\mathcal{F}$ and $\mathcal{G}$ are stably s-P-$\mathcal{K}$-equivalent.
(2) $\mathscr{L}_{\mathcal{F}}(\Sigma_*(\mathcal{F}))$ is s-P-Legendre stable if and only if $\mathcal{F}$ is an infinitesimally P-$\mathcal{K}$-versal unfolding of $\overline{f} = \mathcal{F}|_{\mathbb{R}^k \times \{0\} \times \mathbb{R}}$.
(3) $\mathscr{L}_{\mathcal{F}}(\Sigma_*(\mathcal{F}))$ and $\mathscr{L}_{\mathcal{G}}(\Sigma_*(\mathcal{G}))$ are S.P$^+$-Legendrian equivalent if and only if $\mathcal{F}$ and $\mathcal{G}$ are stably s-S.P$^+$-$\mathcal{K}$-equivalent.

(4) *$\mathscr{L}_{\mathcal{F}}(\Sigma_*(\mathcal{F}))$ is $S.P^+$-Legendre stable if and only if $\mathcal{F}$ is an infinitesimally $S.P^+$-$\mathcal{K}$-versal unfolding of $\overline{f} = \mathcal{F}|_{\mathbb{R}^k \times \{0\} \times \mathbb{R}}$.*

§5. Graph-like Legendrian unfoldings

In this section we explain the theory of graph-like Legendrian unfoldings. Graph-like Legendrian unfoldings belong to a special class of big Legendrian submanifolds. A big Legendrian submanifold $i : \mathscr{L} \subset PT^*(\mathbb{R}^n \times \mathbb{R})$ is said to be a *graph-like Legendrian unfolding* if $\mathscr{L} \subset J^1_{GA}(\mathbb{R}^n, \mathbb{R})$. We call $W(\mathscr{L}) = \overline{\pi}(\mathscr{L})$ a *graph-like wave front* of $\mathscr{L}$, where $\overline{\pi} : J^1_{GA}(\mathbb{R}^n, \mathbb{R}) \to \mathbb{R}^n \times \mathbb{R}$ is the canonical projection. We define a mapping $\Pi : J^1_{GA}(\mathbb{R}^n, \mathbb{R}) \to T^*\mathbb{R}^n$ by $\Pi(x, t, p) = (x, p)$, where $(x, t, p) = (x_1, \dots, x_n, t, p_1, \dots, p_n)$ and the canonical contact form on $J^1_{GA}(\mathbb{R}^n, \mathbb{R})$ is given by $\theta = dt - \sum_{i=1}^n p_i dx_i$. Here, $T^*\mathbb{R}^n$ is a symplectic manifold with the canonical symplectic structure $\omega = \sum_{i=1}^n dp_i \wedge dx_i$ (cf. [1]). Then we have the following proposition.

Proposition 5.1 ([19]). *For a graph-like Legendrian unfolding $\mathscr{L} \subset J^1_{GA}(\mathbb{R}^n, \mathbb{R})$, $z \in \mathscr{L}$ is a singular point of $\overline{\pi}|_{\mathscr{L}} : \mathscr{L} \to \mathbb{R}^n \times \mathbb{R}$ if and only if it is a singular point of $\pi_1 \circ \overline{\pi}|_{\mathscr{L}} : \mathscr{L} \to \mathbb{R}^n$. Moreover, $\Pi|_{\mathscr{L}} : \mathscr{L} \to T^*\mathbb{R}^n$ is immersive, so that $\Pi(\mathscr{L})$ is a Lagrangian submanifold in $T^*\mathbb{R}^n$.*

We have the following corollary of Proposition 5.1.

Corollary 5.2 ([19]). *For a graph-like Legendrian unfolding $\mathscr{L} \subset J^1_{GA}(\mathbb{R}^n, \mathbb{R})$, $\Delta_{\mathscr{L}}$ is the empty set, so that the discriminant of the family of momentary fronts is $C_{\mathscr{L}} \cup M_{\mathscr{L}}$.*

Since $\mathscr{L}$ is a big Legendrian submanifold in $PT^*(\mathbb{R}^n \times \mathbb{R})$, it has a generating family $\mathcal{F} : (\mathbb{R}^k \times (\mathbb{R}^n \times \mathbb{R}), 0) \to (\mathbb{R}, 0)$ at least locally. Since $\mathscr{L} \subset J^1_{GA}(\mathbb{R}^n, \mathbb{R}) = U_\tau \subset PT^*(\mathbb{R}^n \times \mathbb{R})$, it satisfies the condition $(\partial \mathcal{F}/\partial t)(0) \neq 0$. Let $\mathcal{F} : (\mathbb{R}^k \times (\mathbb{R}^n \times \mathbb{R}), 0) \to (\mathbb{R}, 0)$ be a big Morse family of hypersurfaces. We say that $\mathcal{F}$ is a *graph-like Morse family of hypersurfaces* if $(\partial \mathcal{F}/\partial t)(0) \neq 0$. It is easy to show that the corresponding big Legendrian submanifold germ is a graph-like Legendrian unfolding. Of course, all graph-like Legendrian unfolding germs can be constructed by the above way. We say that $\mathcal{F}$ is a *graph-like generating family* of $\mathscr{L}_{\mathcal{F}}(\Sigma_*(\mathcal{F}))$. We remark that the notion of graph-like Legendrian unfoldings and corresponding generating families have been introduced in [14] to describe the perestroikas of wave fronts given as the solutions for general eikonal equations. In this case, there is an additional condition. We say that $\mathcal{F} : (\mathbb{R}^k \times (\mathbb{R}^n \times \mathbb{R}), 0) \to (\mathbb{R}, 0)$ is *non-degenerate* if $\mathcal{F}$ satisfies the conditions $(\partial \mathcal{F}/\partial t)(0) \neq 0$ and $\Delta^* \mathcal{F}|_{\mathbb{R}^k \times \mathbb{R}^n \times \{0\}}$ is a submersion germ. In this case we call $\mathcal{F}$ a *non-degenerate graph-like generating family.* We have the following proposition.

Proposition 5.3 ([24]). *Let $\mathcal{F} : (\mathbb{R}^k \times (\mathbb{R}^n \times \mathbb{R}), 0) \to (\mathbb{R}, 0)$ be a graph-like Morse family of hypersurfaces. Then $\mathcal{F}$ is non-degenerate if and only if $\pi_2 \circ \overline{\pi}|_{\mathscr{L}_{\mathcal{F}}(\Sigma_*(\mathcal{F}))}$ is submersive.*

We say that a graph-like Legendrian unfolding $\mathscr{L} \subset J^1_{GA}(\mathbb{R}^n, \mathbb{R})$ is *non-degenerate* if $\pi_2 \circ \overline{\pi}|\mathscr{L}$ is submersive. Non-degeneracy was assumed for general graph-like Legendrian unfoldings when the notion of graph-like Legendrian unfoldings was introduced in [14]. However, during the last two decades, we have clarified the situation and non-degeneracy is now defined as above.

We can consider the following more restrictive class of graph-like generating families: Let $\mathcal{F}$ be a graph-like Morse family of hypersurfaces. By the implicit function theorem, there exists a function $F : (\mathbb{R}^k \times \mathbb{R}^n, 0) \to (\mathbb{R}, 0)$ such that

$$\langle \mathcal{F}(q, x, t) \rangle_{\mathcal{E}_{k+n+1}} = \langle F(q, x) - t \rangle_{\mathcal{E}_{k+n+1}}.$$

Then we have the following proposition.

Proposition 5.4 ([24]). *Let $\mathcal{F} : (\mathbb{R}^k \times (\mathbb{R}^n \times \mathbb{R}), 0) \to (\mathbb{R}, 0)$ and $F : (\mathbb{R}^k \times \mathbb{R}^n, 0) \to (\mathbb{R}, 0)$ be function germs such that $\langle \mathcal{F}(q, x, t) \rangle_{\mathcal{E}_{k+n+1}} = \langle F(q, x) - t \rangle_{\mathcal{E}_{k+n+1}}$. Then $\mathcal{F}$ is a graph-like Morse family of hypersurfaces if and only if F is a Morse family of functions.*

We now consider the case $\mathcal{F}(q, x, t) = \lambda(q, x, t)(F(q, x) - t)$. In this case,

$$\Sigma_*(\mathcal{F}) = \{(q, x, F(q, x)) \in (\mathbb{R}^k \times (\mathbb{R}^n \times \mathbb{R}), 0) \mid (q, x) \in C(F)\},$$

where $C(F) = \Delta F^{-1}(0)$. Moreover, we have the Lagrangian submanifold germ $L(F)(C(F)) \subset T^*\mathbb{R}^n$, where

$$L(F)(q, x) = \left(x, \frac{\partial F}{\partial x_1}(q, x), \dots, \frac{\partial F}{\partial x_n}(q, x) \right).$$

Since $\mathcal{F}$ is a graph-like Morse family of hypersurfaces, we have a big Legendrian submanifold germ $\mathscr{L}_{\mathcal{F}}(\Sigma_*(\mathcal{F})) \subset J^1_{GA}(\mathbb{R}^n, \mathbb{R})$, where $\mathscr{L}_{\mathcal{F}} : (\Sigma_*(\mathcal{F}), 0) \to J^1_{GA}(\mathbb{R}^n, \mathbb{R}) \cong T^*\mathbb{R}^n \times \mathbb{R}$ is defined by

$$\mathscr{L}_{\mathcal{F}}(q, x, t) = \left(x, t, -\frac{\dfrac{\partial \mathcal{F}}{\partial x_1}(q, x, t)}{\dfrac{\partial \mathcal{F}}{\partial t}(q, x, t)}, \dots, -\frac{\dfrac{\partial \mathcal{F}}{\partial x_n}(q, x, t)}{\dfrac{\partial \mathcal{F}}{\partial t}(q, x, t)}, \right).$$

We also define a map germ $\mathfrak{L}_F : (C(F), 0) \to J^1_{GA}(\mathbb{R}^n, \mathbb{R})$ by

$$\mathfrak{L}_F(q,x) = \left(x, F(q,x), \frac{\partial F}{\partial x_1}(q,x), \ldots, \frac{\partial F}{\partial x_n}(q,x)\right).$$

Since $\partial\mathcal{F}/\partial x_i = \partial\lambda/\partial x_i(F-t)+\lambda\partial F/\partial x_i$ and $\partial\mathcal{F}/\partial t = \partial\lambda/\partial t(F-t)-\lambda$, we have $\partial\mathcal{F}/\partial x_i(q,x,t) = \lambda(q,x,t)\partial F/\partial x_i(q,x,t)$ and $\partial\mathcal{F}/\partial t(q,x,t) = -\lambda(q,x,t)$ for $(q,x,t) \in \Sigma_*(\mathcal{F})$. It follows that $\mathfrak{L}_F(C(F)) = \mathscr{L}_{\mathcal{F}}(\Sigma_*(\mathcal{F}))$. By definition, we have $\Pi(\mathscr{L}_{\mathcal{F}}(\Sigma_*(\mathcal{F}))) = \Pi(\mathfrak{L}_F(C(F))) = L(F)(C(F))$. The graph-like wave front of $\mathscr{L}_{\mathcal{F}}(\Sigma_*(\mathcal{F})) = \mathfrak{L}_F(C(F))$ is the graph of $F|_{C(F)}$. This is the reason why we call it a graph-like Legendrian unfolding. For a non-degenerate graph-like Morse family of hypersurfaces, we have the following proposition.

Proposition 5.5 ([24]). *With the same notations as Proposition 5.4, $\mathcal{F}$ is a non-degenerate graph-like Morse family of hypersurfaces if and only if F is a Morse family of hypersurfaces. In this case, F is also a Morse family of functions such that*

$$\left(\frac{\partial F}{\partial x_1}(0), \ldots, \frac{\partial F}{\partial x_n}(0)\right) \neq \mathbf{0}.$$

The momentary front for a fixed $t \in (\mathbb{R}, 0)$ is $W_t(\mathscr{L}) = \pi_1(\pi_2^{-1}(t) \cap W(\mathscr{L}))$. We define $\mathscr{L}_t = \mathscr{L} \cap (\pi_2 \circ \overline{\pi})^{-1}(t) = \mathscr{L} \cap (T^*\mathbb{R}^n \times \{t\})$ under the canonical identification $J^1_{GA}(\mathbb{R}^n, \mathbb{R}) \cong T^*\mathbb{R}^n \times \mathbb{R}$. Then $\Pi(\mathscr{L}) \subset T^*\mathbb{R}^n$ and $\widetilde{\pi} \circ \Pi(\mathscr{L}_t) \subset PT^*\mathbb{R}^n$, where $\widetilde{\pi} : T^*\mathbb{R}^n \to PT^*(\mathbb{R}^n)$ is the canonical projection. We also have the canonical projections $\pi : T^*\mathbb{R}^n \to \mathbb{R}^n$ and $\varpi : PT^*\mathbb{R}^n \to \mathbb{R}^n$ such that $\pi_1 \circ \overline{\pi} = \pi \circ \Pi$ and $\varpi \circ \widetilde{\pi} = \pi$. Then we have the following proposition.

Proposition 5.6 ([24]). *Let $\mathscr{L} \subset J^1_{GA}(\mathbb{R}^n, \mathbb{R})$ be a non-degenerate graph-like Legendrian unfolding. Then $\widetilde{\pi} \circ \Pi(\mathscr{L}_t)$ is a Legendrian submanifold in $PT^*(\mathbb{R}^n)$.*

The momentary front $W_t(\mathscr{L})$ of a big Legendrian submanifold $\mathscr{L} \subset PT^*(\mathbb{R}^n \times \mathbb{R})$ is not necessarily a wave front of a Legendrian submanifold in the ordinary sense, generally. However, for a non-degenerate Legendrian unfolding in $J^1_{GA}(\mathbb{R}^n, \mathbb{R})$, we have the following corollary.

Corollary 5.7. [24] *Let $\mathscr{L} \subset J^1_{GA}(\mathbb{R}^n, \mathbb{R})$ be a non-degenerate graph-like Legendrian unfolding. Then the momentary front $W_t(\mathscr{L})$ is the wave front set of the Legendrian submanifold $\widetilde{\pi} \circ \Pi(\mathscr{L}_t) \subset PT^*(\mathbb{R}^n)$. Moreover, the caustic C_L is the caustic of the Lagrangian submanifold $\Pi(L) \subset T^*\mathbb{R}^n$. In other words, $W_t(\mathscr{L}) = \varpi(\widetilde{\pi} \circ \Pi(\mathscr{L}_t))$ and $C_{\mathscr{L}}$ is the singular value set of $\pi|\Pi(\mathscr{L})$.*

§6. $S.P^+$-Legendrian equivalence among graph-like Legendrian unfoldings

In this section we describe the properties of $S.P^+$-Legendrian equivalence among graph-like Legendrian unfoldings. For a graph-like Morse family of hypersurfaces $\mathcal{F}(q,x,t) = \lambda(q,x,t)(F(q,x) - t)$, $\mathcal{F}(q,x,t)$ and $\overline{F}(q,x,t) = F(q,x) - t$ are s-$S.P^+$-$\mathcal{K}$-equivalent, so that we consider $\overline{F}(q,x,t) = F(q,x) - t$ as a graph-like Morse family. Moreover, by Proposition 5.4, $F(q,x)$ is a Morse family of functions. We now suppose that $F(q,x)$ is a Morse family of functions. Consider the graph-like Morse family of hypersurfaces $\overline{F}(q,x,t) = F(q,x) - t$ which is not necessarily non-degenerate. Then we have $\mathscr{L}_{\overline{F}}(\Sigma_*(\overline{F})) = \mathfrak{L}_F(C(F))$. We also denote that $\overline{f}(q,t) = f(q) - t$ for any $f \in \mathfrak{M}_k$. We can represent the extended tangent space of $\overline{f} : (\mathbb{R}^k \times \mathbb{R}, 0) \to (\mathbb{R}, 0)$ relative to $S.P^+$-$\mathcal{K}$ by

$$T_e(S.P^+\text{-}\mathcal{K})(\overline{f}) = \left\langle \frac{\partial f}{\partial q_1}(q), \dots, \frac{\partial f}{\partial q_k}(q), f(q) - t \right\rangle_{\mathcal{E}_{k+1}} + \langle 1 \rangle_{\mathbb{R}}.$$

For a unfolding $\overline{F} : (\mathbb{R}^k \times \mathbb{R}^n \times \mathbb{R}, 0) \to (\mathbb{R}, 0)$ of $\overline{f}$, $\overline{F}$ is infinitesimally $S.P^+$-$\mathcal{K}$-versal unfolding of $\overline{f}$ if and only if

$$\mathcal{E}_{k+1} = T_e(S.P^+\text{-}\mathcal{K})(\overline{f}) + \left\langle \frac{\partial F}{\partial x_1}|_{\mathbb{R}^k \times \mathbb{R} \times \{0\}}, \dots, \frac{\partial F}{\partial x_n}|_{\mathbb{R}^k \times \mathbb{R} \times \{0\}} \right\rangle_{\mathbb{R}}.$$

We compare the equivalence relations between Lagrangian submanifold germs and induced graph-like Legendrian unfoldings. As a consequence, we give a relationship between caustics and graph-like wave fronts.

Theorem 6.1 ([19, 25]). *Let $\mathcal{F} : (\mathbb{R}^k \times \mathbb{R}^n \times \mathbb{R}, 0) \to (\mathbb{R}, 0)$ and $\mathcal{G} : (\mathbb{R}^{k'} \times \mathbb{R}^n \times \mathbb{R}, 0) \to (\mathbb{R}, 0)$ be graph-like Morse families of hypersurfaces of the forms $\mathcal{F}(q,x,t) = \lambda(q,x,t)(F(q,x) - t)$ and $\mathcal{G}(q',x,t) = \mu(q',x,t)(G(q',x) - t)$. Then Lagrangian submanifold germs $L(F)(C(F))$ and $L(G)(C(G))$ are Lagrangian equivalent if and only if the graph-like Legendrian unfoldings $\mathscr{L}_{\mathcal{F}}(\Sigma_*(\mathcal{F}))$ and $\mathscr{L}_{\mathcal{G}}(\Sigma_*(\mathcal{G}))$ are $S.P^+$-Legendrian equivalent.*

Proof. By Theorem 2.2, if $L(F)(C(F))$ and $L(G)(C(G))$ are Lagrangian equivalent, then F and G are stably P-$\mathcal{R}^+$-equivalent. In this case we may assume that $k = k'$ and F and G are P-$\mathcal{R}^+$-equivalent, so that there exist a diffeomorphism germ $\Phi : (\mathbb{R}^k \times \mathbb{R}^n, 0) \to (\mathbb{R}^k \times \mathbb{R}^n, 0)$ of the form $\Phi(q,x) = (\phi_1(q,x), \phi(x))$ and a function $\alpha : (\mathbb{R}^n, 0) \to \mathbb{R}$ such that $G(q,x) = F \circ \Phi(q,x) + \alpha(x)$. Then we define a diffeomorphism germ

$\widetilde{\Phi} : (\mathbb{R}^k \times \mathbb{R}^n \times \mathbb{R}, 0) \to (\mathbb{R}^k \times \mathbb{R}^n \times \mathbb{R}, 0)$ by $\widetilde{\Phi}(q, x, t) = (\phi_1(q, x), \phi(x), t - \alpha(x))$. It follows that $G(q, x) - t = F \circ \Phi(q, x) + t - \alpha(x)$. This means that $\mathcal{G}$ and $\mathcal{F}$ are s-$S.P^+$-$\mathcal{K}$-equivalent. By Theorem 4.2, $\mathscr{L}_{\mathcal{F}}(\Sigma_*(\mathcal{F}))$ and $\mathscr{L}_{\mathcal{G}}(\Sigma_*(\mathcal{G}))$ are $S.P^+$-Legendrian equivalent. For the converse assertion, we assume that $\mathscr{L}_{\mathcal{F}}(\Sigma_*(\mathcal{F}))$ and $\mathscr{L}_{\mathcal{G}}(\Sigma_*(\mathcal{G}))$ are $S.P^+$-Legendrian equivalent. By the assumption, there exists a diffeomorphism germ $\Phi : (\mathbb{R}^n \times \mathbb{R}, 0) \to (\mathbb{R}^n \times \mathbb{R}, 0)$ of the form $\Phi(x, t) = (\phi_1(x), t + \alpha(x))$ such that $\widehat{\Phi}(\mathscr{L}_{\mathcal{F}}(\Sigma_*(\mathcal{F}))) = \mathscr{L}_{\mathcal{G}}(\Sigma_*(\mathcal{G}))$. Then we have $\Phi^{-1}(x, t) = (\phi_1^{-1}(x), t - \alpha(\phi_1^{-1}(x)))$, so that the Jacobi matrix is

$$J_{\Phi(x,t)}\Phi^{-1} = \begin{pmatrix} \dfrac{\partial \phi_1^{-1}}{\partial x}(\phi_1(x)) & 0 \\ -\dfrac{\partial \alpha \circ \phi_1^{-1}}{\partial x}(\phi_1(x)) & 1 \end{pmatrix}.$$

It follows that

$$\widehat{\Phi}((x, t), [\xi : \tau]) = \left(\Phi(x, t), \left[\xi \cdot \frac{\partial \phi_1^{-1}}{\partial x}(\phi_1(x)) - \tau \frac{\partial \alpha \circ \phi_1^{-1}}{\partial x}(\phi_1(x)) : \tau \right] \right).$$

Since $\tau \neq 0$,

$$\begin{aligned} &\left[\xi \cdot \frac{\partial \phi_1^{-1}}{\partial x}(\phi_1(x)) - \frac{\partial \alpha \circ \phi_1^{-1}}{\partial x}(\phi_1(x)) : \tau \right] \\ &\quad = \left[-\frac{\xi}{\tau} \cdot \frac{\partial \phi_1^{-1}}{\partial x}(\phi_1(x)) + \frac{\partial \alpha \circ \phi_1^{-1}}{\partial x}(\phi_1(x)) : -1 \right]. \end{aligned}$$

We consider the graph-like affine coordinates $((x, t), p) \in J^1_{GA}(\mathbb{R}^n, \mathbb{R})$, where $p = -\dfrac{\xi}{\tau}$. Then we have $\widehat{\Phi}(J^1_{GA}(\mathbb{R}^n, \mathbb{R})) = J^1_{GA}(\mathbb{R}^n, \mathbb{R})$ and

$$\widehat{\Phi}((x, t), p) = \left(\phi_1(x), t + \alpha(x), p \cdot \frac{\partial \phi_1^{-1}}{\partial x}(\phi_1(x)) + \frac{\partial \alpha \circ \phi_1^{-1}}{\partial x}(\phi_1(x)) \right).$$

We now define a map $\widetilde{\phi_1} : T^*\mathbb{R}^n \to T^*\mathbb{R}^n$ by

$$\widetilde{\phi_1}(x, p) = \left(\phi_1(x), p \cdot \frac{\partial \phi_1^{-1}}{\partial x}(\phi_1(x)) + \frac{\partial \alpha \circ \phi_1^{-1}}{\partial x}(\phi_1(x)) \right).$$

Since $\widehat{\Phi}$ is a contact diffeomorphism germ, there exists a function germ $\mu : J^1_{GA}(\mathbb{R}^n, \mathbb{R}) \to \mathbb{R}$ with $\mu(x, t, p) \neq 0$ such that $\widehat{\Phi}^*\theta = \mu\theta$. Therefore, we have

$$dt + d\alpha - \widetilde{\phi_1}^*(p \cdot dx) = \mu(dt - p \cdot dx) = \mu dt - \mu(p \cdot dx),$$

so that $\mu \equiv 1$. It follows that $-p \cdot dx = d\alpha - \widetilde{\phi_1}^*(p \cdot dx)$. Thus we have

$$\widetilde{\phi_1}^*(\omega) = \widetilde{\phi_1}^*(d(p \cdot dx)) = d\widetilde{\phi_1}^*(p \cdot dx) = d(p \cdot dx) = \omega.$$

This means that $\widetilde{\phi_1}$ is a symplectic diffeomorphism germ (i.e. Lagrangian diffeomorphism germ). Since $\Pi \circ \widehat{\Phi}|J^1_{GA}(\mathbb{R}^n, \mathbb{R}) = \widetilde{\phi_1} \circ \Pi|J^1_{GA}(\mathbb{R}^n, \mathbb{R})$, we have

$$\begin{aligned} L(G)(C(G)) &= \Pi(\mathscr{L}_{\mathcal{G}}(\Sigma_*(\mathcal{G}))) = \Pi \circ \widehat{\Phi}(\mathscr{L}_{\mathcal{F}}(\Sigma_*(\mathcal{F}))) \\ &= \widetilde{\phi_1}(\Pi(\mathscr{L}_{\mathcal{F}}(\Sigma_*(\mathcal{F}))) = \widetilde{\phi_1}(L(F)(C(F))). \end{aligned}$$

This completes the proof. □

By definition, the set of Legendrian singular points of a graph-like Legendrian unfolding $\mathscr{L}_{\mathcal{F}}(\Sigma_*(\mathcal{F}))$ coincides with the set of singular points of $\pi \circ L(F)$. Therefore the singularities of graph-like wave fronts of $\mathscr{L}_{\mathcal{F}}(\Sigma_*(\mathcal{F}))$ lie on the caustics of $L(F)$. It follows that we can apply Proposition 4.1 to $S.P^+$-Legendrian equivalence.

Corollary 6.2 ([25]). *Suppose that $\overline{\pi}|_{\mathscr{L}_{\mathcal{F}}(\Sigma_*(\mathcal{F}))}$, $\overline{\pi}|_{\mathscr{L}_{\mathcal{G}}(\Sigma_*(\mathcal{G}))}$ are proper map germs and the both sets of Legendrian singular points of graph-like Legendrian unfoldings $\mathscr{L}_{\mathcal{F}}(\Sigma_*(\mathcal{F}))$, $\mathscr{L}_{\mathcal{G}}(\Sigma_*(\mathcal{G}))$ are no-where dense respectively. Then the following conditions are equivalent:*
(1) *Lagrangian submanifold germs $L(F)(C(F))$, $L(G)(C(G))$ are Lagrangian equivalent,*
(2) *graph-like wave fronts $W(\mathscr{L}_{\mathcal{F}}(\Sigma_*(\mathcal{F})))$, $W(\mathscr{L}_{\mathcal{G}}(\Sigma_*(\mathcal{G})))$ are $S.P^+$-diffeomorphic.*

Moreover, we have the following direct corollary of Theorem 6.1.

Corollary 6.3 ([21]). *Suppose that $\mathcal{F}(q, x, t) = \lambda(q, x, t)(F(q, x) - t)$ is a graph-like Morse family of hypersurfaces. Then $\mathscr{L}_{\mathcal{F}}(\Sigma_*(\mathcal{F}))$ is $S.P^+$-Legendrian stable if and only if $L(F)(C(F))$ is Lagrangian stable.*

If a Lagrangian submanifold germ $L(F)(C(F))$ is Lagrangian stable, then $\overline{\pi}|_{\mathscr{L}_{\mathcal{F}}(\Sigma_*(\mathcal{F}))}$ is a proper map germ and the regular set of this map germ is dense. Hence we can apply Proposition 4.1 to our situation and obtain the following theorem on the relations among graph-like Legendrian unfoldings and Lagrangian singularities.

Theorem 6.4 ([24]). *Let $\mathcal{F} : (\mathbb{R}^k \times \mathbb{R}^n \times \mathbb{R}, 0) \to (\mathbb{R}, 0)$ and $\mathcal{G} : (\mathbb{R}^{k'} \times \mathbb{R}^n \times \mathbb{R}, 0) \to (\mathbb{R}, 0)$ be graph-like Morse families of hypersurfaces of the forms $\mathcal{F}(q, x, t) = \lambda(q, x, t)(F(q, x) - t)$ and $\mathcal{G}(q', x, t) = \mu(q', x, t)(G(q', x) - t)$ such that $\mathscr{L}_{\mathcal{F}}(\Sigma_*(\mathcal{F}))$ and $\mathscr{L}_{\mathcal{G}}(\Sigma_*(\mathcal{G}))$ are $S.P^+$-Legendrian stable. Then the following conditions are equivalent:*

(1) $\mathscr{L}_{\mathcal{F}}(\Sigma_*(\mathcal{F}))$ *and* $\mathscr{L}_{\mathcal{G}}(\Sigma_*(\mathcal{G}))$ *are* $S.P^+$*-Legendrian equivalent,*
(2) $\mathcal{F}$ *and* $\mathcal{G}$ *are stably* s*-*$S.P^+$*-*$\mathcal{K}$*-equivalent,*
(3) $\overline{f}(q,t) = F(q,0) - t$ *and* $\overline{g}(q',t) = G(q',0) - t$ *are stably* $S.P$*-*$\mathcal{K}$*-equivalent,*
(4) $f(q) = F(q,0)$ *and* $g(q') = G(q',0)$ *are stably* $\mathcal{R}$*-equivalent,*
(5) $F(q,x)$ *and* $G(q',x)$ *are stably* P*-*$\mathcal{R}^+$*-equivalent,*
(6) $L(F)(C(F))$ *and* $L(G)(C(G))$ *are Lagrangian equivalent,*
(7) $W(\mathscr{L}_{\mathcal{F}}(\Sigma_*(\mathcal{F})))$ *and* $W(\mathscr{L}_{\mathcal{G}}(\Sigma_*(\mathcal{G})))$ *are* $S.P^+$*-diffeomorphic.*

Remark 6.5. (i) The above theorem was shown in [24].
(ii) By Corollary 6.3, the assumption of the above theorem is equivalent to the condition that $L(F)(C(F))$ and $L(G)(C(G))$ are Lagrangian stable.
(iii) If $k = k'$ and $q = q'$ in the above theorem, we can remove the word "stably" in conditions (2), (3), (4) and (5).
(iv) By Theorem 6.1, conditions (1), (2), (5) and (6) are always equivalent without any assumptions. This fact was not known when I wrote the survey paper [24]. After I wrote the paper Theorem 6.1 has been shown in [25]. Therefore, this assertion is a new result.
(v) Conditions (3) and (4) are equivalent without any assumptions.
(vi) Equivalency for (2) and (3) (respectively, (4) and (5)), we need the assumption that the $S.P^+$-Legendrian stability (respectively, the Lagrangian stability). Fortunately, these two stability are equivalent by Corollary 6.3.
(vii) The $S.P^+$-Legendrian stability of $\mathscr{L}_{\mathcal{F}}(\Sigma_*(\mathcal{F}))$ is a generic condition for $n \leq 5$.
(viii) By Proposition 4.1 and Corollary 6.2, the conditions (1), (6) and (7) are equivalent generically for an arbitrary dimension n without the assumption on the $S.P^+$-Legendrian stability.

On the other hand, we consider another geometric condition on the generating families. For a function germ $f : (\mathbb{R}^k, 0) \to (\mathbb{R}, 0)$, the *level set foliation germ* of f is defined to be $\mathscr{F}_f = \{f^{-1}(c) \mid c \in (\mathbb{R}, 0)\ \}$. For function germs $f, g : (\mathbb{R}^k, 0) \to (\mathbb{R}, 0)$, we say that the level set foliation germs $\mathscr{F}_f$ and $\mathscr{F}_g$ are *strictly diffeomorphic* if there exists a diffeomorphism germ $\psi : (\mathbb{R}^k, 0) \to (\mathbb{R}^k, 0)$ such that $\psi(f^{-1}(c)) = g^{-1}(c)$ as a set germ for any $c \in (\mathbb{R}, 0)$. Then we have the following proposition.

Proposition 6.6. *For function germs* $f, g : (\mathbb{R}^k, 0) \to (\mathbb{R}, 0)$*, the level set foliation germs* $\mathscr{F}_f$ *and* $\mathscr{F}_g$ *are strictly diffeomorphic if and only if* f *and* g *are* $\mathcal{R}$*-equivalent.*

Proof. By definition, if f and g are $\mathcal{R}$-equivalent, then $\mathscr{F}_f$ and $\mathscr{F}_g$ are strictly diffeomorphic. If $\mathscr{F}_f$ and $\mathscr{F}_g$ are strictly diffeomorphic, then there exists a diffeomorphism germ $\psi : (\mathbb{R}^k, 0) \to (\mathbb{R}^k, 0)$ such that $\psi(f^{-1}(c)) = g^{-1}(c)$ as a set germ for any $c \in (\mathbb{R}, 0)$. We consider $\psi \times 1_{\mathbb{R}} : (\mathbb{R}^k \times \mathbb{R}, 0) \to (\mathbb{R}^k \times \mathbb{R}, 0)$ which is a diffeomorphism germ. For any $(q, f(q)) \in \overline{f}^{-1}(0)$, we have $(\psi \times 1_{\mathbb{R}})(q, f(q)) = (\psi(q), f(q))$. If we set $c = f(q)$, then $\psi(q) \in g^{-1}(c) = \overline{g}^{-1}(0) \cap (\mathbb{R}^k \times \{c\})$, so that $(\psi \times 1_{\mathbb{R}})(q, f(q)) \in \overline{g}^{-1}(0)$. Therefore $(\psi \times 1_{\mathbb{R}})(\overline{f}^{-1}(0)) = \overline{g}^{-1}(0)$ as set germs. For any $q \in (\mathbb{R}^k, 0)$, we have $(q, f(q)) \in \overline{f}^{-1}(0)$. Then we have

$$0 = \overline{g}((\psi \times 1_{\mathbb{R}})(q, f(q)) = \overline{g}(\psi(q), f(q)) = g \circ \psi(q) - f(q),$$

so that $g \circ \psi = f$. This completes the proof. □

For function germs $f : (\mathbb{R}^k, 0) \to (\mathbb{R}, 0)$ and $g : (\mathbb{R}^{k'}, 0) \to (\mathbb{R}, 0)$, we say that the level set foliation germs $\mathscr{F}_f$ and $\mathscr{F}_g$ are *stably strictly diffeomorphic* if they become strictly diffeomorphic after the addition to the arguments q_i of new arguments q'_i and to functions f, g of non-degenerate quadratic forms. Thus, we have the following proposition.

Proposition 6.7. *For function germs $f : (\mathbb{R}^k, 0) \to (\mathbb{R}, 0)$ and $g : (\mathbb{R}^{k'}, 0) \to (\mathbb{R}, 0)$, the following conditions are equivalent:*
(1) f and g are stably $\mathcal{R}$-equivalent,
(2) $\mathscr{F}_f$ and $\mathscr{F}_g$ are stably strictly diffeomorphic,
(3) $\overline{f}$ and $\overline{g}$ are stably $S.P$-$\mathcal{K}$-equivalent.

As a corollary of Theorem 6.4 and Proposition 6.7, we have the following theorem.

Theorem 6.8. *With the same assumptions as those in Theorem 6.4, the following condition is equivalent to (1) $\sim$ (7) in Theorem 6.4:*
(8) $\mathscr{F}_f$ and $\mathscr{F}_g$ are stably strictly diffeomorphic.

We consider another geometric property of graph-like Legendrian unfoldings. Let $(\mathscr{L}, p)$ be a graph-like Legendrian unfolding germ. We consider a representative $\widetilde{\mathscr{L}}$ of $(\mathscr{L}, p)$ on $\overline{\pi}^{-1}(W)$, where $W \subset \mathbb{R}^n \times \mathbb{R}$ is an open neighborhood of $\overline{\pi}(p) \in \mathbb{R}^n \times \mathbb{R}$. We now show that $W(\widetilde{\mathscr{L}}) \cap W \cap (\{\pi_1 \circ \overline{\pi}(p)\} \times \mathbb{R})$ is a discrete set. Suppose that there exists a sequence of points $\{u_i\}_{i=1}^{\infty} \subset U$ such that $\lim_{i\to\infty} u_i = u_0$ and $\overline{\pi}(\mathcal{L}(u_i)) \in W(\widetilde{\mathscr{L}}) \cap W \cap (\{\pi_1 \circ \overline{\pi}(p)\} \times \mathbb{R})$ for any $i \in \mathbb{N}$. Then $\overrightarrow{\overline{\pi}(p)\overline{\pi}(\mathcal{L}(u_i))}$ is parallel to the vector $\partial/\partial t$. If necessary we can choose a subsequence of $\{u_i\}_{i=1}^{\infty}$, we may suppose that

$$\lim_{i\to\infty} \frac{\overrightarrow{\overline{\pi}(p)\overline{\pi}(\mathcal{L}(u_i))}}{\|\overrightarrow{\overline{\pi}(p)\overline{\pi}(\mathcal{L}(u_i))}\|}$$

exists. Therefore $\partial/\partial t$ and $\nu(u_0)$ are orthogonal. This contradicts to the fact that $\nu(u)$ is given by $(p_1(u), \ldots, p_n(u), -1)$ (i.e., $W(\mathscr{L})$ is a graph-like wave front). It follows that $W(\widetilde{\mathscr{L}}) \cap W \cap (\{x\} \times \mathbb{R})$ is a finite set for $x \in \pi_1(W)$ for sufficiently small neighborhood W of $\overline{\pi}(p)$. We now define

$$\mathrm{Max}(W(\widetilde{\mathscr{L}}) \cap W) = \bigcup_{x \in \pi_1(W)} \{\max(W(\widetilde{\mathscr{L}}) \cap (\{x\} \times \mathbb{R}))\},$$

$$\mathrm{mini}(W(\widetilde{\mathscr{L}}) \cap W) = \bigcup_{x \in \pi_1(W)} \{\mathrm{mini}(W(\widetilde{\mathscr{L}}) \cap (\{x\} \times \mathbb{R}))\}.$$

We denote that the germs of the above sets as $(\mathrm{Max}(W(\mathscr{L})), p)$ and $(\min(W(\mathscr{L})), p)$ respectively. We call $(\mathrm{Max}(W(\mathscr{L})), p)$ a *local maximum graph* and $(\min(W(\mathscr{L})), p)$ a *local minimum graph* of the graph-like wave front $W(\mathscr{L})$ respectively. Let $\Phi : (\mathbb{R}^n \times \mathbb{R}, 0) \to (\mathbb{R}^n \times \mathbb{R}, 0)$ be an $S.P^+$-diffeomorphism defined by $\Phi(x,t) = (\phi_1(x), t + \alpha(x))$. Then there exist neighborhoods $U_1, U_2 \subset \mathbb{R}^n \times \mathbb{R}$ of the origin and a diffeomorphism $\widetilde{\Phi} : U_1 \to U_2$ of the form $\widetilde{\Phi}(x,t) = (\widetilde{\phi}_1(x), t + \widetilde{\alpha}(x))$, which is a representative of the map germ Φ. If $t_1 \geq t_2$, then $t_1 + \widetilde{\alpha}(x) \geq t_2 + \widetilde{\alpha}(x)$ for any $x \in \pi_1(U_1)$. Therefore we have the following lemma.

Lemma 6.9. *Let $\Phi : (\mathbb{R}^n \times \mathbb{R}, q_1) \to (\mathbb{R}^n \times \mathbb{R}, q_2)$ be an $S.P^+$-diffeomorphism. Then we have*

$$\Phi(\mathrm{Max}(W(\mathscr{L}))) = \mathrm{Max}(\Phi(W(\mathscr{L}))),\ \Phi(\min(W(\mathscr{L}))) = \min(\Phi(W(\mathscr{L})))$$

as set germs.

We have the following corollary of Theorem 6.1 and Lemma 6.9.

Corollary 6.10. *Let $(\mathscr{L}_1, p_1)$ and $(\mathscr{L}_2, p_2)$ be graph-like Legendrian unfolding germs. If $(\Pi(\mathscr{L}_1), \Pi(p_1))$ and $(\Pi(\mathscr{L}_2), \Pi(p_2))$ are Lagrangian equivalent, then there exists a diffeomorphism germ $\Phi : (\mathbb{R}^n \times \mathbb{R}, \overline{\pi}(p_1)) \to (\mathbb{R}^n \times \mathbb{R}, \overline{\pi}(p_2))$ of the form $\Phi(x,t) = (\phi_1(x), t + \alpha(x))$ such that $\Phi(\mathrm{Max}(W(\mathscr{L}_1))) = \mathrm{Max}(W(\mathscr{L}_2))$ and $\Phi(\min(W(\mathscr{L}_1))) = \min(W(\mathscr{L}_2))$ as set germs.*

The following standard examples clarify the difference between the equivalence relations among graph-like Legendrian unfoldings.

Example 6.11. It is known that one of the germs in the list of 2-parameter $\mathcal{R}^+$-versal unfoldings is the cusp (cf. [6]). The normal form is given by

$$F(q, x_1, x_2) = \mp q^4 \mp x_2 q^2 - x_1 q.$$

It is P-$\mathcal{R}^+$-equivalent to $F_1(q, x_1, x_2) = \mp q^4 \mp x_2(q^2+1) - x_1 q$. Then we have

$$C(F_1) = \{(q, \mp(4q^3 + 2qx_2), x_2) \in \mathbb{R}^3 \mid (q, x_2) \in (\mathbb{R}^2, 0)\}.$$

Since $\partial F_1/\partial x_1 = -q$ and $\partial F_1/\partial x_2 = \mp(q^2+1)$,

$$L(F_1)(C(F_1)) = \{(\mp(4q^3 + 2qx_2), x_2, -q, \mp(q^2+1)) \mid (q, x_2) \in (\mathbb{R}^2, 0)\}$$

are Lagrangian submanifold germs of $T^*\mathbb{R}^2$. If we set $u = \mp q$ and $v = x_2$, then we have Lagrangian embeddings $\mathcal{L}_1^{\pm} : U \to T^*\mathbb{R}^2 \equiv \mathbb{R}^2 \times (\mathbb{R}^2)^*$ defined by

$$\mathcal{L}_1^{\pm}(u, v) = ((4u^3 + 2uv, v), (\pm u, \mp(u^2+1))),$$

where $U \subset \mathbb{R}^2$ is an open subset. Therefore, $L_1^{\pm} = \mathcal{L}_1^{\pm}(U)$ are Lagrangian submanifolds in $T^*\mathbb{R}^2$. Moreover, if we consider graph-like Morse families of hypersurfaces defined by $\overline{F}(q, x_1, x_2, t) = \mp q^4 \mp x_2(q^2+1) - x_1 q - t$, then the corresponding graph-like Legendrian unfoldings are given by mappings $\mathfrak{L}_1^{\pm} : U \to J^1_{GA}(\mathbb{R}^2, \mathbb{R})$ where

$$\mathfrak{L}_1^{\pm}(u, v) = ((4u^3 + 2uv, v), \pm(3u^4 + u^2 v - v), (\pm u, \mp(u^2+1))).$$

Then $\mathscr{L}_1^{\pm} = \mathfrak{L}_1^{\pm}(U)$ are graph-like Legendrian unfoldings such that $\Pi(\mathscr{L}_1^{\pm}) = L_1^{\pm}$.

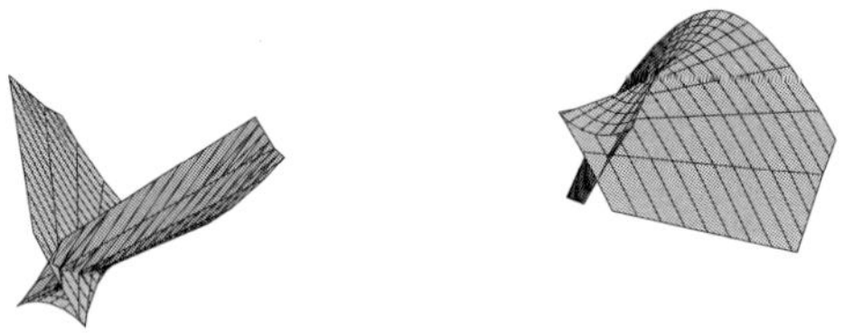

$W(\mathscr{L}_1^+)$ $W(\mathscr{L}_1^-)$

Fig.4: Graph-like wave fronts.

We remark that both the graph-like wave fronts are swallowtails (cf. Fig. 4) at $(u, v) = (0, 0)$. We observe that $(\mathrm{Max}(W(\mathscr{L}_1^+)), 0)$ is a graph of continuous function but $(\mathrm{Max}(W(\mathscr{L}_1^-)), 0)$ is not (cf. Fig.5), so that these are not diffeomorphic as set germs.

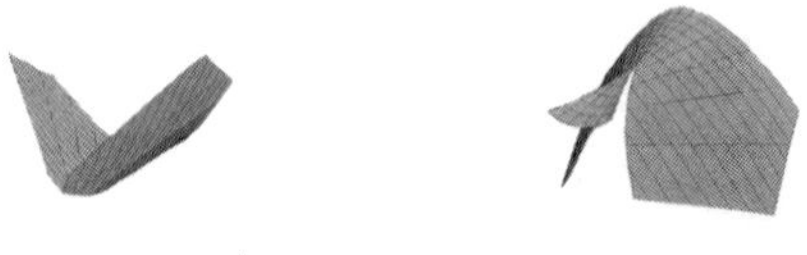

Fig.5: $\mathrm{Max}\, W(\mathscr{L}_1^+)$ $\mathrm{Max}\, W(\mathscr{L}_1^-)$

By Corollary 6.10, the germs of L^+ and L^- at the origin are not Lagrangian equivalent. Since both the caustics of L^+ and L^- at the origin are the ordinary cusp, these are diffeomorphic as set germs (cf. Fig.6).

On the other hand, we consider the bifurcation of the family of momentary fronts for graph-like Legendrian unfoldings. If we consider a diffeomorphism germ $\Phi : (\mathbb{R}^2 \times \mathbb{R}, 0) \to (\mathbb{R}^2 \times \mathbb{R}, 0)$ defined by $\Phi(x,t) = (x, -t)$, then we can show $\Phi(W(\mathscr{L}_1^+)) = W(\mathscr{L}_1^-)$. Thus, $W(\mathscr{L}_1^+)$ and $W(\mathscr{L}_1^-)$ are (s,t)-diffeomorphic but not $S.P^+$-diffeomorphic. As we mentioned above, L_1^+ and L_1^- are not Lagrangian equivalent. The bifurcations of $\{W_t(\mathscr{L}_1^\pm)\}_{t\in(\mathbb{R},0)}$ are depicted in Fig.7. We can observe that

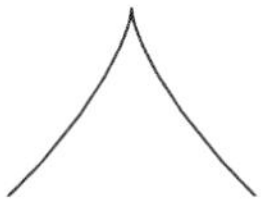

$C_{L_1^\pm}$

Fig.6: The caustic.

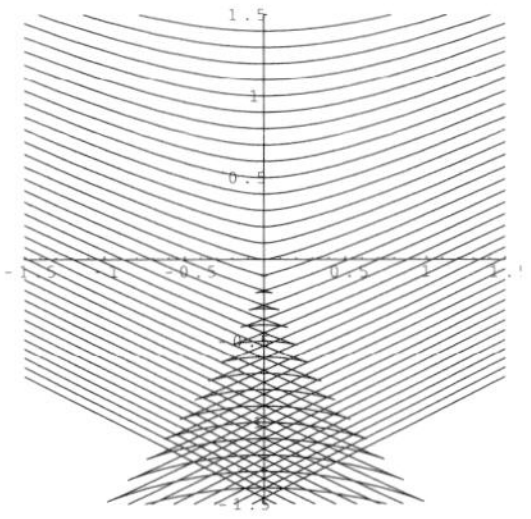

$\{W_t(\mathscr{L}_1^\pm)\}_{t\in(\mathbb{R},0)}$

Fig.7: The bifurcations of momentary fronts.

both the caustics are ordinary cusps. Therefore, these are examples of Lagrangian submanifold germs such that those caustics are diffeomorphic but these are not Lagrangian equivalent.

We also consider $\mathcal{R}^+$-versal unfoldings $F_2(q, x_1, x_2) = \mp q^4 \mp x_2(q^2 - 1) - x_1 q$ which are P-$\mathcal{R}^+$-equivalent to $F(q, x_1, x_2)$. By the calculation similar to the above, we have embeddings $\mathcal{L}_2^\pm : U \to T^*\mathbb{R}^2 \equiv \mathbb{R}^2 \times (\mathbb{R}^2)^*$ defined by

$$\mathcal{L}_2^\pm(u,v) = ((4u^3 + 2uv, v), (\pm u, \mp(u^2 - 1))),$$

where $U \subset \mathbb{R}^2$ is an open subset. Then $L_2^\pm = \mathcal{L}_2^\pm(U)$ are Lagrangian submanifolds. Moreover, we have the corresponding graph-like Legendrian unfoldings defined by mappings $\mathfrak{L}_2^\pm : U \to J^1_{GA}(\mathbb{R}^2, \mathbb{R})$ where

$$\mathfrak{L}_2^\pm(u,v) = ((4u^3 + 2uv, v), \pm(3u^4 + u^2 v + v), (\pm u, \mp(u^2 - 1))).$$

By the same reasons as the above case, L_2^+ and L_2^- are not Lagrangian equivalent (cf. Fig.8). However, if we consider diffeomorphism germ $\Phi^\pm : (\mathbb{R}^2 \times \mathbb{R}, 0) \to (\mathbb{R}^2 \times \mathbb{R}, 0)$ defined by $\Phi^\pm(x_1, x_2, t) = (x_1, x_2, t \pm 2x_2)$, then we have $\Phi^\pm(W(\mathscr{L}_1^\pm)) = W(\mathscr{L}_2^\pm)$ as set germs. By Corollary 6.2,

$W(\mathscr{L}_2^+)$ $W(\mathscr{L}_2^-)$

Fig.8: Graph-like wave fronts.

L_1^+ and L_2^+ (respectively, L_1^- and L_2^-) are Lagrangian equivalent. The pictures of $W(\mathscr{L}_2^\pm)$ are similar to those of $W(\mathscr{L}_1^\pm)$. Moreover, the caustics are the same as the above (Fig.9). However, the bifurcations of momentary fronts $\{W_t(\mathscr{L}_2^\pm)\}_{t\in(\mathbb{R},0)}$ are different from $\{W_t(\mathscr{L}_1^\pm)\}_{t\in(\mathbb{R},0)}$ (cf. Fig.7 and Fig.10). Actually, we can apply the criterion in [11] and show that $W(\mathscr{L}_1^\pm)$ and $W(\mathscr{L}_2^\pm)$ are not $S.P$-diffeomorphic as set germs.

$C_{L_2^\pm}$

Fig.9: The caustic.

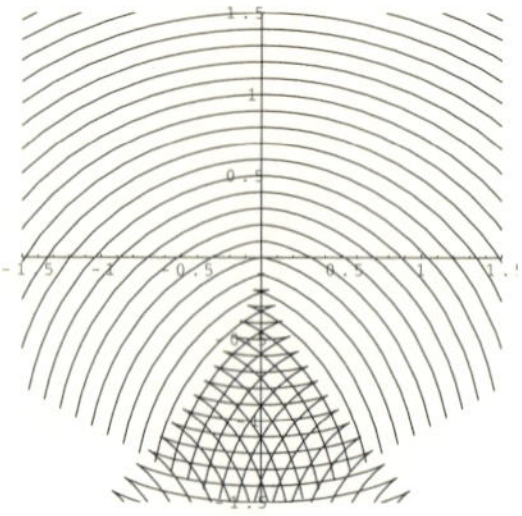

$\{W_t(\mathscr{L}_2^\pm)\}_{t\in(\mathbb{R},0)}$

Fig.10: The bifurcation of momentary fronts.

§7. s-P-Legendrian equivalence among graph-like Legendrian unfoldings

In §6 we have given a brief survey on $S.P^+$-Legendrian equivalence among graph-like Legendrian unfoldings. One of the main consequences is that $S.P^+$-Legendrian equivalence among graph-like Legendrian unfoldings is equivalent to Lagrangian equivalence among induced Lagrangian submanifolds. This fact can be considered as a geometric interpretation of Lagrangian equivalence. On the other hand, s-P-Legendrian equivalence is weaker than $S.P^+$-Legendrian equivalence among graph-like Legendrian unfoldings. Therefore, Lagrangian equivalence is stronger than s-P-Legendrian equivalence. In this section we explain detailed properties of s-P-Legendrian equivalence among graph-like Legendrian unfoldings as an application of the results in [13, 35], which might be new results. We also use a graph-like Morse family of hypersurfaces of the form $\mathcal{F}(q,x,t) = \lambda(q,x,t)(F(q,x) - t)$. Since $\mathcal{F}(q,x,t)$ and $\overline{F}(q,x,t) = F(q,x) - t$ are s-$S.P$-$\mathcal{K}$-equivalent, we consider $\overline{F}(q,x,t) = F(q,x) - t$ as a graph-like Morse family. Moreover, by Proposition 5.4, $F(q,x)$ is a Morse family of functions. We now suppose that $F(q,x)$ is a Morse family of functions. Consider the graph-like Morse family of hypersurfaces $\overline{F}(q,x,t) = F(q,x) - t$ which is not necessarily non-degenerate. Then we have $\mathscr{L}_{\overline{F}}(\Sigma_*(\overline{F})) = \mathfrak{L}_F(C(F))$. We also denote that $\overline{f}(q,t) = f(q) - t$ for any $f \in \mathfrak{M}_k$. For function germs $f, g : (\mathbb{R}^k, 0) \to (\mathbb{R}, 0)$, we say that f, g are *$\mathcal{A}$-equivalent* if there exist diffeomorphism germs $\phi : (\mathbb{R}^k, 0) \to (\mathbb{R}^k, 0)$ and $\psi : (\mathbb{R}, 0) \to (\mathbb{R}, 0)$ such that $\psi \circ f = g \circ \phi$. Moreover, let $F, G : (\mathbb{R}^k \times \mathbb{R}^n, 0) \to (\mathbb{R}, 0)$ be function germs, we say that F, G are *P-$\mathcal{A}$-equivalent* if there exist diffeomorphism germs $\Phi : (\mathbb{R}^k \times \mathbb{R}^n, 0) \to (\mathbb{R}^k \times \mathbb{R}^n, 0)$ of the form $\Phi(q,x) = (\phi_1(q,x), \phi_2(x))$ and $\Psi : (\mathbb{R} \times \mathbb{R}^n, 0) \to (\mathbb{R} \times \mathbb{R}^n, 0)$ of the form $\Psi(t,x) = (\psi(t,x), x)$ such that

$$\Psi(F(q,x), x) = (G \circ \Phi(q,x), x)$$

for any $(q,x) \in (\mathbb{R}^k \times \mathbb{R}^n, 0)$. We remark that if F, G are P-$\mathcal{A}$-equivalent, then $f = F|_{\mathbb{R}^k \times \{0\}}, g = G|_{\mathbb{R}^k \times \{0\}}$ are $\mathcal{A}$-equivalent. Then we have the following proposition.

Proposition 7.1. *Let $F, G : (\mathbb{R}^k \times \mathbb{R}^n, 0) \to (\mathbb{R}, 0)$ be function germs. Then $\overline{F}(q,x,t) = F(q,x) - t, \overline{G}(q,x,t) = G(q,x) - t$ are s-P-$\mathcal{K}$-equivalent if and only if F, G are P-$\mathcal{A}$-equivalent.*

Proof. Suppose that $\overline{F}, \overline{G}$ are s-P-$\mathcal{K}$-equivalent. Then there exists a diffeomorphism germ $\overline{\Psi} : (\mathbb{R}^k \times (\mathbb{R} \times \mathbb{R}^n), 0) \to (\mathbb{R}^k \times (\mathbb{R} \times \mathbb{R}^n), 0)$ of the form $\overline{\Psi}(q,t,x) = (\overline{\psi}(q,t,x), \psi_1(t,x), \psi_2(x))$ such that $\langle \overline{F} \circ \overline{\Psi} \rangle_{\mathcal{E}_{k+1+n}} = \langle \overline{G} \rangle_{\mathcal{E}_{k+1+n}}$. It follows that $\overline{\Psi}(G^{-1}(0)) = F^{-1}(0)$. By definition, we have

$$\overline{F}^{-1}(0) = \{(q, F(q,x), x) \mid (q,x) \in (\mathbb{R}^k \times \mathbb{R}^n, 0)\},$$
$$\overline{G}^{-1}(0) = \{(q, G(q,x), x) \mid (q,x) \in (\mathbb{R}^k \times \mathbb{R}^n, 0)\}.$$

Therefore, we have

$$\overline{\Psi}(q, G(q,x), x) = (\overline{\psi}(q, G(q,x), x), \psi_1(G(q,x), x), \psi_2(x)) = (\overline{q}, F(\overline{q}, \overline{x}), \overline{x}).$$

Hence, we have $\overline{q} = (\overline{\psi}(q, G(q,x), x), \overline{x} = \psi_2(x)$ and

$$\psi_1(G(q,x), x) = F(\overline{q}, \overline{x}) = F(\overline{\psi}(q, G(q,x), x), \psi_2(x)).$$

If we define $\Phi : (\mathbb{R}^k \times \mathbb{R}^n, 0) \to (\mathbb{R}^k \times \mathbb{R}^n, 0)$ by

$$\Phi(q,x) = \overline{\psi}(q, G(q,x), x), \psi_2(x)),$$

then Φ is a diffeomorphism germ. Moreover, we define $\Psi : (\mathbb{R}\times\mathbb{R}^n, 0) \to (\mathbb{R}\times\mathbb{R}^n, 0)$ by $\Psi(t,x) = (\psi_1(t,x), x)$. Then the above equality means that $\Psi(G(q,x),x) = (F\circ\Phi(q,x),x)$, so that F, G are P-$\mathcal{A}$-equivalent.

Suppose that there exist diffeomorphism germs $\Phi : (\mathbb{R}^k\times\mathbb{R}^n, 0) \to (\mathbb{R}^k\times\mathbb{R}^n, 0)$ of the form $\Phi(q,x) = (\phi_1(q,x), \phi_2(x))$ and $\Psi : (\mathbb{R}\times\mathbb{R}^n, 0) \to (\mathbb{R}\times\mathbb{R}^n, 0)$ of the form $\Psi(t,x) = (\psi(t,x), x)$ such that

$$\Psi(F(q,x),x) = (G\circ\Phi(q,x),x)$$

for any $(q,x)\in(\mathbb{R}^k\times\mathbb{R}^n,0)$. We define $\overline{\Psi} : (\mathbb{R}^k\times(\mathbb{R}\times\mathbb{R}^n),0) \to (\mathbb{R}^k\times(\mathbb{R}\times\mathbb{R}^n),0)$ by $\overline{\Psi}(q,t,x) = (\phi_1(q,x),\psi(t,x),\phi_2(x))$. Then $\overline{\Psi}$ is a diffeomorphism germ. Since $\psi(F(q,x),x) = G(\phi_1(q,x),\phi_2(x))$, we have

$$\begin{aligned}\overline{\Psi}(q,F(q,x),x) &= (\phi_1(q,x),\psi(F(q,x),x),\phi_2(x))\\ &= (\phi_1(q,x),G(\phi_1(q,x),\phi_2(x)),\phi_2(x)),\end{aligned}$$

so that $\overline{\Psi}(\overline{F}^{-1}(0)) = \overline{G}^{-1}(0)$ as set germs. Thus $\overline{F}^{-1}(0) = (\overline{G}\circ\overline{\Psi})^{-1}(0)$. Since $\overline{F},\overline{G}$ are submersion germs, we have $\langle\overline{F}\rangle_{\mathcal{E}_{k+1+n}} = \langle\overline{G}\circ\overline{\Psi}\rangle_{\mathcal{E}_{k+1+n}}$. This completes the proof. □

We have the following simple corollary.

Corollary 7.2. *For function germs $f, g : (\mathbb{R}^k,0)\to(\mathbb{R},0)$, f,g are $\mathcal{A}$-equivalent if and only if $\overline{f}(q,t) = f(q)-t, \overline{g}(q,t) = g(q)-t$ are P-$\mathcal{K}$-equivalent.*

For $(q,t,x)\in(\mathbb{R}^k\times\mathbb{R}\times\mathbb{R}^n,0)$ and $(q',t,x)\in(\mathbb{R}^{k'}\times\mathbb{R}\times\mathbb{R}^n,0)$, let $\mathcal{F}(q,x,t) = \lambda(q,x,t)(F(q,x)-t)$ and $\mathcal{G}(q',x,t) = \mu(q',x,t)(G(q',x)-t)$ be graph-like Morse families of hypersurfaces. By Theorem 4.2 and Corollary 7.2, we have the following theorem.

Theorem 7.3. *The graph-like Legendrian unfoldings $\mathscr{L}_{\mathcal{F}}(\Sigma_*(\mathcal{F}))$, $\mathscr{L}_{\mathcal{G}}(\Sigma_*(\mathcal{G}))$ are s-P-Legendrian equivalent if and only if $F(q,x), G(q',x)$ are stably P-$\mathcal{A}$-equivalent.*

The definition of *stably P-$\mathcal{A}$-equivalence* is similar to the definition of stably P-$\mathcal{R}^+$-equivalence, so that we omit to give the definition here.

We now consider the stability of graph-like Legendrian unfoldings relative to *s-P*-Legendrian equivalence. Theorem 4.2 asserts that the graph-like Legendrian unfolding $\mathscr{L}_{\mathcal{F}}(\Sigma_*(\mathcal{F}))$ is *s-P*-Legendrian stable if and only if $\mathcal{F}$ is an infinitesimally *P-$\mathcal{K}$*-versal unfolding of $\mathcal{F}|_{\mathbb{R}^k\times\{0\}\times\mathbb{R}}$. Here, we have $\mathcal{F}(q,x,t) = \lambda(q,x,t)(F(q,x)-t)$. We can represent the extended tangent space of $\overline{f} : (\mathbb{R}^k\times\mathbb{R},0)\to(\mathbb{R},0)$ relative to *P-$\mathcal{K}$* by

$$T_e(P\text{-}\mathcal{K})(\overline{f}) = \left\langle \frac{\partial f}{\partial q_1}(q),\ldots,\frac{\partial f}{\partial q_k}(q), f(q)-t\right\rangle_{\mathcal{E}_{k+1}} + \langle 1\rangle_{\mathcal{E}_1}.$$

In this case the unfolding $\overline{F}(q,x,t)=F(q,x)-t$ of $\overline{f}(q,t)$ is an infinitesimally P-$\mathcal{K}$-versal unfolding of $\overline{f}(q,t)$ if and only if

$$\mathcal{E}_{k+1}=T_e(P\text{-}\mathcal{K})(\overline{f})+\left\langle \frac{\partial F}{\partial x_1}|_{\mathbb{R}^k\times\{0\}\times\mathbb{R}},\dots,\frac{\partial F}{\partial x_n}|_{\mathbb{R}^k\times\{0\}\times\mathbb{R}}\right\rangle_{\mathbb{R}}.$$

Moreover, we now define P-$\mathcal{K}$-versal unfolding of $\overline{f}$ (cf. [13]) as follows: For a map germ $\psi:(\mathbb{R}^m,0)\to(\mathbb{R}^n,0)$, we define an m-dimensional unfolding $\psi^*F:(\mathbb{R}^k\times\mathbb{R}^m\times\mathbb{R},0)\to(\mathbb{R},0)$ of $\overline{f}$ by $\psi^*\overline{F}(q,y,t)=\overline{F}(q,\psi(y),t)$, which we call an *induced unfolding* of $\overline{F}$ by ψ. We say that $\overline{F}$ is a *P-$\mathcal{K}$-versal unfolding* of $\overline{f}$ if for any unfolding $\overline{G}:(\mathbb{R}^k\times\mathbb{R}^m\times\mathbb{R},0)\to(\mathbb{R},0)$ of $\overline{f}$, there exists a P-$\mathcal{K}$-morphism from $\overline{G}$ to $\overline{F}$. Here, a *P-$\mathcal{K}$-morphism* from $\overline{G}$ to $\overline{F}$ is $(\psi,\widetilde{\Phi},\widetilde{\phi},\lambda)$ where $\psi:(\mathbb{R}^m,0)\to(\mathbb{R}^n,0)$ is a map-germ, $\widetilde{\Phi}:(\mathbb{R}^k\times\mathbb{R}^m\times\mathbb{R},0)\to(\mathbb{R}^k\times\mathbb{R}^m\times\mathbb{R},0)$ is a diffeomorphism germ of the form $\widetilde{\Phi}(q,u,t)=(\phi_1(q,u,t),u,\widetilde{\phi}(u,t))$ and $\lambda(q,u,t)\in\mathcal{E}_{k+m+1}$ is a function germ such that $\widetilde{\Phi}(q,0,t)=(q,0,t)$, $\lambda(q,0,t)=1$ and $\psi^*\overline{F}(q,u,t)=\lambda(q,u,t)\overline{G}\circ\widetilde{\Phi}(q,u,t)$. We have the following theorem (cf. [8, 13]).

Theorem 7.4. *An unfolding $\overline{F}:(\mathbb{R}^k\times\mathbb{R}^n\times\mathbb{R},0)\to(\mathbb{R},0)$ is a P-$\mathcal{K}$-versal unfolding of $\overline{f}:(\mathbb{R}^k\times\mathbb{R},0)\to(\mathbb{R},0)$ if and only if it is an infinitesimally P-$\mathcal{K}$-versal unfolding of $\overline{f}$.*

We consider the stability of the unfolding $\overline{F}$ of $\overline{f}$ relative to P-$\mathcal{K}$-equivalence. We say that $\overline{F}$ is *homotopically P-$\mathcal{K}$-stable* if for any one-parameter family of functions $\mathcal{F}:(\mathbb{R}^k\times(\mathbb{R}^n\times\mathbb{R})\times\mathbb{R},0)\to(\mathbb{R},0)$ with $\mathcal{F}(q,x,0,t)=\overline{F}(q,x,t)$, there is a a P-$\mathcal{K}$-morphism from $\mathcal{F}$ to $\overline{F}$ as unfoldings of $\overline{f}$. Here, we remark that $\mathcal{F}(q,x,s,t)$ can be regarded as an unfolding of $\overline{f}$ with the parameter $(x,s)\in\mathbb{R}^n\times\mathbb{R}$. By definition, if $\overline{F}$ is a P-$\mathcal{K}$-versal unfolding of $\overline{f}$, then it is homotopically P-$\mathcal{K}$-stable. Moreover, suppose that $\overline{F}$ is homotopically P-$\mathcal{K}$-stable. For any $h(q,t)\in\mathcal{E}_{k+1}$, we consider a one-parameter family of function germ $\mathcal{F}(q,x,s,t)=\overline{F}(q,x,t)+sh(q,t)$. Then there exist a map-germ $\psi:(\mathbb{R}^n\times\mathbb{R},0)\to(\mathbb{R}^n,0)$, a diffeomorphism germ $\widetilde{\Phi}:(\mathbb{R}^k\times(\mathbb{R}^n\times\mathbb{R})\times\mathbb{R},0)\to(\mathbb{R}^k\times(\mathbb{R}^n\times\mathbb{R})\times\mathbb{R},0)$ of the form $\widetilde{\Phi}(q,x,s,t)=(\phi_1(q,x,s,t),x,s,\widetilde{\phi}(x,s,t))$ and $\lambda(q,x,s,t)\in\mathcal{E}_{k+n+1+1}$ is a function germ such that $\widetilde{\Phi}(q,0,0,t)=(q,0,0,t)$, $\lambda(q,0,0,t)=1$ and

$$\begin{aligned}\overline{F}(q,\psi(x,s),t)&=\lambda(q,x,s,t)\mathcal{F}\circ\widetilde{\Phi}(q,x,s,t)\\&=\lambda(q,x,s,t)(\overline{F}(\phi_1(q,x,s,t),x,\widetilde{\phi}(x,s,t))\\&\qquad+sh(\phi_1(q,x,s,t),\widetilde{\phi}(x,s,t)).\end{aligned}$$

Differentiating with respect to s at $(x,s)=(0,0)$, we have

$$\sum_{i=1}^{n}\frac{\partial\overline{F}}{\partial x_i}(q,0,t)\frac{\partial\psi_i}{\partial s}(0,0)=\frac{\partial\lambda}{\partial s}(q,0,0,t)\overline{F}(q,0,t)$$
$$+\sum_{j=1}^{k}\frac{\partial\overline{F}}{\partial q_j}(q,0,t)\frac{\partial(\phi_1)_j}{\partial s}(q,0,0,t)+\frac{\partial\overline{F}}{\partial t}(q,0,t)\frac{\partial\widetilde{\phi}}{\partial s}(0,0,t)+h(q,t).$$

Therefore, we have

$$h(q,t)\in T_e(P\text{-}\mathcal{K})(\overline{f})+\left\langle\frac{\partial F}{\partial x_1}|_{\mathbb{R}^k\times\{0\}\times\mathbb{R}},\dots,\frac{\partial F}{\partial x_n}|_{\mathbb{R}^k\times\{0\}\times\mathbb{R}}\right\rangle_{\mathbb{R}},$$

so that

$$\mathcal{E}_{k+1}=T_e(P\text{-}\mathcal{K})(\overline{f})+\left\langle\frac{\partial F}{\partial x_1}|_{\mathbb{R}^k\times\{0\}\times\mathbb{R}},\dots,\frac{\partial F}{\partial x_n}|_{\mathbb{R}^k\times\{0\}\times\mathbb{R}}\right\rangle_{\mathbb{R}}.$$

By Theorem 7.4, we have shown the following proposition.

Proposition 7.5. *An unfolding $\overline{F}:(\mathbb{R}^k\times\mathbb{R}^n\times\mathbb{R},0)\to(\mathbb{R},0)$ of $\overline{f}:(\mathbb{R}^k\times\mathbb{R},0)\to(\mathbb{R},0)$ is homotopically P-$\mathcal{K}$-stable if and only if it is a P-$\mathcal{K}$-versal unfolding of $\overline{f}$.*

On the other hand, Wassermann [35] investigated stability and versality of unfoldings of function germs relative to $\mathcal{A}$-equivalence. We say that $F(q,x)$ is an *infinitesimally $\mathcal{A}$-versal unfolding* of $f(q)$ if

$$\mathcal{E}_k=T_e(\mathcal{A})(f)+\left\langle\frac{\partial F}{\partial x_1}|_{\mathbb{R}^k\times\{0\}},\dots,\frac{\partial F}{\partial x_n}|_{\mathbb{R}^k\times\{0\}}\right\rangle_{\mathbb{R}},$$

where

$$T_e(\mathcal{A})(f)=J_f+f^*(\mathcal{E}_1)\text{ and }f^*(\mathcal{E}_1)=\{h\circ f\in\mathcal{E}_k\mid h\in\mathcal{E}_1\}.$$

We also define $\mathcal{A}$-versal unfoldings. An unfolding $F:(\mathbb{R}^k\times\mathbb{R}^n,0)\to(\mathbb{R},0)$ of $f:(\mathbb{R}^k,0)\to(\mathbb{R},0)$ is an *$\mathcal{A}$-versal unfolding* if for any unfolding $G:(\mathbb{R}^k\times\mathbb{R}^m,0)\to(\mathbb{R},0)$ of f, there exists an $\mathcal{A}$-morphism from G to F. Here, an *$\mathcal{A}$-morphism* from G to F is (ψ,Φ,ϕ) where $\psi:(\mathbb{R}^m,0)\to(\mathbb{R}^n,0)$ is a map-germ, $\Phi:(\mathbb{R}^k\times\mathbb{R}^m,0)\to(\mathbb{R}^k\times\mathbb{R}^m,0)$ is a diffeomorphism germ of the form $\Phi(q,u)=(\phi_1(q,u),u)$ and $\phi:(\mathbb{R}\times\mathbb{R}^m,0)\to(\mathbb{R},0)$ is a function germ such that $\Phi(q,0)=(q,0)$, $\phi(y,0)=y$ and $\phi(\psi^*F(q,u),u)=G\circ\Phi(q,u)$. We have the following theorem.

Theorem 7.6 ([35]). *An unfolding $F : (\mathbb{R}^k \times \mathbb{R}^n, 0) \to (\mathbb{R}, 0)$ is an $\mathcal{A}$-versal unfolding of $f : (\mathbb{R}^k, 0) \to (\mathbb{R}, 0)$ if and only if it is an infinitesimally $\mathcal{A}$-versal unfolding of f.*

We say that F is *homotopically $\mathcal{A}$-stable* if for any one-parameter family of functions $\mathcal{F} : (\mathbb{R}^k \times (\mathbb{R}^n \times \mathbb{R}), 0) \to (\mathbb{R}, 0)$ with $\mathcal{F}(q, x, 0) = F(q, x)$, there is a an $\mathcal{A}$-morphism from $\mathcal{F}$ to F as unfoldings of f. Here, we remark that $\mathcal{F}(q, x, s)$ can be regarded as an unfolding of f with the parameter $(x, s) \in \mathbb{R}^n \times \mathbb{R}$. By definition, if F is an $\mathcal{A}$-versal unfolding of f, then it is homotopically $\mathcal{A}$-stable. Moreover, suppose that F is homotopically $\mathcal{A}$-stable. For any $h(q) \in \mathcal{E}_k$, we consider a one-parameter family of function germ $\mathcal{F}(q, x, s) = F(q, x) + sh(q)$. Then there exist a map-germ $\psi : (\mathbb{R}^n \times \mathbb{R}, 0) \to (\mathbb{R}^n, 0)$, a diffeomorphism germ $\Phi : (\mathbb{R}^k \times (\mathbb{R}^n \times \mathbb{R}), 0) \to (\mathbb{R}^k \times (\mathbb{R}^n \times \mathbb{R}), 0)$ of the form $\Phi(q, x, s) = (\phi_1(q, x, s), x, s)$ and a function germ $\phi(y, x, s) \in \mathfrak{M}_{1+n+1}$ such that $\Phi(q, 0, 0) = (q, 0, 0)$, $\phi(y, 0, 0) = y$ and

$$\phi(F(q, \psi(x, s)), x, s) = \mathcal{F} \circ \Phi(q, x, s) = F(\phi_1(q, x, s), x) + sh(\phi_1(q, x, s)).$$

Differentiating with respect to s at $(x, s) = (0, 0)$, we have

$$\begin{aligned} &\sum_{i=1}^{n} \frac{\partial F}{\partial x_i}(q, 0) \frac{\partial \psi_i}{\partial s}(0, 0) + \frac{\partial \phi}{\partial s}(f(q), 0, 0) \\ &\quad = \sum_{j=1}^{k} \frac{\partial F}{\partial q_j}(q, 0, t) \frac{\partial (\phi_1)_j}{\partial s}(q, 0, 0) + h(q, t). \end{aligned}$$

Therefore, we have

$$h(q) \in T_e(\mathcal{A})(f) + \left\langle \frac{\partial F}{\partial x_1}|_{\mathbb{R}^k \times \{0\} \times \mathbb{R}}, \dots, \frac{\partial F}{\partial x_n}|_{\mathbb{R}^k \times \{0\} \times \mathbb{R}} \right\rangle_{\mathbb{R}},$$

so that

$$\mathcal{E}_k = T_e(\mathcal{A})(f) + \left\langle \frac{\partial F}{\partial x_1}|_{\mathbb{R}^k \times \{0\} \times \mathbb{R}}, \dots, \frac{\partial F}{\partial x_n}|_{\mathbb{R}^k \times \{0\} \times \mathbb{R}} \right\rangle_{\mathbb{R}}.$$

By Theorem 7.6, we have shown the following proposition.

Proposition 7.7 ([35]). *An unfolding $F : (\mathbb{R}^k \times \mathbb{R}^n, 0) \to (\mathbb{R}, 0)$ of $f : (\mathbb{R}^k, 0) \to (\mathbb{R}, 0)$ is homotopically $\mathcal{A}$-stable if and only if it is a $\mathcal{A}$-versal unfolding of f.*

We also have the following proposition.

Proposition 7.8. *For an unfolding* $F : (\mathbb{R}^k \times \mathbb{R}^n, 0) \to (\mathbb{R}, 0)$ *of* $f : (\mathbb{R}^k, 0) \to (\mathbb{R}, 0)$, *the following conditions are equivalent:*
(1) $\overline{F} : (\mathbb{R}^k \times \mathbb{R}^n \times \mathbb{R}, 0) \to (\mathbb{R}, 0)$ *is homotopically* P-$\mathcal{K}$-*stable,*
(2) $F : (\mathbb{R}^k \times \mathbb{R}^n, 0) \to (\mathbb{R}, 0)$ *is homotopically* $\mathcal{A}$-*stable.*

Proof. Suppose that F is homotopically $\mathcal{A}$-stable. For any one-parameter family of functions $\mathcal{G} : (\mathbb{R}^k \times (\mathbb{R}^n \times \mathbb{R}) \times \mathbb{R}, 0) \to (\mathbb{R}, 0)$ with $\mathcal{G}(q, x, 0, t) = \overline{F}(q, x, t)$, we have $\partial \mathcal{G}/\partial t(0) \neq 0$, so that there exists a function germ $G(q, x, s)$ and $\mu(q, x, s, t)$ such that

$$\mu(0) \neq 0 \text{ and } \mathcal{G}(q, x, s, t) = \mu(q, x, s, t)(G(q, x, s) - t).$$

It follows that $F(q, x) - t = \mathcal{G}(q, x, t, 0) = \mu(q, x, 0, t)(G(q, x, 0) - t)$. By Proposition 7.1, $F(q, x)$ and $G(q, x, 0)$ are P-$\mathcal{A}$-equivalent. Thus $G(q, x, 0)$ is a homotopically $\mathcal{A}$-stable unfolding. Then there exist a map-germ $\psi : (\mathbb{R}^n \times \mathbb{R}, 0) \to (\mathbb{R}^n, 0)$, a diffeomorphism germ $\Phi : (\mathbb{R}^k \times (\mathbb{R}^n \times \mathbb{R}), 0) \to (\mathbb{R}^k \times (\mathbb{R}^n \times \mathbb{R}), 0)$ of the form $\Phi(q, x, s) = (\phi_1(q, x, s), x, s)$ and a function germ $\phi(t, x, s) \in \mathfrak{M}_{1+n+1}$ such that $\Phi(q, 0, 0) = (q, 0, 0)$, $\phi(t, 0, 0) = t$ and

$$\phi(G(q, \psi(x, s), 0), x, s) = G(\Phi(q, x, s)) = G(\phi_1(q, x, s), x, s).$$

We now define a diffeomorphism germ $\widetilde{\Phi} : (\mathbb{R}^k \times \mathbb{R}^n \times \mathbb{R} \times \mathbb{R}, 0) \to (\mathbb{R}^k \times \mathbb{R}^n \times \mathbb{R} \times \mathbb{R}, 0)$ by $\widetilde{\Phi}(q, x, s, t) = (\phi_1(q, x, s), x, s, \phi(t, x, s))$. The above equality means that

$$\widetilde{\Phi}(q, x, s, G(q, \psi(x, s), 0)) = (\phi_1(q, x, s), x, s, G(\phi_1(q, x, s), x, s)).$$

If we denote that $G(q, x, 0) = G_0(q, x)$, then $\psi^* G_0(q, x, s) = G(q, \psi(x, s), 0)$, so that we have $\widetilde{\Phi}(\overline{\psi^* G_0}^{-1}(0)) = \overline{G}^{-1}(0)$. Therefore, there exists a function germ $\lambda(q, x, s, t) \in \mathcal{E}_{k+n+1+1}$ with $\lambda(0) \neq 0$ such that

$$\overline{G} \circ \widetilde{\Phi}(q, x, s, t) = \lambda(q, x, s, t)\overline{\psi^* G_0}(q, x, s, t).$$

Here, $\widetilde{\Phi}(q, 0, 0, t) = (q, 0, 0, t)$ and

$$\begin{aligned} G_0(q, 0) - t &= \overline{G} \circ \widetilde{\Phi}(q, 0, 0, t) \\ &= \lambda(q, 0, 0, t)\overline{\psi^* G_0}(q, 0, 0, t) = \lambda(q, 0, 0, t)(G_0(q, 0) - t), \end{aligned}$$

so that $\lambda(q, 0, 0, t) = 1$. This means that $\overline{G}(q, x, t)$ is homotopically P-$\mathcal{K}$-stable. Therefore, $\overline{F}$ is homotopically P-$\mathcal{K}$-stable.

For the converse assertion, suppose that $\overline{F}$ is homotopically P-$\mathcal{K}$-stable. Let $G(q, x, s) \in \mathfrak{M}_{k+n+1}$ be a function germ with $G(q, x, 0) =$

$F(q,x)$. Then we have $\overline{G}(q,x,s,t) = G(q,x,s) - t$. Since $\overline{F}$ is homotopically P-$\mathcal{K}$-stable, there exist a map-germ $\psi : (\mathbb{R}^n \times \mathbb{R}, 0) \to (\mathbb{R}^n, 0)$, a diffeomorphism germ $\widetilde{\Phi} : (\mathbb{R}^k \times (\mathbb{R}^n \times \mathbb{R}) \times \mathbb{R}, 0) \to (\mathbb{R}^k \times (\mathbb{R}^n \times \mathbb{R}) \times \mathbb{R}, 0)$ of the form $\widetilde{\Phi}(q,x,s,t) = (\phi_1(q,x,s,t), x, s, \widetilde{\phi}(x,s,t))$ and $\lambda(q,x,s,t) \in \mathcal{E}_{k+n+1+1}$ is a function germ such that $\widetilde{\Phi}(q,0,0,t) = (q,0,0,t)$, $\lambda(q,0,0,t) = 1$ and

$$\begin{aligned} \overline{F}(q,\psi(x,s),t) &= \lambda(q,x,s,t)\overline{G} \circ \widetilde{\Phi}(q,x,s,t) \\ &= \lambda(q,x,s,t)(G(\phi_1(q,x,s,t),x,s) - \widetilde{\phi}(x,s,t)). \end{aligned}$$

It follows that $\overline{\psi^* F}^{-1}(0) = \widetilde{\Phi}^{-1}(\overline{G}^{-1}(0))$. Here, we have

$$\begin{aligned} \overline{\psi^* F}^{-1}(0) &= \{(q,x,s,\psi^* F(q,x,s)) \mid (q,x,s) \in \mathbb{R}^k \times \mathbb{R}^n \times \mathbb{R}\}, \\ \overline{G}^{-1}(0) &= \{(q,x,s,G(q,x,s)) \mid (q,x,s) \in \mathbb{R}^k \times \mathbb{R}^n \times \mathbb{R}\}. \end{aligned}$$

By definition, we have

$$\begin{aligned} &\widetilde{\Phi}(q,x,s,\psi^* F(q,x,s)) \\ &\quad = (\phi_1(q,x,s,\psi^* F(q,x,s)), x, s, \widetilde{\phi}(x,s,\psi^* F(q,x,s))). \end{aligned}$$

Therefore, if we put $\overline{q} = \phi_1(q,x,s,\psi^* F(q,x,s))$, $\overline{x} = x, \overline{s} = s$, then

$$\widetilde{\phi}(x,s,\psi^* F(q,x,s)) = G(\overline{q},\overline{x},\overline{s}) = G(\phi_1(q,x,s,\psi^* F(q,x,s)),x,s).$$

We define $\phi : (\mathbb{R} \times (\mathbb{R}^n \times \mathbb{R}), 0) \to (\mathbb{R} \times (\mathbb{R}^n \times \mathbb{R}), 0)$ by

$$\phi(t,x,s) = (\widetilde{\phi}(x,s,t), x, s).$$

Then the above equality means that

$$\phi(\psi^* F(q,x,s),x,s) = G(\phi_1(q,x,s,\psi^* F(q,x,s)),x,s).$$

Since $\widetilde{\Phi}(q,0,0,t) = (q,0,0,t)$, we have $\phi_1(q,0,0,\psi^* F(q,0,0)) = q$ and $\phi(t,0,0) = \widetilde{\phi}(0,0,t) = t$, so that F is homotopically $\mathcal{A}$-stable. This completes the proof. □

As a consequence of the above arguments, we have the following theorem.

Theorem 7.9. *Let $\mathcal{F}(q,x,t) = \lambda(q,x,t)(F(q,x) - t)$ be a graph-like Morse family of hyper surfaces. Then the following conditions are equivalent:*
(1) *The graph-like Legendrian unfolding $\mathscr{L}_{\mathcal{F}}(\Sigma_*(\mathcal{F}))$ is s-P-Legendrian stable,*

(2) $\overline{F}(q,x,t) = F(q,x) - t$ *is a* P-$\mathcal{K}$*-versal unfolding of* $\overline{f}(q) = \overline{F}(q,0) = F(q,0) - t$,
(3) $F(q,x)$ *is an* $\mathcal{A}$*-versal unfolding of* $f(q) = F(q,0)$.

Proof. By Theorem 4.2, (1) and (2) are equivalent. By Propositions 7.5, 7.7 and 7.8, (2) and (3) are equivalent. □

By the above arguments, $\mathcal{A}$-equivalence among function germs is an important notion for the study of s-P-Legendrian equivalence among graph-like Legendrian unfoldings. We consider geometric characterization for $\mathcal{A}$-equivalence among function germs. For function germs $f, g : (\mathbb{R}^k, 0) \to (\mathbb{R}, 0)$, we say that the level set foliation germs $\mathscr{F}_f$ and $\mathscr{F}_g$ are *diffeomorphic* if there exist diffeomorphism germs $\psi : (\mathbb{R}^k, 0) \to (\mathbb{R}^k, 0)$ and $\phi : (\mathbb{R}, 0) \to (\mathbb{R}, 0)$ such that $\psi(f^{-1}(c)) = g^{-1}(\phi(c))$ as a set germ for any $c \in (\mathbb{R}, 0)$. Then we have the following proposition.

Proposition 7.10. *For function germs* $f, g : (\mathbb{R}^k, 0) \to (\mathbb{R}, 0)$, *the level set foliation germs* $\mathscr{F}_f, \mathscr{F}_g$ *are diffeomorphic if and only if* f, g *are* $\mathcal{A}$*-equivalent.*

Proof. By definition, if f and g are $\mathcal{A}$-equivalent, then $\mathscr{F}_f$ and $\mathscr{F}_g$ are diffeomorphic. If $\mathscr{F}_f$ and $\mathscr{F}_g$ are strictly diffeomorphic, then there exist diffeomorphism germs $\psi : (\mathbb{R}^k, 0) \to (\mathbb{R}^k, 0)$ and $\phi : (\mathbb{R}, 0) \to (\mathbb{R}, 0)$ such that $\psi(f^{-1}(c)) = g^{-1}(\phi(c)) = (\phi^{-1} \circ g)^{-1}(c)$ as a set germ for any $c \in (\mathbb{R}, 0)$. This means that $\mathscr{F}_f$ and $\mathscr{F}_{\phi^{-1} \circ g}$ are strictly diffeomorphic. By Proposition 6.6, f and $\phi^{-1} \circ g$ are $\mathcal{R}$-equivalent, so that f and g are $\mathcal{A}$-equivalent. This completes the proof. □

For function germs $f : (\mathbb{R}^k, 0) \to (\mathbb{R}, 0)$ and $g : (\mathbb{R}^{k'}, 0) \to (\mathbb{R}, 0)$, we say that the level set foliation germs $\mathscr{F}_f$ and $\mathscr{F}_g$ are *stably diffeomorphic* if they become strictly diffeomorphic after the addition to the arguments q_i of new arguments q'_i and to functions f, g of non-degenerate quadratic forms. Then we have the following classification theorem.

Theorem 7.11. *Let* $\mathcal{F} : (\mathbb{R}^k \times \mathbb{R}^n \times \mathbb{R}, 0) \to (\mathbb{R}, 0)$ *and* $\mathcal{G} : (\mathbb{R}^{k'} \times \mathbb{R}^n \times \mathbb{R}, 0) \to (\mathbb{R}, 0)$ *be graph-like Morse families of hypersurfaces of the forms* $\mathcal{F}(q,x,t) = \lambda(q,x,t)(F(q,x) - t)$ *and* $\mathcal{G}(q',x,t) = \mu(q',x,t)(G(q',x) - t)$ *such that* $\mathscr{L}_{\mathcal{F}}(\Sigma_*(\mathcal{F}))$ *and* $\mathscr{L}_{\mathcal{G}}(\Sigma_*(\mathcal{G}))$ *are* s-P*-Legendrian stable. Then the following conditions are equivalent:*
(1) $\mathscr{L}_{\mathcal{F}}(\Sigma_*(\mathcal{F}))$ *and* $\mathscr{L}_{\mathcal{G}}(\Sigma_*(\mathcal{G}))$ *are* s-P*-Legendrian equivalent,*
(2) $\mathcal{F}$ *and* $\mathcal{G}$ *are stably* s-P-$\mathcal{K}$*-equivalent,*
(3) $\overline{f}(q,t) = F(q,0) - t$ *and* $\overline{g}(q',t) = G(q',0) - t$ *are stably* P-$\mathcal{K}$*-equivalent,*
(4) $f(q) = F(q,0)$ *and* $g(q') = G(q',0)$ *are stably* $\mathcal{A}$*-equivalent,*
(5) $F(q,x)$ *and* $G(q',x)$ *are stably* P-$\mathcal{A}$*-equivalent,*

(6) *$W(\mathscr{L}_{\mathcal{F}}(\Sigma_*(\mathcal{F})))$ and $W(\mathscr{L}_{\mathcal{G}}(\Sigma_*(\mathcal{G})))$ are s-P-diffeomorphic,*
(7) *$\mathscr{F}_f$ and $\mathscr{F}_g$ are stably diffeomorphic.*

Proof. By Theorem 4.2, (1) and (2) are equivalent. By Proposition 7.1, (2) and (5) are equivalent. By Corollary 7.2, (3) and (4) are equivalent. By Proposition 7.10, (4) and (7) are equivalent. By definition, (1) implies (6). By Proposition 4.1, (6) implies (1). By definition, (2) implies (3). By the uniqueness of P-$\mathcal{K}$-versal unfoldings, (3) implies (2). This completes the proof. □

Remark 7.12. (i) If $k = k'$ and $q = q'$ in the above theorem, we can remove the word "stably" in conditions (2), (3), (4), (5) and (7).
(ii) By Theorem 4.2 and Proposition 7.1, conditions (1), (2) and (5) are always equivalent without any assumptions.
(iii) By Corollary 7.2 and Proposition 7.10, conditions (3), (4) and (7) are equivalent without any assumptions.
(iv) By Proposition 4.1, conditions (1) and (6) are equivalent generically for an arbitrary dimension n without the assumption on s-P-Legendrian stability.

§8. Applications

In this section we explain some applications of the theory of wave front propagations. In [24] we explained some applications on both of general theory of wave front propagations and the theory of graph-like Legendrian unfoldings. Here, we only give two important cases that the notion of graph-like Legendrian unfoldings are essentially needed.

8.1. Stability of Caustics due to Jänich and Wassermann

Following Thom, Jänich [27] and Wassermann [36] considered the propagation of wave fronts on a manifold depending on the choice of a Hamiltonian on the contangent bundle of the manifold. Let $H : T^*\mathbb{R}^n \setminus 0 \to \mathbb{R}$ be a smooth function, which is called a *Hamiltonian function*, where 0 is the zero-section of $T^*\mathbb{R}^n$. We suppose H to be everywhere positive and positively homogeneous of degree one (i.e. $H(x, \lambda\xi) = \lambda H(x, \xi)$ of any $\lambda > 0$, $(x, \xi) \in T^*\mathbb{R}^n \setminus 0$). If we adopt the canonical coordinates $x_1, \dots, x_n, p_1, \dots p_n$ of $T^*\mathbb{R}^n \cong \mathbb{R}^n \times (\mathbb{R}^*)^n$, then $(x, \xi) = (x_1, \dots, x_n, p_1, \dots, p_n)$. We have a vector field X_H on $T^*\mathbb{R}^n \setminus 0$ associate to H defined by

$$X_H = \sum_{i=1}^{n} \left(\frac{\partial H}{\partial p_i} \frac{\partial}{\partial x_i} - \frac{\partial H}{\partial x_i} \frac{\partial}{\partial p_i} \right).$$

The vector field X_H is determined by the relation $\omega(X_H, Y) = dH(Y)$ for $Y \in TT^*\mathbb{R}^n$, where $\omega = \sum_{i=1}^n dp_i \wedge dx_i$. Since H is positive and positively homogeneous one, we have

$$\sum_{i=1}^n p_i \frac{\partial H}{\partial p_i} = H.$$

Therefore, for any point $(x, \xi) \in H^{-1}(1)$,

$$\sum_{i=1}^n p_i \frac{\partial H}{\partial p_i}(x, \xi) = H(x, \xi) = 1,$$

so that $\mathrm{grad}_p H(x, \xi) \neq \mathbf{0}$. This means that $H^{-1}(1)$ is a regular hypersurface in $T^*\mathbb{R}^n$. Since $dH(X_H) = \omega(X_H, X_H) = 0$, $X_H|_{H^{-1}(1)}$ is tangent to the hypersurface $H^{-1}(1)$. Let $\pi : T^*\mathbb{R}^n \to \mathbb{R}^n$ be the projection of the cotangent bundle. Since $\mathrm{grad}_p H(x, p) \neq \mathbf{0}$ on $H^{-1}(1)$, $\pi|_{H^{-1}(1)} : H^{-1}(1) \to \mathbb{R}^n$ is also a fibre bundle whose fibre is diffeomorphic to an $(n-1)$-sphere. The image under π of the flow lines of $X_H|_{H^{-1}(1)}$ are called the *ray* of H. The flow of the vector field $X_H|_{H^{-1}(1)}$ on $H^{-1}(1)$ induces a map $\rho : \mathbb{R}_+ \times H^{-1}(1) \to H^{-1}(1)$ on at least a neighborhood of $\{0\} \times H^{-1}(1) \subset \mathbb{R}_+ \times H^{-1}(1)$. Then we define a map $\exp : \mathbb{R}_+ \times H^{-1}(1) \to \mathbb{R}^n$ by $\exp = \pi \circ \rho$, which is also defined on at least a neighborhood of $\{0\} \times H^{-1}(1)$. Let V_0 be a co-oriented hypersurface in $\mathbb{R}^n$. We consider V_0 as an *initial wave front.* At any $x \in X_0$ the co-oriented tangent space of V_0 at x defines an element $\overline{\xi}(x) \in T_x^*\mathbb{R}^n \setminus 0$ such that $\mathrm{Ker}\, \overline{\xi}(x) = T_x V_0$ and the direction is compatible with the co-orientation of V_0 (i.e. the positive co-normal vector of V_0 at x). ince H is positive and positively homogeneous degree one, we have $H(x, \overline{\xi}(x)) = \eta(x) > 0$ and $H(x, \overline{\xi}(x)/\eta(x)) = H(x, \overline{\xi}(x))/\eta(x) = 1$. If we put $\xi(x) = \overline{\xi}(x)/\eta(x)$, then $H(x, \xi(x)) = 1$. Thus we have an $(n-1)$-dimensional submanifold $\ell(V_0) = \{(x, \xi(x)) \mid x \in V_0\} \subset H^{-1}(1)$. We now consider the Liouville form $\alpha = \sum_{i=1}^n p_i dx_i$ on $T^*\mathbb{R}^n$. Then $\omega = d\alpha$. Since $\mathrm{Ker}\, \xi(x) = T_x V_0$, we have $\alpha|_{\ell(V_0)} = \sum_{i=1}^n p_i dx_i|_{\ell(V_0)} = 0$, so that $\omega|_{\ell(V_0)} = 0$. This means that $\ell(V_0)$ is an isotropic submanifold of the symplectic structure ω. Moreover, if X_H is tangent to $\ell(V_0)$, then we have $0 = \sum_{i=1}^n p_i dx_i(X_H) = \sum_{i=1}^n p_i(\partial H/\partial p_i) = H(x, \xi(x)) = 1$. This is a contradiction, so that X_H is not tangent to $\ell(V_0)$. If $\exp(t, \xi(x))$ is defined for all $x \in V_0$ for some fixed $t > 0$, we call $V_t = \{\exp(t, \xi(x)) \mid x \in V_0\}$ the *wave front* at time t. We also call the canonical map $V_0 \to V_t$ defined by $x \mapsto \exp(t, \xi(x))$ the *ray map* at time t.

Remark 8.1. The restriction of the Liouville form α on $H^{-1}(1)$ defines a contact structure on $H^{-1}(1)$. Moreover, the projection $\pi|_{H^{-1}(1)}$:

$H^{-1}(1) \to \mathbb{R}^n$ is a Legendrian fibration. Since $\alpha|_{\ell(V_0)} = 0$, $\ell(V_0)$ is a Legendrian submanifold of $H^{-1}(1)$. By definition,

$$\ell(V_t) = \{\rho(t, (x, \xi(x))) \mid x \in V_0\}$$

is a Legendrian submanifold of $H^{-1}(1)$, so that $V_t = \pi(\ell(V_t))$ is the wave front in the sense of §2.

The singular value set of the ray map is called the *caustic points* at time t. We denote the set of caustic points at time t by C_t:

$$C_t = \{\exp(t, \xi(x)) \mid \operatorname{rank} d(\exp)_x\,(t, \xi(x)) < n - 1\}.$$

The set of all the caustic points at all time t in the time interval during which the propagation is considered is called the *caustic* of the propagation.

In [27, 36] Jänich and Wassermann investigated the stability of germs of such caustics. They described the caustics as bifurcation sets in the theory of unfoldings.

Let $t_0 \in \mathbb{R}_+$ and $\xi_0 \in H^{-1}(1)$ be given, such that $\exp(t_0, \xi_0)$ is defined, and such that (t_0, ξ_0) is a regular point of the map

$$(\pi, \exp) : \mathbb{R}_+ \times H^{-1}(1) \to \mathbb{R}^n \times \mathbb{R}^n; (t, \xi) \mapsto (\pi(\xi), \exp(t, \xi)).$$

We set $x_0 = \pi(\xi_0)$ and $u_0 = \exp(t_0, \xi_0)$. Under the above assumptions $(\pi, \exp)$ is a local diffeomorphism at (t_0, ξ_0). Therefore, there exists a local inverse $s : X \times U \to \mathbb{R}_+ \times H^{-1}(1)$ of $(\pi, \exp)$ such that $s(x_0, u_0) = (t_0, \xi_0)$, where X is a neighborhood of x_0 and U is a neighborhood of u_0 in $\mathbb{R}^n$ respectively. We define a function $\tau = \pi_{\mathbb{R}_+} \circ s : X \times U \to \mathbb{R}_+$, where $\pi_{\mathbb{R}_+} : \mathbb{R}_+ \times H^{-1}(1) \to \mathbb{R}_+$ is the canonical projection. τ is called a *ray length function* associated to (t_0, ξ_0). We remark that for given H the germ of τ at (x_0, u_0) depends only on (t_0, ξ_0), not on the choice of s We say that X and U are *sufficiently small* of $s(X \times U)$ which never contains both (t, ξ) and $(t, -\xi)$. Suppose that $x_0 \in V_0$ and $\xi(x_0) = \xi_0$. Given $\varepsilon > 0$, we say that V_0 and ε are *sufficiently small for* s if $(t_0 - \varepsilon, t_0 + \varepsilon) \times \ell(V_0) \subset s(X \times U)$. With these definitions and assumptions, Jänich has shown the following theorem.

Theorem 8.2 (Jänich [27]). *Let (t_0, ξ_0) be a regular point of $(\pi, \exp)$, let $s : X \times U \to \mathbb{R}_+ \times H^{-1}(1)$ be a local inverse to $(\pi, \exp)$ near (t_0, ξ_0), with associated ray-length function τ, and let $V_0 \subset \mathbb{R}^n$ be a normally oriented hypersurface such that $x_0 = \pi(\xi_0) \in V_0$ and $\xi(x_0) = \xi_0$. We define a function $F : V_0 \times U \to \mathbb{R}$ by $F = (\tau - t_0)|_{V_0 \times U}$. Suppose that X and U are sufficiently small, and let $\varepsilon > 0$ be given such that V_0 and*

ε are sufficiently small for s. Then for $t_0 - \varepsilon < t < t_0 + \varepsilon$ we have the following:
(a) $V_t = \{u \in U \mid \exists x \in V_0$ *with* $F(x,u) = t - t_0$ *and* $d_x F(x,u) = 0\}$ *and*
(b) $C_t = \{u \in U \mid \exists x \in V_0$ *with* $F(x,u) = t - t_0,\ d_x F(x,u) = 0$ *and* $d_x^2 F(x,u)$ *degenerate*$\}$,
where $d_x F$ is the differential of F with respect to the first variable (u fixed) and $d_x^2 F$ is the Hessian of F with respect to the first variable.

Remark 8.3. Since F can take values outside the interval $(-\varepsilon, \varepsilon)$, the *full caustic set* (briefly, *full caustic*) $C = \bigcup_{t_0-\varepsilon<t<t_0+\varepsilon} C_t$ is not necessarily be equal to the full bifurcation set

$$B_F = \{u \in U \mid \exists x \in V_0 \text{ with } d_x F(x,u) = 0 \text{ and } d_x^2 F(x,u) \text{ degenerate}\}$$

of F generally. However, for sufficiently small representative F' of the germ F, we have $B_{F'} \subset C \subset B_F$. Moreover, for sufficiently small representative of the germ of V_0 and sufficiently small time interval about t_0, the caustic C' satisfies $C' \subset B_{F'} \subset C$. Therefore, a knowledge of the germ F gives us all information about the local generation of the full caustic.

By assertion (a) in Theorem 8.2, for $t_0 - \varepsilon < t < t_0 + \varepsilon$, we have

$$\ell(V_t) = \{\pi_{H^{-1}(1)} \circ s(x,u) \in H^{-1}(1) \mid F(x,u) = t - t_0 \text{ and } d_x F(x,u) = 0\},$$

where $\pi_{H^{-1}(1)} : \mathbb{R}_+ \times H^{-1}(1) \to H^{-1}(1)$ is the canonical projection. Moreover,

$$L(V_0; (t_0, \xi_0), \varepsilon) = \bigcup_{t_0-\varepsilon<t<t_0+\varepsilon} \ell(V_t)$$

is a Lagrangian submanifold in $H^{-1}(1) \subset T^*\mathbb{R}^n$. Since

$$T_{\xi_0} L(V_0; (t_0, \xi_0), \varepsilon) = T_{\xi_0} \ell(V_t) \oplus \langle X_H \rangle_{\mathbb{R}},$$

$\operatorname{rank} d(\pi|_L)_{\xi_0} < n$ if and only if $\operatorname{rank} d(\pi|_{\ell(V_t)})_{\xi_0} = d(\exp)_{x_0} < n-1$, so that we have $C = C_{L(V_0;(t_0,\xi_0),\varepsilon)}$. Here $C_{L(V_0;(t_0,\xi_0),\varepsilon)}$ is the caustic of the Lagrangian submanifold $L(V_0; (t_0, \xi_0), \varepsilon)$ defined in §2. It follows that $\ell(V_t)$, $(t_0 - \varepsilon, t_0 + \varepsilon)$ is considered to be a momentary front of a graph-like Legendrian unfolding. We define a set germ $\mathscr{L}(V_0; (t_0, \xi_0), \varepsilon) \subset \mathbb{R}_+ \times H^{-1}(1)$ by

$$\{s(x,u) \mid t_0 - \varepsilon, t < t_0 + \varepsilon, F(x,u) = t - t_0 \text{ and } d_x F(x,u) = 0\},$$

which is a graph-like Legendrian unfolding with a generating family $\mathcal{F}(x,u,t) = F(x,u) - (t - t_0)$.

In [27] Jänich considered stability of the caustic in terms of germs of F and of τ. The caustic C is said to be an *$\mathcal{A}$-universal caustic* at time t_0 if (t_0, ξ_0) is a regular point of $(\pi, \exp)$ and if, for a ray-length function $\tau : X \times U \to \mathbb{R}$ associated to $(t_0, \xi(x_0))$, the germ at (x_0, u_0) of the function $F = (\tau - t_0)|_{(V_0 \cap X)\times U}$, considered as an unfolding of the germ of $F(x, u_0)$ at $x_0 \in V_0$ is infinitesimally $\mathcal{A}$-versal. In [27] Jänich said that C is *universal* if it is $\mathcal{A}$-universal. However, we can consider $\mathcal{R}^+$-equivalence instead of $\mathcal{A}$-equivalence here. We also say that C is an *$\mathcal{R}^+$-universal caustic* at time t_0 if the germ of $F(x, u_0)$ at $x_0 \in V_0$ is infinitesimally $\mathcal{R}^+$-versal.

Universality of the caustic is a very strong stability condition, for, roughly speaking, a universal caustic will survive small perturbations not only of the initial wave front but also of the Hamiltonian. However, Jänich conjectured that the stability of the caustic under perturbations of the initial wave front is sufficient to assure the universality of the caustic. In [36] Wassermann has shown that the conjecture of Jänich is true. He considered a general framework as follows: Let $\tau : (\mathbb{R}^m \times \mathbb{R}^n, 0) \to (\mathbb{R}, 0)$ be a function germ and let $\iota : (\mathbb{R}^k, 0) \to (\mathbb{R}^m, 0)$ be a map germ. We say that the pair (τ, ι) is *P-$\mathcal{A}$-stable under perturbations of ι* if the following holds: Given any open neighborhood X of $\mathbb{R}^n$, and any open neighborhood U of 0 in $\mathbb{R}^n$, and any representative $\tau' : X \times U \to \mathbb{R}$ of τ, and given any open neighborhood V of 0 in $\mathbb{R}^k$ and any representative $\iota' : V \to X$ of ι, there is a neighborhood N of ι' in $C^\infty(V, X)$ (in the weak C^∞-topology) such that for every $\kappa \in N$ there are a point $q_0 \in V$ and a point $x_0 \in U$ such that the germ of $\tau(\kappa(q), x)$ at (q_0, x_0) is P-$\mathcal{A}$-equivalent to the germ $\tau'(\iota'(q), x)$ at 0. Wassermann has shown the following theorem.

Theorem 8.4 ([36]). *Let $\tau : (\mathbb{R}^m \times \mathbb{R}^n, 0) \to (\mathbb{R}, 0)$ be a function germ such that $\tau|_{\mathbb{R}^m \times \{0\}}$ is submersive. Let $\iota : (\mathbb{R}^k, 0) \to (\mathbb{R}^m, 0)$ be a map germ and define $F : (\mathbb{R}^k \times \mathbb{R}^n, 0) \to (\mathbb{R}, 0)$ by $F(q, x) = \tau(\iota(q), x)$. Then (τ, ι) is P-$\mathcal{A}$-stable under perturbations of ι if and only if F is an infinitesimally $\mathcal{A}$-versal unfolding of $f(q) = F(q, 0)$.*

Remark 8.5. In [36] Wassermann said that the pair (τ, ι) is *r-stable for caustics* if it is P-$\mathcal{A}$-stable under perturbations of ι. However, we can change P-$\mathcal{A}$-equivalence to P-$\mathcal{R}^+$-equivalence. We say that the pair (τ, ι) is *P-$\mathcal{R}^+$-stable under perturbations of ι* if we change P-$\mathcal{A}$-equivalence to P-$\mathcal{R}^+$-equivalence in the above definition. By exactly the same arguments as in the proof of Theorem 8.4, we have the following theorem.

Theorem 8.6. *Let $\tau : (\mathbb{R}^m \times \mathbb{R}^n, 0) \to (\mathbb{R}, 0)$ be a function germ such that $\tau|\mathbb{R}^m \times \{0\}$ is submersive. Let $\iota : (\mathbb{R}^k, 0) \to (\mathbb{R}^m, 0)$ be a* *map germ and define* $F : (\mathbb{R}^k \times \mathbb{R}^n, 0) \to (\mathbb{R}, 0)$ *by* $F(q, x) = \tau(\iota(q), x)$. *Then* (τ, ι) *is* P-$\mathcal{R}^+$*-stable under perturbations of* ι *if and only if* F *is an infinitesimally* $\mathcal{R}^+$*-versal unfolding of* $f(q) = F(q, 0)$.

We now apply the definition in the above general framework to our situation. In this case ι will be the germ of an embedding. Since the space of embedding $\mathrm{Emb}\,(V, X)$ is an open subset of $C^\infty(V, X)$, we have no problems to change ι to be an embedding. Suppose that V_0 is a normally oriented hypersurface in $\mathbb{R}^n$, $x_0 \in V_0$ and $t_0 > 0$. Let $\xi_0 \in T^*_{x_0}\mathbb{R}^n$ be given by the normally oriented tangent space to V_0 at x_0. We say that V_0 at x_0 *produces an* $\mathcal{A}$*-stable caustics at* t_0 (respectively, *produces an* $\mathcal{R}^+$*-stable caustics at* t_0), if (t_0, ξ_0) is a regular point of $(\pi, \exp)$ and if for some ray length function $\tau : X \times U \to \mathbb{R}$ associated to (t_0, ξ_0), and for some open neighborhood V of 0 in $\mathbb{R}^n$ and some embedding $\iota : V \to X$ whose image is contained in V_0 and such that $\iota(0) = x_0$, and in some choice of coordinates near $x_0 \in X$ and near $u_0 = \exp\,(t_0, \xi_0) \in U$, the germ of the pair (τ, ι) at $((x_0, u_0), 0)$ is P-$\mathcal{A}$-stable (respectively, P-$\mathcal{R}^+$-stable) under perturbations of ι.

Remark 8.7. In [36] Wassermann said that V_0 *at* x_0 *produces a stable caustic at* t_0 if it produces an $\mathcal{A}$-stable caustic at time t_0. However, we also consider $\mathcal{R}^+$-equivalence, so that we have to distinguish these two cases.

In [36] the conjecture of Jänich [27] was solved affirmatively by Wassermann as a corollary of Theorem 8.4.

Theorem 8.8 ([36]). *With the same assumptions as the above paragraph,* V_0 *at* x_0 *produces an* $\mathcal{A}$*-universal caustic if and only if* V_0 *at* x_0 *produces an* $\mathcal{A}$*-stable caustic at* t_0.

Of course, we have the following theorem for $\mathcal{R}^+$-universal caustics as a corollary of Theorem 8.6.

Theorem 8.9. *With the same assumptions as the above paragraph,* V_0 *at* x_0 *produces an* $\mathcal{R}^+$*-universal caustic if and only if* V_0 *at* x_0 *produces an* $\mathcal{R}^+$*-stable caustic at* t_0.

Moreover, we say that V_0 at x_0 *produces a Lagrangian stable Lagrangian submanifold at time* t_0 if (t_0, ξ_0) is a regular point of $(\pi, \exp)$ and if for some ray length function $\tau : X \times U \to \mathbb{R}$ associated to (t_0, ξ_0), and for some open neighborhood V of 0 in $\mathbb{R}^n$ and some embedding $\iota : V \to X$ whose image is contained in V_0 and such that $\iota(0) = x_0$, and in some choice of coordinates near $x_0 \in X$ and near $u_0 = \exp\,(t_0, \xi_0) \in U$, there exists a neighborhood N of ι in $\mathrm{Emb}\,(V, X)$ such that for every $\kappa \in N$ there are a point $q_0 \in V$ and $x \in X$ such that the germ of the Lagrangian submanifold $L(\kappa(V);(t_0,\xi_0),\varepsilon)$ is Lagrangian equivalent to $L(\iota(V);(t_0,\xi_0),\varepsilon)$. We also say that V_0 at x_0 *produces an $S.P^+$-Legendrian stable graph-like Legendrian unfolding at t_0* (respectively, *produces an s-P-Legendrian stable graph-like Legendrian unfolding at t_0*) if (t_0,ξ_0) is a regular point of $(\pi,\exp)$ and if for some ray length function $\tau : X\times U\to\mathbb{R}$ associated to (t_0,ξ_0), and for some open neighborhood V of 0 in $\mathbb{R}^n$ and some embedding $\iota : V\to X$ whose image is contained in V_0 and such that $\iota(0)=x_0$, and in some choice of coordinates near $x_0\in X$ and near $u_0=\exp(t_0,\xi_0)\in U$, there exists a neighborhood N of ι in $\mathrm{Emb}(V,X)$ such that for every $\kappa\in N$ there are a point $q_0\in V$ and $x\in X$ such that the germ of the graph-like Legendrian unfolding $\mathscr{L}(\kappa(V);(t_0,\xi_0),\varepsilon)$ is $S.P^+$-Legendrian equivalent (respectively, s-P-Legendrian equivalent) to $\mathscr{L}(\iota(V);(t_0,\xi_0),\varepsilon)$. Then we have the following theorem as a corollary of Theorems 2.2, 4.2, Corollary 6.3 and Theorem 8.9.

Theorem 8.10. *With the same assumptions as the above paragraph, the following conditions are equivalent*:
(1) *V_0 at x_0 produces a Lagrangian stable Lagrangian submanifold at t_0*,
(2) *V_0 at x_0 produces an $\mathcal{R}^+$-stable caustic at t_0*,
(3) *V_0 at x_0 produces an $\mathcal{R}^+$-universal caustic*,
(4) *$L(V_0;(t_0,\xi_0),\varepsilon)$ is Lagrangian stable*,
(5) *$\mathscr{L}(V_0;(t_0,\xi_0),\varepsilon)$ is $S.P^+$-Legendrian stable*,
(6) *V_0 at x_0 produces an $S.P^+$-Legendrian stable graph-like Legendrian unfolding at t_0*.

On the other hand, Jänich and Wassermann considered $\mathcal{A}$-versality of unfoldings instead of $\mathcal{R}^+$-versality. It has been considered that there might be no corresponding geometric equivalence to P-$\mathcal{A}$-equivalence among generating families. However, from the view point of the theory of graph-like Legendrian unfoldings, we have the following theorem as a corollary of Theorems 4.2 and 8.8.

Theorem 8.11. *With the same assumptions as the above paragraph, the following conditions are equivalent*:
(1) *V_0 at x_0 produces an $\mathcal{A}$-stable caustic at t_0*,
(2) *V_0 at x_0 produces an $\mathcal{A}$-universal caustic*,
(3) *$\mathscr{L}(V_0;(t_0,\xi_0),\varepsilon)$ is s-P-Legendrian stable*,
(6) *V_0 at x_0 produces an s-P-Legendrian stable graph-like Legendrian unfolding at t_0*.

Example 8.12. As a special case, we have parallels of hypersurfaces in the Euclidean space. In this case we induce the metric on

$T_x^*\mathbb{R}^n$ by $\langle dx_i, dx_j\rangle = \delta_{ij}$, therefore $T_x^*\mathbb{R}^n$ can be canonically identified to the Euclidean n-space. Thus, for $(x,\xi) \in T^*\mathbb{R}^n$, we may regard that $\xi \in \mathbb{R}^n \cong T_x^*\mathbb{R}^n$ with the above identification. In this case we consider the Hamiltonian function $H : T^*\mathbb{R}^n \setminus 0 \to \mathbb{R}$ defined by $H(x,\xi) = \sqrt{\sum_{i=1}^n p_i^2} = \|\xi\|$ for the canonical coordinates $(x,\xi) = (x_1,\dots,x_n,p_1,\dots,p_n)$. It follows that

$$\frac{\partial H}{\partial x_i} = 0,\ \frac{\partial H}{\partial p_i} = \frac{p_i}{\sqrt{\sum_{i=1}^n p_i^2}},$$

so that the corresponding system of ODE for the Hamiltonian vector field is given by

$$(*) \quad \begin{cases} \dfrac{dx_i}{dt} = \dfrac{p_i}{\sqrt{\sum_{i=1}^n p_i^2}}, \\ \dfrac{dp_i}{dt} = 0. \end{cases}$$

For $(x,\xi) \in H^{-1}(1) \cong \mathbb{R}^n \times S^{n-1}$, we solve $(*)$ with the initial data $x(0) = x$ and $\xi(0) = \xi$. Then the solution is given by

$$x(t) = t\xi + x,\ \xi(t) = \xi,$$

so that the flow map $\rho : \mathbb{R}_+ \times H^{-1}(1) \to H^{-1}(1)$ is given by $\rho(t,(x,\xi)) = (t\xi + x, \xi)$. Therefore the exponential map is

$$\exp(t,(x,\xi)) = \pi \circ \rho(t,(x,\xi)) = t\xi + x.$$

Let V_0 be an initial front. We assume that V_0 is parametrized by an embedding $\iota : U \to \mathbb{R}^n$ such that $\iota(U) = V_0$, $\iota(\mathbf{0}) = x_0$, $\iota(u) = (x_1(u),\dots,x_n(u))$ and $u = (u_1,\dots,u_{n-1})$. Since V_0 is normally oriented, we have a unit normal vector field $\boldsymbol{n}(u)$ along V_0 in $\mathbb{R}^n$. Then we choose a one-form $(\iota(u), \xi(\iota(u))) \in T^*\mathbb{R}^n$ such that $\operatorname{Ker}\xi(\iota(u)) = T_{\iota(u)}V_0$ and $H(\iota(u), \xi(\iota(u))) = 1$. This means that $\xi(\iota(u)) = \pm\boldsymbol{n}(u)$, so that we choose $\xi(\iota(u)) = \boldsymbol{n}(u)$. For a fixed $t \in \mathbb{R}_+$, the ray map $V_0 \to V_t$ is $\iota(u) \mapsto \exp(t, \xi(\iota(u)) = \iota(u) + t\xi(\iota(u)) = \iota(u) + t\boldsymbol{n}(u)$. Thus we have

$$V_t = \{\iota(u) + t\boldsymbol{n}(u) \in \mathbb{R}^n \mid u \in U\},$$

which is called a *parallel* of V_0 in the classical differential geometry (cf. [17, 26]).

Then the map $(\pi, \exp) : \mathbb{R}_+ \times H^{-1}(1) \to \mathbb{R}^n \times \mathbb{R}^n$ is given by $(\pi,\exp)(t,(x,\xi)) = (x, x+t\xi)$, so that it is a diffeomorphism onto an open set W in $\mathbb{R}^n \times \mathbb{R}^n$. Therefore, we have the inverse mapping $s : W \to \mathbb{R}_+ \times H^{-1}(1)$ and the ray length function is $\tau(x,v) = \pi_{\mathbb{R}_+} \circ s(x,v)$. If we write $s(x,v) = (t,(x,\xi))$, then we have $v = x + t\xi$, so that $t = \|x - v\|$. This means that $\tau(x,v)=\|x-v\|$. Therefore we consider a *distance function* $F:U\times\mathbb{R}^n\to\mathbb{R}$ on V_0 defined by $F(u,v)=\|\iota(u)-v\|=\tau(\iota(u),v)$. If we consider the extended distance function $\widetilde{F}(u,v,t)=F(u,v)-t$, then the discriminant set

$$D_{\widetilde{F}}=\left\{(v,t)\ \middle|\ \exists u\in U \text{ s.t. } \widetilde{F}(u,v,t)=\frac{\partial\widetilde{F}}{\partial u_i}(u,v,t)=0, i=1,\dots,u_{n-1}\right\}$$

of $\widetilde{F}$ is the graph-like wave front whose momentary fronts are $\{V_t\}_{t\in\mathbb{R}_+}$.

On the other hand, we can use the distance squared function $D:U\times\mathbb{R}^n\to\mathbb{R}$ defined by $D(u,v)=\|\iota(u)-v\|^2=\langle\iota(u)-v,\iota(u)-v\rangle$ instead of the distance function F. In this case, calculations are rather easier than the case when we adopt the distance function F. Actually the (full) bifurcation set B_D is the caustic C and it is also called the *evolute* (or, the *focal set*) of V_0 (cf. [17]). For $n=2$, V_0 is a regular curve. In this case, V_0 is parametrized by an immersion $\boldsymbol{\gamma}:I\to\mathbb{R}^2$ from an open interval I such that $\boldsymbol{t}(s)=\boldsymbol{\gamma}'(s)$ is a unit vector. Then we have the Frenet frame $\{\boldsymbol{t}(s),\boldsymbol{n}(s)\}$ along the curve $\boldsymbol{\gamma}$, where $\boldsymbol{n}(s)$ is the unit normal vector defined by the anti-clockwise $\pi/2$-rotation of $\boldsymbol{t}(s)$. Then we have the Frenet formulae:

$$\begin{cases} \boldsymbol{t}'(s)= \kappa(s)\boldsymbol{n}(s),\\ \boldsymbol{n}'(s)=-\kappa(s)\boldsymbol{t}(s),\end{cases}$$

where $\kappa(s)=\langle\boldsymbol{t}'(s),\boldsymbol{n}(s)\rangle$ is the *curvature* of $\boldsymbol{\gamma}$. In this case the parallel is

$$V_t=\{\boldsymbol{\gamma}(s)+t\boldsymbol{n}(s)\mid s\in I\}$$

and the evolute of $\boldsymbol{\gamma}$ is

$$C=\left\{\boldsymbol{\gamma}(s)+\frac{1}{\kappa(s)}\boldsymbol{n}(s)\ \middle|\ \kappa(s)\neq 0\right\}.$$

The evolute of $\boldsymbol{\gamma}$ is known to be the bifurcation set of the diastase squared function $D:I\times\mathbb{R}^2\to\mathbb{R}$. In fact, we have

$$\frac{\partial D}{\partial s}(s,\boldsymbol{v})=2\langle\boldsymbol{t}(s),\boldsymbol{\gamma}(s)-\boldsymbol{v}\rangle,\ \frac{\partial^2 D}{\partial s^2}(s,\boldsymbol{v})=2(\langle\kappa(s)\boldsymbol{n}(s),\boldsymbol{\gamma}(s)-\boldsymbol{v}\rangle+1),$$

so that $B_D=C$. Here, the equation $\langle\boldsymbol{t}(s),\boldsymbol{\gamma}(s)-\boldsymbol{v}\rangle=0$ defines a normal line of $\boldsymbol{\gamma}$ at $\boldsymbol{\gamma}(s)$. Therefore, the evolute is the envelope of the normal lines along $\boldsymbol{\gamma}$. Moreover, the singular point of the evolute is $s\in I$ such that $\kappa'(s)=0$, which is called a *vertex* of $\boldsymbol{\gamma}$ in the classical differential geometry (cf. Fig.2 and Fig.3)

8.2. Caustics of world hyper-sheets in Lorentz-Minkowski space-time

In the Lorentz-Minkowski space-time, a world hyper-sheet is a timelike hypersurface formed by a one-parameter family of spacelike submanifolds. In the theory of relativity, we do not have the notion of time constant, so that everything that is moving depends on the time. Therefore, we consider world sheets. Although we have the notion of world sheets with general codimension, we stick to the case when the codimension one, that are called world hyper-sheets in the Lorentz-Minkowski space-time. Recently, Bousso and Randall introduced the notion of caustics of world hyper-sheets in order to define the notion of holographic domains in the space-time. Here, we give a mathematical framework for describing the caustics of world hyper-sheets in the Lorentz-Minkowski space-time.

We now introduce some basic notions on the $(n+1)$-dimensional Lorentz-Minkowski space-time. For basic concepts and properties, see [32]. Let $\mathbb{R}^{n+1} = \{(x_0, x_1, \dots, x_n) \mid x_i \in \mathbb{R}\ (i = 0, 1, \dots, n)\ \}$ be an $(n+1)$-dimensional cartesian space. For any $\boldsymbol{x} = (x_0, x_1, \dots, x_n)$, $\boldsymbol{y} = (y_0, y_1, \dots, y_n) \in \mathbb{R}^{n+1}$, the *pseudo scalar product* of $\boldsymbol{x}$ and $\boldsymbol{y}$ is defined to be $\langle \boldsymbol{x}, \boldsymbol{y} \rangle = -x_0 y_0 + \sum_{i=1}^{n} x_i y_i$. We call $(\mathbb{R}^{n+1}, \langle , \rangle)$ the $(n+1)$-*dimensional Minkowski space-time* (or briefly, the *Lorentz-Minkowski* $(n+1)$-*space*). We write $\mathbb{R}^{n+1}_1$ instead of $(\mathbb{R}^{n+1}, \langle , \rangle)$. We say that a non-zero vector $\boldsymbol{x} \in \mathbb{R}^{n+1}_1$ is *spacelike, lightlike or timelike* if $\langle \boldsymbol{x}, \boldsymbol{x} \rangle > 0$, $\langle \boldsymbol{x}, \boldsymbol{x} \rangle = 0$ or $\langle \boldsymbol{x}, \boldsymbol{x} \rangle < 0$, respectively. The norm of the vector $\boldsymbol{x} \in \mathbb{R}^{n+1}_1$ is defined to be $\|\boldsymbol{x}\| = \sqrt{|\langle \boldsymbol{x}, \boldsymbol{x} \rangle|}$. We have the canonical projection $\pi : \mathbb{R}^{n+1}_1 \to \mathbb{R}^n$ defined by $\pi(x_0, x_1, \dots, x_n) = (x_1, \dots, x_n)$. Here we identify $\{\mathbf{0}\} \times \mathbb{R}^n$ with $\mathbb{R}^n$ and it is considered as the Euclidean n-space whose scalar product is induced by the pseudo scalar product $\langle , \rangle$. For a vector $\boldsymbol{v} \in \mathbb{R}^{n+1}_1$ and a real number c, we define a *hyperplane with pseudo normal* $\boldsymbol{v}$ by

$$HP(\boldsymbol{v}, c) = \{\boldsymbol{x} \in \mathbb{R}^{n+1}_1 \mid \langle \boldsymbol{x}, \boldsymbol{v} \rangle = c\ \}.$$

We call $HP(\boldsymbol{v}, c)$ a *spacelike hyperplane*, a *timelike hyperplane* or a *lightlike hyperplane* if $\boldsymbol{v}$ is timelike, spacelike or lightlike, respectively. We now define

$$LC(\boldsymbol{\lambda}) = \{\boldsymbol{x} = (x_0, x_1, \dots, x_n) \in \mathbb{R}^{n+1}_1 \mid \langle \boldsymbol{x} - \boldsymbol{\lambda}, \boldsymbol{x} - \boldsymbol{\lambda} \rangle = 0\}$$

and we call it *the lightcone* with the vertex $\boldsymbol{\lambda} \in \mathbb{R}^{n+1}_1$. We write $LC^* = LC(\mathbf{0}) \setminus \{\mathbf{0}\}$, which is called an *open lightcone* at the origin.

For any $\boldsymbol{x}_1, \boldsymbol{x}_2, \ldots, \boldsymbol{x}_n \in \mathbb{R}^{n+1}_1$, we define a vector $\boldsymbol{x}_1 \wedge \boldsymbol{x}_2 \wedge \cdots \wedge \boldsymbol{x}_n$ by

$$\boldsymbol{x}_1 \wedge \boldsymbol{x}_2 \wedge \cdots \wedge \boldsymbol{x}_n = \begin{vmatrix} -\boldsymbol{e}_0 & \boldsymbol{e}_1 & \cdots & \boldsymbol{e}_n \\ x_0^1 & x_1^1 & \cdots & x_n^1 \\ x_0^2 & x_1^2 & \cdots & x_n^2 \\ \vdots & \vdots & \cdots & \vdots \\ x_0^n & x_1^n & \cdots & x_n^n \end{vmatrix},$$

where $\boldsymbol{e}_i = (0, \ldots, 0, \overset{i}{1}, 0, \ldots, 0)$ and $\boldsymbol{x}_i = (x_0^i, x_1^i, \ldots, x_n^i)$. We can easily show that $\boldsymbol{x}_1 \wedge \boldsymbol{x}_2 \wedge \cdots \wedge \boldsymbol{x}_n$ is pseudo orthogonal to any $\boldsymbol{x}_i$ $(i = 1, \ldots, n)$.

We briefly review the basic geometrical framework for the study of world hyper-sheets in the $(n+1)$-dimensional Lorentz-Minkowski spacetime in [22]. Let $\mathbb{R}^{n+1}_1$ be a time-oriented space (cf. [32]). We choose $\boldsymbol{e}_0 = (1, 0, \ldots, 0)$ as the future timelike vector field. A world hypersheet is defined to be a timelike hypersurface foliated by codimension one spacelike submanifolds. Here, we only investigate the local situation, so that we consider a one-parameter family of spacelike submanifolds. Let $\boldsymbol{X} : U \times I \to \mathbb{R}^{n+1}_1$ be a timelike embedding, where $U \subset \mathbb{R}^{n-1}$ is an open subset and I is an open interval. We write $W = \boldsymbol{X}(U \times I)$ and identify W and $U \times I$ through the embedding $\boldsymbol{X}$. The embedding $\boldsymbol{X}$ is said to be *timelike* if the tangent space T_pW of W is a timelike hyperplane at any point $p \in W$. We write that $\mathcal{S}_t = \boldsymbol{X}(U \times \{t\})$ for each $t \in I$. We have a foliation of W defined by $\mathcal{S} = \{\mathcal{S}_t\}_{t \in I}$. We say that $W = \boldsymbol{X}(U \times I)$ (or, $(W, \mathcal{S})$) is a *world hyper-sheet* if W is a time-orientable timelike hypersurface and each $\mathcal{S}_t$ is spacelike. Here, we say that $\mathcal{S}_t$ is *spacelike* if the tangent space $T_p\mathcal{S}_t$ consists only spacelike vectors (i.e. spacelike subspace of $\mathbb{R}^{n+1}_1$) for any point $p \in \mathcal{S}_t$. Each $\mathcal{S}_t$ is called a *momentary space* of W. For any $p = \boldsymbol{X}(\overline{u}, t) \in W \subset \mathbb{R}^{n+1}_1$, we have

$$T_pW = \langle \boldsymbol{X}_{u_1}(\overline{u}, t), \ldots, \boldsymbol{X}_{u_{n-1}}(\overline{u}, t), \boldsymbol{X}_t(\overline{u}, t) \rangle_{\mathbb{R}},$$

where we write $(\overline{u}, t) = (u_1, \ldots, u_{n-1}, t) \in U \times I$, $\boldsymbol{X}_t = \partial \boldsymbol{X}/\partial t$ and $\boldsymbol{X}_{u_j} = \partial \boldsymbol{X}/\partial u_j$. We also have

$$T_p\mathcal{S}_t = \langle \boldsymbol{X}_{u_1}(\overline{u}, t), \ldots, \boldsymbol{X}_{u_{n-1}}(\overline{u}, t) \rangle_{\mathbb{R}}.$$

Since W is time-orientable, there exists a timelike vector field $\boldsymbol{v}(\overline{u}, t)$ on W [32, Lemma 32] Moreover, we can choose that $\boldsymbol{v}$ is *future directed* which means that $\langle \boldsymbol{v}(\overline{u}, t), \boldsymbol{e}_0 \rangle < 0$. Since $\operatorname{codim} W = 1$, we have $\operatorname{codim} \mathcal{S}_t = 2$. Moreover, $\mathcal{S}_t$ is spacelike, so that we can apply the method developed in [18]. We consider the unit normal spacelike vector of W

defined by

$$\boldsymbol{n}^S(\overline{u},t) = \frac{\boldsymbol{X}_{u_1}(\overline{u},t)\wedge\cdots\wedge\boldsymbol{X}_{u_{n-1}}(\overline{u},t)\wedge\boldsymbol{X}_t(\overline{u},t)}{\|\boldsymbol{X}_{u_1}(\overline{u},t)\wedge\cdots\wedge\boldsymbol{X}_{u_{n-1}}(\overline{u},t)\wedge\boldsymbol{X}_t(\overline{u},t)\|}.$$

For any $t\in I$, Let $N_p(\mathcal{S}_t)$ be the pseudo-normal space of $\mathcal{S}_t$ at $p=\boldsymbol{X}(\overline{u},t)$ in $\mathbb{R}^{n+1}_1$. Since $\mathcal{S}_t$ is a codimension one in W, $N_p(\mathcal{S}_t)$ is a two dimensional Lorentz space. There exists a unique timelike unit vector field $\boldsymbol{n}^T(\overline{u},t)\in N_p(\mathcal{S}_t)\cap T_pW$ such that it is future directed (i.e. $\langle\boldsymbol{n}^T(\overline{u},t),\boldsymbol{e}_0\rangle<0$). We now define maps $\mathbb{LG}^{\pm}(\mathcal{S}_t):\mathcal{S}_t\to LC^*$ by $\mathbb{LG}^{\pm}(\mathcal{S}_t)(p)=\boldsymbol{n}^T(\overline{u},t)\pm\boldsymbol{n}^S(\overline{u},t)$, where $p=\boldsymbol{X}(\overline{u},t)$. We call each one of $\mathbb{LG}^{\pm}(\mathcal{S}_t)$ a *momentary lightcone Gauss map.* These maps lead us to the notion of curvatures (cf. [22]). We have linear maps $d\mathbb{LG}^{\pm}(\mathcal{S}_t)_p:T_p\mathcal{S}_t\to T_{\widetilde{p}}LC^*\subset T_{\widetilde{p}}\mathbb{R}^{n+1}_1$, where $p=\boldsymbol{X}(\overline{u},t)$ and $\widetilde{p}=\boldsymbol{n}^T(\overline{u},t)\pm\boldsymbol{n}^S(\overline{u},t)$. With the identification $T_{\widetilde{p}}\mathbb{R}^{n+1}_1\equiv\mathbb{R}^{n+1}_1\equiv T_p\mathbb{R}^{n+1}_1$, we have the canonical decomposition $T_p\mathbb{R}^{n+1}=T_p\mathcal{S}_t\oplus N_p(\mathcal{S}_t)$. Let $\Pi^t:T_p\mathbb{R}^{n+1}=T_p\mathcal{S}_t\oplus N_p(\mathcal{S}_t)\to T_p\mathcal{S}_t$ be the canonical projection. Then we have linear transformations

$$S^{\pm}_{\ell}(\mathcal{S}_t)_p=-\Pi^t\circ d\mathbb{LG}^{\pm}(\mathcal{S}_t)_p:T_p\mathcal{S}_t\to T_p\mathcal{S}_t.$$

Each one of the above mappings is called a *momentary lightcone shape operator* of $\mathcal{S}_t$ at $p=\boldsymbol{X}(\overline{u},t)$. Let $\{\kappa^{\pm}_i(\mathcal{S}_t)(p)\}_{i=1}^{n-1}$ be the set of eigenvalues of $S^{\pm}_{\ell}(\mathcal{S}_t)_p$, which are called *momentary lightcone principal curvatures* of $\mathcal{S}_t$ at $p=\boldsymbol{X}(\overline{u},t)$. Then *momentary lightcone Gauss-Kronecker curvatures* of $\mathcal{S}_t$ at $p=\boldsymbol{X}(\overline{u},t)$ are defined to be

$$K^{\pm}_{\ell}(\mathcal{S}_t)(p)=\det S^{\pm}_{\ell}(\mathcal{S}_t)_p.$$

We obtain now the lightcone Weingarten formulae. Since $\mathcal{S}_t$ is a spacelike submanifold, we have a Riemannian metric (the *first fundamental form*) on $\mathcal{S}_t$ defined by $ds^2=\sum_{i=1}^{n-1}g_{ij}du_idu_j$, where $g_{ij}(\overline{u},t)=\langle\boldsymbol{X}_{u_i}(\overline{u},t),\boldsymbol{X}_{u_j}(\overline{u},t)\rangle$ for any $(\overline{u},t)\in U\times I$. *Lightcone second fundamental invariants* are defined to be

$$h_{ij}[\pm](\overline{u},t)=\langle-(\boldsymbol{n}^T\pm\boldsymbol{n}^S)_{u_i}(\overline{u},t),\boldsymbol{X}_{u_j}(\overline{u},t)\rangle$$

for any $(\overline{u},t)\in U\times I$. The following lightcone Weingarten formulae are given as special cases of the formulae in [18]:

(a) $(\boldsymbol{n}^T\pm\boldsymbol{n}^S)_{u_i}=\langle\boldsymbol{n}^S,\boldsymbol{n}^T_{u_i}\rangle(\boldsymbol{n}^T\pm\boldsymbol{n}^S)-\sum_{j=1}^{n-1}h^j_i[\pm]\boldsymbol{X}_{u_j}$

(b) $\Pi^t\circ(\boldsymbol{n}^T+\boldsymbol{n}^S)_{u_i}=-\sum_{j=1}^{n-1}h^j_i[\pm]\boldsymbol{X}_{u_j}$.

Here $\left(h^j_i[\pm]\right)=(h_{ik}[\pm])\left(g^{kj}\right)$ and $\left(g^{kj}\right)=(g_{kj})^{-1}$.

It follows that the momentary lightcone principal curvatures are the eigenvalues of $\left(h_i^j[\pm]\right)$.

On the other hand, as an application of the theory of Legendrian unfoldings, a geometric framework for the study of caustics of world hyper-sheets in the Lorentz-Minkowski space-time has been constructed in [23]. We give a brief survey here. We define hypersurfaces $\mathbb{LH}^{\pm}_{\mathcal{S}_{t_0}} : U \times \{t_0\} \times \mathbb{R} \to \mathbb{R}^{n+1}_1$ by

$$\mathbb{LH}^{\pm}_{\mathcal{S}_{t_0}}(p,\mu) = \mathbb{LH}^{\pm}_{\mathcal{S}_{t_0}}(\overline{u}, t_0, \mu) = \boldsymbol{X}(\overline{u}, t_0) + \mu\mathbb{LG}^{\pm}(\mathcal{S}_{t_0})(\overline{u}, t_0),$$

where $p = \boldsymbol{X}(\overline{u}, t_0)$. We call $\mathbb{LH}^{\pm}_{\mathcal{S}_{t_0}}$ *light sheets* along $\mathcal{S}_{t_0}$. A hypersurface $H \subset \mathbb{R}^{n+1}_1$ is, generally, called a *lightlike hypersurface* if it is tangent to a lightcone at any point. The light sheet along $\mathcal{S}_{t_0}$ is a lightlike hypersurface. We also define $\mathbb{LH}^{\pm}_W : U \times I \times \mathbb{R} \to \mathbb{R}^{n+1}_1 \times I$ by

$$\mathbb{LH}^{\pm}_W(\overline{u}, t, \mu) = (\mathbb{LH}^{\pm}_{\mathcal{S}_t}(\overline{u}, t, \mu), t),$$

which are called *unfolded light sheets* of $(W, \mathcal{S})$.

We introduce the notion of Lorentz distance-squared functions on a world hyper-sheet, which is useful for the study of singularities of light sheets. We define a family of functions $G : W \times \mathbb{R}^{n+1}_1 \to \mathbb{R}$ on $W = \boldsymbol{X}(U \times I)$ by

$$G(p, \boldsymbol{\lambda}) = G(\overline{u}, t, \boldsymbol{\lambda}) = \langle \boldsymbol{X}(\overline{u}, t) - \boldsymbol{\lambda}, \boldsymbol{X}(\overline{u}, t) - \boldsymbol{\lambda}\rangle,$$

where $p = \boldsymbol{X}(\overline{u}, t)$. We call G a *Lorentz distance-squared function* on the world hyper-sheet $(W, \mathcal{S})$. For any fixed $(t_0, \boldsymbol{\lambda}_0) \in I \times \mathbb{R}^{n+1}_1$, we write $g(\overline{u}) = G_{(t_0,\boldsymbol{\lambda}_0)}(\overline{u}) = G(\overline{u}, t_0, \boldsymbol{\lambda}_0)$ and have the following proposition.

Proposition 8.13 ([23]). *Let $\mathcal{S}_{t_0}$ be a momentary space of $(W, \mathcal{S})$ and $G : W \times \mathbb{R}^{n+1}_1 \to \mathbb{R}$ the Lorentz distance-squared function on $(W, \mathcal{S})$. Suppose that $p_0 = \boldsymbol{X}(\overline{u}_0, t_0) \neq \boldsymbol{\lambda}_0$. Then we have the following:*
(1) $g(\overline{u}_0) = \partial g/\partial u_i(\overline{u}_0) = 0$ $(i = 1, \dots, n-1)$ if and only if $p_0 - \boldsymbol{\lambda}_0 = \mu\mathbb{LG}^{\pm}(\mathcal{S}_{t_0})(p_0)$ for some $\mu \in \mathbb{R} \setminus \{0\}$.
(2) $g(\overline{u}_0) = \partial g/\partial u_i(\overline{u}_0) = \det \mathcal{H}(g)(\overline{u}_0) = 0$ $(i = 1, \dots, n-1)$ if and only if

$$p_0 - \boldsymbol{\lambda}_0 = \mu\mathbb{LG}^{\pm}(\mathcal{S}_{t_0})(p_0)$$

for $\mu \in \mathbb{R} \setminus \{0\}$ such that $-1/\mu$ is one of the non-zero momentary lightcone principal curvatures $\{\kappa_i^{\pm}(\mathcal{S}_t)(p)\}_{i=1}^{n-1}$.
Here, $\det \mathcal{H}(g)(\overline{u}_0)$ *is the determinant of the Hessian matrix of g at $\overline{u}_0$.*

Inspired by the above result, we define a set $\mathbb{LF}^{\pm}_{\mathcal{S}_{t_0}}$ by

$$\bigcup_{i=1}^{n-1}\left\{\boldsymbol{X}(u,t_0)+\frac{1}{\kappa_i^{\pm}(\mathcal{S}_t)(p)}\mathbb{LG}^{\pm}(\mathcal{S}_{t_0})(p)\mid u\in U, p=\boldsymbol{X}(u,t_0)\right\},$$

which are called *lightlike focal sets* of $\mathcal{S}_{t_0}$. Moreover, *unfolded lightcone focal sets* of $(W,\mathcal{S})$ are defined to be

$$\mathbb{LF}^{\pm}_{(W,\mathcal{S})}=\bigcup_{t\in I}\mathbb{LF}^{\pm}_{\mathcal{S}_t}\times\{t\}\subset\mathbb{R}^{n+1}_1\times I.$$

Each one of $\mathbb{LF}^{\pm}_{(W,\mathcal{S})}$ is the critical value set of $\mathbb{LH}^{\pm}_W$, respectively.

We consider the relationship between the contact of a one parameter family of submanifolds with a submanifold and $S.P$-$\mathcal{K}$-equivalence among functions (cf. [13]). Let $U_i\subset\mathbb{R}^r$, $(i=1,2)$ be open sets and $g_i:(U_i\times I,(\overline{u}_i,t_i))\to(\mathbb{R}^n,\boldsymbol{y}_i)$ immersion germs. We define $\overline{g}_i:(U_i\times I,(\overline{u}_i,t_i))\to(\mathbb{R}^n\times I,(\boldsymbol{y}_i,t_i))$ by $\overline{g}_i(\overline{u},t)=(g_i(\overline{u}),t)$. We write that $(\overline{Y}_i,(\boldsymbol{y}_i,t_i))=(\overline{g}_i(U_i\times I),(\boldsymbol{y}_i,t_i))$. Let $f_i:(\mathbb{R}^n,\boldsymbol{y}_i)\to(\mathbb{R},0)$ be submersion germs and write that $(V(f_i),\boldsymbol{y}_i)=(f_i^{-1}(0),\boldsymbol{y}_i)$. We say that *the contact of* $\overline{Y}_1$ *with the trivial family of* $V(f_1)$ *at* $(\boldsymbol{y}_1,t_1)$ is of the *same type in the strict sense* as *the contact of* $\overline{Y}_2$ *with the trivial family of* $V(f_2)$ *at* $(\boldsymbol{y}_2,t_2)$ if there is a diffeomorphism germ $\Phi:(\mathbb{R}^n\times I,(\boldsymbol{y}_1,t_1))\to(\mathbb{R}^n\times I,(\boldsymbol{y}_2,t_2))$ of the form $\Phi(\boldsymbol{y},t)=(\phi_1(\boldsymbol{y},t),t+(t_2-t_1))$ such that $\Phi(\overline{Y}_1)=\overline{Y}_2$ and $\Phi(V(f_1)\times I)=V(f_2)\times I$. In this case we write $SK(\overline{Y}_1,V(f_1)\times I;(\boldsymbol{y}_1,t_1))=SK(\overline{Y}_2,V(f_2)\times I;(\boldsymbol{y}_2,t_2))$. In [23] we claimed that the following proposition holds which is analogous to Montaldi's theorem on contact between submanifolds in [31]:

Proposition 8.14. *With the same notations as in the above paragraphs,*

$$SK(\overline{Y}_1,V(f_1)\times I;(\boldsymbol{y}_1,t_1))=SK(\overline{Y}_2,V(f_2)\times I;(\boldsymbol{y}_2,t_2))$$

if and only if $f_1\circ g_1$ *and* $f_2\circ g_2$ *are* $S.P$-$\mathcal{K}$*-equivalent* [*i.e. there exists a diffeomorphism germ* $\Psi:(U_1\times I,(\overline{u}_1,t_1))\to(U_2\times I,(\overline{u}_2,t_2))$ *of the form* $\Psi(\overline{u},t)=(\psi_1(\overline{u},t),t-(t_2-t_1))$ *and a function germ* $\lambda:(U_1\times I,(\overline{u}_1,t_1))\to\mathbb{R}$ *with* $\lambda(\overline{u}_1,t_1)\neq 0$ *such that* $(f_2\circ g_2)\circ\Psi(\overline{u},t)=\lambda(\overline{u},t)f_1\circ g_1(\overline{u},t)$].

On the other hand, we also consider a little weaker version of the above definition of the contact of a one parameter family of submanifolds with a submanifold. We say that *the contact of* $\overline{Y}_1$ *with the trivial family of* $V(f_1)$ *at* $(\boldsymbol{y}_1,t_1)$ is of the *same type* as *the contact of* $\overline{Y}_2$ *with the trivial family of* $V(f_2)$ *at* $(\boldsymbol{y}_2,t_2)$ if there is a diffeomorphism germ $\Phi : (\mathbb{R}^n \times I, (\boldsymbol{y}_1, t_1)) \to (\mathbb{R}^n \times I, (\boldsymbol{y}_2, t_2))$ of the form $\Phi(\boldsymbol{y}, t) = (\phi_1(\boldsymbol{y}, t), \phi_2(t))$ such that $\Phi(\overline{Y}_1) = \overline{Y}_2$ and $\Phi(V(f_1) \times I) = V(f_2) \times I$. In this case we write $PK(\overline{Y}_1, V(f_1) \times I; (\boldsymbol{y}_1, t_1)) = PK(\overline{Y}_2, V(f_2) \times I; (\boldsymbol{y}_2, t_2))$. We also claim that the following proposition holds which is analogous to Montaldi's theorem [31] and we omit to give the proof here.

Proposition 8.15. *With the same notations as in the above paragraphs,*

$$PK(\overline{Y}_1, V(f_1) \times I; (\boldsymbol{y}_1, t_1)) = PK(\overline{Y}_2, V(f_2) \times I; (\boldsymbol{y}_2, t_2))$$

if and only if $f_1 \circ g_1$ and $f_2 \circ g_2$ are P-$\mathcal{K}$-equivalent [i.e. there exists a diffeomorphism germ $\Psi : (U_1 \times I, (\overline{u}_1, t_1)) \to (U_2 \times I, (\overline{u}_2, t_2))$ of the form $\Psi(\overline{u}, t) = (\psi_1(\overline{u}, t), \psi_2(t))$ and a function germ $\lambda : (U_1 \times I, (\overline{u}_1, t_1)) \to \mathbb{R}$ with $\lambda(\overline{u}_1, t_1) \neq 0$ such that $(f_2 \circ g_2) \circ \Psi(\overline{u}, t) = \lambda(\overline{u}, t) f_1 \circ g_1(\overline{u}, t)$].

We now consider a function $\mathfrak{g}_{\boldsymbol{\lambda}} : \mathbb{R}^{n+1}_1 \to \mathbb{R}$ defined by $\mathfrak{g}_{\boldsymbol{\lambda}}(\boldsymbol{x}) = \langle \boldsymbol{x} - \boldsymbol{\lambda}, \boldsymbol{x} - \boldsymbol{\lambda} \rangle$, where $\boldsymbol{\lambda} \in \mathbb{R}^{n+1}_1 \setminus W$. For any $\boldsymbol{\lambda}_0 \in \mathbb{R}^{n+1}_1$, we have a lightcone $\mathfrak{g}^{-1}_{\boldsymbol{\lambda}_0}(0) = LC(\boldsymbol{\lambda}_0)$. Moreover, we consider the lightlike vectors $\boldsymbol{\lambda}^{\pm}_0 = \mathbb{LH}^{\pm}_{\mathcal{S}_{t_0}}(p_0, \mu_0)$, where $p_0 = \boldsymbol{X}(\overline{u}_0, t_0)$. Then we have

$$\mathfrak{g}_{\boldsymbol{\lambda}^{\pm}_0} \circ \boldsymbol{X}(\overline{u}_0, t_0) = G((u_0, t_0), \mathbb{LH}^{\pm}_{\mathcal{S}_{t_0}}(p_0, \mu_0)) = 0.$$

By Proposition 8.13, we also have relations that

$$\frac{\partial \mathfrak{g}_{\boldsymbol{\lambda}^{\pm}_0} \circ \boldsymbol{X}}{\partial u_i}(\overline{u}_0, t_0) = \frac{\partial G}{\partial u_i}((u_0, t_0), \mathbb{LH}^{\pm}_{\mathcal{S}_{t_0}}(p_0, \mu_0)) = 0.$$

for $i = 1, \dots, n-1$. These relations mean that the lightcones $\mathfrak{g}^{-1}_{\boldsymbol{\lambda}^{\pm}_0}(0) = LC(\boldsymbol{\lambda}^{\pm}_0)$ are tangent to $\mathcal{S}_{t_0} = \boldsymbol{X}(U \times \{t_0\})$ at $p_0 = \boldsymbol{X}(\overline{u}_0, t_0)$. Each one of the lightcones $LC(\boldsymbol{\lambda}^{\pm}_0)$ is said to be a *tangent lightcone* of $\mathcal{S}_{t_0} = \boldsymbol{X}(U \times \{t_0\})$ at $p_0 = \boldsymbol{X}(\overline{u}_0, t_0)$, which we write $TLC(\mathcal{S}_{t_0}, \boldsymbol{\lambda}^{\pm}_0)$, where $\boldsymbol{\lambda}^{\pm}_0 = \mathbb{LH}^{\pm}_{\mathcal{S}_{t_0}}(p_0, \mu_0)$. Then we have the following simple lemma.

Lemma 8.16. *Let $\boldsymbol{X} : U \times I \to \mathbb{R}^{n+1}_1$ be a world hyper-sheet. Consider two points $p_i = \boldsymbol{X}(\overline{u}_i, t_0)$, $(i = 1, 2)$. Then*

$$\mathbb{LH}^{\pm}_{\mathcal{S}_{t_0}}(p_1, \mu_1) = \mathbb{LH}^{\pm}_{\mathcal{S}_{t_0}}(p_2, \mu_2)$$

if and only if

$$TLC(\mathcal{S}_{t_0}, \mathbb{LH}^{\pm}_{\mathcal{S}_{t_0}}(p_1, \mu_1)) = TLP(\mathcal{S}_{t_0}, \mathbb{LH}^{\pm}_{\mathcal{S}_{t_0}}(p_2, \mu_2)).$$

As a consequence, we have tools for the study of the contact between momentary spaces and families of lightcones. We write that $g_{\boldsymbol{\lambda}}(\overline{u},t) = G(\overline{u},t,\boldsymbol{\lambda})$. Then we have $g_{\boldsymbol{\lambda}}(\overline{u},t) = \mathfrak{g}_{\boldsymbol{\lambda}} \circ \boldsymbol{X}(\overline{u},t)$, so that we have the following proposition as a corollary of Proposition 8.14.

Proposition 8.17. *Let* $\boldsymbol{X}_i : (U \times I, (\overline{u}_i, t_0)) \to (\mathbb{R}^{n+1}_1, p_i)$, $(i = 1,2)$, *be world hypersheet germs and* $\boldsymbol{\lambda}_i^{\pm} = \mathbb{LH}^{\pm}_{\mathcal{S}_{t_0}}(p_i, \mu_i)$ *and* $W_i = \boldsymbol{X}_i(U \times I)$. *Then the following conditions are equivalent*:
(1) $SK(\overline{W}_1, TLC(\mathcal{S}_{t_0}, \boldsymbol{\lambda}_1^{\pm}) \times I; (p_1, t_0))$
$= SK(\overline{W}_2, TLC(\mathcal{S}_{t_0}, \boldsymbol{\lambda}_2^{\pm}) \times I; (p_2, t_0))$,
(2) $g_{1,\boldsymbol{\lambda}_1^{\pm}}$ *and* $g_{2,\boldsymbol{\lambda}_2^{\pm}}$ *are S.P-$\mathcal{K}$-equivalent.*
Here, $g_{i,\boldsymbol{\lambda}_i^{\pm}}(\overline{u},t) = G_i(\overline{u},t,\boldsymbol{\lambda}_i^{\pm}) = \langle \boldsymbol{X}_i(\overline{u},t) - \boldsymbol{\lambda}_i^{\pm}, \boldsymbol{X}_i(\overline{u},t) - \boldsymbol{\lambda}_i^{\pm} \rangle$, $(i = 1,2)$.

We also have the following proposition as a corollary of Proposition 8.15.

Proposition 8.18. *With the same notations as those in Proposition 8.17, the following conditions are equivalent*:
(1) $PK(\overline{W}_1, TLC(\mathcal{S}_{t_0}, \boldsymbol{\lambda}_1^{\pm}) \times I; (p_1, t_0))$
$= PK(\overline{W}_2, TLC(\mathcal{S}_{t_0}, \boldsymbol{\lambda}_2^{\pm}) \times I; (p_2, t_0))$,
(2) $g_{1,\boldsymbol{\lambda}_1^{\pm}}$ *and* $g_{2,\boldsymbol{\lambda}_2^{\pm}}$ *are P-$\mathcal{K}$-equivalent.*

We can investigate unfolded lightcone focal sets of world hypersheets as an application of the theory of graph-like Legendrian unfoldings. We have shown the following key-proposition in [23].

Proposition 8.19 ([23])**.** *Let* $G : U \times I \times (\mathbb{R}^{n+1}_1 \setminus W) \to \mathbb{R}$ *be a Lorentz distance-squared function on a world hyper-sheet* $(W, \mathcal{S})$. *For any point* $(\overline{u}_0, t_0, \boldsymbol{\lambda}_0) \in \Sigma_*(G)$, G *is a non-degenerate graph-like Morse family of hypersurfaces around* $(\overline{u}_0, t_0, \boldsymbol{\lambda}_0)$.

By Proposition 8.13, we have

$$\Sigma_*(G) = \{(\overline{u}, t, \mathbb{LH}^{\pm}_{\mathcal{S}_t}(p,\mu)) \in U \times I \times \mathbb{R}^{n+1}_1 \mid p = \boldsymbol{X}(\overline{u},t), \mu \in \mathbb{R} \setminus \{0\}\}.$$

We define a map $\mathscr{L}_G : \Sigma_*(G) \to J^1(\mathbb{R}^{n+1}_1, I)$ by

$$\mathscr{L}_G(\overline{u}, t, \mathbb{LH}^{\pm}_{\mathcal{S}_t}(p,\mu)) = \left(\mathbb{LH}^{\pm}_{\mathcal{S}_t}(p,\mu), t, \frac{2}{\langle \boldsymbol{X}_t(\overline{u},t), \boldsymbol{n}^T(\overline{u},t)\rangle} \overline{\mathbb{LG}^{\pm}(\mathcal{S}_t)(\overline{u},t)} \right),$$

where we define $\overline{\boldsymbol{x}} = (-x_0, x_1, \ldots x_n)$ for $\boldsymbol{x} = (x_0, x_1, \ldots, x_n) \in \mathbb{R}^{n+1}_1$. By the construction of the graph-like Legendrian unfolding from a graph-like Morse family of hypersurfaces, $\mathscr{L}_G(\Sigma_*(G))$ is a graph-like Legendrian unfolding in $J^1(\mathbb{R}^{n+1}_1, I)$. Therefore, the graph-like wave front

$W(\mathscr{L}_G(\Sigma_*(G)))$ is equal to

$$\{(\mathbb{LH}^{\pm}_{\mathcal{S}_t}(p,\mu),t) \in \mathbb{R}^{n+1}_1 \times I \mid p = \boldsymbol{X}(\overline{u},t), (\overline{u},t) \in U \times I, \mu \in \mathbb{R} \setminus \{0\}\}.$$

This means that $W(\mathscr{L}_G(\Sigma_*(G))) = \mathbb{LH}^+_W(U \times I \times (\mathbb{R} \setminus \{0\})) \cup \mathbb{LH}^-_W(U \times I \times (\mathbb{R} \setminus \{0\}))$. By Proposition 8.13, the singular set of $W(\mathscr{L}_G(\Sigma_*(G)))$ is the union of the critical value sets of $\mathbb{LH}^{\pm}_W$ which is the union of unfolded lightcone focal sets $\mathbb{LF}^+_W \cup \mathbb{LF}^-_W$. Therefore, we have shown the following proposition.

Proposition 8.20 ([23]). *Let $(W, \mathcal{S})$ be a world hyper-sheet in $\mathbb{R}^{n+1}_1$ and $G : W \times (\mathbb{R}^{n+1}_1 \setminus W) \to \mathbb{R}$ the Lorentz distance squared function. Then we have the graph-like Legendrian unfolding $\mathscr{L}_G(\Sigma_*(G)) \subset J^1(\mathbb{R}^{n+1}_1, I)$ such that*

$$W(\mathscr{L}_G(\Sigma_*(G))) = \mathbb{LH}^+_W(U \times I \times (\mathbb{R} \setminus \{0\})) \cup \mathbb{LH}^-_W(U \times I \times (\mathbb{R} \setminus \{0\})).$$

We write $\mathbb{LH}^{\pm}_{(W,\mathcal{S})} = \mathbb{LH}^{\pm}_W(U \times I \times (\mathbb{R} \setminus \{0\}))$. We also call $\mathbb{LH}^+_{(W,\mathcal{S})} \cup \mathbb{LH}^-_{(W,\mathcal{S})}$ an *unfolded light sheet* of $(W, \mathcal{S})$. On the other hand, we have the corresponding Lagrangian submanifold $\Pi(\mathscr{L}_G(\Sigma_*(G))) \subset T^*\mathbb{R}^{n+1}_1$. We now consider the natural question what are the caustic $C_{\mathscr{L}_G(\Sigma_*(G))}$ and the Maxwell set $M_{\mathscr{L}_G(\Sigma_*(G))}$? Moreover, are there any meanings of $C_{\mathscr{L}_G(\Sigma_*(G))}$ and $M_{\mathscr{L}_G(\Sigma_*(G))}$ in physics?

In [4, 5] Bousso and Randall gave an idea of caustics of world hyper-sheets in order to define the notion of holographic domains. The family of light sheets $\{\mathbb{LH}^{\pm}_{\mathcal{S}_t}(U \times \{t\}) \times \mathbb{R}\}_{t \in J}$ sweeps out a region in $\mathbb{R}^{n+1}_1$. A *caustic* of a world hyper-sheet is the union of the sets of critical values of light sheets along momentary spaces $\{\mathcal{S}_t\}_{t \in I}$. A *holographic domain* of the world hyper-sheet is the region where the light-sheets sweep out until *caustics*. So this means that the boundary of the holographic domain consists the caustic of the world hyper-sheet. The set of critical values of the light sheet of a momentary space is the lightlike focal set of the momentary space. Therefore the notion of caustics in the sense of Bousso-Randall is formulated as follows: *Caustics of a world sheet* $(W, \mathcal{S})$ are defined to be

$$C^{\pm}(W, \mathcal{S}) = \bigcup_{t \in I} \mathbb{LF}^{\pm}_{\mathcal{S}_t} = \pi_1(\mathbb{LF}^{\pm}_{(W,\mathcal{S})}),$$

where $\pi_1 : \mathbb{R}^{n+1}_1 \times I \to \mathbb{R}^{n+1}_1$ is the canonical projection. We call $C^{\pm}(W, \mathcal{S})$ *BR-caustics* of $(W, \mathcal{S})$ (cf. [23]). We write that $C(W, \mathcal{S}) = \pi_1(\mathbb{LF}^+_W \cup \mathbb{LF}^-_W)$ and call it a *total BR-caustic* of $(W, \mathcal{S})$. By definition, we have $\Sigma(W(\mathscr{L}_G(\Sigma_*(G))) = \mathbb{LF}^+_{(W,\mathcal{S})} \cup \mathbb{LF}^-_{(W,\mathcal{S})}$, so that we have the following proposition.

Proposition 8.21 ([23]). *Let $(W, \mathcal{S})$ be a world hyper-sheet in $\mathbb{R}^{n+1}_1$ and $G : U \times I \times (\mathbb{R}^{n+1}_1 \setminus W) \to \mathbb{R}$ the Lorentz distance squared function. Then we have $C(W, \mathcal{S}) = C_{\mathscr{L}_G(\Sigma_*(G))}$.*

In [4, 5] the authors did not consider the Maxwell set of a world hyper-sheet. However, the notion of Maxwell sets plays an important role in the cosmology which has been called a *crease set* by Penrose (cf. [33, 34]). Actually, the topological shape of the event horizon is determined by the crease set of light sheets. Here, we write $M(W, S) = M_{\mathscr{L}_G(\Sigma_*(G))}$ and call it a *BR-Maxwell set* of the world sheet $(W, \mathcal{S})$.

Let $\boldsymbol{X}_i : (U_i \times I_i, (\overline{u}_i, t_i)) \to (\mathbb{R}^{n+1}_1, p_i)$, $(i = 1, 2)$ be germs of timelike embeddings such that $(W_i, \mathcal{S}_i)$ are world hyper-sheet germs, where $W_i = \boldsymbol{X}_i(U)$. For $\boldsymbol{\lambda}_i = \mathbb{LH}^+_{\mathcal{S}_i}(p_i, \overline{u}_i)$ or $\boldsymbol{\lambda}_i = \mathbb{LH}^-_{\mathcal{S}_i}(p_i, \overline{u}_i)$, let $G_i : (U_i \times I_i \times (\mathbb{R}^{n+1}_1 \setminus W_i), (\overline{u}_i, t_i, \boldsymbol{\lambda}_i)) \to \mathbb{R}$ be Lorentz distance squared function germs. We also write that $g_{i,\boldsymbol{\lambda}_i}(\overline{u}, t) = G_i(\overline{u}, t, \boldsymbol{\lambda}_i)$. Since

$$W(\mathscr{L}_{G_i}(\Sigma_*(G_i))) = \mathbb{LH}^+_{(W_i,\mathcal{S}_i)} \cup \mathbb{LH}^-_{(W_i,\mathcal{S}_i)},$$

we can apply Theorem 6.1 and Corollary 6.2 to our case. Then we have the following theorem.

Theorem 8.22. *Suppose that $\overline{\pi}|_{\mathscr{L}_{G_i}(\Sigma_*(G_i))}$ is a proper map germ and the singular set of the map germ is nowhere dense for each $i = 1, 2$, respectively. Then the following conditions are equivalent:*
(1) $(\mathbb{LH}^+_{(W_1,\mathcal{S}_1)} \cup \mathbb{LH}^-_{(W_1,\mathcal{S}_1)}, \boldsymbol{\lambda}_1), (\mathbb{LH}^+_{(W_2,\mathcal{S}_2)} \cup \mathbb{LH}^-_{(W_2,\mathcal{S}_2)}, \boldsymbol{\lambda}_2)$ *are $S.P^+$-diffeomorphic,*
(2) $\mathscr{L}_{G_1}(\Sigma_*(G_1))$, $\mathscr{L}_{G_2}(\Sigma_*(G_2))$ *are $S.P^+$-Legendrian equivalent,*
(3) $\Pi(\mathscr{L}_{G_1}(\Sigma_*(G_1)))$, $\Pi(\mathscr{L}_{G_2}(\Sigma_*(G_2))$ *are Lagrangian equivalent.*

We remark that conditions (2) and (3) are equivalent without any assumptions (cf. Theorem 6.1). Moreover, if we assume that $\mathscr{L}_{G_i}(\Sigma_*(G_i))$ is $S.P^+$-Legendrian stable, then we can apply Proposition 8.14 and Theorem 6.4 and show the following theorem.

Theorem 8.23. *Suppose that $\mathscr{L}_{G_i}(\Sigma_*(G_i))$ is $S.P^+$-Legendrian stable for each $i = 1, 2$, respectively. Then the following conditions are equivalent:*
(1) $(\mathbb{LH}^+_{(W_1,\mathcal{S}_1)} \cup \mathbb{LH}^-_{(W_1,\mathcal{S}_1)}, \boldsymbol{\lambda}_1), (\mathbb{LH}^+_{(W_2,\mathcal{S}_2)} \cup \mathbb{LH}^-_{(W_2,\mathcal{S}_2)}, \boldsymbol{\lambda}_2)$ *are $S.P^+$-diffeomorphic,*
(2) $\mathscr{L}_{G_1}(\Sigma_*(G_1))$, $\mathscr{L}_{G_2}(\Sigma_*(G_2))$ *are $S.P^+$-Legendrian equivalent,*
(3) $\Pi(\mathscr{L}_{G_1}(\Sigma_*(G_1)))$, $\Pi(\mathscr{L}_{G_2}(\Sigma_*(G_2))$ *are Lagrangian equivalent,*
(4) $g_{1,\boldsymbol{\lambda}_1}$, $g_{2,\boldsymbol{\lambda}_2}$ *are $S.P$-$\mathcal{K}$-equivalent,*
(5) $SK(\overline{W}_1, TLC(\mathcal{S}_{t_0}, \boldsymbol{\lambda}_1) \times I; (p_1, t_0))$
$$= SK(\overline{W}_2, TLC(\mathcal{S}_{t_0}, \boldsymbol{\lambda}_2) \times I; (p_2, t_0)).$$

For s-P-Legendrian equivalence, we have the following theorem as a corollary of Theorem 7.11 and Proposition 8.18.

Theorem 8.24. *Suppose that $\mathscr{L}_{G_i}(\Sigma_*(G_i))$ is s-P-Legendrian stable for each $i = 1, 2$, respectively. Then the following conditions are equivalent:*
(1) $(\mathbb{LH}^+_{(W_1,\mathcal{S}_1)} \cup \mathbb{LH}^-_{(W_1,\mathcal{S}_1)}, \boldsymbol{\lambda}_1), (\mathbb{LH}^+_{(W_2,\mathcal{S}_2)} \cup \mathbb{LH}^-_{(W_2,\mathcal{S}_2)}, \boldsymbol{\lambda}_2)$ *are s-P-diffeomorphic,*
(2) $\mathscr{L}_{G_1}(\Sigma_*(G_1))$, $\mathscr{L}_{G_2}(\Sigma_*(G_2))$ *are s-P-Legendrian equivalent,*
(3) $g_{1,\boldsymbol{\lambda}_1}$, $g_{2,\boldsymbol{\lambda}_2}$ *are P-$\mathcal{K}$-equivalent,*
(4) $PK(\overline{W}_1, TLC(\mathcal{S}_{t_0}, \boldsymbol{\lambda}_1) \times I; (p_1, t_0))$
$$= PK(\overline{W}_2, TLC(\mathcal{S}_{t_0}, \boldsymbol{\lambda}_2) \times I; (p_2, t_0)).$$

Moreover, $S.P^+$-Legendrian equivalence among graph-like Legendrian unfoldings implies s-P-Legendrian equivalence. By Proposition 8.19, the caustic and the Maxwell set of $\mathscr{L}_G(\Sigma_*(G))$ are the BR-caustic and the BR-Mawxell set. Since s-P-Legendrian equivalence among graph-like Legendrian unfoldings preserves both the diffeomorphism types of caustics and Maxwell sets, we have the following proposition.

Proposition 8.25. *If $\Pi(\mathscr{L}_{G_1}(\Sigma_*(G_1)))$ and $\Pi(\mathscr{L}_{G_2}(\Sigma_*(G_2)))$ are Lagrangian equivalent, then $\mathscr{L}_{G_1}(\Sigma_*(G_1))$ and $\mathscr{L}_{G_2}(\Sigma_*(G_2))$ are s-P-Legendrian equivalent. It follows that total BR-caustics*

$$C(W_1, \mathcal{S}_1),\ C(W_2, \mathcal{S}_2)$$

and BR-Maxwell sets

$$M(W_1, \mathcal{S}_1),\ M(W_2, \mathcal{S}_2)$$

are diffeomorphic as set germs, respectively.

References

[1] V. I. Arnol'd, S. M. Gusein-Zade and A. N. Varchenko, Singularities of Differentiable Maps vol. I, Birkhäuser, 1986.
[2] V. I. Arnol'd, Contact geometry and wave propagation, Monograph. Enseignement Math., **34**, (1989).
[3] V. I. Arnol'd, Singularities of caustics and wave fronts, Math. Appl., **62**, Kluwer, Dordrecht, 1990.
[4] R. Bousso, The holographic principle, REVIEWS OF MODERN PHYSICS, **74** (2002), 825–874.
[5] R. Bousso and L. Randall, Holographic domains of anti-de Sitter space, Journal of High Energy Physics, **04** (2002), 057.

[6] TH. Bröcker, Differentiable Germs and Catastrophes, London Mathematical Society Lecture Note Series, **17**, Cambridge University Press, 1975.

[7] J. W. Bruce, Wavefronts and parallels in Euclidean space, Math. Proc. Cambridge Philos. Soc., **93** (1983) 323–333.

[8] J. Damon, The unfolding and determinacy theorems for subgroups of $\mathcal{A}$ and $\mathcal{K}$, Memoirs of A.M.S., **50** No. **306**, (1984).

[9] J. J. Duistermaat, Osculating Integrals, Lagrange Immersions and Unfolding of Singularitites, Comm. Pure and Applied Math., **XXVII** (1974), 207–281.

[10] V. Goryunov and V. M. Zakalyukin, Lagrangian and Legendrian Singularities, Real and Complex Singularities, Trends in Mathematics, 169–185, Birkhäuger, 2006.

[11] A. Hayakawa, G. Ishikawa, S. Izumiya and K. Yamaguchi, Classification of generic integral diagrams and first order ordinary differential equations, International Journal of Mathematics, **5** (1994), 447–489.

[12] L. Hörmander, Fourier Integral Operators, I. Acta. Math., **128** (1972), 79–183.

[13] S. Izumiya, Generic bifurcations of varieties. manuscripta math., **46** (1984), 137–164.

[14] S. Izumiya, Perestroikas of optical wave fronts and graphlike Legendrian unfoldings, J. Differential Geom., **38** (1993), 485–500.

[15] S. Izumiya, Completely integrable holonomic systems of first-order differential equations, Proc. Royal Soc. Edinburgh, **125A** (1995), 567–586.

[16] S. Izumiya, D-H. Pei and T. Sano, Singularities of Hyperbolic Gauss maps. Proc. London Math. Soc., **86** (2003), 485–512.

[17] S. Izumiya, Differential Geometry from the viewpoint of Lagrangian or Legendrian singularity theory. in *Singularity Theory, Proceedings of the 2005 Marseille Singularity School and Conference, by D. Chéniot et al.* World Scientific, (2007) 241–275.

[18] S. Izumiya and M.C. Romero Fuster, The lightlike flat geometry on spacelike submanifolds of codimension two in Minkowski space. Selecta Math. (N.S.), **13** (2007), no. 1, 23–55.

[19] S. Izumiya and M. Takahashi, Spacelike parallels and evolutes in Minkowski pseudo-spheres, Journal of Geometry and Physics, **57** (2007), 1569–1600.

[20] S. Izumiya and M. Takahashi, Caustics and wave front propagations: Applications to differential geometry, Banach Center Publications. Geometry and topology of caustics, **82** (2008) 125–142.

[21] S. Izumiya and M. Takahashi, Pedal foliations and Gauss maps of hypersurfaces in Euclidean space, Journal of Singularities, **6** (2012) 84–97.

[22] S. Izumiya, Geometry of world sheets in Lorentz-Minkowski space, RIMS Kôkyûroku Bessatsu, **B55** (2016), 89–109.

[23] S. Izumiya, *Caustics of world hyper-sheets in the Minkowski space-time*, to appear in Contemporary Mathematics, AMS (2016).

[24] S. Izumiya, The theory of graph-like Legendrian unfoldings and its applications, Journal of Singularities, **12** (2015), 53–79.

[25] S. Izumiya, Geometric interpretation of Lagrangian equivalence, Canad. Math. Bull., **59** (2016), 806–812.
[26] S. Izumiya, M. C. Romero Fuster, M. A. Soares Ruas and F. Tari, Differential Geometry from a Singularity Theory Viewpoint. World Scientific, (2015)
[27] K. Jänich, Caustics and catastrophes, Math. Ann., **209**, (1974), 161–180.
[28] J. Martinet, Singularities of Smooth Functions and Maps, London Math. Soc. Lecture Note Series, Cambridge University Press, **58** (1982).
[29] V. P. Malsov and M. V. Fedoruik, Semi-classical approximation in quantum mechanics, D. Reidel Publishing Company, Dordrecht, 1981.
[30] J. Mather, Stability of C^∞-mappings IV. Classification of stable germs by R-algebras, Publications Mathématiques, Institut de Hautes Études Scientifiques, **37**, 223–248 (1969).
[31] J. A. Montaldi, On contact between submanifolds, Michigan Math. J., **33** (1986), 81–85.
[32] B. O'Neill, Semi-Riemannian Geometry, Academic Press, New York, 1983.
[33] R. Penrose, Null Hypersurface Initial Data for Classical Fields of Arbitrary Spin and for General Relativity, General Relativity and Gravitation, **12** (1963), 225–264.
[34] M. Siino and T. Koike, Topological classification of black holes: generic Maxwell set and crease set of a horizon, International Journal of Modern Physics D, **20**, (2011), 1095.
[35] G. Wassermann, Stability of Unfoldings, Lecture Notes in Mathematics, **393** (1974).
[36] G. Wassermann, Stability of Caustics, Math. Ann., **216**, 43–50 (1975).
[37] V. M. Zakalyukin, Lagrangian and Legendrian singularities, Funct. Anal. Appl., **10** (1976), 23–31.
[38] V. M. Zakalyukin, Reconstructions of fronts and caustics depending one parameter, Funct. Anal. Appl., **10** (1976), 139–140.
[39] V. M. Zakalyukin, Reconstructions of fronts and caustics depending one parameter and versality of mappings, J. Sov. Math., **27** (1984), 2713–2735.
[40] V. M. Zakalyukin, Envelope of Families of Wave Fronts and Control Theory, Proc. Steklov Inst. Math., **209** (1995), 114–123.

Department of Mathematics, Hokkaido University,
Sapporo 060-0810, Japan
E-mail address: `izumiya@math.sci.hokudai.ac.jp`

Advanced Studies in Pure Mathematics 78, 2018
Singularities in Generic Geometry
pp. 163–182

Asymptotic directions on a surface in a 4-dimensional metric Lie group

Pierre Bayard and Federico Sánchez-Bringas

Abstract.

In this paper we introduce the notion of asymptotic directions of a surface in a 4-dimensional riemannian manifold, and study the special case of a surface in a 4-dimensional metric Lie group. It appears that this notion depends on the left invariant metric in general.

§1. Introduction

The aim of this paper is to introduce the notion of asymptotic directions of a surface in a general 4-dimensional riemannian manifold, and to study their first properties.

If M is a surface in a 4-dimensional riemannian manifold $\tilde{M}$, we define the asymptotic directions of M at a point x as the directions of T_xM which annulate a natural quadratic form $\delta : T_xM \to \mathbb{R}$: this quadratic form is constructed from the generalized Gauss map $\varphi : M \to \Lambda^2 T\tilde{M}$ of the surface by the formula

$$\delta(X) := -\frac{1}{2}\nabla_X\varphi \wedge \nabla_X\varphi$$

for all $X \in T_xM$, where ∇ denotes the Levi-Civita connection on $\Lambda^2 T\tilde{M}$; it measures the complexity of the first variation of the tangent planes of the surface at x. The construction is explained in details in the first section of the paper. We show that these directions may be equivalently defined in terms of the second fundamental form, or in terms of the curvature ellipses of the surface; it appears that these directions

Received May 26, 2016.
Revised February 7, 2017.
2010 *Mathematics Subject Classification.* Primary 53B20; Secondary 53B25.

are also the directions of higher contact of the surface with some 3-dimensional totally geodesic spaces. So, the definition extends to a general 4-dimensional riemannian manifold the notion of asymptotic directions for a surface in $\mathbb{R}^4$.

We then study the special case of a surface in a Lie group equipped with a left invariant metric. The case of a surface which is also a subgroup is particularly interesting, since the asymptotic directions satisfy a purely algebraic equation in that case. We then give an example showing that the notion of asymptotic directions generally depends on the invariant metric on the group; this is in contrast with the notion of asymptotic directions on a surface in $\mathbb{R}^4$, or more generally in a Lie group if we consider only metrics which are bi-invariant. We then finish the paper with the special case of a surface in a 3-dimensional subgroup of a metric Lie group, and especially study the case of a surface in the 4-dimensional hyperbolic space $\mathbb{H}^4$, which is regarded as a Lie group with a left invariant metric.

We quote some related papers: the notion of asymptotic directions of a surface in $\mathbb{R}^4$ has been widely studied, especially in relation to the theory of singularities, see e.g. [5, 7]. It has been extended to other pseudo-euclidian 4-dimensional spaces in [2, 3, 4].

The outline of the paper is as follows: we introduce the generalized Gauss map and the asymptotic directions of a surface in a 4-dimensional riemannian manifold in Section 2; we also show in the same section that the principal properties of the asymptotic directions of a surface in $\mathbb{R}^4$ hold in this general setting. We then study in Section 3 the notion of asymptotic directions on a surface in a 4-dimensional metric Lie group, especially when the surface is also a subgroup, or belongs to a 3-dimensional subgroup. We finally recall the fundamental equations of Gauss, Codazzi and Ricci in a short appendix, at the end of the paper.

§2. Asymptotic directions on a surface in a 4-dimensional riemannian manifold

We assume here that $\widetilde{M}$ is a 4-dimensional riemannian manifold, with a given orientation, and denote by $\widetilde{\nabla}$ its Levi-Civita connection. We suppose that $F : M \to \widetilde{M}$ is the immersion of an oriented surface into $\widetilde{M}$. The purpose of the section is to define in that context the notion of asymptotic directions of M into $\widetilde{M}$, and study their first properties.

2.1. The generalized Gauss map

We consider, in a neighborhood $\mathcal{U}$ of $x_o \in M$, a positively oriented and orthonormal frame (e_1, e_2) of TM. The tangent planes of M in $\mathcal{U}$

may be represented by $e_1 \wedge e_2$, a local section of $F^*\Lambda^2 T\tilde{M}$, the bundle of the bivectors of $T\tilde{M}$, induced on M : we set

$$\begin{array}{rccl} \varphi : & \mathcal{U} & \to & F^*\Lambda^2 T\tilde{M} \\ & x & \mapsto & e_1 \wedge e_2\,(x); \end{array}$$

this map is a natural generalization of the Gauss map. The connection ∇ on $T\tilde{M}$ naturally induces a connection on $F^*\Lambda^2 T\tilde{M}$, still denoted by ∇.

Lemma 2.1. *For $X \in \Gamma(T\mathcal{U})$, $|X| = 1$, we have the formula*

$$\nabla_X \varphi = B(X,X) \wedge X^\perp + X \wedge B(X, X^\perp), \tag{1}$$

where $B : TM \times TM \to E$ is the second fundamental form of M in $\tilde{M}$ (E denotes the normal bundle of M in $\tilde{M}$) and $X^\perp$ is the tangent vector obtained from X by a rotation of angle $+\pi/2$ in TM.

Proof. We assume that (e_1, e_2) is a positively oriented and orthonormal frame in $\mathcal{U} \subset M$ such that $\nabla^M e_1 = \nabla^M e_2 = 0$ at x_o (∇^M is the Levi-Civita connection in M), and we compute

$$\begin{aligned} \nabla_X \varphi &= \{\nabla_X e_1\} \wedge e_2 + e_1 \wedge \{\nabla_X e_2\} \\ &= B(X, e_1) \wedge e_2 + e_1 \wedge B(X, e_2), \end{aligned}$$

since $B(X, e_i) = \nabla_X e_i$ at x_o. We finally choose (e_1, e_2) such that $e_1 = X$ and $e_2 = X^\perp$ at x_o. Q.E.D.

Remark 1. *Formula* (1) *has the following interpretation: at $x_o \in M$, $\nabla_X \varphi$ represents the infinitesimal rotation of the tangent planes of M in the direction X in the 4-space $T_{x_o}\tilde{M}$, and* (1) *is its decomposition in infinitesimal rotations in two 3-spaces: the first term $B(X,X) \wedge X^\perp$ represents an infinitesimal rotation in the 3-space $T_{x_o}M \oplus B(X,X)$, around the tangent direction $X^\perp$, and the second term $X \wedge B(X, X^\perp)$ an infinitesimal rotation in the 3-space $T_{x_o}M \oplus B(X, X^\perp)$, around the tangent direction X.*

2.2. Asymptotic directions

We note that $\Lambda^4 T_{x_o}\tilde{M} \simeq \mathbb{R}$, since $\tilde{M}$ is 4-dimensional (by fixing a positively oriented and orthonormal frame of $T_{x_o}\tilde{M}$). This allows the following definition:

Definition 2.1. *Let us consider the quadratic map*

$$\begin{array}{rccl} \delta : & T_{x_o}M & \to & \mathbb{R} \\ & X & \mapsto & -\dfrac{1}{2}\nabla_X \varphi \wedge \nabla_X \varphi. \end{array} \tag{2}$$

We will say that $X \in T_{x_o}M$, $X \neq 0$, *defines an asymptotic direction of* M *at* x_o *if* $\delta(X) = 0$.

Remark 2. *By Lemma 2.1, for all* $X \in T_{x_o}M$, $|X| = 1$,

$$\delta(X) = X \wedge X^{\perp} \wedge B(X, X) \wedge B(X, X^{\perp}), \tag{3}$$

and X *is an asymptotic direction if and only if*

$$B(X, X) \wedge B(X, X^{\perp}) = 0.$$

An asymptotic direction has the following interpretation: by Remark 1, $\nabla_X \varphi$ *represents an infinitesimal rotation of the tangent planes of* M *in the direction* X, *and* (1) *is its decomposition in two infinitesimal rotations in 3-spaces; we readily see that* X *is an asymptotic direction of* M *if and only if the two 3-spaces in the decomposition* (1) *coincide: the infinitesimal rotation of the tangent planes of* M *in the direction* X *takes place in a 3-dimensional space* $\subset T_{x_o}\tilde{M}$, *instead of in the whole 4-space* $T_{x_o}\tilde{M}$; *this 3-dimensional space is an osculating 3-space of* M *in the direction* X. *We will precise this fact in the next section.*

Remark 3. *The notion of asymptotic directions does not really depend on the metric but rather on the associated connection: this notion is in fact defined for a surface in a general 4-dimensional manifold equipped with a linear connection. For example, in* $\mathbb{R}^4$, *the notion of asymptotic lines does not depend on the choice of the metric invariant by translation: we have a unique notion in* $\mathbb{R}^4$, $\mathbb{R}^{3,1}$ *or* $\mathbb{R}^{2,2}$, *since the Levi-Civita connection is the canonical connection in all these cases. However, we will see that in a general Lie group we may have different notions of asymptotic lines, since different left invariant metrics give different Levi-Civita connections in general.*

Similarly to the euclidean case $\mathbb{R}^4$ [5], we obtain the following:

Proposition 2.2. *If* $\tilde{R}$ *is the curvature tensor of* $\tilde{M}$ *and* (e_1, e_2) *and* (n_1, n_2) *are positively oriented and orthonormal bases of the tangent and the normal planes of* M *at* x_o, *we set*

$$\tilde{K}_N := \langle \tilde{R}(e_1, e_2)(n_2), n_1 \rangle.$$

The trace of δ *is*

$$tr_g \delta = K_N - \tilde{K}_N$$

where $K_N = \langle R^N(e_1, e_2)(n_2), n_1 \rangle$ *is the normal curvature of* M.

We moreover define

$$\Delta := det_g \delta.$$

By definition, M has an asymptotic direction at x_o if and only if $\Delta \leq 0$; it has two distinct asymptotic directions if and only if $\Delta < 0$.

Proof. We set $B_i = \langle B, n_i \rangle$, $i = 1, 2$. Let $(X, X^{\perp})$ be an orthonormal basis of TM. The Ricci equation (26) in the appendix reads

$$(4) \quad K_N - \tilde{K}_N = \langle B(X, B^*(X^{\perp}, n_2)), n_1 \rangle - \langle B(X^{\perp}, B^*(X, n_2)), n_1 \rangle.$$

We compute the right-hand side terms: since $(X, X^{\perp})$ is an orthonormal basis of TM, we have

$$\begin{aligned} B^*(X^{\perp}, n_2) &= \langle B^*(X^{\perp}, n_2), X \rangle X + \langle B^*(X^{\perp}, n_2), X^{\perp} \rangle X^{\perp} \\ &= \langle B(X^{\perp}, X), n_2 \rangle X + \langle B(X^{\perp}, X^{\perp}), n_2 \rangle X^{\perp} \\ &= B_2(X^{\perp}, X) X + B_2(X^{\perp}, X^{\perp}) X^{\perp} \end{aligned}$$

and thus

$$\begin{aligned} \langle B(X, B^*(X^{\perp}, n_2)), n_1 \rangle &= B_1(X, B^*(X^{\perp}, n_2)) \\ &= B_1(X, X) B_2(X^{\perp}, X) \\ &\quad + B_1(X, X^{\perp}) B_2(X^{\perp}, X^{\perp}). \end{aligned}$$

Similarly,

$$\begin{aligned} B^*(X, n_2) &= \langle B^*(X, n_2), X \rangle X + \langle B^*(X, n_2), X^{\perp} \rangle X^{\perp} \\ &= \langle B(X, X), n_2 \rangle X + \langle B(X, X^{\perp}), n_2 \rangle X^{\perp} \\ &= B_2(X, X) X + B_2(X, X^{\perp}) X^{\perp} \end{aligned}$$

and

$$\begin{aligned} \langle B(X^{\perp}, B^*(X, n_2)), n_1 \rangle &= B_1(X^{\perp}, B^*(X, n_2)) \\ &= B_1(X^{\perp}, X) B_2(X, X) \\ &\quad + B_1(X^{\perp}, X^{\perp}) B_2(X, X^{\perp}). \end{aligned}$$

The Ricci equation (4) thus reads

$$\begin{aligned} K_N - \tilde{K}_N &= B_1(X, X^{\perp}) \left(B_2(X^{\perp}, X^{\perp}) - B_2(X, X) \right) \\ &\quad + B_2(X^{\perp}, X) \left(B_1(X, X) - B_1(X^{\perp}, X^{\perp}) \right). \end{aligned}$$

On the other hand, we note that

$$\begin{aligned}\delta(X) &= X\wedge X^{\perp}\wedge B(X,X)\wedge B(X,X^{\perp})\\ &= X\wedge X^{\perp}\wedge (B_1(X,X)n_1+B_2(X,X)n_2)\wedge (B_1(X,X^{\perp})n_1\\ &\quad +B_2(X,X^{\perp})n_2)\\ &= X\wedge X^{\perp}\wedge n_1\wedge n_2(B_1(X,X)B_2(X,X^{\perp})\\ &\quad -B_2(X,X)B_1(X,X^{\perp}))\\ &\simeq B_1(X,X)B_2(X,X^{\perp})-B_2(X,X)B_1(X,X^{\perp}),\end{aligned}$$

and similarly

$$\delta(X^{\perp})=-B_1(X^{\perp},X^{\perp})B_2(X^{\perp},X)+B_2(X^{\perp},X^{\perp})B_1(X^{\perp},X).$$

We thus obtain

$$tr_g\delta=\delta(X)+\delta(X^{\perp})=K_N-\tilde{K}_N.$$

Q.E.D.

Remark 4. *Since the metric g is positive definite, we have*

$$(tr_g\delta)^2\geq 4\ det_g\delta,$$

and thus the inequality

$$(K_N-\tilde{K}_N)^2\geq 4\Delta.$$

We may also interpret the invariants of δ in terms of the curvature ellipse at x_o, which is an ellipse in the plane normal to the surface at x_o : the curvature ellipse of M at x_o is classically defined as

$$\mathcal{E}_{x_o}:=\{B(X,X),\ X\in T_{x_o}M,\ |X|=1\}\qquad\subset\ E_{x_o}.$$

Since $B(X,X^{\perp})$ is tangent to the ellipse at $B(X,X)$, the direction X is an asymptotic direction of M at x_o if and only if the line $(0,B(X,X))\subset E_{x_o}$ is tangent to the ellipse $\mathcal{E}_{x_o}$. Moreover, the sign of Δ has the following interpretation: $\Delta>0$ if and only if the null vector $0\in E_{x_o}$ belongs to the interior of the ellipse $\mathcal{E}_{x_o}$, $\Delta=0$ if and only if 0 belongs to $\mathcal{E}_{x_o}$, and $\Delta<0$ if and only 0 is exterior to the ellipse $\mathcal{E}_{x_o}$. We omit the proofs since they are similar to the case $\tilde{M}=\mathbb{R}^4$.

2.3. Asymptotic directions and height functions

If ν is a vector in $T_{x_o}\tilde{M}$, we may define the height function on a neighborhood $\mathcal{U}$ of x_o in M by

$$\begin{array}{rrcl} h_\nu : & \mathcal{U} & \to & \mathbb{R} \\ & x & \mapsto & \langle \nu, \exp_{x_o}^{-1}(x)\rangle, \end{array}$$

where $\exp_{x_o}$ is the riemannian exponential map at x_o (this is a local diffeomorphism between a neighborhood of 0 in $T_{x_o}\tilde{M}$ and a neighborhood of x_o in $\tilde{M}$). The function h_ν represents the "height" of $\mathcal{U} \subset M$ with respect to the totally geodesic 3-dimensional submanifold $exp_{x_o}(\nu^\perp) \subset \tilde{M}$. We have the following:

Lemma 2.3. *$dh_\nu = 0$ at x_o if and only if ν belongs to the normal plane E_{x_o}. In that case, the hessian of h_ν is*

$$\nabla^M dh_\nu = B_\nu, \tag{5}$$

where B_ν is the quadratic form on $T_{x_o}M$ such that

$$B_\nu(X,Y) = \langle B(X,Y), \nu\rangle$$

for all $X, Y \in T_{x_o}M$. Here ∇^M stands for the Levi-Civita connection on M.

Proof. Let us set, for x belonging to a neighborhood $\tilde{\mathcal{U}}$ of x_o in $\tilde{M}$,

$$\tilde{h}_\nu(x) = \langle \nu, \exp_{x_o}^{-1}(x)\rangle;$$

the function h_ν is the restriction of $\tilde{h}_\nu$ to $\mathcal{U} \subset M$. Let $\gamma : (-\varepsilon, \varepsilon) \to \tilde{M}$ be the geodesic of $\tilde{M}$ such that $\gamma(0) = x_o$ and $\gamma'(0) = v \in T_{x_o}M$. By definition of the exponential map, $\gamma(t) = \exp_{x_o}(tv)$. Thus

$$\tilde{h}_\nu(\gamma(t)) = t\langle \nu, v\rangle$$

and

$$dh_\nu(v) = d\tilde{h}_\nu(v) = \frac{d}{dt}_{|t=0} \tilde{h}_\nu(\gamma(t)) = \langle \nu, v\rangle.$$

Thus $dh_\nu = 0$ at x_o if and only if $\langle \nu, v\rangle = 0$ for all $v \in T_{x_o}M$, that is ν belongs to E_{x_o}. We now assume that u and v are vector fields defined on $\tilde{\mathcal{U}}$, whose restrictions to $\mathcal{U}$ are tangent to M. By definition,

$$\nabla^M dh_\nu(u,v) = \partial_u\{dh_\nu(v)\} - dh_\nu(\nabla^M_u v)$$

and

$$\nabla d\tilde{h}_\nu(u,v) = \partial_u\{d\tilde{h}_\nu(v)\} - d\tilde{h}_\nu(\nabla_u v),$$

and thus

$$\begin{aligned}\nabla^M dh_\nu(u,v) - \nabla d\tilde{h}_\nu(u,v) &= -d\tilde{h}_\nu(\nabla^M_u v - \nabla_u v)\\ &= d\tilde{h}_\nu(B(u,v))\\ &= \langle \nu, B(u,v)\rangle\\ &= B_\nu(u,v).\end{aligned}$$

Now, if γ is a geodesic of $\tilde{M}$,

$$\frac{d^2}{dt^2}\,\tilde{h}_\nu(\gamma(t)) = \nabla d\tilde{h}_\nu(\gamma'(t),\gamma'(t)).$$

For $\gamma(t) = \exp_{x_o}(tv)$ we get $\tilde{h}_\nu(\gamma(t)) = t\langle\nu, v\rangle$, and deduce that $\nabla d\tilde{h}_\nu = 0$. Identity (5) follows. Q.E.D.

The following proposition shows that an asymptotic direction corresponds to a direction of higher contact between M and some totally geodesic 3-dimensional submanifold:

Proposition 2.4. *M admits an asymptotic direction X at x_o if and only if there exists $\nu \in E_{x_o}$, $|\nu| = 1$, such that $X \in Ker\ \nabla^M dh_\nu$.*

Proof. We suppose that ν is a unit vector belonging to E_{x_o}; by Lemma 2.3, $\nabla^M dh_\nu$ is degenerate if and only if so is B_ν, i.e. there exists $X \in T_{x_o}M$, $|X| = 1$, such that $B_\nu(X,.) : T_{x_o}M \to \mathbb{R}$ is the null form. But the later is equivalent to

$$B_\nu(X,X) = B_\nu(X,X^\perp) = 0,$$

i.e. the normal vectors $B(X,X)$ and $B(X,X^\perp)$ are both orthogonal to ν. We thus get the following: there exists a unit vector $\nu \in E_{x_o}$ such that $\nabla^M dh_\nu$ is degenerate in the direction X if and only if $B(X,X)$ and $B(X,X^\perp)$ are colinear, that is if and only if X is an asymptotic direction of M at x_o. Q.E.D.

The direction ν appearing in the proposition is traditionally called *a binormal direction* of the surface M at the point x_o; see [7] for surfaces in $\mathbb{R}^4$.

Remark 5. *We may interpret Proposition 2.4 as follows: if X is an asymptotic direction of M at x_o, then, for some open subset $\mathcal{V} \subset T_{x_o}M \oplus \mathbb{R}B(X,X) \subset T_{x_o}\tilde{M}$ containing 0, the totally geodesic 3-dimensional manifold $\exp_{x_o}(\mathcal{V})$ has a contact with M of order ≥ 2 in the direction X.*

§3. Asymptotic directions on a surface in a metric Lie group

We suppose here that $\tilde{M}$ is a Lie group G, and denote by $\mathcal{G}$ its Lie algebra: $\mathcal{G}$ is the space of the left invariant vector fields on G, equipped with the Lie bracket $[.,.]$ and is identified to the linear space tangent to G at the identity. We consider the Maurer-Cartan form $\omega \in \Omega^1(G, \mathcal{G})$ defined by

$$\omega_g(v) = L_{g^{-1}}{}_*(v) \qquad \in \mathcal{G}$$

for all $v \in T_gG$, where $L_{g^{-1}}$ denotes the left multiplication by g^{-1} on G and $L_{g^{-1}}{}_* : T_gG \to \mathcal{G}$ is its differential. This form induces a bundle isomorphism

$$\begin{array}{rcl} TG & \to & G \times \mathcal{G} \\ (g, v) & \mapsto & (g, \omega_g(v)). \end{array} \tag{6}$$

We note that a vector field $X \in \Gamma(TG)$ is left invariant if $\omega(X) : G \to \mathcal{G}$ is a constant map. We consider the canonical connection ∇^o on G defined by

$$\omega(\nabla^o_X Y) = \partial_X \{\omega(Y)\}$$

for all $X, Y \in \Gamma(TG)$, where ∂_X stands for the usual derivative in the direction X; it is left invariant, and such that $\nabla^o_X Y = 0$ if X, Y are left invariant vector fields.

3.1. Left invariant metrics and Levi-Civita connections

We now assume that a left invariant metric $\langle .,. \rangle$ is given on G, and denote by ∇ its Levi-Civita connection. Since ∇ is also left invariant, there exists a left invariant tensor Γ belonging to $T^*G \otimes T^*G \otimes TG$ such that, for all $X, Y \in \Gamma(TG)$,

$$\nabla_X Y = \nabla^o_X Y + \Gamma(X, Y). \tag{7}$$

Since Γ is left invariant, we may alternatively consider Γ as a bilinear map

$$\begin{array}{rrcl} \Gamma : & \mathcal{G} \times \mathcal{G} & \to & \mathcal{G} \\ & (X, Y) & \mapsto & \Gamma(X, Y); \end{array}$$

it is such that, for all $X, Y \in \mathcal{G}$

$$\nabla_X Y = \Gamma(X, Y). \tag{8}$$

By the Koszul formula, Γ is determined by the metric as follows: for all $X, Y, Z \in \mathcal{G}$,

$$\langle \Gamma(X, Y), Z \rangle = \frac{1}{2}\langle [X, Y], Z \rangle + \frac{1}{2}\langle [Z, X], Y \rangle - \frac{1}{2}\langle [Y, Z], X \rangle. \tag{9}$$

Since ∇ is without torsion, we have, for all $X, Y \in \mathcal{G}$,

$$\Gamma(X,Y) - \Gamma(Y,X) = [X,Y]. \tag{10}$$

If we consider Γ as a map

$$\begin{array}{rcl} \mathcal{G} & \to & \Lambda^2\mathcal{G} \subset End(\mathcal{G}) \\ X & \mapsto & \Gamma(X) : \ Y \mapsto \Gamma(X,Y) \end{array}$$

(note that $\Gamma(X) : \mathcal{G} \to \mathcal{G}$ is skew-symmetric since ∇ is compatible with the metric), the curvature of ∇ is given by

$$R(X,Y) = [\Gamma(X), \Gamma(Y)] - \Gamma([X,Y]) \quad \in \Lambda^2\mathcal{G} \tag{11}$$

for all $X, Y \in \mathcal{G}$, where the first bracket in the right-hand side stands for the commutator of the endomorphisms.

3.2. A basic example: the Lie group $\mathbb{H}^n$

Here we briefly describe the group structure on $\mathbb{H}^n$, and refer to [6] for further details. Let us set

$$\mathbb{H}^n = \{a = (a', a_n) \ \in \mathbb{R}^n : \ a_n > 0\},$$

and, for $a \in \mathbb{H}^n$, *the similarity* of $\mathbb{R}^{n-1}$ (by a similarity we mean an homothety composed by a translation)

$$\begin{array}{rrcl} \varphi_a : & \mathbb{R}^{n-1} & \to & \mathbb{R}^{n-1} \\ & x & \mapsto & a_n x + a'. \end{array}$$

The similarities of $\mathbb{R}^{n-1}$ naturally form a group under composition, and the bijection

$$\begin{array}{rrcl} \varphi : & \mathbb{H}^n & \to & \{\text{similarities } \mathbb{R}^{n-1} \to \mathbb{R}^{n-1}\} \\ & a & \mapsto & \varphi_a \end{array}$$

induces a group structure on $\mathbb{H}^n$: it is such that

$$ab = (a_n b' + a', a_n b_n) \tag{12}$$

for all $a, b \in \mathbb{H}^n$; the identity element is $e = (0, 1) \in \mathbb{H}^n$. Let us denote by $(e_1, e_2, \ldots, e_n)$ the canonical basis of $T_e\mathbb{H}^n = \mathbb{R}^n$ and keep the same letters to denote the corresponding left invariant vector fields on $\mathbb{H}^n$. The Lie bracket may be easily seen to be given by

$$[e_i, e_j] = 0 \quad \text{and} \quad [e_n, e_i] = e_i$$

for $i, j = 1, \ldots, n-1$. This may also be written in the form

$$[X, Y] = l(X)Y - l(Y)X$$

for all $X, Y \in \mathbb{R}^n$, where $l : \mathbb{R}^n \to \mathbb{R}$ is the linear form such that $l(e_i) = 0$ if $i \leq n-1$ and $l(e_n) = 1$. This property implies that every left invariant metric on $\mathbb{H}^n$ has constant negative curvature $-|l|^2$ [8, 6]. We now suppose that a left invariant metric $\langle ., . \rangle$ is given on $\mathbb{H}^n$ and define

$$g_{ij} = \langle e_i, e_j \rangle$$

for all $i, j = 1, \ldots, n$; the structure constants Γ_{ij}^k, $1 \leq i, j, k \leq n$ associated to the Levi-Civita connection are defined by

$$\Gamma(e_i, e_j) = \sum_{k=1}^{n} \Gamma_{ij}^k e_k$$

for all $i, j = 1, \ldots, n$, and are easily computed using the Koszul formula (9): we have

$$\Gamma_{ij}^k = g^{kn} g_{ij} - \delta_{ik}\delta_{jn} \tag{13}$$

for all $i, j, k = 1, \ldots, n$, where $(g^{ij})_{1 \leq i,j \leq n} = {(g_{ij})_{1 \leq i,j \leq n}}^{-1}$.

3.3. The Gauss map and the quadratic form δ of a surface in a metric Lie group

Since the Lie algebra $\mathcal{G}$ of the group is supposed to be equipped with a scalar product, so is $\Lambda^2\mathcal{G}$. Let us still denote this scalar product by $\langle ., . \rangle$. There is an other symmetric and bilinear form on $\Lambda^2\mathcal{G}$

$$\begin{array}{rccc} \wedge : & \Lambda^2\mathcal{G} \times \Lambda^2\mathcal{G} & \to & \mathbb{R} \\ & (\eta, \eta') & \mapsto & \eta \wedge \eta', \end{array}$$

where a fixed positively oriented and orthonormal basis of $\mathcal{G}$ is used to identify $\Lambda^4\mathcal{G}$ to $\mathbb{R}$. The Hodge operator

$$\begin{array}{rccc} * : & \Lambda^2\mathcal{G} & \to & \Lambda^2\mathcal{G} \\ & \eta & \mapsto & *\eta \end{array}$$

is the symmetric operator on $\Lambda^2\mathcal{G}$ associated to the symmetric bilinear form $\wedge$: it is defined by the relation

$$\eta \wedge \eta' = \langle *\eta, \eta' \rangle \tag{14}$$

for all $\eta, \eta' \in \Lambda^2\mathcal{G}$. We set

$$\mathcal{Q} := \{\eta \in \Lambda^2\mathcal{G} : \ \eta \wedge \eta = 0 \ \text{ and } \ \langle \eta, \eta \rangle = 1\}.$$

$\mathcal{Q}$ is the set of the oriented 2-planes in $\mathcal{G}$. It is well known that $\mathcal{Q}$ is naturally isometric to the product of spheres $S^2(\sqrt{2}/2) \times S^2(\sqrt{2}/2)$.

We now use the trivialization $TG \simeq G \times \mathcal{G}$ given in (6) to identify vector fields on G with smooth maps $G \to \mathcal{G}$, and sections of $\Lambda^2 TG$ with smooth maps $G \to \Lambda^2\mathcal{G}$. If M is an oriented surface immersed in G, and (e_1, e_2) is a positively oriented and orthonormal frame of M defined in some open set $\mathcal{U} \subset M$, identifying the vector fields e_1 and e_2 to maps $\mathcal{U} \to \mathcal{G}$, the generalized Gauss map introduced in Section 2.1 reads

$$\begin{array}{rcccl} \varphi: & M & \to & \mathcal{Q} & \subset \Lambda^2\mathcal{G} \\ & x & \mapsto & e_1 \wedge e_2\ (x). & \end{array} \tag{15}$$

If $\varphi = e_1 \wedge e_2$ belongs to $\Lambda^2\mathcal{G}$, we set, for all $X \in \mathcal{G}$,

$$\Gamma(X)(\varphi) = \Gamma(X)(e_1) \wedge e_2 + e_1 \wedge \Gamma(X)(e_2). \tag{16}$$

This formula extends in fact $\Gamma(X)$ to a skew-symmetric operator $\Lambda^2\mathcal{G} \to \Lambda^2\mathcal{G}$. If M is a surface in G, the quadratic form δ defined in Section 2.2 reads as follows: for $X \in T_{x_o}M$,

$$\begin{aligned} \delta(X) &= -\frac{1}{2}\nabla_X\varphi \wedge \nabla_X\varphi \\ &= -\frac{1}{2}\left\{\partial_X\varphi + \Gamma(X)(\varphi)\right\} \wedge \left\{\partial_X\varphi + \Gamma(X)(\varphi)\right\}. \end{aligned}$$

Note that it is distinct to the form $\delta^o(X) = -\frac{1}{2}\partial_X\varphi \wedge \partial_X\varphi$ (this form is associated to the canonical connection of G, rather than to the Levi-Civita connection).

3.4. Asymptotic directions on a 2-dimensional subgroup

Our purpose here is to study the special case of a 2-dimensional Lie subgroup in G; this is a surface which admits a pair of left invariant and orthonormal vector fields. We are principally concerned with the question of the existence of the asymptotic directions on such a surface, and their dependence on the left invariant metric. This is indeed the simplest case in a Lie group since we will see that it reduces to an algebraic problem; it moreover furnishes the first basic examples.

We suppose that H is a 2-dimensional Lie subgroup of G, and denote by $\mathcal{H}$ its Lie algebra: there exist two orthonormal and left invariant vector fields u_1, u_2 on H; since $\varphi = \pm u_1 \wedge u_2$ we get $\partial_X \varphi = 0$ and the form δ of H simply reads

$$\delta(X) = -\frac{1}{2}\Gamma(X)(\varphi) \wedge \Gamma(X)(\varphi)$$

for all $X \in \mathcal{H}$. By definition, the tangent direction X is an asymptotic direction if $\delta(X) = 0$, which is an algebraic condition since Γ is moreover determined by the algebraic relation (9).

3.4.1. *A first negative result: the case of a bi-invariant metric.*

Proposition 3.1. *If the metric on G is bi-invariant, then the quadratic form δ on H is identically zero: all the tangent directions of H are asymptotic directions.*

We note that this result generalizes the trivial case of a plane through 0 in $\mathbb{R}^4$.

Proof. Since the metric is left invariant and H is a subgroup, it is sufficient to show that $\delta = 0$ at the origin $e \in H$. Since the metric is moreover bi-invariant, then $\Gamma(X, Y) = \frac{1}{2}[X, Y]$ (see e.g. [1] p. 61). Let us fix an orthonormal basis (u_1, u_2) of T_eH. Then, for all $X \in T_eH$,

$$\begin{aligned} \delta(X) &= u_1 \wedge u_2 \wedge \Gamma(X)(u_1) \wedge \Gamma(X)(u_2) \\ &= \frac{1}{4}\, u_1 \wedge u_2 \wedge [X, u_1] \wedge [X, u_2]. \end{aligned}$$

Writing $X = X_1 u_1 + X_2 u_2$, we see that $\delta(X) = 0$. Q.E.D.

3.4.2. *An intermediate case: the group* $\mathbb{H}^4$. We consider here the group $\mathbb{H}^4$ with a left invariant metric $\langle ., . \rangle$. The metric is not bi-invariant, since the group $\mathbb{H}^4$ is not unimodular. We keep the notation of Section 3.2, and introduce the vector $U_o \in T_e\mathbb{H}^4$ such that

$$l(X) = \langle U_o, X \rangle$$

for all $X \in T_e\mathbb{H}^4$. By the Koszul formula (9) the Levi-Civita connection is easily seen to be given by the map

$$\Gamma(X)(Y) = -\langle Y, U_o \rangle X + \langle X, Y \rangle U_o \tag{17}$$

for all $X, Y \in T_e\mathbb{H}^4$. We note that an arbitrary linear plane in $T_e\mathbb{H}^4$ is also a sub-algebra of the Lie algebra of $\mathbb{H}^4$, and thus generates a 2-dimensional subgroup. The following proposition shows that the 2-dimensional subgroups are umbilic surfaces:

Proposition 3.2. *Let H be a 2-dimensional subgroup of $\mathbb{H}^4$. The second fundamental form of H in $\mathbb{H}^4$ is given by*

$$B(X,Y) = \langle X,Y\rangle U_o^\perp$$

for all $X,Y \in T_eH$, where $U_o^\perp$ denotes the orthogonal projection of U_o onto the plane normal to H at e. Moreover, H is a totally geodesic surface if and only if $U_o^\perp = 0$ that is U_o belongs to T_eH.

In particular, on a 2-dimensional subgroup of $\mathbb{H}^4$ the quadratic differential δ vanishes identically.

Proof. Let us extend $X, Y \in T_eH$ to left invariant vector fields on H. The Levi-Civita connection of $\mathbb{H}^4$ on left invariant vector fields is given by (17); the first result follows since the second fundamental form is by definition the component normal to the surface of the covariant derivative. Finally, $\delta = 0$ by formula (3). Q.E.D.

The previous discussion contains the special case of the 4-dimensional hyperbolic space with its usual metric, if the left invariant metric is given at e by the canonical metric in $T_e\mathbb{H}^4 = \mathbb{R}^4$ (ie $g_{ij} = \delta_{ij}$ in Section 3.2). Note that in that case the vector U_o is the last vector of the canonical basis e_4.

3.4.3. *A first positive example: the group $\mathbb{H}^2 \times \mathbb{H}^2$.* We consider

$$\mathbb{H}^2 \times \mathbb{H}^2 = \{(x_1,x_3),\ x_1 \in \mathbb{R},\ x_3 > 0\} \times \{(x_2,x_4),\ x_2 \in \mathbb{R},\ x_4 > 0\}$$

with the product

$$\begin{aligned}((x_1,x_3),(x_2,x_4)) \cdot ((x_1',x_3'),(x_2',x_4')) &= ((x_3x_1' + x_1, x_3x_3'),\\ &\quad (x_4x_2' + x_2, x_4x_4'));\end{aligned}$$

this is the natural structure on the product (recall that the group structure on $\mathbb{H}^2$ is given by (12), with $n = 2$). We also consider the subgroup

$$H = \{((x_1,1),(x_2,1)),\ x_1,x_2 \in \mathbb{R}\},$$

and denote by

$$e_1 = ((1,0),(0,0)) \quad \text{and} \quad e_2 = ((0,0),(1,0))$$

the natural basis of $T_eH \subset T_e(\mathbb{H}^2 \times \mathbb{H}^2)$. We moreover suppose that a left invariant metric is given on $\mathbb{H}^2 \times \mathbb{H}^2$.

Proposition 3.3. *The subgroup H has two asymptotic directions at every point; moreover the two asymptotic directions depend on the left invariant metric $\langle .,. \rangle$. Especially, e_1, e_2 are the two asymptotic directions at $e \in H$ if and only if the left invariant metric is such that $\langle e_1, e_2\rangle = 0$.*

Proof. Let us also consider the vectors

$$e_3 = ((0,1),(0,0)) \quad \text{and} \quad e_4 = ((0,0),(0,1)).$$

The Lie algebra structure of $\mathbb{H}^2 \times \mathbb{H}^2$ is given by

$$[e_3, e_1] = -[e_1, e_3] = e_1, \; [e_4, e_2] = -[e_2, e_4] = e_2$$

and the other brackets are zero. The Koszul formula (9) implies that Γ is given by the following formula: for $i, j \in \{1, 2\}$,

$$\Gamma(e_i, e_j) = \frac{1}{2} g_{ij} \sum_{k=1}^{4} \left(g^{k(i+2)} + g^{k(j+2)} \right) e_k. \tag{18}$$

Since the metric is left invariant and H is a subgroup, we only have to do the computations at the origin $e \in H$. Let us fix an orthonormal basis (u_1, u_2) of T_eH. Then, for all $X \in T_eH$,

$$\delta(X) = u_1 \wedge u_2 \wedge \Gamma(X)(u_1) \wedge \Gamma(X)(u_2).$$

Since $u_1 \wedge u_2$ is proportional to $e_1 \wedge e_2$, and $\Gamma(X)(u_1) \wedge \Gamma(X)(u_2)$ is proportional to $\Gamma(X)(e_1) \wedge \Gamma(X)(e_2)$, $\delta(X) = 0$ if and only if

$$\Gamma(X)(e_1) \wedge \Gamma(X)(e_2) = \sum_{1 \le i < j \le 4} c_{ij}(X)\; e_i \wedge e_j$$

is such that $c_{34}(X) = 0$. A straightforward computation using (18) then shows that, for $X = X_1 e_1 + X_2 e_2$,

$$c_{34}(X) = \frac{1}{2} \left(g^{33} g^{44} - (g^{34})^2 \right) \left(g_{12} g_{11} X_1^2 + 2 g_{11} g_{22} X_1 X_2 + g_{12} g_{22} X_2^2 \right).$$

Thus $\delta(X) = 0$ if and only if

$$g_{12} g_{11} X_1^2 + 2 g_{11} g_{22} X_1 X_2 + g_{12} g_{22} X_2^2 = 0. \tag{19}$$

The discriminant of this quadratic form is

$$g_{11} g_{22} \left(g_{12}^2 - g_{11} g_{22} \right) < 0,$$

which proves that there always exist two distinct asymptotic directions; by (19) they clearly depend on the metric, and e_1, e_2 are these asymptotic directions if and only if $g_{12} = \langle e_1, e_2 \rangle = 0$. Q.E.D.

Remark 6. *Note that, by contrast, the notion of asymptotic directions for a general surface in $\mathbb{R}^4$ does not depend on the metric since the Levi-Civita connection of a left invariant metric always coincide with the canonical connection in that case (the Koszul formula (9) with $[.,.] = 0$ yields $\Gamma = 0$, whatever the invariant metric is).*

3.5. Asymptotic directions on a surface in a 3-dimensional subgroup

We consider here a 4-dimensional metric Lie group G, and a 3-dimensional subgroup H of G. Our purpose is to study the asymptotic directions of a surface belonging to H. Let $\mathcal{G}$ and $\mathcal{H}$ denote the Lie algebras of G and H. We fix $(\underline{u}_1, \underline{u}_2, \underline{u}_3, \underline{u}_4)$ an orthonormal basis of $\mathcal{G}$ such that $(\underline{u}_1, \underline{u}_2, \underline{u}_3)$ is a basis of $\mathcal{H}$. If M is a surface in H, then its Gauss map is of the form

$$\varphi = \sum_{1\leq i<j\leq 3} a_{ij}\ \underline{u}_i \wedge \underline{u}_j \qquad \in \Lambda^2\mathcal{H}$$

where a_{ij}, $1 \leq i < j \leq 3$ are smooth functions on M, and, for $X \in TM$,

$$\partial_X\varphi = \sum_{1\leq i<j\leq 3} \partial_X a_{ij}\ \underline{u}_i \wedge \underline{u}_j \qquad \in \Lambda^2\mathcal{H}.$$

Thus $\partial_X\varphi \wedge \partial_X\varphi = 0$ and

$$\delta(X) = -\partial_X\varphi \wedge \Gamma(X)\varphi - \frac{1}{2}\Gamma(X)\varphi \wedge \Gamma(X)\varphi \tag{20}$$

for all $X \in TM$. We note that $\Lambda^2\mathcal{H} \simeq \mathcal{H}$ since $\dim\mathcal{H} = 3$, and the Gauss map φ is equivalent to a map

$$\begin{array}{rccl} N: & M & \to & S^2 \subset \mathcal{H} \\ & x & \mapsto & N(x), \end{array}$$

where $N(x)$ is a unit vector normal to M at x, translated to the Lie algebra $\mathcal{H}$ by left multiplication: precisely, if $* : \Lambda^2\mathcal{G} \to \Lambda^2\mathcal{G}$ stands for the Hodge operator, we have

$$\varphi = *(N \wedge \underline{u}_4).$$

We note that, as it is usual, $dN : TM \to TS^2$ may be regarded as an operator of TM; it is related to ∇N by the formula

$$\nabla_X N = dN(X) + \Gamma(X)N \tag{21}$$

for all $X \in TM$ (see (7)).

3.5.1. *The group* $\mathbb{R}^4$. If $G = \mathbb{R}^4$, then $H = \mathbb{R}^3 \subset \mathbb{R}^4$, $\Gamma = 0$, $\delta = 0$, and we recover the well known fact that all the tangent directions of a surface in $\mathbb{R}^3 \subset \mathbb{R}^4$ are asymptotic directions.

3.5.2. *The group $\mathbb{H}^4$.* In that case, the situation is different:

Proposition 3.4. *Let G be the group $\mathbb{H}^4$, equipped with a left invariant metric, and let H be a 3-dimensional subgroup of G. Recall that $U_o \in \mathcal{G}$ is such that (17) holds. If M is a surface belonging to H, then we have the following:*

(1) *If U_o belongs to $\mathcal{H}$ then all the directions in TM are asymptotic directions of M.*

(2) *If U_o does not belong to $\mathcal{H}$, then $X \in TM$ is an asymptotic direction of M if and only if it is a principal direction of M, i.e. is such that*

$$\nabla_X N = \lambda X \tag{22}$$

for some λ belonging to $\mathbb{R}$.

Remark 7. *The connection ∇ in (22) is the Levi-Civita connection of G; let us note that ∇N also coincides with the covariant derivative of N with respect to the Levi-Civita connection of H, since, by (21) and (17), it is in fact tangent to H. Thus $\nabla N : TM \to TM$ is a symmetric operator and (22) implies that the asymptotic directions are two orthogonal directions in the case (2) (if moreover ∇N is not an homothety of TM).*

Remark 8. *If $\mathbb{H}^4$ is equipped with the left invariant metric such that the canonical basis (e_1, e_2, e_3, e_4) of $T_e\mathbb{H}^4 = \mathbb{R}^4$ is orthonormal (see Section 3.2), then $\mathbb{H}^4$ is the usual hyperbolic space, and a 3-dimensional sub-algebra $\mathcal{H}$ of $T_e\mathbb{H}^4$ may be assumed to be generated by vectors of the form $e_1, e_2, \lambda e_3 + \mu e_4$, with $\lambda, \mu \neq (0,0)$. In that case $U_o = e_4$, and U_o belongs to $\mathcal{H}$ if and only if $\lambda = 0$. Thus, by the proposition, denoting by H the subgroup generated by $\mathcal{H}$, if $\lambda = 0$ then all the tangent directions of a surface in H are asymptotic directions, and if $\lambda \neq 0$ then the asymptotic directions of a surface in H coincide with its principal directions.*

Proof. We first note that, for all $X \in TM$,

$$\Gamma(X)\varphi = p(U_o) \wedge X^\perp \tag{23}$$

where $p(U_o)$ is the orthogonal projection of U_o onto the normal plane of M. Indeed, if $\varphi = e_1 \wedge e_2$ and $X = X_1 e_1 + X_2 e_2$, a straightforward computation using (16) and (17) gives

$$\begin{aligned} \Gamma(X)\varphi &= X_1 \left(U_o - \langle e_1, U_o\rangle e_1\right) \wedge e_2 - X_2 \left(U_o - \langle e_2, U_o\rangle e_2\right) \wedge e_1 \\ &= X_1\, p(U_o) \wedge e_2 - X_2\, p(U_o) \wedge e_1, \end{aligned}$$

since

$$p(U_o) = U_o - \langle e_1, U_o\rangle e_1 - \langle e_2, U_o\rangle e_2.$$

This gives (23) since

$$X_1 e_2 - X_2 e_1 = X^\perp.$$

Thus $\Gamma(X)\varphi \wedge \Gamma(X)\varphi = 0$ and (20) reads

$$\delta(X) = -\partial_X\varphi \wedge p(U_o) \wedge X^\perp.$$

Since $\partial_X\varphi = *(\partial_X N \wedge \underline{u}_4)$ and writing $p(U_o) = \alpha N + \beta\underline{u}_4$, with $\alpha, \beta \in \mathbb{R}$, we get

$$\delta(X) \quad = \quad -\alpha * (\partial_X N \wedge \underline{u}_4) \wedge N \wedge X^\perp - \beta * (\partial_X N \wedge \underline{u}_4) \wedge \underline{u}_4 \wedge X^\perp.$$

The first right-hand side term is zero since the bivectors $*(\partial_X N \wedge \underline{u}_4)$ and $N \wedge X^\perp$ belong to $\Lambda^2\mathcal{H}$ and $\dim\mathcal{H} = 3$. For the second term, we note that by the very definition (14) of the Hodge operator $*$ and since $** = id_{\Lambda^2\mathcal{G}}$, we have

$$*\eta \wedge \eta' \simeq \langle \eta, \eta' \rangle$$

for all $\eta, \eta' \in \Lambda^2\mathcal{G}$, which implies in particular

$$\begin{aligned} *(\partial_X N \wedge \underline{u}_4) \wedge \underline{u}_4 \wedge X^\perp &= \langle \partial_X N \wedge \underline{u}_4, \underline{u}_4 \wedge X^\perp \rangle \\ &= -\langle \partial_X N, X^\perp \rangle. \end{aligned}$$

Observing that

$$\nabla_X N - \partial_X N = \Gamma(X)N = -\langle N, U_o\rangle X$$

(by (21) and (17)), we finally get

$$\delta(X) = \beta\langle \nabla_X N, X^\perp \rangle \tag{24}$$

for all $X \in TM$.

If $\beta = \langle U_o, \underline{u}_4\rangle$ is zero, then $\delta = 0$, which gives the first claim of the proposition. If now $\beta \neq 0$, then $\delta(X) = 0$ if and only if $\nabla_X N$ is orthogonal to $X^\perp$, i.e. is colinear to X; this is the second claim of the proposition. Q.E.D.

§Appendix A. The fundamental equations

We recall here the equations of Gauss, Ricci and Codazzi for an immersion of a submanifold M into a riemannian manifold $\tilde{M}$: let us denote by $\tilde{R}$ the curvature tensor of $\tilde{M}$, R^T and R^N the curvature tensors of the connections on TM and on the normal bundle E, $B : TM \times TM \to$

E the second fundamental form and $B^* : TM \times E \to TM$ the bilinear map such that for all $X, Y \in \Gamma(TM)$ and $N \in \Gamma(E)$

$$\langle B(X,Y), N\rangle = \langle Y, B^*(X,N)\rangle;$$

then we have, for all $X, Y, Z \in \Gamma(TM)$ and $N \in \Gamma(E)$,

(1) the Gauss equation

$$(\tilde{R}(X,Y)Z)^T = R^T(X,Y)Z - B^*(X, B(Y,Z)) + B^*(Y, B(X,Z)), \tag{25}$$

(2) the Ricci equation

$$(\tilde{R}(X,Y)N)^N = R^N(X,Y)N - B(X, B^*(Y,N)) + B(Y, B^*(X,N)), \tag{26}$$

(3) the Codazzi equation

$$(\tilde{R}(X,Y)Z)^N = \tilde{\nabla}_X B(Y,Z) - \tilde{\nabla}_Y B(X,Z); \tag{27}$$

in the last equation, $\tilde{\nabla}$ denotes the natural connection on $T^*M \otimes T^*M \otimes E$.

Acknowledgments: P. Bayard was supported by the project PAPIIT IA105116, UNAM, and F. Sánchez-Bringas was supported by the project PAPIIT IN118217, UNAM.

References

[1] A. Arvanitogeorgos, *An introduction to Lie groups and the geometry of homogeneous spaces*, Student Math. Library 22, AMS (2003).

[2] P. Bayard, V. Patty, F. Sánchez-Bringas, *On lorentzian surfaces in* $\mathbb{R}^{2,2}$, to appear in J. Geom. Physics (2016).

[3] P. Bayard and F. Sánchez-Bringas, *Geometric invariants and principal configurations on spacelike surfaces immersed in* $\mathbb{R}^{3,1}$, Proc. Roy. Soc. Edimb. **140 A** (2010) 1141–1160.

[4] S. Izumiya, J.J. Nuno Ballesteros, M.C. Romero Fuster, *Global properties of codimension two spacelike submanifolds in Minkowski space*, Adv. Geom. **10** (2010) 51–75.

[5] J. A. Little, *On the singularities of submanifolds of higher dimensional Euclidean spaces*, Annali Mat. Pura et Appl. **83:4A** (1969) 261–336.

[6] W. Meeks & J. Perez, *Constant mean curvature surfaces in metric Lie groups*, in "Geometric Analysis: Partial Differential Equations and Surfaces", Contemp. Math. **570** (2012) 25–110.

[7] D.K.H. Mochida, M.C. Romero Fuster, M.A.S. Ruas, *The geometry of surfaces in 4-space from a contact viewpoint*, Geom. Dedicata **54** (1995) 323–332.

[8] J. Milnor, *Curvatures of left invariant metrics on Lie groups*, Adv. Math. **21** (1976) 293–329.

Facultad de Ciencias, UNAM, México
E-mail address: bayard@ciencias.unam.mx, sanchez@unam.mx

Advanced Studies in Pure Mathematics 78, 2018
Singularities in Generic Geometry
pp. 183–200

Singularity analysis of lightlike hypersurfaces of partially null curves

Xiupeng Cui[a] and Donghe Pei[b,*]

Abstract.

We have gotten singularity classifications of lightlike hypersurfaces of a pseudo null curve in $\mathbb{R}^4_2$[6]. This paper is to characterize singularities of lightlike hypersurfaces of a partially null curve in the same space and give an example of such curves.

§1. Introduction

The notions of partially and pseudo null curves are derived from null curves, also called lightlike curves. There widely exist null curves in Minkowski spacetime. About half a century ago, null curves were researched from the view point of differential geometry [3]. In 1985, W. B. Bonnor further investigated curves with lightlike normals [4]. Until 1995, J. Walrave gave the definitions of partially and pseudo null curves [24].

A pseudo null curve is not a lightlike curve, but its tangent curve is a lightlike curve. A partially null curve is not a lightlike curve, nor is its tangent curve. Normally, partially null curves are curves with lightlike binormals [21].

Additionally, M. Petrović-Torgašev, K. İlarslan, and E. Nešović ([21], 2005) give the Frenet equations of pseudo null and partially null curves in $\mathbb{R}^4_2$ and classify all such curves with constant curvatures. Thereafter, pseudo and partially null curves have been widely concerned and many good results have been obtained from the view point of differential geometry [1, 8, 9, 10, 11, 20, 23, 25]. Pseudo null Bertrand curves, pseudo

---

Received March 31, 2016.
Revised August 4, 2016.
2010 *Mathematics Subject Classification.* Primary 53A35; Secondary 58C27, 58C28.
Key words and phrases. partially null curve, semi-Euclidean 4-space with index 2, lightlike hypersurface, Gaussian surface.

null Mannheim curves, the inextensible flows and the position vector of partially null curves are considered, respectively in [8], [9], [23] and [25]. And the relations are gotten in [10] between pseudo and partially null rectifying curves and centrodes (Darboux vectors), which play some important roles in mechanics, kinematics as well as in differential geometry. Moreover, the involute-evolute of the pseudo null curve is studied in [20], and they prove that there is no involute of pseudo null curves in Minkowski 3-space. On the other hand, the research about submanifolds in semi-Riemannian spaces have been hot issues in recent years from the view point of singularity theory and differential geometry. There appeared many good achievements [2, 12, 13, 14, 15, 16, 17, 18, 19, 22].

We have researched pseudo null timelike curves with lightlike frames given in [21], that are curves with lightlike principal normal vectors, i.e. $\|\gamma''\| = 0$. In this paper, we focus on partially null timelike curves, that are curves with lightlike binormals. However, we find it difficult to construct the lightlike frame in [21]. For example if we take γ''' as a lightlike binormal vector, then γ'' is also lightlike. Therefore, we construct a frame without lightlike vectors and naturally extend our research to the case of $\|\gamma''\| \neq 0$. Take $\boldsymbol{n}_1 = \gamma''/\|\gamma''\|$ as the unit principal normal vector. When $\boldsymbol{n}_1$ is spacelike, γ has two lightlike binormal vectors which is a partially null curve. We also consider the case that $\boldsymbol{n}_1$ is timelike. The current study is inspired by the report of S. Izumiya and T. Sato [18]. We focus on the singularity analysis of lightlike hypersurfaces of partially null curves.

The paper is organized as follows: Section 2 summarizes the required formalism of the basic notions concerning the semi-Euclidean 4-space with index 2 and gives the main results about geometric information of singularities of lightlike hypersurfaces, which can measure the the order of the contact between a partially null curve and a lightcone. Section 3 introduces the one parameter family of lightcone Gaussian indicatrices named lightcone Gaussian surfaces from the view point of differential geometry. Section 4 constructs Lorentz distance-squared functions to characterize the contact relations between partially null curves and the lightcone. Section 5 gives the proof of the main result, i.e. Theorem 1, through the methods of the classical unfolding theory in singularity theory. Finally, in Section 6 we give an example to illustrate the singularities of lightlike hypersurfaces and some properties of the lightcone Gaussian surfaces.

We assume throughout the paper that all manifolds and maps are C^∞ unless explicitly stated otherwise.

§2. The basic concepts and Main Results

The semi-Euclidean four space with index two $(\mathbb{R}^4_2, \langle , \rangle)$ is the vector space $\mathbb{R}^4$ endowed with the metric induced by the pseudo-scalar product $\langle \boldsymbol{x}, \boldsymbol{y} \rangle = -x^0y^0 - x^1y^1 + x^2y^2 + x^3y^3$, for any vectors $\boldsymbol{x} = (x^0, x^1, x^2, x^3)$, $\boldsymbol{y} = (y^0, y^1, y^2, y^3)$ in $\mathbb{R}^4$. The non-zero vector $\boldsymbol{x} \in \mathbb{R}^4_2$ is called *spacelike, lightlike or timelike* if $\langle \boldsymbol{x}, \boldsymbol{x} \rangle > 0$, $\langle \boldsymbol{x}, \boldsymbol{x} \rangle = 0$ or $\langle \boldsymbol{x}, \boldsymbol{x} \rangle < 0$ respectively. The *norm* of the vector $\boldsymbol{x} \in \mathbb{R}^4_2$ is defined as $\|\boldsymbol{x}\| = \sqrt{|\langle \boldsymbol{x}, \boldsymbol{x} \rangle|}$. The signature of a vector $\boldsymbol{x} \in \mathbb{R}^4_2 \backslash \{\boldsymbol{0}\}$ is defined as

$$\operatorname{sign}(\boldsymbol{x}) = \begin{cases} 1 & \boldsymbol{x} \text{ is spacelike} \\ 0 & \boldsymbol{x} \text{ is lightlike} \\ -1 & \boldsymbol{x} \text{ is timelike.} \end{cases}$$

We call $NC_\alpha = \{\boldsymbol{x} = (x^0, x^1, x^2, x^3) \in \mathbb{R}^4_2 | \langle \boldsymbol{x} - \boldsymbol{\alpha}, \boldsymbol{x} - \boldsymbol{\alpha} \rangle = 0\}$ a *lightcone with vertex* $\boldsymbol{\alpha}$, and denote $NC^* = NC_0 \backslash \{\boldsymbol{0}\}$.

For any $\boldsymbol{x}_1, \boldsymbol{x}_2, \boldsymbol{x}_3 \in \mathbb{R}^4_2$, we define the vector $\boldsymbol{x}_1 \wedge \boldsymbol{x}_2 \wedge \boldsymbol{x}_3$ as

$$\boldsymbol{x}_1 \wedge \boldsymbol{x}_2 \wedge \boldsymbol{x}_3 = \begin{vmatrix} -\boldsymbol{e}_0 & -\boldsymbol{e}_1 & \boldsymbol{e}_2 & \boldsymbol{e}_3 \\ x_1^0 & x_1^1 & x_1^2 & x_1^3 \\ x_2^0 & x_2^1 & x_2^2 & x_2^3 \\ x_3^0 & x_3^1 & x_3^2 & x_3^3 \end{vmatrix},$$

where $\boldsymbol{x}_i = (x_i^0, x_i^1, x_i^2, x_i^3)$ and $\{\boldsymbol{e}_0, \boldsymbol{e}_1, \boldsymbol{e}_2, \boldsymbol{e}_3\}$ is the canonical basis of $\mathbb{R}^4_2$. Obviously,

$$\langle \boldsymbol{x}, \boldsymbol{x}_1 \wedge \boldsymbol{x}_2 \wedge \boldsymbol{x}_3 \rangle = \det(\boldsymbol{x}, \boldsymbol{x}_1, \boldsymbol{x}_2, \boldsymbol{x}_3),$$

so that $\boldsymbol{x}_1 \wedge \boldsymbol{x}_2 \wedge \boldsymbol{x}_3$ is pseudo orthogonal to any $\boldsymbol{x}_i (i = 1, 2, 3)$.

Let $\gamma : I \longrightarrow \mathbb{R}^4_2$ be a smooth regular curve (i.e.,$\gamma'(t) \neq \boldsymbol{0}$), where I is an open interval. For any $t \in I$, the curve γ is called *spacelike, lightlike or timelike* if the velocity of the curve is $\langle \dot\gamma(t), \dot\gamma(t) \rangle > 0$, $\langle \dot\gamma(t), \dot\gamma(t) \rangle = 0$ or $\langle \dot\gamma(t), \dot\gamma(t) \rangle < 0$ respectively.

Let $\gamma : I \longrightarrow \mathbb{R}^4_2$ be a unit speed timelike curve, parameterized by the arclength parameter s, i.e. $\langle \gamma'(s), \gamma'(s) \rangle = -1$. If γ'' is a spacelike vector, we can choose two lightlike binormal vectors such that γ is a partially null curve. In [21], the authors have given a frame of partially null curve, which contains two transversal lightlike vectors. The tangent and the principal normal vector fields are defined respectively by

$$T(s) = \gamma'(s), \;\; N(s) = \frac{\gamma''(s)}{\|\gamma''(s)\|}.$$

The first and second binormal vector fields are taken from the subspace $\{T, N\}^\perp$, denoted respectively by B_1 and B_2. Then the lightlike frame

$\{T, N, B_1, B_2\}$[21] associated with the partially null curve satisfies

$$\langle T,T\rangle = -\langle N,N\rangle = -1,\ \langle B_1,B_1\rangle = \langle B_2,B_2\rangle = 0,\ \langle B_1,B_2\rangle = -1,$$
$$\langle N,B_1\rangle = \langle N,B_2\rangle = \langle T,N\rangle = \langle T,B_1\rangle = \langle T,B_2\rangle = 0.$$

The two transversal lightlike vectors can be substituted by a spacelike vector and a timelike vector

$$\frac{B_1+B_2}{\sqrt{2}} \ \textit{and}\ \frac{B_1-B_2}{\sqrt{2}}.$$

Therefore, for a general situation, if $\|\gamma''(s)\| \neq 0$ (i.e. γ'' is a spacelike vector or a timelike vector), we can construct, without loss of generality, a pseudo-orthogonal frame without lightlike vectors. Denote $\boldsymbol{t}(s) = T(s)$, $\boldsymbol{n}_1(s) = N(s)$. Take $k_1(s) = \|\gamma''(s)\|$ as a curvature function. As γ is not a pseudo null curve, $k_1(s) \neq 0$. Then take $\boldsymbol{n}_2(s) = (\delta_1 k_1(s)\boldsymbol{t}(s) - \boldsymbol{n}_1'(s))/k_2(s)$, where $k_2(s) = \|\delta_1 k_1(s)\boldsymbol{t}(s) - \boldsymbol{n}_1'(s)\|$ and $\delta_i = \mathrm{sign}(\boldsymbol{n}_i(s))$ $(i = 1, 2, 3)$. $\boldsymbol{n}_3(s)$ is defined as

$$\boldsymbol{n}_3(s) = \frac{\boldsymbol{t}(s)\wedge\boldsymbol{n}_1(s)\wedge\boldsymbol{n}_2(s)}{\|\boldsymbol{t}(s)\wedge\boldsymbol{n}_1(s)\wedge\boldsymbol{n}_2(s)\|}.$$

So we define a pseudo-orthogonal frame $F = \{\boldsymbol{t}(s), \boldsymbol{n}_1(s), \boldsymbol{n}_2(s), \boldsymbol{n}_3(s)\}$ of $\mathbb{R}^4_2$ which is a positively oriented 4-tuple of vectors satisfying

$$\text{(1)}\quad \begin{aligned} &\langle \boldsymbol{t},\boldsymbol{t}\rangle = -1,\ \langle \boldsymbol{n}_i,\boldsymbol{n}_i\rangle = \delta_i,\\ &\langle \boldsymbol{t},\boldsymbol{n}_1\rangle = \langle \boldsymbol{t},\boldsymbol{n}_2\rangle = \langle \boldsymbol{t},\boldsymbol{n}_3\rangle = \langle \boldsymbol{n}_1,\boldsymbol{n}_2\rangle = \langle \boldsymbol{n}_1,\boldsymbol{n}_3\rangle = \langle \boldsymbol{n}_2,\boldsymbol{n}_3\rangle = 0,\end{aligned}$$

where $\delta_1\delta_2\delta_3 = -1$ and $\delta_1 + \delta_2 + \delta_3 = 1$.

The Frenet formula of γ with respect to the frame F is as follows

$$\text{(2)}\quad \begin{cases} \boldsymbol{t}'(s) = k_1(s)\boldsymbol{n}_1(s)\\ \boldsymbol{n}_1'(s) = \delta_1 k_1(s)\boldsymbol{t}(s) - k_2(s)\boldsymbol{n}_2(s)\\ \boldsymbol{n}_2'(s) = -\delta_3 k_2(s)\boldsymbol{n}_1(s) - k_3(s)\boldsymbol{n}_3(s)\\ \boldsymbol{n}_3'(s) = -\delta_1 k_3(s)\boldsymbol{n}_2(s), \end{cases}$$

where $k_2(s) = -\delta_2\langle \boldsymbol{n}_1'(s), \boldsymbol{n}_2(s)\rangle$, $k_3(s) = -\delta_3\langle \boldsymbol{n}_2'(s), \boldsymbol{n}_3(s)\rangle$.

Remark 1. *γ is a partially null curve when $\delta_1 = 1$. We can take $\boldsymbol{n}_2 \pm \boldsymbol{n}_3$ as the two lightlike binormal vectors. For the sake of completeness and unification, we take the pseudo-orthogonal frame $F = \{\boldsymbol{t}, \boldsymbol{n}_1, \boldsymbol{n}_2, \boldsymbol{n}_3\}$ without lightlike vectors, and naturally extend our results to the case of $\|\gamma''\| \neq 0$.*

Remark 2. *We focus on $k_2(s) \neq 0$. Otherwise,*

$$\boldsymbol{n}_1'(s) \equiv \delta_1 k_1(s)\boldsymbol{t}(s).$$

It means that γ is locally a plane curve.

We define $\mathfrak{ng}_\gamma : U \longrightarrow NC^*$, where $U = I \times [0, 2\pi)$, by

$$\begin{aligned} \mathfrak{ng}_\gamma(s,\theta) = &\left(\frac{1-\delta_1}{2} + \frac{1-\delta_2}{2}\cos\theta + \frac{1-\delta_3}{2}\cos\theta\right)\boldsymbol{n}_1(s) \\ &+ \left(\frac{1-\delta_1}{2}\cos\theta + \frac{1-\delta_2}{2} + \frac{1-\delta_3}{2}\sin\theta\right)\boldsymbol{n}_2(s) \\ &+ \left(\frac{1-\delta_1}{2}\sin\theta + \frac{1-\delta_2}{2}\sin\theta + \frac{1-\delta_3}{2}\right)\boldsymbol{n}_3(s), \end{aligned} \tag{3}$$

it is called the *lightcone Gaussian surface* of γ.

Remark 3. *If $\delta_1 = \delta_3 = 1$, $\delta_2 = -1$, $\mathfrak{ng}_\gamma(s,\theta) = \cos\theta\boldsymbol{n}_1(s) + \boldsymbol{n}_2(s) + \sin\theta\boldsymbol{n}_3(s)$, that is a surface on the lightcone.*

We define the *lightlike hypersurface along γ*

$$\mathfrak{nh}_\gamma : U \times \mathbb{R} \longrightarrow \mathbb{R}^4_2$$

by $\mathfrak{nh}_\gamma(s,\theta,t) = \gamma(s) + t\mathfrak{ng}_\gamma(s,\theta)$. If we fix θ_0, the lightlike hypersurface is just a lightlike ruled surface along γ.

We also define a new important function of the timelike curve in $\mathbb{R}^4_2$ by

$$\begin{aligned} \eta(s) = \Big(&k_1k_2(k_1'' + \delta_3 k_1 k_2^2) - k_1'(2k_1'k_2 + k_1k_2') \\ &\mp k_1k_2k_3\sqrt{\delta_2 k_1^2 k_2^2 + \delta_1 (k_1')^2}\,\Big)(s). \end{aligned} \tag{4}$$

Let $F : \mathbb{R}^4_2 \longrightarrow \mathbb{R}$ be a submersion and $\gamma : I \longrightarrow \mathbb{R}^4_2$ be a timelike curve. We say that γ and $F^{-1}(0)$ have *k-point contact* for $t = t_0$ if the function $h(t) = F \circ \gamma(t)$ satisfies $h(t_0) = h'(t_0) = \cdots = h^{(k-1)}(t_0) = 0$, $h^{(k)}(t_0) \neq 0$. We also say that γ and $F^{-1}(0)$ have *at least k-point contact* for $t = t_0$ if the function $h(t) = F \circ \gamma(t)$ satisfies $h(t_0) = h'(t_0) = \cdots = h^{(k-1)}(t_0) = 0$. For any fixed $\boldsymbol{v}_0 \in \mathbb{R}^4_2$, we have a model surface $NC_{\boldsymbol{v}_0}$. It is a lightcone with vertex $\boldsymbol{v}_0$. We now consider the following conditions

(A 1) The number of points p of $\gamma(I)$ where the model surface at p having five-point contact with the curve γ is finite.

(A 2) There is no point p of $\gamma(I)$ where the model surface at p having greater than or equal to six-point contact with the curve γ.

Here, we present the main results in this paper.

Theorem 1. *Let $\gamma : I \longrightarrow \mathbb{R}^4_2$ be a timelike curve with $\|\gamma''(s)\| \neq 0$. Let $\boldsymbol{v}_0 = \mathfrak{nh}_\gamma(s_0, \theta_0, t_0)$, we have the following:*
(1) NC_{v_0} and γ have at least 2-point contact at s_0.
(2) NC_{v_0} and γ have 3-point contact at s_0 if and only if there exists $\theta_0 \in [0, 2\pi)$ such that

$$\varphi(s_0, \theta_0) \neq 0 \text{ and } \gamma(s_0) - \boldsymbol{v}_0 = \frac{1}{k_1(s_0)(\frac{\delta_1 - 1}{2} + \frac{\delta_1 + 1}{2}\cos\theta_0)} \mathfrak{ng}_\gamma(s_0, \theta_0),$$

where $\varphi(s,\theta) = k_1'(s)(\frac{\delta_1-1}{2} + \frac{\delta_1+1}{2}\cos\theta) - k_1(s)k_2(s)(\frac{1-\delta_1}{2}\cos\theta + \frac{\delta_2-1}{2} + \frac{1-\delta_3}{2}\sin\theta)$. Under this condition, the lightlike hypersurface $\mathfrak{nh}_\gamma$ at $\boldsymbol{v}_0$ is locally diffeomorphic to $C(2,3) \times \mathbb{R}^2$ and the lightlike focal set $\mathfrak{nf}_\gamma$ is non-singular.
(3) NC_{v_0} and γ have 4-point contact at s_0 if and only if there exists $\theta_0 = \theta(s_0) \in [0, 2\pi)$ such that $\varphi(s_0, \theta(s_0)) = 0$, $\eta(s_0) \neq 0$ and

$$\gamma(s_0) - \boldsymbol{v}_0 = \frac{1}{k_1(s_0)(\frac{\delta_1 - 1}{2} + \frac{\delta_1 + 1}{2}\cos\theta_0)} \mathfrak{ng}_\gamma(s_0, \theta(s_0)).$$

Under this condition, the lightlike hypersurface $\mathfrak{nh}_\gamma$ at $\boldsymbol{v}_0$ is locally diffeomorphic to $SW \times \mathbb{R}$, the lightlike focal set $\mathfrak{nf}_\gamma$ is locally diffeomorphic to $C(2,3,4) \times \mathbb{R}$ and the singular value set of $\mathfrak{nf}_\gamma$ is a regular curve.
(4) NC_{v_0} and γ have 5-point contact at s_0 if and only if there exists $\theta(s_0) \in [0, 2\pi)$ such that $\varphi(s_0, \theta(s_0)) = \eta(s_0) = 0$, $\eta'(s_0) \neq 0$ and

$$\gamma(s_0) - \boldsymbol{v}_0 = \frac{1}{k_1(s_0)(\frac{\delta_1 - 1}{2} + \frac{\delta_1 + 1}{2}\cos\theta_0)} \mathfrak{ng}_\gamma(s_0, \theta(s_0)).$$

Under this condition, the lightlike hypersurface $\mathfrak{nh}_\gamma$ at $\boldsymbol{v}_0$ is locally diffeomorphic to BF, the lightlike focal set $\mathfrak{nf}_\gamma$ is locally diffeomorphic to $C(BF) \times \mathbb{R}$ and the singular value set of $\mathfrak{nf}_\gamma$ is locally diffeomorphic to the $C(2,3,4,5)$-cusp.

We respectively call

$$C(2,3) = \{(x^1, x^2) | x^1 = u^2, x^2 = u^3\},$$
$$C(2,3,4) = \{(x^1, x^2, x^3) | x^1 = u^2, x^2 = u^3, x^3 = u^4\},$$
$$C(2,3,4,5) = \{(x^1, x^2, x^3, x^4) | x^1 = u^2, x^2 = u^3, x^3 = u^4, x^4 = u^5\}$$

$(2,3)$-*cusp*, $(2,3,4)$-*cusp*, $(2,3,4,5)$-*cusp*.

And we respectively call $SW = \{(x^1, x^2, x^3) | x^1 = 3u^4 + u^2 v, x^2 = 4u^3 + 2uv, x^3 = v\}$, $BF = \{(x^1, x^2, x^3, x^4) | x^1 = 5u^4 + 3vu^2 + 2wu, x^2 = 4u^5 + 2vu^3 + wu^2, x^3 = u, x^4 = v\}$, $C(BF) = \{(x^1, x^2, x^3, x^4) | x^1 =$

$6u^5 + u^3v, x^2 = 25u^4 + 9u^2v, x^3 = 10u^3 + 3uv, x^4 = v\}$ *swallowtail, butterfly, c-butterfly* (i.e., the singular value set of the butterflies). One can see Figure 1, Figure 2 and Figure 3. We will give the proof of Theorem 1 in §5.

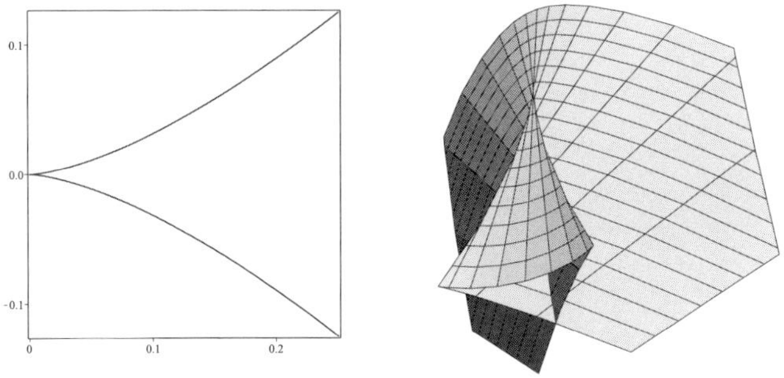

Fig. 1. (2,3)-cusp and Swallowtail.

§3. Lightcone Gaussian Surface

In this section, we analyse the lightcone Gaussian surface from the view point of differential geometry. And we obtain that a lightcone Gaussian surface is locally either a regular Lorentz surface, or a 1-lightlike surface.

Let $\boldsymbol{x} : U \longrightarrow NC^*$ be an embedding of an open subset $U \subset \mathbb{R}^2$. We denote $M = \boldsymbol{x}(U)$ and identify M and U through the embedding $\boldsymbol{x}$. Denote TM and T_pM the tangent bundle and the tangent space of M at $p \in M$. M is called a *Lorentz surface* if T_pM is a Lorentz plane for any point $p \in M$.

If $\langle \cdot, \cdot \rangle$ is degenerate on TM, we say that M is a *lightlike submanifold* of NC^*. Next, we introduce some basic notions about lightlike submanifolds (see [7]). Denote by $\mathcal{F}(M)$ the algebra of smooth functions on M and by $\Gamma(E)$ the $\mathcal{F}(M)$ module of smooth sections of a vector bundle E (same notation for any other vector bundle) over M. For a degenerate tensor field $\langle \cdot, \cdot \rangle$ on M, there exists locally a vector field $\xi \in \Gamma(TM)$ such that $\langle \xi, \boldsymbol{X} \rangle = 0$ for any $\boldsymbol{X} \in \Gamma(TM)$. Then for each tangent space T_pM we have $T_pM^\perp = \{\boldsymbol{u} \in T_pNC^* | \langle \boldsymbol{u}, \boldsymbol{v} \rangle = 0 \ \forall \boldsymbol{v} \in T_pM\}$, which is a degenerate 1-dimensional subspace of T_pNC^*. The *radical subspace* of T_pM (denoted as $\mathrm{Rad}T_pM$) is defined by

$$\mathrm{Rad}T_pM = \{\xi_p \in T_pM | \langle \xi_p, \boldsymbol{X} \rangle = 0 \ \forall \boldsymbol{X} \in T_pM\}.$$

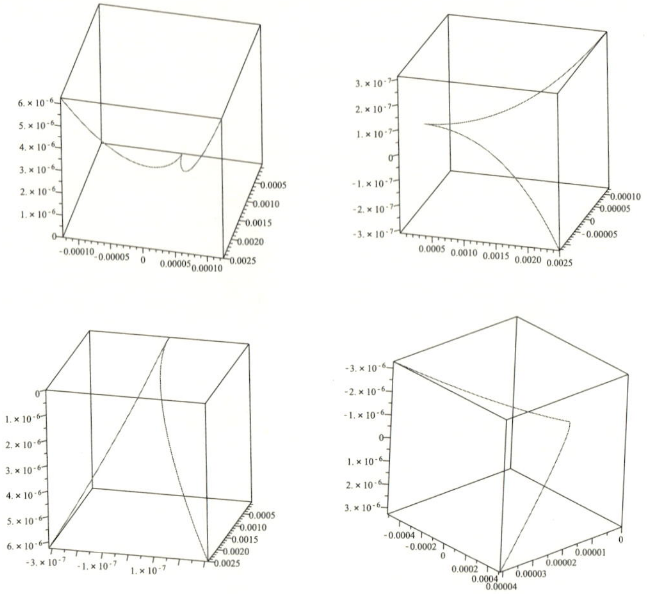

Fig. 2. Projection of a (2,3,4,5)-cusp respectively on $x^1x^2x^3$-space, $x^1x^2x^4$-space, $x^1x^3x^4$-space, $x^2x^3x^4$-space.

The dimension of $\mathrm{Rad}T_pM = T_pM \cap T_pM^{\perp}$ depends on $p \in M$. The submanifold M of NC^* is said to be a 1-*lightlike submanifold* if the mapping

$$\begin{aligned}\mathrm{Rad}TM : M &\longrightarrow TM\\ p &\longmapsto \mathrm{Rad}T_pM\end{aligned}$$

defines a smooth distribution of rank 1 on M.

For the lightcone Gaussian surface $\mathfrak{ng}_\gamma$, we have the following results.

Proposition 1. *Let $\mathfrak{ng}_\gamma$ be the lightcone Gaussian surface of $\boldsymbol{\gamma}$ with $\|\boldsymbol{\gamma}''(s)\| \neq 0$.*
(1) *If $\boldsymbol{n}_1$ is a timelike vector, $\mathfrak{ng}_\gamma$ is a regular surface.*
(2) *If $\boldsymbol{n}_1$ is a spacelike vector, the singular set of $\mathfrak{ng}_\gamma$ is*

$$\{(\boldsymbol{n}_2 + \boldsymbol{n}_3)(s_0), \pm(\boldsymbol{n}_2 - \boldsymbol{n}_3)(s_0) | k_3(s_0) = 0\}.$$

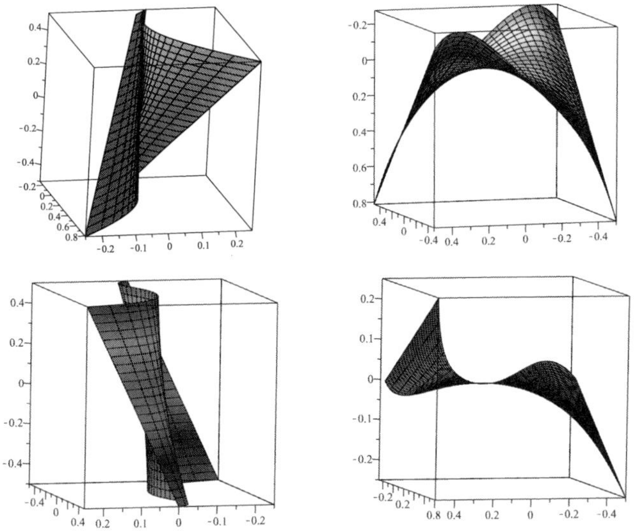

Fig. 3. When $v = 0$, projection of a butterfly respectively on $x^1x^2x^3$-space, x^1x^2w-space, x^1x^3w-space, x^2x^3w-space.

Proof. If $\delta_1 = -1$,

$$\frac{\partial \mathfrak{ng}_\gamma}{\partial \theta} = -\sin\theta \boldsymbol{n}_2 + \cos\theta \boldsymbol{n}_3,$$
$$\frac{\partial \mathfrak{ng}_\gamma}{\partial s} = -k_1 \boldsymbol{t} - k_2\cos\theta \boldsymbol{n}_1 + (k_3\sin\theta - k_2)\boldsymbol{n}_2 - k_3\cos\theta \boldsymbol{n}_3.$$

As $k_1(s) \neq 0$, the above two vectors are definitely linearly independent. It means $\mathfrak{ng}_\gamma$ is a regular surface. If $\delta_1 \neq -1$,

$$\frac{\partial \mathfrak{ng}_\gamma}{\partial \theta} = -\sin\theta \boldsymbol{n}_1 + \frac{1-\delta_3}{2}\cos\theta \boldsymbol{n}_2 + \frac{1+\delta_3}{2}\cos\theta \boldsymbol{n}_3,$$
$$\begin{aligned}\frac{\partial \mathfrak{ng}_\gamma}{\partial s} =& \frac{1+\delta_1}{2}k_1\cos\theta \boldsymbol{t} + \left(-\frac{1-\delta_2}{2}k_2 + \frac{1-\delta_3}{2}k_2\sin\theta\right)\boldsymbol{n}_1 \\ &+ \left(\frac{\delta_2-1}{2}k_3\sin\theta - k_2\cos\theta - \frac{1-\delta_3}{2}k_3\right)\boldsymbol{n}_2 \\ &+ \left(-\frac{1-\delta_2}{2}k_3 - \frac{1-\delta_3}{2}k_3\sin\theta\right)\boldsymbol{n}_3.\end{aligned}$$

$\partial \mathfrak{ng}_\gamma/\partial\theta$ and $\partial \mathfrak{ng}_\gamma/\partial s$ are linearly dependent if and only if $\cos\theta_0 = 0$ and $k_3(s_0) = 0$. Therefore, the singular set of $\mathfrak{ng}_\gamma$ is given by

$$\{(\boldsymbol{n}_2 + \boldsymbol{n}_3)(s_0), \pm(\boldsymbol{n}_2 - \boldsymbol{n}_3)(s_0)|k_3(s_0) = 0\}.$$

Q.E.D.

Proposition 2. *Let $\mathfrak{ng}_\gamma$ be the lightcone Gaussian surface of γ with $\|\gamma''(s)\| \neq 0$.*
(1) If $\boldsymbol{n}_1$ is a timelike vector, $\mathfrak{ng}_\gamma$ is a Lorentz surface.
(2) If $\boldsymbol{n}_1$ is a spacelike vector, except the singular parts $\mathfrak{ng}_\gamma$ is a Lorentz surface in the local neighborhood of (s, θ_0), where $\theta_0 \neq \pi/2, 3\pi/2$. Otherwise, if $\theta_0 = \pi/2$ or $3\pi/2$, it is a 1-lightlike surface.

Proof. At regular parts, $\mathfrak{ng}_\gamma$ can be locally generated by this two vectors $\partial \mathfrak{ng}_\gamma/\partial\theta$ and $\partial \mathfrak{ng}_\gamma/\partial s$. Obviously, $\partial \mathfrak{ng}_\gamma/\partial\theta$ is a spacelike vector. Let

$$\boldsymbol{\iota} = P - \left\langle P, \frac{\partial \mathfrak{ng}_\gamma}{\partial\theta}\right\rangle \frac{\partial \mathfrak{ng}_\gamma}{\partial\theta},$$

where

$$P = \left(\frac{1+\delta_3}{2}k_3 + \frac{1-\delta_3}{2}k_2\right)\frac{\partial \mathfrak{ng}_\gamma}{\partial\theta} + \left(\frac{1+\delta_2}{2} + \frac{1-\delta_2}{2}\cos\theta\right)\frac{\partial \mathfrak{ng}_\gamma}{\partial s}.$$

Then $\langle\boldsymbol{\iota}, \partial \mathfrak{ng}_\gamma/\partial\theta\rangle = 0$ and

$$\langle\boldsymbol{\iota}, \boldsymbol{\iota}\rangle = \frac{1-\delta_1}{2}(-k_1^2) + \frac{1-\delta_2}{2}(-k_1^2\cos^4\theta) + \frac{1-\delta_3}{2}(-k_1^2\cos^2\theta).$$

Thus, $\mathfrak{ng}_\gamma$ can also be locally generated by $\partial \mathfrak{ng}_\gamma/\partial\theta$ and $\boldsymbol{\iota}$ at regular parts.

When $\delta_1 = -1$, $\langle\boldsymbol{\iota}, \boldsymbol{\iota}\rangle = -k_1^2 < 0$. It means $\mathfrak{ng}_\gamma$ is a Lorentz surface. When $\delta_1 \neq -1$, $\langle\boldsymbol{\iota}, \boldsymbol{\iota}\rangle \leq 0$. It means $\mathfrak{ng}_\gamma$ is a Lorentz surface (1-lightlike surface) in the local neighborhood of (s, θ_0), where

$$\theta_0 \neq \frac{\pi}{2}, \frac{3}{2}\pi \ \left(\theta_0 = \frac{\pi}{2} \ or \ \frac{3}{2}\pi\right).$$

This completes the proof. Q.E.D.

§4. A Family of Lorentz Distance-Squared Functions

In this section we introduce one very useful family of functions on a partially null curve. For a partially null curve $\boldsymbol{\gamma}$, we define the function

$$G : I \times \mathbb{R}^4_2 \longrightarrow \mathbb{R},\ G(s, \boldsymbol{v}) = \langle\boldsymbol{\gamma}(s) - \boldsymbol{v}, \boldsymbol{\gamma}(s) - \boldsymbol{v}\rangle.$$

This function is called the *Lorentz distance-squared function* of $\boldsymbol{\gamma}$. We use the notation $g_v(s) = G(s, \boldsymbol{v})$ for any fixed vector $\boldsymbol{v}$ in $\mathbb{R}^4_2$. They describe the contact between $\boldsymbol{\gamma}(s)$ and a lightcone. As we study this family of functions, it will become clear how singularities and the corresponding catastrophes arise.

By using Eqs. (2) and by making tedious calculations, we can state Proposition 3.

Proposition 3. *Let* $\gamma : I \longrightarrow \mathbb{R}^4_2$ *be a timelike curve with* $\|\gamma''(s)\| \neq 0$. *Suppose that* $\boldsymbol{\gamma}(s_0) \neq \boldsymbol{v}_0$, *then we have the following.*
(1) $g_{v_0}(s_0) = g'_{v_0}(s_0) = 0$ *if and only if there exist* $\theta_0 \in [0, 2\pi)$ *and* $\mu \in \mathbb{R} \setminus \{0\}$ *such that* $\boldsymbol{\gamma}(s_0) - \boldsymbol{v}_0 = \mu \mathfrak{ng}_\gamma(s_0, \theta_0)$.
(2) $g_{v_0}(s_0) = g'_{v_0}(s_0) = g''_{v_0}(s_0) = 0$ *if and only if there exists* $\theta_0 \in [0, 2\pi)$ *such that*

$$\boldsymbol{\gamma}(s_0) - \boldsymbol{v}_0 = \frac{1}{k_1(s_0)(\frac{\delta_1 - 1}{2} + \frac{\delta_1 + 1}{2} \cos\theta_0)} \mathfrak{ng}_\gamma(s_0, \theta(s_0)).$$

(3) $g_{v_0}(s_0) = g'_{v_0}(s_0) = g''_{v_0}(s_0) = g'''_{v_0}(s_0) = 0$ *if and only if there exists* $\theta_0 \in [0, 2\pi)$ *such that*

$$\boldsymbol{\gamma}(s_0) - \boldsymbol{v}_0 = \frac{1}{k_1(s_0)(\frac{\delta_1 - 1}{2} + \frac{\delta_1 + 1}{2} \cos\theta_0)} \mathfrak{ng}_\gamma(s_0, \theta(s_0))$$

and $\varphi(s_0, \theta_0) = 0$, *where* $\varphi(s, \theta) = k'_1(s)(\frac{\delta_1 - 1}{2} + \frac{\delta_1 + 1}{2} \cos\theta) - k_1(s)k_2(s)$ $(\frac{1-\delta_1}{2} \cos\theta + \frac{\delta_2 - 1}{2} + \frac{1-\delta_3}{2} \sin\theta)$. *So we can write* $\theta_0 = \theta(s_0)$.
(4) $g_{v_0}(s_0) = g'_{v_0}(s_0) = g''_{v_0}(s_0) = g'''_{v_0}(s_0) = g^{(4)}_{v_0}(s_0) = 0$ *if and only if there exists* $\theta(s_0) \in [0, 2\pi)$ *such that*

$$\boldsymbol{\gamma}(s_0) - \boldsymbol{v}_0 = \frac{1}{k_1(s_0)(\frac{\delta_1 - 1}{2} + \frac{\delta_1 + 1}{2} \cos\theta_0)} \mathfrak{ng}_\gamma(s_0, \theta(s_0))$$

and $\varphi(s_0, \theta(s_0)) = \eta(s_0) = 0$.
(5) $g_{v_0}(s_0) = g'_{v_0}(s_0) = g''_{v_0}(s_0) = g'''_{v_0}(s_0) = g^{(4)}_{v_0}(s_0) = g^{(5)}_{v_0}(s_0) = 0$ *if and only if there exists* $\theta(s_0) \in [0, 2\pi)$ *such that*

$$\boldsymbol{\gamma}(s_0) - \boldsymbol{v}_0 = \frac{1}{k_1(s_0)(\frac{\delta_1 - 1}{2} + \frac{\delta_1 + 1}{2} \cos\theta_0)} \mathfrak{ng}_\gamma(s_0, \theta(s_0))$$

and $\varphi(s_0, \theta(s_0)) = \eta(s_0) = \eta'(s_0) = 0$.

The above proposition also states that the discriminant set of the Lorentz distance-squared function G is given by

$$D_G = \mathfrak{nh}_\gamma(U \times \mathbb{R}) = \{\boldsymbol{v} = \boldsymbol{\gamma}(s) + \mu \mathfrak{ng}_\gamma(s, \theta) \mid (s, \theta) \in U, \mu \in \mathbb{R}\},$$

which is the image of the lightlike hypersurface along γ. Therefore, a singular point of the lightlike hypersurface is the point $\boldsymbol{v}_0 = \gamma(s_0) + \mu_0 \mathfrak{ng}_\gamma(s_0, \theta_0)$, where $\mu_0 = (k_1(s_0)(\frac{\delta_1 - 1}{2} + \frac{\delta_1+1}{2}\cos\theta_0))^{-1}$.

We define $\mathfrak{nf}_\gamma : U \longrightarrow \mathbb{R}^4_2$ as

$$\mathfrak{nf}_\gamma(s, \theta) = \gamma(s) + \mu_0 \mathfrak{ng}_\gamma(s, \theta),$$

we call it the *lightlike focal set of* γ. By definition, the lightlike focal set is the singular value set of the lightlike hypersurface $\mathfrak{nh}_\gamma$.

§5. Proof of the Main Results

In this section we classify singularities of the lightlike hypersurface along γ as an application of the unfolding theory of functions. Detailed descriptions could be found in [5]. Let

$$F : (\mathbb{R} \times \mathbb{R}^r, (s_0, \boldsymbol{x}_0)) \longrightarrow \mathbb{R}$$

be a function germ. We call F an *r-parameter unfolding of f*, if $f(s) = F_{x_0}(s, \boldsymbol{x}_0)$. We say f has A_k-singularity at s_0, if $f^{(p)}(s_0) = 0$ for all $1 \le p \le k$ and $f^{(k+1)}(s_0) \neq 0$. Let F be an r-parameter unfolding of f, where f has A_k-singularity ($k \ge 1$) at s_0. We denote the $(k-1)$-jet of the partial derivative $\partial F/\partial x^i$ at s_0 as

$$j^{(k-1)}\left(\frac{\partial F}{\partial x^i}(s, \boldsymbol{x}_0)\right)(s_0) = \sum_{j=1}^{k-1} \alpha_{ji}(s - s_0)^j, \ (i = 1, \ldots, r).$$

If the rank of $k \times r$ matrix $(\alpha_{0i}, \alpha_{ji})$ is k ($k \le r$), then F is called a *versal unfolding* of f, where $\alpha_{0i} = \partial F/\partial x^i(s_0, \boldsymbol{x}_0)$.

Inspired by the proposition in the previous section, we have:

$$D^l_F = \left\{\boldsymbol{x} \in \mathbb{R}^r \mid \exists\, s \in \mathbb{R}, F(s, \boldsymbol{x}) = \frac{\partial F}{\partial s}(s, \boldsymbol{x}) = \cdots = \frac{\partial^l F}{\partial s^l}(s, \boldsymbol{x}) = 0\right\},$$

which is called a *discriminant set with order l*. Therefore, we have the following proposition.

Proposition 4. *For a timelike curve* γ *with* $\|\gamma''(s)\| \neq 0$,

$$D_G = D^1_G = \mathfrak{nh}_\gamma(U \times \mathbb{R}), \ D^2_G = \mathfrak{nf}_\gamma(U)$$

and

$$D^3_G \textit{ is the singular value set of } \mathfrak{nf}_\gamma.$$

Then we have the following classification theorem as Corollary 7.7 in [18].

Theorem 2. *Let* $F : (\mathbb{R} \times \mathbb{R}^r, (s_0, \boldsymbol{x}_0)) \longrightarrow \mathbb{R}$ *be an r-parameter unfolding of f with* A_k*-singularity at* s_0. *Suppose F is a versal unfolding of f, then we have the following assertions*:
(*a*) *If* $k = 1$, *then* D_F *is locally diffeomorphic to* $\{0\} \times \mathbb{R}^{r-1}$ *and* $D_F^2 = \emptyset$.
(*b*) *If* $k = 2$, *then* D_F *is locally diffeomorphic to* $C(2,3) \times \mathbb{R}^{r-2}$, D_F^2 *is locally diffeomorphic to* $\{\mathbf{0}\} \times \mathbb{R}^{r-2}$ *and* $D_F^3 = \emptyset$.
(*c*) *If* $k = 3$, *then* D_F *is locally diffeomorphic to* $SW \times \mathbb{R}^{r-3}$, D_F^2 *is locally diffeomorphic to* $C(2,3,4) \times \mathbb{R}^{r-3}$, D_F^3 *is locally diffeomorphic to* $\{\mathbf{0}\} \times \mathbb{R}^{r-3}$ *and* $D_F^4 = \emptyset$.
(*d*) *If* $k = 4$, *then* D_F *is locally diffeomorphic to* $BF \times \mathbb{R}^{r-4}$, D_F^2 *is locally diffeomorphic to* $C(BF) \times \mathbb{R}^{r-4}$, D_F^3 *is locally diffeomorphic to* $C(2,3,4,5) \times \mathbb{R}^{r-4}$, D_F^4 *is locally diffeomorphic to* $\{\mathbf{0}\} \times \mathbb{R}^{r-4}$ *and* $D_F^5 = \emptyset$.

For the proof of Theorem 1 we have the following fundamental proposition in this paper.

Proposition 5. *If* $g(s)$ *has* A_k*-singularity* $(k = 1, 2, 3, 4)$ *at* s_0, *then G is a versal unfolding of g.*

Proof. By definition,

$$G(s, \boldsymbol{v}) = -(x^0(s) - v^0)^2 - (x^1(s) - v^1)^2 + (x^2(s) - v^2)^2 + (x^3(s) - v^3)^2,$$

where $\gamma(s) = (x^0(s), x^1(s), x^2(s), x^3(s))$ and $\boldsymbol{v} = (v^0, v^1, v^2, v^3)$. For a fixed $\boldsymbol{v}_0 = (v_0^0, v_0^1, v_0^2, v_0^3)$, the 3-*jet* of $\partial G/\partial v^i(s, \boldsymbol{v}_0)$ at s_0 is

$$j^{(3)} \frac{\partial G}{\partial v^i}(s_0)$$
$$= \begin{cases} 2(x^i)'(s - s_0) + (x^i)''(s - s_0)^2 + \frac{(x^i)'''}{3}(s - s_0)^3 & i = 0, 1 \\ -2(x^i)'(s - s_0) - (x^i)''(s - s_0)^2 - \frac{(x^i)'''}{3}(s - s_0)^3 & i = 2, 3 \end{cases}.$$

The condition for versatility can be checked as follows.
(1) When g has A_1-singularity at s_0, we require the 1×4 matrix

$$A_1 = (2(x^0 - v^0), 2(x^1 - v^1), -2(x^2 - v^2), -2(x^3 - v^3))$$

to have rank 1, which it always does since $\boldsymbol{v_0} \neq \gamma(s_0)$.
(2) When g has A_2-singularity at s_0, we require 2×4 matrix

$$A_2 = \begin{pmatrix} 2(x^0 - v^0) & 2(x^1 - v^1) & -2(x^2 - v^2) & -2(x^3 - v^3) \\ 2(x^0)' & 2(x^1)' & -2(x^2)' & -2(x^3)' \end{pmatrix}$$

to have rank 2. Otherwise, if $\mathrm{rank} A_2 = 1$, it means that $\gamma(s_0) - \boldsymbol{v}_0$ and $\boldsymbol{t}(s_0)$ are linearly dependent. This contradicts with the fact that $\{\boldsymbol{t}(s), \boldsymbol{n}_1(s), \boldsymbol{n}_2(s), \boldsymbol{n}_3(s)\}$ is the pseudo-orthogonal frame of γ.

(3) When g has A_3-singularity at s_0, we require 3×4 matrix

$$A_3 = \begin{pmatrix} 2(x^0 - v^0) & 2(x^1 - v^1) & -2(x^2 - v^2) & -2(x^3 - v^3) \\ 2(x^0)' & 2(x^1)' & -2(x^2)' & -2(x^3)' \\ (x^0)'' & (x^1)'' & -(x^2)'' & -(x^3)'' \end{pmatrix}$$

to have rank 3. Otherwise, if $\mathrm{rank} A_3 = 2$, it means that $\gamma''(s_0)$ can be generated by $\gamma(s_0) - \boldsymbol{v}_0$ and $\boldsymbol{t}(s_0)$. Through a straightforward calculation, we can easily show that it is a contradiction.

(4) When g has A_4-singularity at s_0, we require 4×4 matrix

$$A_4 = \begin{pmatrix} 2(x^0 - v^0) & 2(x^1 - v^1) & -2(x^2 - v^2) & -2(x^3 - v^3) \\ 2(x^0)' & 2(x^1)' & -2(x^2)' & -2(x^3)' \\ (x^0)'' & (x^1)'' & -(x^2)'' & -(x^3)'' \\ \frac{1}{3}(x^0)''' & \frac{1}{3}(x^1)''' & -\frac{1}{3}(x^2)''' & -\frac{1}{3}(x^3)''' \end{pmatrix}$$

to have rank 4.

In fact

$$\begin{aligned} \det A_4 &= \frac{4}{3}\det(\gamma(s_0) - \boldsymbol{v}_0, \gamma'(s_0), \gamma''(s_0), \gamma'''(s_0)) \\ &= -\frac{4k_1(s_0)k_2(s_0)}{3(\frac{\delta_1 - 1}{2} + \frac{\delta_1 + 1}{2}\cos\theta_0)}\left(\frac{\delta_3 + 1}{2}\sin\theta_0 + \frac{\delta_3 - 1}{2}\right). \end{aligned}$$

When $\delta_3 = -1$

$$\det A_4 = \frac{4k_1(s_0)k_2(s_0)}{3(\frac{\delta_1 - 1}{2} + \frac{\delta_1 + 1}{2}\cos\theta_0)} \neq 0.$$

When $\delta_3 = 1$ and $\delta_1 = -1$, we have $\det A_4 \neq 0$ under the condition that $k_1'(s_0) \neq \pm(k_1k_2)(s_0)$. If $k_1'(s_0) = \pm(k_1k_2)(s_0)$, then $k_1''(s_0) = k_1(s_0)k_2^2(s_0) \pm k_1(s_0)k_2'(s_0)$ because $\eta(s_0) = 0$. Then $\eta'(s_0) = 0$. This contradicts with the assumption that g has A_4-singularity at s_0. When $\delta_3 = 1$ and $\delta_2 = -1$, the proof is the same. Here, it is omitted. Therefore, $\mathrm{rank} A_4 = 4$.

In summary, G is a versal unfolding of g. This completes the proof.
Q.E.D.

We now give the proofs of Theorem 1.

Proof of Theorem 1 Let $\gamma : I \longrightarrow \mathbb{R}^4_2$ be a timelike regular curve with $\|\gamma''(s)\| \neq 0$. As $\boldsymbol{v}_0 = \mathfrak{nh}_\gamma(s_0, \theta_0, t_0)$, we give a function $\mathfrak{G} : \mathbb{R}^4_2 \longrightarrow \mathbb{R}$, by $\mathfrak{G}(\boldsymbol{u}) = \langle \boldsymbol{u} - \boldsymbol{v}_0, \boldsymbol{u} - \boldsymbol{v}_0 \rangle$, then we assume that $g_{v_0}(s) = \mathfrak{G}(\gamma(s))$. Because $\mathfrak{G}^{-1}(0) = NC_{v_0}$ and 0 is a regular value of $\mathfrak{G}$, γ and NC_{v_0} have $(k+1)$-point contact for s_0 if and only if $g_{v_0}(s)$ has A_k-singularity at s_0.

Thus γ and NC_{v_0} have at least 2-point contact for s_0 if and only if $g_{v_0}(s_0) = g'_{v_0}(s_0) = 0$. By Proposition 4, we have

$$D_G^1 = \mathfrak{nh}_\gamma(U \times \mathbb{R}),\ D_G^2 = \mathfrak{nf}_\gamma(U)$$

and

$$D_G^3 \text{ is the singular value set of } \mathfrak{nf}_\gamma.$$

By combining Proposition 3, Theorem 2, and Proposition 5, we get the results. Q.E.D.

For the proof of the generic properties, one can see [6] that are omitted here.

§6. Example

As an application and an illustration of the main result (Theorem 1), we give an example of a partially null curve in this section.

Example 1. *Let γ be a unit speed timelike curve of $\mathbb{R}_2^4$ defined by $\gamma(s) = (\sqrt{2}e^s, s, e^s \cos s, e^s \sin s)$ with respect to arclength parameter s and satisfying $\|\gamma''(s)\| \neq 0$. The tangent vector $\boldsymbol{t}(s)$ is given by*

$$\boldsymbol{t}(s) = (\sqrt{2}e^s, 1, e^s \cos s - e^s \sin s, e^s \sin s + e^s \cos s).$$

And

$$\begin{aligned}
\boldsymbol{n}_1(s) =&(1, 0, -\sqrt{2}\sin s, \sqrt{2}\cos s),\\
\boldsymbol{n}_2(s) =&\frac{1}{\sqrt{e^s+1}}(\sqrt{2}e^{2s}, e^s, e^{2s}\cos s - e^{2s}\sin s + \cos s,\\
&e^{2s}\cos s + e^{2s}\sin s + \sin s),\\
\boldsymbol{n}_3(s) =&\frac{1}{\sqrt{e^{2s}+1}}(\sqrt{2}, -e^s, -\sin s, \cos s),\\
k_1(s) =&\sqrt{2}e^s,\\
k_2(s) =&\sqrt{2e^{2s}+2},\\
k_3(s) =&\frac{-e^{3s}\cos s + e^{3s}\sin s + 2e^{2s} - e^s\cos s + 1}{\sqrt{e^s+1}(e^{2s}+1)^{3/2}}.
\end{aligned}$$

Obviously, $\boldsymbol{n}_2$ and $\boldsymbol{n}_3$ can be substituted by two transversal lightlike vectors $\boldsymbol{n}_2 + \boldsymbol{n}_3$ and $\boldsymbol{n}_2 - \boldsymbol{n}_3$. Accordingly, γ is a partially null curve.

In this example $\mathfrak{ng}_\gamma(s,\theta) = \cos\theta\boldsymbol{n}_1(s) + \sin\theta\boldsymbol{n}_2(s) + \boldsymbol{n}_3(s)$. By maple, $k_3(s) \neq 0$ for any $s \in \mathbb{R}$. Thus it is a regular surface. Moreover, $\mathfrak{ng}_\gamma$ is a Lorentz surface in the local neighborhood of (s, θ_0), where $\theta_0 \neq \pi/2, 3\pi/2$. Otherwise, if $\theta_0 = \pi/2$ or $3\pi/2$, it is a 1-lightlike surface.

The important functions associated with γ are as follows

$$\varphi(s,\theta) = \sqrt{2}e^s\cos\theta - 2e^s\sqrt{e^{2s}+1}\sin\theta,$$

$$\eta(s) = -2e^{2s}\sqrt{2e^{2s}+2}(1+2e^{2s}) - \frac{4e^{2s}(3e^{2s}+2)}{\sqrt{2e^{2s}+2}} + \frac{2e^s\sqrt{2e^{2s}(2e^{2s}+3)}(-e^{3s}\cos s + e^{3s}\sin s + 2e^{2s} - e^s\cos s + 1)}{\sqrt{e^s+1}(e^{2s}+1)}.$$

By maple, we find $\eta(s) \neq 0$ for any $s \in \mathbb{R}$ and $\varphi(s,\theta) = 0$ if and only if $\tan\theta = 1/\sqrt{2e^{2s}+2}$.

Furthermore, the vector parametric equations of the lightlike hypersurface $\mathfrak{nh}_\gamma$ are given by

$$\{\mathfrak{nh}_{\gamma 1}, \mathfrak{nh}_{\gamma 2}, \mathfrak{nh}_{\gamma 3}, \mathfrak{nh}_{\gamma 4}\},$$

where

$$\mathfrak{nh}_{\gamma 1}(s,\theta,t) = \sqrt{2}e^s + t\Big(\cos\theta + \frac{\sqrt{2}+\sqrt{2}e^{2s}\sin\theta}{\sqrt{1+e^{2s}}}\Big),$$

$$\mathfrak{nh}_{\gamma 2}(s,\theta,t) = s + \frac{te^s(\sin\theta - 1)}{\sqrt{1+e^{2s}}},$$

$$\mathfrak{nh}_{\gamma 3}(s,\theta,t) = e^s\cos s + t\Big(-\sqrt{2}\sin s\cos\theta + \frac{\sin\theta(e^{2s}\cos s - e^{2s}\sin s + \cos s) - \sin s}{\sqrt{e^{2s}+1}}\Big),$$

$$\mathfrak{nh}_{\gamma 4}(s,\theta,t) = e^s\sin s + t\Big(\sqrt{2}\cos s\cos\theta + \frac{\sin\theta(e^{2s}\sin s + e^{2s}\cos s + \sin s) + \cos s}{\sqrt{e^{2s}+1}}\Big).$$

We take $\boldsymbol{v}_0 = \mathfrak{nh}_\gamma(s_0,\theta_0,t_0)$, where $s_0 = 0$, $\theta_0 = \arctan(1/\sqrt{4})$ and $t_0 = -(\sqrt{2}\cos(\arctan(1/\sqrt{4})))^{-1}$. So $\varphi(s_0,\theta_0) = 0$ and $\eta(s_0) \neq 0$. By Theorem 1, $\mathfrak{nh}_\gamma$ at $\boldsymbol{v}_0$ is locally diffeomorphic to SW $\times \mathbb{R}$, see Figure 1. In general, for any

$$\boldsymbol{v} = \mathfrak{nh}_\gamma(s, \arctan(1/\sqrt{2e^{2s}+2}), -(\sqrt{2}\cos(\arctan(1/\sqrt{4})))^{-1}),$$

we have $\varphi(s,\arctan(1/\sqrt{2e^{2s}+2})) = 0$ and $\eta(s) \neq 0$. Accordingly, $\mathfrak{nh}_\gamma$ is locally diffeomorphic to SW $\times \mathbb{R}$ at $\boldsymbol{v}$.

Acknowledgements

The authors would like to thank G. Ishikawa, S. Izumiya, K. Saji, M. Takahashi, M. Yamamoto and T. Yamamoto for their great help in the conference named Singularities in Generic Geometry and its Applications, Kobe-Kyoto 2015. The authors are grateful to the referee for careful reading and helpful comments. The second author are partially supported by NSF of China No.11671070. The work was partially supported by Science and Technology Foundation of The Education Department of Jilin Province during the "13th Five-Year Plan" of China No. JJKH20181016KJ.

References

[1] A. T. Ali, R. López, M. Turgut, K-type partially null and pseudo null slant helices in Minkowski 4-space, Math. Commun., **17** (2012), 93–103.

[2] P. Bayard, F. Sánchez-Bringas, Geometric invariants and principal configurations on spacelike surfaces immersed in $\mathbb{R}^{3,1}$, Proc. Roy. Soc. Edinburgh Sect. A, **140** (2010), 1141–1160.

[3] W. B. Bonnor, Null curves in a Minkowski space-time, Tensor, **20** (1969), 229–242.

[4] W. B. Bonnor, Curves with null normals in Minkowski space-time, A random walk in relativity and cosmology, **1** (1985), 33–47.

[5] J. W. Bruce, P. J. Giblin, Curves and Singularities: A geometrical introduction to singularity theory, Second edition, Cambridge University Press: Cambridge, 1992.

[6] X. Cui, D. Pei, Singularities of lightlike hypersurfaces of pseudo null curves, J. Funct. Spaces, (2015), Art. ID 502789.

[7] K. L. Duggal, B. Aurel, Lightlike submanifolds of semi-Riemannian manifolds and applications, Mathematics and its Applications, Netherlands, **364**, Dordrecht; Kluwer Academic Publishers Group, Springer, 1996.

[8] İ. Gök, S. Kaya Nurkan, K. İlarslan, On pseudo null Bertrand curves in Minkowski space-time, Kyungpook Math. J., **54** (2014), 685–697.

[9] M. Grbović, K. İlarslan, E. Nešović, On null and pseudo null Mannheim curves in Minkowski 3-space, J. Geom., **105** (2014), 177–183.

[10] K. İlarslan, E. Nešović, Some characterizations of null, pseudo null and partially null rectifying curves in Minkowski space-time, Taiwanese J. Math., **12** (2008), 1035–1048.

[11] K. İlarslan, E. Nešović, Some characterizations of pseudo null and partially null osculating curves in Minkowski space-time, Int. Electron. J. Geom., **4** (2011), 1–12.

[12] S. Izumiya, D. Pei, M. C. Romero-Fuster, Spacelike surfaces in anti de Sitter four-space from a contact viewpoint, Proc. Steklov Inst. Math., **267** (2009), 156–173.

[13] S. Izumiya, D. Pei, M. C. Romero-Fuster, M. Takahashi, The horospherical geometry of submanifolds in hyperbolic space, J. London Math. Soc., **71** (2005), 779–800.
[14] S. Izumiya, D. Pei, T. Sano, E. Torii, Evolutes of hyperbolic plane curves, Acta Math. Sin. (Engl. Ser.), **20** (2004), 543–550.
[15] S. Izumiya, M. C. Romero-Fuster, K. Saji, Flat lightlike hypersurfaces in Lorentz-Minkowski 4-space, J. Geom. Phys., **59** (2009), 1528–1546.
[16] S. Izumiya, K. Saji, The mandala of Legendrian dualities for pseudo-spheres in Lorentz-Minkowski space and "flat" spacelike surfaces, J. Singul., **2** (2010), 92–127.
[17] S. Izumiya, K. Saji, M. Takahashi, Horospherical flat surfaces in hyperbolic 3-space, J. Math. Soc. Japan, **62** (2010), 789–849.
[18] S. Izumiya, T. Sato, Lightlike hypersurfaces along spacelike submanifolds in Minkowski space-time, J. Geom. Phys., **71** (2013), 30–52.
[19] L. F. Martins, J. J. Nuño-Ballesteros, Contact properties of surfaces in $\mathbb{R}^3$ with corank 1 singularities, Tohoku Math. J., **67** (2015), 105–124.
[20] U. Ozturk, E. B. Koc Ozturk, K. İlarslan, On the involute-evolute of the pseudonull curve in Minkowski 3-space, J. Appl. Math., (2013), Art. ID 651495.
[21] M. Petrović-Torgašev, K. İlarslan, E. Nešović, On partially null and pseudo null curves in the semi-Euclidean space $\mathbb{R}^4_2$, J. Geom., **84** (2005), 106–116.
[22] F. Tari, Umbilics of surfaces in the Minkowski 3-space, J. Math. Soc. Japan, **65** (2013), 723–731.
[23] A. Uçum, H. A. Erdem, K. İlarslan, Inextensible flows of partially null and pseudo null curves in semi-Euclidean 4-space with index 2, Preprint.
[24] J. Walrave, Curves and surfaces in Minkowski space, Ph.D Dissertation, K. U. Leuven, Fac. of Science, Leuven, 1995.
[25] S. Yılmaz, E. Özyılmaz, M. Turgut, Ş. Nizamoğlu, Position vector of a partially null curve derived from a vector differential equation, Int. J. Comput. Math. Sci., **3** (2009), 71–73.

[a] *School of Mathematics and Statistics, Changchun University of Technology, Changchun, 130012, P.R.China*
E-mail address: cuixp606@nenu.edu.cn

[b] *School of Mathematics and Statistics, Northeast Normal University, Changchun, 130024, P.R.China*
E-mail address: peidh340@nenu.edu.cn

Advanced Studies in Pure Mathematics 78, 2018
Singularities in Generic Geometry
pp. 201–219

Triply periodic zero mean curvature surfaces in Lorentz-Minkowski 3-space

Shoichi Fujimori

Abstract.

We construct triply periodic zero mean curvature surfaces of mixed type in the Lorentz-Minkowski 3-space $\mathbb{L}^3$, with the same topology as the triply periodic minimal surfaces in the Euclidean 3-space $\mathbb{R}^3$, called Schwarz rPD surfaces.

§1. Introduction

Zero mean curvature surfaces in the Lorentz-Minkowski 3-space $\mathbb{L}^3$ are smooth surfaces with mean curvature zero wherever the mean curvature is defined. Having the mean curvature defined at all points is not expected, because these surfaces can change causal type, meaning that some parts may have spacelike tangent planes and other parts may have timelike tangent planes, with lightlike tangent planes at boundary points between these parts. An interesting aspect of these surfaces is precisely that they change causal type, often resulting in interesting singular and topological behaviors, and these surfaces have been investigated in [1, 5, 6, 7, 9]. One of the main tools for the construction of such surfaces is based on the fact that fold singularities of spacelike maximal surfaces have real analytic extensions to timelike minimal surfaces [5].

In contrast to minimal surfaces in the Euclidean 3-space $\mathbb{R}^3$, the only known triply periodic zero mean curvature surfaces in $\mathbb{L}^3$ were those in a 1-parameter family constructed in [9], while there are many known triply periodic minimal surfaces in $\mathbb{R}^3$, see for example [4, 11, 12, 16].

Received July 7, 2016.
Revised March 8, 2017.
2010 *Mathematics Subject Classification.* Primary 53A10; Secondary 53A35, 53C50.
Key words and phrases. zero mean curvature, triply periodic surface, fold singularity.
The author was partially supported by the Grant-in-Aid for Young Scientists (B) No. 25800047 from Japan Society for the Promotion of Science (JSPS), as well as the Joint Project Grant between the Austrian Science Fund (FWF) and JSPS I1671-N26.

This motivates us to broaden our knowledge of triply periodic zero mean curvature surfaces in $\mathbb{L}^3$, and in this paper we construct a new 1-parameter family of triply periodic zero mean curvature surfaces in $\mathbb{L}^3$ based on the conjugate surfaces of the triply periodic minimal surfaces in $\mathbb{R}^3$ called the Schwarz H surfaces. This is the main original result of this paper, and the family here is interesting because it exhibits both a change of causal type, and also a greater complexity than the previously known examples. The method we use to construct the family is essentially the same as in [9], but the surfaces here are less symmetric, and so the construction is more involved. It is expected that by using the method in this paper one could construct families of surfaces with more complicated topologies, based on the data of triply periodic minimal surfaces in $\mathbb{R}^3$ constructed in [11, 12]. It is also expected that the family of surfaces we construct in this paper is a prototype for the study of the moduli space of triply periodic zero mean curvature surfaces.

We remark that the surfaces constructed in this paper have the same topology and symmetry as Schwarz rPD minimal surfaces, not Schwarz H surfaces. (As for the symbols “rPD” and “H”, see Remark A.4 in Appendix A.)

§2. Preliminaries

We denote by $\mathbb{L}^3$ the Lorentz-Minkowsiki 3-space with indefinite metric $\langle\ ,\ \rangle = -dx_0^2 + dx_1^2 + dx_2^2$. Let M be a Riemann surface. A conformal immersion $f : M \to \mathbb{L}^3$ is called a *spacelike surface* if the induced metric $ds^2 = \langle df, df\rangle$ is positive definite on M. A spacelike surface $f : M \to \mathbb{L}^3$ is called *maximal* if its mean curvature vanishes identically. In [18] a notion of *maxface* was introduced as a maximal surface with certain kind of singularities. More precisely, $f : M \to \mathbb{L}^3$ is called a *maxface* if there exists an open dense subset W of M such that the restriction $f|_W$ of f to W gives a conformal maximal immersion and $df(p) \neq 0$ for all $p \in M$.

For maxfaces, a similar representation formula to Weierstrass representation for minimal surfaces in $\mathbb{R}^3$ is known.

Theorem 2.1 (Weierstrass-type representation [14, 18]). *Let (g, η) be a pair of a meromorphic function g and a holomorphic differential η on a Riemann surface M so that $(1 + |g|^2)^2\eta\bar{\eta}$ gives a Riemannian metric on M. We set*

$$\Phi = \begin{pmatrix} -2g\eta \\ (1+g^2)\eta \\ i(1-g^2)\eta \end{pmatrix}, \tag{2.1}$$

where $i = \sqrt{-1}$. *Then*

$$f = \mathrm{Re} \int_{z_0}^{z} \Phi : M \to \mathbb{L}^3 \qquad (z_0 \in M) \tag{2.2}$$

defines a maxface. The singular set $S(f)$ *of* f *is given by*

$$S(f) = \{p \in M \ ; \ |g(p)| = 1\}.$$

Moreover, f *is single-valued on* M *if and only if*

$$\mathrm{Re} \oint_{\ell} \Phi = \mathbf{0} \tag{2.3}$$

for any closed curve ℓ *on* M. *Conversely, any maxface can be obtained in this manner.*

The pair (g, η) in Theorem 2.1 is called the *Weierstrass data* of f.

Remark 2.2. The first fundamental form ds^2 and the second fundamental form $\mathrm{I\!I}$ of the surface (2.2) are given by

$$ds^2 = \left(1 - |g|^2\right)^2 \eta\bar{\eta}, \qquad \mathrm{I\!I} = -\eta dg - \overline{\eta dg}.$$

Moreover, $g|_{M \setminus S(f)} : M \setminus S(f) \to (\mathbb{C} \cup \{\infty\}) \setminus \{|z| = 1\}$ coincides with the composition of the Gauss map

$$G|_{M \setminus S(f)} : M \setminus S(f) \to H^2 = \{x \in \mathbb{L}^3 \ ; \ \langle x, x \rangle = -1\}$$

of the maximal surface and the stereographic projection

$$\sigma : H^2 \ni (x_0, x_1, x_2) \mapsto \frac{x_1 + i x_2}{1 - x_0} \in \mathbb{C} \cup \{\infty\},$$

that is, $g|_{M \setminus S(f)} = \sigma \circ G|_{M \setminus S(f)}$. So we call g the Gauss map of the maxface.

Generic singularities of maxfaces are classified in [10]. Moreover several criteria for singular points of maxfaces by using their Weierstrass data are given in [8, 10, 18].

Definition 2.3 (Fold singular point [5]). Let $f : M \to \mathbb{L}^3$ be a maxface with Weierstrass data (g, η). We denote by $S(f)$ the singular set of f, that is, $S(f) = \{p \in M \ ; \ |g(p)| = 1\}$.

(1) A singular point $p \in S(f)$ of f is called *non-degenerate* if dg does not vanish at p.

(2) A regular curve $\hat{\gamma}$ on M is called a *non-degenerate fold singularity* if it consists of non-degenerate singular points such that

$$\mathrm{Re}\left(\frac{dg}{g^2\eta}\right)$$

vanish identically along the curve $\hat{\gamma}$. Each point on the non-degenerate fold singularity is called a *fold singular point.*

Remark 2.4. Let Σ be a smooth 2-manifold and $f : \Sigma \to \mathbb{L}^3$ a smooth map. A singular point $p \in \Sigma$ of f is called *fold singularity* if there exist a local coordinate system $(U;\varphi)$ ccentered at $p \in \Sigma$ and a diffeomorphism ψ of $\mathbb{L}^3$ such that $(\psi \circ f \circ \varphi^{-1})(u,v) = (u, v^2, 0)$.

In [5, Lemma 2.17], it is shown that a non-degenerate fold singularity of a maxface is indeed a fold singularity.

A regular curve $\gamma : (a,b) \to \mathbb{L}^3$ is called *null* or *isotropic* if $\gamma'(t) = (d\gamma/dt)(t)$ is a lightlike vector for all $t \in (a,b)$.

Definition 2.5 (Non-degenerate null curve [5]). A null curve $\gamma : (a,b) \to \mathbb{L}^3$ is called *degenerate* or *non-degenerate* at $t = c$ if $\gamma''(c)$ is or is not proportional to the velocity vector $\gamma'(c)$, respectively. If γ is non-degenerate at each $t \in (a,b)$, it is called a *non-degenerate* null curve.

Theorem 2.6 (Analytic extension of maxface [5]). *Let $f : M \to \mathbb{L}^3$ be a maxface which has non-degenerate fold singularities along a singular curve $\hat{\gamma} : (a,b) \to M$. Then $\gamma = f \circ \hat{\gamma}$ is a non-degenerate null curve and the image of the map*

$$f^*(u,v) = \frac{\gamma(u+v) + \gamma(u-v)}{2} \tag{2.4}$$

is real analytically connected to the image of f along γ as a timelike minimal immersion. Conversely, any real analytic immersion with mean curvature, whereever well-defined, equal to zero which changes type across a non-degenerate null curve is obtained as a real analytic extension of non-degenerate fold singularities of a maxface.

We call an immersion in $\mathbb{L}^3$ with mean curvature, whereever well-defined, equal to zero a *zero mean curvature* (*ZMC*) *surface.*

§3. Schwarz H-type ZMC surfaces

For a constant $a \in (0,\infty)$, we set M_a a Riemann surface of genus 3 defined by the hyperelliptic curve

$$w^2 = z(z^3 + a^3)(z^3 + a^{-3}).$$

Consider the family $\{f_a\}_{0<a<\infty}$ of maxfaces

$$f_a = \begin{pmatrix} x_0 \\ x_1 \\ x_2 \end{pmatrix} = \mathrm{Re} \int \begin{pmatrix} -2g \\ 1+g^2 \\ i(1-g^2) \end{pmatrix} \eta \tag{3.1}$$

in $\mathbb{L}^3$ with the Weierstrass data

$$g = z, \qquad \eta = i\frac{dz}{w}. \tag{3.2}$$

The singular set of f_a is $\{|z| = 1\}$. It is easy to verify that each singular point is fold singularity.

We define a $\mathbb{C}^3$-valued 1 form Φ_a on M_a by

$$\Phi_a = \begin{pmatrix} -2g \\ 1+g^2 \\ i(1-g^2) \end{pmatrix} \eta.$$

Direct computations show the following lemma.

Lemma 3.1 (Symmetries of the surface). *Define anti-holomorphic maps $\psi_j : M_a \to M_a$ $(j = 1, 2, 3)$ as follows:*

$$\begin{aligned} \psi_1(z, w) &= (\bar{z}, \bar{w}), \\ \psi_2(z, w) &= (e^{2\pi i/3}\bar{z}, e^{\pi i/3}\bar{w}), \\ \psi_3(z, w) &= \left(\frac{1}{\bar{z}}, \frac{\bar{w}}{\bar{z}}\right). \end{aligned}$$

Then we have the following:

$$\begin{aligned} \psi_1^*\Phi_a &= \begin{pmatrix} -1 & 0 & 0 \\ 0 & -1 & 0 \\ 0 & 0 & 1 \end{pmatrix} \overline{\Phi}_a, \\ \psi_2^*\Phi_a &= \begin{pmatrix} 1 & 0 & 0 \\ 0 & -\cos(\pi/3) & \sin(\pi/3) \\ 0 & \sin(\pi/3) & \cos(\pi/3) \end{pmatrix} \overline{\Phi}_a, \\ \psi_3^*\Phi_a &= \overline{\Phi}_a. \end{aligned}$$

By the above lemma, we can consider

$$\Omega_a^{max} = \{f_a(z) \ ; \ |z| \leq 1, \ 0 \leq \arg z \leq \pi/3\}$$

as the fundamental piece of the maxface, that is, the entire maxface consists of pieces each of which is congruent to Ω_a^{max}.

Lemma 3.2. *In Ω_a^{max}, the images of $\{0 \le |z| \le 1,\ \arg z = 0\}$ and $\{a \le |z| \le 1,\ \arg z = \pi/3\}$ by f_a are straight lines, and the image of $\{a \le |z| \le 1,\ \arg z = \pi/3\}$ by f_a is a curve in some timelike plane.*

Proof. Consider the Hopf differential

$$Q = \eta dg = i\frac{dz^2}{w}$$

of f_a. If $z = t$ $(0 \le t \le 1)$, then $Q \in i\mathbb{R}$. Also, if $z = e^{\pi i/3}t$ $(0 \le t \le 1)$, then

$$Q = \frac{-dt^2}{\sqrt{t(t^3 - a^3)(t^3 - a^{-3})}} \in \begin{cases} \mathbb{R} & (0 \le t \le a) \\ i\mathbb{R} & (a \le t \le 1) \end{cases}$$

this completes the proof. Q.E.D.

Next we consider the singular curve γ of f_a. The singular curve is the image of $z = e^{it}$ $(0 \le t \le \pi/3)$. Hence we can write

$$\gamma(s) = \int_0^s \begin{pmatrix} 1 \\ -\cos t \\ -\sin t \end{pmatrix} \xi(t)dt, \quad \xi(t) = \frac{2}{\sqrt{2\cos 3t + a^3 + a^{-3}}} \quad \left(0 \le s \le \frac{\pi}{3}\right)$$

by a direct computation. Thus if we set

$$f_a^*(u,v) = \frac{1}{2}\big(\gamma(u+v) + \gamma(u-v)\big), \tag{3.3}$$

then f_a^* is a timelike minimal surface such that $\{v = 0\}$ corresponds to the fold singularities and f_a^* is the analytic extension of the maximal surface f_a.

Arguments similar to those in [9] show the following two lemmas.

Lemma 3.3 ([9, Lemma 3.1]). *$f_a^*(u,v)$ is an immersion on $(u,v) \in \mathbb{R} \times (0,\pi)$.*

Lemma 3.4 ([9, Lemma 3.2]). *$f_a^*(0,v)$ $(0 < v < \pi)$ is a straight line parallel to x_2-axis, and $f_a^*(\pi/3,v)$ $(0 < v < \pi)$ is a straight line parallel to $x_0 = x_1 + \sqrt{3}x_2 = 0$.*

Moreover, since $f_a^*(u,\pi+v) = f_a^*(u,\pi-v)$ holds, we have the following lemma.

Lemma 3.5. *$f_a^*(u,\pi)$ $(u \in \mathbb{R})$ corresponds to fold singularities.*

We set $\Omega_a^{min} = \{f_a^*(u,v)\ ;\ 0 \le u \le \pi/3,\ 0 \le v \le \pi\}$.

Remark 3.6. For the Schwarz D-type ZMC surface in [9], $f_a^*(u,\pi/2)$ is a straight line parallel to x_0-axis, but we do not have such a symmetry in this case.

We set

$$\sigma(s) = f_a^*(s,\pi) = \frac{1}{2}\big(\gamma(s+\pi) + \gamma(s-\pi)\big) \qquad (0 \le s \le \pi/3)$$

to further extend analytically from $f_a^*(u,\pi)$ to spacelike surface. Then we have

$$\sigma'(s) = \begin{pmatrix} 1 \\ \cos s \\ \sin s \end{pmatrix} \hat{\xi}(s),$$

where

$$\hat{\xi}(s) = \xi(s+\pi) = \xi(s-\pi) = \frac{2}{a^3 + a^{-3} - 2\cos 3s}.$$

A direct computation shows the following lemma.

Lemma 3.7. *The following equation*

$$\sigma'(s) = A\gamma'\left(\frac{\pi}{3} - s\right)$$

holds, where

$$A = \begin{pmatrix} 1 & 0 & 0 \\ 0 & -\cos(\pi/3) & -\sin(\pi/3) \\ 0 & -\sin(\pi/3) & \cos(\pi/3) \end{pmatrix}.$$

By this lemma, we have

$$\sigma(s) = A\gamma\left(\frac{\pi}{3} - s\right) + \boldsymbol{c},$$

where

$$\boldsymbol{c} = \sigma(0) - A\gamma(\pi/3) = f_a^*(0,\pi) - Af_a^*(\pi/3, 0) \in \mathbb{L}^3.$$

Thus we have the following proposition (See Fig. 3.1).

Proposition 3.8. *We denote by* $\hat{f}_a$ *the spacelike extension from* $\sigma(s)$. *Then we have*

$$\hat{f}_a(z) = -Af_a(z) + \boldsymbol{c} \qquad (|z| \le 1,\ \ 0 \le \arg z \le \pi/3).$$

We set $\hat{\Omega}_a^{max} = \{\hat{f}_a(z)\ ;\ |z| \le 1,\ 0 \le \arg z \le \pi/3\}$. Then the boundary of

$$\Omega_a^{max} \cup \Omega_a^{min} \cup \hat{\Omega}_a^{max} \tag{3.4}$$

consists of two planar curves and two straight lines. See Fig. 3.1. Now we extend this piece (3.4) by reflections with respect to planar curves, then six copies of (3.4) look like "twisted" equilateral triangular catenoid, see Fig. 3.2. This triangular catenoid is homeomorphic and has the same symmetry to the half of rPD family (in $\mathbb{R}^3$) as in Example A.5. Therefore the ZMC surface we obtain by extending (3.4) by reflections infinitely many times is triply periodic. Though the triply periodic ZMC surface looks like embedded for any $a \in (0,1)$, we leave the study of embeddedness of this family for another occasion. See Fig. 3.2.

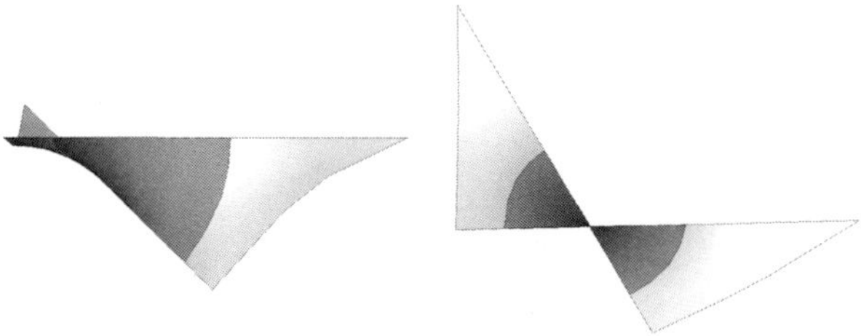

$a = 0.5$

Fig. 3.1. The piece $\Omega_a^{max} \cup \Omega_a^{min} \cup \hat{\Omega}_a^{max}$ in different view points. The spacelike parts are indicated by grey shades, and the timelike part is indicated by black shade.

We call the 1 parameter family of this triply periodic ZMC surface by *Schwarz H-type ZMC surfaces.*

Remark 3.9. The 1 parameter family of the conjugate surface of the maxface we have considered in this section, that is, the maxface with the Weierstrass data $g = z$, $\eta = dz/w$, have conelike singularities, and the half of the fundamental piece looks like "twisted" equilateral triangular Lorentzian catenoid. See Fig. 3.3. Hence by extending these surfaces by reflections with respect to boundary straight lines, we have triply periodic maxfaces with conelike singularities.

§4. Limits of Schwarz H-type ZMC surfaces

In this section we consider the limits of Schwarz H-type ZMC surfaces. As $a \to 0$, the surface, with rescaled by $\sqrt{a^3 + a^{-3}}$, converges to the helicoid by the same arguments as in [9, Remark 3.6].

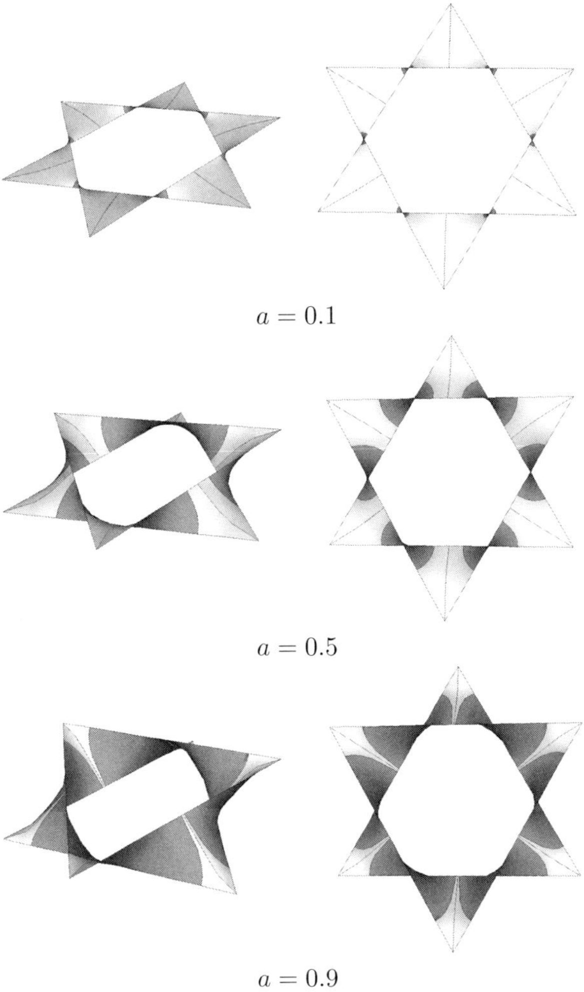

Fig. 3.2. Schwarz H-type ZMC surfaces.

Next we consider the limit as $a \to 1$. Since the hyperelliptic curve $w^2 = z(z^3 + a^3)(z^3 + a^{-3})$ converges to

$$w^2 = z(z^3 + 1)^2,$$

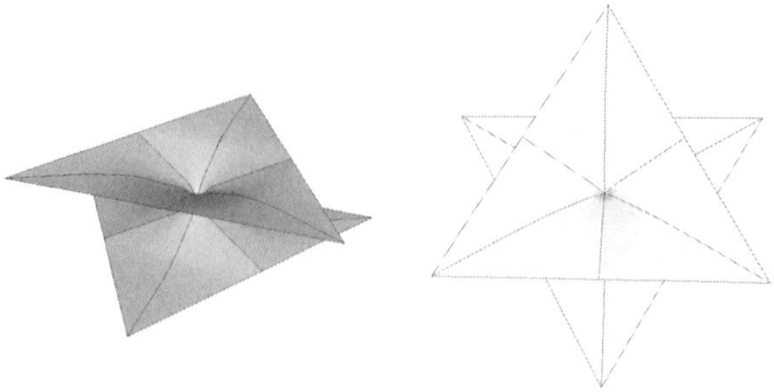

$a = 0.1$

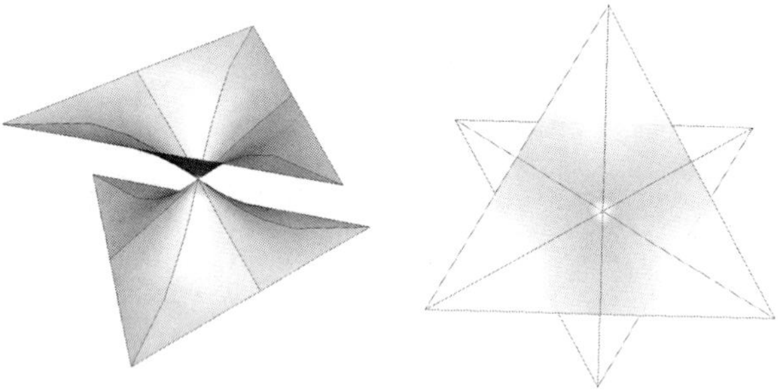

$a = 0.5$

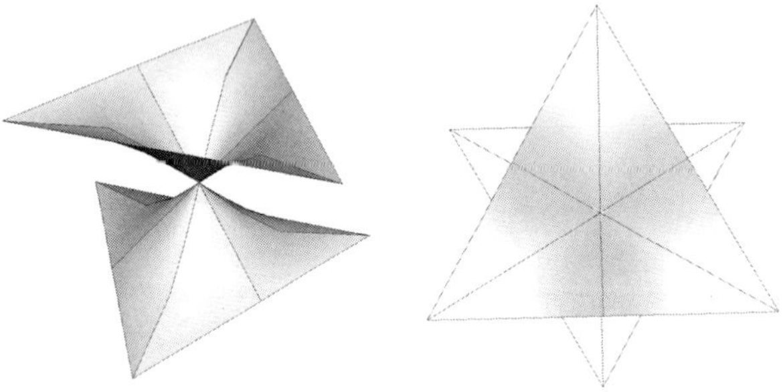

$a = 0.9$

Fig. 3.3. The conjugate surfaces of Schwarz H-type surface.

the Riemann surface M_a converges to a Riemann surface with six nodes at

$$z = e^{\pi i/3}, \quad -1, \quad e^{-\pi i/3}$$

and two branch points at

$$z = 0, \quad \infty.$$

This Riemann surface is of genus zero with six nodal singular points. Hence the maxface f_a converges to

$$f_a \to \pm \operatorname{Re} \int \begin{pmatrix} -2z \\ 1+z^2 \\ i(1-z^2) \end{pmatrix} \frac{i\,dz}{\sqrt{z}(z^3+1)}. \tag{4.1}$$

Let ζ be a branch of $\zeta^2 = z$. Then ζ is a coordinates of this Riemann surface, with six nodes at $\zeta = \pm e^{\pm \pi i/6}$, $\pm i$, and the right hand side of (4.1) becomes

$$\pm 2 \operatorname{Re} \int \begin{pmatrix} -2\zeta^2 \\ 1+\zeta^4 \\ i(1-\zeta^4) \end{pmatrix} \frac{i\,d\zeta}{\zeta^6+1}.$$

This surface coincides with the Karcher-type maxface with $k = 3$, which is a maxface obtained by the conjugate of the maxface with the Weierstrass data as in Example A.7. See Fig. 4.1.

Remark 4.1. In contrast to the Karcher tower in $\mathbb{R}^3$, the Karcher-type maxface is single-valued on M_k for any $k \geq 2$. Moreover, it is easy to verify that each singular point is fold singularity.

For $k = 2$, the image of the analytic extension of the maxface to ZMC surface coincides with the entire graph

$$x_0 = \log \frac{\cosh x_1}{\cosh x_2} \tag{4.2}$$

which is called the Scherk-type ZMC surface.

§Appendix A. Minimal surfaces in $\mathbb{R}^3$

Here we review several examples of minimal surfaces in $\mathbb{R}^3$ which are related to ZMC surfaces we constructed in this paper. For the detail of these examples, see for example [3, 11, 12, 13].

Theorem A.1 (Weierstrass representation [15]). *Let (g, η) be a pair of a meromorphic function g and a holomorphic differential η on a Riemann surface M so that*

$$(1+|g|^2)^2 \eta\bar{\eta} \tag{A.1}$$

gives a Riemannian metric on M. We set

$$\Phi = \begin{pmatrix} (1-g^2)\eta \\ i(1+g^2)\eta \\ 2g\eta \end{pmatrix}. \tag{A.2}$$

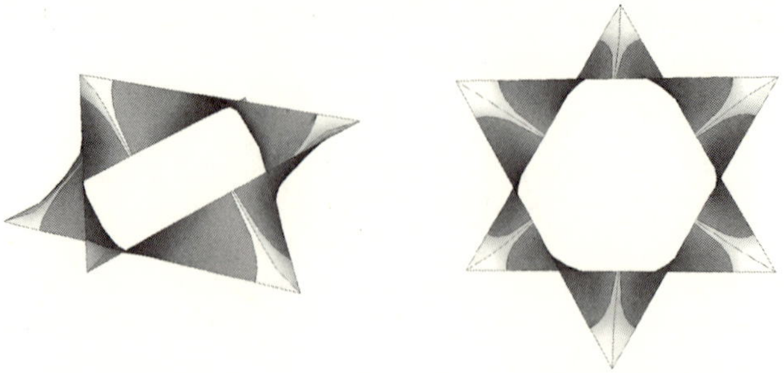

$a = 0.9$

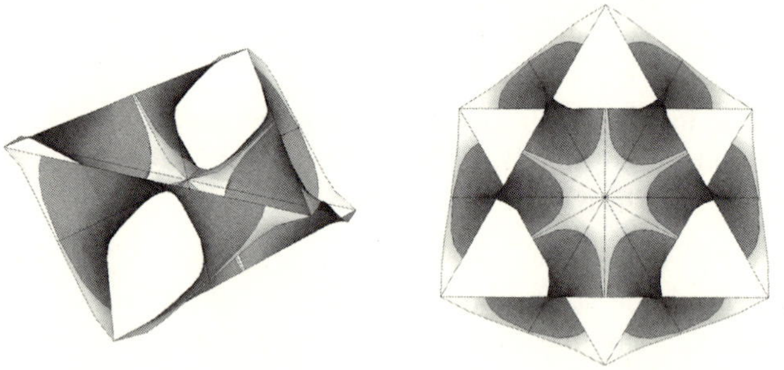

$a = 0.9$ (the same surface as above with different lattice)

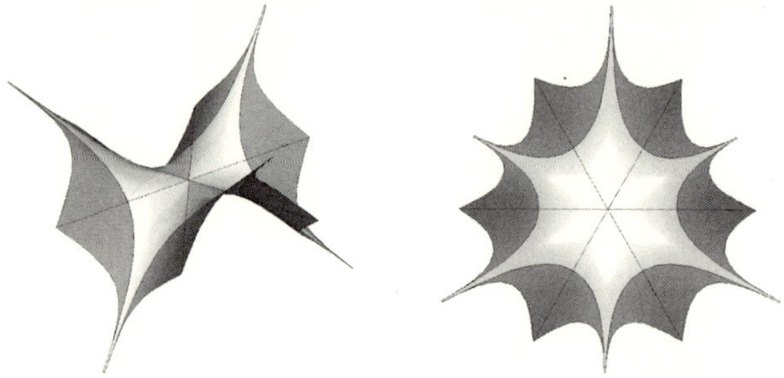

$a \to 1$

Fig. 4.1. The limit of Schwarz H-type ZMC surface as $a \to 1$.

Then

$$f = \mathrm{Re} \int_{z_0}^{z} \Phi : M \to \mathbb{R}^3 \qquad (z_0 \in M) \tag{A.3}$$

defines a conformal minimal immersion. Moreover, f is single-valued on M if and only if

$$\mathrm{Re} \oint_{\ell} \Phi = \mathbf{0} \tag{A.4}$$

for any closed curve ℓ on M. Conversely, any minimal surface can be obtained in this manner.

The pair $(g,\,\eta)$ in Theorem A.1 is called the *Weierstrass data* of f.

Remark A.2. To verify the periodicity of surfaces, consider the following map.

$$\mathrm{Per}(f) = \left\{ \mathrm{Re} \oint_{\ell} \Phi \ ; \ \ell \in H_1(M, \mathbb{Z}) \right\}. \tag{A.5}$$

The periodicity can be determined in the following way:

- If $\mathrm{Per}(f) = \{\mathbf{0}\}$, that is, f satisfies the condition (A.4) for any closed curve ℓ on M, then $f : M \to \mathbb{R}^3$ is well-defined on M, that is, f is *single-valued in* $\mathbb{R}^3$.
- If there exists only one direction $\boldsymbol{v} \in \mathbb{R}^3 \setminus \{\mathbf{0}\}$ such that
$$\mathrm{Per}(f) \subset \Lambda_1 = \{n\boldsymbol{v} \ ; \ n \in \mathbb{Z}\},$$
then f is *singly periodic*. In this case, f is single-valued in $\mathbb{R}^3/\Lambda_1 \approx \mathbb{R}^2 \times S^1$. (A surface invariant under *screw-motions* $\Lambda_1 + R$, where R is a rotation around an axis in the direction of Λ_1, is also singly periodic. See, for example, [2] and the references therein.)
- If there exist two linearly independent vectors $\boldsymbol{v}_1, \boldsymbol{v}_2 \in \mathbb{R}^3$ (with $\mathrm{span}\{\boldsymbol{v}_1, \boldsymbol{v}_2\}$ uniquely determined) such that
$$\mathrm{Per}(f) \subset \Lambda_2 = \left\{ \sum_{j=1}^{2} n_j \boldsymbol{v}_j \ ; \ n_j \in \mathbb{Z} \right\},$$
then f is *doubly periodic*. In this case, f is single-valued in $\mathbb{R}^3/\Lambda_2 \approx T^2 \times \mathbb{R}$.
- If there exist three linearly independent vectors $\boldsymbol{v}_1, \boldsymbol{v}_2, \boldsymbol{v}_3 \in \mathbb{R}^3$ such that
$$\mathrm{Per}(f) \subset \Lambda_3 = \left\{ \sum_{j=1}^{3} n_j \boldsymbol{v}_j \ ; \ n_j \in \mathbb{Z} \right\},$$
then f is *triply periodic*. In this case, f is single-valued in $\mathbb{R}^3/\Lambda_3 \approx T^3$.

Remark A.3. The first fundamental form ds^2 and the second fundamental form $\mathrm{I\!I}$ of the surface (A.3) are given by

$$ds^2 = \left(1 + |g|^2\right)^2 \eta\bar{\eta}, \qquad \mathrm{I\!I} = -\eta dg - \overline{\eta dg}.$$

Moreover, $g : M \to \mathbb{C} \cup \{\infty\}$ coincides with the composition of the Gauss map $G : M \to S^2$ of the minimal surface and the stereographic projection $\sigma : S^2 \to \mathbb{C} \cup \{\infty\}$, that is, $g = \sigma \circ G$. So we call g the Gauss map of the minimal surface.

Remark A.4 (A historical remark about triply periodic minimal surfaces). The first examples of triply periodic minimal surfaces in $\mathbb{R}^3$ are found by H. A. Schwarz in the 19th century [17]. Then in 1970, a NASA scientist A. Schoen found many more examples, and he named three of Schwarz' examples P surface, D surface, and H family, because they have the symmetry related to those of the *primitive* cubic lattice, *diamond* crystal structure, and *hexagonal* crystal structure, respectively [16].

In 1989, H. Karcher found a 1-parameter family of triply periodic minimal surfaces [12]. Since a half of the fundamental piece of the surface looks like *twisted* (equilateral) *trianglar* catenoid (see Fig. A.1), he named the family TT, but since the family contains both Schwarz P and D surfaces, the family is now called rPD family. See for example [4].

Example A.5 (Schwarz rPD family). For a constant $a \in (0, \infty)$, we set M_a a Riemann surface of genus 3 defined by the hyperelliptic curve

$$w^2 = z(z^3 - a^3)(z^3 + a^{-3}).$$

We define the Weierstrass data

$$g = z, \qquad \eta = \frac{dz}{w}. \tag{A.6}$$

Then

$$f_a = \begin{pmatrix} x_1 \\ x_2 \\ x_3 \end{pmatrix} = \operatorname{Re} \int \begin{pmatrix} 1 - g^2 \\ i(1 + g^2) \\ 2g \end{pmatrix} \eta \tag{A.7}$$

gives a 1-parameter family $\{f_a\}_{0<a<\infty}$ of embedded triply periodic minimal surfaces in $\mathbb{R}^3$. This family is called the *Schwarz rPD family.* When $a = 1/\sqrt{2}$, the surface coincides with Schwarz P surface, and when $a = \sqrt{2}$, the surface coincides with Schwarz D surface.

As we mentioned in Remark A.4, a half of the fundamental piece of rPD surface looks like "twisted" equilateral triangular catenoid. See Fig. A.1.

Fig. A.2 shows the relation between Schwarz P and rPD for $a = 1/\sqrt{2}$.

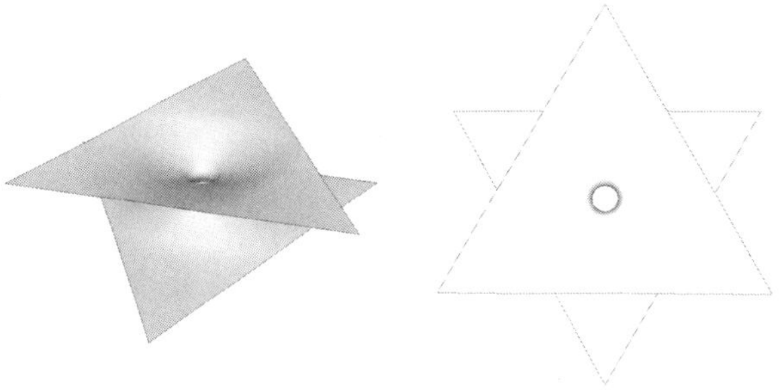

$a = 0.1$

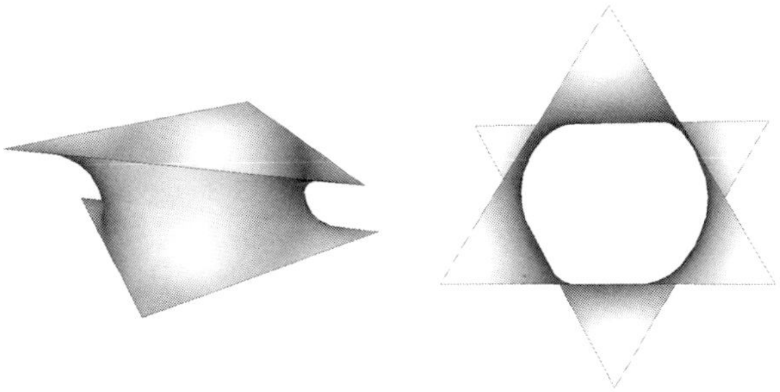

$a = 1.0$

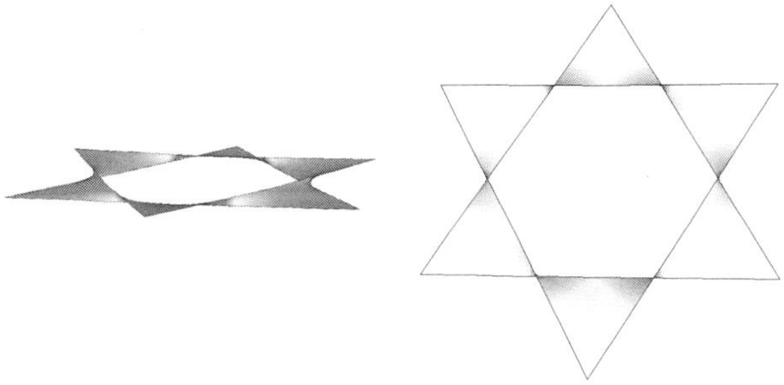

$a = 10.0$

Fig. A.1. Schwarz rPD surfaces.

Example A.6 (Schwarz H family). For a constant $a \in (0, \infty)$, we set M_a a Riemann surface of genus 3 defined by the hyperelliptic curve

$$w^2 = z(z^3 + a^3)(z^3 + a^{-3}).$$

Then the family $\{f_a\}_{0<a<1}$ of minimal surfaces (A.7) with the Weierstrass data (A.6) is a family of embedded triply periodic minimal surfaces in $\mathbb{R}^3$. This family is called the *Schwarz H family*. As $a \to 0$, f_a, with

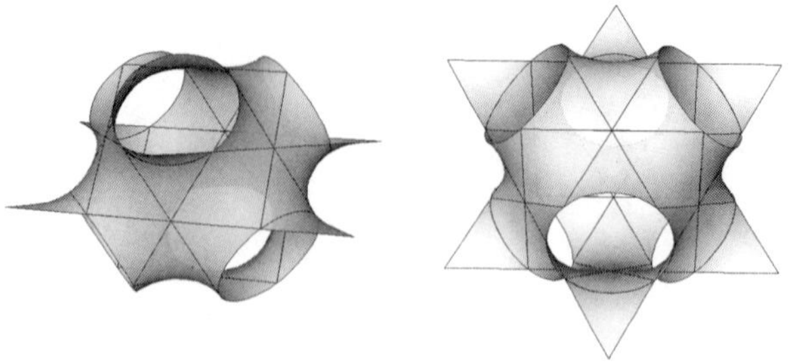

$a = 1/\sqrt{2}$

Fig. A.2. Relation between Schwarz P and rPD.

rescaled by $\sqrt{a^3 + a^{-3}}$, converges to catenoid. Also, as $a \to 1$, f_a converges to Karcher tower with $k = 3$ (see Example A.7).

A half of the fundamental piece of Schwarz H surface looks like "non-twisted " equilateral triangular catenoid. See Fig. A.3.

Fig. A.4 shows the conjugate surface of Schwarz H surface. Each vertex of the hexagon in the right hand side figure lies in the straight line parallel to x_3-axis. Hence after reflections with respect to these lines, we see that the surface has self-intersections.

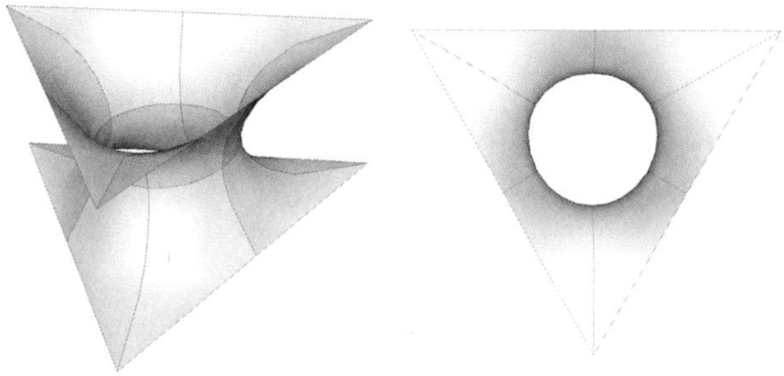

Fig. A.3. Schwarz H surface ($a = 0.5$).

Example A.7 (Karcher tower). For an integer $k \geq 2$, we define M_k by

$$M_k = (\mathbb{C} \cup \{\infty\}) \setminus \{z \in \mathbb{C}\,;\ z^{2k} = -1\}. \tag{A.8}$$

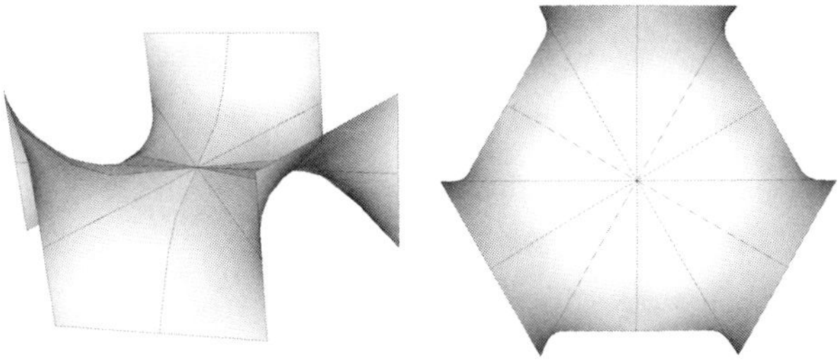

Fig. A.4. The comjugate surface of Schwarz H surface ($a = 0.5$).

Then the minimal surface (A.7) with the Weierstrass data

$$g = z^{k-1}, \qquad \eta = \frac{dz}{z^{2k}+1}$$

is embedded singly periodic with $2k$ ends in $\mathbb{R}^3$. This minimal surface is called the *Karcher tower*. When $k = 2$, this surface coincides with the Scherk tower (the conjugate surface of doubly periodic Scherk surface). See Fig. A.5.

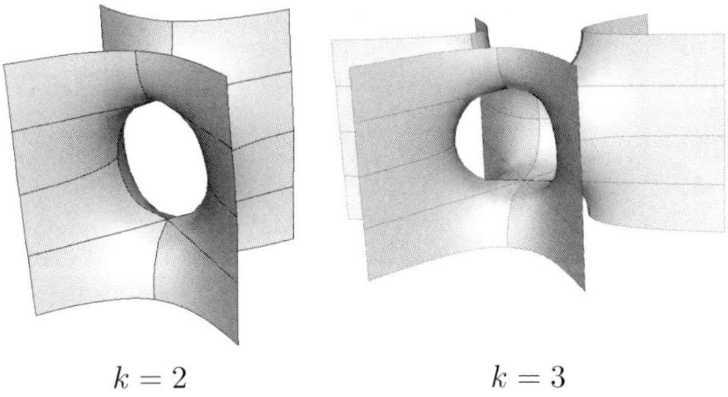

Fig. A.5. The Karcher tower.

References

[1] S. Akamine, *Causal characters of zero mean curvature surfaces of Riemann-type in the Lorentz-Minkowski 3-space*, Kyushu J. Math. **71** (2017), 211–249.

[2] M. Callahan, D. Hoffman and W. Meeks III, *The structure of singly-periodic minimal surfaces*, Invent. Math. **99** (1990), 455–481.

[3] N. Ejiri, S. Fujimori and T. Shoda, *A remark on limits of triply periodic minimal surfaces of genus 3*, Topology Appl. **196** (2015), 880–903.

[4] W. Fischer and E. Koch, *Crystallographic aspects of minimal surfaces*, Journal de Physique Colloques **51** (1990), C7-131–C7-147.

[5] S. Fujimori, Y. W. Kim, S.-E. Koh, W. Rossman, H. Shin, M. Umehara, K. Yamada and S.-D. Yang, *Zero mean curvature surfaces in Lorentz-Minkowski 3-space and 2-dimensional fluid mechanics*, Math. J. Okayama Univ. **57** (2015), 173–200.

[6] S. Fujimori, Y. Kawakami, M. Kokubu, W. Rossman, M. Umehara and K. Yamada, *Analytic extension of Jorge-Meeks type maximal surfaces in Lorentz-Minkowski 3-space*, Osaka J. Math. **54** (2017), 249–272.

[7] S. Fujimori, Y. Kawakami, M. Kokubu, W. Rossman, M. Umehara and K. Yamada, *Entire zero mean curvature graphs of mixed type in Lorentz-Minkowski 3-space*, Quart. J. Math. **67** (2016), 801–837.

[8] S. Fujimori, W. Rossman, M. Umehara, K. Yamada and S.-D. Yang, *New maximal surfaces in Minkowski 3-space with arbitrary genus and their cousins in de Sitter 3-space*, Result. Math. **56** (2009), 41–82.

[9] S. Fujimori, W. Rossman, M. Umehara, K. Yamada and S.-D. Yang, *Embedded triply periodic zero mean curvature surfaces of mixed type in Lorentz-Minkowski 3-space*, Michigan Math. J. **63** (2014), 189–207.

[10] S. Fujimori, K. Saji, M. Umehara and K. Yamada, *Singularities of maximal surfaces*, Math. Z. **259** (2008), 827–848.

[11] S. Fujimori and M. Weber, *Triply periodic minimal surfaces bounded by vertical symmetry planes*, Manuscripta Math. **129** (2009), 29–53.

[12] H. Karcher, *The Triply Periodic Minimal Surfaces of Alan Schoen and their Constant Mean Curvature Companions*, Manuscripta Math. **64** (1989), 291–357.

[13] H. Karcher, *Construction of Minimal Surfaces*, Surveys in Geometry, 1–96, University of Tokyo, 1989.

[14] O. Kobayashi, *Maximal surfaces in the 3-dimensional Minkowski space* $\mathbb{L}^3$, Tokyo J. Math. **6** (1983), 297–309.

[15] R. Osserman, *Global properties of minimal surfaces in* E^3 *and* E^n, Ann. of Math. (2) **80** (1964), 340–364.

[16] A. Schoen, *Infinite Periodic Minimal Surfaces without Self-Intersections*, Technical Note D-5541, NASA, Cambridge, Mass. 1970.

[17] H. A. Schwarz, *Gesammelte Mathematische Abhandlungen*, Springer, 1890.

[18] M. Umehara and K. Yamada, *Maximal surfaces with singularities in Minkowski space*, Hokkaido Math. J. **35** (2006), 13–40.

Department of Mathematics
Okayama University
Tsushima-naka 3-1-1
Okayama 700-8530
Japan
E-mail address: fujimori@math.okayama-u.ac.jp

Advanced Studies in Pure Mathematics 78, 2018
Singularities in Generic Geometry
pp. 221–249

Hamiltonian systems on submanifolds

T. Fukuda and S. Janeczko

Abstract.

A constraint submanifold in a symplectic space after P.A.M. Dirac is determined locally by geometric restriction of the symplectic form to the constraint. The natural symplectic invariant associated to this restriction is the space of Hamiltonian vector fields which uniquely restrict to the solvable Hamiltonian ones on a constraint. By investigation of solvability of generalized Hamiltonian systems we characterize the constraint invariants and find them explicitly in the generic cases. Moreover the Poisson-Lie algebra on submanifold is described and an example of the Hamiltonian vector fields on the 2-sphere in symplectic space was constructed.

§1. Introduction.

Let (M, ω) be a symplectic 2n-dimensional manifold, endowed with the nondegenerate, closed two-form ω. By the vector bundle morphism $\beta : TM \ni u \mapsto \omega(u, \cdot) \in T^*M$ we introduce the canonical symplectic structure $\dot{\omega}$ on TM, namely the pullback of the Liouville symplectic form $d\theta$ defined on the cotangent bundle T^*M, $\dot{\omega} = \beta^* d\theta$. A vector field $X : M \to TM$ is said to be Hamiltonian if the form $\omega(X, \cdot)$ is closed and exact. A function $H : M \to \mathbb{R}$ is called Hamiltonian function for X if $\omega(X, \cdot) = -dH(\cdot)$. If X is Hamiltonian, then its image $X(M) \subset TM$ is a Lagrangian submanifold of $(TM, \dot{\omega})$ generated by H (cf. [15]). In local Darboux coordinates, $M \cong \mathbb{R}^{2n}$, $\omega = \sum_{i=1}^n dy_i \wedge dx_i$, and $\dot{\omega} = \beta^* d\theta = \sum_{i=1}^n (d\dot{y}_i \wedge dx_i - d\dot{x}_i \wedge dy_i)$, where $(q, \dot{q}) = ((x, y), (\dot{x}, \dot{y}))$ are coordinates on $T\mathbb{R}^{2n} \equiv \mathbb{R}^{2n} \times \mathbb{R}^{2n}$.

Received June 2, 2016.
Revised October 12, 2016.
2010 *Mathematics Subject Classification.* Primary 53D05, 51N10; Secondary 53D22, 70H05, 15A04 .
Key words and phrases. Symplectic manifold, Hamiltonian systems, Symplectic constraints.
The second author was partially supported by NCN grant no. DEC-2013/11/B/ST1/03080.

In what follows a smooth submanifold $N \subset TM$ is called *Hamiltonian system* if N is Lagrangian, i.e. $\dot{\omega} |_N = 0$ and $dimN = 2n$. If $\tau |_N : N \to M$ is singular, where τ is a tangent bundle projection, we also call N an *implicit Hamiltonian system.*

Fundamental property of a differential system which we investigate in this paper is its local solvability. Let M be a smooth manifold, not necessarily a symplectic manifold. A submanifold N of TM is called an implicit differential system. A point $(q, \dot{q}) \in N \subset TM$ is called a *solvable point* of N if there exists a smooth curve $\gamma : (-\varepsilon, \varepsilon) \to M$, $\gamma(0) = q$ such that its tangent lifting $\dot{\gamma}(t)$ belongs to N. N is called a *solvable manifold* if N consists of solvable points only. N is called *smoothly solvable* if it consists smoothly solvable points, i.e. around each $v \in N$ there exists a smooth family $\alpha : U \times (-\epsilon, \epsilon) \ni (\bar{v}, t) \mapsto M$ of smooth solutions of N for a neighbourhood U of v in N such that $\dot{\alpha}_{\bar{v}}(0) = \bar{v}$. A submanifold of N is called a *solvable submanifold of* N (*smoothly solvable*) if it is a solvable (*smoothly solvable* respectively) submanifold of TM.

If $\tau |_N$ is a diffeomorphism, then N is smoothly solvable vector field on M. If $\tau |_N$ is singular, then N may not be solvable in the critical points of $\tau |_N$. The simplest representative example of such manifold is given by $N = \{(q, \dot{q}) \in T\mathbb{R} : q = (\dot{q} - a)^2\}$. For $a \neq 0$, N is not solvable at $(0, a)$ and this is a singular point of $\tau |_N$. In general case of any submanifold of TM the necessary and sufficient conditions for a manifold $N \subset TM$ to be solvable are found in [4, 9].

Let $v = (q, \dot{q}) \in N$ be a solvable point of N. Then there exists a smooth curve $\gamma : (-\varepsilon, \varepsilon) \to M$ as above. Thus an immediate necessary condition for a point $v = (q, \dot{q}) \in N$ to be solvable is that (cf. [4])

$$\dot{q} \in d(\tau |_N)_v(T_v N), \tag{1}$$

where $d(\tau |_N)_v$ is the tangent mapping to $\tau |_N$ at v. In what follows we will call this condition *tangential solvability condition.*

We can ask whether this condition is also a sufficient condition for a submanifold N to be solvable. Although the answer for this question is negative, there is a wide class of submanifolds of TM for which the tangential solvability condition is also sufficient. An example of the submanifold N for which the tangential solvability condition is fulfilled but N is not solvable is given in [4].

1.1. Implicit Hamiltonian systems.

Now let M be a symplectic manifold. A smooth submanifold $N \subset TM$ is called *Hamiltonian system* if N is Lagrangian, i.e. $\dot{\omega} |_N = 0$ and $dimN = 2n$. If $\tau |_N \to M$ is singular, where τ is atangent bundle projection, we also call N an *implicit Hamiltonian system.*

In this work we concentrate only on Hamiltonian systems and symplectic invariants connected to their solvability. As the solvability is a local property investigated globally on a manifold, then we will use the local coordinate systems and replace manifolds by their Euclidean representatives.

Let $N \subset (T\mathbb{R}^{2n}, \dot{\omega})$ be a Hamiltonian system. Suppose that

$$corankd(\tau\,|_N)_v = k$$

for some $v \in N$. Then there exists an open neighborhood $\mathcal{O}$ of v in $T\mathbb{R}^{2n}$ and a smooth function $F : \mathbb{R}^{2n}\times\mathbb{R}^k \ni (q,\lambda) \mapsto F(q,\lambda) \in \mathbb{R}$ defined on an open neighborhood of $(q_0, 0)$ in $\mathbb{R}^{2n}\times\mathbb{R}^k$, $q_0 = \tau(v)$ such that

$$N \cap \mathcal{O} = \left\{(q,\dot{q}); \exists_{\lambda\in\mathbb{R}^k}, \dot{x}_i = \frac{\partial F}{\partial y_i}(q,\lambda), \dot{y}_j = -\frac{\partial F}{\partial x_j}(q,\lambda), 0 = \frac{\partial F}{\partial \lambda_l}(q,\lambda)\right\},$$

where $1 \leq i, j \leq n, \quad 1 \leq l \leq k$, and

$$(2)\quad rank\left(\frac{\partial^2 F}{\partial x_j \partial \lambda_l}, \frac{\partial^2 F}{\partial y_i \partial \lambda_l}, \frac{\partial^2 F}{\partial \lambda_s \partial \lambda_l}\right)(q_0, 0) = k, \qquad \frac{\partial^2 F}{\partial \lambda_s \partial \lambda_r}(q_0, 0) = 0,$$

where $1 \leq s, r \leq k$. And F is called a *generating family* of $N \cap \mathcal{O}$.

If N is a Hamiltonian system generated by a generating family $F : \mathbb{R}^{2n}\times\mathbb{R}^k \to \mathbb{R}$, then the tangential solvability condition for N is equivalent to the existence of a solution $\mu = (\mu_1, \ldots, \mu_n) \in \mathbb{R}^k$ of the following linear equation (cf. [4]),

$$(3)\qquad \sum_{j=1}^{k} \frac{\partial^2 F}{\partial \lambda_i \partial \lambda_j}(q,\lambda)\mu_j = \left\{\frac{\partial F}{\partial \lambda_i}, F\right\}(q,\lambda), \quad i = 1, \ldots, k$$

for each $(q,\lambda) \in \mathbb{R}^{2n} \times \mathbb{R}^k$, where $\{.,.\}$ denotes the Poisson bracket on $\mathbb{R}^{2n}$ induced by ω.

By the Cramer's rule equation (3) is equivalent to

$$\sum_{j=1}^{k} \left[\frac{\partial^2 F}{\partial \lambda_i \partial \lambda_j}\right] \left\{\frac{\partial F}{\partial \lambda_j}, F\right\} |_{\{detH=0\}\cap C_F} \equiv 0,$$

where $[\frac{\partial^2 F}{\partial \lambda_i \partial \lambda_j}]$ is a cofactor matrix of $H(x,y,\lambda) = (\frac{\partial^2 F}{\partial \lambda_i \partial \lambda_j}(x,y,\lambda))$ and C_F is a critical manifold defined by

$$C_F = \left\{(x,y,\lambda); \frac{\partial F}{\partial \lambda_i}(x,y,\lambda) = 0, \quad 0 \leq i \leq k\right\}.$$

Smooth solvability of N is implied by the condition that the linear equation (3) has a smooth solution $(\mu_1(x,y,\lambda),\dots,\mu_k(x,y,\lambda))$ on the critical manifold C_F.

The natural problems concerning solvability phenomena of implicit Hamiltonian systems are formulated as follows,

a) *find conditions to be posed on a smooth generating family* $F : \mathbb{R}^{2n}\times\mathbb{R}^k\to\mathbb{R}$, *so that the linear equation* (3) *has a smooth solution on* C_F.

b) *specify the insolvability area in general implicit Hamiltonian systems in particular those defined by constraints in the symplectic space.*

c) *determine the Poisson-Lie algebras induced by smooth submanifolds of symplectic space.*

Point *a*) is already considered in [4]. The point *b*) needs an extra conditions on regions of L fulfilling tangential solvability condition to be finally solvable, and the point *c*) provides constructions of function algebras which are equipped with the Poisson structure. This is the subject of research in the rest of the paper.

1.2. Solvability conditions.

Let $E_s(k,k)$ denote the space of $k\times k$ symmetric matrices of real numbers. For each integer $r \geq 0$ let S_r denote the subset of $E_s(k,k)$ consisting of all symmetric matrices of rank r. Then S_r is a submanifold of $E_s(k,k)$ of codimension $(k-r)(k-r+1)/2$. Now we have a smooth mapping $\hat{H} : \mathbb{R}^{2n}\times\mathbb{R}^k \to E_s(k,k)$ defined by

$$\hat{H}(x,y,\lambda) = \left(\frac{\partial^2 F}{\partial\lambda_i\partial\lambda_j}(x,y,\lambda)\right). \tag{4}$$

Definition 1. *An implicit Hamiltonian system*

$$N = \left\{(x,y,\frac{\partial F}{\partial y}(x,y,\lambda),-\frac{\partial F}{\partial x}(x,y,\lambda)) \in T\mathbb{R}^{2n} : (x,y,\lambda)\in C_F\right\}$$

generated by a generating family $F : \mathbb{R}^{2n}\times\mathbb{R}^k \to \mathbb{R}$ *is called generic if the map* $\hat{H} : R^{2n}\times\mathbb{R}^k \to E_s(k,k)$ *is transversal to all* S_r, $r = 0,\cdots,k-1$, *at every point of* C_F.

Now we can formulate the main result we will use in this paper (cf. [4, 11]).

Theorem 1. ([4]) *The generic implicit Hamiltonian system* $N \subset T\mathbb{R}^{2n}$ *is smoothly solvable if and only if it satisfies the tangential solvability condition.*

Using the results concerning solvability of general implicit differential systems in [4] we can get the corresponding results concerning solvability of implicit Hamiltonian systems. Now our function-matrix $\hat{H}\mid_{\{(x,y,\lambda)\in C_F\}}: C_F \to E_s(k,k)$ corresponds to matrix $A(x)$ in [4]. Let $\mathcal{O}_{C_F,0}$ denote the ring of germs at $0 \in C_F$ of real analytic functions on C_F. Then we get the following result.

Theorem 2. ([4]) *Let* $F : (\mathbb{R}^{2n}\times\mathbb{R}^k, 0)\to\mathbb{R}$ *be a real analytic function-germ. Suppose that the implicit Hamiltonian system* N *generated by* F *fulfills the tangential solvability condition. If the ideal* $< det(\hat{H}\mid_{C_F})(x,y,\lambda) >$ *in* $\mathcal{O}_{C_F,0}$ *has the property of zeros (i.e. any function vanishing on the variety defined by this ideal belongs to it), then the germ at* $(0,0)$ *of* N *is smoothly solvable.*

1.3. Generalized Hamiltonian systems.

Let K be a submanifold of $\mathbb{R}^{2n}$ and $h : K \to \mathbb{R}$ be a smooth function on K. The notion of generalized Hamiltonian system (generalized Hamiltonian dynamics) was introduced by P.A.M. Dirac in [2]. It is defined as a sub-bundle of $T\mathbb{R}^{2n}$ over K, being a Lagrangian submanifold L of $(T\mathbb{R}^{2n}, \dot{\omega})$,(cf. [10])

$$L = \{v \in T\mathbb{R}^{2n} : \omega(v,u) = -dh(u) \quad \forall_{u\in TK}\}. \tag{5}$$

In local coordinates which we use in the setting, the generalized Hamiltonian system (5) can be written by linear in λ as generating family $F : \mathbb{R}^{2n} \times \mathbb{R}^k \to \mathbb{R}$,

$$F(x,y,\lambda) = \sum_{\ell=1}^{k} a_\ell(x,y)\lambda_\ell + b(x,y), \tag{6}$$

where K, being a complete intersection, is defined by an ideal $I_K =< a_1,\ldots,a_k >$ having property of zeros with analytic generators $a_i, 1 \le i \le k$. K is a zero-level set of the mapping $a : (x,y) \mapsto (a_1(x,y),\ldots,a_k(x,y))$, $K = \{(x,y) \in \mathbb{R}^{2n} : a_i(x,y) = 0, i = 1,\ldots,k\}$, and $b(x,y)$ is an arbitrary smooth extension of the function $h : K \to \mathbb{R}$ and the rank condition (2) is fulfilled. In what follows we consider the smooth K and b identified with h.

Generalized Hamiltonian systems are not generic in the sense of Definition 1. For such systems the necessary tangential solvability condition is also sufficient. The aim of this paper is to investigate conditions on subvarieties of symplectic space on which the solvable generalized Hamiltonian systems may exist. We find conditions that L is smoothly solvable under some properties of K and general function on K.

Let us notice that the tangential solvability condition for generalized Hamiltonian system is reformulated after (3) as the system of equations fulfilled in the smoothly solvable points of L,

$$\left\{\frac{\partial F}{\partial \lambda_i}, F\right\}(x, y, \lambda) = 0 \text{ for } (x, y, \lambda) \in C_F. \tag{7}$$

Concerning the solvability of the generalized Hamiltonian system L, we have already the following basic result proved in [4]. L is smoothly solvable if (7) is fulfilled on $K \times \mathbb{R}^k$ which is a very strong condition expressed in the following,

Theorem 3. ([4]) *A generalized Hamiltonian system* $L \subset (T\mathbb{R}^{2n}, \dot{\omega})$ *generated by the generating family* (6) *is smoothly solvable if and only if*

$$\{a_i, a_\ell\} = 0 \quad \text{and} \quad \{b, a_\ell\} = 0, \quad 1 \le i, \ell \le k,$$

$$\text{on} \quad K = \{(x, y) \in \mathbb{R}^{2n} : a_i(x, y) = 0, \quad 1 \le i \le k\},$$

and $1 \le k \le n$. If $k = n$, then $b \equiv 0$.

Solvability property of L defines K to be an involutive, coisotropic submanifold of $(\mathbb{R}^{2n}, \omega)$, i.e. geometrically $T_qK \supset (T_qK)^\omega = \{u \in T_q\mathbb{R}^{2n} : \omega(u, v) = 0, \forall_{v \in T_qK}\}$, and b restricts to those functions who are constant on leaves of the characteristic foliation of coisotropic K, (cf. [13]).

Remark 1. *If* $\dim K < n$ *and* K *is isotropic, i.e.* $(TK)^\omega \supset TK$, *then* TK *is solvable submanifold of* L *with* $b \equiv 0$. *In this case* L *can not be completely solvable Hamiltonian system. If* $\dim K = n$, *and* $TK = L$ *is solvable with* $b \equiv 0$, *then* K *is Lagrangian.*

Corollary 4. *Let* L *be a generalized Hamiltonian system over the submanifold* $K \subset \mathbb{R}^{2n}$ *and its generating family* F *fulfills the tangential integrability condition. Then* K *is a coisotropic submanifold of* $(\mathbb{R}^{2n}, \omega)$ *and* L *is smoothly solvable.*

In what follows we investigate the case when L is not smoothly solvable. We clarify the properties of such L with respect to the structure of non-solvable part of it and symplectic invariant properties of constraints. The regions of solvability on L may be identified by analysis of (7) under some assumptions on K.

§2. Solvability on even dimensional submanifolds.

The generalized Hamiltonian system L is given by an immersion

$$\phi : C_F \to L \subset (T\mathbb{R}^{2n}, \dot{\omega})$$

defined by

$$\phi(x,y,\lambda)=\left(x,y,\frac{\partial F}{\partial y}(x,y,\lambda),-\frac{\partial F}{\partial x}(x,y,\lambda)\right),\quad (x,y,\lambda)\in C_F.$$

Since $\frac{\partial F}{\partial \lambda_\ell}(x,y,\lambda)=a_\ell(x,y)$, we have $C_F=K\times\mathbb{R}^k$. Then L can be written as

$$L=\phi(C_F)=\left\{\left(x,y,\frac{\partial F}{\partial y}(x,y,\lambda),-\frac{\partial F}{\partial x}(x,y,\lambda)\right)\in T\mathbb{R}^{2n}:(x,y,\lambda)\in K\times\mathbb{R}^k\right\}.$$

We find conditions for a submanifold or domain of L to be smoothly solvable. Thus the traditionally solvable Hamiltonian system exists on a submanifold K in the case where the generating family does not satisfy the involutivity condition in Theorem 3, i.e. $\{a_i,a_\ell\}=0$ and $\{b,a_\ell\}=0$ on K, $1\le i,\ \ell\le k$.

Consider the $k\times k$ skew-symmetric matrix $A(x,y)=(\{a_i,a_j\}(x,y))$ and the linear equation

$$\sum_{j=1}^{k}\{a_i,a_j\}(x,y)\lambda_j=\{b,a_i\}(x,y),\quad i=1,\ldots,k. \tag{8}$$

Set

$$\widetilde{S}_F=\{(x,y,\lambda)\in C_F:\sum_{j=1}^{k}\{a_i,a_j\}(x,y)\lambda_j=\{b,a_i\}(x,y),\quad i=1,\ldots,k\}$$

and $S_F=\phi(\widetilde{S}_F)\subset L$.

Comparing to the general implicit Hamiltonian systems (cf. [4]) we can easily see that the following three properties still hold in the present irregular generalized Hamiltonian case. Thus before we proceed to the more specified cases we formulate the following Lemmas.

Lemma 1.
1) *$\phi:C_F\to L$ is a local diffeomorphism.*
2) *Let $Q\subset L$ be a submanifold. Then the following three conditions are equivalent,*

(a) *a submanifold Q of L is smoothly solvable*
(b) *there exists a smooth vector field ξ tangent to Q such that*

$$d\tau(\xi(x,y,\dot{x},\dot{y}))=\sum_{i=1}^{n}\dot{x}_i\frac{\partial}{\partial x_i}+\sum_{i=1}^{n}\dot{y}_i\frac{\partial}{\partial y_i},$$

(c) *there exists a smooth vector field* $\widetilde{\xi}$ *tangent to* $\widetilde{Q} = \phi^{-1}(Q)$ *such that*

$$d\widetilde{\pi}(\widetilde{\xi}(x,y,\lambda)) = \sum_{i=1}^{n} \frac{\partial F}{\partial y_i}(x,y,\lambda)\frac{\partial}{\partial x_i} - \sum_{i=1}^{n} \frac{\partial F}{\partial x_i}(x,y,\lambda)\frac{\partial}{\partial y_i},$$

where $\widetilde{\pi} : \mathbb{R}^{2n} \times \mathbb{R}^k \to \mathbb{R}^{2n}$, $\quad \widetilde{\pi}(x,y,\lambda) = (x,y)$.

Lemma 2. 1) *For* $(x,y,\lambda) \in C_F$, *the vector field*

$$\sum_{i=1}^{n} \frac{\partial F}{\partial y_i}(x,y,\lambda)\frac{\partial}{\partial x_i} - \sum_{i=1}^{n} \frac{\partial F}{\partial x_i}(x,y,\lambda)\frac{\partial}{\partial y_i}$$

is tangent to K *if and only if equation* (8) *is fulfilled.*

2) *Equivalently, for a point* $(x,y,\dot{x},\dot{y}) \in L$, *the vector field*

$$\sum_{i=1}^{n} \dot{x}_i \frac{\partial}{\partial x_i} + \sum_{i=1}^{n} \dot{y}_i \frac{\partial}{\partial y_i}$$

is tangent to K *at* (x,y) *if and only if* $(x,y,\dot{x},\dot{y}) \in S_F$.

Proof. Since K is defined by equations $a_1(x,y) = 0, \ldots, a_k(x,y) = 0$, then the vector field

$$\sum_{i=1}^{n} \frac{\partial F}{\partial y_i}(x,y,\lambda)\frac{\partial}{\partial x_i} - \sum_{i=1}^{n} \frac{\partial F}{\partial x_i}(x,y,\lambda)\frac{\partial}{\partial y_i}$$

is tangent to K if and only if

$$\left(\sum_{i=1}^{n} \frac{\partial F}{\partial y_i}(x,y,\lambda)\frac{\partial}{\partial x_i} - \sum_{i=1}^{n} \frac{\partial F}{\partial x_i}(x,y,\lambda)\frac{\partial}{\partial y_i}\right)(a_j(x,y)) = 0, \quad j = 1,\ldots,k,$$

which holds if and only if $\{F, a_j\}(x,y,\lambda) = 0, \quad j = 1,\ldots,k$. Inserting (6) the last equality holds if and only if

$$\sum_{i=1}^{k} \{a_i, a_j\}(x,y)\lambda_i + \{b, a_j\}(x,y) = 0, \quad j = 1,\ldots,k,$$

which gives an equation (8) and completes the proof of Lemma 2 □

Lemma 3. *Let* $(x_0, y_0, \dot{x}_0, \dot{y}_0) \in L$ *and let*

$$(x_0, y_0, \lambda_0) = \phi^{-1}(x_0, y_0, \dot{x}_0, \dot{y}_0) \in C_F.$$

If $(x_0, y_0, \dot{x}_0, \dot{y}_0)$ is a solvable point of L, then $\lambda_0 = (\lambda_{01}, \dots, \lambda_{0k})$ is a solution of the linear equation $\sum_{j=1}^{k} \{a_i, a_j\}(x_0, y_0)\lambda_j = \{b, a_i\}(x_0, y_0)$, $i = 1, \dots, k$, which means that

$$(x_0, y_0, \lambda_0) \in \widetilde{S}_F \quad \textit{and} \quad (x_0, y_0, \dot{x}_0, \dot{y}_0) \in S_F.$$

Consequently any solvable submanifold of L is a subset of $S_F = TK \cap L$.

Proof. Since $(x_0, y_0, \dot{x}_0, \dot{y}_0) \in L$ is a solvable point of L, there exists a smooth curve $\gamma(t) = (x(t), y(t)) \in \mathbb{R}^{2n}$, $-\epsilon < t < \epsilon$ such that $(\gamma(t), \dot{\gamma}(t)) \in L$, $-\epsilon < t < \epsilon$, and $(\gamma(0), \dot{\gamma}(0)) = (x_0, y_0, \dot{x}_0, \dot{y}_0)$. Let $\widetilde{\gamma} : (\epsilon, \epsilon) \to C_F$ be the curve defined by $\widetilde{\gamma}(t) = (x(t), y(t), \lambda(t))$, then $\phi(\widetilde{\gamma}(t)) = (\gamma(t), \dot{\gamma}(t))$.

Since $(x_0, y_0, \lambda_0) = \phi^{-1}(x_0, y_0, \dot{x}_0, \dot{y}_0)$, we have $\lambda(0) = \lambda_0$. Since $\widetilde{\gamma}(t) \in C_F$, $\quad -\epsilon < t < \epsilon$, we see that

$$\frac{d\widetilde{\gamma}}{dt}(0) = \dot{x}_0 \frac{\partial}{\partial x} + \dot{y}_0 \frac{\partial}{\partial y} + \frac{d\lambda}{dt}(0) \frac{\partial}{\partial \lambda}$$

is tangent to L. Since L is contained in $T\mathbb{R}^{2n} |_K$ and K is defined by $a_1(x, y) = 0, \dots, a_k(x, y) = 0$, we have

$$\left(\dot{x}_0 \frac{\partial}{\partial x} + \dot{y}_0 \frac{\partial}{\partial y} + \frac{d\lambda}{dt}(0) \frac{\partial}{\partial \lambda} \right) (a_j) = 0, \quad j = 1, \dots, k.$$

And

$$0 = \frac{\partial F}{\partial y}(x_0, y_0, \lambda_0) \frac{\partial a_j}{\partial x}(0) - \frac{\partial F}{\partial x}(x_0, y_0, \lambda_0) \frac{\partial a_j}{\partial y}(0) = \{F, a_j\}(x_0, y_0, \lambda_0).$$

And using the form (6) of F we have

$$\sum_{i=1}^{k} \{a_i, a_j\}(x_0, y_0)\lambda_{0i} + \{b, a_j\}(x_0, y_0) = 0, \quad j = 1, \dots, k.$$

Thus $\lambda_0 = (\lambda_{01}, \dots, \lambda_{0k})$ is a solution of the system of linear equations

$$\sum_{j=1}^{k} \left(\{a_i, a_j\}(x_0, y_0) \right) \lambda_j = \{b, a_i\}(x_0, y_0), 1 \leq i \leq k.$$

This completes the proof of Lemma 3. □

First we have the following introductory result.

Proposition 4.
1) *If a submanifold Q of L is a solvable submanifold of the generalized Hamiltonian system L, then it is a solvable submanifold of the tangent bundle TK of K.*

2) *If the linear equation* (8) *has a smooth solution* $(\lambda_1(x,y),\cdots,\lambda_k(x,y))$ *defined on* K*, then the image* $G_\lambda = \phi(\widetilde{G}_\lambda)$ *by* ϕ *of the graph of this solution*

$$\widetilde{G}_\lambda = \{(x,y,\lambda_1(x,y),\ldots,\lambda_k(x,y)) : (x,y) \in K\}$$

is a smoothly solvable submanifold of L*.*

Proof. Part 1) is immediate by Lemma 3.
For part 2) suppose that the linear equation (8) has a smooth solution $\lambda(x,y) = (\lambda_1(x,y),\ldots,\lambda_k(x,y))$ defined on K. Consider the image

$$G_\lambda = \phi(\widetilde{G}_\lambda)$$

by ϕ of the graph $\widetilde{G}_\lambda = \{(x,y,\lambda_1(x,y),\ldots,\lambda_k(x,y)) \mid (x,y) \in K\}$ of the solution $(\lambda_1(x,y),\ldots,\lambda_k(x,y))$.

Since $\lambda(x,y)$ is a solution of the linear equation (8), from Lemma 2, we see that the vector field

$$d\widetilde{\pi}\left(\frac{\partial F}{\partial y}(x,y,\lambda(x,y))\frac{\partial}{\partial x} - \frac{\partial F}{\partial x}(x,y,\lambda(x,y))\frac{\partial}{\partial y}\right)$$

is tangent to K. Since $\lambda(x,y)$ is smooth, then this vector field depends smoothly on (x,y). Since $\widetilde{\pi}\mid_{\widetilde{G}_\lambda} : \widetilde{G}_\lambda \to K$ is a diffeomorphism then there exists a smooth vector field $\widetilde{\xi}$ tangent to $\widetilde{G}_\lambda$ such that

$$d\widetilde{\pi}(\widetilde{\xi}(x,y,\lambda(x,y))) = \frac{\partial F}{\partial y}(x,y,\lambda(x,y))\frac{\partial}{\partial x} - \frac{\partial F}{\partial x}(x,y,\lambda(x,y))\frac{\partial}{\partial y}.$$

Then, from Lemma 1. 2), the image $G_\lambda = \phi(\widetilde{G}_\lambda)$ is a smoothly solvable submanifold of L. This completes the proof of Proposition 4. □

Remark 2. *In Proposition 4. 1), in order to check that* Q *is smoothly solvable, it is enough to check that* Q *is a submanifold of* TK *and that* Q *is smoothly solvable as an implicit differential system, to which one can apply results of* [4].

We see that the Proposition 4. 1) is a direct consequence of Lemmas 2 and 3. Situation diametrically opposite to that in Theorem 3 is in the case if

$$\det(\{a_\ell, a_m\}(x,y)) \neq 0. \tag{9}$$

Under this condition we have

Proposition 5. *Let L be a generalized Hamiltonian system generated by a generating family* (6). *Suppose that*

$$k \quad \textit{is even and} \quad \det(\{a_\ell, a_m\}(x,y)) \neq 0, \quad \text{on } K.$$

Then S_F is a smoothly solvable submanifold of L and it is the maximal solvable submanifold of L in the sense that any other smoothly solvable submanifold of L is a submanifold of S_F. Moreover, the projection $\tau_{|S_F} : S_F \to K$ is a diffeomorphism and has no singular points. Consequently, S_F is a unique smoothly solvable submanifold of L such that $\tau(S_F) = K$.

Proof. Consider the $k \times k$ matrix $(\{a_\ell, a_m\}(x,y))$ and the linear equation

$$(10) \qquad \sum_{m=1}^{k} \{a_\ell, a_m\}(x,y)\lambda_m = \{b, a_\ell\}(x,y), \quad 1 \leq \ell \leq k.$$

Since $\det(\{a_\ell, a_m\}(x,y)) \neq 0$ on K, the linear equation (10) has a unique smooth solution $\lambda(x,y) = (\lambda_1(x,y), \ldots, \lambda_k(x,y))$ on K. Then we have

$$\widetilde{S}_F = \{(x,y,\lambda) \in \mathbb{R}^{2n} \times \mathbb{R}^k \mid \lambda = \lambda(x,y), \quad (x,y) \in K\}.$$

Thus $\widetilde{S}_F$ is the graph of the map $\lambda : K \to \mathbb{R}^k$. Therefore the projection map $\widetilde{\pi} \mid_{\widetilde{S}_F} : \widetilde{S}_F \to K$ is a submersion and so is $\tau_{|S_F} : S_F \to K$. Moreover, from Lemma 2, S_F is an implicit differential system as a submanifold of TK. Thus S_F is a smoothly solvable implicit differential system and it is a smoothly solvable submanifold of L. Now the maximality of S_F follows from Lemma 3. This completes the proof of Proposition 5. □

Remark 3. If b is a pre-Hamiltonian function defined on K for the generalized Hamiltonian system L, then the corresponding Hamiltonian function for the solvable Hamiltonian vector field in the restricted symplectic space $(K, \omega \mid_K)$ is defined by

$$\hat{F}(x,y) = b(x,y) + \sum_{i=1}^{k} \lambda_i(x,y) a_i(x,y),$$

where $\lambda(x,y)$ is a unique smooth solution of the equation (10) and

$$\frac{\partial \hat{F}}{\partial y}(x,y)\frac{\partial}{\partial x} - \frac{\partial \hat{F}}{\partial x}(x,y)\frac{\partial}{\partial y} \mid_K$$

is a smooth section of TK.

§3. Solvability over constant rank constraints.

Let K be a submanifold of (M, ω). The constant rank condition of matrix $A(q)$ at all points of K is related to the special cases of submanifolds of M.

K is said to be coisotropic if $(T_qK)^\omega \subset T_qK$ at each $q \in K$ is isotropic if $T_qK \subset (T_qK)^\omega$ at each $q \in K$. $det A(q)$ is vanishing on K in both these cases. K is said to be symplectic if $T_qK \cap (T_qK)^\omega = 0$ at each $q \in K$.

Let us denote the intersection $V_q = T_qK \cap (T_qK)^\omega$ and we assume $dim V_q = l$ is constant at each $q \in K$. V_q is a kernel of $A(q)$. The two form induced on the quotient space $(T_qK)^\omega / V_q$ is nondegenerated for $k = 1, \ldots, 2n-1$. $dim(T_qK)^\omega/V_q = k - l$ and there is a natural relation for the kernel dimension, $l \leq max\{k, 2n-k\}$. Obviously $k-l$ is an even number. We easily find that $rank A(q) = k - l, l \leq n,\ \ l \leq 2n-k$. The kernel $N_q = Ker A(q)$ gives an intersection of skew-conormal fibre of K with the tangent space TK. The constant rank condition of A along K implies that $V = \bigcup_{q \in K} V_q$ is a distribution on K, i.e. a smooth field of linear subspaces of TK. This is the characteristic distribution of $\omega\mid_K$. V is defined by the generating function

$$F(x, y, \lambda) = \sum_{i=1}^{k} \lambda_i a_i(x, y), (x, y) = q \in K.$$

$corank A(x, y) \leq 2n - k$, $rank A(x, y) \geq 2k - 2n$, for $k > n$

$$V_q = \left\{ \sum_{i=1}^{k} \lambda_i \left(\frac{\partial a_i}{\partial y}(x, y) \frac{\partial}{\partial x} - \frac{\partial a_i}{\partial x}(x, y) \frac{\partial}{\partial y} \right) \right\},$$

where $\lambda \in Ker A(x, y), (x, y) \in K$.

Proposition 6. *V is an integrable distribution of TK and it is a solvable submanifold of L with $b \equiv 0$.*

Let $\lambda^j(x, y)$, $j = 1, \ldots l$ be $l-$independent smooth sections of the fibre bundle $Ker A(x, y)$ over K, then we can re-define the defining generators a_i taking instead the new l functions, $c^j(x, y) \in I_K$, where I_K is an ideal which consists of all functions vanishing on K (cf. p. 5),

$$c^j(x, y) = \sum_{i=1}^{k} \lambda_i^j(x, y) a_i(x, y), \quad j = 1, \ldots, l.$$

We can easily check that

$$\{c^j, a_i\}\mid_K = 0, \quad j = 1, \ldots, l, i = 1, \ldots, k$$

and

$$\{c^j, c^s\} \mid_K = 0, \quad j = 1, \ldots, l, s = 1, \ldots, l.$$

After re-numeration of $a_1, \ldots, a_k$ assume that $c^1, \ldots, c^l, a_{l+1}, \ldots, a_k$ are independent and define K. Thus the matrix A reduces to the maximal rank sub-matrix $(\{a_i, a_j\})_{l+1 \leq i,j \leq k}$.

Thus the problem reduces to the coisotropic submanifold $C \in \mathbb{R}^{2n}$ defined by $c^j = 0$ with b preserving fibers of V on C. The rest of functions a_i define K as a section of the foliation defined by integral surfaces of V.

§4. Solvable domains in generalized Hamiltonian systems.

When k is odd we have $\det A(x, y) = 0$ everywhere. As a result corresponding to Proposition 5, we have

Proposition 7. *Let $L \subset T\mathbb{R}^{2n}$ be a generalized Hamiltonian system generated by a generating family* (6). *Suppose that k is odd and the rank of $(\{a_\ell, a_m\}(x, y))$ is constant and equal to $k - 1$. Suppose also that the linear equation* (8) *has a smooth solution $\lambda(x, y) = (\lambda_1(x, y), \ldots, \lambda_k(x, y))$ on K. Then*

1) *S_F is a smoothly solvable submanifold of L and it is the maximal solvable submanifold in the sense that any other smoothly solvable submanifold of L is a submanifold of S_F.*

2) *Moreover, S_F is a line bundle over K with the submersion map $\tau_{|S_F} : S_F \to K$.*

To prove Proposition 7, we use the following more general theorem which we will prove later.

Theorem 5. *Let $L \subset T\mathbb{R}^{2n}$ be a generalized Hamiltonian system generated by a generating family* (6). *Let Q be a submanifold of L such that the projection $\tau \mid_Q: Q \to K$ is a submersion. Then Q is smoothly solvable if and only if $Q \subset S_F$.*

Proof of Proposition 7. Let $L \subset T\mathbb{R}^{2n}$ be an implicit Hamiltonian system generated by a Morse family (6). Suppose that k is odd and the rank of $(\{a_i, a_j\}(x, y))$ is constant and equal to $k - 1$. Suppose also that the linear equation (8) has a smooth solution $\lambda(x, y) = (\lambda_1(x, y), \ldots, \lambda_k(x, y))$ on K.

Since the matrix $(\{a_i, a_j\}(x, y))$ depends smoothly on $(x, y) \in K$ and has a constant rank $k - 1$, the kernel set

$$\widetilde{K}_F = \{(x, y, \lambda) \in C_F \mid (\{a_i, a_j\}(x, y))\, \lambda = 0\}$$

is a smooth line bundle over K and we see that

$$\widetilde{S}_F = \{(x, y, \lambda(x, y) + \lambda) \mid (x, y) \in K, (x, y, \lambda) \in \widetilde{K}_F\}.$$

Therefore $\widetilde{S}_F$ is also a smooth line bundle over K and so is $S_F = \phi(\widetilde{S}_F)$. Thus, S_F is a smooth submanifold of L and the projection $\pi : S_F \to K$ is a submersion. From Theorem 5, $S_F = \phi(\widetilde{S}_F)$ is a smoothly solvable submanifold of L. The maximality of S_F follows from Lemma 3. This completes the proof of Proposition 7. □

Proof of Theorem 5. Let $L \subset T\mathbb{R}^{2n}$ be an implicit Hamiltonian system generated by a Morse family (6). Suppose that M is a submanifold of L such that the projection $\tau\,|_M \colon M \to K$ is a submersion.

If M is smoothly solvable, then, from Lemma 3, we have $M \subset S_F$.

Conversely, suppose that $M \subset S_F$. Let

$$(x_0, y_0, \dot{x}_0, \dot{y}_0) \in M \quad \text{and} \quad (x_0, y_0, \lambda_0) = \phi^{-1}(x_0, y_0, \dot{x}_0, \dot{y}_0).$$

Since

$$(x_0, y_0, \dot{x}_0, \dot{y}_0) \in S_F \quad \text{and} \quad (x_0, y_0, \lambda_0) \in \widetilde{S}_F,$$

from the definition of S_F and from Lemma 2, the vector

$$\dot{x}_0 \frac{\partial}{\partial x} + \dot{y}_0 \frac{\partial}{\partial y} = \frac{\partial F}{\partial y}(x_0, y_0, \lambda_0)\frac{\partial}{\partial x} - \frac{\partial F}{\partial x}(x_0, y_0, \lambda_0)\frac{\partial}{\partial y}$$

is tangent to K at (x_0, y_0) and smoothly depends on $(x_0, y_0, \dot{x}_0, \dot{y}_0) \in M$. Since $\tau\,|_M \colon M \to K$ is a submersion, there exists a smooth vector field ξ tangent to M such that

$$d\tau(\xi(x_0, y_0, \dot{x}_0, \dot{y}_0)) = \dot{x}_0 \frac{\partial}{\partial x} + \dot{y}_0 \frac{\partial}{\partial y}, \quad \forall (x_0, y_0, \dot{x}_0, \dot{y}_0) \in M.$$

Thus, from Lemma 1, M is smoothly solvable. This completes the proof of Theorem 5. □

As a direct corollary of Theorem 5, we have the following Proposition which is a generalization of Proposition 7.

Proposition 8. *Let $L \subset T\mathbb{R}^{2n}$ be a generalized Hamiltonian system generated by* (6). *Suppose that the linear equation* (8) *has a smooth solution on K,*

$$\lambda(x, y) = (\lambda_1(x, y), \ldots, \lambda_k(x, y)).$$

Suppose also that the kernel set

$$\widetilde{K}_F = \ker(\{a_i, a_j\}) = \{(x, y, \lambda) \in C \times \mathbb{R}^k \mid (\{a_i, a_j\}(x, y))\, \lambda = 0\}$$

contains an $m-$ dimensional smooth vector subbundle $\widetilde{K}$ of the vector bundle $K \times \mathbb{R}^k$ over K. Then

$$R_F = \left\{ \left(x, y, \frac{\partial F}{\partial y}(x, y, \lambda(x, y) + \lambda), -\frac{\partial F}{\partial x}(x, y, \lambda(x, y) + \lambda) \right) \mid (x, y, \lambda) \in \widetilde{K} \right\}$$

is a $(2n - k + m)$ dimensional smoothly solvable submanifold of L.

The condition in Proposition 8 that the kernel set $\widetilde{K}_F$ contains an m dimensional smooth vector subbundle is not generic condition if $m > 0$ for k even, and if $m > 1$ for k odd. Because generically if k is even, $\det(\{a_\ell, a_m\}(x, y)) \neq 0$ almost everywhere, and if k is odd, $\operatorname{rank}(\{a_\ell, a_m\}(x, y)) = k - 1$ almost everywhere. Note that smooth vector subbundles of $\widetilde{K}_F$ are not unique in general: If $\bar{K}$ is a smooth vector subbundle of $\widetilde{K}_F$, then any smooth vector subbundle of $\bar{K}$ is also a smooth vector subbundle of $\widetilde{K}_F$. For k even we define

$$K_{reg} = \{(x, y) \in K \mid \det(\{a_\ell, a_m\}(x, y)) \neq 0\},$$

for k odd we have

$$K_{k-1} = \{(x, y) \in K \mid \operatorname{rank}(\{a_\ell, a_m\}(x, y)) = k - 1\}.$$

In generic situation, we have

Proposition 9. *Suppose that k is even and* $\det(\{a_\ell, a_m\}(x, y)) \neq 0$ *almost everywhere but* $\det(\{a_\ell, a_m\}(0, 0)) = 0$. *Then $L_F \cap \tilde{\pi}^{-1}(K_{reg})$ is smoothly solvable implicit differential system of TK_{reg}. Moreover there exists a smoothly solvable differential system Q such that $\tilde{\pi}(Q) = K$ if and only if the linear equation (8) has a smooth solution on K. Such a smoothly solvable differential system Q is unique and it has the properties that $Q \cap \pi^{-1}(K_{reg}) = L_F \cap \tau^{-1}(K_{reg})$ and that $\tau_Q : Q \to K$ is a diffeomorphism.*

Proof. The fact that $L \cap \tau^{-1}(K_{reg})$ is a smoothly solvable implicit differential system of TK_{reg} is a direct corollary of Theorem 5. Now suppose that the linear equation (8) has a smooth solution

$$(\lambda_1(x, y), \ldots, \lambda_k(x, y))$$

on K. Then, by Proposition 4. 2), the image $G_\lambda = \phi(\widetilde{G}_\lambda)$ of the graph $\widetilde{G}_\lambda$ of the solution $(\lambda_1(x, y), \ldots, \lambda_k(x, y))$ is a smoothly solvable submanifold of L. Take G_λ as M we seek. Then, by Theorem 5, $M \cap TK_{reg} = G_\lambda \cap TK_{reg}$ and $S_F \cap TC_{reg}$ must coincide. Since K_{reg} is dense in K, the uniqueness of such M follows.

Conversely suppose that there exists a smoothly solvable differentiable system M such that $\pi(M) = K$. Then, again by Proposition 5, $M \cap TK_{reg}$ must coincide with $S_F \cap TK_{reg}$. Consider the inverse image $\widetilde{M} = \phi^{-1}(M) \subset C_F \subset K \times \mathbb{R}^k$. Since, by Proposition 5, $\widetilde{S}_F \cap (K_{reg} \times \mathbb{R}^k)$ is the graph of a smooth solution $\lambda : K_{reg} \to \mathbb{R}^k$ of the linear equation (8), $\widetilde{M} \cap (K_{reg} \times \mathbb{R}^k)$ must coincide with the graph of this smooth solution $\lambda(x, y)$, $(x, y) \in K_{reg}$. Since K_{reg} is dense in K and $\widetilde{M}$ is a smooth submanifold such that $\widetilde{\pi}(\widetilde{M}) = K$, $\lambda(x, y)$ can be extended to a smooth solution defined on K of the linear equation. Thus the linear equation has a smooth solution on K. This completes the proof of Proposition 9 □.

Remark 4. *In the case where k is odd we can have a similar result. However, when k is odd and the rank of the matrix* $(\{a_i, a_j\}(0, 0))$ *is less than* $k - 1$,

1) *There is a question, in a generic situation, whether the kernel set*

$$\widetilde{K}_F = \ker(\{a_i, a_j\}) = \{(x, y, \lambda) \in K \times \mathbb{R}^k \mid (\{a_i, a_j\}(x, y))\, \lambda = 0\}$$

contains or not a smooth line bundle over K appeared in Proposition 8.

2) *Moreover when k is odd, we can not apply our condition for the linear equation to have a smooth solution. Since* $\det(\{a_i, a_j\}(x, y)) = 0$, *the product of the matrix* $(\{a_i, a_j\}(x, y))$ *and its cofactor matrix is always the zero matrix. Thus we can not apply our method.*

Theorem 5 and Propositions 7, 5 and 8 are obtained by reducing the fibers of the bundle $\tau : L \to K$. However reducing the base space K, we obtain

Proposition 10. *Suppose that L is not smoothly solvable. Let* $g_1, \ldots, g_s : \mathbb{R}^{2n} \to \mathbb{R}$ *be smooth functions such that the Jacobian matrix of the map* $(a, g) = (a_1, \ldots, a_k, g_1, \ldots, g_s) : \mathbb{R}^{2n} \to \mathbb{R}^{k+s}$ *has the maximal rank* $k + s$. *Let* $C_g \subset K$ *be a submanifold defined by*

$$C_g = \{(x, y) \in K \mid g_1(x, y) = \cdots = g_s(x, y) = 0\}.$$

Then $\phi(C_g \times \mathbb{R}^k)(\subset L_F)$ *is smoothly solvable if and only if*

$$\{a_\ell, a_m\} = \{b, a_m\} = 0, \{a_\ell, g_t\} = \{b, g_t\} = 0 \qquad on \quad C_g,$$

$$1 \le \ell, m \le k, \quad 1 \le t \le s.$$

Proof. This Proposition can be proved in the same way as it was done for Theorem 3. Let us consider the Morse family (6). Then we have

$$\frac{\partial F}{\partial \lambda_\ell}(x,y,\lambda) = a_\ell(x,y).$$

Set

$$\begin{aligned}&C_{F,g} = \{(x,y,\lambda) \in C_F \mid g_1(x,y) = \cdots = g_s(x,y) = 0\} = C_g \times \mathbb{R}^k,\\ &L_{F,g} = \phi(C_{F,g}).\end{aligned}$$

Now $L_{F,g}$ is smoothly solvable if and only if there exists a smooth tangent vector filed ξ on $L_{F,g} = \phi(C_{F,g})$ such that

$$d\pi(\xi(x,y,\dot{x},\dot{y})) = \sum_{i=1}^{n} \dot{x}_i \frac{\partial}{\partial x_i} + \dot{y}_i \frac{\partial}{\partial y_i}$$

where $\pi : T\mathbb{R}^{2n} \to \mathbb{R}^{2n}$ is the projection of the tangent bundle, then there are smooth functions $\mu_\ell(x,y,\lambda), \ell = 1,\ldots,k$, such that the vector field

$$\widetilde{\xi}(x,y,\lambda) = \sum_{i=1}^{n} \frac{\partial F}{\partial y_i}(x,y,\lambda)\frac{\partial}{\partial x_i} - \frac{\partial F}{\partial x_i}(x,y,\lambda)\frac{\partial}{\partial y_i} + \sum_{\ell=1}^{k} \mu_\ell(x,y,\lambda)\frac{\partial}{\partial \lambda_\ell}$$

is tangent to $C_{F,g} = C_g \times \mathbb{R}^k$ if and only if

$$\sum_{i=1}^{n} \frac{\partial F}{\partial y_i}(x,y,\lambda)\frac{\partial}{\partial x_i} - \frac{\partial F}{\partial x_i}(x,y,\lambda)\frac{\partial}{\partial y_i} \quad \text{is tangent to} \quad C_{F,g}.$$

Then

$$\left(\sum_{i=1}^{n} \frac{\partial F}{\partial y_i}\frac{\partial}{\partial x_i} - \frac{\partial F}{\partial x_i}\frac{\partial}{\partial y_i}\right) a_\ell = 0 \quad \text{on} \quad C_{F,g}, \quad 1 \le \ell \le k,$$

$$\left(\sum_{i=1}^{n} \frac{\partial F}{\partial y_i}\frac{\partial}{\partial x_i} - \frac{\partial F}{\partial x_i}\frac{\partial}{\partial y_i}\right) g_t = 0 \quad \text{on} \quad C_{F,g}, \quad 1 \le t \le s,$$

and

$$\{F, a_\ell\} = \sum_{i=1}^{k} \{a_i, a_\ell\}\lambda_i + \{b, a_\ell\} = 0 \quad \text{on} \quad C_F, \quad 1 \le \ell \le k.$$

$$\{F, g_t\} = \sum_{i=1}^{k} \{a_i, g_t\}\lambda_i + \{b, g_t\} = 0 \quad \text{on} \quad C_{F,g}, \quad 1 \le t \le s.$$

Differentiating the equalities with respect to λ_i, we have

$$\{a_i, a_\ell\} = \{a_i, g_t\} = 0, \quad \text{and then} \quad \{b, a_\ell\} = \{b, g_t\} = 0,$$

$$\text{on} \quad C_{F,g}, \quad 1 \le \ell \le k, 1 \le t \le s.$$

Conversely, if

$$\{a_i, a_\ell\} = \{a_i, g_t\} = 0, \quad \text{then} \quad \{b, a_\ell\} = \{b, g_t\} = 0,$$

$$\text{on} \quad C_{F,g}, \quad 1 \le \ell \le k, 1 \le t \le s,$$

then $L_{F,g} = \phi(C_{F,g})$ is smoothly solvable. This completes the proof of Proposition 10. □

Example 1. *Consider the following function.*

$$F(x, y, \lambda) = \sum_{i=1}^{k} x_i\lambda_i + b_1(y_1, \ldots, y_k)b_2(x_{k+1}, \ldots, x_m),$$

$$k + 1 \le m \le n, \quad b_1(0) = b_2(0) = 0, \quad b_1, b_2 \quad \textit{are not constantly} \quad 0.$$

Then

$$\{a_\ell, a_m\} = \{x_\ell, x_m\} = 0, \quad 1 \le \ell, m \le k.$$

However

$$\{a_\ell, b\} = \{x_\ell, b\} = -\frac{\partial b_1}{\partial y_\ell} \cdot b_2 \ne 0$$

$$\textit{on} \quad K = \{a_1 = \cdots = a_k = 0\} = \{x_1 = \cdots = x_k = 0\}.$$

Thus L itself is not smoothly solvable. Now consider functions

$$g_1(x, y) = x_{k+1}, \ldots, g_s(x, y) = x_{k+s} = x_m, \quad \textit{where} \quad s = m - k,$$

and set

$$S = \{(x, y) \in \mathbb{R}^{2n} \mid a_1(x, y) = \cdots = a_k(x, y) = g_1(x, y) = \cdots = g_s(x, y) = 0\}.$$

Then

$$\{a_\ell, b\} = -\frac{\partial b_1}{\partial y_\ell} b_2(x_{x+1}, \ldots, x_m) = 0,$$

$$\{a_\ell, g_t\} = \{x_\ell, x_{k+t}\} = 0, \quad \{b, g_t\} = \{b, x_{k+t}\} = 0,$$

$$1 \le \ell \le k, \quad 1 \le t \le s = m - k,$$

$$\textit{on} \quad S = \{a_1 = \cdots = a_k = g_1 = \cdots = g_s = 0\}.$$

Then, by Proposition 10, $L_F \cap (S \times \mathbb{R}^{2n})$ is smoothly solvable.

§5. Poisson-Lie algebras on submanifolds.

The Poisson-Lie algebra is an associative algebra equipped with a Poisson bracket, which is a Lie bracket. A Poisson-Lie algebra structure on an associative, commutative algebra A is defined by the bilinear skew-symmetric mapping $\{.,.\} : A \times A \to A$, which satisfies the Leibnitz rule $\{f, gh\} = \{f, g\}h + g\{f, h\}$, $f, g, h \in A$ and is called the Poisson bracket.

Poisson structures on manifolds are the basic mathematical structures of mechanics. The representative one is the algebra of all smooth functions on the phase space under ordinary multiplication and the Lie structure induced by the Poisson bracket usually defined by the symplectic form. For the implicit Hamiltonian systems, defined by singular mappings, the Poisson-Lie algebra is formed by the solvable implicit Hamiltonian systems [6]. In this section we search for Poisson-Lie algebras associated to generalized Hamiltonian systems.

Let Q be a submanifold of L_F. If $\pi|_Q : Q \to K$ is a diffeomorphism, then Q is smoothly solvable. We showed that $\pi|_Q : Q \to K$ is a diffeomorphism if and only if there exists a smooth solution $\lambda(x, y)$ of (8) such that

$$Q = \phi_F\Big(\big\{(x, y, \lambda(x, y)) \mid (x, y) \in K\big\}\Big) = \phi_F(\text{ the graph of } \lambda(x, y)).$$

Let us define

$$\{a_1, \cdots, a_k\}_K^{\perp} = \{h \in \mathcal{E}_{x,y} \mid \{h, a_i\} = 0 \quad \text{on } K\},$$

where $\mathcal{E}_{x,y}$ is the $\mathbb{R}$-algebra of germs of smooth functions on $\mathbb{R}^{2n}$ with a fixed base point in K.

If $h \in \{a_1, \cdots, a_k\}_K^{\perp}$, then the corresponding Hamiltonian vector field X_h is tangent to K.

Theorem 6. *Equation* (8) *has a smooth solution defined on* K *if and only if*

$$b \in \langle a_1, \cdots, a_k \rangle_{\mathcal{E}_{x,y}} + \{a_1, \cdots, a_k\}_K^{\perp}.$$

Proof. Suppose that (8) has a smooth solution $\lambda(x, y)$ defined on K;

$$\Big(\{a_\ell, a_m\}(x, y)\Big) \begin{pmatrix} \lambda_1(x, y) \\ \vdots \\ \lambda_k(x, y) \end{pmatrix} = \begin{pmatrix} \{b, a_1\}(x, y) \\ \vdots \\ \{b, a_k\}(x, y) \end{pmatrix}, \quad (x, y) \in K.$$

Let's consider a function $h(x,y) = b(x,y) - \sum_{m=1}^{k} \lambda_m(x,y)a_m(x,y)$. Then

$$\begin{pmatrix} \{h,a_1\}(x,y) \\ \vdots \\ \{h,a_k\}(x,y) \end{pmatrix} = \begin{pmatrix} \{b,a_1\}(x,y) \\ \vdots \\ \{b,a_k\}(x,y) \end{pmatrix} - \Big(\{a_\ell,a_m\}(x,y)\Big) \begin{pmatrix} \lambda_1(x,y) \\ \vdots \\ \lambda_k(x,y) \end{pmatrix}$$

is vanishing on K. In the above calculations we have $\{a_\ell, \lambda_m\}(x,y)$ $a_m(x,y) = 0$ on K. Thus $h \in \{a_1, \cdots, a_k\}_K^\perp$ and $b(x,y) = \sum_{m=1}^{k} \lambda_m(x,y)$ $a_m(x,y) + h(x,y)$. Hence

$$b \in \langle a_1, \cdots, a_k \rangle_{\mathcal{E}_{x,y}} + \{a_1, \cdots, a_k\}_K^\perp.$$

Conversely suppose tha $b \in \langle a_1, \cdots, a_k \rangle_{\mathcal{E}_{x,y}} + \{a_1, \cdots, a_k\}_K^\perp$. Then $b(x,y)$ has the form

$$b(x,y) = \sum_{m=1}^{k} \mu_m(x,y)a_m(x,y) + h(x,y), \quad \mu_m \in \mathcal{E}_{x,y}, \quad h \in \{a_1, \cdots, a_k\}_K^\perp.$$

Then

$$\begin{pmatrix} \{b,a_1\}(x,y) \\ \vdots \\ \{b,a_k\}(x,y) \end{pmatrix} = -\Big(\{a_\ell,a_m\}(x,y)\Big) \begin{pmatrix} \mu_1(x,y) \\ \vdots \\ \mu_k(x,y) \end{pmatrix} + \begin{pmatrix} \{h,a_1\}(x,y) \\ \vdots \\ \{h,a_k\}(x,y) \end{pmatrix} = -\Big(\{a_\ell,a_m\}(x,y)\Big) \begin{pmatrix} \mu_1(x,y) \\ \vdots \\ \mu_k(x,y) \end{pmatrix}$$

on K since $h \in \{a_1, \cdots, a_k\}_K^\perp$. Thus $-\mu(x,y) = -(\mu_1(x,y), \cdots, \mu_k(x,y)$ is a smooth solution of (8) defined on K. □

Now we introduce the following notation:
$\mathcal{S}_{a,b} = \{\lambda(x,y) \in \mathcal{E}_{x,y} \mid \text{a smooth solution of (8) defined on } K\} \subset \mathcal{E}_{x,y}$,
$F_{a,b,\lambda}(x,y) = \sum_{i=1}^{k} a_i(x,y)\lambda_i(x,y) + b(x,y), \quad \lambda = (\lambda_1, \cdots, \lambda_k) \in \mathcal{S}_{a,b}$,
$\mathcal{H}_{a,K} = \{F_{a,b,\lambda}(x,y) \mid \lambda(x,y) \in \mathcal{S}_{a,b},\ b \in \langle a_1, \cdots, a_k \rangle_{\mathcal{E}_{x,y}} + \{a_1, \cdots, a_k\}_K^\perp\}$,

$$M_{F_{a,b,\lambda}} = \phi_F\Big(\big\{(x,y,\lambda(x,y)) \mid (x,y) \in K\big\}\Big), \quad \lambda = (\lambda_1, \cdots, \lambda_k) \in \mathcal{S}_{a,b}.$$

Proposition 11. *If $F_{a,b,\lambda} \in \mathcal{H}_{a,K}$, then the Hamiltonian vector field $X_{F_{a,b,\lambda}}$ is tangent to K and $M_{F_{a,b,\lambda}}$ is smoothly solvable.*

Proof. Let $F_{a,b,\lambda} \in \mathcal{H}_{a,K}$. $\lambda(x,y)$ is a smooth solution of (8) defined on K and $F_{a,b,\lambda}$ has the form

$$F_{a,b,\lambda}(x,y) = \sum_{m=1}^{k} a_m(x,y)\lambda_m(x,y) + b(x,y).$$

Since $\lambda(x,y)$ is a smooth solution of (8) defined on K, we have

$$\begin{aligned}\{F_{a,b,\lambda}, a_\ell\}(x,y) &= \sum_{m=1}^{k} \{a_m, a_\ell\}(x,y)\lambda_m(x,y) + \{b, a_\ell\}(x,y) \\ &= -\sum_{m=1}^{k} \{a_\ell, a_m\}(x,y)\lambda_m(x,y) + \{b, a_\ell\}(x,y) = 0\end{aligned}$$

on K. Thus $\{F_{a,b,\lambda}, a_\ell\}(x,y) = 0$ on K. Hence $X_{F_{a,b,\lambda}}$ is tangent to K and $M_{F_{a,b,\lambda}}$ is smoothly solvable. □

Theorem 7.
1) $\mathcal{H}_{a,K} = \{a_1, \cdots, a_k\}_K^{\perp}$.
2) $\mathcal{H}_{a,K} = \{a_1, \cdots, a_k\}_K^{\perp}$ *is a Poisson-Lie algebra with respect to* ω ; *if* $F_{a,b,\lambda}, F_{a,b',\lambda'} \in \mathcal{H}_{a,K}$, *then* $\{F_{a,b,\lambda}, F_{a,b',\lambda'}\} \in \mathcal{H}_{a,K}$, *and equivalently if* $h, h' \in \{a_1, \cdots, a_k\}_K^{\perp}$, *then* $\{h, h'\} \in \{a_1, \cdots, a_k\}_K^{\perp}$.

Proof. 1) Let $F_{a,b,\lambda} \in \mathcal{H}_{a,K}$. Then as seen on the last line of the proof of Proposition 11, we have $\{F_{a,b,\lambda}, a_\ell\}(x,y) = 0$ on K for $1 \le \ell \le k$. Therefore $F_{a,b,\lambda} \in \{a_1, \cdots, a_k\}_K^{\perp}$ and $\mathcal{H}_{a,K} \subset \{a_1, \cdots, a_k\}_K^{\perp}$.

Conversely let $h \in \{a_1, \cdots, a_k\}_K^{\perp}$. For any k-tuple $\lambda_1, \cdots, \lambda_k \in \mathcal{E}_{x,y}$ set

$$b(x,y) = \sum_{m=1}^{k} -a_m(x,y)\lambda_m(x,y) + h(x,y). \tag{11}$$

Then we see that $\lambda(x,y) = (\lambda_1(x,y), \cdots, \lambda_k(x,y))$ is a smooth solution of (8) defined on K and that $F_{a,b,\lambda} = h$. Thus $h \in \mathcal{H}_{a,K}$.

$$\begin{aligned}\begin{pmatrix} \{b, a_1\}(x,y) \\ \vdots \\ \{b, a_k\}(x,y) \end{pmatrix} &= -\Big(\{a_m, a_\ell\}(x,y)\Big) \begin{pmatrix} \lambda_1(x,y) \\ \vdots \\ \lambda_k(x,y) \end{pmatrix} + \\ \begin{pmatrix} \{h, a_1\}(x,y) \\ \vdots \\ \{h, a_k\}(x,y) \end{pmatrix} &= \Big(\{a_\ell, a_m\}(x,y)\Big) \begin{pmatrix} \lambda_1(x,y) \\ \vdots \\ \lambda_k(x,y) \end{pmatrix}\end{aligned}$$

on K since $h \in \{a_1, \cdots, a_k\}_K^{\perp}$. Thus $\lambda(x,y)$ is a smooth solution of (8) and $F_{a,b,\lambda}(x,y) \in \mathcal{H}_{a,K}$. Then by the definition of $F_{a,b,\lambda}(x,y)$ and (16) we have,

$$F_{a,b,\lambda}(x,y) = \sum_{m=1}^{k} a_m(x,y)\lambda_m(x,y) + b(x,y) = h(x,y).$$

Thus $h(x,y) \in \mathcal{H}_{a,K}$ and $\{a_1, \cdots, a_k\}_K^{\perp} \subset \mathcal{H}_{a,K}$. This complete the proof of 1).

2) Suppose that $h, h' \in \{a_1, \cdots, a_k\}_K^{\perp}$. Then the Hamiltonian vector fields X_h and $X_{h'}$ are both tangent to K. Then $X_{\{h,h'\}} = [X_h, X_{h'}]$ is also tangent to K. Thus $\{h, h'\} \in \{a_1, \cdots, a_k\}_K^{\perp}$. □

Definition 2. *We say that two map-germs $(a_1, \cdots, a_k)$ and $(\bar{a}_1, \cdots, \bar{a}_k)$ are* **symplectic$\mathcal{K}$-equivalent** *if there exist a symplectic diffeomorphism germ $\varphi : (\mathbb{R}^{2n}, 0) \to (\mathbb{R}^{2n}, 0)$ and a family of regular matrices $G(x,y) \in Gl(k, \mathbb{R})$ smoothly depending on (x,y) such that*

$$\begin{pmatrix} \bar{a}_1(x,y) \\ \vdots \\ \bar{a}_k(x,y) \end{pmatrix} = G(x,y) \begin{pmatrix} a_1 \circ \varphi(x,y) \\ \vdots \\ a_k \circ \varphi(x,y) \end{pmatrix}. \tag{12}$$

Proposition 12. *Suppose that $(a_1, \cdots, a_k)$ and $(\bar{a}_1, \cdots, \bar{a}_k)$ are symplectic $\mathcal{K}$-equivalent. Then*

$$\{a_1, \cdots, a_k\}_K^{\perp} \cong \{\bar{a}_1, \cdots, \bar{a}_k\}_{\varphi^{-1}(K)}^{\perp}$$

as the Poisson-Lie algebras.

Proof. If their symplectic $\mathcal{K}$-equivalence relation is given by (12), then the isomorphism is given by

$$\varphi^* : \{a_1, \cdots, a_k\}_K^{\perp} \to \{\bar{a}_1, \cdots, \bar{a}_k\}_{\varphi^{-1}(K)}^{\perp}$$

and for $h, h' \in \{a_1, \cdots, a_k\}_K^{\perp}$ we have $\{h \circ \varphi, h' \circ \varphi\} = \{h, h'\} \circ \varphi$. □

Proposition 13. *Let $k \leq n$. If*

$$\mathrm{rank}\Big(\{a_i, a_j\}(x,y)\Big) = 0 \quad \text{constantly on } \mathbb{R}^{2n},$$

then $(a_1, \cdots, a_k)$ is symplectic $\mathcal{K}$-equivalent to the projection map-germ

$$p(x,y) = (y_1, \cdots, y_k)$$

and

$$\mathcal{H}_{a,K} \cong \langle y_1, \cdots, y_k \rangle_{\mathcal{E}_{x,y}}^2 + \mathcal{E}_{x_{k+1}, \cdots, x_n, y}$$

Proof. Since rank $(\{a_i, a_j\}(x,y)) = 0$ constantly on $\mathbb{R}^{2n}$, by Darboux Theorem there exists a symplectic coordinate systems $\xi_1, \ldots, \xi_n, \eta_1, \ldots, \eta_n$ such that $a_i = \eta_i, i = 1, \ldots, k$. Hence $(a_1, \ldots, a_k)$ is symplectic $\mathcal{K}$-equivalent to the projection map-germ $p(x,y) = (y_1, \ldots, y_k)$. This completes the proof. □

Example 2. *Let $k = 2r$. Let $a = (a_1, \cdots, a_k) : (\mathbb{R}^{2n}, (0,0)) \to (\mathbb{R}^k, 0)$.*

$$\operatorname{rank}\Big(\{a_i, a_j\}(0,0)\Big) = k.$$

Then $(a_1, \cdots, a_k)$ is symplectic $\mathcal{K}$-equivalent to the projection map-germ

$$p(x,y) = (y_1, \cdots, y_r, x_1, \cdots, x_r)$$

and

$$\mathcal{H}_{a,K} \cong \langle x_1, \cdots, x_{r+s}, y_1, \cdots, y_r \rangle^2_{\mathcal{E}_{x,y}} + \mathcal{E}_{x_{r+1}, \cdots, x_n, y_{r+s+1}, \cdots, y_n}$$

Example 3. *Let $k = 2r + s$. If*

$$\operatorname{rank}\Big(\{a_i, a_j\}(x,y)\Big) = 2r \quad \textit{constantly on } \mathbb{R}^{2n},$$

then $(a_1, \cdots, a_k)$ is symplectic $\mathcal{K}$-equivalent to the projection map-germ

$$p(x,y) = (y_1, \cdots, y_r, x_1, \cdots, x_{r+s})$$

and

$$\mathcal{H}_{a,K} \cong \langle x_1, \cdots, x_{r+s}, y_1, \cdots, y_r \rangle^2_{\mathcal{E}_{x,y}} + \mathcal{E}_{x_{r+1}, \cdots, x_n, y_{r+s+1}, \cdots, y_n}.$$

§6. Hamiltonian vector fields on S^2.

Let K be a submanifold of $(\mathbb{R}^{2n}, \omega)$. First we will prove the following general property.

Proposition 14. *Let U be an open subset of $\mathbb{R}^{2n}$. Suppose that $\omega\,|_{T(K\cap U)}$ is a non-singular 2 form on $K \cap U$. Let $h \in C^\infty(U)$ such that X_h is tangent to $K \cap U$. Then*

$$X_h\,|_{K\cap U} = X_{\omega|_{T(K\cap U)}, h|_{K\cap U}}, \tag{13}$$

where $X_{\omega_{K\cap U}, f}$ denotes the Hamiltonian vector field on the symplectic manifold $(K \cap U, \omega\,|_{T(K\cap U)})$ generated by a function $f \in C^\infty(K \cap U)$.

Proof. Since the problem is a local problem, it suffices to prove the proposition in the case where U is a small open ball. Since $\omega\mid_{K\cap U}$ is non-singular on $K\cap U$, by the Darboux-Givental theorem, we may assume that

$$a_i(x,y)=y_i,\quad a_{r+i}(x,y)=x_i,\quad i\le r,\quad (x,y)\in U.$$

Then $(x_{r+1},\cdots,x_n,y_{r+1},\cdots,y_n)$ is a coordinate system on $K\cap U$ and

$$\omega\mid_{K\cap U}=dy_{r+1}\wedge dx_{r+1}+\cdots+dy_n\wedge dx_n.$$

Let $h\in C^\infty(U)$ such that X_h is tangent to $K\cap U$. Then

$$\begin{aligned}
X_h\mid_{K\cap U} &= \left(\sum_{i=1}^{n}\frac{\partial h}{\partial y_i}\frac{\partial}{\partial x_i}-\frac{\partial h}{\partial x_i}\frac{\partial}{\partial y_i}\right)\mid_{K\cap U}\\
&= \sum_{i=r+1}^{n}\frac{\partial h}{\partial y_i}\mid_{K\cap U}\frac{\partial}{\partial x_i}-\frac{\partial h}{\partial x_i}\mid_{K\cap U}\frac{\partial}{\partial y_i},\\
&\qquad\qquad\qquad\text{since } X_h \text{ is tangent to } K\cap U,\\
&= \sum_{i=r+1}^{n}\frac{\partial h\mid_{K\cap U}}{\partial y_i}\frac{\partial}{\partial x_i}-\frac{\partial h\mid_{K\cap U}}{\partial x_i}\frac{\partial}{\partial y_i}\\
&= X_{\omega_{K\cap U},h\mid_{K\cap U}}.
\end{aligned}$$

This completes the proof of Proposition 14. □

Let us consider the two-dimensional case. If f is a smooth function on a 2-dimensional symplectic manifold M, then an integral curve of Hamiltonian vector field X_f passing through a point $p\in M$ is contained in the level set $f^{-1}(f(p))=\{q\in M: f(q)=f(p)\}$ which is a curve in a generic situation. For a proper Morse function we have the following straightforward result.

Proposition 15. *Suppose that f is a proper Morse function on a 2-dimensional symplectic manifold M. Let $c\in f(M)$ and let Γ be a connected component of $f^{-1}(c)$.*
1) If Γ does not contain critical points of f, then Γ is the orbit of a periodic solution of X_f.
2) If Γ contains one and only one critical point p of f, then

$$\Gamma=\begin{cases}\{p\}, & \text{if } p \text{ is a critical point of } f \text{ with index } 0 \text{ or } 2,\\ \{p\}\cup W_s(p)\cup W_u(p), & \text{if } p \text{ is a critical point of } f \text{ with index } 1,\end{cases}$$

where $W_s(p)$ and $W_u(p)$ are the stable and unstable manifolds of a stationary point p with index 1 respectively.

Let us consider the sphere $K = S^2$ in the symplectic space $(\mathbb{R}^4, \omega)$. We assume $\omega = dy_1 \wedge dx_1 + dy_2 \wedge dx_2$, and

$$(14) \qquad a_1(x,y) = y_2, \quad a_2(x,y) = x_1^2 + x_2^2 + y_1^2 + y_2^2 - 1.$$

Thus

$$K \quad = \quad S^2 = \{(x,y) \in \mathbb{R}^4 \mid y_2 = 0, \quad x_1^2 + x_2^2 + y_1^2 = 1\}.$$

We see that $\{a_1, a_2\} = 2x_2$ and a point $(x_1, x_2, y_1, 0) \in S^2$ is a singular point of the restricted symplectic structure $\omega\mid_{TS^2}$ if and only if $x_2 = 0$:

$$(15) \qquad \Sigma(\omega\mid_{TS^2}) = S^1 = \{(x_1, x_2, y_1, 0) \mid x_2 = 0, \ x_1^2 + y_1^2 = 1\}.$$

Now we consider a Hamiltonian function on $\mathbb{R}^4$

$$(16) \qquad F_b(x,y) = a_1(x,y)\lambda_1(x,y) + a_2(x,y)\lambda_2(x,y) + b(x,y)$$

together with

$$(17) \qquad b(x,y) = x_2^2\left(1 + 2x_1^2\right)$$

and the Hamiltonian vector field X_{F_b} on $\mathbb{R}^4$

$$(18) \qquad X_{F_b} = \sum_{i=1,2} \frac{\partial F_b}{\partial y_i}(x,y)\frac{\partial}{\partial x_i} - \frac{\partial F_b}{\partial x_i}(x,y)\frac{\partial}{\partial y_i}.$$

generated by $F_b(x,y)$. In (16)

$$\begin{aligned} \lambda_1(x,y) \quad &= \quad 4x_1y_1x_2 + 2y_2\left(1 + 2x_1^2\right), \\ (19) \qquad \lambda_2(x,y) \quad &= \quad -(1 + 2x_1^2) \end{aligned}$$

is a unique smooth solution of the structure equation

$$(20) \qquad \begin{pmatrix} 0 & \{a_1, a_2\} \\ \{a_2, a_1\} & 0 \end{pmatrix} \begin{pmatrix} \lambda_1 \\ \lambda_2 \end{pmatrix} = \begin{pmatrix} \{b, a_1\} \\ \{b, a_2\} \end{pmatrix},$$

which is precisely

$$(21) \quad \begin{pmatrix} 0 & 2x_2 \\ -2x_2 & 0 \end{pmatrix} \begin{pmatrix} \lambda_1 \\ \lambda_2 \end{pmatrix} = \begin{pmatrix} -2x_2\left(1 + 2x_1^2\right) \\ -8x_1y_1x_2^2 - 4x_2y_2\left(1 + 2x_1^2\right) \end{pmatrix},$$

since

$$\{a_1, a_2\} = 2x_2, \ \{b, a_1\} = -2x_2(1+2x_1^2), \ \{b, a_2\} = -8x_1y_1x_2^2 - 4x_2y_2(1+2x_1^2).$$

Thus we have got that X_{F_b} is tangent to S^2 as well as to $\Sigma(\omega\mid_{TS^2})$. $X_{F_b}\mid_{\Sigma(\omega\mid_{TS^2})}$ has no stationary points and its integral curves move anti-clockwise in the (x_1, y_1)-plane. Therefore, in order to understand the phase portrait of $X_{F_b}\mid_{S^2}$, it suffices to understand the phase portrait of $X_{F_b}\mid_{S^2-\Sigma(\omega\mid_{TS^2})}$. Let

$$(22)\qquad U_+ = \{(x,y)\in\mathbb{R}^4 : x_2 > 0\}, \qquad U_- = \{(x,y)\in\mathbb{R}^4 : x_2 < 0\}.$$

Then

$$S^2\cap(U_+\cup U_-) = S^2 - \Sigma(\omega\mid_{TS^2}).$$

Thus in order to understand the phase portrait of $X_{F_b}\mid_{S^2}$, it suffices to understand the phase portrait of $X_{F_b}\mid_{S^2\cap U_\pm}$. Since $\omega\mid_{T(S^2\cap U_\pm)}$ is nonsingular, by Proposition 14, we see that

$$(23)\qquad X_{F_b|_{S^2\cap U_\pm}} = X_{\omega|_{T(S^2\cap U_\pm)}, F_b|_{S^2\cap U_\pm}}.$$

Now we apply Proposition 15. In our case M is $S^2\cap U_\pm$ and the function is $F_b\mid_{S^2\cap U_\pm}$. Note that $F_b\mid_{S^2\cap U_\pm}$ is a proper function. We adapt

$$(x_1, y_1), \quad (\text{with} \quad x_1^2 + y_1^2 < 1),$$

as a coordinate system on $S^2\cap U_\pm$. Then, we see that

$$(24)\qquad \omega\mid_{T(S^2\cap U_\pm)} = dy_1\wedge dx_1.$$

Since

$$F_b(x,y) = a_1(x,y)\lambda_1(x,y) + a_2(x,y)\lambda_2(x,y) + b(x,y)$$

we have

$$F_b\mid_{S^2\cap U_\pm} = b\mid_{S^2\cap U_\pm} = x_2^2(1+2x_1^2)\mid_{S^2\cap U_\pm}.$$

Therefore

$$(25)\qquad F_b\mid_{S^2\cap U_\pm} = b\mid_{S^2\cap U_\pm} = (1-x_1^2-y_1^2)(1+2x_1^2),$$

where $x_2^2 = 1-x_1^2-y_1^2$ on S^2. From (23), (24) and (25) we see that the solutions of the Hamiltonian systems $X_{F_b|_{S^2\cap U_+}}$ and $X_{F_b|_{S^2\cap U_-}}$ are the same in x_1, y_1 coordinates. Hence, from now on we investigate only the case $X_{F_b|_{S^2\cap U_+}}$.

Proposition 16. 1) *The function $F_b\mid_{S^2\cap U_+} = b\mid_{S^2\cap U_+}$ is a Morse function with three critical points $(x_1, y_1) = (0,0)$ and $(x_1, y_1) = (\pm\frac{1}{2}, 0)$.* 2) *The point $(0,0)$ is a saddle point (index 1) and $b(0,0) = 1$. The points $(\pm\frac{1}{2}, 0)$ are the maximum points of $b\mid_{U_+}$ and $b(\pm\frac{1}{2}, 0) = 9/8$.*

3) *Thus the phase portrait of* $X_{F_b|_{S^2\cap U_+}}$ *is as follows.*

$$(b_{U_+})^{-1}(c) = \begin{cases} \emptyset, & \textit{for } c > \frac{9}{8}, \\ \{(\frac{1}{2},0),(-\frac{1}{2},0)\} \textit{ the two stationary points,} & \textit{for } c = \frac{9}{8}, \\ \textit{the disjoint sum of two periodic orbits,} & \textit{for } \frac{9}{8} > c > 1, \\ \{(0,0)\} \cup W_s(0,0) \cup W_u(0,0), & \textit{for } c = 1, \\ \textit{a periodic orbit,} & \textit{for } 1 > c > 0, \end{cases}$$

where all the periodic solutions move anti-clockwise in (x_1, y_1)*-plane, (see Fig.1).*

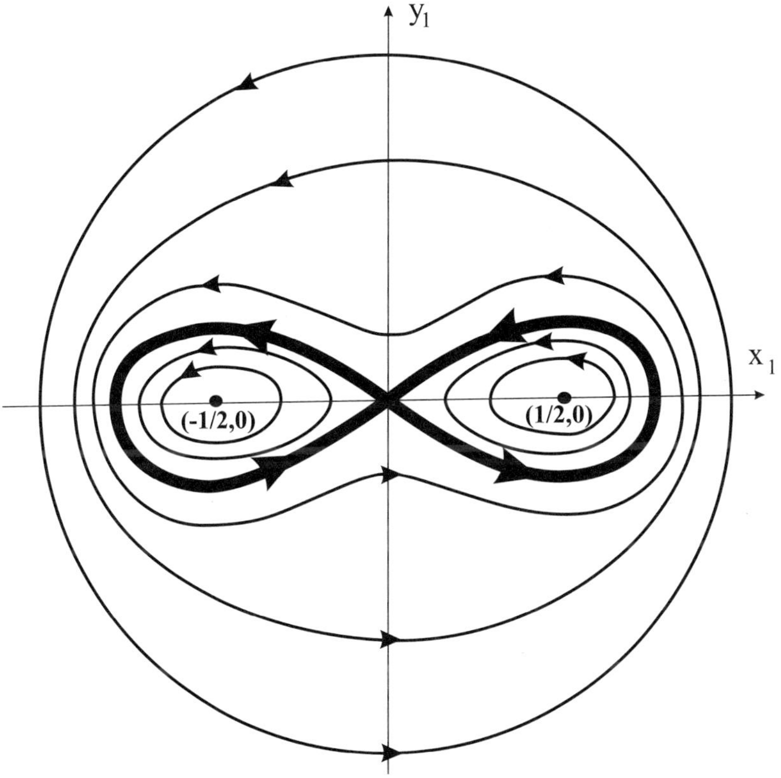

Fig. 1. Flow of $d\pi(X_{F_b}\,|_{S^2}), \pi(x_1, x_2, y_1) = (x_1, y_1)$

Proof. First we search for critical points of $F_b\,|_{S^2\cap U_\pm} = b\,|_{S^2\cap U_\pm} = (1 - x_1^2 - y_1^2)(1 + 2x_1^2)$. Since

$$\begin{aligned}\frac{\partial b\,|_{S^2\cap U}}{\partial y_1}(x_1, y_1) &= -2y_1(1 + 2x_1^2)\\ \frac{\partial b\,|_{S^2\cap U}}{\partial x_1}(x_1, y_1) &= -2x_1((1 + 2x_1^2) + 4x_1(1 - x_1^2 - y_1^2)\\ &= 2x_1(1 - 4x_1^2 - 2y_1^2),\end{aligned}$$

the critical points of $F_b\,|_{S^2\cap U} = b\,|_{S^2\cap U}$ are

$$(x_1, y_1) = (0, 0) \quad \text{and} \quad (x_1, y_1) = \left(\pm\frac{1}{2}, 0\right) \tag{26}$$

and the critical values are

$$F_b\,|_{S^2\cap U}\,(0, 0) = 1 \quad \text{and} \quad F_b\,|_{S^2\cap U}\left(\pm\frac{1}{2}, 0\right) = \frac{9}{8}. \tag{27}$$

The Hessian matrices at the critical points are

$$\begin{pmatrix} 2 & 0 \\ 0 & -2 \end{pmatrix} \ \text{at} \ (0, 0), \qquad \begin{pmatrix} -4 & 0 \\ 0 & -3 \end{pmatrix} \ \text{at} \ \left(\pm\frac{1}{2}, 0\right). \tag{28}$$

Thus $(0, 0)$ is a saddle point and $F_b\,|_{S^2\cap U} = b\,|_{S^2\cap U}$ takes maximal (actually maximum) values $\frac{9}{8}$ at $(\pm\frac{1}{2}, 0)$. This proves 1) and 2). Therefore the shape of the graph of $b\,|_{S^2\cap U}$ looks like an island having two mountains with the same hight and a mountain pass (a saddle point) between them. Every periodic solution inherits its orientation from those of integral curves on $\Sigma(\omega\,|_{TS^2})$ which move anti-clockwise in the (x_1, y_1)-plane. This proves 3). This completes the proof of Proposition 16. □

Acknowledgements. The authors wish to thank to the referee for helpful comments.

References

[1] V.I. Arnold, *Lagrangian submanifolds with singularities, asymptotic rays and open swallowtail*, Funct. Anal. Appl. Vol. 15, (4), (1981), 1–14.

[2] P.A.M. Dirac, *Generalized Hamiltonian Dynamics*, Canadian J. Math. **2**, (1950), 129–148.

[3] T. Fukuda, *Local topological properties of differentiable mappings I*, Invent. Math. **65**, (1981), 227–250.

[4] T. Fukuda, S. Janeczko, *Singularities of implicit differential systems and their integrability*, Banach Center Publications, **65**, (2004), 23–47.

[5] T. Fukuda, S. Janeczko, *Global properties of integrable implicit Hamiltonian systems*, Proc. of the 2005 Marseille Singularity School and Conference, World Scientific, (2007), 593–611.

[6] T. Fukuda, S. Janeczko, *On the Poisson algebra of a singular map*, Journal of Geometry and Physics 86, (2014), 194–202.

[7] M. Glubitsky, D. Tischler, *An example of moduli for singular symplectic forms*, Invent. Math., (1977), 38:3, p. 219225.

[8] S. Janeczko, *Constrained Lagrangian submanifolds over singular constraining varieties and discriminant varieties*, Ann. Inst. Henri Poincare, Phys. theorique, **46**, No. 1, (1987), 1–26.

[9] S. Janeczko, *On implicit Lagrangian differential systems*, Annales Polonici Mathematici, **LXXIV**, (2000), 133–141.

[10] S. Janeczko, F. Pelletier, *Singularities of implicit differential systems and Maximum Principle*, Banach Center Publications, **62,** (2004), 117–132.

[11] J.N. Mather, *Solutions of generic linear equations*, Dynamical Systems, (1972), 185–193.

[12] R.B. Melrose, *Equivalence of glancing hypersurfaces*, Invent. Math. Vol. 37, No. 3, (1976), 165–192.

[13] M.R. Menzio, W.M. Tulczyjew, *Infinitesimal symplectic relations and generalized Hamiltonian dynamics*, Ann. Inst. H. Poincaré, XXVIII, No. 4, (1978), 349–367.

[14] R. Thom, *Sur les équations différentielles multiforms et leurs intégrales singuliérs*, Colloque E. Cartan, Paris, 1971.

[15] A. Weinstein, *Lectures on Symplectic Manifolds*, CBMS Regional Conf. Ser. in Math., 29, AMS Providence, R.I. 1977.

(T. Fukuda) *Department of Mathematics College of Humanities and Sciences, Nihon University, Sakurajousui*
3-25-40, Setagaya-ku, 156-8550 Tokyo, Japan,
E-mail address: fukuda@math.chs.nihon-u.ac.jp

(S. Janeczko) *Instytut Matematyczny PAN,*
ul. Śniadeckich 8, 00-950 Warszawa, Poland
and Wydział Matematyki i Nauk Informacyjnych, Politechnika Warszawska,
Pl. Politechniki 1, 00-661 Warszawa, Poland
E-mail address: janeczko@mini.pw.edu.pl

Advanced Studies in Pure Mathematics 78, 2018
Singularities in Generic Geometry
pp. 251–272

Equidistants and their duals for families of plane curves

Peter Giblin and Graham Reeve

Abstract.

We consider the local geometry of a generic 1-parameter family of smooth curves in the real plane for which one member of the family has parallel tangents at two inflexion points. We study the equidistants of this family, that is the loci of points at a fixed ratio along chords joining points with parallel tangents, as a 2-parameter family depending on the value of the fixed ratio and on the parameter in the family of curves. Codimension 2 singularities of type 'gull' arise in this way and are in general versally unfolded by the two parameters. We also calculate the family of duals of the equidistants; here it is necessary to view them as bifurcation sets of bigerms and they evolve through 'moth' and 'nib' singularities also encountered in 1-parameter families of symmetry sets in the plane. Finally we show that certain sub-families of the 2-parameter family of equidistants can be classified by reduction to a normal form.

§1. Introduction

A generic smooth, closed plane curve C will not possess two inflexion points at which the tangents are parallel, but a generic 1-parameter family of plane curves $\{C_\varepsilon\}$ can be expected to contain isolated members with this property. In a previous article [6], the authors investigated some affinely invariant constructions—equidistants and centre symmetry sets—based on such a 1-parameter family of plane curves. Besides the parameter ε in the curve family there is an additional parameter λ inherent in the equidistants: for each λ we take all chords joining pairs of points of C_ε at which the tangents are parallel and construct the locus of points at a fixed ratio $\lambda : 1 - \lambda$ along these chords.

Received May 19, 2016.
Revised November 11, 2016.
2010 *Mathematics Subject Classification.* 58K05, 57R45, 53A04.

In [6] we were not able to consider this 2-parameter family of equidistants as a whole, but in the present article we give a method for doing this, describing the evolution of equidistants by unfoldings of singularities of maps from the plane to the plane. The most degenerate case is that of a 'gull singularity', in the language of [5], and in general this is versally unfolded by the parameters λ, ε. In addition to this we are able to analyse the inflexions, that is the *dual* structure of the equidistants, using an approach via multi-local germs of mappings reminiscent of the classification of 1-parameter families of symmetry sets in [3].

The classification of maps from the plane to the plane has another natural application, in the study of projections of smooth surfaces in $\mathbb{R}^3$ to the plane, where the critical values of the projection form the apparent contour (also called outline or profile) of the surface for a given direction of projection. Many of the same singularities, namely cusp, swallowtail, lips, beaks, and including the gull singularity, occur in that context, and the duals exhibit the same singularities—for example the dual of a gull is also a gull. In our situation the duals have a completely different structure and are described by means of Maxwell sets of bigerms of functions (called 'moth' and 'nib' in [3]).

All our results are local in nature and the article is organized as follows. In §2 we give the local form of the family of curves being studied, the same as in [6]. In §3 we explain our first method of studying the whole family of equidistants and show how they evolve with varying λ, ε. In §4 we study the *duals* of the equidistants, that is we capture their inflexions and bitangents, features which are not preserved by the methods of §3. In §5 we find the loci along which the geometry of the equidistants, including the inflexions, changes, and give illustrations of the results. Finally in §6 we adopt a different approach to the problem, via reduction of families to a normal form. There is an Appendix filling in some details from §6.

§2. Families of curves

Consider a family of plane curves, one member of the family having parallel but not identical tangents at inflexion points. By means of a family of affine maps of the plane to itself (compare [6]) this situation is modelled locally by two local families of curves $y = f(x, \varepsilon)$, $y = g(x, \varepsilon)$,

$$\begin{aligned} f(x,\varepsilon) &= x^3 f_1(x,\varepsilon) \\ &= f_{30}x^3 + f_{40}x^4 + f_{31}x^3\varepsilon + f_{50}x^5 + f_{41}x^4\varepsilon + f_{32}x^3\varepsilon^2 + \ldots, \\ g(x,\varepsilon) &= 1 + xg_1(\varepsilon) + x^3 g_2(x,\varepsilon) \\ &= 1 + g_{11}x\varepsilon + g_{30}x^3 + g_{12}x\varepsilon^2 + g_{40}x^4 + g_{31}x^3\varepsilon + g_{13}x\varepsilon^3 + \ldots, \end{aligned} \tag{1}$$

where (at least) $f_{30}, g_{30}, f_{30} - g_{30}, g_{11}$ are nonzero. For all ε, the first curve has a horizontal inflexion at the origin, and the second curve has an inflexion at $(0, 1)$.

We shall consider two situations; see Figure 1.

Case 1 Here, f_{30}, g_{30} have the same sign, say positive, when we write $f_{30} = a_3^2$, $g_{30} = b_3^2$ where $a_3 > 0$, $b_3 > 0$.

Case 2 Here, f_{30}, g_{30} have opposite signs, and we write $f_{30} = a_3^2$, $g_{30} = -b_3^2$ where $a_3 > 0$, $b_3 > 0$.

Note In general we do not need to assume $a_3 \neq b_3$ in what follows, even though $a_3 = b_3$ implies a greater degree of 'similarity' between the two inflexions. We shall note below when $a_3 \neq b_3$ is required.

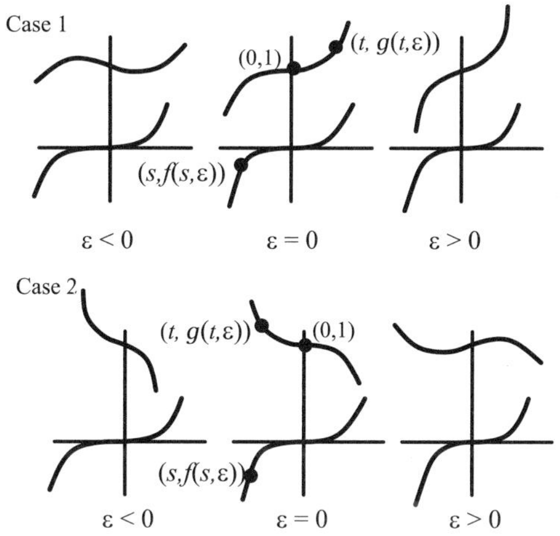

Fig. 1. Schematic representation of the two cases considered here. Case 1 ($f_{30}g_{30} > 0$ in (1)) has the inflexions 'oriented the same way' or 'of the same sign' and Case 2 ($f_{30}g_{30} < 0$) 'oriented opposite ways' or 'of opposite signs'. The sign given for ε assumes (without loss of generality) that $g_{11} > 0$ in the notation of (1), that is the upper inflexional tangent turns anticlockwise as ε increases. For Case 2 there are no pairs of parallel tangents at parameter points s, t for $\varepsilon < 0$, and for $\varepsilon > 0$ the ranges of values of s, t providing parallel tangents are bounded.

We are interested in the *equidistants* which are the points which are at a fixed ratio along the chord joining a pair $(s, f(s, \varepsilon))$ and $(t, g(t, \varepsilon))$ at which the tangents to the two curves are parallel (we do not include pairs where both belong to the same local curve). Thus the equidistant, for a fixed λ, consists of points

$$(1-\lambda)(s, f(s,\varepsilon)) + \lambda(t, g(t,\varepsilon)),$$

subject to the parallel tangency condition.

Definition 2.1. (i) When considering values of λ close to some fixed $\lambda_0 \neq 0, 1$ we write $\lambda = \lambda_0 + \alpha$.
(ii) For Case 1 it will be necessary to separate out two *special values* of λ_0, namely $\lambda_0 = \dfrac{b_3}{b_3 \pm a_3}$. (For the $-$ sign we require $a_3 \neq b_3$.) These were also discussed in [6]. For Case 2 there are no special values of λ_0.

§3. Equidistants

For any pair of smooth plane curves varying in a 1-parameter family, $\gamma_1(s,\varepsilon)$ and $\gamma_2(t,\varepsilon)$, we can consider the "λ-point map", defined by

$$(s,t,\lambda,\varepsilon) \mapsto (1-\lambda)\gamma_1(s,\varepsilon) + \lambda\gamma_2(t,\varepsilon) \in \mathbb{R}^2.$$

For fixed λ, ε the critical set consists of points $(s,t,\lambda,\varepsilon)$ where the tangent lines to the two curves at $\gamma_1(s,\varepsilon)$ and $\gamma_2(t,\varepsilon)$ are parallel. Thus for fixed λ, ε the discriminant of this mapping, that is the set of its critical values, is exactly the equidistant for those values of λ, ε. Extending the map to

$$(s,t,\lambda,\varepsilon) \mapsto ((1-\lambda)\gamma_1(s,\varepsilon) + \lambda\gamma_2(t,\varepsilon), \lambda, \varepsilon) \in \mathbb{R}^4,$$

the discriminant is the union of all equidistants of all members of the family.

For our family this becomes

$$(s,t,\lambda,\varepsilon) \mapsto ((1-\lambda)s + \lambda t,\ (1-\lambda)f(s,\varepsilon) + \lambda g(t,\varepsilon)). \tag{2}$$

In order to compare this more easily with standard forms we write $u = (1-\lambda)s + \lambda t$ and solve for t:

$$t = \frac{u-(1-\lambda)s}{\lambda},$$

where we shall avoid working near $\lambda = 0$. We shall also write $\lambda = \lambda_0 + \alpha$ for a fixed $\lambda_0 \neq 0, 1$. Then the above family will be written as follows:

$$\mathcal{H}(s,u,\alpha,\varepsilon) = (u,\ H(s,u,\alpha,\varepsilon)), \tag{3}$$

where H is a real-valued function. We shall use the following abbreviations:

$$H_0(s,u) = H(s,u,0,0); \quad \mathcal{H}_0(s,u) = \mathcal{H}(s,u,0,0) = (u, H_0(s,u)).$$

We proceed to examine the map $\mathcal{H}_0 : \mathbb{R}^2, (0,0) \to \mathbb{R}^2$ and its unfolding $\mathcal{H} : \mathbb{R}^2 \times \mathbb{R}^2, (0,0,0,0) \to \mathbb{R}^2$ and prove the following proposition, where we assume as always that $\lambda \neq 0, 1$.

Proposition 3.1. *For Case 1,*
(i) *if* $\lambda_0 \neq \dfrac{b_3}{b_3 \pm a_3}$, *then the map* $\mathcal{H}_0$ *has the type 'beaks', and is versally unfolded by the* ε *parameter alone;*
(ii) *if* $\lambda_0 = \dfrac{b_3}{b_3 \pm a_3}$ *(for the* $-$ *sign we require* $a_3 \neq b_3$*) then the map* $\mathcal{H}_0$ *has type 'gull'*[1] *and is versally unfolded by the parameters* α, ε *for generic values of the coefficients. The precise conditions are given below in* (6).
For Case 2, and any λ,
(iii) *the map* $\mathcal{H}_0$ *has the type 'lips', and is versally unfolded by the* ε *parameter alone.*

Proof (i) Then $H_0(s,u) = \lambda_0 + c_{30}s^3 + c_{21}s^2u + c_{12}su^2 + c_{03}u^3 +$ higher terms, where

$$c_{30} = -\frac{(1-\lambda_0)(\lambda_0(b_3+a_3)-b_3)(\lambda_0(b_3-a_3)-b_3)}{\lambda_0^2} \neq 0.$$

It is then easy to remove the s^2u term by substituting $s = s_1 + ku$ where $k = -c_{21}/(3c_{30})$. Using $\mathcal{A}$-equivalence on the map $\mathcal{H}_0$, the term in u^3 can be removed too. The result in the present situation is

$$\mathcal{H}_0(s_1,u) =$$
$$\left(u,\ \lambda_0 + c_{30}s_1^3 + \frac{3(1-\lambda_0)a_3^2b_3^2}{(\lambda_0(b_3-a_3)-b_3)(\lambda_0(b_3+a_3)-b_3)}s_1u^2 + \ldots,\right)$$

where both coefficients are nonzero. This is essentially a normal form for a (3-$\mathcal{A}$-determined) *lips/beaks singularity*, where beaks occurs if and only

---

[1]See [5]. In [8], Clint McCrory describes three types of 'gulls' [plural] singularity which arise from local projections of surfaces to the plane, one of 'elliptic type' and two of 'hyperbolic type'. As local maps from the plane to the plane these are all $\mathcal{A}$-equivalent but the patterns of inflexions differ. We shall see in the next section that the pattern of inflexions on the equidistants is quite different from this.

if the coefficients of s_1^3 and s_1u^2 have opposite sign. But the product of these coefficients is $-\dfrac{3(1-\lambda_0)^2a_3^2b_3^2}{\lambda_0^2}$, which is always negative. Hence the mapping $\mathcal{H}_0$ has a beaks singularity at $s_1 = u = 0$. (The lips singularity occurs for Case 2, for any value of λ.)

The function $H(s_1, u, \alpha, \varepsilon)$ when the same reduction is made has the form

(4)

$$\lambda_0 + \alpha - (1-\lambda_0)g_{11}s_1\varepsilon - \frac{a_3^2\lambda_0^2 g_{11}}{(\lambda_0(b_3-a_3)-b_3)(\lambda_0(b_3+a_3)-b_3)}u\varepsilon + \ldots.$$

This means that the 'initial speed' $\dfrac{\partial H}{\partial \alpha}$ provides, up to this order, only the constant 1, while the 'initial speed' $\dfrac{\partial H}{\partial \varepsilon}$ provides (since $g_{11} \neq 0$) the term s_1 which is needed for a versal unfolding of the beaks singularity. Thus changing λ to values close to λ_0 provides only a trivial unfolding but changing ε to values close to 0 does give a versal unfolding of the beaks singularity, as observed in [6].

(ii) We take the case $\lambda_0 = \dfrac{b_3}{b_3 - a_3}$.

Remark 3.2. Clearly this requires $a_3 \neq b_3$. However all the corresponding calculations for the other special value $\lambda_0 = \dfrac{b_3}{b_3+a_3}$ are obtained by the formal substitution of $-b_3$ for b_3 throughout, and at that special value it is permissible to have $a_3 = b_3$ in which case $\lambda_0 = \frac{1}{2}$.

With $\lambda_0 = b_3/(b_3 - a_3)$ the s^3 term is absent from H_0 and there is a standard technique (see for example [9]) for reducing the 3-jet of H_0 to a multiple of s^2u, using $\mathcal{A}$-equivalence on $\mathcal{H}_0$. We can always remove powers of u using a left-equivalence; we then replace s by $s_1 - c_{12}u/(2c_{21})$. This gives the 4-jet up to $\mathcal{A}$-equivalence

$$j^4H_0(s_1, u) = \lambda_0 + c_{21}s_1^2u + c_{40}s_1^4 + c_{31}s_1^3u + c_{22}s_1^2u^2 + c_{13}s_1u^3,$$

where

$$c_{21} = 3a_3^2 \neq 0, \quad c_{40} = \frac{a_3(a_3^3g_{40} - b_3^3f_{40})}{b_3^3(b_3 - a_3)}$$

and for certain coefficients c_{31}, c_{22}, c_{13}. We therefore assume that the coefficient c_{40} of s_1^4 is nonzero (compare [6, Prop. 4.1]). A second substitution $s_2 = s_1 + \frac{1}{6a_3^2}(c_{31}s_1^2 + c_{22}s_1u + c_{13}u^2)$ (which is then solved for s_1 as a function of s_2, u) removes the degree 4 terms other than s_1^4 and the 5-jet of $H_0(s_2,u))$ becomes

$$\lambda_0 + 3a_3^2 s_2^2 u + c_{40}s_2^4 + c_{50}s_2^5 + \text{ other degree 5 terms}$$

where

$$c_{50} = \frac{4b_3^6 f_{40}^2 - 4a_3^6 g_{40}^2 + 3a_3^6 b_3^2 g_{50} - 3a_3^2 b_3^6 f_{50}}{3a_3 b_3^6 (b_3 - a_3)}. \tag{5}$$

The condition that c_{50} is nonzero also arose in [6, Prop. 4.2] as the additional condition that a function on a swallowtail surface yielded the transition on equidistants at a special point. We shall now assume that c_{40} and c_{50} are both nonzero. The conditions $c_{40} \neq 0$, $c_{50} \neq 0$ can be written in the more suggestive forms (as in [6])

$$\frac{f_{40}}{a_3^3} \neq \frac{g_{40}}{b_3^3}, \quad \frac{4f_{40}^2}{a_3^6} - \frac{3f_{50}}{a_3^4} \neq \frac{4g_{40}^2}{b_3^6} - \frac{3g_{50}}{b_3^4} \tag{6}$$

in which the two curves are separated on the two sides.

A further substitution $s_3 = s_3(s_2,u)$ of a similar kind removes the degree 5 terms other than s_2^5 but does not affect c_{40} or c_{50}, resulting in the normal form up to $\mathcal{A}$-equivalence

$$\mathcal{H}_0(s_3,u) = (u,\ \lambda_0 + c_{21}s_3^2 u + c_{40}s_3^4 + c_{50}s_3^5). \tag{7}$$

All three coefficients c_{ij} are nonzero, so that this jet is 5-$\mathcal{A}$-determined and occurs in Rieger's list [9, Table 1] as 11_5, and is also the 'gull singularity' of [5].

Applying the same transformations to the function H we obtain

$$\lambda_0 + \alpha + \frac{a_3 g_{11}}{b_3 - a_3} s_3\varepsilon + \ldots + \frac{2a_3^2(b_3 - a_3)}{b_3} s_3^3\alpha + \ldots + ks_3^3\varepsilon + \ldots,$$

where the terms exhibited are the significant ones in the initial speeds $(0, \partial H/\partial\alpha)$ and $(0, \partial H/\partial\varepsilon)$, and k is an expression involving the coefficients of f, g encountered so far and in addition f_{31} and g_{31}. These terms guarantee that unfolding terms $(0, s_3)$ and $(0, s_3^3)$ are provided by the initial speeds and since in the normal form (7) other terms in s_3, u up to degee 5 are in the extended tangent space of $\mathcal{H}_0$, it follows that α and ε versally unfold the 'gull' singularity of $\mathcal{H}_0$ at $s_3 = u = 0$.

The proof of (iii) is similar to (i) except that there are no special values of λ here. □

§4. Duals of the equidistants

4.1. Inflexions

We should like to study the duals of the equidistants in order to decide the pattern of inflexions (not preserved by $\mathcal{A}$-equivalence). As a start we derive here the condition for an equidistant to have an inflexion. The following discussion applies both to Case 1 and to Case 2.

An equidistant has the form, for fixed λ and ε,

$$\{((1-\lambda)s+\lambda t,\quad (1-\lambda)f(s,\varepsilon)+\lambda g(t,\varepsilon))\},$$

subject to the parallel tangent condition $f_s = g_t$. For the purpose of calculation assume that the condition $f_s = g_t$ is solved as $t = T(s)$ so that $f_s(s,\varepsilon) \equiv g_t(T(s),\varepsilon)$. Similar arguments apply if s is a function of t and one or the other holds except for $f_{ss} = g_{tt} = 0$: inflexions on both curves, which happens for parallel tangents only at $s = t = \varepsilon = 0$. Note that the tangent to the equidistant is parallel to the tangents to the two given curves. In fact, for fixed λ and ε, writing

$$\delta(s) = ((1-\lambda)s+\lambda t,\quad (1-\lambda)f(s,\varepsilon)+\lambda g(T(s),\varepsilon))$$

the condition for an inflexion is that $\delta'(s)$ and $\delta''(s)$ are parallel vectors. The derivatives are

$$\delta' = ((1-\lambda)+\lambda T')(1,f_s) \text{ and } \delta'' = (\lambda T'', (1-\lambda)f_{ss}+\lambda g_{tt}T'^2+g_tT'').$$

Computing T' and T'' from $f_s(s,\varepsilon) \equiv g_t(T(s),\varepsilon)$ we find $T' = f_{ss}/g_{tt}$ and $T'' = (f_{sss}g_{tt}^2 - g_{ttt}f_{ss}^2)/g_{tt}^3$. Finally substituting these into the condition for δ' and δ'' to be parallel, and clearing denominators, gives the following condition (besides $f_s = g_t$):

$$((1-\lambda)g_{tt}+\lambda f_{ss})f_{ss}g_{tt} = 0. \tag{8}$$

However the bracket is in fact exactly the condition for the equidistant to be *singular*, that is $\delta' = 0$ so the condition reduces to $f_{ss} = 0$ or $g_{tt} = 0$. (Both occur simultaneously only for $s = t = \varepsilon = 0$.) Extending the calculation we find that the inflexion is ordinary (δ''' not parallel to δ') if and only if the corresponding derivative f_{sss} or g_{ttt} is nonzero, and this will be the case for small values of s, t, ε since a_3 and b_3 are nonzero. We deduce the following.

Proposition 4.1. *The equidistant has an inflexion at the (nonsingular) point corresponding to (s,t) where the tangents are parallel if and only if $f_{ss} = 0$ or $g_{tt} = 0$, that is one of the two curves has an inflexion. In our case this means that, for any ε, inflexions occur exactly for $s = 0$ and for $t = 0$ and all these inflexions are ordinary inflexions.* □

Corollary 4.2. *Away from any singular points of the equidistant, for Case 1 (inflexions 'facing the same way') there are exactly two inflexions on the equidistant, both ordinary, and for Case 2 (inflexions 'facing opposite ways') there are four.*

Also, the tangent to an equidistant at an inflexion point will either be parallel to the x-axis or to the line $y = g_{11}x$.

The second statement of the corollary follows because for a pair of points $\mathbf{p} = (s, f(s, \varepsilon))$ and $\mathbf{q} = (t, g(t, \varepsilon))$ with parallel tangents, the tangent at the corresponding point $(1 - \lambda)\mathbf{p} + \lambda\mathbf{q}$ to any equidistant is parallel to those tangents. The tangent at the the origin $s = 0$ to the curve $y = f(x, \varepsilon)$ is horizontal and the tangent at $(0, 1)$ $(t = 0)$ to the curve $y = g(x, \varepsilon)$ has slope g_{11}. □

4.2. The family of duals

We now explain how to study the duals of the equidistants by means of a suitable map. For fixed ε and $\lambda \neq 0, 1$ let

$$F(s, t, u, v) = ((1 - \lambda)(s, f(s, \varepsilon)) + \lambda(t, g(t, \varepsilon))) \cdot (u, 1) - v, \tag{9}$$

where $\cdot$ is the euclidean scalar product of vectors. Then $F_s = 0$ means that the tangent to the curve $\{(s, f(s, \varepsilon))\}$ is perpendicular to $(u, 1)$ and $F_t = 0$ similarly for the curve $\{(t, g(t, \varepsilon))\}$, while $F = 0$ means that $v = \mathbf{p} \cdot (u, 1)$ where $\mathbf{p}$ is the equidistant point for the value λ corresponding to parameter values s, t. Since the tangent to the equidistant has the equation $(\mathbf{x} - \mathbf{p}) \cdot (u, 1) = 0$ we can use (u, v) to parametrize the *dual* of the equidistant. For fixed λ and ε the set of points

$$\{(u, v) : \exists s, t \text{ such that } F = F_s = F_t = 0\}$$

is the dual of the equidistant for that λ and ε. This is the discriminant set of a family of functions of two variables s, t. However this description is not quite satisfactory since $F(s, t, 0, 0)$ has type $D_4^{\pm}$ and hence λ and ε cannot versally unfold the singularity.

To get round this problem we split F into two parts, $F = (F_1, F_2)$ and regard it as a *bigerm*, of singularity type A_2^2 at $u = v = \varepsilon = 0$ (and any $\lambda \neq 0, 1$), as follows. Let, again for fixed λ,

$$\begin{aligned} F_1(s, u, v, \varepsilon) &= (1 - \lambda)su + (1 - \lambda)f(s, \varepsilon), \\ F_2(t, u, v, \varepsilon) &= -\lambda tu - \lambda g(t, \varepsilon) + v. \end{aligned}$$

Thus, $F_1(s, 0, 0, 0) = a_3^2 s^3 +$ higher terms, and $F_2(t, 0, 0, 0) = \pm b_3^2 t^3 +$ higher terms, where the sign is $+$ for Case 1 and $-$ for Case 2. All these are of type A_2 since a_3 and b_3 are nonzero. We have the following.

Proposition 4.3. *For fixed* λ, ε *the dual of the equidistant is the* levels bifurcation set *or* Maxwell set

$$\mathcal{B} = \{(u,v) : \exists s,t \text{ such that } F_{1s} = F_{2t} = 0, F_1 = F_2\}.$$

Furthermore, for any $\lambda \neq 0, 1$, *the parameters* u, v, ε *in* F *give a multiversal unfolding of the* A_2^2 *singularity of this bigerm at* $(s,t) = (0,0)$.

Note that in our situation we do not need to specify $s \neq t$ since they are parameters on two separate curve pieces.

Proof of the last statement. The 'initial speeds', evaluated at $u = v = \varepsilon = 0$, are

$$\begin{aligned} F_u &= ((1-\lambda)s, -\lambda t),\ \ F_v = (0,1) \\ \text{and } F_\varepsilon &= ((1-\lambda)O(s^3), -\lambda g_{11}(t + O(t^3))), \end{aligned}$$

which give the requisite terms $(s,0), (0,1)$ (or $(1,0)$) and $(0,t)$ for a multiversal unfolding of the A_2^2 singularity, since $g_{11} \neq 0$. □

This result implies that, for any λ, including special values, the levels bifurcation set $\mathcal{B}$ is locally diffeomorphic to the levels bifurcation set $\mathcal{B}_G$ of the standard multiversal unfolding of an A_2^2 singularity, namely (see [3]):

$$G_1(x,p,q,r) = x^3 + px, \quad G_2(y,p,q,r) = y^3 + qy + r,$$

$$\begin{aligned} \text{with } \mathcal{B}_G &= \{(p,q,r) : G_{1x} = G_{2y} = G_1 - G_2 = 0\} \\ &= \{(-3x^2, -3y^2, 2(x^3 - y^3))\}. \end{aligned}$$

To understand the evolution of the dual equidistants for a fixed λ and ε passing through 0 we therefore need to identify the function ε on $\mathcal{B}$. For a stable function the evolution will be, up to local diffeomorphism, a standard 'nib' or 'moth' transition from [3], and this occurs when the plane $\varepsilon = 0$ is transverse to the limiting singular strata of the surface $\mathcal{B}$, namely the limiting tangent lines to the cuspidal edges and the double curve. The 'nib' is illustrated in Figure 3 along the ε-axis and the 'moth' is the 4-cusped curve on the right of Figure 4, which shrinks to a point and disappears in the transition.

It is clear that we can solve the equations $F_{1s} = F_{2t} = 0, F_1 = F_2$ for u and v, and the remaining equation is then simply $f_s(s,\varepsilon) = g_t(t,\varepsilon)$ which is the condition for parallel tangents of the original two inflexional curves $y = f(x,\varepsilon)$ and $y = g(x,\varepsilon)$. This equation is solved locally by say $\varepsilon = E(s,t)$, such a smooth solution being guaranteed by $g_{11} \neq 0$. The lowest terms of (u, v, ε) as functions of s and t, say $(U(s,t), V(s,t), E(s,t))$, are as follows, for any λ:

$$(U, V, E) =$$

$$\left(-3a_3^2 s^2 + ..., \lambda + 2a_3^2(\lambda - 1)s^3 \mp 2b_3^2 \lambda t^3 + ..., \frac{3a_3^2}{g_{11}} s^2 \pm \frac{3b_3^2}{g_{11}} t^2 + ...\right). \tag{10}$$

Here the upper sign is for Case 1 and the lower sign for Case 2. Note that $U(0,t) \equiv 0$ since $U = -f_s(s, \varepsilon)$ before we substitute for ε.

The cuspidal edges on $\mathcal{B}$ correspond to inflexions on the equidistants and these occur for $s = 0$ and $t = 0$ only (see Proposition 4.1). Putting $s = 0$ or $t = 0$ in (10) then gives the limiting tangent vectors to the cuspidal edges on $\mathcal{B}$ as $(0,0,1)$ and $(1,0,1)$, both of which are transverse to the plane $\varepsilon = 0$. We therefore need to examine the limiting tangent vector to the double curve on $\mathcal{B}$. Calculation shows that the relation between s and t on the double curve of $\mathcal{B}$ is $D(s,t) = 0$ where

$$D(s,t) = a_3^2(1 - \lambda)s^3 \pm b_3^2 \lambda t^3 + \dots,$$

so that $t = ks + \dots$ where $k^3 = \pm a_3^2(\lambda - 1)/b_3^2\lambda$. The tangent vector to the image of this curve on $\mathcal{B}$ then has the form $(ps^3 + \dots,\ qs^5 + \dots, \mp 18a_3^2 b_3^2(\lambda k + 1 - \lambda)s^3/g_{11} + \dots)$, where $p \neq 0$. The limit as $s \to 0$ is transverse to the plane $\varepsilon = 0$ unless the coefficient of s^3 in the third component is 0. This is equivalent to

$$\left(\frac{\lambda - 1}{\lambda}\right)^2 = \pm\frac{a_3^2}{b_3^2}; \text{ for the upper sign + this is } \lambda = \frac{b_3}{b_3 \pm a_3}.$$

For the upper sign (Case 1), these are the special values of λ. For the lower sign (Case 2) there are no special values of λ. We deduce the following.

Proposition 4.4. *For Case 1 ($f_{30} = a_3^2, g_{30} = b_3^2$ in (1)) the function ε on the set $\mathcal{B}$ is stable provided λ is not one of the special values $b_3/(b_3 \pm a_3)$, and the transition on dual equidistants as ε passes through 0 is then a 'nib' transition.*

For Case 2 ($f_{30} = a_3^2, g_{30} = -b_3^2$ in (1)), the function ε on $\mathcal{B}$ is always stable and the transition on the dual equidistants as ε passes through 0 is a 'moth' transition.

Remark 4.5. At a special value of λ in Case 1 we can examine the situation less formally by direct calculation. The transition is almost identical to a nib transition, except that for $\varepsilon = 0$ the two cusps with a common tangent are no longer both ordinary cusps. The outer one is ordinary and the inner one is rhamphoid. In terms of the standard A_2^2 surface $\mathcal{B}_G$, parametrized by $(x, y) \mapsto (u, v, w) = (x^2, y^2, x^3 - y^3)$, a function whose level sets model the transition at a special value of λ is $u - v + w^2$, while stable functions which model the nib/moth are respectively $2u - v$ and $u + v$. There is an additional condition for the non-stable function ε to be equivalent to $u - v + w^2$ and that is the second condition of (6) above.

§5. Equidistants and duals together

Our aim in this section is to describe the equidistants and their duals simultaneously, that is to include inflexions in the 'clock diagram' of the equidistants, in the (ε, α)-plane where $\lambda = \lambda_0 + \alpha$ for a fixed λ_0. The most interesting situation is Case 1, with λ_0 a special value, and we shall start with that.

Let $f_{30} = a_3^2$, $g_{30} = b_3^2$ in (1) (Case 1) and $\lambda = b_3/(b_3 - a_3) + \alpha$, and consider a neighbourhood of the origin in the (ε, α)-plane. (The calculations for the other special value $b_3/(b_3 + a_3)$ are similar.) There are some loci in the (ε, α)-plane which help us to understand the geometrical structure of the equidistants.

(S) The set of points (ε, α) for which the equidistant has a swallowtail singularity (the dual a double inflexion or undulation). Crossing this locus, the number of cusps on the equidistant (inflexions on the dual) changes by 2, and the number of self-intersections on the equidistant (double tangents on the dual) changes by 1.

(T) The set of points (ε, α) for which the equidistant, and hence also the dual, has a self-tangency or tacnode. Crossing this locus the number of self-intersections of the equidistant, or of the dual, changes by 2.

The calculations to identify these loci are straightforward but tedious and we state the results, as follows. We shall assume without loss of generality that $g_{11} > 0$.

Proposition 5.1. *Assume that the conditions of* (6) *both hold.*

(i) *The swallowtail locus* S *has the form*

$$\varepsilon = \frac{a_3^3 b_3^3 (a_3 - b_3)^6}{2 g_{11} (a_3^3 g_{40} - b_3^3 f_{40})^2} \alpha^3 + \dots .$$

(ii) *The self-tangency locus* T *has the form*

$$\varepsilon = \frac{3 a_3^4 b_3^4 (a_3 - b_3)^4}{g_{11} B} \alpha^2 + \dots ,$$

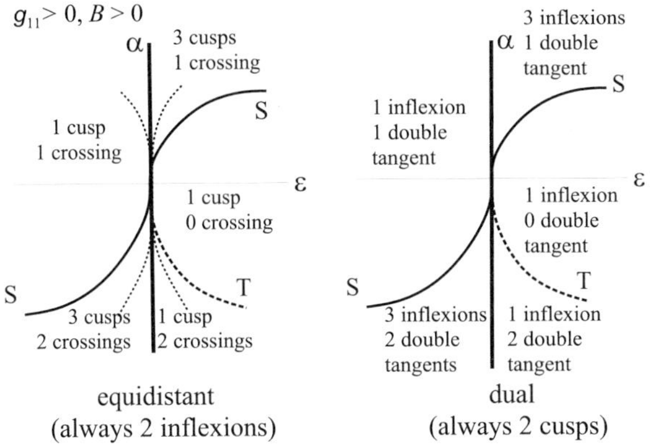

Fig. 2. Case 1, $\lambda = b_3/(b_3 - a_3) + \alpha$. The swallowtail (S) and self-tangency (T) loci for the equidistants, as in Proposition 5.1, showing how the number of self-intersections and the number of cusps on the equidistant changes around the (ε, α) diagram. The thin dashed lines in the left-hand diagram represent the coincidence of a self-intersection and an inflexion, as in Proposition 5.2. The quantity B is given by (11). If $B < 0$ then the figure is reflected in the origin, so that S remains essentially unaltered and T moves to the second quadrant.

where

$$B = 4b_3^6 f_{40}^2 - 4a_3^6 g_{40}^2 + 3a_3^6 b_3^2 g_{50} - 3a_3^2 b_3^6 f_{50} \tag{11}$$

(*the same as the numerator of* c_{50} *in* (5)), ε *has the sign of* B *and* α *has the sign of* $-B$.
(iii) *In addition, crossing the* α *axis,* $\varepsilon = 0$, *where beaks transitions occur on the equidistant, the number of cusps changes by two but although there is a self-tangency on the equidistant the number of self-intersections does not change.* □

These loci are illustrated in Figure 2 and the equidistants and duals themselves in Figures 3, 4.

There are two other loci in the (ε, α)-plane, where $\lambda = \lambda_0 + \alpha$ and λ_0 is a special value, which affect the configuration of the equidistants in a minor way. Firstly, an inflexion, which always corresponds to $s = 0$ or $t = 0$, can occur at the same place as a self-intersection on the equidistant. Crossing this locus an inflexion migrates from one sides

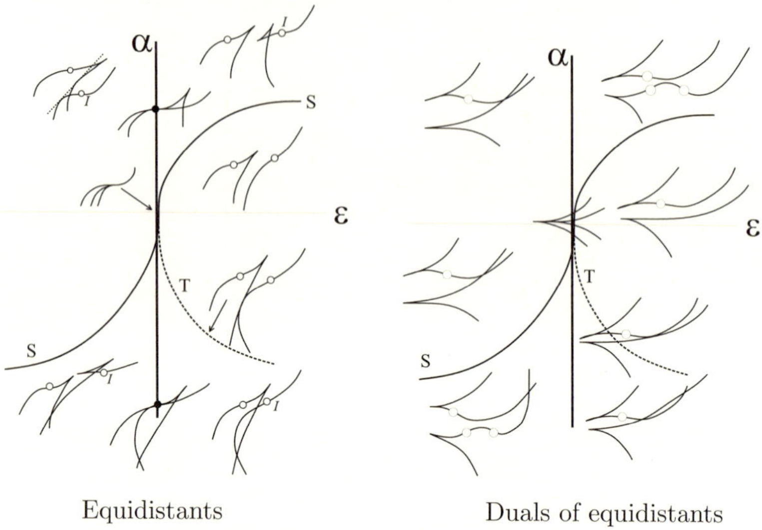

Fig. 3. (As with Figure 2 we take $g_{11} > 0, B > 0$, B given by (11).) Left: for Case 1, $\lambda = b_3/(b_3 - a_3) + \alpha$, α and ε small, this shows the evolution of the equidistants of the family of curves studied in this article. Inflexions are marked by a circle and those labelled I migrate across the nearby self-intersection as $\varepsilon \to 0$. For $\varepsilon < 0$ the inflexions on the equidistants are horizontal and for $\varepsilon > 0$ they are parallel and of positive slope g_{11} (see Corollary 4.2). One bitangent line is indicated at top left. The swallowtail locus S and self-tangency locus T are as in Figure 2. Right: a necessarily more schematic indication of the duals, since there is no canonical coordinate system in the dual plane. The transition along the ε axis is called a 'nib' in [3].

of a self-intersection to the other. It occurs only for Case 1 (there are no special values of λ for Case 2) and is given by (i) in the following proposition. Secondly, and also for Case 1, the same event can occur on the dual equidistant, which means that, on the equidistant itself, there is a bitangent line which, for one tangency, is the (limiting) tangent at a cusp. Crossing this locus a bitangent line migrates through a cuspidal tangent on the equidistant. See (ii) below.

Proposition 5.2. *Suppose as before that* $a_3^3 g_{40} - b_3^3 f_{40} \neq 0$. *Then* (i) *the locus in the* (ε, α) *plane where an inflexion and a self-intersection* *coincide on the equidistant has the form*

$$\varepsilon = \pm \frac{a_3^2 b_3^2 (a_3 - b_3)^8}{3g_{11}(a_3^3 g_{40} - b_3^3 f_{40})^2} \alpha^4 + \dots,$$

where the sign is $+$ *for inflexions corresponding to* $t = 0$ *and* $-$ *for inflexions corresponding to* $s = 0$.
(ii) *the locus in the* (ε, α) *plane where the dual has an inflexion coinciding with a self-intersection has the form*

$$\varepsilon = -\frac{2a_3^3 b_3^3 (a_3 - b_3)^6}{g_{11}(a_3^3 g_{40} - b_3^3 f_{40})^2} \alpha^3 + \dots.$$

Assuming as usual that $g_{11} > 0$, *this locus lies locally in the second and fourth quadrants of the* (ε, α) *plane where* ε *and* α *have opposite signs.* □

The migration of inflexions on the equidistant is noted on Figure 3. We do not attempt to include the second migration (ii) on the same figure but an idea of what is happening on the equidistant itself is in Figure 5.

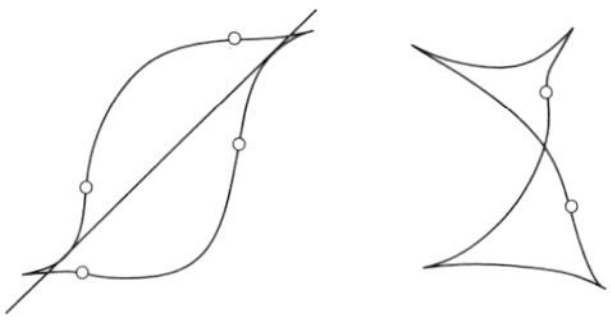

Fig. 4. Case 2, the equidistant (left, a lips) and dual (right) for $\varepsilon > 0$, with a bitangent line indicated corresponding to the self-intersection on the dual. The dual is called a 'moth' as in [3] and it disappears as $\varepsilon \to 0$. Note that this lips necessarily has four inflexions, as in Corollary 4.2.

§6. Reduction to a normal form preserving the ε-fibration

In Proposition 3.1 it was shown that the family $\mathcal{H}$ given by (3) is $\mathcal{A}$-equivalent to a versal unfolding of a lips, beaks or gull singularity. In this section the same family is considered using a finer notion of equivalence which we call (α, ε)-$\mathcal{A}$ equivalence. As before, $\lambda = \lambda_0 + \alpha$ and λ_0 is either a general value or a special value, but never 0 or 1. Restricting

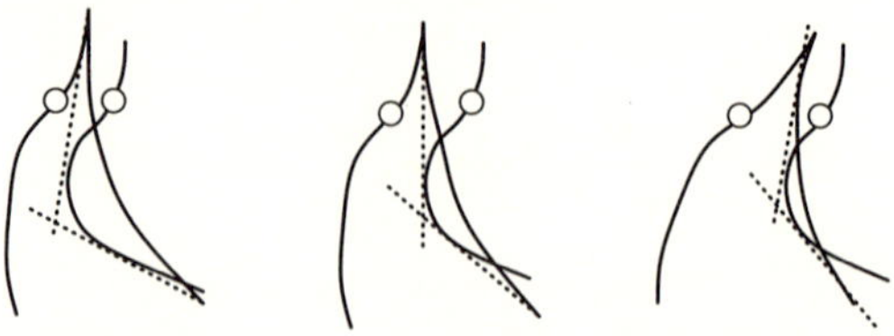

Fig. 5. Migration of an inflexion on the *dual* equidistant across a self-intersection produces this change on the equidistant, where the configuration is from the bottom right of the equidistants diagram in Figure 3. A similar change occurs in the top left but with a single bitangent line.

the permissible changes in the parameters so that ε can be replaced by a function in ε only (not involving α) preserves the fibration over ε. In particular, this preserves how the families of equidistants evolve as α varies for fixed values of ε near zero. It does not, however, preserve the families of equidistants as ε varies, for a fixed α, such as $\alpha = 0$. The advantage of the method here is that it is possible to reduce to a *normal form*.

Definition 6.1. *Two germs of families* $\mathcal{H}_i : \mathbb{R}^2 \times \mathbb{R}^2, \mathbf{0} \to \mathbb{R}^2$, $i = 1, 2$, *of the variables* s, u *and with parameters* α, ε *are called* (α, ε)-$\mathcal{A}$ *equivalent if there exists a diffeomorphism germs* $\theta : \mathbb{R}^2 \times \mathbb{R}^2 \to \mathbb{R}^2 \times \mathbb{R}^2$, *of the form*

$$\theta : (s, u, \alpha, \varepsilon) \mapsto (\theta_1(s, u, \alpha, \varepsilon), \theta_2(s, u, \alpha, \varepsilon), A(\alpha, \varepsilon), E(\varepsilon))$$

and a diffeomorphism $\phi : \mathbb{R}^2 \times \mathbb{R}^2 \to \mathbb{R}^2 \times \mathbb{R}^2$, *of the form* $(\phi_1, \phi_2) \times \mathrm{id}$, *which is the identity on the last two coordinates, such that* $\phi \circ (\mathcal{H}_1 \times \mathrm{id}) = (\mathcal{H}_2 \times \mathrm{id}) \circ \theta + c(\alpha, \varepsilon)$ *for a smooth germ* $c(\alpha, \varepsilon)$. *(Here* $\mathcal{H}_i \times \mathrm{id}$ *is the identity on the last two coordinates* (α, ε).*)*

$$\begin{array}{ccc} \mathbb{R}^2 \times \mathbb{R}^2 & \xrightarrow{\mathcal{H}_1 \times \mathrm{id}} & \mathbb{R}^2 \times \mathbb{R}^2 \\ \downarrow \theta & & \downarrow \phi \\ \mathbb{R}^2 \times \mathbb{R}^2 & \xrightarrow{\mathcal{H}_2 \times \mathrm{id}} & \mathbb{R}^2 \times \mathbb{R}^2 \end{array}$$

In other words,

$$\begin{aligned}&(\phi_1(\mathcal{H}_1(s,u,\alpha,\varepsilon),\alpha,\varepsilon),\phi_2(\mathcal{H}_1(s,u,\alpha,\varepsilon)))\\ =\;&\mathcal{H}_2(\theta_1(s,u,\alpha,\varepsilon),\theta_2(s,u,\alpha,\varepsilon),A(\alpha,\varepsilon),E(\varepsilon))+c'(\alpha,\varepsilon)\end{aligned}$$

where c' is the first two components of c. (The last two components must be of the form $(\alpha - A, \varepsilon - E)$.) The key requirement is of course that the equivalence preserves, up to local diffeomorphism, the critical point set and its image, the set of critical values, of a family $\mathcal{H}$ for α and ε close to 0, and also preserves the projection onto the ε-axis.

Theorem 6.2. *At $\alpha = \varepsilon = 0$ the generating family $\mathcal{H}$ is (α, ε)-$\mathcal{A}$ equivalent to one of the following versal deformations in variables $(u, s) \in \mathbb{R}^2$ and parameters $(\alpha, \varepsilon) \in \mathbb{R}^2$:*

Case 1 $f_{30}g_{30} > 0$	**Beaks**	$\lambda_0 \neq \frac{b_3}{b_3 \pm a_3}$ $\lambda_0 \neq 0, 1$	$\mathcal{H} = (u, s^3 - su^2 + \varepsilon s)$
	Gull	$\lambda_0 = \frac{b_3}{b_3 \pm a_3}$	$\mathcal{H} =$ $(u, s^5 + s^4 + s^2u + \alpha s^3 + \varepsilon s)$
Case 2 $f_{30}g_{30} < 0$	**Lips**	$\lambda_0 \neq 0, 1$	$\mathcal{H} = (u, s^3 + su^2 + \varepsilon s)$

The classification follows on from the reduction given in §3 and the observation that the component of H that is linear in s and u does not depend on α (see equation (4) above). The complete proof depends upon a special version of the versality theorem which holds for (α, ε)-$\mathcal{A}$ versality; see the Appendix (Lemma A.1). □

The local singularities for the two parallel inflexions case studied in this article fit into the general adjacency diagram for equidistants as follows:

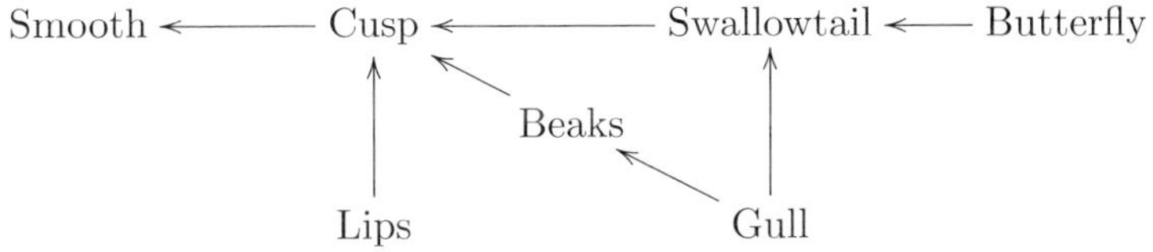

Note that the normal forms for lips and beaks in Proposition 6.2 do not contain α, that is they do not depend on the particular λ-equidistant where $\lambda = \lambda_0 + \alpha$ is close to a non-special value λ_0. Along the chord between a pair of parallel inflexions, which we call a *supercaustic chord*, every ε-family of equidistants, for a fixed non-special value of λ, undergoes either a beaks or lips bifurcation depending on whether the inflexions have the same or opposite sign. (For inflexions of the same sign, this happens when crossing the α-axis at constant $\alpha \neq 0$ in Figure 3; for inflexions of opposite sign the 'lips' of Figure 4 shrinks to a point and vanishes.) It is well-known that such bifurcations are not possible for Legendrian curves such as the λ-equidistants, that is where λ varies but ε remains fixed, since they alter the topology of the curves (see for example [1, p.60]).

In the case of inflexions of the same sign, where $f_{30}g_{30}$ is positive, the supercaustic chord contains two special values of λ_0 where the more degenerate gull singularity occurs with normal form as given in Proposition 6.2. Since (α, ε)-$\mathcal{A}$ equivalence preserves the fibration over ε, the normal form of the gull singularity preserves how the λ-equidistants bifurcate for fixed values of ε near zero ($\lambda = \lambda_0 + \alpha$ where λ_0 is a special value).

Using the normal form for gull in the table, we can calculate the critical set of $\mathcal{H}$ and hence the critical locus (set of critical values) which is the 'big equidistant'. From that a clock diagram can be drawn and this will correctly depict the bifurcations of the equidistants but only projection to the ε-axis can be relied on as representing the bifurcations for a fixed ε and α close to 0. Of course this clock diagram does not include any information about inflexions of the equidistants, such information not being preserved by diffeomorphisms. It is not difficult to calculate the swallowtail locus S and the self-tangency locus T in the (ε, α)-plane; these come to

$$\begin{aligned} S : \varepsilon &= -4s^3 - 15s^4, \alpha = -4s - 10s^2, \text{ so } \varepsilon = \tfrac{1}{16}\alpha^3 + \ldots; \\ T : \varepsilon &= s^4, \alpha = -2s^2, \text{ so } \varepsilon = \tfrac{1}{4}\alpha^2, \ \alpha \leq 0. \end{aligned}$$

A sketch of the resulting clock diagram is in Figure 6.

As previously mentioned, gull singularities also occur as projections of smooth surfaces in $\mathbb{R}^3$ to the plane (see for example [3, 5, 8]). In such projections, other singularities of the same codimension as the gull include butterfly and goose singularities. It is interesting to note that whilst butterfly and gull singularities both occur in one-parameter families of equidistants (only the latter in the context of our article), goose singularities are absent from the list. Goose singularities occur as the closure of the intersection of lips and beaks strata, and for equidistants these occur separately depending on the sign of $f_{30}g_{30}$. Since we assume our original curves have ordinary inflexions, so that f_{30} and g_{30} are nonzero, goose singularities do not occur.

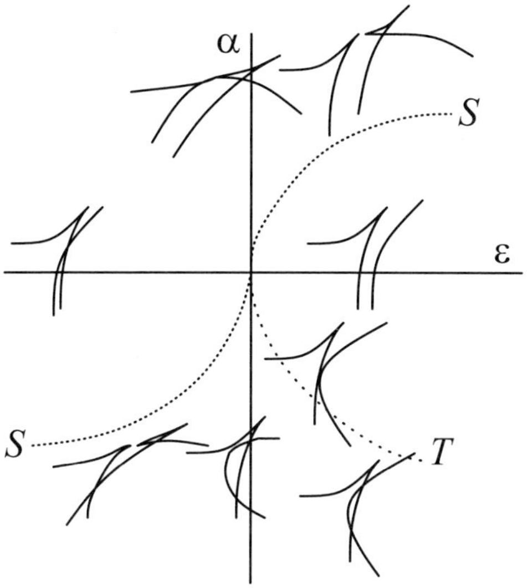

Fig. 6. The clock diagram for a gull singularity derived from the normal form in Theorem 6.2. Here S is the swallowtail set and T is the self-tangency set.

§Appendix A. Sketch of the proof of Theorem 6.2

Denote by Ω the space of map germs $F = (F_1, F_2) : \mathbb{R}^2 \times \mathbb{R}^2, 0 \to \mathbb{R}^2, 0$ in variables $x \in \mathbb{R}^2$, parameters α and ε, with the property that the linear terms in $x = (x_1, x_2)$ do not depend on α. That is, for each $i, j \in \{1, 2\}$ we have $\left.\dfrac{\partial^2 F_i}{\partial x_j \partial \alpha}\right|_{x=0} \equiv 0$.

Lemma A.1. *An infinitesimally (α, ε)-$\mathcal{A}$ versal germ $F \in \Omega$ is (α, ε)-$\mathcal{A}$ versal.*

Proof (This proof is modelled on [7]; see also [2, p.151].) Let F be an infinitesimal (α, ε)-$\mathcal{A}$ versal deformation of the germ $f : \mathbb{R}^2, 0 \to \mathbb{R}^2, 0$ and let G be a 1-parameter deformation of F (parameter β):

$$G(x, \alpha, \varepsilon, 0) \equiv F(x, \alpha, \varepsilon), \quad F(x, 0, 0) \equiv f(x)$$

$$G : (\mathbb{R}^2 \times \mathbb{R} \times \mathbb{R} \times \mathbb{R}, 0) \to (\mathbb{R}^2, 0).$$

We may consider G as an 3-parameter deformation of the germ of a map in x with parameters $\alpha, \varepsilon, \beta \in \mathbb{R}$. The result follows from the following lemma.

Lemma A.2. *The deformation G of f is (α,ε)-$\mathcal{A}$ equivalent to one induced from F.*

Proof Following [7, p.8–9] the key step in the argument is to solve the following "homological equation" for the unknown map germs Ξ_1, Ξ_2, L_1, L_2 and A.

$$\begin{aligned}&\frac{\partial G}{\partial \beta}+\frac{\partial G}{\partial \alpha}\Xi_1(\alpha,\varepsilon,\beta)+\frac{\partial G}{\partial \varepsilon}\Xi_2(\varepsilon,\beta)+\sum_{i=1}^{2}\frac{\partial G}{\partial x_i}L_i(x,\alpha,\varepsilon,\beta)\\ =\;& A(G(\alpha,\varepsilon,\beta,x),\alpha,\varepsilon,\beta),\end{aligned}$$

where for us it is important that Ξ_2 is a function of ε,β only, that is independent of α.

For any germ $s:\mathbb{R}^2,0\to\mathbb{R}^2,0$ there exists a decomposition signifying infinitesimal (α,ε)-$\mathcal{A}$ versality (here and below terms involving derivatives with respect to $x=(x_1,x_2)$ stand for the sum of two terms):

$$s(x)=\frac{\partial f}{\partial x}h(x)+k(f(x))+\frac{\partial F}{\partial \alpha}\xi+\frac{\partial F}{\partial \varepsilon}\nu,\ \text{where } \xi,\nu\in\mathbb{R}.$$

Consequently for every germ $S(x,\alpha,\varepsilon,\beta)$ there exists a decomposition:

$$\begin{aligned}S(x,\alpha,\varepsilon,\beta)\;=\;&\frac{\partial G}{\partial x}h(x)+K(G,\alpha,\varepsilon,\beta)+\frac{\partial G}{\partial \alpha}\xi+\frac{\partial G}{\partial \varepsilon}\nu\\ &+[\beta\sigma_0(x,\alpha,\beta,\varepsilon)+\alpha\sigma_1(x,\alpha)+\varepsilon\sigma_2(x,\alpha,\varepsilon)]\end{aligned}$$

Decompose σ_0,σ_1 and σ_2 using the same procedure:

$$\begin{aligned}S(x,\alpha,\varepsilon,\beta)\;=\;&\frac{\partial G}{\partial x}(h(x)+\beta h_0(x)+\alpha h_1(x)+\varepsilon h_2(x))\\ &+\widetilde{K}(G(\alpha,\varepsilon,\beta,x),\alpha,\varepsilon,\beta)\\ &+\frac{\partial G}{\partial \alpha}(\xi+\beta\xi_0+\alpha\xi_1+\varepsilon\xi_2)\\ &+\frac{\partial G}{\partial \varepsilon}(\nu+\beta\nu_0+\alpha\nu_1+\varepsilon\nu_2)\\ &+[\beta^2\sigma_{00}(x,\alpha,\beta,\varepsilon)+\beta\alpha\sigma_{01}(x,\alpha,\varepsilon)+\alpha^2\sigma_{11}(x,\alpha,\varepsilon)\\ &\quad+\beta\varepsilon\sigma_{02}(x,\varepsilon)+\alpha\varepsilon\sigma_{12}(x,\varepsilon)+\varepsilon^2\sigma_{22}(x,\varepsilon)]\end{aligned}$$

We have now obtained a better decomposition where the part in the square bracket is now of the second order in α,ε and β. The coefficients of the other terms will form the linear parts in α,ε,β of L_i, A, Ξ_1, Ξ_2 respectively.

Notice that if we consider mappings $S \in \Omega$ (i.e. with the property that no linear part in the variables contain α) then ν_1 will be zero and the coefficient of $\partial G/\partial\varepsilon$ will be a function in ε and β only. Continuing in this way we obtain the decomposition

$$\begin{aligned} S(x,\alpha,\varepsilon,\beta) &= \frac{\partial G}{\partial x}L(x,\alpha,\varepsilon,\beta) \\ &\quad + K(G(x,\alpha,\varepsilon,\beta),\alpha,\varepsilon,\beta) + \frac{\partial G}{\partial \alpha}\Xi_1(\alpha,\varepsilon,\beta) + \frac{\partial G}{\partial \varepsilon}\Xi_2(\varepsilon,\beta) \end{aligned} \tag{12}$$

at the level of a formal power series. Note that Ξ_2 does not depend on α. The preparation theorem (see [2]) shows that the decomposition (12) exists for convergent series and in the C^∞ case, where it is necessary to apply the preparation theorem to the $\mathcal{E}_{x,\alpha,\varepsilon,\beta}$ module $(\mathcal{E}_{x,\alpha,\varepsilon,\beta})^2/\{\frac{\partial G}{\partial x} + K(G(x,\alpha,\varepsilon,\beta),\alpha,\varepsilon,\beta)\}$, to the map $(y,\alpha,\varepsilon,\beta) \to (\alpha,\varepsilon,\beta)$ and to the generators $\frac{\partial G}{\partial \alpha}$ and $\frac{\partial G}{\partial \varepsilon}$. The decomposition (12) for $S = -\frac{\partial G}{\partial \beta}, K = -A$ provides the desired solution of the homological equation and the lemma is proved. □

References

[1] V.I.Arnol'd, *Singularities of Caustics and Wave Fronts*, Math. Appl. Soviet Ser., vol. 62, Kluwer, Dordrecht, (1990).

[2] V.I. Arnold, S.M. Gusein-Zade and A.N. Varchenko, *Singularities of Differentiable Maps Vol* **1**, Birkhäuser, (1988).

[3] J.W.Bruce and P.J.Giblin, 'Growth, motion and one-parameter families of symmetry sets', *Proc. Royal Soc. Edinburgh* 104A (1986), 179–204.

[4] J.W.Bruce and P.J.Giblin, 'Outlines and their duals', *Proc. London Math. Soc.* 50 (1985), 552–570.

[5] T. Gaffney, 'The structure of $T\mathcal{A}(f)$, classification and an application to differential geometry', *Proc. Arcata Conf. on Singularities 1981, AMS Proc. of Symposia in Pure Math.* 40 (1983), 409–427.

[6] Peter Giblin and Graham Reeve, 'Centre symmetry sets of families of plane curves', *Demonstratio Mathematica* 48 (2015), 167–192.

[7] Jean Martinet, 'Déploiements versels des applications differentiables et classification des applications stables', *Springer Lecture Notes in Mathematics* 535 (1976), 1–44.

[8] C. McCrory, *Profiles of surfaces*, unpublished preprint, University of Warwick 1980; see `https://warwick.ac.uk/fac/sci/maths/research/preprints/1980/` or `https://sites.google.com/site/clintonmccrory/home/vita`

[9] J. Rieger, 'Families of maps from the plane to the plane', *J. London Math. Soc.* 36 (1987), 351–369.

Peter Giblin
Department of Mathematical Sciences, The University of Liverpool, Liverpool L69 7ZL, UK
E-mail address: pjgiblin@liv.ac.uk

Graham Reeve
Department of Mathematics and Computer Science, Liverpool Hope University, Liverpol L16 9JD, UK
E-mail address: reeveg@hope.ac.uk

Advanced Studies in Pure Mathematics 78, 2018
Singularities in Generic Geometry
pp. 273–292

Rectifying developable surfaces of framed base curves and framed helices

Shun'ichi Honda

Abstract.

We study the rectifying developable surface of a framed base curve and a framed helix in the Euclidean space. A framed base curve is a smooth curve with a moving frame which may have singular points. By using the curvature of a framed base curve, we investigate the rectifying developable surface and a framed helix. Moreover, we introduce two new invariants of a framed base curve, which characterize singularities of the rectifying developable surface and a framed helix.

§1. Introduction

There are several articles concerning singularities of the tangent developable surface and the focal developable surface of a space curve with singular points ([5, 6, 7, 8]). In [6, 7, 8] Ishikawa investigated relationships between singularities of the tangent developable surface and the type (a_1, a_2, a_3) of a space curve. In [5] the author and Takahashi introduced relationships between singularities of the focal developable surface of a framed base curve and invariants, that is, the curvature of a framed curve. On the other hand, Izumiya, Katsumi and Yamasaki introduced the rectifying developable surface of a regular space curve in [9]. They showed relationships between singularities of the rectifying developable surface of a regular space curve and geometric invariants of the curve which are deeply related to the order of contact with a helix. A regular space curve γ is always a geodesic of its rectifying developable surface. In this sense, the rectifying developable surface is an interesting subject.

Received June 30, 2016.

Revised March 22, 2017.

2010 *Mathematics Subject Classification.* Primary 53A05; Secondary 53A04, 58K05.

Key words and phrases. Framed base curve, rectifying developable surface, framed helix, singularities.

In this paper we consider the rectifying developable surface of a space curve with singular points, and a helix which may have singular points. In order to define these notions, we apply the theory of framed base curves under a certain condition. A framed base curve is a smooth space curve with a moving frame which may have singular points, see [4] and Appendix A. In Section 3, we define the rectifying developable surface of a framed base curve under a certain condition. That is a natural generalization of the rectifying developable surface of a regular space curve in [9]. By using two new invariants, we give basic properties of the rectifying developable surface (cf. Proposition 3.2 and Theorem 3.3). Moreover, we define a framed helix and consider relationships between the rectifying developable surface of a framed base curve and a framed helix in Section 4. In Section 5, we introduce the notion of support functions of a framed base curve. By using this function, we give relationships between singularities of the rectifying developable surface and invariants of a framed base curve. The proof of Theorem 3.3 is given in Section 5. We give some examples of the rectifying developable surface of a framed base curve and a framed helix in Section 6.

All maps and manifolds considered here are differential of class C^∞.

Acknowledgements. The author would like to express sincere gratitude to Professor Izumiya and the referee for helpful comments.

§2. Basic notions

Let $\mathbb{R}^3$ be the 3-dimensional Euclidean space equipped with the canonical inner product $\langle \boldsymbol{a}, \boldsymbol{b}\rangle = a_1b_1+a_2b_2+a_3b_3$, where $\boldsymbol{a} = (a_1, a_2, a_3)$, $\boldsymbol{b} = (b_1, b_2, b_3) \in \mathbb{R}^3$. The norm of $\boldsymbol{a}$ is given by $\|\boldsymbol{a}\| = \sqrt{\langle \boldsymbol{a}, \boldsymbol{a}\rangle}$. We define the vector product of $\boldsymbol{a}$ and $\boldsymbol{b}$ by

$$\boldsymbol{a} \times \boldsymbol{b} = \begin{vmatrix} a_1 & a_2 & a_3 \\ b_1 & b_2 & b_3 \\ \boldsymbol{e}_1 & \boldsymbol{e}_2 & \boldsymbol{e}_3 \end{vmatrix} = \begin{vmatrix} \boldsymbol{e}_1 & \boldsymbol{e}_2 & \boldsymbol{e}_3 \\ a_1 & a_2 & a_3 \\ b_1 & b_2 & b_3 \end{vmatrix},$$

where $\boldsymbol{e}_1, \boldsymbol{e}_2, \boldsymbol{e}_3$ are the canonical basis on $\mathbb{R}^3$.

We quickly review some basic concepts on classical differential geometry of regular space curves in $\mathbb{R}^3$. Let I be an interval. Suppose that $\boldsymbol{\gamma} : I \to \mathbb{R}^3$ is a regular space curve with linearly independent condition, that is, $\dot{\boldsymbol{\gamma}}(t)$ and $\ddot{\boldsymbol{\gamma}}(t)$ are linearly independent for all $t \in I$, where $\dot{\boldsymbol{\gamma}}(t) = (d\boldsymbol{\gamma}/dt)(t)$ and $\ddot{\boldsymbol{\gamma}}(t) = (d^2\boldsymbol{\gamma}/dt^2)(t)$. Then we have an orthonormal frame

$$\{T(t), N(t), B(t)\} = \left\{ \frac{\dot{\boldsymbol{\gamma}}(t)}{\|\dot{\boldsymbol{\gamma}}(t)\|}, \frac{(\dot{\boldsymbol{\gamma}}(t) \times \ddot{\boldsymbol{\gamma}}(t)) \times \dot{\boldsymbol{\gamma}}(t)}{\|(\dot{\boldsymbol{\gamma}}(t) \times \ddot{\boldsymbol{\gamma}}(t)) \times \dot{\boldsymbol{\gamma}}(t)\|}, \frac{\dot{\boldsymbol{\gamma}}(t) \times \ddot{\boldsymbol{\gamma}}(t)}{\|\dot{\boldsymbol{\gamma}}(t) \times \ddot{\boldsymbol{\gamma}}(t)\|} \right\}$$

along $\gamma(t)$, which is called the *Frenet frame* along $\gamma(t)$. Then we have the following Frenet-Serret formula:

$$\begin{pmatrix} \dot{T}(t) \\ \dot{N}(t) \\ \dot{B}(t) \end{pmatrix} = \begin{pmatrix} 0 & \|\dot{\gamma}(t)\|\kappa(t) & 0 \\ -\|\dot{\gamma}(t)\|\kappa(t) & 0 & \|\dot{\gamma}(t)\|\tau(t) \\ 0 & -\|\dot{\gamma}(t)\|\tau(t) & 0 \end{pmatrix} \begin{pmatrix} T(t) \\ N(t) \\ B(t) \end{pmatrix},$$

where

$$\kappa(t) = \frac{\|\dot{\gamma}(t) \times \ddot{\gamma}(t)\|}{\|\dot{\gamma}(t)\|^3}, \quad \tau(t) = \frac{\det(\dot{\gamma}(t), \ddot{\gamma}(t), \dddot{\gamma}(t))}{\|\dot{\gamma}(t) \times \ddot{\gamma}(t)\|^2},$$

where $\dddot{\gamma}(t) = (d^3\gamma/dt^3)(t)$. We call $\kappa(t)$ a *curvature* and $\tau(t)$ a *torsion* of $\gamma(t)$. Note that the curvature $\kappa(t)$ and the torsion $\tau(t)$ are independent of a choice of parametrization. For any regular space curve $\gamma : I \to \mathbb{R}^3$, we define a vector $D(t) = \tau(t)T(t) + \kappa(t)B(t)$ and we call it a *Darboux vector* along $\gamma(t)$ (cf. [9, 11]). Since $\kappa(t) > 0$, we also define a spherical Darboux vector $\overline{D} : I \to S^2$ by

$$\overline{D}(t) = \frac{\tau(t)T(t) + \kappa(t)B(t)}{\sqrt{\kappa^2(t) + \tau^2(t)}}.$$

and the rectifying developable surface $RD_\gamma : I \times \mathbb{R} \to \mathbb{R}^3$ of $\gamma(t)$ by

$$RD_\gamma(t, u) = \gamma(t) + u\overline{D}(t) = \gamma(t) + u\frac{\tau(t)T(t) + \kappa(t)B(t)}{\sqrt{\kappa^2(t) + \tau^2(t)}}.$$

By a direct calculation, we have

$$\begin{aligned} &\frac{\partial RD_\gamma}{\partial t}(t, u) \times \frac{\partial RD_\gamma}{\partial u}(t, u) \\ &\quad = -\left(\frac{\kappa(t)\|\dot{\gamma}(t)\|}{\sqrt{\kappa^2(t) + \tau^2(t)}} + u\frac{\kappa(t)\dot{\tau}(t) - \dot{\kappa}(t)\tau(t)}{\kappa^2(t) + \tau^2(t)} \right) N(t). \end{aligned}$$

Therefore, $(t_0, u_0) \in I \times \mathbb{R}$ is a singular point of RD_γ if and only if

$$\frac{\kappa(t_0)\|\dot{\gamma}(t_0)\|}{\sqrt{\kappa^2(t_0) + \tau^2(t_0)}} + u_0\frac{\kappa(t_0)\dot{\tau}(t_0) - \dot{\kappa}(t_0)\tau(t_0)}{\kappa^2(t_0) + \tau^2(t_0)} = 0.$$

If (t_0, u_0) is a singular point of RD_γ, then we have $u_0 \neq 0$, that is, RD_γ has no singular value on the base curve $\gamma(t)$. Izumiya, Katsumi and Yamasaki investigated the rectifying developable surfaces of a regular space curve in [9].

In this paper we do not assume that $\gamma : I \to \mathbb{R}^3$ is a regular curve with linearly independent condition, so that γ may have singular points.

If γ has a singular point, we can not construct the Frenet frame along $\gamma(t)$. However, we can define the Frenet type frame along $\gamma(t)$ under a certain condition.

Definition 2.1. We say that $\gamma : I \to \mathbb{R}^3$ is a *Frenet type framed base curve* if there exist a regular spherical curve $\mathcal{T} : I \to S^2$ and a smooth function $\alpha : I \to \mathbb{R}$ such that $\dot{\gamma}(t) = \alpha(t)\mathcal{T}(t)$ for all $t \in I$. Then we call $\mathcal{T}(t)$ a *unit tangent vector* and $\alpha(t)$ a *speed function* of $\gamma(t)$.

Clearly, t_0 is a singular point of γ if and only if $\alpha(t_0) = 0$. We define a *unit principal normal vector* $\mathcal{N}(t) = \dot{\mathcal{T}}(t)/\|\dot{\mathcal{T}}(t)\|$ and a *unit binormal vector* $\mathcal{B}(t) = \mathcal{T}(t) \times \mathcal{N}(t)$ of $\gamma(t)$. Then we have an orthonormal frame $\{\mathcal{T}(t), \mathcal{N}(t), \mathcal{B}(t)\}$ along $\gamma(t)$, which is called the *Frenet type frame* along $\gamma(t)$. Then we have the following Frenet-Serret type formula:

$$\begin{pmatrix} \dot{\mathcal{T}}(t) \\ \dot{\mathcal{N}}(t) \\ \dot{\mathcal{B}}(t) \end{pmatrix} = \begin{pmatrix} 0 & \kappa(t) & 0 \\ -\kappa(t) & 0 & \tau(t) \\ 0 & -\tau(t) & 0 \end{pmatrix} \begin{pmatrix} \mathcal{T}(t) \\ \mathcal{N}(t) \\ \mathcal{B}(t) \end{pmatrix},$$

where

$$\kappa(t) = \|\dot{\mathcal{T}}(t)\|, \ \tau(t) = \frac{\det(\mathcal{T}(t), \dot{\mathcal{T}}(t), \ddot{\mathcal{T}}(t))}{\|\dot{\mathcal{T}}(t)\|^2}.$$

We call $\kappa(t)$ a *curvature* and $\tau(t)$ a *torsion* of γ. Note that the curvature $\kappa(t)$ and the torsion $\tau(t)$ are depend on a choice of parametrization.

We define a vector $\mathcal{D}(t)$ along $\gamma(t)$ by

$$\mathcal{D}(t) = \tau(t)\mathcal{T}(t) + \kappa(t)\mathcal{B}(t),$$

which is called a *Darboux type vector* along $\gamma(t)$. By using the Darboux type vector, the Frenet-Serret type formula is rewritten as follows:

$$\begin{cases} \dot{\mathcal{T}}(t) = \mathcal{D}(t) \times \mathcal{T}(t), \\ \dot{\mathcal{N}}(t) = \mathcal{D}(t) \times \mathcal{N}(t), \\ \dot{\mathcal{B}}(t) = \mathcal{D}(t) \times \mathcal{B}(t). \end{cases}$$

Thus the Darboux type vector plays an important role for the study of framed base curves. Since $\kappa(t) > 0$, we can define a *spherical Darboux-type vector* by

$$\overline{\mathcal{D}}(t) = \frac{\tau(t)\mathcal{T}(t) + \kappa(t)\mathcal{B}(t)}{\sqrt{\kappa^2(t) + \tau^2(t)}}.$$

Remark 2.2. Since $\mathcal{T}(t)$ is a regular curve, we uniquely obtain the unit principal normal vector $\mathcal{N}(t)$ and the unit binormal vector $\mathcal{N}(t)$.

Therefore, $\kappa(t)$, $\tau(t)$ and $\overline{\mathcal{D}}(t)$ is uniquely determined with respect to $\mathcal{T}(t)$. On the other hand, we can easily check that $\overline{\mathcal{D}}(t)$ is a spherical dual of $\mathcal{N}(t)$ (cf. [12]).

Example 2.3. Above objects are natural generalizations of corresponding notions for regular space curves. In fact, let $\gamma : I \to \mathbb{R}^3$ be a regular space curve with linearly independent condition. If we take $\mathcal{T}(t) = T(t)$ and $\alpha(t) = \|\dot{\gamma}(t)\|$, then $\mathcal{N}(t) = N(t)$, $\mathcal{B}(t) = B(t)$ and the curvature $\kappa(t)$ (respectively, $\tau(t)$) as a framed base curve coincides with the curvature $\kappa(t)$ (respectively, the torsion $\tau(t)$) in the sense of classical differential geometry. Therefore, we have $\mathcal{D}(t) = D(t)$ and $\overline{\mathcal{D}}(t) = \overline{D}(t)$.

We can easily check that $\gamma : I \to \mathbb{R}^3$ is a framed base curve (cf. [4] and Appendix A). More precisely, $(\gamma, \mathcal{N}, \mathcal{B}) : I \to \mathbb{R}^3 \times \Delta \subset \mathbb{R}^3 \times S^2 \times S^2$ is a framed curve with the curvature of the framed curve $(\tau(t), -\kappa(t), 0, \alpha(t))$. This is the reason why we call γ the Frenet type framed base curve. In [4], the author and Takahashi have shown the existence and the uniqueness for framed curves. Since Remark 2.2, the speed function $\alpha(t)$, the curvature $\kappa(t)$ and the torsion $\tau(t)$ are invariants of the pair $(\gamma, \mathcal{T})$.

§3. Rectifying developable surfaces

In this section, we consider the rectifying developable surface of a Frenet type framed base curve. Let $\gamma : I \to \mathbb{R}^3$ be a Frenet type framed base curve with unit tangent vector $\mathcal{T}(t)$.

Definition 3.1. We define a map $\mathcal{RD}_\gamma : I \times \mathbb{R} \to \mathbb{R}^3$ by

$$\mathcal{RD}_\gamma(t, u) = \gamma(t) + u\overline{\mathcal{D}}(t) = \gamma(t) + u\frac{\tau(t)\mathcal{T}(t) + \kappa(t)\mathcal{B}(t)}{\sqrt{\kappa^2(t) + \tau^2(t)}}.$$

We call $\mathcal{RD}_\gamma$ the *rectifying developable surface* of Frenet type framed curve γ.

Since Example 2.3, Definition 3.1 is a natural generalization of the rectifying developable surface of a regular space curve in [9]. The rectifying developable surface $\mathcal{RD}_\gamma(t, u)$ is a ruled surface and we have

$$\dot{\overline{\mathcal{D}}}(t) = \left(\frac{\kappa(t)\dot{\tau}(t) - \dot{\kappa}(t)\tau(t)}{\kappa^2(t) + \tau^2(t)}\right)\frac{\kappa(t)\mathcal{T}(t) - \tau(t)\mathcal{B}(t)}{\sqrt{\kappa^2(t) + \tau^2(t)}},$$

so that we have

$$\begin{aligned}\det\left(\dot{\gamma}, \overline{\mathcal{D}}, \dot{\overline{\mathcal{D}}}\right) &= \det\left(\alpha\mathcal{T}, \frac{\tau\mathcal{T} + \kappa\mathcal{B}}{\sqrt{\kappa^2 + \tau^2}}, \left(\frac{\kappa\dot{\tau} - \dot{\kappa}\tau}{\kappa^2 + \tau^2}\right)\frac{\kappa\mathcal{T} - \tau\mathcal{B}}{\sqrt{\kappa^2 + \tau^2}}\right)\\ &= 0\end{aligned}$$

for all $t \in I$. This means that $\mathcal{RD}_\gamma$ is a developable surface (cf. [10]). Moreover, we introduce two invariants $\delta(t)$, $\sigma(t)$ as follows:

$$\delta(t) = \frac{\kappa(t)\dot{\tau}(t) - \dot{\kappa}(t)\tau(t)}{\kappa^2(t) + \tau^2(t)},$$

$$\sigma(t) = \frac{\alpha(t)\tau(t)}{\sqrt{\kappa^2(t) + \tau^2(t)}} - \frac{d}{dt}\left(\frac{\alpha(t)\kappa(t)}{\delta(t)\sqrt{\kappa^2(t) + \tau^2(t)}}\right), \text{ (when } \delta(t) \neq 0).$$

We remark that $\delta(t)$ corresponds to $(d/dt)(\tau/\kappa)(t)$ which is investigated in [9]. By a direct calculation, $\delta(t) = 0$ if and only if $\dot{\overline{\mathcal{D}}}(t) = \mathbf{0}$. We can also calculate that

$$\frac{\partial \mathcal{RD}_\gamma}{\partial t}(t,u) \times \frac{\partial \mathcal{RD}_\gamma}{\partial u}(t,u) = -\left(\frac{\alpha(t)\kappa(t)}{\sqrt{\kappa^2(t) + \tau^2(t)}} + u\delta(t)\right)\mathcal{N}(t).$$

Therefore, $(t_0, u_0) \in I \times \mathbb{R}$ is a singular point of $\mathcal{RD}_\gamma$ if and only if

$$\frac{\alpha(t_0)\kappa(t_0)}{\sqrt{\kappa^2(t_0) + \tau^2(t_0)}} + u_0 \frac{\kappa(t_0)\dot{\tau}(t_0) - \dot{\kappa}(t_0)\tau(t_0)}{\kappa^2(t_0) + \tau^2(t_0)} = 0.$$

Since $\dot{\mathcal{N}}(t) \neq \mathbf{0}$ for all $t \in I$, $\mathcal{RD}_\gamma$ is a wave front (cf. [1, 2]).

Proposition 3.2. *Let $\gamma : I \to \mathbb{R}^3$ be a Frenet type framed base curve with $\mathcal{T}(t)$. Then we have the following:*

(A) *The following are equivalent:*
  (1) *$\mathcal{RD}_\gamma$ is a cylinder,*
  (2) *$\delta(t) = 0$ for all $t \in I$.*

(B) *If $\delta(t) \neq 0$ for all $t \in I$, then the following are equivalent:*
  (3) *$\mathcal{RD}_\gamma$ is a conical surface,*
  (4) *$\sigma(t) = 0$ for all $t \in I$.*

Proof. (A) By definition, $\mathcal{RD}_\gamma$ is a cylinder if and only if $\overline{\mathcal{D}}(t)$ is a constant. Since

$$\dot{\overline{\mathcal{D}}}(t) = \delta(t)\frac{\kappa(t)\mathcal{T}(t) - \tau(t)\mathcal{B}(t)}{\sqrt{\kappa^2(t) + \tau^2(t)}},$$

$\overline{\mathcal{D}}(t)$ is a constant if and only if $\delta(t) = 0$ for all $t \in I$.

(B) We consider the striction curve $\boldsymbol{\sigma}(t)$ defined by

$$\boldsymbol{\sigma}(t) = \gamma(t) - \frac{\langle \dot{\gamma}(t), \dot{\overline{\mathcal{D}}}(t)\rangle}{\langle \dot{\overline{\mathcal{D}}}(t), \dot{\overline{\mathcal{D}}}(t)\rangle}\overline{\mathcal{D}}(t) = \gamma(t) - \frac{\alpha(t)\kappa(t)}{\delta(t)\sqrt{\kappa^2(t) + \tau^2(t)}}\overline{\mathcal{D}}(t).$$

Then (B)-(3) is equivalent to the condition $\dot{\boldsymbol{\sigma}}(t) = 0$ for all $t \in I$. We can calculate that

$$\begin{aligned}
\dot{\boldsymbol{\sigma}} &= \dot{\gamma} - \frac{d}{dt}\left(\frac{\alpha\kappa}{\delta\sqrt{\kappa^2+\tau^2}}\right)\overline{\mathcal{D}} - \frac{\alpha\kappa}{\delta\sqrt{\kappa^2+\tau^2}}\dot{\overline{\mathcal{D}}} \\
&= \alpha\mathcal{T} - \frac{d}{dt}\left(\frac{\alpha\kappa}{\delta\sqrt{\kappa^2+\tau^2}}\right)\overline{\mathcal{D}} - \frac{\alpha\kappa}{\sqrt{\kappa^2+\tau^2}}\frac{\kappa\mathcal{T}-\tau\mathcal{B}}{\sqrt{\kappa^2+\tau^2}} \\
&= \left(\frac{\alpha\tau}{\sqrt{\kappa^2+\tau^2}} - \frac{d}{dt}\left(\frac{\alpha\kappa}{\delta\sqrt{\kappa^2+\tau^2}}\right)\right)\frac{\tau\mathcal{T}+\kappa\mathcal{B}}{\sqrt{\kappa^2+\tau^2}} \\
&= \sigma\overline{\mathcal{D}}.
\end{aligned}$$

It follows that (B)-(3) and (B)-(4) are equivalent. □

We give relationships between singularities of the rectifying developable surface of a framed base curve and such invariants.

Theorem 3.3. *Let $\gamma : I \to \mathbb{R}^3$ be a Frenet type framed base curve with $\mathcal{T}(t)$. Then we have the following:*

(1) *(t_0, u_0) is a regular point of $\mathcal{RD}_\gamma$ if and only if*

$$\frac{\alpha(t_0)\kappa(t_0)}{\sqrt{\kappa^2(t_0)+\tau^2(t_0)}} + u_0\delta(t_0) \neq 0.$$

(2) *Suppose that (t_0, u_0) is a singular point of $\mathcal{RD}_\gamma$, then the rectifying developable surface $\mathcal{RD}_\gamma$ is locally diffeomorphic to the cuspidal edge* ***ce*** *at (t_0, u_0) if*

(i) *$\delta(t_0) \neq 0$, $\sigma(t_0) \neq 0$ and*

$$u_0 = -\frac{\alpha(t_0)\kappa(t_0)}{\delta(t_0)\sqrt{\kappa^2(t_0)+\tau^2(t_0)}},$$

or

(ii) *$\delta(t_0) = \alpha(t_0) = 0$, $\dot{\delta}(t_0) \neq 0$ and*

$$u_0 \neq -\dot{\alpha}(t_0)\kappa(t_0)\frac{\sqrt{\kappa^2(t_0)+\tau^2(t_0)}}{\kappa(t_0)\ddot{\tau}(t_0)-\ddot{\kappa}(t_0)\tau(t_0)},$$

or

(iii) *$\delta(t_0) = \alpha(t_0) = 0$ and $\dot{\alpha}(t_0) \neq 0$.*

(3) *Suppose that (t_0, u_0) is a singular point of $\mathcal{RD}_\gamma$, then the rectifying developable surface $\mathcal{RD}_\gamma$ is locally diffeomorphic to the swallowtail* ***sw*** *at (t_0, u_0) if $\delta(t_0) \neq 0$, $\sigma(t_0) = 0$, $\dot{\sigma}(t_0) \neq 0$ and*

$$u_0 = -\frac{\alpha(t_0)\kappa(t_0)}{\delta(t_0)\sqrt{\kappa^2(t_0)+\tau^2(t_0)}}.$$

Here, $\boldsymbol{ce} : (\mathbb{R}^2, 0) \to (\mathbb{R}^3, 0);\ (u, v) \mapsto (u, v^2, v^3)$ *is the cuspidal edge,* $\boldsymbol{sw} : (\mathbb{R}^2, 0) \to (\mathbb{R}^3, 0);\ (u, v) \mapsto (3u^4 + u^2 v, 4u^3 + 2uv, v)$ *is the swallowtail.*

Remark 3.4. Suppose that t_0 is a singular point of γ. Then $(t_0, 0)$ is a singular point of $\mathcal{RD}_\gamma$.

Remark 3.5. γ is always a geodesic of the rectifying developable surface $\mathcal{RD}_\gamma(t)$, away from singular points of γ (cf. [9]). Therefore, γ is the geodesic which formed to stride over singular points of $\mathcal{RD}_\gamma$.

§4. Framed helices

In this section, we define a helix which may have singular points. Let $\gamma : I \to \mathbb{R}^3$ be a Frenet type framed base curve with $\mathcal{T}(t)$.

Definition 4.1. We say that $\gamma : I \to \mathbb{R}^3$ is a *framed helix* if there exist constants $\boldsymbol{v} \in S^2$ and $C \in \mathbb{R}$ such that $\mathcal{T}(t) \cdot \boldsymbol{v} = C$ for all $t \in I$.

Definition 4.1 means that the tangent line of γ makes a constant angle with a fixed direction. In this sense, a framed helix is a natural generalization of a regular helix. The invariant $\delta(t)$ characterize a framed helix. In fact, we can prove the following Proposition.

Proposition 4.2. *Let* $\gamma : I \to \mathbb{R}^3$ *be a Frenet type framed base curve with* $\mathcal{T}(t)$*. Then the following are equivalent:*

(1) γ *is a framed helix.*
(2) $\delta(t) = 0$ *for all* $t \in I$.

Proof. Suppose that $\gamma(t)$ is a framed helix. Here we put $\boldsymbol{v} = a(t)\mathcal{T}(t) + b(t)\mathcal{N}(t) + c(t)\mathcal{B}(t)$, where $a(t)$, $b(t)$ and $c(t)$ are smooth functions. By the assumption,

$$\langle \boldsymbol{v}, \mathcal{T}(t) \rangle = a(t) = C. \tag{1}$$

Moreover, taking the derivative of the both sides of the formula (1), we have

$$-b(t)\kappa(t)\mathcal{T}(t) + (C\kappa(t) + \dot{b}(t) - c(t)\tau(t))\mathcal{N}(t) + (\dot{c}(t) + b(t)\tau(t))\,\mathcal{B}(t) = \mathbf{0}.$$

Then we have $b(t) = 0$, $c(t) = C_1$ and $C = C_1(\tau(t)/\kappa(t))$, where C_1 is a constant. On the other hand, since

$$1 = \|\boldsymbol{v}\|^2 = C_1^2 \left(\frac{\tau^2(t)}{\kappa^2(t)} + 1 \right),$$

$C_1 \neq 0$. Thus, $C/C_1 = \tau(t)/\kappa(t)$. We remark that $\delta(t) = 0$ if and only if $(d/dt)(\tau/\kappa)(t) = 0$. Hence, we have $\delta(t) = 0$ for all $t \in I$.

Conversely, suppose that $\delta(t) = 0$ for all $t \in I$. We put a constant vector $\boldsymbol{v} = (\tau(t)/\kappa(t))\mathcal{T}(t) + \mathcal{B}(t)$ and $\overline{\boldsymbol{v}} = \boldsymbol{v}/\|\boldsymbol{v}\|$. Then

$$\langle \overline{\boldsymbol{v}}, \mathcal{T}(t) \rangle = \frac{\frac{\tau(t)}{\kappa(t)}}{\sqrt{\frac{\tau^2(t)}{\kappa^2(t)} + 1}},$$

that is, $\langle \overline{\boldsymbol{v}}, \mathcal{T}(t) \rangle$ is a constant. This means that $\boldsymbol{\gamma}(t)$ is a framed helix. □

Corollary 4.3. *Let $\boldsymbol{\gamma} : I \to \mathbb{R}^3$ be a Frenet type framed base curve with $\mathcal{T}(t)$. Then the following are equivalent:*

(1) *$\mathcal{RD}_{\boldsymbol{\gamma}}$ is a cylinder,*
(2) *$\delta(t) = 0$ for all $t \in I$,*
(3) *$\boldsymbol{\gamma}$ is a framed helix.*

We recall the notion of the contact between framed curves, see [4]. Let $(\boldsymbol{\gamma}, \boldsymbol{\nu}_1, \boldsymbol{\nu}_2) : I \to \mathbb{R}^3 \times \Delta; t \mapsto (\boldsymbol{\gamma}(t), \boldsymbol{\nu}_1(t), \boldsymbol{\nu}_2(t))$ and $(\widetilde{\boldsymbol{\gamma}}, \widetilde{\boldsymbol{\nu}}_1, \widetilde{\boldsymbol{\nu}}_2) : \widetilde{I} \to \mathbb{R}^3 \times \Delta; u \mapsto (\widetilde{\boldsymbol{\gamma}}(u), \widetilde{\boldsymbol{\nu}}_1(u), \widetilde{\boldsymbol{\nu}}_2(u))$ be framed curves, respectively. Let k be a natural number. We denote the curvatures $\mathcal{F}(t) = (\ell(t), m(t), n(t), \alpha(t))$ and $\widetilde{\mathcal{F}}(u) = (\widetilde{\ell}(u), \widetilde{m}(u), \widetilde{n}(u), \widetilde{\alpha}(u))$ for convenience. We say that $(\boldsymbol{\gamma}, \boldsymbol{\nu}_1, \boldsymbol{\nu}_2)$ *and* $(\widetilde{\boldsymbol{\gamma}}, \widetilde{\boldsymbol{\nu}}_1, \widetilde{\boldsymbol{\nu}}_2)$ *have k-th order contact at $t = t_0$, $u = u_0$* if

$$\frac{d^i}{dt^i}(\boldsymbol{\gamma}, \boldsymbol{\nu}_1, \boldsymbol{\nu}_2)(t_0) = \frac{d^i}{du^i}(\widetilde{\boldsymbol{\gamma}}, \widetilde{\boldsymbol{\nu}}_1, \widetilde{\boldsymbol{\nu}}_2)(u_0),$$

$$\frac{d^k}{dt^k}(\boldsymbol{\gamma}, \boldsymbol{\nu}_1, \boldsymbol{\nu}_2)(t_0) \neq \frac{d^k}{du^k}(\widetilde{\boldsymbol{\gamma}}, \widetilde{\boldsymbol{\nu}_1}, \widetilde{\boldsymbol{\nu}}_2)(u_0)$$

for $i = 0, 1, ..., k-1$. Moreover, we say that $(\boldsymbol{\gamma}, \boldsymbol{\nu}_1, \boldsymbol{\nu}_2)$ *and* $(\widetilde{\boldsymbol{\gamma}}, \widetilde{\boldsymbol{\nu}_1}, \widetilde{\boldsymbol{\nu}_2})$ *have at least k-th order contact at $t = t_0$, $u = u_0$* if

$$\frac{d^i}{dt^i}(\boldsymbol{\gamma}, \boldsymbol{\nu}_1, \boldsymbol{\nu}_2)(t_0) = \frac{d^i}{du^i}(\widetilde{\boldsymbol{\gamma}}, \widetilde{\boldsymbol{\nu}_1}, \widetilde{\boldsymbol{\nu}_2})(u_0),$$

for $i = 0, 1, ..., k-1$.

In general, we may assume that $(\boldsymbol{\gamma}, \boldsymbol{\nu}_1, \boldsymbol{\nu}_2)$ and $(\widetilde{\boldsymbol{\gamma}}, \widetilde{\boldsymbol{\nu}}_1, \widetilde{\boldsymbol{\nu}}_2)$ have at least first order contact at any point $t = t_0$, $u = u_0$, up to congruence as framed curves.

Theorem 4.4. ([4]) *If $(\boldsymbol{\gamma}, \boldsymbol{\nu}_1, \boldsymbol{\nu}_2)$ and $(\widetilde{\boldsymbol{\gamma}}, \widetilde{\boldsymbol{\nu}}_1, \widetilde{\boldsymbol{\nu}}_2)$ have at least $(k+1)$-th order contact at $t = t_0$, $u = u_0$ then*

$$\frac{d^i}{dt^i}\mathcal{F}(t_0) = \frac{d^i}{du^i}\widetilde{\mathcal{F}}(u_0), \tag{2}$$

for $i = 0, 1, ..., k-1$. *Conversely, if the condition* (2) *hold, then* $(\boldsymbol{\gamma}, \boldsymbol{\nu}_1, \boldsymbol{\nu}_2)$ *and* $(\widetilde{\gamma}, \widetilde{\boldsymbol{\nu}}_1, \widetilde{\boldsymbol{\nu}}_2)$ *have at least* $(k+1)$*-th order contact at* $t = t_0$, $u = u_0$, *up to congruence as framed curves.*

By Theorem 4.4, we can show the following propositions:

Proposition 4.5. *If* $(\gamma, \mathcal{N}, \mathcal{B})(t_0)$ *and* $(\widetilde{\gamma}, \widetilde{\mathcal{N}}, \widetilde{\mathcal{B}})(u_0)$ *have at least* $(k+2)$*-th order contact as framed curves, then* $\delta^{(p)}(t_0) = \widetilde{\delta}^{(p)}(u_0)$ (for $0 \leq p \leq k-1$), *where*

$$\delta^{(p)}(t_0) = (d^p\delta/dt^p)(t_0) \text{ and } \widetilde{\delta}^{(p)}(u_0) = (d^p\widetilde{\delta}/du^p)(u_0).$$

Proposition 4.6. *Let* $\gamma : I \to \mathbb{R}^3$ *be a Frenet type framed base curve with* $\mathcal{T}(t)$. *Then there exists a framed curve* $(\widetilde{\gamma}, \widetilde{\mathcal{N}}, \widetilde{\mathcal{B}}) : I \to \mathbb{R}^3 \times \Delta$ *such that* $\widetilde{\gamma}(t)$ *is a framed helix, and* $(\gamma, \mathcal{N}, \mathcal{B})$ *and* $(\widetilde{\gamma}, \widetilde{\mathcal{N}}, \widetilde{\mathcal{B}})$ *have at least second order contact as framed curves at a point* $t_0 \in I$.

Proof. Choose any fixed value $t = t_0$ of the parameter. We consider a new curvature as a framed curve

$$(\widetilde{\tau}(t), -\widetilde{\kappa}(t), 0, \widetilde{\alpha}(t)) = ((\tau(t_0)/\kappa(t_0))\kappa(t), -\kappa(t), 0, \alpha(t)).$$

Since the existence and the uniqueness of framed curves, there exists a framed curve $(\widetilde{\gamma}, \widetilde{\mathcal{N}}, \widetilde{\mathcal{B}})$ with $(\widetilde{\tau}(t), -\widetilde{\kappa}(t), 0, \widetilde{\alpha}(t))$. Moreover, by Theorem 4.4 and an appropriate Euclid transformation, $(\gamma, \mathcal{N}, \mathcal{B})$ and $(\widetilde{\gamma}, \widetilde{\mathcal{N}}, \widetilde{\mathcal{B}})$ have at least second order contact as framed curves at $t_0 \in I$. On the other hand, by a direct calculation, we have $\widetilde{\delta}(t) = 0$ for all $t \in I$. Thus, $\widetilde{\gamma}(t)$ is a framed helix. □

§5. Support functions

For a Frenet type framed base curve $\gamma : I \to \mathbb{R}^3$, we define a function $G : I \times \mathbb{R}^3 \to \mathbb{R}$ by $G(t, \boldsymbol{x}) = \langle \boldsymbol{x} - \gamma(t), \mathcal{N}(t) \rangle$. We call G a *support function of* γ with respect to the unit principal normal vector $\mathcal{N}(t)$. We denote that $g_{\boldsymbol{x}_0}(t) = G(t, \boldsymbol{x}_0)$ for any $\boldsymbol{x}_0 \in \mathbb{R}^3$. Then we have the following proposition.

Proposition 5.1. *For a support function* $g_{\boldsymbol{x}_0}(t) = \langle \boldsymbol{x}_0 - \gamma(t), \mathcal{N}(t) \rangle$, *we have the following*:

(1) $g_{\boldsymbol{x}_0}(t_0) = 0$ *if and only if there exist* u, $v \in \mathbb{R}$ *such that* $\boldsymbol{x}_0 - \gamma(t_0) = u\mathcal{T}(t_0) + v\mathcal{B}(t_0)$.

(2) $g_{\boldsymbol{x}_0}(t_0) = \dot{g}_{\boldsymbol{x}_0}(t_0) = 0$ *if and only if there exists* $u \in \mathbb{R}$ *such that*

$$\boldsymbol{x}_0 - \gamma(t_0) = u\frac{\tau(t_0)\mathcal{T}(t_0) + \kappa(t_0)\mathcal{B}(t_0)}{\sqrt{\kappa^2(t_0) + \tau^2(t_0)}}.$$

(A) *Suppose that* $\delta(t_0)\neq 0$. *Then we have the following:*
(3) $g_{\boldsymbol{x}_0}(t_0)=\dot{g}_{\boldsymbol{x}_0}(t_0)=\ddot{g}_{\boldsymbol{x}_0}(t_0)=0$ *if and only if*

$$(*)\ \boldsymbol{x}_0-\boldsymbol{\gamma}(t_0)=-\frac{\alpha(t_0)\kappa(t_0)}{\delta(t_0)\sqrt{\kappa^2(t_0)+\tau^2(t_0)}}\frac{\tau(t_0)\mathcal{T}(t_0)+\kappa(t_0)\mathcal{B}(t_0)}{\sqrt{\kappa^2(t_0)+\tau^2(t_0)}}.$$

(4) $g_{\boldsymbol{x}_0}(t_0)=\dot{g}_{\boldsymbol{x}_0}(t_0)=\ddot{g}_{\boldsymbol{x}_0}(t_0)=g^{(3)}_{\boldsymbol{x}_0}(t_0)=0$ *if and only if* $\sigma(t_0)=0$ *and* $(*)$.
(5) $g_{\boldsymbol{x}_0}(t_0)=\dot{g}_{\boldsymbol{x}_0}(t_0)=\ddot{g}_{\boldsymbol{x}_0}(t_0)=g^{(3)}_{\boldsymbol{x}_0}(t_0)=g^{(4)}_{\boldsymbol{x}_0}(t_0)=0$ *if and only if* $\sigma(t_0)=0$, $\dot{\sigma}(t_0)=0$ *and* $(*)$.

(B) *Suppose that* $\delta(t_0)=0$. *Then we have the following:*
(6) $g_{\boldsymbol{x}_0}(t_0)=\dot{g}_{\boldsymbol{x}_0}(t_0)=\ddot{g}_{\boldsymbol{x}_0}(t_0)=0$ *if and only if* $\alpha(t_0)=0$ *and there exists* $u\in\mathbb{R}$ *such that*

$$\boldsymbol{x}_0-\boldsymbol{\gamma}(t_0)=u\frac{\tau(t_0)\mathcal{T}(t_0)+\kappa(t_0)\mathcal{B}(t_0)}{\sqrt{\kappa^2(t_0)+\tau^2(t_0)}}.$$

(7) $g_{\boldsymbol{x}_0}(t_0)=\dot{g}_{\boldsymbol{x}_0}(t_0)=\ddot{g}_{\boldsymbol{x}_0}(t_0)=g^{(3)}_{\boldsymbol{x}_0}(t_0)=0$ *if and only if one of the following conditions holds:*
(a) $\dot{\delta}(t_0)\neq 0$, $\alpha(t_0)=0$ *and*

$$\boldsymbol{x}_0-\boldsymbol{\gamma}(t_0)=-\dot{\alpha}(t_0)\kappa(t_0)\frac{\tau(t_0)\mathcal{T}(t_0)+\kappa(t_0)\mathcal{B}(t_0)}{\kappa(t_0)\ddot{\tau}(t_0)-\ddot{\kappa}(t_0)\tau(t_0)}.$$

(b) $\dot{\delta}(t_0)=0$, $\alpha(t_0)=\dot{\alpha}(t_0)=0$ *and there exists* $u\in\mathbb{R}$ *such that*

$$\boldsymbol{x}_0-\boldsymbol{\gamma}(t_0)=u\frac{\tau(t_0)\mathcal{T}(t_0)+\kappa(t_0)\mathcal{B}(t_0)}{\sqrt{\kappa^2(t_0)+\tau^2(t_0)}}.$$

Proof. Since $g_{\boldsymbol{x}_0}(t)=\langle \boldsymbol{x}_0-\boldsymbol{\gamma}(t),\mathcal{N}(t)\rangle$, we have the following calculations:

$$(\alpha)\ g_{\boldsymbol{x}_0}=\langle \boldsymbol{x}_0-\boldsymbol{\gamma},\mathcal{N}\rangle,$$
$$(\beta)\ \dot{g}_{\boldsymbol{x}_0}=\langle \boldsymbol{x}_0-\boldsymbol{\gamma},-\kappa\mathcal{T}+\tau\mathcal{B}\rangle,$$
$$(\gamma)\ \ddot{g}_{\boldsymbol{x}_0}=\alpha\kappa+\langle \boldsymbol{x}_0-\boldsymbol{\gamma},-\dot{\kappa}\mathcal{T}-(\kappa^2+\tau^2)\mathcal{N}+\dot{\tau}\mathcal{B}\rangle,$$
$$\begin{aligned}(\delta)\ g^{(3)}_{\boldsymbol{x}_0}&=2\alpha\dot{\kappa}+\dot{\alpha}\kappa+\langle \boldsymbol{x}_0-\boldsymbol{\gamma},(\kappa(\kappa^2+\tau^2)-\ddot{\kappa})\mathcal{T}\\&\quad-3(\kappa\dot{\kappa}+\tau\dot{\tau})\mathcal{N}+(-\tau(\kappa^2+\tau^2)+\ddot{\tau})\mathcal{B}\rangle,\end{aligned}$$
$$\begin{aligned}(\epsilon)\ g^{(4)}_{\boldsymbol{x}_0}&=\ddot{\alpha}\kappa+3\dot{\alpha}\dot{\kappa}+3\alpha\ddot{\kappa}-\alpha\kappa(\kappa^2+\tau^2)\\&\quad+\langle \boldsymbol{x}_0-\boldsymbol{\gamma},(\dot{\kappa}(6\kappa^2+\tau^2)+5\kappa\tau\dot{\tau}-\dddot{\kappa})\mathcal{T}\\&\quad+((\kappa^2+\tau^2)^2-4(\kappa\ddot{\kappa}+\tau\ddot{\tau})-3(\dot{\kappa}^2+\dot{\tau}^2))\mathcal{N}\\&\quad+(-\dot{\tau}(\kappa^2+6\tau^2)-5\kappa\dot{\kappa}\tau+\dddot{\tau})\mathcal{B}\rangle.\end{aligned}$$

By definition and (α), (1) follows.

By (β), $g_{\boldsymbol{x}_0}(t_0) = \dot{g_{\boldsymbol{x}_0}}(t_0) = 0$ if and only if there exist $u, v \in \mathbb{R}$ such that $\boldsymbol{x}_0 - \boldsymbol{\gamma}(t_0) = u\mathcal{T}(t_0) + v\mathcal{B}(t_0)$ and $-\kappa(t_0)u + \tau(t_0)v = 0$. Since $\kappa(t_0) > 0$, we have

$$u = v\frac{\tau(t_0)}{\kappa(t_0)},$$

so that there exists $w \in \mathbb{R}$ such that

$$\boldsymbol{x}_0 - \boldsymbol{\gamma}(t_0) = w\frac{\tau(t_0)\mathcal{T}(t_0) + \kappa(t_0)\mathcal{B}(t_0)}{\sqrt{\kappa^2(t_0) + \tau^2(t_0)}}.$$

Therefore (2) holds.

By (γ), $g_{\boldsymbol{x}_0}(t_0) = \dot{g_{\boldsymbol{x}_0}}(t_0) = \ddot{g_{\boldsymbol{x}_0}}(t_0) = 0$ if and only if there exists $u \in \mathbb{R}$ such that

$$\boldsymbol{x}_0 - \boldsymbol{\gamma}(t_0) = u\frac{\tau(t_0)\mathcal{T}(t_0) + \kappa(t_0)\mathcal{B}(t_0)}{\sqrt{\kappa^2(t_0) + \tau^2(t_0)}}$$

and

$$\alpha(t_0)\kappa(t_0) + u\frac{\kappa(t_0)\dot{\tau}(t_0) - \dot{\kappa}(t_0)\tau(t_0)}{\sqrt{\kappa^2(t_0) + \tau^2(t_0)}} = 0.$$

It follows that

$$\frac{\alpha(t_0)\kappa(t_0)}{\sqrt{\kappa^2(t_0) + \tau^2(t_0)}} + u\frac{\kappa(t_0)\dot{\tau}(t_0) - \dot{\kappa}(t_0)\tau(t_0)}{\kappa^2(t_0) + \tau^2(t_0)} = 0.$$

Thus,

$$\delta(t_0) = \frac{\kappa(t_0)\dot{\tau}(t_0) - \dot{\kappa}(t_0)\tau(t_0)}{\kappa^2(t_0) + \tau^2(t_0)} \neq 0 \text{ and } u = -\frac{\alpha(t_0)\kappa(t_0)}{\delta(t_0)\sqrt{\kappa^2(t_0) + \tau^2(t_0)}},$$

or $\delta(t_0) = 0$ and $\alpha(t_0) = 0$. This completes the proof of (A)-(3), and (B)-(6).

Suppose that $\delta(t_0) \neq 0$. By (δ),

$$g_{\boldsymbol{x}_0}(t_0) = \dot{g_{\boldsymbol{x}_0}}(t_0) = \ddot{g_{\boldsymbol{x}_0}}(t_0) = g^{(3)}_{\boldsymbol{x}_0}(t_0) = 0$$

if and only if

$$\boldsymbol{x}_0 - \boldsymbol{\gamma}(t_0) = -\frac{\alpha(t_0)\kappa(t_0)}{\delta(t_0)\sqrt{\kappa^2(t_0) + \tau^2(t_0)}}\frac{\tau(t_0)\mathcal{T}(t_0) + \kappa(t_0)\mathcal{B}(t_0)}{\sqrt{\kappa^2(t_0) + \tau^2(t_0)}}$$

and

$$2\alpha(t_0)\dot{\kappa}(t_0) + \dot{\alpha}(t_0)\kappa(t_0) - \frac{\alpha(t_0)\kappa(t_0)}{\delta(t_0)}\left(\frac{\kappa(t_0)\ddot{\tau}(t_0) - \ddot{\kappa}(t_0)\tau(t_0)}{\kappa^2(t_0) + \tau^2(t_0)}\right) = 0.$$

We can rewrite $\sigma(t)$:

$$\sigma = -\sqrt{\kappa^2+\tau^2}\left(2\alpha\dot{\kappa}+\dot{\alpha}\kappa-\frac{\alpha\kappa}{\delta}\left(\frac{\kappa\ddot{\tau}-\ddot{\kappa}\tau}{\kappa^2+\tau^2}\right)\right).$$

Therefore, (A)-(3) holds. By the similar arguments to the above, we have (A)-(5).

Suppose that $\delta(t_0)=0$. Then by (δ), $g_{\boldsymbol{x}_0}(t_0)=\dot{g}_{\boldsymbol{x}_0}(t_0)=\ddot{g}_{\boldsymbol{x}_0}(t_0)=g^{(3)}_{\boldsymbol{x}_0}(t_0)=0$ if and only if $\alpha(t_0)=0$, there exists $u\in\mathbb{R}$ such that

$$\boldsymbol{x}_0-\gamma(t_0)=u\frac{\tau(t_0)\mathcal{T}(t_0)+\kappa(t_0)\mathcal{B}(t_0)}{\sqrt{\kappa^2(t_0)+\tau^2(t_0)}}$$

and

$$\dot{\alpha}(t_0)\kappa(t_0)\sqrt{\kappa^2(t_0)+\tau^2(t_0)}+u\left(\kappa(t_0)\ddot{\tau}(t_0)-\ddot{\kappa}(t_0)\tau(t_0)\right)=0.$$

It follows that

$$\kappa(t_0)\ddot{\tau}(t_0)-\ddot{\kappa}(t_0)\tau(t_0)\neq 0 \text{ and } u=-\dot{\alpha}(t_0)\kappa(t_0)\frac{\sqrt{\kappa^2(t_0)+\tau^2(t_0)}}{\kappa(t_0)\ddot{\tau}(t_0)-\ddot{\kappa}(t_0)\tau(t_0)},$$

or

$$\kappa(t_0)\ddot{\tau}(t_0)-\ddot{\kappa}(t_0)\tau(t_0)=0 \text{ and } \dot{\alpha}(t_0)=0.$$

Therefore, we have (B)-(7)-(a) and (B)-(7)-(b). This completes the proof. □

In order to prove Theorem 3.3, we use some general results on the singularity theory for families of function germs. Detailed descriptions are found in the book [3]. Let $\mathbb{R}^r$ be the r-dimensional Euclidean space with coordinates $(x_1,x_2,\dots,x_r)$ and $F:(\mathbb{R}\times\mathbb{R}^r,(t_0,\boldsymbol{x_0}))\to\mathbb{R}$ be a function germ. We call F an *r-parameter unfolding* of f, where $f(t)=F(t,\boldsymbol{x}_0)$. We say that f has the *A_k-singularity* at t_0 if $f^{(p)}(t_0)=0$ for all $1\le p\le k$, and $f^{(k+1)}(t_0)\neq 0$. Let F be an unfolding of f and f has the A_k-singularity ($k\ge 1$) at t_0. We write the $(k-1)$-jet of the partial derivative $\frac{\partial F}{\partial x_i}$ at t_0 by $j^{(k-1)}(\frac{\partial F}{\partial x_i}(t,\boldsymbol{x}_0))(t_0)=\sum_{j=0}^{k-1}\alpha_{ji}(t-t_0)^j$ for $i=1,\dots,r$. Then F is called an $\mathcal{R}$-versal unfolding if the $k\times r$ matrix of coefficients $(\alpha_{ji})_{j=0,\dots,k-1;i=1,\dots,r}$ has rank k ($k\le r$). We introduce an important set concerning the unfoldings relative to the above notions. The *discriminant set* of F is defined to be

$$D_F=\left\{\boldsymbol{x}\in\mathbb{R}^r \mid \text{there exists } s \text{ such that } F=\frac{\partial F}{\partial t}=0 \text{ at } (s,\boldsymbol{x})\right\}.$$

Then we have the following classification (cf. [3]).

Theorem 5.2. *Let $F : (\mathbb{R} \times \mathbb{R}^r, (t, \boldsymbol{x}_0)) \to \mathbb{R}$ be an r-parameter unfolding of f which has the A_k-singularity at t_0. Suppose that F is an $\mathcal{R}$-versal unfolding.*

(1) *If $k = 2$, then D_F is locally diffeomorphic to the cuspidal edge* ***ce****.*
(2) *If $k = 3$, then D_F is locally diffeomorphic to the swallowtail* ***sw****.*

For the proof of Theorem 3.3, we have the following proposition.

Proposition 5.3. *Let $\gamma : I \to \mathbb{R}^3$ be a Frenet type framed base curve with $\mathcal{T}(t)$. If $g_{\boldsymbol{x}_0}$ has the A_k-singularity $(k = 2, 3)$ at t_0, then G is an $\mathcal{R}$-versal unfolding of $g_{\boldsymbol{x}_0}$. Here, we assume that $\delta(t_0) \neq 0$ for $k = 3$.*

Proof. We write that $\boldsymbol{x} = (x_1, x_2, x_3)$, $\gamma(t) = (\gamma_1(t), \gamma_2(t), \gamma_3(t))$ and $\mathcal{N}(t) = (n_1(t), n_2(t), n_3(t))$. Then we have

$$G(t, \boldsymbol{x}) = n_1(t)(x_1 - \gamma_1(t)) + n_2(t)(x_2 - \gamma_2(t)) + n_3(t)(x_3 - \gamma_3(t)),$$

so that

$$\frac{\partial G}{\partial x_i}(t, \boldsymbol{x}) = n_i(t), \ (i = 1, 2, 3).$$

Therefore the 2-jet is

$$j^2 \frac{\partial G}{\partial x_i}(t_0, \boldsymbol{x}_0) = n_i(t_0) + \dot{n}_i(t_0)(t - t_0) + \frac{1}{2}\ddot{n}_i(t_0)(t - t_0)^2.$$

We consider the following matrix:

$$A = \begin{pmatrix} n_1(t_0) & n_2(t_0) & n_3(t_0) \\ \dot{n}_1(t_0) & \dot{n}_2(t_0) & \dot{n}_3(t_0) \\ \ddot{n}_1(t_0) & \ddot{n}_2(t_0) & \ddot{n}_3(t_0) \end{pmatrix} = \begin{pmatrix} \mathcal{N}(t_0) \\ \dot{\mathcal{N}}(t_0) \\ \ddot{\mathcal{N}}(t_0) \end{pmatrix}.$$

By the Frenet-Serret type formula, we have

$$\begin{aligned} \dot{\mathcal{N}}(t_0) &= -\kappa(t_0)\mathcal{T}(t_0) + \tau(t_0)\mathcal{B}(t_0), \\ \ddot{\mathcal{N}}(t_0) &= -\dot{\kappa}(t_0)\mathcal{T}(t_0) - (\kappa^2(t_0) + \tau^2(t_0))\mathcal{N}(t_0) + \dot{\tau}(t_0)\mathcal{B}(t_0). \end{aligned}$$

Since $\{\mathcal{T}(t_0), \mathcal{N}(t_0), \mathcal{B}(t_0)\}$ is an orthonormal basis of $\mathbb{R}^3$, the rank of

$$A = \begin{pmatrix} \mathcal{N}(t_0) \\ -\kappa(t_0)\mathcal{T}(t_0) + \tau(t_0)\mathcal{B}(t_0) \\ -\dot{\kappa}(t_0)\mathcal{T}(t_0) - (\kappa^2(t_0) + \tau^2(t_0))\mathcal{N}(t_0) + \dot{\tau}(t_0)\mathcal{B}(t_0) \end{pmatrix}$$

is equal to the rank of

$$\begin{pmatrix} 0 & 1 & 0 \\ -\kappa(t_0) & 0 & \tau(t_0) \\ -\dot{\kappa}(t_0) & -(\kappa^2(t_0) + \tau^2(t_0)) & \dot{\tau}(t_0) \end{pmatrix}.$$

Therefore, rank $A = 3$ if and only if

$$0 \neq \kappa(t_0)\dot{\tau}(t_0) - \dot{\kappa}(t_0)\tau(t_0).$$

The last condition is equivalent to the condition $\delta(t_0) \neq 0$. Moreover, the rank of

$$\begin{pmatrix} \mathcal{N}(t_0) \\ \dot{\mathcal{N}}(t_0) \end{pmatrix} = \begin{pmatrix} \mathcal{N}(t_0) \\ -\kappa(t_0)\mathcal{T}(t_0) + \tau(t_0)\mathcal{B}(t_0) \end{pmatrix}$$

is always two.

If $g_{\boldsymbol{x}_0}$ has the A_k-singularity ($k = 2, 3$) at t_0, then G is the $\mathcal{R}$-versal unfolding of $g_{\boldsymbol{x}_0}$. This completes the proof. □

Proof of Theorem 3.3. By a straightforward calculation, we have

$$\frac{\partial \mathcal{RD}_\gamma}{\partial t}(t,u) \times \frac{\partial \mathcal{RD}_\gamma}{\partial u}(t,u) = -\left(\frac{\alpha(t)\kappa(t)}{\sqrt{\kappa^2(t) + \tau^2(t)}} + u\delta(t) \right) \mathcal{N}(t).$$

Therefore, (t_0, u_0) is a regular point of $\mathcal{RD}_\gamma$ if and only if

$$\frac{\alpha(t_0)\kappa(t_0)}{\sqrt{\kappa^2(t_0) + \tau^2(t_0)}} + u_0\delta(t_0) \neq 0.$$

This completes the proof of (1).

By Proposition 5.1-(2), the discriminant set $\mathcal{D}_G$ of the support function G of γ with respect to $\mathcal{N}(t)$ is the image of the rectifying developable surface of γ.

Suppose that $\delta(t_0) \neq 0$. It follows from Proposition 5.1-(A)-(3), (4) and (5) that $g_{\boldsymbol{x}_0}$ has the A_2-type singularity (respectively, the A_3-type singularity) at $t = t_0$ if and only if

$$(^{**})\ u_0 = -\frac{\alpha(t_0)\kappa(t_0)}{\delta(t_0)\sqrt{\kappa^2(t_0) + \tau^2(t_0)}}$$

and $\sigma(t_0) \neq 0$ (respectively, $(^{**})$, $\sigma(t_0) = 0$ and $\dot{\sigma}(t_0) \neq 0$). By Theorem 5.2 and Proposition 5.3, we have (2)-(i) and (3).

Suppose that $\delta(t_0) = 0$. It follows from Proposition 5.1 (B)-(6) and (7) that $g_{\boldsymbol{x}_0}$ has the A_3-type singularity if and only if $\alpha(t_0) = 0$, $\dot{\alpha}(t_0) \neq 0$ and $\dot{\delta}(t_0) = 0$, or $\alpha(t_0) = 0$, $\dot{\delta}(t_0) \neq 0$ and

$$\dot{\alpha}(t_0)\kappa(t_0)\sqrt{\kappa^2(t_0) + \tau^2(t_0)} + u_0(\kappa(t_0)\ddot{\tau}(t_0) - \ddot{\kappa}(t_0)\tau(t_0)) \neq 0.$$

By Theorem 5.2 and Proposition 5.3, we have the assertion (2)-(β). This completes the proof. □

§6. Examples

We give examples to understand the phenomena for rectifying developable surfaces of framed base curves and framed helices.

Example 6.1 (The astroid). The *astroid* $\gamma : [0, 2\pi) \to \mathbb{R}^3$ is defined by $\gamma(t) = (\cos^3 t, \sin^3 t, \cos 2t)$. See Fig.1. Then

$$\mathcal{T}(t) = \frac{1}{5}(-3\cos t, 3\sin t, -4)$$

gives the unit tangent vector and $\alpha(t) = 5\cos t \sin t$ is a speed function. By a direct calculation, we have

$$\mathcal{N}(t) = (\sin t, \cos t, 0)\,,\ \mathcal{B}(t) = \frac{1}{5}(4\cos t, -4\sin t, -3)\,,$$

$\kappa(t) = 3/5$ and $\tau(t) = 4/5$. Since $\delta(t) \equiv 0$ and Corollary 4.3, γ is a framed helix. For the astroid γ, the rectifying developable surface is given by $\mathcal{RD}_\gamma(t, u) = \left(\cos^3 t, \sin^3 t, -u + \cos 2t\right)$. By Theorem 3.3 (2)-(iii), we have the cuspidal edge singularities at $t = 0, \pi/2, \pi, 3\pi/2$ (Fig.2).

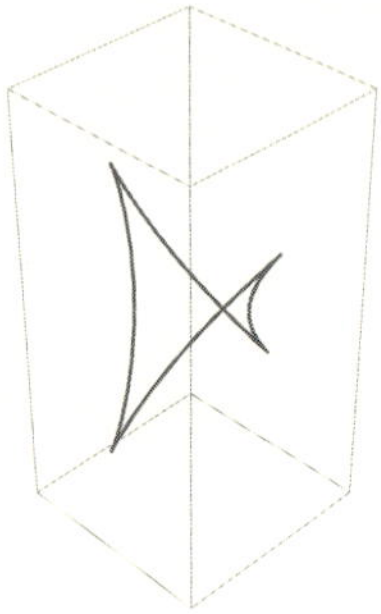

Fig. 1. γ of Example 6.1

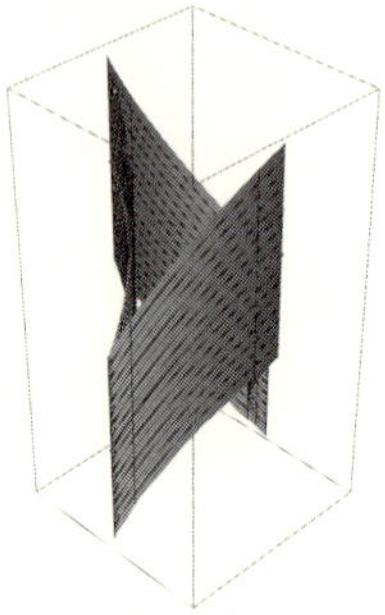

Fig. 2. γ and $\mathcal{RD}_\gamma$ of Example 6.1

Example 6.2 (The spherical nephroid (cf. [13])). The *spherical nephroid* $\gamma : [0, 2\pi) \to S^2 \subset \mathbb{R}^3$ is defined by

$$\gamma(t) = \left(\frac{3}{4}\cos t - \frac{1}{4}\cos 3t, \frac{3}{4}\sin t - \frac{1}{4}\sin 3t, \frac{\sqrt{3}}{2}\cos t\right).$$

See Fig.3. Then

$$\mathcal{T}(t) = \frac{1}{2}\left(\sqrt{3}\cos 2t, \sqrt{3}\sin 2t, -1\right)$$

gives the unit tangent vector and $\alpha(t) = \sqrt{3}\sin(t)$ is a speed function. By a direct calculation, we have

$$\mathcal{N}(t) = (-\sin 2t, \cos 2t, 0), \ \mathcal{B}(t) = \frac{1}{2}\left(\cos 2t, \sin 2t, \sqrt{3}\right),$$

$\kappa(t) = \sqrt{3}$ and $\tau(t) = -1$. Since $\delta(t) \equiv 0$ and Corollary 4.3, γ is a framed helix. For the spherical nephroid, the rectifying developable surface is given by

$$\mathcal{RD}_\gamma(t,u) = \left(\frac{3}{4}\cos t - \frac{1}{4}\cos 3t, \frac{3}{4}\sin t - \frac{1}{4}\sin 3t, u + \frac{\sqrt{3}}{2}\cos t\right).$$

By Theorem 3.3 (2)-(iii), we have the cuspidal edge singularities at $t = 0, \pi$ (Fig.4).

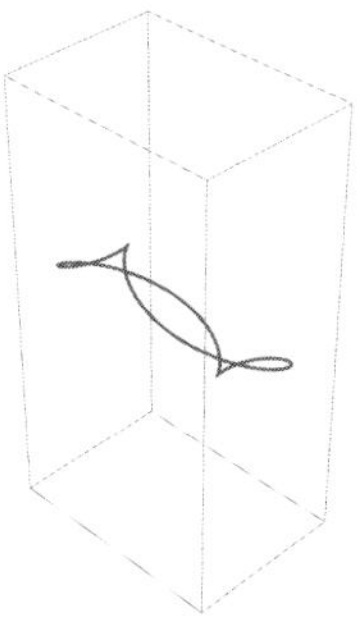

Fig. 3. γ of Example 6.2

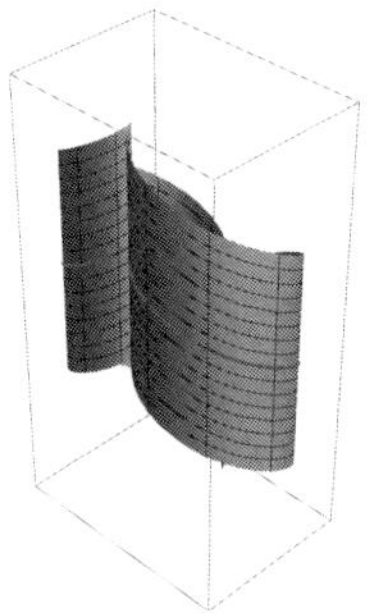

Fig. 4. γ and $\mathcal{RD}_\gamma$ of Example 6.2

Example 6.3 ((2, 3, 5)-type)**.** Let $\gamma : \mathbb{R} \to \mathbb{R}^3$ be

$$\gamma(t) = \left(\frac{1}{2}t^2, \frac{1}{3}t^3, \frac{1}{5}t^5\right).$$

See Fig.5. We say that γ is of type (2, 3, 5). Then

$$\mathcal{T}(t) = \frac{1}{\sqrt{1+t^2+t^6}}\left(1, t, t^3\right)$$

gives the unit tangent vector and $\alpha(t) = t\sqrt{1+t^2+t^6}$ is a speed function. By a direct calculation, we have

$$\kappa(t) = \frac{\sqrt{1+9t^4+4t^6}}{1+t^2+t^6}, \ \tau(t) = \frac{6t\sqrt{1+t^2+t^6}}{1+9t^4+4t^6}.$$

Since $\delta(0) = 6$, $\sigma(0) = 1/6$ and $\alpha(0) = 0$ and Theorem 3.3 (2)-(i), the rectifying developable surface $\mathcal{RD}(t,u)$ is locally diffeomorphic to the cuspidal edge $\boldsymbol{ce}$ at $(0,0)$ (Fig.6).

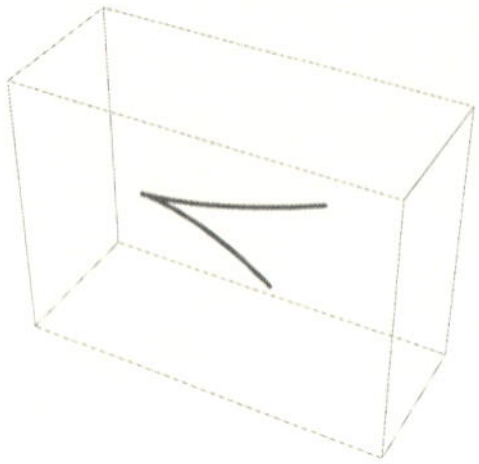

Fig. 5. γ of Example 6.3

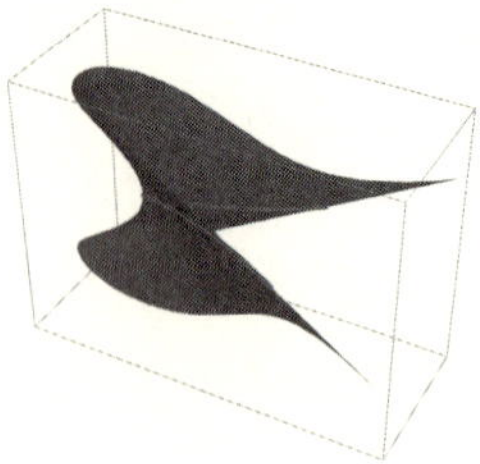

Fig. 6. γ and $\mathcal{RD}_\gamma$ of Example 6.3

§Appendix A. Framed base curves in the Euclidean space

We define a set

$$\begin{aligned}\Delta &= \{(\boldsymbol{\nu}_1,\boldsymbol{\nu}_2) \in \mathbb{R}^3 \times \mathbb{R}^3 \mid \langle \boldsymbol{\nu}_1,\boldsymbol{\nu}_1\rangle = \langle \boldsymbol{\nu}_2,\boldsymbol{\nu}_2\rangle = 1, \langle \boldsymbol{\nu}_1,\boldsymbol{\nu}_2\rangle = 0\}\\ &= \{(\boldsymbol{\nu}_1,\boldsymbol{\nu}_2) \in S^2 \times S^2 \mid \langle \boldsymbol{\nu}_1,\boldsymbol{\nu}_2\rangle = 0\}.\end{aligned}$$

Definition A.1. We say that $(\gamma,\boldsymbol{\nu}_1,\boldsymbol{\nu}_2) : I \to \mathbb{R}^3\times\Delta \subset \mathbb{R}^3\times S^2\times S^2$ is a *framed curve* if $\langle \dot{\gamma}(t),\boldsymbol{\nu}_1(t)\rangle = 0$ and $\langle \dot{\gamma}(t),\boldsymbol{\nu}_2(t)\rangle = 0$ for all $t \in I$. We also say that $\gamma : I \to \mathbb{R}^3$ is a *framed base curve* if there exists $(\boldsymbol{\nu}_1,\boldsymbol{\nu}_2) : I \to \Delta$ such that $(\gamma,\boldsymbol{\nu}_1,\boldsymbol{\nu}_2)$ is a framed curve.

Then we have the Frenet-Serret type formula of the framed curve γ. We define $\boldsymbol{\mu}(t) = \boldsymbol{\nu}_1(t) \times \boldsymbol{\nu}_2(t)$ and call $\{\boldsymbol{\nu}_1(t),\boldsymbol{\nu}_2(t),\boldsymbol{\mu}(t)\}$ a *moving frame along the framed base curve* $\gamma(t)$. By standard arguments, we have the Frenet-Serret type formulae as follows:

Proposition A.2. ([4]) *Let $(\gamma,\boldsymbol{\nu}_1,\boldsymbol{\nu}_2) : I \to \mathbb{R}^3 \times \Delta$ be a framed curve. Then we have*

$$\begin{pmatrix} \dot{\boldsymbol{\nu}}_1(t) \\ \dot{\boldsymbol{\nu}}_2(t) \\ \dot{\boldsymbol{\mu}}(t) \end{pmatrix} = \begin{pmatrix} 0 & \ell(t) & m(t) \\ -\ell(t) & 0 & n(t) \\ -m(t) & -n(t) & 0 \end{pmatrix} \begin{pmatrix} \boldsymbol{\nu}_1(t) \\ \boldsymbol{\nu}_2(t) \\ \boldsymbol{\mu}(t) \end{pmatrix},$$

where $\ell(t) = \langle \dot{\boldsymbol{\nu}}_1(t),\boldsymbol{\nu}_2(t)\rangle$, $m(t) = \langle \dot{\boldsymbol{\nu}}_1(t),\boldsymbol{\mu}(t)\rangle$ and $n(t) = \langle \dot{\boldsymbol{\nu}}_2(t),\boldsymbol{\mu}(t)\rangle$. Moreover, there exists a smooth function $\alpha(t)$ such that $\dot{\gamma}(t) = \alpha(t)\boldsymbol{\mu}(t)$.

The quadruplet (ℓ,m,n,α) is an important invariant of a framed curve. We call (ℓ,m,n,α) the *curvature* of the framed curve. Note that t_0 is a singular point of γ if and only if $\alpha(t_0) = 0$.

Definition A.3. Let $(\gamma, \boldsymbol{\nu}_1, \boldsymbol{\nu}_2)$ and $(\widetilde{\gamma}, \widetilde{\boldsymbol{\nu}_1}, \widetilde{\boldsymbol{\nu}_2}) : I \to \mathbb{R}^3 \times \Delta$ be framed curves. We say that $(\gamma, \boldsymbol{\nu}_1, \boldsymbol{\nu}_2)$ and $(\widetilde{\gamma}, \widetilde{\boldsymbol{\nu}_1}, \widetilde{\boldsymbol{\nu}_2})$ are *congruent* as framed curves if there exists a rotation $A \in SO(3)$ and a translation $\boldsymbol{a} \in \mathbb{R}^3$ such that $\widetilde{\gamma}(t) = A(\gamma(t)) + \boldsymbol{a}$, $\widetilde{\boldsymbol{\nu}_1}(t) = A(\boldsymbol{\nu}_1(t))$ and $\widetilde{\boldsymbol{\nu}_2}(t) = A(\boldsymbol{\nu}_2(t))$ for all $t \in I$.

We have shown the existence and the uniqueness for framed curves similarly to the case of regular space curves in [4].

Theorem A.4 (The Existence Theorem, [4]). *Let $(\ell, m, n, \alpha) : I \to \mathbb{R}^4$ be a smooth mapping. There exists a framed curve $(\gamma, \boldsymbol{\nu}_1, \boldsymbol{\nu}_2) : I \to \mathbb{R}^3 \times \Delta$ whose curvature is (ℓ, m, n, α).*

Theorem A.5 (The Uniqueness Theorem, [4]). *Let $(\gamma, \boldsymbol{\nu}_1, \boldsymbol{\nu}_2)$ and $(\widetilde{\gamma}, \widetilde{\boldsymbol{\nu}_1}, \widetilde{\boldsymbol{\nu}_2}) : I \to \mathbb{R}^3 \times \Delta$ be framed curves with curvatures (ℓ, m, n, α) and $(\widetilde{\ell}, \widetilde{m}, \widetilde{n}, \widetilde{\alpha})$. If $(\ell, m, n, \alpha) = (\widetilde{\ell}, \widetilde{m}, \widetilde{n}, \widetilde{\alpha})$, then $(\gamma, \boldsymbol{\nu}_1, \boldsymbol{\nu}_2)$ and $(\widetilde{\gamma}, \widetilde{\boldsymbol{\nu}_1}, \widetilde{\boldsymbol{\nu}_2})$ are congruent as framed curves.*

References

[1] V. I. Arnol'd, *Singularities of Caustics and Wave fronts*, *Mathematics and its Applications* **62**, Kluwer Academic Publishers, 1990.

[2] V. I. Arnol'd, S. M. Gusein-Zade and A. N. Varchenko, *Singularities of Differentiable Maps Vol. I*, Birkhäuser, 1986.

[3] J. W. Bruce and P. J. Giblin, *Curves and singularities. A geometrical introduction to singularity theory. Second edition*, Cambridge University Press, Cambridge, 1992.

[4] S. Honda and M. Takahashi, Framed curves in the Euclidean space, *Adv. Geom.* **16** (2016), 265–276.

[5] S. Honda and M. Takahashi, Evolutes and focal surfaces of framed immersions in the Euclidean space, to appear in *Proc. Roy. Soc.* Edinburgh Sect. A.

[6] G. Ishikawa, Determinacy of envelope of the osculating hyperplanes to a curve, *Bull. London Math. Soc.* **25** (1993), 603–610.

[7] G. Ishikawa, Developable of a curve and its determinacy relative to the osculation-type, *Quart. J. Math. Oxford Ser.* (2) **46** (1995), 437–451.

[8] G. Ishikawa, Topological classification of the tangent developable of space curves, *J. London Math. Soc.* **62** (2000), 583–598.

[9] S. Izumiya, H. Katsumi and T. Yamasaki, The rectifying developable and the spherical Darboux image of a space curve, *Banach Center Publications* **50** (1999), 137–149.

[10] S. Izumiya and N. Takeuchi, Geometry of ruled surfaces, *Applicable Math. in the Golden Age* (2003), 305–338.

[11] J. Koenderink, *Solid Shape*, MIT Press, 1990.

[12] I. R. Porteous, Some remarks on duality in S^3, *Banach Center Publications* **50** (1999), 217–226.

[13] M. Takahashi, Legendre curves in the unit spherical bundle over the unit sphere and evolutes, *Contemp. Math.* **675** (2016), 337–355.

Department of Mathematics, Hokkaido University,
Sapporo 060-0810, Japan.
E-mail address: s-honda@math.sci.hokudai.ac.jp

Advanced Studies in Pure Mathematics 78, 2018
Singularities in Generic Geometry
pp. 293–311

On keen Heegaard splittings

Ayako Ido, Yeonhee Jang and Tsuyoshi Kobayashi

Abstract.

In this paper, we introduce a new concept of *strongly keen* for Heegaard splittings, and show that, for any integers $n \geq 2$ and $g \geq 3$, there exists a strongly keen Heegaard splitting of genus g whose Hempel distance is n.

§1. Introduction

The *curve complex* $\mathcal{C}(S)$ of a compact surface S introduced by Harvey[4] has been used to prove many deep results in 3-dimentional topology. In particular, Hempel [5] defined the *Hempel distance* for a Heegaard splitting $V_1 \cup_S V_2$ by $d(S) = d_S(\mathcal{D}(V_1), \mathcal{D}(V_2)) = \min\{d_S(x, y) \mid x \in \mathcal{D}(V_1), y \in \mathcal{D}(V_2)\}$, where d_S is the simplicial distance of $\mathcal{C}(S)$ (for the definition, see Section 2), and $\mathcal{D}(V_i)$ is the disk complex of the handlebody V_i $(i = 1, 2)$. There have been many works on Hempel distance. For example, some authors showed that the existence of high distance Heegaard splittings (see [1, 3, 5], for example). Moreover, it is also shown that there exist Heegaard splittings of Hempel distance exactly n for various integers n (see [2, 6, 7, 11, 12], for example). Here we note that the pair (x, y) in the above definition that realizes $d(S)$ may not be unique. Hence it may be natural to settle: we say that a Heegaard splitting $V_1 \cup_S V_2$ is *keen* if its Hempel distance is realized by a unique pair of elements of $\mathcal{D}(V_1)$ and $\mathcal{D}(V_2)$. Namely, $V_1 \cup_S V_2$ is keen if it satisfies the following.

- If $d_S(a, b) = d_S(a', b') = d_S(\mathcal{D}(V_1), \mathcal{D}(V_2))$ for $a, a' \in \mathcal{D}(V_1)$ and $b, b' \in \mathcal{D}(V_2)$, then $a = a'$ and $b = b'$.

In Proposition 3.1, we give necessary conditions for a Heegaard splitting to be keen. We note that these show that Heegaard splittings given in

Received May 10, 2016.
Revised January 24, 2017.
2010 *Mathematics Subject Classification.* 57M50.
Key words and phrases. Heegaard splitting, curve complex, distance.

[6, 7, 11] are not keen (Remark 3.2). We also note that Proposition 3.1 shows that every genus-2 Heegaard splitting with Hempel distance $n\,(\geq 1)$ is not keen.

By the way, for a keen Heegaard splitting $V_1 \cup_S V_2$, the geodesics joining the unique pair of elements of $\mathcal{D}(V_1)$ and $\mathcal{D}(V_2)$ may not be unique (see Remark 4.15). We say that a Heegaard splitting $V_1 \cup_S V_2$ is *strongly keen* if the geodesics joining the pair of elements of $D(V_1)$ and $D(V_2)$ are unique. The main result of this paper gives the existence of strongly keen Heegaard splitting with Hempel distance n for each $g \geq 3$ and $n \geq 2$ as follows.

Theorem 1.1. *For any integers $n \geq 2$ and $g \geq 3$, there exists a 3-manifold with a strongly keen genus-g Heegaard splitting of Hempel distance n.*

§2. Preliminaries

Let S be a compact connected orientable surface. A simple closed curve in S is *essential* if it does not bound a disk in S and is not parallel to a component of ∂S. An arc properly embedded in S is *essential* if it does not co-bound a disk in S together with an arc on ∂S.

Heegaard splittings

A connected 3-manifold C is a *compression-body* if there exists a closed (possibly empty) surface F and a 0-handle B such that C is obtained from $(F \times [0,1]) \cup B$ by adding 1-handles to $F \times \{1\} \cup \partial B$. The subsurface of ∂C corresponding to $F \times \{0\}$ is denoted by $\partial_- C$, and $\partial_+ C$ denotes the subsurface $\partial C \setminus \partial_- C$ of ∂C. A compression-body C is called a *handlebody* if $\partial_- C = \emptyset$.

Let M be a closed orientable 3-manifold. We say that $V_1 \cup_S V_2$ is a *Heegaard splitting* of M if V_1 and V_2 are handlebodies in M such that $V_1 \cup V_2 = M$ and $V_1 \cap V_2 = \partial V_1 = \partial V_2 = S$. The genus of S is called the *genus* of the Heegaard splitting $V_1 \cup_S V_2$. Alternatively, given a Heegaard splitting $V_1 \cup_S V_2$ of M, we may regard that there is a homeomorphism $f : \partial V_2 \to \partial V_1$ such that M is obtained from V_1 and V_2 by identifying ∂V_1 and ∂V_2 via f. When we take this viewpoint, we will denote the Heegaard splitting by the expression $V_1 \cup_f V_2$.

Curve complexes

Let S be a compact connected orientable surface with genus g and p boundary components, where $3g + b > 4$. We call such surfaces *non-sporadic*. The *curve complex* $\mathcal{C}(S)$ is defined as follows: each vertex of $\mathcal{C}(S)$ is the isotopy class of an essential simple closed curve on S, and a collection of $k + 1$ vertices forms a k-simplex of $\mathcal{C}(S)$ if they can be realized by mutually disjoint curves in S. The *arc-and-curve complex* $\mathcal{AC}(S)$ is defined similarly, as follows: each vertex of $\mathcal{AC}(S)$ is the isotopy class of an essential properly embedded arc or an essential simple closed curve on S, and a collection of $k+1$ vertices forms a k-simplex of $\mathcal{AC}(S)$ if they can be realized by mutually disjoint arcs or simple closed curves in S. The symbol $\mathcal{C}^0(S)$ (resp. $\mathcal{AC}^0(S)$) denotes the 0-skeleton of $\mathcal{C}(S)$ (resp. $\mathcal{AC}(S)$). Throughout this paper, for a vertex $x \in \mathcal{C}^0(S)$ we often abuse notation and use x to represent (the isotopy class of) a geometric representative of x, and we assume that any pair of geometric representatives has minimal intersections.

For two vertices a, b of $\mathcal{C}(S)$, we define the *distance* $d_{\mathcal{C}(S)}(a, b)$ between a and b, which will be denoted by $d_S(a, b)$ in brief, as the minimal number of 1-simplexes of a simplicial path in $\mathcal{C}(S)$ joining a and b. For a subset A of $\mathcal{C}^0(S)$, we define $\mathrm{diam}_S(A) :=$ the diameter of A in $\mathcal{C}(S)$. Similarly, we can define $d_{\mathcal{AC}(S)}(a, b)$ for $a, b \in \mathcal{AC}^0(S)$ and $\mathrm{diam}_{\mathcal{AC}(S)}(A)$ for $A \subset \mathcal{AC}^0(S)$.

For a sequence $a_0, a_1, \ldots, a_n$ of vertices in $\mathcal{C}(S)$ with $a_i \cap a_{i+1} = \emptyset$ in S $(i = 0, 1, \ldots, n-1)$, we denote by $[a_0, a_1, \ldots, a_n]$ the path in $\mathcal{C}(S)$ with vertices $a_0, a_1, \ldots, a_n$ in this order. We say that a path $[a_0, a_1, \ldots, a_n]$ is a *geodesic* if $n = d_S(a_0, a_n)$.

Let C be a compression-body. A disk D properly embedded in C is *essential* if ∂D is an essential simple closed curve in $\partial_+ C$. Then the *disk complex* $\mathcal{D}(C)$ is the subset of $\mathcal{C}^0(\partial_+ C)$ consisting of the vertices with representatives bounding essential disks of C.

For a genus-$g(\geq 2)$ Heegaard splitting $V_1 \cup_S V_2$, the *Hempel distance* of $V_1 \cup_S V_2$ is defined by $d_S(\mathcal{D}(V_1), \mathcal{D}(V_2)) = \min\{d_S(x, y) \mid x \in \mathcal{D}(V_1), y \in \mathcal{D}(V_2)\}$.

Subsurface projection maps

For a set Y, let $\mathcal{P}(Y)$ denote the set consisting of the finite subsets of Y. Let S be a compact connected orientable surface, and let X be a subsurface of S. We suppose that both S and X are non-sporadic, and each component of ∂X is either contained in ∂S or essential in S. Let $\pi_A : \mathcal{C}^0(S) \to \mathcal{P}(\mathcal{AC}^0(X))$ and $\pi_0 : \mathcal{P}(\mathcal{AC}^0(X)) \to \mathcal{P}(\mathcal{C}^0(X))$ be maps defined as follows: for $\alpha \in \mathcal{C}^0(S)$, take a representative of α so that $|\alpha \cap X|$ is minimal, where $|\cdot|$ is the number of connected components. Then

- $\pi_A(\alpha)$ is the set of all isotopy classes of the components of $\alpha \cap X$,
- $\pi_0(\{\alpha_1, \ldots, \alpha_n\})$ is the union, for all $i = 1, \ldots, n$, of the set of all isotopy classes of the components of $\partial N(\alpha_i \cup \partial X)$ which are

essential in X, where $N(\alpha_i \cup \partial X)$ is a regular neighborhood of $\alpha_i \cup \partial X$ in X.

We call the composition $\pi_0 \circ \pi_A$ the *subsurface projection* and denote it by π_X. We say that α *misses* X (resp. α *cuts* X) if $\alpha \cap X = \emptyset$ (resp. $\alpha \cap X \neq \emptyset$). The following lemma can be proved by using [10, Lemma 2.2].

Lemma 2.1. *Let* $A \in \mathcal{P}(\mathcal{AC}^0(X))$ *and* $n \in \mathbb{N}$. *If* $\mathrm{diam}_{\mathcal{AC}(X)}(A) \leq n$, *then* $\mathrm{diam}_X(\pi_0(A)) \leq 2n$.

The following lemma is proved by using the above lemma.

Lemma 2.2. ([6, Lemma 2.1]) *Let* $[\alpha_0, \alpha_1, \ldots, \alpha_n]$ *be a path in* $\mathcal{C}(S)$ *such that every* α_i *cuts* X. *Then* $\mathrm{diam}_X(\pi_X(\alpha_0) \cup \pi_X(\alpha_n)) \leq 2n$.

Maps induced on curve complexes

Let Y, Z be non-sporadic surfaces. Suppose that there exists an embedding $\varphi : Y \to Z$ such that for each component l of ∂Y either $\varphi(l) \subset \partial Z$ or $\varphi(l)$ is essential in Z. Note that φ naturally induces maps $\mathcal{C}^0(Y) \to \mathcal{C}^0(Z)$ and $\mathcal{P}(\mathcal{C}^0(Y)) \to \mathcal{P}(\mathcal{C}^0(Z))$. Throughout this paper, under this setting, we abuse notation and use φ to denote these maps.

Let S be a non-sporadic closed surface.

Lemma 2.3. *Let* X *be a non-sporadic subsurface of* S *such that each component of* ∂X *is essential in* S. *Let* $\alpha, \beta \in \mathcal{C}^0(S)$ *such that* α, β *cut* X. *For any* $k \in \mathbb{N}$, *there exists a homeomorphism* $h : S \to S$ *such that* $h|_{S \setminus X} = \mathrm{id}_{S \setminus X}$ *and that* $\mathrm{diam}_X(\pi_X(\alpha) \cup \pi_X(h(\beta))) > k$.

Proof. Let γ be an element of $\pi_X(\beta)$. Take and fix a pseudo-Anosov homeomorphism $f : X \to X$ such that $f|_{\partial X} = \mathrm{id}_{\partial X}$. Then, by [9, Proposition 4.6], there is a positive integer n such that

$$d_X(\gamma, f^n(\gamma)) > k + \mathrm{diam}_X(\pi_X(\alpha) \cup \pi_X(\beta)).$$

Let $h : S \to S$ be the extension of f^n. Then

$$\begin{aligned}
& k + \mathrm{diam}_X(\pi_X(\alpha) \cup \pi_X(\beta)) \\
< \ & d_X(\gamma, h(\gamma)) \\
\leq \ & \mathrm{diam}_X(\pi_X(\beta) \cup \pi_X(\alpha)) + \mathrm{diam}_X(\pi_X(\alpha) \cup h(\pi_X(\beta))) \\
= \ & \mathrm{diam}_X(\pi_X(\beta) \cup \pi_X(\alpha)) + \mathrm{diam}_X(\pi_X(\alpha) \cup \pi_X(h(\beta)))
\end{aligned}$$

and hence we have $\mathrm{diam}_X(\pi_X(\alpha) \cup \pi_X(h(\beta))) > k$. Q.E.D.

The following two lemmas can be proved by using arguments in the proof of [6, Propositions 4.1, 4.4].

Lemma 2.4. *Let $[\alpha_0, \alpha_1, \ldots, \alpha_n]$ and $[\beta_0, \beta_1, \ldots, \beta_m]$ be geodesics in $\mathcal{C}(S)$. Suppose that α_n and β_0 are non-separating on S, and let $X = \mathrm{Cl}(S \setminus N(\alpha_n))$. Let $h : S \to S$ be a homeomorphism such that*

- $h(\beta_0) = \alpha_n$, *and*
- $\mathrm{diam}_X(\pi_X(\alpha_0) \cup \pi_X(h(\beta_m))) > 2(n+m)$.

Then $[\alpha_0, \alpha_1, \ldots, \alpha_n(= h(\beta_0)), h(\beta_1), \ldots, h(\beta_m)]$ is a geodesic in $\mathcal{C}(S)$.

Moreover, every geodesic connecting α_0 and $h(\beta_m)$ passes through α_n. In fact, for any geodesic $[\gamma_0, \gamma_1, \ldots, \gamma_{n+m}]$ in $\mathcal{C}(S)$ such that $\gamma_0 = \alpha_0$ and $\gamma_{n+m} = h(\beta_m)$, we have $\gamma_n = \alpha_n$.

Lemma 2.5. *Suppose that the genus of S is greater than 2. Let $[\alpha_0, \alpha_1, \ldots, \alpha_n]$ and $[\beta_0, \beta_1, \ldots, \beta_m]$ be geodesics in $\mathcal{C}(S)$. Suppose that $\alpha_{n-1} \cup \alpha_n$ and $\beta_0 \cup \beta_1$ are non-separating on S, and let $X = \mathrm{Cl}(S \setminus N(\alpha_{n-1} \cup \alpha_n))$. Let $h : S \to S$ be a homeomorphism such that*

- $h(\beta_0) = \alpha_{n-1}$, $h(\beta_1) = \alpha_n$, *and*
- $\mathrm{diam}_X(\pi_X(\alpha_0) \cup \pi_X(h(\beta_m))) > 2(n+m-1)$.

Then $[\alpha_0, \alpha_1, \ldots, \alpha_{n-1}(= h(\beta_0)), \alpha_n(= h(\beta_1)), h(\beta_2), \ldots, h(\beta_m)]$ is a geodesic in $\mathcal{C}(S)$.

Moreover, every geodesic connecting α_0 and $h(\beta_m)$ passes through α_{n-1} or α_n. In fact, for any geodesic $[\gamma_0, \gamma_1, \ldots, \gamma_{n+m-1}]$ in $\mathcal{C}(S)$ such that $\gamma_0 = \alpha_0$ and $\gamma_{n+m-1} = h(\beta_m)$, we have $\gamma_{n-1} = \alpha_{n-1}$ or $\gamma_n = \alpha_n$.

Remark 2.6. Note that, in Lemmas 2.4 and 2.5, since S is closed and non-sporadic (that is, the genus of S is greater than 1) in Lemma 2.4 and the genus of S is greater than 2 in Lemma 2.5, the subsurfaces denoted by X are non-sporadic.

§3. Keen Heegaard splittings

Recall that a Heegaard splitting $V_1 \cup_S V_2$ is called *keen* if its Hempel distance is realized by a unique pair of elements of $\mathcal{D}(V_1)$ and $\mathcal{D}(V_2)$.

Proposition 3.1. *Let $V_1 \cup_S V_2$ be a genus-$g(\geq 2)$ Heegaard splitting with Hempel distance $n(\geq 1)$. Let $[l_0, l_1, \ldots, l_n]$ be a geodesic in $\mathcal{C}(S)$ such that $l_0 \in \mathcal{D}(V_1)$ and $l_n \in \mathcal{D}(V_2)$. If $V_1 \cup_S V_2$ is keen, then the following holds.*

(1) *l_0 and l_n are non-separating on S.*
(2) *l_1 and l_{n-1} are non-separating on S.*
(3) *$l_0 \cup l_1$ and $l_{n-1} \cup l_n$ are separating on S.*

Proof. (1) Assume on the contrary that either l_0 or l_n is separating on S. Without loss of generality, we may assume that l_0 is separating on S. Let D_0 be a disk properly embedded in V_1 such that $\partial D_0 = l_0$.

Let $V_1^{(1)}$ be the component of $V_1 \setminus D_0$ that contains l_1, and let $V_1^{(2)}$ be the other component. It is easy to see that there is an essential disk D_0' properly embedded in $V_1^{(2)}$ such that $D_0' \cap D_0 = \emptyset$. Then $l_0' := \partial D_0'$ is also disjoint from l_1, and hence, $[l_0', l_1, \ldots, l_n]$ is a geodesic in $\mathcal{C}(S)$. Hence, we have $d_S(l_0', l_n) = d_S(\mathcal{D}(V_1), \mathcal{D}(V_2))$, where l_0' is an element of $\mathcal{D}(V_1)$ different from l_0, a contradiction.

(2) Assume on the contrary that either l_1 or l_{n-1}, say l_1, is separating on S. Let $S^{(1)}$ be the component of $S \setminus l_1$ that contains l_0. Since l_0 is non-separating on S by (1) and l_1 is separating on S, we can see that l_0 is non-separating on $S^{(1)}$. Then there exists an essential simple closed curve l^* on $S^{(1)}$ such that l^* intersects l_0 transversely in one point. Let D_0 be a disk properly embedded in V_1 such that $\partial D_0 = l_0$, and let D_0^+ and D_0^- be the components of $\mathrm{Cl}(\partial N(D_0) \setminus \partial V_1)$, where $N(D_0)$ is a regular neighborhood of D_0 in V_1. Take the subarc of l^* lying outside of the product region $N(D_0)$ between D_0^+ and D_0^-, and let D_0'' be the disk in V_1 obtained from $D_0^+ \cup D_0^-$ by adding a band along the subarc of l^*. Then $l_0'' := \partial D_0''$ is also disjoint from l_1, and hence, $[l_0'', l_1, \ldots, l_n]$ is a geodesic in $\mathcal{C}(S)$. Hence, we have $d_S(l_0'', l_n) = d_S(\mathcal{D}(V_1), \mathcal{D}(V_2))$, where l_0'' is an element of $\mathcal{D}(V_1)$ different from l_0, a contradiction.

(3) Assume on the contrary that either $l_0 \cup l_1$ or $l_{n-1} \cup l_n$, say $l_0 \cup l_1$, is non-separating on S. Then there exists an essential simple closed curve l^* on S such that l^* intersects l_0 transversely in one point and $l^* \cap l_1 = \emptyset$. We can lead to a contradiction by the arguments in (2). Q.E.D.

Remark 3.2. (1) By Proposition 3.1, we see that every genus-2 Heegaard splitting with Hempel distance $n\,(\geq 1)$ is not keen. In fact, if a genus-2 Heegaard splitting $V_1 \cup_S V_2$ is keen, and $[l_0, l_1, \ldots, l_n]$ is a path that realizes the Hempel distance, then by (1) and (2) of Proposition 3.1, we see that $l_0 \cup l_1$ cuts S into four punctured sphere, contradicting (3) of Proposition 3.1. Hence, if a genus-g Heegaard splitting with Hempel distance $n\,(\geq 1)$ is keen, then $g \geq 3$.

(2) Heegaard splittings given in [6, 7, 11] are not keen, since their Hempel distances are realized by pairs of separating elements.

§4. Proof of Theorem 1.1 when $n \geq 4$

Let n and g be integers with $n \geq 4$ and $g \geq 3$. Let S be a closed connected orientable surface of genus g. Let l_0 and l_1 be non-separating simple closed curves on S such that $l_0 \cap l_1 = \emptyset$, $l_0 \cup l_1$ is separating and l_0, l_1 are not parallel on S. Let $F_1 = \mathrm{Cl}(S \setminus N(l_1))$. Choose and fix an integer $k \in \{2, 3, \ldots, n-2\}$. Let $[l_1', l_2', \ldots, l_k']$ and $[l_1'', l_2'', \ldots, l_{n-k}'']$ be geodesics in $\mathcal{C}(S)$ such that l_1', l_k', l_1'' and l_{n-k}'' are non-separating on S. (For the existence of such geodesics, see [6] or the proof of Proposition 4.14 below for example.) By Lemma 2.3, there exist homeomorphisms $h_1 : S \to S$ and $h_2 : S \to S$ such that

- $h_1(l_1') = l_1$,
- $h_2(l_1'') = l_1$,
- $\mathrm{diam}_{F_1}(\pi_{F_1}(l_0) \cup \pi_{F_1}(h_1(l_k'))) \geq 4n + 16$, and
- $\mathrm{diam}_{F_1}(\pi_{F_1}(l_0) \cup \pi_{F_1}(h_2(l_{n-k}''))) \geq 4n + 16$.

Note that $\pi_{F_1}(l_0) = \{l_0\}$ since $l_0 \cap l_1 = \emptyset$. By Lemma 2.4, $[l_0, l_1 (= h_1(l_1')), h_1(l_2'), \ldots, h_1(l_k')]$ and $[l_0, l_1 (= h_2(l_1'')), h_2(l_2''), \ldots, h_2(l_{n-k}'')]$ are geodesics in $\mathcal{C}(S)$. Let $F_k = \mathrm{Cl}(S \setminus N(h_1(l_k')))$. By Lemma 2.3, there exists a homeomorphism $h_3 : S \to S$ such that

- $h_3(h_2(l_{n-k}'')) = h_1(l_k')$, and
- $\mathrm{diam}_{F_k}(\pi_{F_k}(l_0) \cup \pi_{F_k}(h_3(l_0))) > 2n$.

Let $l_i = h_1(l_i')$ for $i \in \{2, \ldots, k\}$, $l_i = h_3(h_2(l_{n-i}''))$ for $i \in \{k+1, \ldots, n-1\}$, and $l_n = h_3(l_0)$. By Lemma 2.4, $[l_0, l_1, \ldots, l_n]$ is a geodesic in $\mathcal{C}(S)$. Moreover, by the construction of the geodesic, the following are satisfied.

(G1) l_0, l_1, l_{n-1} and l_n are non-separating on S,
(G2) $l_0 \cup l_1$ and $l_{n-1} \cup l_n$ are separating on S,
(G3) $\mathrm{diam}_{F_1}(\pi_{F_1}(l_0) \cup \pi_{F_1}(l_k)) \geq 4n + 16$,
(G4) $\mathrm{diam}_{F_{n-1}}(\pi_{F_{n-1}}(l_k) \cup \pi_{F_{n-1}}(l_n)) \geq 4n + 16$, and
(G5) $\mathrm{diam}_{F_k}(\pi_{F_k}(l_0) \cup \pi_{F_k}(l_n)) > 2n$,

where $F_{n-1} = \mathrm{Cl}(S \setminus N(l_{n-1}))$.

Let C_1 and C_2 be copies of the compression-body obtained by adding a 1-handle to $F \times [0, 1]$, where F is a closed connected orientable surface of genus $g - 1$. Let D_1 (resp. D_2) be the non-separating essential disk properly embedded in C_1 (resp. C_2) corresponding to the co-core of the 1-handle. We may assume that $\partial_+ C_1 = S$ and $\partial D_1 = l_0$. Choose a homeomorphism $f : \partial_+ C_2 \to \partial_+ C_1$ such that $f(\partial D_2) = l_n$.

Let H_1 and H_2 be copies of the handlebody of genus $g - 1$. In the remainder of this section, we identify ∂H_i and $\partial_- C_i$ ($i = 1, 2$) so that we obtain a keen Heegaard splitting of genus g whose Hempel distance is n.

For each $i = 1, 2$, let $C_i' = \mathrm{Cl}(C_i \setminus N(D_i))$ and $X_i = \partial C_i' \cap \partial_+ C_i$. Note that C_i' is homeomorphic to $\partial_- C_i \times [0, 1]$. Let $\varphi_i : C_i' \to \partial_- C_i \times [0, 1]$ be a homeomorphism such that $\varphi_i(\partial C_i' \setminus \partial_- C_i) = \partial_- C_i \times \{1\}$ and $\varphi_i(\partial_- C_i) = \partial_- C_i \times \{0\}$, and let $\psi_i : \partial_- C_i \times \{1\} \to \partial_- C_i \times \{0\}$ be the natural homeomorphism. Let $P_i : X_i \to \partial_- C_i$ be the composition of the inclusion map $X_i \to \partial C_i' \setminus \partial_- C_i$ and the map $\left(\varphi_i|_{\partial_- C_i}\right)^{-1} \circ \psi_i \circ \left(\varphi_i|_{\partial C_i' \setminus \partial_- C_i}\right) : \partial C_i' \setminus \partial_- C_i \to \partial_- C_i$.

It is clear that l_1 represents an essential simple closed curve on X_1. Since l_1 is non-separating on S, $P_1(l_1)$ is an essential simple closed curve on $\partial_- C_1$. By [5, Theorem 2.7] and its proof (see also [1, Theorem 2.4]), there exists a homeomorphism $f_1 : \partial H_1 \to \partial_- C_1$ such that

$$d_{\partial_- C_1}(f_1(\mathcal{D}(H_1)), P_1(l_1)) \geq 2. \tag{1}$$

Let $V_1 = C_1 \cup_{f_1} H_1$, that is, V_1 is the manifold obtained from C_1 and H_1 by identifying $\partial_- C_1$ and ∂H_1 via f_1. Note that V_1 is a handlebody.

Claim 4.1. *l_1 intersects every element of $\mathcal{D}(V_1) \setminus \{l_0\}$.*

Proof. Assume on the contrary that there exists an element a of $\mathcal{D}(V_1) \setminus \{l_0\}$ such that $a \cap l_1 = \emptyset$. Let D_a be a disk in V_1 bounded by a, and recall that l_0 bounds the disk D_1 in C_1, and hence, in V_1 (see Fig. 1). We may assume that $|D_a \cap D_1| = |D_a \cap N(D_1)|$ and is minimal. By using innermost disk arguments, we see that $D_a \cap D_1$ has no loop components. Let Δ be a disk properly embedded in $C_1' \cup_{f_1} H_1$ defined as follows.

- If $D_a \cap D_1 = \emptyset$, let $\Delta = D_a$.
- If $D_a \cap D_1 \neq \emptyset$, let Δ be the closure of a component of $D_a \setminus N(D_1)$ that is outermost in D_a.

Since $a \cap l_1 = \emptyset$, the disk Δ is disjoint from l_1. Since l_0, l_1 are non-separating and $l_0 \cup l_1$ is separating on S by the condition (G2), and $a \neq l_0$, we see that Δ is essential in $C_1' \cup_{f_1} H_1$.

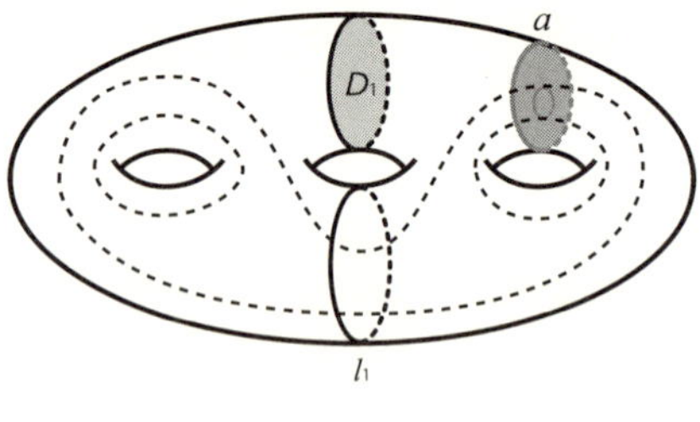

Fig. 1

Since C_1' is homeomorphic to $\partial_- C_1 \times [0, 1]$, we may assume that Δ is obtained by gluing a vertical annulus in C_1' and an essential disk Δ' in H_1 via f_1, after boundary compressions and isotopies toward $\partial_- C_1$ if necessary. This together with $\Delta \cap l_1 = \emptyset$ implies that $d_{\partial_- C_1}(f_1(\partial \Delta'), P_1(l_1)) \leq 1$. Since $f_1(\partial \Delta') \in f_1(\mathcal{D}(H_1))$, we have $d_{\partial_- C_1}(f_1(\mathcal{D}(H_1)), P_1(l_1)) \leq 1$, a contradiction to the inequality (1). Q.E.D.

Let $\pi_{F_1} = \pi_0 \circ \pi_A : \mathcal{C}^0(S) \to \mathcal{P}(\mathcal{AC}^0(F_1)) \to \mathcal{P}(\mathcal{C}^0(F_1))$ be the subsurface projection introduced in Section 2. Recall that $\pi_{F_1}(l_0) = \{l_0\}$ since $l_0 \cap l_1 = \emptyset$.

Claim 4.2. *For any element $a \in \mathcal{D}(V_1)$, we have $\pi_{F_1}(a) \neq \emptyset$, and $\mathrm{diam}_{F_1}(l_0 \cup \pi_{F_1}(a)) \leq 4$.*

Proof. Note that, by Claim 4.1, we immediately have $\pi_{F_1}(a) \neq \emptyset$. If $a = l_0$ or $a \cap l_0 = \emptyset$, that is, $d_S(l_0, a) \leq 1$, then we have $\mathrm{diam}_{F_1}(l_0 \cup \pi_{F_1}(a)) \leq 2$ by Lemma 2.2. Hence, we suppose that $a \neq l_0$ and $a \cap l_0 \neq \emptyset$ in the following.

Let D_a be a disk in V_1 bounded by a, and recall that l_0 bounds the disk D_1 in V_1. Here, we may assume that $|a \cap l_1| = |a \cap N(l_1)|$ and is minimal. We may also assume that $|D_a \cap D_1| = |D_a \cap N(D_1)|$ and is minimal. Let Δ be the closure of a component of $D_a \setminus N(D_1)$ that is outermost in D_a. If $\Delta \cap l_1 = \emptyset$, then we can lead to a contradiction by arguments in the proof of Claim 4.1. Hence, $\Delta \cap l_1 \neq \emptyset$. Since $l_0 \cup l_1$ is separating on S by the condition (G2), there exists a component γ of $\mathrm{Cl}(\partial\Delta \setminus (N(D_1) \cup N(l_1)))$ such that $\partial\gamma \subset \partial N(l_1)$. It is clear that γ is an essential arc on F_1. Note that γ is disjoint from l_0, that is, $d_{\mathcal{AC}(F_1)}(l_0, \gamma) = 1$, since $l_0 \cap \Delta = \emptyset$ and γ is a subarc of $\partial\Delta$. Since $\gamma \in \pi_A(a)$, we have $d_{\mathcal{AC}(F_1)}(l_0, \pi_A(a)) \leq d_{\mathcal{AC}(F_1)}(l_0, \gamma) = 1$. Hence,

$$\begin{aligned}\mathrm{diam}_{\mathcal{AC}(F_1)}(l_0 \cup \pi_A(a)) &\leq d_{\mathcal{AC}(F_1)}(l_0, \pi_A(a)) + \mathrm{diam}_{\mathcal{AC}(F_1)}(\pi_A(a)) \\ &\leq 1 + 1 = 2.\end{aligned}$$

By Lemma 2.1, we have $\mathrm{diam}_{F_1}(l_0 \cup \pi_{F_1}(a)) \leq 4$. Q.E.D.

Lemma 4.3. *$d_S(\mathcal{D}(V_1), l_n) = n$.*

Proof. Since $l_0 \in \mathcal{D}(V_1)$, we have $d_S(\mathcal{D}(V_1), l_n) \leq n$. To prove $d_S(\mathcal{D}(V_1), l_n) = n$, assume on the contrary that $d_S(\mathcal{D}(V_1), l_n) < n$. Then there exists a geodesic $[m_0, m_1, \ldots, m_p]$ in $\mathcal{C}(S)$ such that $p < n$, $m_0 \in \mathcal{D}(V_1)$ and $m_p = l_n$.

Claim 4.4. *$m_i = l_1$ for some $i \in \{0, 1, \ldots, p\}$.*

Proof. Assume on the contrary that $m_i \neq l_1$ for every $i \in \{0, 1, \ldots, p\}$. Namely, every m_i cuts F_1. By Lemma 2.2, we have

$$\mathrm{diam}_{F_1}(\pi_{F_1}(m_0) \cup \pi_{F_1}(m_p)) \leq 2p. \tag{2}$$

Similarly, we have

$$\mathrm{diam}_{F_1}(\pi_{F_1}(l_n) \cup \pi_{F_1}(l_k)) \leq 2(n-k). \tag{3}$$

By the triangle inequality, we have

$$\begin{aligned}\operatorname{diam}_{F_1}(\pi_{F_1}(l_0)\cup\pi_{F_1}(l_k)) &\le \operatorname{diam}_{F_1}(\pi_{F_1}(l_0)\cup\pi_{F_1}(m_0))\\ &\quad+\operatorname{diam}_{F_1}(\pi_{F_1}(m_0)\cup\pi_{F_1}(m_p))\\ &\quad+\operatorname{diam}_{F_1}(\pi_{F_1}(l_n)\cup\pi_{F_1}(l_k)).\end{aligned}\tag{4}$$

By the inequalities (2), (3), (4) and Claim 4.2, we obtain

$$\begin{aligned}\operatorname{diam}_{F_1}(\pi_{F_1}(l_0)\cup\pi_{F_1}(l_k)) &\le 4+2p+2(n-k)\\ &< 4+2n+2n,\end{aligned}\tag{5}$$

which contradicts the condition (G3). Q.E.D.

By Claim 4.4, we have $d_S(m_i, m_p) = d_S(l_1, l_n)$. Since $[m_0, m_1, \ldots, m_p]$ and $[l_0, l_1, \ldots, l_n]$ are geodesics, $d_S(m_i, m_p) = p-i$ and $d_S(l_1, l_n) = n-1 > p-1$. Hence, $p-i > p-1$, which implies $i = 0$, that is, $m_0 = l_1$. This contradicts Claim 4.1. Hence, we have $d_S(\mathcal{D}(V_1), l_n) = n$. Q.E.D.

Note that $f^{-1}(l_{n-1})$ represents an essential simple closed curve on X_2. Since $f^{-1}(l_{n-1})$ is non-separating on $\partial_+ C_2$ by the condition (G1), $P_2(f^{-1}(l_{n-1}))$ is an essential simple closed curve on $\partial_- C_2$. By [5, Theorem 2.7] and its proof (see also [1, Theorem 2.4]), there exists a homeomorphism $f_2 : \partial H_2 \to \partial_- C_2$ such that

$$d_{\partial_- C_2}(f_2(\mathcal{D}(H_2)), P_2(f^{-1}(l_{n-1}))) \ge 2.\tag{6}$$

Let $V_2 = C_2 \cup_{f_2} H_2$. Then $V_1 \cup_f V_2$ is a genus-g Heegaard splitting.

Claims 4.5, 4.6 and Lemma 4.7 below can be proved by the arguments similar to those for Claims 4.1, 4.2 and Lemma 4.3, respectively.

Claim 4.5. *l_{n-1} intersects every element of $f(\mathcal{D}(V_2)) \setminus \{l_n\}$.*

Claim 4.6. *For any element $a \in f(\mathcal{D}(V_2))$, we have $\pi_{F_{n-1}}(a) \neq \emptyset$, and $\operatorname{diam}_{F_{n-1}}(l_n \cup \pi_{F_{n-1}}(a)) \le 4$.*

Lemma 4.7. *$d_S(f(\mathcal{D}(V_2)), l_0) = n$.*

Claim 4.8. (1) $\operatorname{diam}_{F_1}(\pi_{F_1}(f(\mathcal{D}(V_2)))) \le 12$.
(2) $\operatorname{diam}_{F_{n-1}}(\pi_{F_{n-1}}(\mathcal{D}(V_1))) \le 12$.

Proof. By Lemma 4.3, we have $d_S(\mathcal{D}(V_1), l_{n-1}) = n-1 \ge 3$. Hence, by [8, Theorem 1], $\operatorname{diam}_{F_{n-1}}(\pi_{F_{n-1}}(\mathcal{D}(V_1))) \le 12$. Similarly, we have $\operatorname{diam}_{F_1}(\pi_{F_1}(f(\mathcal{D}(V_2)))) \le 12$ by Lemma 4.7 and [8]. Q.E.D.

Lemma 4.9. *$d_S(\mathcal{D}(V_1), f(\mathcal{D}(V_2))) = n$. Namely, the Hempel distance of the Heegaard splitting $V_1 \cup_f V_2$ is n.*

Proof. Since $l_0 \in \mathcal{D}(V_1)$ and $l_n \in f(\mathcal{D}(V_2))$, we have

$$d_S(\mathcal{D}(V_1), f(\mathcal{D}(V_2))) \leq n.$$

Let $[m_0, m_1, \ldots, m_p]$ be a geodesic in $\mathcal{C}(S)$ such that $m_0 \in \mathcal{D}(V_1)$, $m_p \in f(\mathcal{D}(V_2))$ and $p \leq n$.

Claim 4.10. *$m_i = l_1$ for some $i \in \{0, 1, \ldots, p\}$.*

Proof. Assume on the contrary that $m_i \neq l_1$ for every $i \in \{0, 1, \ldots, p\}$. Namely, every m_i cuts F_1. By Lemma 2.2, we have

$$\operatorname{diam}_{F_1}(\pi_{F_1}(m_0) \cup \pi_{F_1}(m_p)) \leq 2p. \tag{7}$$

Recall that $k \in \{2, 3, \ldots, n-2\}$. Similarly, we have

$$\operatorname{diam}_{F_1}(\pi_{F_1}(l_n) \cup \pi_{F_1}(l_k)) \leq 2(n-k). \tag{8}$$

By the triangle inequality, we have

$$\begin{aligned} \operatorname{diam}_{F_1}(\pi_{F_1}(l_0) \cup \pi_{F_1}(l_k)) \leq\ & \operatorname{diam}_{F_1}(\pi_{F_1}(l_0) \cup \pi_{F_1}(m_0)) \\ & + \operatorname{diam}_{F_1}(\pi_{F_1}(m_0) \cup \pi_{F_1}(m_p)) \\ & + \operatorname{diam}_{F_1}(\pi_{F_1}(m_p) \cup \pi_{F_1}(l_n)) \\ & + \operatorname{diam}_{F_1}(\pi_{F_1}(l_n) \cup \pi_{F_1}(l_k)). \end{aligned} \tag{9}$$

By the inequalities (7), (8), (9) together with Claims 4.2 and 4.8, we obtain

$$\begin{aligned} \operatorname{diam}_{F_1}(\pi_{F_1}(l_0) \cup \pi_{F_1}(l_k)) & \leq 4 + 2p + 12 + 2(n-k) \\ & < 4 + 2n + 12 + 2n, \end{aligned} \tag{10}$$

which contradicts the condition (G3). Q.E.D.

The following claim can be proved similarly.

Claim 4.11. *$m_j = l_{n-1}$ for some $j \in \{0, 1, \ldots, p\}$.*

Note that $l_1 \notin \mathcal{D}(V_1)$ by Claim 4.1. Note also that $l_1 \notin f(\mathcal{D}(V_2))$ since, otherwise, we have $d_S(f(\mathcal{D}(V_2)), l_0) \leq d_S(l_1, l_0) = 1$, which contradicts Lemma 4.7. Since $m_0 \in \mathcal{D}(V_1)$ and $m_p \in f(\mathcal{D}(V_2))$ by the assumption, we have $m_i (= l_1) \neq m_0$ and $m_i (= l_1) \neq m_p$, which implies $1 \leq i \leq p-1$. Similarly, we have $1 \leq j \leq p-1$. Hence, we have

$$|i - j| \leq (p-1) - 1 = p - 2. \tag{11}$$

On the other hand, by Claims 4.10 and 4.11, we have

$$|i - j| = d_S(m_i, m_j) = d_S(l_1, l_{n-1}) = n - 2,$$

which together with the inequality (11) implies $p = n$. Hence,

$$d_S(\mathcal{D}(V_1), f(\mathcal{D}(V_2))) = n.$$

Q.E.D.

Lemma 4.12. *The Heegaard splitting $V_1 \cup_f V_2$ is keen.*

Proof. Let $[m_0, m_1, \ldots, m_n]$ be a geodesic in $\mathcal{C}(S)$ such that $m_0 \in \mathcal{D}(V_1)$ and $m_n \in f(\mathcal{D}(V_2))$. By the proof of Lemma 4.9, we have $m_1 = l_1$ and $m_{n-1} = l_{n-1}$. By Claims 4.1 and 4.5, we have $m_0 = l_0$ and $m_n = l_n$. Q.E.D.

In Claim 4.13 and Proposition 4.14, we show that the existence of strongly keen Heegaard splitting.

Claim 4.13. *In the above construction, if the following conditions are satisfied, then the Heegaard splitting constructed from the geodesic $[l_0, l_1, \ldots, l_n]$ is strongly keen.*

- *The geodesic $[l'_1, l'_2, \ldots, l'_k]$ (resp. $[l''_1, l''_2, \ldots, l''_{n-k}]$) is the unique geodesic from l'_1 to l'_k (resp. l''_1 to l''_{n-k}).*

Proof. By the proof of Lemma 4.12, $m_i = l_i$ holds for $i = 0$, 1, $n-1$ and n. Moreover, by the condition (G5) and Lemma 2.4, we have $m_k = l_k$. Hence, if the geodesics $[l'_1, l'_2, \ldots, l'_k]$ (resp. $[l''_1, l''_2, \ldots, l''_{n-k}]$) is the unique geodesic connecting l'_1 and l'_k (resp. l''_1 and l''_{n-k}), then we obtain the desired result. Q.E.D.

Hence the next proposition completes the proof of Theorem 1.1.

Proposition 4.14. *Let S be a closed, non-sporadic surface. For each p, there exists a geodesic $[\alpha_0, \alpha_1, \ldots, \alpha_p]$ in $\mathcal{C}(S)$ such that each α_i ($i = 0, 1, \ldots, p$) is non-separating on S and $[\alpha_0, \alpha_1, \ldots, \alpha_p]$ is the unique geodesic connecting α_0 and α_p.*

Proof. Let α_0 and α_1 be non-separating simple closed curve on S such that $\alpha_0 \cap \alpha_1 = \emptyset$, and let $X_1 = \mathrm{Cl}(S \setminus N(\alpha_1))$. Let α'_2 be a non-separating simple closed curve on S disjoint from α_1. By Lemma 2.3, there exists a homeomorphism $g_1 : S \to S$ such that $g_1(\alpha_1) = \alpha_1$ and $\mathrm{diam}_{X_1}(\pi_{X_1}(\alpha_0) \cup \pi_{X_1}(g_1(\alpha'_2))) > 4$. Let $\alpha_2 = g_1(\alpha'_2)$. By Lemma 2.4, $[\alpha_0, \alpha_1, \alpha_2]$ is a geodesic in $\mathcal{C}(S)$. Moreover, by Lemma 2.4, $[\alpha_0, \alpha_1, \alpha_2]$ is the unique geodesic connecting α_0 and α_2.

For any positive integer p, we repeat this process to construct a geodesic $[\alpha_0, \alpha_1, \ldots, \alpha_p]$ inductively as follows. Suppose we have constructed a geodesic $[\alpha_0, \alpha_1, \ldots, \alpha_i]$ for $i < p$ such that

- α_i is non-separating on S, and

- $[\alpha_0, \alpha_1, \ldots, \alpha_i]$ is the unique geodesic connecting α_0 and α_i.

Let $X_i = \mathrm{Cl}(S \backslash N(\alpha_i))$. Let α'_{i+1} be a non-separating simple closed curve on S disjoint from α_i. By Lemma 2.3, there exists a homeomorphism $g_i : S \to S$ such that $g_i(\alpha_i) = \alpha_i$ and $\mathrm{diam}_{X_i}(\pi_{X_i}(\alpha_0) \cup \pi_{X_i}(g_i(\alpha'_{i+1}))) > 2(i+1)$. Let $\alpha_{i+1} = g_i(\alpha'_{i+1})$. By Lemma 2.4, $[\alpha_0, \alpha_1, \ldots, \alpha_{i+1}]$ is a geodesic in $\mathcal{C}(S)$. Moreover, every geodesic connecting α_0 and α_{i+1} passes through α_i. Since $[\alpha_0, \alpha_1, \ldots, \alpha_i]$ is the unique geodesic connecting α_0 and α_i, we have that $[\alpha_0, \alpha_1, \ldots, \alpha_{i+1}]$ is the unique geodesic connecting α_0 and α_{i+1}. Hence, we obtain a geodesic $[\alpha_0, \alpha_1, \ldots, \alpha_p]$ such that every α_i $(i = 0, 1, \ldots, p)$ is non-separating on S and $[\alpha_0, \alpha_1, \ldots, \alpha_p]$ is the unique geodesic connecting α_0 and α_p. Q.E.D.

Remark 4.15. There exists a keen Heegaard splitting which is not strongly keen. For example, in the construction at the beginning of this section, let $k = 3$ and take l'_1 and l'_3 such that l'_1 and l'_3 intersect transversely in one point. Let l'_2 be an essential simple closed curve disjoint from $l'_1 \cup l'_3$. Then $[l'_1, l'_2, l'_3]$ is a geodesic in $\mathcal{C}(S)$. We can apply the arguments up to Lemma 4.12 to obtain a geodesic $[l_0, l_1, \ldots, l_n]$ and obtain a keen Heegaard splitting $V_1 \cup_f V_2$. Since l_1 and l_3 intersect transversely in one point, there exists an essential simple closed curve l^*_2 that is disjoint from $l_1 \cup l_3$ and different from l_2. Then $[l_0, l_1, l^*_2, l_3, \ldots, l_n]$ is a geodesic realizing the Hempel distance of $V_1 \cup_f V_2$ which is different from $[l_0, l_1, l_2, l_3, \ldots, l_n]$. Hence, $V_1 \cup_f V_2$ is not strongly keen.

§5. Proof of Theorem 1.1 when $n = 2$

Let $n = 2$ and g be an integer with $g \geq 3$. Let S be a closed connected orientable surface of genus g. Let l_0 and l_1 be non-separating simple closed curves on S such that $l_0 \cup l_1$ is separating on S and l_0, l_1 are not parallel on S. By Lemma 2.3, there exists a homeomorphism $h : S \to S$ such that $h(l_1) = l_1$ and

$$d_{F_1}(l_0, h(l_0)) > 12,$$

where $F_1 = \mathrm{Cl}(S \setminus N(l_1))$. Let $l_2 = h(l_0)$. By Lemma 2.4, $[l_0, l_1, l_2]$ is a geodesic in $\mathcal{C}(S)$.

Let C_1 and C_2 be copies of the compression-body obtained by adding a 1-handle to $F \times [0, 1]$, where F is a closed connected orientable surface of genus $g - 1$. Let D_1 and D_2 be the non-separating essential disk properly embedded in C_1 and C_2 corresponding to the co-cores of the 1-handles, respectively. We may assume that $\partial_+ C_1 = S$ and $\partial D_1 = l_0$. Choose a homeomorphism $f : \partial_+ C_2 \to \partial_+ C_1$ such that $f(\partial D_2) = l_2$.

Let H_i, C_i', X_i, P_i $(i = 1, 2)$ be as in Section 4. Note that l_1 is non-separating on S, and hence, $P_1(l_1)$ and $P_2(f^{-1}(l_1))$ are essential simple closed curves on $\partial_- C_1$ and $\partial_- C_2$, respectively. By [5, Theorem 2.7] and its proof (see also [1, Theorem 2.4]), there exist homeomorphisms $f_1 : \partial H_1 \to \partial_- C_1$ and $f_2 : \partial H_2 \to \partial_- C_2$ such that $d_{\partial_- C_1}(f_1(\mathcal{D}(H_1)), P_1(l_1)) \geq 2$ and $d_{\partial_- C_2}(f_2(\mathcal{D}(H_2)), P_2(f^{-1}(l_1))) \geq 2$, respectively. Let $V_i = C_i \cup_{f_i} H_i$ $(i = 1, 2)$. Then, $V_1 \cup_f V_2$ is a genus-g Heegaard splitting. By the arguments similar to those for Claims 4.1, 4.2, 4.5 and 4.6, we obtain the following.

Claim 5.1. (1) *l_1 intersects every element of $\mathcal{D}(V_1) \setminus \{l_0\}$ and every element of $f(\mathcal{D}(V_2)) \setminus \{l_2\}$.*

(2) *For any element $a \in \mathcal{D}(V_1)$, we have $\pi_{F_1}(a) \neq \emptyset$, and* $\mathrm{diam}_{F_1}(l_0 \cup \pi_{F_1}(a)) \leq 4$.

(3) *For any element $a \in f(\mathcal{D}(V_2))$, we have $\pi_{F_1}(a) \neq \emptyset$, and* $\mathrm{diam}_{F_1}(l_2 \cup \pi_{F_1}(a)) \leq 4$.

Lemma 5.2. *$V_1 \cup_f V_2$ is a strongly keen Heegaard splitting whose Hempel distance is 2.*

Proof. Since $l_0 \in \mathcal{D}(V_1)$ and $l_2 \in f(\mathcal{D}(V_2))$, we have

$$d_S(\mathcal{D}(V_1), f(\mathcal{D}(V_2))) \leq 2.$$

Let $[m_0, m_1, m_2]$ be a geodesic in $\mathcal{C}(S)$ such that $m_0 \in \mathcal{D}(V_1)$ and $m_2 \in f(\mathcal{D}(V_2))$. (Possibly, $m_1 \in \mathcal{D}(V_1)$ or $m_1 \in f(\mathcal{D}(V_2))$.) By Claim 5.1 (1), both m_0 and m_2 cut F_1. If m_1 also cuts F_1, then we have $\mathrm{diam}_{F_1}(\pi_{F_1}(m_0) \cup \pi_{F_1}(m_2)) \leq 4$ by Lemma 2.2, which together with Claim 5.1 (2) and (3) implies that

$$\begin{aligned} d_{F_1}(l_0, l_2) &\leq \mathrm{diam}_{F_1}(l_0 \cup \pi_{F_1}(m_0)) + \mathrm{diam}_{F_1}(\pi_{F_1}(m_0) \cup \pi_{F_1}(m_2)) \\ &\quad + \mathrm{diam}_{F_1}(\pi_{F_1}(m_2) \cup l_2) \\ &\leq 4 + 4 + 4 = 12. \end{aligned}$$

This contradicts the fact that $d_{F_1}(l_0, l_2) > 12$. Hence, m_1 misses F_1, that is, $m_1 = l_1$. By Claim 5.1 (1), we have $m_0 = l_0$ and $m_2 = l_2$, and we obtain the desired result. Q.E.D.

§6. Proof of Theorem 1.1 when $n = 3$

Let $n = 3$ and g be an integer with $g \geq 3$. Let S be a closed connected orientable surface of genus g. Let l_0 and l_1 be non-separating simple closed curves on S such that $l_0 \cup l_1$ is separating on S and l_0, l_1 are not parallel on S. Let l_2' be a simple closed curve on S such that

$l'_2 \cap l_1 = \emptyset$ and $l_1 \cup l'_2$ is non-separating on S. By Lemma 2.3, there exists a homeomorphism $h_1 : S \to S$ such that $h_1(l_1) = l_1$ and

$$d_{F_1}(l_0, h_1(l'_2)) > 8,$$

where $F_1 = \mathrm{Cl}(S \setminus N(l_1))$. Let $l_2 = h_1(l'_2)$. By Lemma 2.4, $[l_0, l_1, l_2]$ is a geodesic in $\mathcal{C}(S)$. Note that there exists a homeomorphism $h_2 : S \to S$ such that $h_2(l_1) = l_2$ and $h_2(l_2) = l_1$, since l_1 and l_2 are non-separating on S. Let $l'_3 = h_2(l_0)$. Note that $[l_1, l_2, l'_3]$ is a geodesic in $\mathcal{C}(S)$.

Let $S' = \mathrm{Cl}(S \setminus N(l_1 \cup l_2))$. Let $\pi_{S'} = \pi_0 \circ \pi_A : \mathcal{C}^0(S) \to \mathcal{P}(\mathcal{AC}^0(S')) \to \mathcal{P}(\mathcal{C}^0(S'))$ be the subsurface projection introduced in Section 2.

Claim 6.1. *There exists a homeomorphism $h : S \to S$ such that $h(l_1) = l_1$, $h(l_2) = l_2$ and* $\mathrm{diam}_{S'}(\pi_{S'}(l_0) \cup \pi_{S'}(h(l'_3))) > 14$.

Proof. Let γ be the closure of a component of $l'_3 \setminus l_1$. Since $l'_3 \cap l_2 = \emptyset$, we have $\gamma \in \pi_A(l'_3)$, and hence, $\pi_0(\gamma) \in \pi_0(\pi_A(l'_3)) = \pi_{S'}(l'_3)$. Note that $\pi_0(\gamma)$ consists of a single simple closed curve or two disjoint simple closed curves on S'. By Lemma 2.3, there exists a homeomorphism $h : S \to S$ such that $h(l_1) = l_1$, $h(l_2) = l_2$ and $d_{S'}(\pi_{S'}(l_0), h(\pi_0(\gamma))) > 14$. This inequality, together with the fact that $h(\pi_0(\gamma)) \in h(\pi_{S'}(l'_3))$, implies

$$\begin{aligned} \mathrm{diam}_{S'}(\pi_{S'}(l_0) \cup \pi_{S'}(h(l'_3))) &= \mathrm{diam}_{S'}(\pi_{S'}(l_0) \cup h(\pi_{S'}(l'_3))) \\ &\geq d_{S'}(\pi_{S'}(l_0), h(\pi_0(\gamma))) \\ &> 14. \end{aligned}$$

Q.E.D.

Let $l_3 = h(l'_3)$. By Lemma 2.5, $[l_0, l_1, l_2, l_3]$ is a geodesic in $\mathcal{C}(S)$. Note that the following hold.

- $d_{F_1}(l_0, l_2) > 8$.
- $d_{F_2}(l_1, l_3) > 8$, where $F_2 = \mathrm{Cl}(S \setminus N(l_2))$, since $d_{F_1}(l_0, l_2) > 8$ and the homeomorphism $h \circ h_2$ sends l_0, l_1, l_2 to l_3, l_2, l_1, respectively.
- $\mathrm{diam}_{S'}(\pi_{S'}(l_0) \cup \pi_{S'}(l_3)) > 14$.

Let C_1 and C_2 be copies of the compression-body obtained by adding a 1-handle to $F \times [0,1]$, where F is a closed connected orientable surface of genus $g-1$. Let D_1 and D_2 be the non-separating essential disk properly embedded in C_1 and C_2 corresponding to the co-cores of the 1-handles, respectively. We may assume that $\partial_+ C_1 = S$ and $\partial D_1 = l_0$. Choose a homeomorphism $f : \partial_+ C_2 \to \partial_+ C_1$ such that $f(\partial D_2) = l_3$.

Let H_i, C'_i, X_i, P_i $(i = 1, 2)$ be as in Section 4. Note that l_1 and l_2 are non-separating on S and not isotopic to l_0 or l_3. Hence, $P_1(l_1)$ and

$P_2(f^{-1}(l_2))$ are essential simple closed curves on $\partial_- C_1$ and $\partial_- C_2$, respectively. By [5, Theorem 2.7] and its proof (see also [1, Theorem 2.4]), there exist homeomorphisms $f_1 : \partial H_1 \to \partial_- C_1$ and $f_2 : \partial H_2 \to \partial_- C_2$ such that $d_{\partial_- C_1}(f_1(\mathcal{D}(H_1)), P_1(l_1)) \geq 2$ and $d_{\partial_- C_2}(f_2(\mathcal{D}(H_2)), P_2(f^{-1}(l_2))) \geq 2$, respectively. Let $V_i = C_i \cup_{f_i} H_i$ $(i = 1, 2)$. Then, $V_1 \cup_f V_2$ is a genus-g Heegaard splitting. By the arguments similar to those for Claims 4.1, 4.2, 4.5 and 4.6, we obtain the following.

Claim 6.2. (1) *l_1 intersects every element of $\mathcal{D}(V_1) \setminus \{l_0\}$, and l_2 intersects every element of $f(\mathcal{D}(V_2)) \setminus \{l_3\}$.*

(2) *For any element $a \in \mathcal{D}(V_1)$, we have $\pi_{F_1}(a) \neq \emptyset$, and* $\mathrm{diam}_{F_1}(l_0 \cup \pi_{F_1}(a)) \leq 4$.

(3) *For any element $a \in f(\mathcal{D}(V_2))$, we have $\pi_{F_2}(a) \neq \emptyset$, and* $\mathrm{diam}_{F_2}(l_3 \cup \pi_{F_2}(a)) \leq 4$.

Lemma 6.3. (1) *For any element $a \in \mathcal{D}(V_1)$, we have $\pi_{S'}(l_0) \neq \emptyset$, $\pi_{S'}(a) \neq \emptyset$, and* $\mathrm{diam}_{S'}(\pi_{S'}(l_0) \cup \pi_{S'}(a)) \leq 4$.

(2) *For any element $a \in f(\mathcal{D}(V_2))$, we have $\pi_{S'}(l_3) \neq \emptyset$, $\pi_{S'}(a) \neq \emptyset$, and* $\mathrm{diam}_{S'}(\pi_{S'}(l_3) \cup \pi_{S'}(a)) \leq 4$.

Proof. We give a proof for (1) only, since (2) can be proved similarly. Suppose that $\pi_{S'}(l_0) = \emptyset$ (resp. $\pi_{S'}(a) = \emptyset$). This means that for each component γ of $l_0 \cap S'$ (resp. $a \cap S'$), each component of $S' \setminus \gamma$ is an annulus. This shows that S' is a sphere with three boundary components, a contradiction. If $a = l_0$ or $a \cap l_0 = \emptyset$, then we have $\mathrm{diam}_{S'}(\pi_{S'}(l_0) \cup \pi_{S'}(a)) \leq 2$ by Lemma 2.2. Hence, we suppose that $a \neq l_0$ and $a \cap l_0 \neq \emptyset$ in the following.

Let D_a be a disk in V_1 bounded by a, and recall l_0 bounds the disk D_1 in V_1. We may assume that $|D_a \cap D_1|$ is minimal. Let Δ be the closure of a component of $D_a \setminus D_1$ that is outermost in D_a. Let $D_1^{(1)}$ and $D_1^{(2)}$ be the components of $D_1 \setminus \Delta$. By the minimality of $|D_a \cap D_1|$, the disks $D_1^{(1)} \cup \Delta$ and $D_1^{(2)} \cup \Delta$ are essential in V_1.

Claim 6.4. *$D_1^{(1)} \cup \Delta$ or $D_1^{(2)} \cup \Delta$, say $D_1^{(1)} \cup \Delta$, is not isotopic to D_1 in V_1.*

Proof. Let m_1 and m_2 be the two simple closed curves obtained from $l_0 (= \partial D_1)$ by a band move along $\Delta \cap \partial V_1$. Suppose both $D_1^{(1)} \cup \Delta$ and $D_1^{(2)} \cup \Delta$ are isotopic to D_1 in V_1. This implies that m_1 and m_2 are parallel in ∂V_1, and hence, they co-bound an annulus, say A, in S. Further, by slight isotopy, we may suppose that $l_0 \cap (m_1 \cup m_2) = \emptyset$. Note that l_0 is retrieved from $m_1 \cup m_2$ by a band move along an arc α such that $|\alpha \cap (\Delta \cap \partial V_1)| = 1$. Since l_0 is essential, $(\mathrm{int}\alpha) \cap A = \emptyset$. This shows that l_0 cuts off a punctured torus from ∂V_1, which contradicts the assumption that l_0 is non-separating on ∂V_1. Q.E.D.

Hence, by Claim 6.2 (1), l_1 intersects $D_1^{(1)} \cup \Delta$. Since $l_1 \cap D_1 = \emptyset$, l_1 intersects $\partial\Delta \setminus D_1$. Since $l_0 \cup l_1$ is separating on S, there is a subarc γ of $\partial\Delta \setminus D_1$ such that $\partial\gamma \subset l_1$. Let γ' be the closure of a component of $\gamma \setminus N(l_1 \cup l_2)$. Then γ' is an element of $\pi_A(a)\,(\subset \mathcal{AC}^0(S'))$. Hence, we have

$$\mathrm{diam}_{\mathcal{AC}(S')}(\gamma' \cup \pi_A(a)) \leq 1.$$

On the other hand, since γ' is disjoint from l_0, we have

$$\mathrm{diam}_{\mathcal{AC}(S')}(\pi_A(l_0) \cup \gamma') \leq 1.$$

By the triangle inequality, we have

$$\begin{aligned}\mathrm{diam}_{\mathcal{AC}(S')}(\pi_A(l_0) \cup \pi_A(a)) &\leq \mathrm{diam}_{\mathcal{AC}(S')}(\pi_A(l_0) \cup \gamma') \\ &\quad + \mathrm{diam}_{\mathcal{AC}(S')}(\gamma' \cup \pi_A(a)) \\ &\leq 1+1=2.\end{aligned}$$

By Lemma 2.1, we have $\mathrm{diam}_{S'}(\pi_{S'}(l_0) \cup \pi_{S'}(a)) \leq 4$. This completes the proof of Lemma 6.3 (1). Q.E.D.

Lemma 6.5. *$V_1 \cup_f V_2$ is a strongly keen Heegaard splitting whose Hempel distance is* 3.

Proof. Since $l_0 \in \mathcal{D}(V_1)$ and $l_3 \in f(\mathcal{D}(V_2))$, we have

$$d_S(\mathcal{D}(V_1), f(\mathcal{D}(V_2))) \leq 3.$$

Let $[m_0, \ldots, m_p]$ be a geodesic in $\mathcal{C}(S)$ such that $m_0 \in \mathcal{D}(V_1)$, $m_p \in f(\mathcal{D}(V_2))$ and $p \leq 3$.

Claim 6.6. *$m_i = l_1$ or $m_i = l_2$ for some $i \in \{0, \ldots, p\}$.*

Proof. Assume on the contrary that $m_i \neq l_1$ and $m_i \neq l_2$ for every $i \in \{0, \ldots, p\}$. Namely, every m_i cuts S'. By Lemma 2.2, we have

$$\mathrm{diam}_{S'}(\pi_{S'}(m_0) \cup \pi_{S'}(m_p)) \leq 2p \leq 6. \tag{12}$$

By the triangle inequality, we have

$$\begin{aligned}\mathrm{diam}_{S'}(\pi_{S'}(l_0) \cup \pi_{S'}(l_3)) &\leq \mathrm{diam}_{S'}(\pi_{S'}(l_0) \cup \pi_{S'}(m_0)) \\ &\quad + \mathrm{diam}_{S'}(\pi_{S'}(m_0) \cup \pi_{S'}(m_p)) \\ &\quad + \mathrm{diam}_{S'}(\pi_{S'}(m_p) \cup \pi_{S'}(l_3)).\end{aligned} \tag{13}$$

By the inequalities (12), (13) together with Lemma 6.3, we obtain

$$\mathrm{diam}_{S'}(\pi_{S'}(l_0) \cup \pi_{S'}(l_3)) \leq 4+6+4=14,$$

which contradicts the inequality $\mathrm{diam}_{S'}(\pi_{S'}(l_0) \cup \pi_{S'}(l_3)) > 14$ (see Claim 6.1). Q.E.D.

Assume that $m_i = l_1$ for some $i \in \{0, \ldots, p\}$. (The case where $m_i = l_2$ for some $i \in \{0, \ldots, p\}$ can be treated similarly.) By Claim 6.2 (1), we have $l_1 \notin \mathcal{D}(V_1)$ and $l_1 \notin f(\mathcal{D}(V_2))$, which imply $i \neq 0$ and $i \neq p$, respectively. Hence, we have $1 \leq i \leq p-1$ and $2 \leq p (\leq 3)$. If $p = 2$, then $m_1 = l_1$ and $m_2 \in f(\mathcal{D}(V_2))$, and hence

$$
\begin{aligned}
d_{F_2}(l_1, l_3) &= d_{F_2}(m_1, l_3) \\
&\leq \mathrm{diam}_{F_2}(m_1 \cup \pi_{F_2}(m_2)) + \mathrm{diam}_{F_2}(\pi_{F_2}(m_2) \cup l_3) \qquad (14) \\
&\leq 2 + 4 = 6,
\end{aligned}
$$

which contradicts the inequality $d_{F_2}(l_1, l_3) > 8$. Hence, $p = 3$, and this implies that the Hempel distance of $V_1 \cup_f V_2$ is 3. Moreover, we have $i = 1$ (that is, $m_1 = l_1$) since, if $i = 2$, then $[l_0, l_1(= m_2), m_3]$ is a path of length 2 from $\mathcal{D}(V_1)$ to $f(\mathcal{D}(V_2))$, a contradiction.

To prove $m_2 = l_2$, assume on the contrary that $m_2 \neq l_2$. Then m_2, as well as $m_1(= l_1)$ and m_3, cuts F_2. By Lemma 2.2 and Claim 6.2 (3),

$$
\begin{aligned}
d_{F_2}(l_1, l_3) &= d_{F_2}(m_1, l_3) \\
&\leq \mathrm{diam}_{F_2}(m_1 \cup \pi_{F_2}(m_3)) + \mathrm{diam}_{F_2}(\pi_{F_2}(m_3) \cup l_3) \qquad (15) \\
&\leq 4 + 4 = 8,
\end{aligned}
$$

which contradicts the inequality $d_{F_2}(l_1, l_3) > 8$. Hence, $m_2 = l_2$.

By Claim 6.2 (1), we have $m_0 = l_0$ and $m_3 = l_3$. Hence, $[l_0, l_1, l_2, l_3]$ is the unique geodesic realizing the Hempel distance. Q.E.D.

§ **Acknowledgement**

We would like to thank Dr Jesse Johnson for many helpful discussions, particularly for teaching us an idea of constructing a unique geodesic path in the curve complex. We would also like to thank the referee for careful reading of the paper and helpful suggestions.

References

[1] A. Abrams and S. Schleimer, *Distances of Heegaard splittings*, Geom. Topology, 9 (2005), 95–119.

[2] J. Berge and M. Scharlemann, *Multiple genus 2 Heegaard splittings: a missed case*, Algebraic and Geometric Topology, 11 (2011), 1781–1792.

[3] T. Evans, *High distance Heegaard splittings of 3-manifolds*, Topology Appl., 153 (2006), 2631–2647.

[4] W. J. Harvey, *Boundary structure of the modular group. In Riemann surfaces and related topics*, Proceedings of the 1978 Stony Brook Conference (State Univ. New York, Stony Brook, N.Y., 1978), pages 245–251, Princeton, N.J., 1981. Princeton Univ. Press.

[5] J. Hempel, *3-manifolds as viewed from the curve complex*, Topology, 40 (2001), 631–657.

[6] A. Ido, Y. Jang and T. Kobayashi, *Heegaard splittings of distance exactly n*, Algebr. Geom. Topol., 14 (2014), 1395–1411.

[7] J. Johnson, *Non-uniqueness of high distance Heegaard splittings*, arXiv:1308.4599.

[8] T. Li, *Images of the disk complex*, Geom. Dedicata., 158 (2012), 121–136.

[9] H. Masur and Y. Minsky, *Geometry of the complex of curves. I. Hyperbolicity*, Invent. Math., 138 (1999), 103–149.

[10] H. Masur and Y. Minsky, *Geometry of the complex of curves. II. Hierarchical structure*, Geom. Funct. Anal., 10 (2000), 902–974.

[11] Ruifeng Qiu, Yanqing Zou and Qilong Guo, *The Heegaard distances cover all non-negative integers*, Pacific J. Math., 275 (2015), 231–255.

[12] M. Yoshizawa, *High distance Heegaard splittings via Dehn twists*, Algebr. Geom. Topol., 14 (2014), 979–1004.

(A. Ido) *Department of Mathematics Education, Aichi University of Education 1 Hirosawa, Igaya-cho, Kariya-shi, 448-8542, Japan*

(Y. Jang, T. Kobayashi) *Department of Mathematics, Nara Women's University Kitauoya Nishimachi, Nara, 630-8506 Japan*

E-mail address, A. Ido: ayakoido@auecc.aichi-edu.ac.jp
E-mail address, Y. Jang: yeonheejang@cc.nara-wu.ac.jp
E-mail address, T. Kobayashi: tsuyoshi@cc.nara-wu.ac.jp

Advanced Studies in Pure Mathematics 78, 2018
Singularities in Generic Geometry
pp. 313–330

Evolutes of curves in the Lorentz-Minkowski plane

S. Izumiya, M. C. Romero Fuster and M. Takahashi

Abstract.

We can use a moving frame, as in the case of regular plane curves in the Euclidean plane, in order to define the arc-length parameter and the Frenet formula for non-lightlike regular curves in the Lorentz-Minkowski plane. This leads naturally to a well defined evolute associated to non-lightlike regular curves without inflection points in the Lorentz-Minkowski plane. However, at a lightlike point the curve shifts between a spacelike and a timelike region and the evolute cannot be defined by using this moving frame. In this paper, we introduce an alternative frame, the *lightcone frame*, that will allow us to associate an evolute to regular curves without inflection points in the Lorentz-Minkowski plane. Moreover, under appropriate conditions, we shall also be able to obtain globally defined evolutes of regular curves with inflection points. We investigate here the geometric properties of the evolute at lightlike points and inflection points.

§1. Introduction

The evolute of a regular plane curve is a classical subject of differential geometry on Euclidean plane which is defined to be the locus of the centres of the osculating circles of the curve (cf. [3, 7, 8]). It is useful to recognize a vertex of a regular plane curve as a singularity (generically, a 3/2 cusp singularity) of the evolute. Recently, the evolutes have been considered in other spaces, such as hyperbolic, de Sitter, anti de Sitter and Minkowski space, as an application of singularity theory, see [4, 9, 10, 11, 13, 14, 15, 16, 17].

Received March 26, 2016.
Revised June 29, 2016.
2010 *Mathematics Subject Classification.* Primary 53A35; Secondary 53D35, 53C50.
Key words and phrases. evolute, inflection point, lightcone frame, Lagrangian singularity, Legendrian singularity.

For a non-lightlike regular curve in the Lorentz-Minkowski plane, we can use a moving frame along the curve and define the arc-length parameter and the Frenet formula. This leads to the definition of the curvature and the evolute of a non-lightlike regular curves without inflection points in the Lorentz-Minkowski plane, see [14] for the definition and properties of the evolute of a non-lightlike regular curves without inflection points. On the other hand, we can consider the caustics of a regular curve, which is defined even at the lightlike points of the curve. Then the caustics of a non-lightlike regular curve without inflection points coincides the evolute.

The lightlike points occur when of the curve moves between spacelike and timelike regions and it can be seen that closed curves in the Lorentz-Minkowski plane must have at least four lightlike points. Hence we can not define the evolute globally by using the standard moving frame. In this paper, we introduce an alternative frame, composed of lightlike vector directions at each point, that we shall call the *lightcone frame*. This allows us to define not only an evolute for the regular curves without inflection points, but also for regular curves with inflection points under certain conditions in the Lorentz-Minkowski plane. We can see that the evolute of a regular curve with lightlike points is a completion of the evolute of a non-lightlike regular curve.

In §2, we introduce the Frenet formula for non-lightlike curves and the evolute of a non-lightlike regular curves without inflection points. In order to consider the lightlike points, we introduce to the lightcone frame in §3. We obtain a kind of a curvature for a regular curve in the Lorentz-Minkowski plane and prove the corresponding existence and uniqueness theorems. In §4, we see that the evolute of a regular curve without inflection points can be regarded not only as a front (a wavefront) but also as a caustic. Furthermore, we describe the behaviour of the evolute at a lightlike point. In §5, we define the evolute of a regular curve with inflection points under appropriate conditions. We show with some examples that the evolutes obtained in this way for the Lorentz-Minkowski geometry happen to be quite different from the corresponding ones in the well known case of the Euclidean geometry.

All maps and manifolds considered here are differentiable of class C^∞.

Acknowledgement. This work started during the visit of the first and third authors to the Universitat de València. We would like to thank Universitat de València for their kind hospitality. S. Izumiya was partially supported by JSPS KAKENHI Grant Number 26287009, M.C. romero-Fuster was partially supported by grant MTM2012-33073 from

Ministerio de Economia y Competitividad (Spain) and M. Takahashi was partially supported by JSPS KAKENHI Grant Number 26400078.

§2. Preliminaries

The *Lorentz-Minkowski plane* $\mathbb{R}^2_1$ is the plane $\mathbb{R}^2$ endowed with the metric induced by the pseudo-scalar product $\langle \boldsymbol{u}, \boldsymbol{v} \rangle = -u_0v_0 + u_1v_1$, where $\boldsymbol{u} = (u_0, u_1)$ and $\boldsymbol{v} = (v_0, v_1)$.

We say that a non-zero vector $\boldsymbol{u} \in \mathbb{R}^2_1$ is *spacelike* if $\langle \boldsymbol{u}, \boldsymbol{u} \rangle > 0$, *lightlike* if $\langle \boldsymbol{u}, \boldsymbol{u} \rangle = 0$, and *timelike* if $\langle \boldsymbol{u}, \boldsymbol{u} \rangle < 0$ respectively. The *norm* of a vector $\boldsymbol{u} = (u_0, u_1) \in \mathbb{R}^2_1$ is defined by $||\boldsymbol{u}|| = \sqrt{|\langle \boldsymbol{u}, \boldsymbol{u} \rangle|}$ and the vector $\boldsymbol{u}^\perp$ is given by $\boldsymbol{u}^\perp = (u_1, u_0)$. By definition, $\langle \boldsymbol{u}, \boldsymbol{u}^\perp \rangle = 0$ and $||\boldsymbol{u}|| = ||\boldsymbol{u}^\perp||$. We have $\boldsymbol{u}^\perp = \pm \boldsymbol{u}$ if and only if $\boldsymbol{u}$ is lightlike, and $\boldsymbol{u}^\perp$ is timelike (respectively, spacelike) if and only if $\boldsymbol{u}$ is spacelike (respectively, timelike).

We have the pseudo-circle in $\mathbb{R}^2_1$ with centre $\boldsymbol{v} \in \mathbb{R}^2_1$ and $a \in \mathbb{R}$,

$$PS(\boldsymbol{v}, a) = \{\boldsymbol{u} \in \mathbb{R}^2_1 \mid \langle \boldsymbol{u} - \boldsymbol{v}, \boldsymbol{u} - \boldsymbol{v} \rangle = a\}.$$

We can classify the pseudo-circles with centre $\boldsymbol{v} \in \mathbb{R}^2_1$ and radius $r > 0$ into the following types:

$$\begin{aligned} S^1_1(\boldsymbol{v}, r) &= \{\boldsymbol{u} \in \mathbb{R}^2_1 \mid \langle \boldsymbol{u} - \boldsymbol{v}, \boldsymbol{u} - \boldsymbol{v} \rangle = r^2\}, \\ LC^*(\boldsymbol{v}, 0) &= \{\boldsymbol{u} \in \mathbb{R}^2_1 \mid \langle \boldsymbol{u} - \boldsymbol{v}, \boldsymbol{u} - \boldsymbol{v} \rangle = 0\}, \\ H^1(\boldsymbol{v}, -r) &= \{\boldsymbol{u} \in \mathbb{R}^2_1 \mid \langle \boldsymbol{u} - \boldsymbol{v}, \boldsymbol{u} - \boldsymbol{v} \rangle = -r^2\}. \end{aligned}$$

We denote by $S^1_1(r), LC^*$ and $H^1(-r)$ the pseudo-circles centred at the origin in $\mathbb{R}^2_1$.

Let $\gamma : I \to \mathbb{R}^2_1$ be a smooth curve, where I is an interval of $\mathbb{R}$. We say that γ is *spacelike* (respectively, *timelike*) if $\dot\gamma(t) = (d\gamma/dt)(t)$ is a spacelike (respectively, timelike) vector for any $t \in I$. Moreover, a point $\gamma(t)$ (or, t) is called a *spacelike* (respectively, *lightlike*, *timelike*) point if $\dot\gamma(t)$ is a spacelike (respectively, lightlike, timelike) vector.

Let $\gamma : I \to \mathbb{R}^2_1$ be a spacelike or a timelike curve. In this case, we may take the arc-length parameter s of γ. It follows that $||\gamma'(s)|| = 1$ for all $s \in I$, where $\gamma'(s) = (d\gamma/ds)(s)$. We denote by $\boldsymbol{t}(s)$ the unit tangent vector and $\boldsymbol{n}(s)$ the unit normal vector to $\gamma(s)$ such that $\{\boldsymbol{t}(s), \boldsymbol{n}(s)\}$ is oriented anti-clockwise. Actually, $\boldsymbol{t}(s) = \gamma'(s)$ and $\boldsymbol{n}(s) = (-1)^{\omega+1}\gamma'(s)^\perp$, where $\omega = 1$ if γ is timelike and $\omega = 2$ if γ is spacelike. Then we have the Frenet formula:

$$\begin{pmatrix} \boldsymbol{t}'(s) \\ \boldsymbol{n}'(s) \end{pmatrix} = \begin{pmatrix} 0 & \kappa(s) \\ \kappa(s) & 0 \end{pmatrix} \begin{pmatrix} \boldsymbol{t}(s) \\ \boldsymbol{n}(s) \end{pmatrix},$$

where $\kappa(s)$ is defined to be the curvature of γ. Thus,

$$\kappa(s) = \frac{\langle \boldsymbol{t}'(s), \boldsymbol{n}(s)\rangle}{\langle \boldsymbol{n}(s), \boldsymbol{n}(s)\rangle} = (-1)^{\omega+1}\langle \boldsymbol{t}'(s), \boldsymbol{n}(s)\rangle = \langle \gamma''(s), \gamma'(s)^{\perp}\rangle.$$

Even if γ is not parametrised by the arc-length and t denotes the parameter, then the unit tangent and the unit normal vectors to $\gamma(t)$ such that $\{\boldsymbol{t}(t), \boldsymbol{n}(t)\}$ is oriented anti-clockwise are given by

$$\boldsymbol{t}(t) = \frac{\dot\gamma(t)}{||\dot\gamma(t)||},\ \boldsymbol{n}(t) = (-1)^{\omega+1}\frac{\dot\gamma(t)^{\perp}}{||\dot\gamma(t)||}.$$

It follows that

$$\begin{pmatrix} \dot{\boldsymbol{t}}(t) \\ \dot{\boldsymbol{n}}(t) \end{pmatrix} = \begin{pmatrix} 0 & ||\dot\gamma(t)||\kappa(t) \\ ||\dot\gamma(t)||\kappa(t) & 0 \end{pmatrix} \begin{pmatrix} \boldsymbol{t}(t) \\ \boldsymbol{n}(t) \end{pmatrix}$$

and the curvature is given by $\kappa(t) = \langle \ddot\gamma(t), \dot\gamma(t)^{\perp}\rangle / ||\dot\gamma(t)||^3$.

We call a point $\gamma(t_0)$ (or, t_0) an *inflection point* if $\langle \ddot\gamma(t_0), \dot\gamma(t_0)^{\perp}\rangle = 0$. An inflection point of a spacelike, or a timelike regular curve γ is a point $\gamma(t)$ such that $\kappa(t) = 0$.

The *evolute* of a curve γ without inflection points is defined to be the curve in $\mathbb{R}^2_1$ given by

$$e(t) = \gamma(t) - \frac{1}{\kappa(t)}\boldsymbol{n}(t). \tag{1}$$

The properties of the evolute of a spacelike or a timelike curve are given in [14].

We cannot consider the evolute (1) at a lightlike point, since the curvature is not well defined at it. In this paper, we introduce another frame and define the evolutes of regular curves, both without inflection points and with inflection points under appropriate conditions, in the Lorentz-Minkowski plane.

§3. Lightcone frame

We denote $\mathbb{L}^+ = (1, 1)$ and $\mathbb{L}^- = (1, -1)$. By definition, $\mathbb{L}^+$ and $\mathbb{L}^-$ are independent lightlike vectors and $\langle \mathbb{L}^+, \mathbb{L}^-\rangle = -2$. We call $\{\mathbb{L}^+, \mathbb{L}^-\}$ a *lightcone frame* on $\mathbb{R}^2_1$.

Let $\gamma : I \to \mathbb{R}^2_1$ be a regular curve (with lightlike points). There exists a smooth function $(\alpha, \beta) : I \to \mathbb{R}^2 \setminus \{0\}$ such that

$$\dot\gamma(t) = \alpha(t)\mathbb{L}^+ + \beta(t)\mathbb{L}^- \tag{2}$$

for all $t \in I$. We say that a *regular curve* γ *with the lightlike tangential data* (α, β) if the condition (2) holds. Then we have $\dot{\gamma}(t)^{\perp} = \alpha(t)\mathbb{L}^{+} - \beta(t)\mathbb{L}^{-}$. Since $\langle \dot{\gamma}(t), \dot{\gamma}(t) \rangle = -4\alpha(t)\beta(t)$, $\gamma(t)$ is a spacelike (respectively, lightlike or timelike) point if and only if $\alpha(t)\beta(t) < 0$ (respectively, $= 0$ or > 0).

Theorem 1. (The Existence Theorem) *Let* $(\alpha, \beta) : I \to \mathbb{R}^2 \setminus \{0\}$ *be a smooth mapping. There exists a regular curve* $\gamma : I \to \mathbb{R}^2_1$ *with the lightlike tangential data* (α, β).

Proof. Let $\gamma : I \to \mathbb{R}^2_1$ be

$$\gamma(t) = \left(\int (\alpha(t) + \beta(t))\, dt, \int (\alpha(t) - \beta(t))\, dt \right).$$

By a direct calculation, γ is a regular curve and satisfies the condition (2). Q.E.D.

Proposition 1. *If* γ *and* $\widetilde{\gamma} : I \to \mathbb{R}^2_1$ *are regular curves with the same lightlike tangential data* (α, β), *then there exists a constant* $c \in \mathbb{R}^2_1$ *such that* $\widetilde{\gamma}(t) = \gamma(t) + c$.

Proof. Since $\dot{\gamma}(t) = \dot{\widetilde{\gamma}}(t)$ for all $t \in I$, we have the result. Q.E.D.

The condition of Proposition 1 seems to be strong. We consider a mild condition for the uniqueness as a Lorentz motion.

Definition 1. Let γ and $\widetilde{\gamma} : I \to \mathbb{R}^2_1$ be regular curves. We say that γ and $\widetilde{\gamma}$ are *congruent through a Lorentz motion* if there exist a matrix A and a constant $c \in \mathbb{R}^2_1$ such that $\widetilde{\gamma}(t) = A(\gamma(t)) + c$ for all $t \in I$, where A is given by

$$A = \begin{pmatrix} \cosh\theta & -\sinh\theta \\ -\sinh\theta & \cosh\theta \end{pmatrix} \text{ or } A = -\begin{pmatrix} \cosh\theta & -\sinh\theta \\ -\sinh\theta & \cosh\theta \end{pmatrix}$$

for some $\theta \in \mathbb{R}$.

Proposition 2. *Let* γ *and* $\widetilde{\gamma} : I \to \mathbb{R}^2_1$ *be regular curves with the lightlike tangential data* (α, β) *and* $(\widetilde{\alpha}, \widetilde{\beta})$ *respectively. Suppose that* γ *and* $\widetilde{\gamma}$ *are congruent through a Lorentz motion, that is, there exist a matrix*

$$A = \begin{pmatrix} \cosh\theta & -\sinh\theta \\ -\sinh\theta & \cosh\theta \end{pmatrix} \left(\text{or, } A = -\begin{pmatrix} \cosh\theta & -\sinh\theta \\ -\sinh\theta & \cosh\theta \end{pmatrix} \right)$$

and a constant $c \in \mathbb{R}^2_1$ *such that* $\widetilde{\gamma}(t) = A(\gamma(t)) + c$. *Then*

$$\widetilde{\alpha}(t) = (\cosh\theta - \sinh\theta)\alpha(t), \; \widetilde{\beta}(t) = (\cosh\theta + \sinh\theta)\beta(t)$$

(*or*, $\widetilde{\alpha}(t) = -(\cosh\theta - \sinh\theta)\alpha(t)$, $\widetilde{\beta}(t) = -(\cosh\theta + \sinh\theta)\beta(t)$).

Proof. Suppose that $\widetilde{\gamma}(t) = A(\gamma(t)) + c$. Since

$$\begin{aligned}\dot{\widetilde{\gamma}}(t) &= A(\dot{\gamma}(t)) = A\left(\alpha(t)\mathbb{L}^+ + \beta(t)\mathbb{L}^-\right) = \alpha(t)A(\mathbb{L}^+) + \beta(t)A(\mathbb{L}^-)\\ &= \alpha(t)(\cosh\theta - \sinh\theta)\mathbb{L}^+ + \beta(t)(\cosh\theta + \sinh\theta)\mathbb{L}^-,\end{aligned}$$

we have the result. Q.E.D.

Note that $\cosh\theta - \sinh\theta = e^{-\theta}$ and $\cosh\theta + \sinh\theta = e^{\theta}$.

Theorem 2. (The Uniqueness Theorem) *Let* γ *and* $\widetilde{\gamma} : I \to \mathbb{R}^2_1$ *be regular curves with the lightlike tangential data* (α, β) *and* $(\widetilde{\alpha}, \widetilde{\beta})$ *respectively. Suppose that the lightlike points of* γ *and* $\widetilde{\gamma}$ *are isolated. If*

$$\alpha(t)\beta(t) = \widetilde{\alpha}(t)\widetilde{\beta}(t)$$

and

$$\dot{\alpha}(t)\beta(t) - \alpha(t)\dot{\beta}(t) = \dot{\widetilde{\alpha}}(t)\widetilde{\beta}(t) - \widetilde{\alpha}(t)\dot{\widetilde{\beta}}(t)$$

for all $t \in I$, *then* γ *and* $\widetilde{\gamma}$ *are congruent through a Lorentz motion.*

Proof. We fix a non-lightlike point $\gamma(t_0)$ of γ and $\widetilde{\gamma}(t_0)$ of $\widetilde{\gamma}$. Then $\alpha(t_0)\beta(t_0) = \widetilde{\alpha}(t_0)\widetilde{\beta}(t_0) > 0$ or < 0. There exists a Lorenz motion, namely, a matrix $A = \begin{pmatrix} \cosh\theta & -\sinh\theta \\ -\sinh\theta & \cosh\theta \end{pmatrix}$ and a constant $c \in \mathbb{R}^2_1$, such that

$$\widetilde{\gamma}(t_0) = \pm A(\gamma(t_0)) + c, \quad \dot{\widetilde{\gamma}}(t_0) = \pm A\dot{\gamma}(t_0).$$

By differentiating $\alpha(t)\beta(t) = \widetilde{\alpha}(t)\widetilde{\beta}(t)$, we have

$$\dot{\alpha}(t)\beta(t) + \alpha(t)\dot{\beta}(t) = \dot{\widetilde{\alpha}}(t)\widetilde{\beta}(t) + \widetilde{\alpha}(t)\dot{\widetilde{\beta}}(t).$$

It follows from the second condition

$$\dot{\alpha}(t)\beta(t) - \alpha(t)\dot{\beta}(t) = \dot{\widetilde{\alpha}}(t)\widetilde{\beta}(t) - \widetilde{\alpha}(t)\dot{\widetilde{\beta}}(t)$$

that $\dot{\alpha}(t)\beta(t) = \dot{\widetilde{\alpha}}(t)\widetilde{\beta}(t)$ and $\alpha(t)\dot{\beta}(t) = \widetilde{\alpha}(t)\dot{\widetilde{\beta}}(t)$. Thus we have

$$\begin{pmatrix} \alpha(t) & \widetilde{\alpha}(t) \\ \dot{\alpha}(t) & \dot{\widetilde{\alpha}}(t) \end{pmatrix} \begin{pmatrix} \beta(t) \\ -\widetilde{\beta}(t) \end{pmatrix} = \begin{pmatrix} 0 \\ 0 \end{pmatrix}.$$

For a non-lightlike point $\gamma(t)$, we have $\alpha(t) \neq 0$ and $\beta(t) \neq 0$. Therefore $\alpha(t)\dot{\widetilde{\alpha}}(t) - \dot{\alpha}(t)\widetilde{\alpha}(t) = 0$ for non-lightlike points. It follows that $(d/dt)(\widetilde{\alpha}(t)/\alpha(t)) = 0$ and hence there is a constant $b \in \mathbb{R}$ such that $\widetilde{\alpha}(t) = b\alpha(t)$. Since $\gamma(t_0)$ is a non-lightlike point and $\widetilde{\alpha}(t_0) = b\alpha(t_0)$, we have $b = \pm e^{-\theta}$. Moreover, $\widetilde{\beta}(t) = (1/b)\beta(t)$ for non-lightlike points. Since lightlike points of γ and $\widetilde{\gamma}$ are isolated, we have $\widetilde{\alpha}(t) = b\alpha(t)$ and $\widetilde{\beta}(t) = (1/b)\beta(t)$ on I. Thus,

$$\widetilde{\alpha}(t) = \pm(\cosh\theta - \sinh\theta)\alpha(t),\ \widetilde{\beta}(t) = \pm(\cosh\theta + \sinh\theta)\beta(t).$$

It follows that $(d/dt)(\widetilde{\gamma}(t) \mp A(\gamma(t))) = 0$. By $\widetilde{\gamma}(t_0) = \pm A(\gamma(t_0)) + c$, we have $\widetilde{\gamma}(t) = \pm A(\gamma(t)) + c$. Therefore, γ and $\widetilde{\gamma}$ are congruent through the Lorentz motion. Q.E.D.

Remark 1. Let $\gamma(t) = (t,t)$ and $\widetilde{\gamma}(t) = (t,-t)$. Since $(\alpha(t),\beta(t)) = (1,0)$ and $(\widetilde{\alpha}(t),\widetilde{\beta}(t)) = (0,1)$, the conditions $\alpha(t)\beta(t) = \widetilde{\alpha}(t)\widetilde{\beta}(t)$ and $\dot{\alpha}(t)\beta(t) - \alpha(t)\dot{\beta}(t) = \dot{\widetilde{\alpha}}(t)\widetilde{\beta}(t) - \widetilde{\alpha}(t)\dot{\widetilde{\beta}}(t)$ in Theorem 2 are satisfied. However, $\mathbb{L}^+$ and $\mathbb{L}^-$ are not congruent through a Lorentz motion by Proposition 2.

§4. Evolutes of regular curves without inflection points

Let $\gamma : I \to \mathbb{R}^2_1$ be a regular curve with the lightlike tangential data (α,β). Since $\langle\ddot{\gamma}(t), \dot{\gamma}(t)^{\perp}\rangle = 2(\dot{\alpha}(t)\beta(t) - \alpha(t)\dot{\beta}(t))$, $\gamma(t_0)$ is an inflection point of γ if and only if

$$\dot{\alpha}(t_0)\beta(t_0) - \alpha(t_0)\dot{\beta}(t_0) = 0. \tag{3}$$

We define an *evolute* $Ev(\gamma) : I \to \mathbb{R}^2_1$ of $\gamma : I \to \mathbb{R}^2_1$ with the lightlike tangential data (α,β) by

$$Ev(\gamma)(t) = \gamma(t) - \frac{2\alpha(t)\beta(t)}{\dot{\alpha}(t)\beta(t) - \alpha(t)\dot{\beta}(t)}\left(\alpha(t)\mathbb{L}^+ - \beta(t)\mathbb{L}^-\right) \tag{4}$$

without inflection points.

Suppose that γ is a spacelike (or, timelike) regular curve. We have the following expression for the curvature κ in terms of the lightlike tangential data (α,β) of γ.

Proposition 3. *Let $\gamma : I \to \mathbb{R}^2_1$ be a spacelike (or, timelike) regular curve with the lightlike tangential data (α,β). The curvature κ of γ is given by*

$$\kappa(t) = \frac{\dot{\alpha}(t)\beta(t) - \alpha(t)\dot{\beta}(t)}{4|\alpha(t)\beta(t)|\sqrt{|\alpha(t)\beta(t)|}}.$$

Proof. Since $\dot{\gamma}(t) = \alpha(t)\mathbb{L}^+ + \beta(t)\mathbb{L}^-$, we have $\ddot{\gamma}(t) = \dot{\alpha}(t)\mathbb{L}^+ + \dot{\beta}(t)\mathbb{L}^-$ and $\dot{\gamma}(t)^{\perp} = \alpha(t)\mathbb{L}^+ - \beta(t)\mathbb{L}^-$. It follows that

$$\kappa(t) = \frac{\langle\ddot{\gamma}(t), \dot{\gamma}(t)^{\perp}\rangle}{||\dot{\gamma}(t)||^3} = \frac{\dot{\alpha}(t)\beta(t) - \alpha(t)\dot{\beta}(t)}{4|\alpha(t)\beta(t)|\sqrt{|\alpha(t)\beta(t)|}}.$$

Q.E.D.

Remark 2. By Proposition 3, the conditions of Theorem 2 say that the curvatures of spacelike (or, timelike) congruent regular curves are the same.

Since

$$\boldsymbol{n}(t) = (-1)^{\omega}\frac{\dot{\gamma}(t)^{\perp}}{||\dot{\gamma}(t)||} = (-1)^{\omega}\frac{\alpha(t)\mathbb{L}^{+} - \beta(t)\mathbb{L}^{-}}{2\sqrt{|\alpha(t)\beta(t)|}}$$

and Proposition 3, the evolute (1) of a regular non-lightlike curve is given by

$$\begin{aligned} e(t) &= \gamma(t) - \frac{1}{\kappa(t)}\boldsymbol{n}(t) \\ &= \gamma(t) + (-1)^{\omega}\frac{2|\alpha(t)\beta(t)|}{\dot{\alpha}(t)\beta(t) - \alpha(t)\dot{\beta}(t)}(\alpha(t)\mathbb{L}^{+} - \beta(t)\mathbb{L}^{-}). \end{aligned}$$

If γ is spacelike (respectively, timelike), then $\omega = 2$ and $\alpha(t)\beta(t) < 0$ (respectively, $\omega = 1$ and $\alpha(t)\beta(t) > 0$). It follows that

$$e(t) = \gamma(t) - \frac{2\alpha(t)\beta(t)}{\dot{\alpha}(t)\beta(t) - \alpha(t)\dot{\beta}(t)}\left(\alpha(t)\mathbb{L}^{+} - \beta(t)\mathbb{L}^{-}\right) = Ev(\gamma)(t).$$

Therefore, the evolute $Ev(\gamma)(t)$ is a generalization of the evolute $e(t)$.

Remark 3. If $\gamma(t_0)$ is a lightlike point of γ, then $\alpha(t_0) = 0$ and $\beta(t_0) \neq 0$, or $\alpha(t_0) \neq 0$ and $\beta(t_0) = 0$. Thus, we have $Ev(\gamma)(t_0) = \gamma(t_0)$.

We see next that the evolute $Ev(\gamma)(t)$ of γ without inflection points can be regarded not only as a front (a wavefront), but also as a caustic.

Let $\gamma : I \to \mathbb{R}^2_1$ be a regular curve with the lightlike tangential data (α, β) and without inflection points. We consider two families of functions:

$$F : I \times \mathbb{R}^2_1 \to \mathbb{R}$$

is given by

$$F(t, \boldsymbol{v}) = \langle \gamma(t) - \boldsymbol{v}, \dot{\gamma}(t)\rangle,$$

and

$$D : I \times \mathbb{R}^2_1 \to \mathbb{R}$$

is given by

$$D(t, \boldsymbol{v}) = \langle \gamma(t) - \boldsymbol{v}, \gamma(t) - \boldsymbol{v}\rangle.$$

Given $\boldsymbol{v} \in \mathbb{R}^2_1$, we denote $f_{\boldsymbol{v}}(t) = F(t, \boldsymbol{v})$ and $d_{\boldsymbol{v}}(t) = D(t, \boldsymbol{v})$.

Proposition 4. (1) $f_{\boldsymbol{v}}(t)=0$ *if and only if there exists* $\lambda \in \mathbb{R}$ *such that* $\boldsymbol{v}=\gamma(t)-\lambda\dot{\gamma}(t)^{\perp}$.

(2) $f_{\boldsymbol{v}}(t)=\dot{f}_{\boldsymbol{v}}(t)=0$ *if and only if*

$$\boldsymbol{v}=\gamma(t)-\Big(2\alpha(t)\beta(t)/(\dot{\alpha}(t)\beta(t)-\alpha(t)\dot{\beta}(t))\Big)\dot{\gamma}(t)^{\perp}.$$

Proof. (1) $\langle\gamma(t)-\boldsymbol{v},\dot{\gamma}(t)\rangle=0$ if and only if there exists $\lambda \in \mathbb{R}$ such that $\gamma(t)-\boldsymbol{v}=\lambda\dot{\gamma}(t)^{\perp}$ if and only if $\boldsymbol{v}=\gamma(t)-\lambda\dot{\gamma}(t)^{\perp}$.

(2) Since $\dot{f}_{\boldsymbol{v}}(t)=\langle\dot{\gamma}(t),\dot{\gamma}(t)\rangle+\langle\gamma(t)-\boldsymbol{v},\ddot{\gamma}(t)\rangle=-4\alpha(t)\beta(t)+2\lambda(\dot{\alpha}(t)\beta(t)-\alpha(t)\dot{\beta}(t))=0$, we have $\lambda=2\alpha(t)\beta(t)/(\dot{\alpha}(t)\beta(t)-\alpha(t)\dot{\beta}(t))$. The converse also holds. Q.E.D.

Clearly, we have the following relationship between $f_{\boldsymbol{v}}$ and $d_{\boldsymbol{v}}$: $\dot{d}_{\boldsymbol{v}}(t)=2f_{\boldsymbol{v}}(t)$. Then, as a consequence of Proposition 4, we obtain the following result.

Proposition 5. (1) $\dot{d}_{\boldsymbol{v}}(t)=0$ *if and only if there exists* $\lambda \in \mathbb{R}$ *such that* $\boldsymbol{v}=\gamma(t)-\lambda\dot{\gamma}(t)^{\perp}$.

(2) $\dot{d}_{\boldsymbol{v}}(t)=\ddot{d}_{\boldsymbol{v}}(t)=0$ *if and only if*

$$\boldsymbol{v}=\gamma(t)-\Big(2\alpha(t)\beta(t)/(\dot{\alpha}(t)\beta(t)-\alpha(t)\dot{\beta}(t))\Big)\dot{\gamma}(t)^{\perp}.$$

We refer to [1, 2, 11, 12, 13, 18] for the definitions of Morse families in the theories of Legendre and Lagrange singularities. In particular, we shall follow the notations in [11, 12, 13].

Proposition 6. *The map* $F: I\times\mathbb{R}^2_1\to\mathbb{R}$ *is a Morse family of hypersurfaces, namely,*

$$\left(F,\frac{\partial F}{\partial t}\right): I\times\mathbb{R}^2_1\to\mathbb{R}\times\mathbb{R}$$

is non-singular.

Proof. We denote $\gamma(t)=(x(t),y(t))$ and $\boldsymbol{v}=(x,y)$. It is enough to show that

$$\operatorname{rank}\begin{pmatrix}\partial F/\partial t & \partial^2F/\partial t^2\\ \partial F/\partial x & \partial^2F/\partial t\partial x\\ \partial F/\partial y & \partial^2F/\partial t\partial y\end{pmatrix}(t,\boldsymbol{v})=2.$$

Since $F(t,\boldsymbol{v})=\langle\gamma(t)-\boldsymbol{v},\dot{\gamma}(t)\rangle=-(x(t)-x)\dot{x}(t)+(y(t)-y)\dot{y}(t)$, we have

$$\frac{\partial F}{\partial x}(t,\boldsymbol{v})=\dot{x}(t),\ \frac{\partial F}{\partial y}(t,\boldsymbol{v})=-\dot{y}(t),$$
$$\frac{\partial^2F}{\partial t\partial x}(t,\boldsymbol{v})=\ddot{x}(t),\ \frac{\partial^2F}{\partial t\partial y}(t,\boldsymbol{v})=-\ddot{y}(t).$$

It follows that $-\dot{x}(t)\ddot{y}(t)+\ddot{x}(t)\dot{y}(t)=-\langle\ddot{\gamma}(t),\dot{\gamma}(t)^{\perp}\rangle\neq 0$. Q.E.D.

The discriminant set of F is given by

$$\Sigma(F)=\left\{(t,\boldsymbol{v})\in I\times\mathbb{R}^2_1 \mid f_{\boldsymbol{v}}(t)=f'_{\boldsymbol{v}}(t)=0\right\}.$$

We consider the projective cotangent bundle $\pi : PT^*\mathbb{R}^2_1\to\mathbb{R}^2_1$ over $\mathbb{R}^2_1$. By Proposition 6, we have that $\Sigma(F)$ is a 1-dimensional submanifold and

$$\mathcal{L}_F:\Sigma(F)\to PT^*\mathbb{R}^2_1;\ (t,\boldsymbol{v})\mapsto\left(\boldsymbol{v},\left[\frac{\partial F}{\partial x}(t,\boldsymbol{v}):\frac{\partial F}{\partial y}(t,\boldsymbol{v})\right]\right)$$

is a Legendre immersion with respect to the canonical contact structure on $PT^*\mathbb{R}^2_1$. Now, it follows from Proposition 4 that $\pi\circ\mathcal{L}_F(\Sigma(F))$ coincides with the evolute of γ. Therefore, we get that the evolute $Ev(\gamma)$ can be interpreted as the front (wavefront) of the $\mathcal{L}_F$.

Proposition 7. *The map $D : I\times\mathbb{R}^2_1\to\mathbb{R}$ is a Morse family of functions, namely,*

$$\frac{\partial D}{\partial t}:I\times\mathbb{R}^2_1\to\mathbb{R}$$

is a non-singular.

Proof. We use the same notations as in the proof of Proposition 6. Since $(\partial D/\partial t)(t,\boldsymbol{v})=2F(t,\boldsymbol{v})$, it is enough to show that the gradient vector of F is non-zero. $(\partial F/\partial x)(t,\boldsymbol{v})=\dot{x}(t),(\partial F/\partial y)(t,\boldsymbol{v})=-\dot{y}(t)$ and since γ is a regular curve, we have the conclusion. Q.E.D.

The catastrophe set and the bifurcation set of D are respectively given by

$$C(D)=\left\{(t,\boldsymbol{v})\in I\times\mathbb{R}^2_1 \mid \dot{d}_{\boldsymbol{v}}(t)=0\right\}$$

and

$$B_D=\left\{\boldsymbol{v}\in\mathbb{R}^2_1 | \text{ there exists } t\in I \text{ such that } (t,\boldsymbol{v})\in C(D),\ \ddot{d}_{\boldsymbol{v}}(t)=0\right\}.$$

We consider the cotangent bundle $\widetilde{\pi}:T^*\mathbb{R}^2_1\to\mathbb{R}^2_1$ over $\mathbb{R}^2_1$. By Proposition 7, $C(D)$ is a smooth 2-dimensional submanifold and

$$L(D):C(D)\to T^*\mathbb{R}^2_1;\ (t,\boldsymbol{v})\mapsto\left(\boldsymbol{v},\frac{\partial D}{\partial x}(t,\boldsymbol{v}),\frac{\partial D}{\partial y}(t,\boldsymbol{v})\right)$$

is a Lagrange immersion with respect to the canonical symplectic structure on $T^*\mathbb{R}^2_1$. By Proposition 5, the critical value set of $\widetilde{\pi}\circ L(D)$ is the bifurcation set of D. Therefore, the evolute $Ev(\gamma)$ is the caustic of $L(D)$.

Example 1. Let $\gamma : [0, 2\pi) \to \mathbb{R}^2_1$ be a circle $\gamma(t) = (r\cos t, r\sin t)$ in the Minkowski plane, where $r > 0$. Since

$$\begin{aligned}\dot\gamma(t) &= (-r\sin t, r\cos t)\\ &= \frac{1}{2}(-r\sin t + r\cos t)\,\mathbb{L}^+ + \frac{1}{2}(-r\sin t - r\cos t)\,\mathbb{L}^-,\end{aligned}$$

we have

$$\alpha(t) = \frac{1}{2}(-r\sin t + r\cos t),\ \beta(t) = \frac{1}{2}(-r\sin t - r\cos t).$$

It follows that the evolute of the circle is given by

$$Ev(\gamma)(t) = \left(r(1 - \sin^2 t + \cos^2 t)\cos t, r(1 + \sin^2 t - \cos^2 t)\sin t\right),$$

see Figure 1.

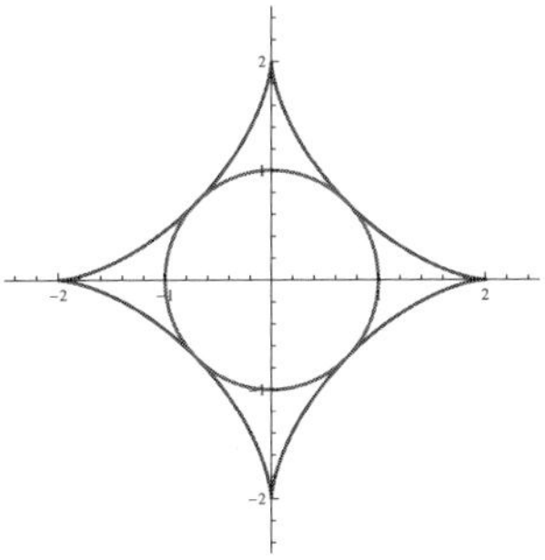

Figure 1. the circle with $r = 1$ and the evolute.

Remark 4. It is worth noting that the evolute of circles in the Euclidean plane is a point. Therefore the evolute in the Lorenz-Minkowski plane is different from the evolute in the Euclidean plane.

A point t (or, $\gamma(t)$) is called a *vertex* for a non-lightlike regular curve γ if $\dot\kappa(t) = 0$. The following result has been given in [14].

Proposition 8. ([14, Proposition 3.2]) *Let $\gamma : I \to \mathbb{R}^2_1$ be a non-lightlike regular curve without inflection points.*

(1) *The evolute of a spacelike (respectively, timelike) curve is a timelike (respectively, spacelike) curve.*

(2) *The evolute of γ is singular precisely at the vertices of γ.*

We consider now the case of lightlike points.

Proposition 9. *Let $\gamma : I \to \mathbb{R}^2_1$ be a regular curve with the lightlike tangential data (α, β) and without inflection points.*

(1) *If $\gamma(t_0)$ is a lightlike point of γ, then $Ev(\gamma)(t_0)$ is also a lightlike point of $Ev(\gamma)$.*

(2) *If $\gamma(t_0)$ is a lightlike point of γ, then $Ev(\gamma)(t_0)$ is a regular point of $Ev(\gamma)$.*

Proof. (1) By definition of the evolute of γ, we have

$$\dot{Ev}(\gamma)(t) = \alpha(t)\mathbb{L}^+ + \beta(t)\mathbb{L}^- - \frac{2\alpha(t)\beta(t)}{\dot\alpha(t)\beta(t) - \alpha(t)\dot\beta(t)}\left(\dot\alpha(t)\mathbb{L}^+ - \dot\beta(t)\mathbb{L}^-\right)$$
$$-\frac{d}{dt}\left(\frac{2\alpha(t)\beta(t)}{\dot\alpha(t)\beta(t) - \alpha(t)\dot\beta(t)}\right)\left(\alpha(t)\mathbb{L}^+ - \beta(t)\mathbb{L}^-\right).$$

Moreover, $(d/dt)(2\alpha(t)\beta(t)/\dot\alpha(t)\beta(t) - \alpha(t)\dot\beta(t))$ is given by

$$2\frac{\dot\alpha^2(t)\beta^2(t) - \alpha^2(t)\dot\beta^2(t) - \alpha(t)\beta(t)(\ddot\alpha(t)\beta(t) - \alpha(t)\ddot\beta(t))}{\left(\dot\alpha(t)\beta(t) - \alpha(t)\dot\beta(t)\right)^2}.$$

If $\alpha(t_0) = 0$ and $\beta(t_0) \neq 0$, then $\dot{Ev}(\gamma)(t_0) = 3\beta(t_0)\mathbb{L}^-$. On the other hand, if $\beta(t_0) = 0$ and $\alpha(t_0) \neq 0$, then $\dot{Ev}(\gamma)(t_0) = 3\alpha(t_0)\mathbb{L}^+$. Hence $Ev(\gamma)(t_0)$ is also a lightlike point of $Ev(\gamma)$.

(2) By the same calculation of (1), $\dot{Ev}(\gamma)(t_0) \neq 0$ at a lightlike point $\gamma(t_0)$ of the curve. Q.E.D.

If we denote $\dot{Ev}(\gamma)(t) = \alpha_{Ev}(t)\mathbb{L}^+ + \beta_{Ev}(t)\mathbb{L}^-$, then $\alpha_{Ev}(t) =$

$$\alpha(t)\left(\frac{-3\dot\alpha^2(t)\beta^2(t) + 3\alpha^2(t)\dot\beta^2(t) + 2\alpha(t)\beta(t)(\ddot\alpha(t)\beta(t) - \alpha(t)\ddot\beta(t))}{(\dot\alpha(t)\beta(t) - \alpha(t)\dot\beta(t))^2}\right),$$

$\beta_{Ev}(t) =$

$$-\beta(t)\left(\frac{-3\dot\alpha^2(t)\beta^2(t) + 3\alpha^2(t)\dot\beta^2(t) + 2\alpha(t)\beta(t)(\ddot\alpha(t)\beta(t) - \alpha(t)\ddot\beta(t))}{(\dot\alpha(t)\beta(t) - \alpha(t)\dot\beta(t))^2}\right).$$

As a corollary of Propositions 8 and 9, we have the following result.

Corollary 1. *Let $\gamma : I \to \mathbb{R}^2_1$ be a regular curve with lightlike tangential data (α, β) and without inflection points.*

(1) *Suppose that $Ev(\gamma)$ is a regular curve. Then γ is a spacelike (respectively, lightlike or timelike) curve if and only if $Ev(\gamma)$ is a timelike (respectively, lightlike or spacelike) curve.*

(2) *The evolute $Ev(\gamma)$ is singular precisely at the vertices of γ.*

The singularities of $d_{\boldsymbol{v}}$ estimate the contact of γ with the pseudo circles. By Proposition 5, the evolute is given by the locus of the centres of the pseudo circles of at least second order contact with γ at t_0. This pseudo circle is given by its centre $\boldsymbol{v} = Ev(\gamma)(t_0)$ and radius $r = ||\gamma(t_0) - \boldsymbol{v}||$, namely,

$$\begin{aligned} &PS(\boldsymbol{v}, \langle \gamma(t_0) - \boldsymbol{v}, \gamma(t_0) - \boldsymbol{v} \rangle) \\ &= \{(x,y) \in \mathbb{R}^2_1 \mid \langle (x,y) - \boldsymbol{v}, (x,y) - \boldsymbol{v} \rangle = \langle \gamma(t_0) - \boldsymbol{v}, \gamma(t_0) - \boldsymbol{v} \rangle \}. \end{aligned}$$

By a direct calculation, we have

$$\langle \gamma(t_0) - \boldsymbol{v}, \gamma(t_0) - \boldsymbol{v} \rangle = 4 \left(\frac{2\alpha(t_0)\beta(t_0)}{\dot{\alpha}(t_0)\beta(t_0) - \alpha(t_0)\dot{\beta}(t_0)} \right)^2 \alpha(t_0)\beta(t_0).$$

Since $\gamma(t_0)$ is a timelike (respectively, lightlike, or spacelike) point of $\gamma(t)$ if and only if $\alpha(t_0)\beta(t_0) > 0$ (respectively, $= 0$ or < 0), the pseudo circle is $S^1_1(\boldsymbol{v}, r)$ (respectively, $LC^*(\boldsymbol{v}, 0)$ or $H^1(\boldsymbol{v}, -r)$), see Figure 2.

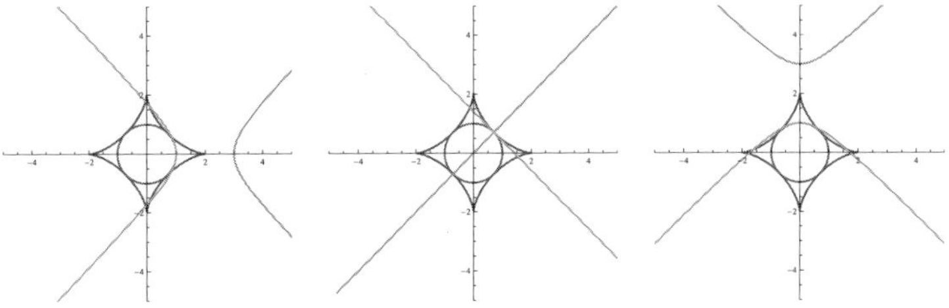

$t = 0$ $\qquad$ $t = \pi/4$ $\qquad$ $t = \pi/2$

Figure 2. The pseudo circles and the evolute of the circle in Example 1.

§5. Evolutes of regular curves with inflection points

In the Euclidean plane, we cannot define the evolutes of regular curves and fronts at their inflection points (cf. [3, 5, 7, 8]). On the other hand, under appropriate conditions in the Euclidean plane, we can define an evolute at the inflection points of a frontal (cf. [6]).

In the Lorentz-Minkowski plane, the lightlike points play the role of the singular points. We may define the evolute of a regular curve at its inflection points under appropriate conditions. It follows that the situation in both cases, the Euclidean geometry and the Lorentz-Minkowski geometry, appears to be quite different.

Let $\gamma : I \to \mathbb{R}^2_1$ be a regular curve with inflection points, having lightlike tangential data (α, β). We may define an evolute under the following existence and uniqueness conditions:

Definition 2. The *evolute* $Ev(\gamma) : I \to \mathbb{R}^2_1$ *of* γ is given by

$$Ev(\gamma)(t) = \gamma(t) + \lambda(t)(\alpha(t)\mathbb{L}^+ - \beta(t)\mathbb{L}^-), \tag{5}$$

if there exists a unique smooth function $\lambda : I \to \mathbb{R}$ such that

$$-2\alpha(t)\beta(t) = \lambda(t)(\dot\alpha(t)\beta(t) - \alpha(t)\dot\beta(t)).$$

In such case, we say that *the evolute* $Ev(\gamma)$ *exists.*

The uniqueness condition is well-known as a topological condition.

Lemma 1. *Suppose that there exists a continuous function* $\lambda : I \to \mathbb{R}$ *such that* $\lambda(t) = -2\alpha(t)\beta(t)/(\dot\alpha(t)\beta(t) - \alpha(t)\dot\beta(t))$ *on* $\Lambda = \{t \in I \mid \dot\alpha(t)\beta(t) - \alpha(t)\dot\beta(t) \neq 0\}$. *Then the function* λ *is a unique if and only if* Λ *is a dense subset of* I.

Remark 5. If the inflection points are isolated, then the condition that Λ is a dense subset of I is satisfied.

In this section, we assume that $\Lambda = \{t \in I \mid \dot\alpha(t)\beta(t) - \alpha(t)\dot\beta(t) \neq 0\}$ is a dense subset of I. Then we have that if such a smooth function λ exists, the uniqueness condition is guaranteed by Lemma 1.

Observe that provided the evolute $Ev(\gamma)$ exists at an inflection point, then this must be a lightlike point of γ. Since γ is a regular curve, the function D is a Morse family of functions. Hence $Ev(\gamma)$ is still a caustic of $L(D)$. However, the function F is not a Morse family of hypersurface.

We can now prove an extension of Proposition 9 including the inflection points case.

Proposition 10. *Let* $\gamma : I \to \mathbb{R}^2_1$ *be a regular curve with the lightlike tangential data* (α, β). *Suppose that the evolute* $Ev(\gamma)$ *exists and* $-2\alpha(t)\beta(t) = \lambda(t)(\dot\alpha(t)\beta(t) - \alpha(t)\dot\beta(t))$.

(1) *If* $\gamma(t_0)$ *is an inflection point of* γ *and a regular point of* $Ev(\gamma)$, *then* $Ev(\gamma)(t_0)$ *is a lightlike point of* $Ev(\gamma)$. *Moreover,* $Ev(\gamma)(t_0)$ *is an inflection point of* $Ev(\gamma)$.

(2) *Suppose that* $\gamma(t_0)$ *is a lightlike point. Then* $Ev(\gamma)(t_0)$ *is a singular point of* $Ev(\gamma)$ *if and only if one of the following condition holds.*

(*i*) $\alpha(t_0) = \lambda(t_0) = 1 - \dot\lambda(t_0) = 0$ *and* $\beta(t_0) \neq 0$,

(*ii*) $\beta(t_0) = \lambda(t_0) = 1 + \dot\lambda(t_0) = 0$ *and* $\alpha(t_0) \neq 0$,

(*iii*) $\alpha(t_0) = \dot\alpha(t_0) = (1 - \dot\lambda(t_0))\beta(t_0) - \lambda(t_0)\dot\beta(t_0) = 0$ *and* $\beta(t_0) \neq 0$,

(*iv*) $\beta(t_0) = \dot\beta(t_0) = (1 + \dot\lambda(t_0))\alpha(t_0) + \lambda(t_0)\dot\alpha(t_0) = 0$ *and* $\alpha(t_0) \neq 0$.

Proof. (1) By differentiating the evolute

$$Ev(\gamma)(t) = \gamma(t) + \lambda(t)(\alpha(t)\mathbb{L}^+ - \beta(t)\mathbb{L}^-),$$

we have

$$\begin{aligned}\frac{d}{dt}Ev(\gamma)(t) &= \Big((1+\dot\lambda(t))\alpha(t)+\lambda(t)\dot\alpha(t)\Big)\,\mathbb{L}^+\\ &\quad+\Big((1-\dot\lambda(t))\beta(t)-\lambda(t)\dot\beta(t)\Big)\,\mathbb{L}^-.\end{aligned}$$

It follows that $\alpha_{Ev}(t) = (1+\dot\lambda(t))\alpha(t)+\lambda(t)\dot\alpha(t)$ and $\beta_{Ev}(t) = (1-\dot\lambda(t))\beta(t)-\lambda(t)\dot\beta(t)$. Since $\gamma(t_0)$ is an inflection point of γ, it holds that $\gamma(t_0)$ is a lightlike point of γ. It follows that $\alpha(t_0)=\dot\alpha(t_0)=0, \beta(t_0)\neq 0$ or $\beta(t_0)=\dot\beta(t_0)=0, \alpha(t_0)\neq 0$. Therefore, we have $\alpha_{Ev}(t_0)=0$ or $\beta_{Ev}(t_0)=0$. If $Ev(\gamma)(t_0)$ is a regular point of $Ev(\gamma)$, then $Ev(\gamma)(t_0)$ is a lightlike point of $Ev(\gamma)$.

By differentiating $-2\alpha(t)\beta(t)=\lambda(t)(\dot\alpha(t)\beta(t)-\alpha(t)\dot\beta(t))$, we have

$$\begin{aligned}&-2(\dot\alpha(t)\beta(t)+\alpha(t)\dot\beta(t))\\ &\quad=\dot\lambda(t)(\dot\alpha(t)\beta(t)-\alpha(t)\dot\beta(t))+\lambda(t)(\ddot\alpha(t)\beta(t)-\alpha(t)\ddot\beta(t)).\end{aligned}$$

Moreover,

$$\begin{aligned}&\dot\alpha_{Ev}(t)\beta_{Ev}(t)-\alpha_{Ev}(t)\dot\beta_{Ev}(t)\\ &\quad=\Big(\ddot\lambda(t)\alpha(t)+(1+2\dot\lambda(t))\dot\alpha(t)+\lambda(t)\ddot\alpha(t)\Big)\,\beta_{Ev}(t)\\ &\qquad-\Big(-\ddot\lambda(t)\beta(t)+(1-2\dot\lambda(t))\dot\beta(t)-\lambda(t)\ddot\beta(t)\Big)\,\alpha_{Ev}(t).\end{aligned}$$

If $\alpha(t_0)=\dot\alpha(t_0)=0, \beta(t_0)\neq 0$, then $\alpha_{Ev}(t_0)=0$ and $\lambda(t_0)\ddot\alpha(t_0)=0$. Also, if $\beta(t_0)=\dot\beta(t_0)=0, \alpha(t_0)\neq 0$, then $\beta_{Ev}(t_0)=0$ and $\lambda(t_0)\ddot\beta(t_0)=0$. Both cases, we have $\dot\alpha_{Ev}(t_0)\beta_{Ev}(t_0)-\alpha_{Ev}(t_0)\dot\beta_{Ev}(t_0)=0$. Hence $Ev(\gamma)(t_0)$ is an inflection points of $Ev(\gamma)$.

(2) Since $\gamma(t_0)$ is a lightlike point of γ, we have $\lambda(t_0)=0$ or $\gamma(t_0)$ is an inflection point of γ. By definition, $Ev(\gamma)(t_0)$ is a singular point of $Ev(\gamma)$ if and only if $\alpha_{Ev}(t_0)=\beta_{Ev}(t_0)=0$.

First we assume that $\lambda(t_0)=0$. If $\alpha(t_0)=0$ and $\beta(t_0)\neq 0$, then $Ev(\gamma)(t_0)$ is a singular point of $Ev(\gamma)$ if and only if $1-\dot\lambda(t_0)=0$. Also if $\beta(t_0)=0$ and $\alpha(t_0)\neq 0$, then $Ev(\gamma)(t_0)$ is a singular point of $Ev(\gamma)$ if and only if $1+\dot\lambda(t_0)=0$.

Next, we assume that $\gamma(t_0)$ is an inflection point of γ. By the proof of (1), $Ev(\gamma)(t_0)$ is a singular point of $Ev(\gamma)$ if and only if $\alpha(t_0)=\dot\alpha(t_0)=0, \beta(t_0)\neq 0$ and $\beta_{Ev}(t_0)=0$, or $\beta(t_0)=\dot\beta(t_0)=0, \alpha(t_0)\neq 0$ and $\alpha_{Ev}(t_0)=0$. This completes the proof. Q.E.D.

Remark 6. We can use the same definition (5) in order to define the evolute of γ with singular points. In this case, α and β vanish simultaneously at the singular points. Moreover, a singular point of γ is also an inflection point of γ.

Example 2. Let $\gamma : \mathbb{R} \to \mathbb{R}^2_1$ be a graph of a smooth function f, that is, $\gamma(t) = (t, f(t))$. Then we have $\alpha(t) = (1+\dot{f}(t))/2, \beta(t) = (1-\dot{f}(t))/2$. It follows that

$$\alpha(t)\beta(t) = \frac{1}{4}(1+\dot{f}(t))(1-\dot{f}(t)),\ \dot{\alpha}(t)\beta(t) - \alpha(t)\dot{\beta}(t) = \frac{\ddot{f}(t)}{2}.$$

Hence if there exists a unique smooth function λ such that

$$-(1+\dot{f}(t))(1-\dot{f}(t)) = \lambda(t)\ddot{f}(t),$$

then we have the evolute $Ev(\gamma)(t) = \gamma(t) + \lambda(t)(\alpha(t)\mathbb{L}^+ - \beta(t)\mathbb{L}^-)$ of $\gamma(t)$.

For example, let $f(t) = t+t^3$. Note that $\gamma(0)$ is an inflection point of γ. Then $\alpha(t) = (2+3t^2)/2, \beta(t) = -(3/2)t^2, \alpha(t)\beta(t) = -3t^2(2+3t^2)/4$ and $\dot{\alpha}(t)\beta(t) - \alpha(t)\dot{\beta}(t) = 6t$. It follows that we have $\lambda(t) = (1/2)t(2+3t^2)$ and the evolute $Ev(\gamma)$ is given by

$$Ev(\gamma)(t) = \left(t + \frac{1}{2}t(2+3t^2)(1+3t^2), 2t + \frac{5}{2}t^3\right),$$

see Figure 3.

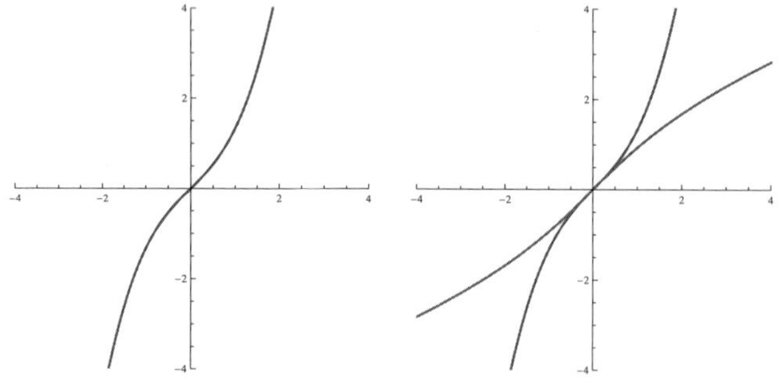

$\gamma(t) = (t, t+t^3)$ $\gamma(t)$ and the evolute $Ev(\gamma)$

Figure 3.

Example 3. Let $\gamma : [0, 2\pi) \to \mathbb{R}, \gamma(t) = (\cos t, \sin t \cos t)$ be an eight figure. Then $\alpha(t) = (\cos 2t - \sin t)/2, \beta(t) = -(\sin t + \cos 2t)/2$, $\alpha(t)\beta(t) = -(\cos 2t - \sin t)(\cos 2t + \sin t)/4$ and $\dot{\alpha}(t)\beta(t) - \alpha(t)\dot{\beta}(t) = \cos t(1+2\sin^2 t)/2$. It follows that we have $\lambda(t) = \cos t(4\cos^2 t - 3)/(1+2\sin^2 t)$ and the evolute $Ev(\gamma)$ is given by $Ev(\gamma)(t) =$

$$\left(\cos t\left(1 + \frac{(4\cos^2 t - 3)\cos 2t}{1+2\sin^2 t}\right), \sin t \cos t\left(1 - \frac{4\cos^2 t - 3}{1+2\sin^2 t}\right)\right),$$

see Figure 4. Note that $\gamma(t)$ for $t = \pi/2$ and $t = 3\pi/2$ are inflection points.

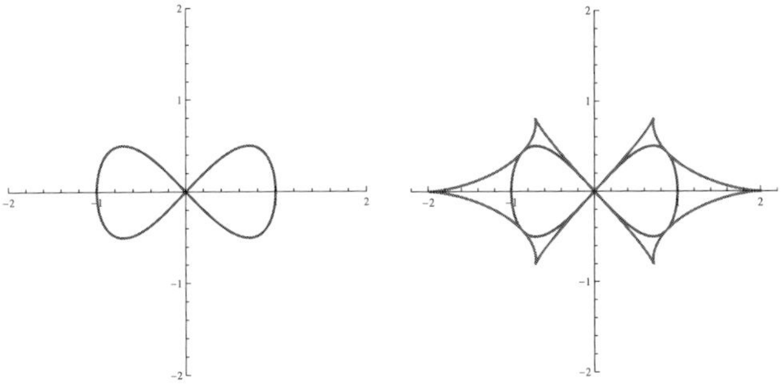

$\gamma(t) = (\cos t, \sin t \cos t)$ $\quad$ $\gamma(t)$ and the evolute $Ev(\gamma)$

Figure 4.

References

[1] V. I. Arnol'd, Singularities of Caustics and Wave Fronts, Mathematics and Its Applications, **62** Kluwer Academic Publishers, 1990.

[2] V. I. Arnol'd, S. M. Gusein-Zade and A. N. Varchenko, Singularities of Differentiable Maps vol. I, Birkhäuser, 1986.

[3] J. W. Bruce and P. J. Giblin, Curves and singularities. A geometrical introduction to singularity theory. Second edition, Cambridge University Press, Cambridge, 1992.

[4] L. Chen and M. Takahashi, Dualities and evolutes of fronts in hyperbolic 2-space and de Sitter 2-space, J. Math. Analysis and Applications. **437** (2016), 133–159.

[5] T. Fukunaga and M. Takahashi, Evolutes of fronts in the Euclidean plane, J. Singul., **10** (2014), 92–107.

[6] T. Fukunaga and M. Takahashi, Evolutes and involutes of frontals in the Euclidean plane, Demonstr. Math., **48** (2015), 147–166.

[7] C. G. Gibson, Elementary geometry of differentiable curves. An undergraduate introduction, Cambridge University Press, Cambridge, 2001.

[8] A. Gray, E. Abbena, and S. Salamon, Modern differential geometry of curves and surfaces with Mathematica. Third edition, Studies in Advanced Mathematics, Chapman and Hall/CRC, Boca Raton, FL, 2006.

[9] R. Hayashi, S. Izumiya, and T. Sato, Duals of curves in hyperbolic space, Note Mat., **33** (2013), 97–106.

[10] S. Izumiya, D. Pei, T. Sano and E. Torii, Evolutes of hyperbolic plane curves, Acta Math. Sin., (Engl. Ser.) **20** (2004), 543–550.

[11] S. Izumiya and M. Takahashi, Spacelike parallels and evolutes in Minkowski pseudo-spheres, Journal of Geometry and Physics, **57** (2007), 1569–1600.

[12] S. Izumiya and M. Takahashi, Caustics and wave front propagations: applications to differential geometry, Geometry and topology of caustics–CAUSTICS '06, Banach Center Publ., **82** (2008), 125–142.

[13] S. Izumiya and M. Takahashi, On caustics of submanifolds and canal hypersurfaces in Euclidean space, Topology Appl., **159** (2012), 501–508.

[14] A. Saloom and F. Tari, Curves in the Minkowski plane and their contact with pseudo-circles, Geom. Dedicata. **159** (2012), 109–124.

[15] S. Tabachnikov, Parametrized plane curves, Minkowski caustics, Minkowski vertices and conservative line fields, Enseign. Math., **43** (1997), 3–26.

[16] M. Takahashi, Legendre curves in the unit spherical bundle over the unit sphere and evolutes, Contemp. Math., **675** (2016), 337–355.

[17] F. Tari, Caustics of surfaces in the Minkowski 3-space, Q. J. Math., **63** (2012), 189–209.

[18] V. M. Zakalyukin, Reconstructions of fronts and caustics depending on a parameter and versality of mappings, J. Soviet Math., **27** (1983), 2713–2735.

Department of Mathematics, Hokkaido University,
Sapporo 060-0810, Japan.
E-mail address: izumiya@math.sci.hokudai.ac.jp

Department de Geometria i Topologia, Universitat de Valencia,
46100 Valencia, Spain.
E-mail address: Carmen.Romero@uv.es

Muroran Institute of Technology,
Muroran 050-8585, Japan.
E-mail address: masatomo@mmm.muroran-it.ac.jp

Advanced Studies in Pure Mathematics 78, 2018
Singularities in Generic Geometry
pp. 331–344

A new method for computing the limiting tangent space of an isolated hypersurface singularity via algebraic local cohomology

Katsusuke Nabeshima and Shinichi Tajima

Abstract.

Limiting tangent hyperplanes associated with isolated hypersurface singularities are considered in the context of symbolic computation. A new effective method is proposed to compute the limiting tangent space of a given hypersurface. The key of the method is the concept of parametric local cohomology systems. The proposed method can provide the decomposition of the limiting tangent space by Milnor numbers of hyperplane sections of a given hypersurface. The resulting algorithm has been implemented in the computer algebra system Risa/Asir. Examples of the computation for some typical cases are given.

§1. Introduction

We introduce a new approach for studying limiting tangent spaces of an isolated hypersurface singularity. The limiting tangent spaces were introduced in 1965 by H. Whitney [21, 22] and have been extensively utilized in various ways in singularity theory, especially in problems that involve Whitney stratifications.

In pioneering works [7, 11] published in 1977 and 1979, J-P. G. Henry, Lê Dũng Tráng and B. Teissier studied the geometry of the limiting tangent space of a complex analytic surface, and they have highlighted in a series of papers the importance of limiting tangent spaces themselves ([8, 9, 10]). In general, limiting tangent spaces encode, as H. Whitney already showed in [21, 22], much more information of singularities than

Received July 4, 2016.
Revised March 1, 2017.
2010 *Mathematics Subject Classification.* 13D45, 32C37, 13J05, 32A27.
Key words and phrases. limiting tangent space, local cohomology.

tangent cones. Note that A. G. Flores [2] recently generalized some results on limiting tangent spaces of [12].

In 1991, D. O'Shea gave computation methods of limiting tangent spaces in [17] by noticing the fact that only few computations of limiting tangent spaces had been made. The methods require Gröbner bases computation that involves elimination of many variables.

In this paper, we propose a new method for computing the limiting tangent space of a hypersurface with an isolated singularity. By utilizing the theory of algebraic local cohomology and the Grothendieck local duality theorem on residues, we have introduced, in previous papers [4, 5, 13, 14, 18], a concept of parametric local cohomology systems and have developed a framework for treating parametric problems in local rings.

Based on some results due to B. Teissier [19, 20], we derive in the present paper a new method of computing limiting tangent spaces by adapting the algorithms described in [14]. One of the advantage of the proposed method lies in the fact that the method is free from Gröbner bases computation and the main body consists of linear algebra computation. We emphasize here the fact that the resulting algorithm can provide a stratification of the limiting tangent space by Milnor numbers of hyperplane sections of a given hypersurface. This is another advantage.

This paper is organized as follows. Section 2 reviews algebraic local cohomology, parametric local cohomology systems, μ-stratifications, a definition of limiting tangent spaces and O'Shea's theorem. Section 3 provides a new method for computing limiting tangent spaces of isolated hypersurface singularities and gives some limiting tangent spaces.

§2. Preliminaries

Here we briefly recall the notions of algebraic local cohomology, parametric local cohomology systems and limiting tangent spaces and fix some notations. For details, we refer the reader to [4, 5, 13, 14, 18] for local cohomology, [7, 11, 17, 21, 22] for limiting tangent spaces. The set of natural numbers $\mathbb{N}$ includes zero. $\mathbb{C}$ is the field of complex numbers.

2.1. Algebraic local cohomology

Let X be an open neighborhood of the origin O of the n-dimensional complex space $\mathbb{C}^n$ with coordinates $x = (x_1, x_2, \ldots, x_n)$ and let $\mathcal{O}_X$ be the sheaf on X of holomorphic functions. Let $H^n_{[O]}(\mathcal{O}_X)$ denote the set of algebraic local cohomology classes, defined by $H^n_{[O]}(\mathcal{O}_X) =$

$\lim_{k\to\infty} \mathrm{Ext}^n_{\mathcal{O}_X}(\mathcal{O}_X/\langle x_1, x_2, \ldots, x_n\rangle^k, \mathcal{O}_X)$, where $\langle x_1, x_2, \ldots, x_n\rangle$ is the maximal ideal generated by $x_1, x_2, \ldots, x_n$.

We represent an algebraic local cohomology class as

$$\sum_\lambda c_\lambda \xi^\lambda = \sum c_\lambda \xi_1^{\lambda_1}\xi_2^{\lambda_2}\cdots\xi_n^{\lambda_n}$$

where $\xi^\lambda = \xi_1^{\lambda_1}\xi_2^{\lambda_2}\cdots\xi_n^{\lambda_n}$ and $\xi_1, \xi_2, \ldots, \xi_n$ correspond to $x_1, x_2, \ldots, x_n$ (see [18]). The multiplication is defined as follows:

$$x^\alpha * \xi^\lambda := \begin{cases} \xi^{\lambda-\alpha}, & \text{if } \lambda_i \geq \alpha_i, i = 1, \ldots, n, \\ 0, & \text{otherwise,} \end{cases}$$

where $x^\alpha = x_1^{\alpha_1}\cdots x_n^{\alpha_n} \in \mathbb{C}[x]$, $\xi^\lambda = \xi_1^{\lambda_1}\cdots\xi_n^{\lambda_n} \in H^n_{[O]}(\mathcal{O}_X)$, $\alpha = (\alpha_1, \ldots, \alpha_n) \in \mathbb{N}^n$, $\lambda = (\lambda_1, \ldots, \lambda_n) \in \mathbb{N}^n$, and $\lambda - \alpha = (\lambda_1 - \alpha_1, \ldots, \lambda_n - \alpha_n)$. We use the symbol " $*$ " to represent the multiplication. The action of monomials on algebraic local cohomology classes is extended to polynomials by linearity. For example, let $f = 2x_1^2x_2 + x_2 \in \mathbb{C}[x_1, x_2]$ and $\psi = 3\xi_1^3\xi_2^2 + \xi_2 \in H^2_{[O]}(\mathcal{O}_X)$, where $X \subset \mathbb{C}^2$ with coordinates (x_1, x_2). Then,

$$\begin{aligned} f * \psi &= 2x_1^2x_2 * \psi + x_2 * \psi \\ &= (2x_1^2x_2 * 3\xi_1^3\xi_2^2 + 2x_1^2x_2 * \xi_2) + (x_2 * 3\xi_1^2\xi_2^2 + 3x_2 * \xi_2) \\ &= 6\xi_1\xi_2 + 0 + 3\xi_1^2\xi_2 + 1 \\ &= 3\xi_1^2\xi_2 + 6\xi_1\xi_2 + 1. \end{aligned}$$

Let f be a holomorphic function defined on X with an isolated singularity at the origin. We define a vector space $H_{J(f)}$ to be the set of algebraic local cohomology classes in $H^n_{[O]}(\mathcal{O}_X)$ that are annihilated by the Jacobi ideal $J(f) = \langle \frac{\partial f}{\partial x_1}, \ldots, \frac{\partial f}{\partial x_n}\rangle$:

$$H_{J(f)} := \left\{ \psi \in H^n_{[O]}(\mathcal{O}_X) \,\middle|\, \frac{\partial f}{\partial x_1} * \psi = \frac{\partial f}{\partial x_2} * \psi = \cdots = \frac{\partial f}{\partial x_n} * \psi = 0 \right\}.$$

It follows from the Grothendieck local duality theorem [4, 5] on residues that the vector space $H_{J(f)}$ is a dual space to the Milnor algebra $\mathcal{O}_{X,O}/J(f)$. Therefore $\dim_{\mathbb{C}}(H_{J(f)})$ is equal to the Milnor number $\mu(f)$ of the singularity.

It is known that, according to a result of M. Artin [1], a defining holomorphic function of any isolated hypersurface singularity can be represented by a polynomial. We can assume therefore that the defining holomorphic function is actually a polynomial.

In our previous works [13, 18], we introduced and implemented algorithms for computing bases of the vector space $H_{J(f)}$ for the case where f is a polynomial.

2.2. Parametric local cohomology systems

We turn to parametric cases of algebraic local cohomology. For details, we refer the reader to [13, 14].

We use the notation t as the abbreviation of m parameters $t_1, \ldots, t_m$. For $g_1, \ldots, g_k$ in $\mathbb{C}[t]$, $\mathbb{V}(g_1, \ldots, g_k)$ denotes the affine variety of $g_1, \ldots, g_k$, i.e., $\mathbb{V}(g_1, \ldots, g_k) := \{t \in \mathbb{C}^m \mid g_1(t) = \cdots = g_k(t) = 0\}$. We consider a finite partition of $\mathbb{C}^m$ into disjoint algebraically constructible subsets of the form $\mathbb{V}(g_1, \ldots, g_k) \setminus \mathbb{V}(g'_1, \ldots, g'_{k'}) \subseteq \mathbb{C}^m$. For simplicity, we call those subsets *strata* and let notations $\mathbb{A}_1, \ldots, \mathbb{A}_q$, $\mathbb{B}_1, \ldots, \mathbb{B}_r$ stand for them.

We define $\mathbb{C}[t]_{\mathbb{A}}$, for a stratum $\mathbb{A} \subseteq \mathbb{C}^m$, as $\mathbb{C}[t]_{\mathbb{A}} = \{\frac{c}{b} | c, b \in \mathbb{C}[t], b(t) \neq 0 \text{ for } t \in \mathbb{A}\}$. Then, for every $\bar{a} \in \mathbb{A}$, we can define the canonical specialization homomorphism $\sigma_{\bar{a}} : \mathbb{C}[t]_{\mathbb{A}}[x] \to \mathbb{C}[x]$ (or $\sigma_{\bar{a}} : \mathbb{C}[t]_{\mathbb{A}}[\xi] \to \mathbb{C}[\xi]$) by putting $t = \bar{a}$. When we say that $\sigma_{\bar{a}}(h_t)$ makes sense for $h_t \in \mathbb{C}(t)[x]$, it has to be understood that $h_t \in \mathbb{C}[t]_{\mathbb{A}}[x]$ for some $\mathbb{A}$ with $\bar{a} \in \mathbb{A}$ where $\mathbb{C}(t)$ is the field of rational functions of t. For instance, let $h_t = t_1 x_1^3 x_2 + \frac{1}{t_2} x_1$ in $\mathbb{C}(t_1, t_2)[x_1, x_2]$ and $(2, 1), (0, \frac{2}{3}) \in \mathbb{C}^2 \setminus \mathbb{V}(t_2)$. Then, $\sigma_{(2,1)}(h_t) = 2x_1^3 x_2 + x_1$ and $\sigma_{(0,\frac{2}{3})}(h_t) = \frac{3}{2} x_1$.

Let h_t be a polynomial in $\mathbb{C}[x]$ with parameters $t = (t_1, t_2, \ldots, t_m)$ which generically has an isolated singularity at the origin O, namely, there exists a Zariski open dense subset $U \subset \mathbb{C}^m$ such that for all t in U, h_t has an isolated singularity at the origin. The following notion is used to describe the parameter dependency of the structure of the vector space $H_{J(h_t)}$.

Definition 2.1. *Let $\mathbb{A}_1, \ldots, \mathbb{A}_q, \mathbb{B}_1, \ldots, \mathbb{B}_r$ be strata in $\mathbb{C}^m$ suth that $\mathbb{A}_1 \cup \cdots \cup \mathbb{A}_q \cup \mathbb{B}_1 \cup \cdots \cup \mathbb{B}_r = \mathbb{C}^m$, $S_1, \ldots, S_q$ subsets of $\mathbb{C}(t)[\xi]$. Set $\mathcal{S} = \{(\mathbb{A}_1, S_1), \ldots, (\mathbb{A}_q, S_q)\}$ and $\mathcal{W} = \{\mathbb{B}_1, \ldots, \mathbb{B}_r\}$. Then, a pair $(\mathcal{S}, \mathcal{W})$ is called a* **parametric local cohomology system (PLCS)** *of $H_{J(h_t)}$ on the parameter space $\mathbb{C}^m$, if for all $i \in \{1, \ldots, q\}$, $S_i \subset \mathbb{C}[t]_{\mathbb{A}_i}[\xi]$ and $\bar{a} \in \mathbb{A}_i$, $\sigma_{\bar{a}}(S_i)$ is a basis of the vector space $H_{J(\sigma_{\bar{a}}(h_t))}$ and for all $j \in \{1, \ldots, r\}$ and $\bar{b} \in \mathbb{B}_j$, $\sigma_{\bar{b}}(h_t)$ does not define an isolated singularity at the origin. We call a pair $(\mathbb{A}_i, S_i)$ a* **segment** *of the PLCS of $H_{J(h_t)}$, for $1 \leq i \leq q$.*

In the papers [13, 14], we have introduced algorithms for computing a PLCS of $H_{J(h_t)}$ on the parameter space $\mathbb{C}^m$, which have been implemented in a computer algebra system Risa/Asir [16].

Definition 2.2. *Let $\mathbb{A}_1, \dots, \mathbb{A}_q$, $\mathbb{B}_1, \dots, \mathbb{B}_r$ strata in $\mathbb{C}^m$ such that $\mathbb{A}_1 \cup \cdots \cup \mathbb{A}_q \cup \mathbb{B}_1 \cup \cdots \cup \mathbb{B}_r = \mathbb{C}^m$ and $\mu_1, \dots, \mu_q$ natural numbers. Set $\mathcal{M} = \{(\mathbb{A}_1, \mu_1), \dots, (\mathbb{A}_q, \mu_q)\}$ and $\mathcal{W} = \{\mathbb{B}_1, \dots, \mathbb{B}_r\}$. Then, a pair $(\mathcal{M}, \mathcal{W})$ is called a* **μ-stratification** *of $\mathbb{C}^m$ for h_t if for all $i \in \{1, \dots, q\}$, μ_i is the Milnor number of h_t at the origin $O \in \mathbb{C}^n$ on $\mathbb{A}_i$, and for all $j \in \{1, \dots, r\}$ and $\bar{b} \in \mathbb{B}_j$, $\sigma_{\bar{b}}(h_t)$ does not define an isolated singularity at the origin. We call a pair $(\mathbb{A}_i, \mu_i)$ a* **segment** *of the μ-stratification of h_t, for $1 \le i \le q$.*

Example 1. *Let us consider $h_t = x_1^3 + t_1 x_1 x_2^3 + x_2^6 + t_2 x_2^4$ where x_1, x_2 are variables and t_1, t_2 are parameters. Then, our* Risa/Asir*-implementation outputs Table 1 as a PLCS of $H_{J(h_t)}$ and Milnor numbers μ.*

stratum	*basis of $H_{J(h_t)}$*	μ
$\mathbb{C}^2 \setminus \mathbb{V}(t_1 t_2)$	$\{1, \xi_1, \xi_2, \xi_1\xi_2, \xi_2^2, -\frac{3}{4}t_1\xi_2^3 + t_2\xi_1\xi_2 + \frac{1}{4}a^2\xi_1^2\}$	6
$\mathbb{V}(t_2) \setminus \mathbb{V}(t_1, t_2)$	$\{1, \xi_1, \xi_2, \xi_1\xi_2, \xi_2^2, \xi_2^3 - \frac{1}{3}a\xi_1^2, \xi_2^4 - \frac{1}{3}a\xi_1^2\xi_2\}$	7
$\mathbb{V}(t_1) \setminus \mathbb{V}(t_1, t_2)$	$\{1, \xi_1, \xi_2, \xi_1\xi_2, \xi_2^2, \xi_1\xi_2^2\}$	6
$\mathbb{V}(t_1, t_2)$	$\{1, \xi_1, \xi_2, \xi_1\xi_2, \xi_2^2, \xi_1\xi_2^2, \xi_2^3, \xi_1\xi_2^3, \xi_2^4, \xi_1\xi_2^4\}$	10

Table 1. A PLCS of $H_{J(h_t)}$ on $\mathbb{C}^2$

Since the dimension of the vector space $H_{J(h_t)}$ is equal to the Milnor number of h_t, μ-stratifications of parametric polynomials can be constructed by computing PLCS of $H_{J(h_t)}$.

2.3. Limiting tangent spaces

Here we recall a definition of the limiting tangent space for a hypersurface with an isolated singularity and O'Shea's method for computing limiting tangent spaces.

Let $f(x)$ be an holomorphic function with an isolated singularity at the origin and $S = \{x \in X | f(x) = 0\}$. If $x \in S - \{O\}$, we let $T(S, x)$ denote the tangent hyperplane to S at x in $\mathbb{C}^n$ translated so that it passes through the origin. If we identify a hyperplane $p_1x_1 + p_2x_2 + \cdots + p_nx_n = 0$ with the conormal vector $[p_1, p_2, \dots, p_n]$ in projective space $\check{\mathbb{P}}^{n-1}$, then we can write the map

$$\mathrm{grad}(f) : x \longrightarrow \left[\frac{\partial f}{\partial x_1}, \frac{\partial f}{\partial x_2}, \dots, \frac{\partial f}{\partial x_n}\right] \in \check{\mathbb{P}}^{n-1}.$$

Consider the graph $\mathrm{graph}(\mathrm{grad}(f)) \subset S \times \check{\mathbb{P}}^{n-1}$ and the closure $\overline{\mathrm{graph}(\mathrm{grad}(f))}$. Projection onto the first factor $S \times \check{\mathbb{P}}^{n-1} \longrightarrow S$ induces the Nash blow up ν.

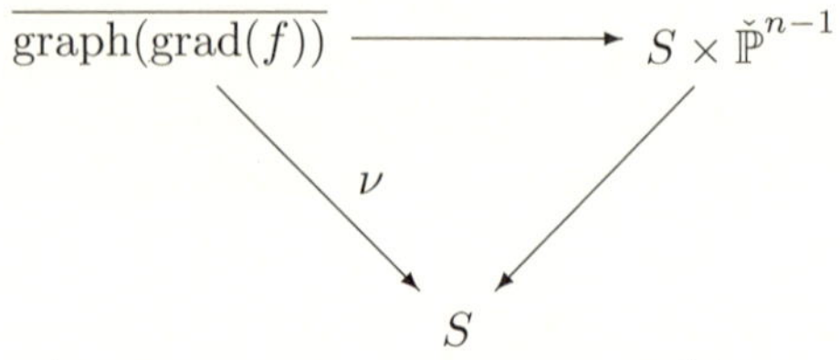

We denote the second factor of the fiber $\nu^{-1}(O)$ as $K(S, O)$. The space $K(S,O) \subset \check{\mathbb{P}}^{n-1}$ is called the **limiting tangent space** of S. We refer [6, 9, 10, 15, 21, 22] for the notion of limiting tangent spaces.

In general, the limiting tangent space of the tangent cone of S is a subset of the limiting tangent space of the hypersurface S, which means in particular that the limiting tangent space has more information than the tangent cone.

D. O'Shea proved the following theorem for computing limiting tangent spaces in [17].

Theorem 1 (D. O'Shea [17]). *Let f be polynomial in $\mathbb{C}[x_1, \ldots, x_n]$ with an isolated singularity at the origin and $S = \{x \in X | f(x) = 0\}$. Let $K(S,O) \subset \check{\mathbb{P}}^{n-1}$ be the limiting tangent space. Let A denote the ideal in $\mathbb{C}[x_1, \ldots, x_n, u, p_1, \ldots, p_n]$ given by setting*

$$A = \left\langle f, p_1 - u\frac{\partial f}{\partial x_1}, \ldots, p_n - u\frac{\partial f}{\partial x_n} \right\rangle.$$

Then, the ideal $\mathbb{I}(K(S,O))$ of $K(S,O)$ in $\check{\mathbb{P}}^{n-1}$ is the radical of the ideal in $\mathbb{C}[p_1, \ldots, p_n]$ given by eliminating u and setting $x_1, \ldots, x_n$ equal to zero. That is,

$$\mathbb{I}(K(S,O)) = \sqrt{A \cap \mathbb{C}[x_1, \ldots, x_n, p_1, \ldots, p_n]/\langle x_1, \ldots, x_n \rangle}.$$

(For an ideal I, $\sqrt{I}$ is a radical ideal of I.)

The theorem above implies that the limiting tangent space can be obtained by Gröbner bases computation of A w.r.t. the elimination term order.

§3. Main results

Here we see some results of B. Teissier and give a new computation method of limiting tangent spaces at the singular point O. The key ingredient of the method is the μ-stratification (i.e., PLCS).

3.1. A new computation method

Let $p = (p_1, p_2, \ldots, p_n)$ be a non-zero vector and let $[p]$ be the corresponding point in the complex projective space $\check{\mathbb{P}}^{n-1}$. We identify the hyperplane

$$H_p = \{x \in \mathbb{C}^n \mid p_1x_1 + p_2x_2 + \cdots + p_nx_n = 0\}$$

with the point $[p]$ in $\check{\mathbb{P}}^{n-1}$.

Let $f(x)$ be a holomorphic function defined on X with an isolated singularity at the origin O and $S = \{x \in X | f(x) = 0\}$. If the restriction $f|_{H_p}$ of f on $H_p \cap X$ has an isolated singularity at the origin of H_p, we define $\mu^{(n-1)}(f|_{H_p})$ to be $\mu(S \cap H_p)$, the Milnor number at the origin of the hyperplane section $S \cap H_p$. Otherwise we define $\mu^{(n-1)}(f|_{H_p}) = \infty$.

Definition 3.1. *Let $f(x)$ be a holomorphic function defined on X with an isolated singularity at the origin. Then, $\mu^{(n-1)}(f)$ is defined by*

$$\mu^{(n-1)}(f) = \min_{[p]\in\check{\mathbb{P}}^{n-1}} \mu(f|_{H_p}).$$

B. Teissier proved the following important theorems [19, 20].

Theorem 2 (B. Teissier). *Let $U = \{[p] \in \check{\mathbb{P}}^{n-1} \mid \mu^{(n-1)}(f|_{H_p}) = \mu^{(n-1)}(f)\}$. Then, U is Zariski open and a dense subset of $\check{\mathbb{P}}^{n-1}$.*

Theorem 3 (B. Teissier). *For $[p] \in \check{\mathbb{P}}^{n-1}$, the following are equivalent:*

(i) $[p] \in K(S, O)$.
(ii) $\mu^{(n-1)}(f|_{H_p}) > \mu^{(n-1)}(f)$.

Let us recall the cell decomposition of the projective space $\check{\mathbb{P}}^{n-1}$ given by

$$\begin{aligned}\check{\mathbb{P}}^{n-1} &= (\check{\mathbb{P}}^{n-1} - \check{\mathbb{P}}^{n-2}) \cup (\check{\mathbb{P}}^{n-2} - \check{\mathbb{P}}^{n-3}) \cup \cdots \cup (\check{\mathbb{P}}^{1} - \check{\mathbb{P}}^{0}) \cup \check{\mathbb{P}}^{0} \\ &\cong \mathbb{C}^{n-1} \cup \mathbb{C}^{n-2} \cup \cdots \cup \mathbb{C} \cup \check{\mathbb{P}}^{0}\end{aligned}$$

where

$$\check{\mathbb{P}}^{n-i} = \{[p] \in \check{\mathbb{P}}^{n-1} \mid p_1 = p_2 = \cdots = p_i = 0\}, \quad i = 1, 2, ..., n-1.$$

Let $\kappa_i = (\check{\mathbb{P}}^{n-i} - \check{\mathbb{P}}^{n-i-1}) \cap K(S, O)$. Then,

$$\kappa_i = \{[p] \in \check{\mathbb{P}}^{n-i} - \check{\mathbb{P}}^{n-i-1} \mid \mu^{(n-1)}(f|_{H_p}) > \mu^{(n-1)}(f)\}.$$

Now, we are ready to describe a method to compute the limiting tangent space $K(S,O)$.

Method 1

Let $f(x_1, x_2, \ldots, x_n)$ be a polynomial with an isolated singularity at the origin.

Step 1: Let us assume $p_1 \neq 0$ and $[p_1, p_2, \ldots, p_n] \in \check{\mathbb{P}}^{n-1}$. Consider the hyperplane $x_1 = s_2x_2 + s_3x_3 + \cdots + s_nx_n$ where $s_i = -\frac{p_i}{p_1}$ and $i \in \{2, \ldots, n\}$. Compute a μ-stratification of $f(s_2x_2 + s_3x_3 + \cdots + s_nx_n, x_2, \ldots, x_n)$ on the cell $\mathbb{C}^{n-1}$ by the method that is described in section 2.2, where $s_2, s_3, \ldots, s_n$ are parameters. After that, compute the union of the strata whose Milnor number is not the minimum (or the open dense subset's one), and compute κ_1 by $s_i = -\frac{p_i}{p_1}$ and $p_1 \neq 0$. Then, $\kappa_1 \subset \check{\mathbb{P}}^{n-1}$ becomes a part of the limiting tangent space $K(S,O)$ by Theorem 2 and Theorem 3.

Step 2: Next, we consider the case $p_1 = 0$, $p_2 \neq 0$ and $x_2 = s_3x_3 + \cdots + s_nx_n$ where $s_i = -\frac{p_i}{p_2}$ and $i \in \{3, \ldots, n\}$. Compute a μ-stratification of $f(x_1, s_3x_3 + \cdots + s_nx_n, x_3, \ldots, x_n)$ on $\mathbb{C}^{n-2}$ where $s_3, \ldots, s_n$ are parameters. Compute the union of the strata whose Milnor number is not $\mu^{(n-1)}(f)$ and compute κ_2 by $s_i = -\frac{p_i}{p_2}, p_1 = 0$ and $p_2 \neq 0$. The set κ_2 becomes a part of the limiting tangent space $K(S,O)$.

Repeat: Repeat the same procedure until $x_n = 0$ i.e., $(p_1, \ldots, p_{n-1}, p_n) = (0, \ldots, 0, p_n)$ with $p_n \neq 0$. Then,

$$K(S,O) = \kappa_1 \cup \kappa_2 \cup \cdots \cup \kappa_n.$$

Remark that an inequality $\mu^{(n)}(f) \geq \mu^{(n-1)}(f)$ holds [20]. Thus, if a number of elements of a basis of algebraic local cohomology classes associated with $f|_{H_p}$ becomes bigger than $\mu(f)$ on a stratum $\mathbb{A}$ as the halfway result in the computation of a PLCS, then we can stop the computation on the stratum $\mathbb{A}$. That is, $\mathbb{A} \subset K(S,O)$. This technique have been implemented in our implementation of μ-stratifications.

Theorem 4. *Method 1 returns the limiting tangent space of S correctly and terminates.*

Proof. Since the algorithm for computing a μ-stratification always terminates by the remark above and the algorithm for computing PLCSs [14], this method terminates. The correctness also follows from the algorithm for computing a μ-stratification, Theorem 2 and Theorem 3. Q.E.D.

We can compute a limiting tangent space of S via PLCSs. We give an example to facilitate the method.

Example 2. *Let us consider* $f(x_1, x_2, x_3) = x_1^2x_3 + x_2^3 + x_3^4 + x_2x_3^3$ *(*Q_{10}*-singularity) with an isolated singularity at the origin.* $(S = \{x \in X | f(x) = 0\})$

Set a hyperplane H_p *with the point* $[p_1, p_2, p_3]$ *in* $\check{\mathbb{P}}^2$*. First, we consider the case* $p_1 \neq 0$*. Set* $x_1 = s_2x_2 + s_3x_3$ *where* $s_2 = -\frac{p_2}{p_1}$ *and* $s_3 = -\frac{p_3}{p_1}$*. Let compute a* μ*-stratification of* $h_{(s_2,s_3)}(x_2, x_3) = f(s_2x_2 + s_3x_3, x_2, x_3)$ *where* s_2 *and* s_3 *are parameters. Then, our implementation for computing* μ*-stratifications, returns Table 2.*

strata	μ
$\mathbb{C}^2 \setminus \mathbb{V}(s_2s_3(4s_2^3 - 27s_3))$, $\mathbb{V}(s_2)\setminus \mathbb{V}(s_2, s_3)$	*4*
$\mathbb{V}(s_3)\setminus \mathbb{V}(s_2, s_3)$, $\mathbb{V}(4s_2^3 - 27s_3)\setminus \mathbb{V}(4s_3^3 - s_3, s_2 - 3s_3)$	*5*
$\mathbb{V}(4s_3^3 - s_3, s_2 - 3s_3)\setminus \mathbb{V}(s_2, s_3)$, $\mathbb{V}(s_2, s_3)$	*6*

Table 2. μ-stratification of $h_{(s_2,s_3)}$

Since the stratum $\mathbb{C}^2 \setminus \mathbb{V}(s_2s_3(4s_2^3-27s_3))$ *is open and dense,* $\mu^{(2)}(f) = 4$*. (Obviously 4 is the minimum.) Thus, the union of strata whose Milnor number is not 4, is*

$$\begin{aligned}&(\mathbb{V}(s_3)\setminus \mathbb{V}(s_2, s_3)) \cup \left(\mathbb{V}(4s_2^3 - 27s_3)\setminus \mathbb{V}(4s_3^3 - s_3, s_2 - 3s_3)\right)\\ &\quad \cup \left(\mathbb{V}(4s_3^3 - s_3, s_2 - 3s_3)\setminus \mathbb{V}(s_2, s_3)\right) \cup \mathbb{V}(s_2, s_3) = \mathbb{V}(s_3(4s_2^3 - 27s_3)).\end{aligned}$$

As $s_2 = -\frac{p_2}{p_1}$, $s_3 = -\frac{p_3}{p_1}$ *and* $p_1 \neq 0$*, the stratum defined by* $s_3(4s_2^3 - 27s_3) = 0$ *on* $\mathbb{C}^2$ *can be written as* $\kappa_1 = \mathbb{V}(p_3(4p_2^3 - 27p_1^2p_3))\setminus \mathbb{V}(p_1, p_2p_3)$*.*

Second, we consider the case $p_1 = 0$ *and* $p_2 \neq 0$*, i.e.,* $x_2 = -\frac{p_3}{p_2}x_3$*. Set* $x_2 = t_3x_3$ *where* $t_3 = -\frac{p_3}{p_2}$*. Let compute a* μ*-stratification of* $h_{(t_3)}(x_1, x_3) = f(x_1, t_3x_3, x_3)$ *where* t_3 *is a parameter. Our implementation returns Table 3.*

strata	μ
$\mathbb{C}^2 \setminus \mathbb{V}(t_3^2 + t^3)$, $\mathbb{V}(t_3 + 1)$	*4*
$\mathbb{V}(t_3)$	*5*

Table 3. μ-stratification of $h_{(t_3)}$

As $\mu^{(2)}(f) = 4$, *we require* $t_3 = -\frac{p_3}{p_1} = 0$ *as a part of* $K(S, O)$. *Thus, in this case,* $p_3 = 0$ *corresponds to the limiting tangent hyperplane* $H_{\{(0,p_2,0)\}} = \{p_2 x_2 | p_2 \neq 0\}$. *Hence, we have* $\kappa_2 = \mathbb{V}(p_1, p_3) \setminus \mathbb{V}(p_1, p_2, p_3)$.

Finally, we consider the case $p_1 = 0$, $p_2 = 0$ *and* $p_3 \neq 0$, *i.e.,* $x_3 = 0$. *Then, the Milnor number of* $f(x_1, x_2, 0) = x_2^3$ *is infinity* (∞). *Thus,* $H_{\{(0,0,p_3)\}} = \{p_3 x_3 | p_3 \neq 0\}$ *is a limiting tangent hyperplane. Hence, we have* $\kappa_3 = \mathbb{V}(p_1, p_2) \setminus \mathbb{V}(p_1, p_2, p_3)$. *Since*

$$(\mathbb{V}(p_3(4p_2^3 - 27p_1^2 p_3)) \setminus \mathbb{V}(p_1)) \cup (\mathbb{V}(p_1, p_3) \setminus \mathbb{V}(p_2)) \cup (\mathbb{V}(p_1, p_2) \setminus \mathbb{V}(p_3))$$
$$= \mathbb{V}(p_3(4p_2^3 - 27p_1^2 p_3)) \setminus \mathbb{V}(p_1, p_2, p_3),$$

$p_3(4p_2^3 - 27p_1^2 p_3) = 0$ *is the limiting tangent space of* S. *(As* $(p_1, p_2, p_3) \neq (0, 0, 0)$, *we omit* $\mathbb{V}(p_1, p_2, p_3)$.*)*

It is easy to see that the limiting tangent space of the tangent cone $x_1^2 x_3 + x_2^3 = 0$ *of the hypersurface* S *is* $4p_2^3 - 27p_1^2 p_3 = 0$.

Let M_S be the set of all Milnor numbers at the origin of hyperplane sections of S: $M_S = \{\mu^{(n-1)}(f|_{H_p}) \in \mathbb{N} \cup \{\infty\} \mid [p] \in \check{\mathbb{P}}^{n-1}\}$. Then, Method 1 can also compute

$$\tau_\mu = \{p \in \check{\mathbb{P}}^{n-1} | \mu^{(n-1)}(f|_{H_p}) = \mu\}, \quad \mu \in M_S.$$

The next example is a parametric case.

Example 3. *Let us consider* $f(x_1, x_2, x_3) = x_1^2 x_2 + x_1 x_3^3 + x_2^2 x_3 + a x_3^5$ *(*S_{12} *singularity) with an isolated singularity at the origin where "a" is a deformation parameter.*

First, set $h_{(s_2,s_3)}(x_2, x_3) = f(s_2 x_2 + s_3 x_3, x_2, x_3)$ *with parameters* s_2 *and* s_3. *Then, we can obtain a* μ*-stratification of* $h_{(s_2,s_3)}$ *by our implementation, that is Table 4.*

strata	μ
$\mathbb{C}^3 \setminus \mathbb{V}(s_3(4s_2 s_3 + 1)a)$, $\mathbb{V}(a) \setminus \mathbb{V}(s_3(4s_2 s_3 + 1), a)$	*4*
$\mathbb{V}(s_3) \setminus \mathbb{V}(s_2(4a - s_2^2)a, s_3)$, $\mathbb{V}(s_2, s_3) \setminus \mathbb{V}(a, s_2, s_3)$ $\mathbb{V}(s_3, a) \setminus \mathbb{V}(a, s_2, s_3)$	*6*
$\mathbb{V}(4s_2 s_3 + 1) \setminus \mathbb{V}(a, 4s_2 s_3 + 1)$, $\mathbb{V}(4s_2 s_3 + 1, a)$	*5*
$\mathbb{V}(4a - s_2^2, s_3) \setminus \mathbb{V}(a, s_2, s_3)$	*7*
$\mathbb{V}(a, s_2, s_3)$	∞

Table 4. μ-stratification of $h_{(s_1,s_2)}$

Since the stratum $\mathbb{C}^3 \setminus \mathbb{V}(s_3(4s_2s_3+1)a)$ *is open and dense,* $\mu^{(2)}(f) = 4$*. Thus, the union of strata whose Milnor number is not 4, is* $\mathbb{V}(s_3(4s_2s_3+1))$*. As* $s_2 = -\frac{p_2}{p_1}$, $s_3 = -\frac{p_3}{p_1}$ *and* $p_1 \neq 0$*, the stratum* $s_3(4s_2s_3+1) = 0$ *on* $\mathbb{C}^3$ *can be written as*

$$\kappa_1 = \mathbb{V}(p_3(4p_2p_3 + p_1^2)) \setminus \mathbb{V}(p_1, p_2p_3).$$

Second, set $h_{(t_3)}(x_1, x_3) = f(x_1, t_3x_3, x_3)$ *with a parameter* t_3*. A* μ*-stratification of* $h_{(t_3)}$ *is Table 5.*

strata	μ
$\mathbb{C}^2 \setminus \mathbb{V}(t_3(10at_3-3)(4at_3-1))$, $\mathbb{V}(10at_3-3)$, $\mathbb{V}(4at_3-1)$	*4*
$\mathbb{V}(t_3)$	∞

Table 5. μ-stratification of $h_{(t_3)}$

As $t_3 = -\frac{p_3}{p_2}$, $p_1 = 0$ *and* $p_2 \neq 0$*, the stratum* $t_3 = 0$ *can be written as* $\kappa_2 = \mathbb{V}(p_1, p_3) \setminus \mathbb{V}(p_1, p_2, p_3)$*.*

Finally, we consider the case $p_1 = 0$*,* $p_2 = 0$ *and* $p_3 \neq 0$*, i.e.,* $x_3 = 0$*. Then, the Milnor number of* $h(x_1, x_2) = f(x_1, x_2, 0) = x_1^2x_2$ *is infinity* (∞)*. Thus,* $\kappa_3 = \mathbb{V}(p_1, p_2) \setminus \mathbb{V}(p_1, p_2, p_3)$*.*

Hence,

$$\kappa_1 \cup \kappa_2 \cup \kappa_3 = \mathbb{V}(p_3(4p_2p_3 + p_1^2)).$$

Therefore, $p_3(4p_2p_3 + p_1^2) = 0$ *is the limiting tangent space of* S *that does not depend on the parameter* a*.*

Notice that, if $a = 0$*, then* $M_S = \{4, 5, 6, \infty\}$*, and if* $a \neq 0$ *then* $M_S = \{4, 5, 6, 7, \infty\}$ *holds. Furthermore,* $\tau_7 = \mathbb{V}(4a - s_2^2, s_3) \setminus \mathbb{V}(a, s_2, s_3)$*. Namely, the* μ*-stratification depends on the deformation parameter "*a*" (see [3]).*

3.2. Comparisons

Here we give the results of benchmark tests and some limiting tangent spaces. Table 6 shows a comparison of the Risa/Asir implementation of Method 1 with our Risa/Asir implementation of O'Shea's method (Theorem 1) in computation time (CPU time). x, y, z are variables and the hyperplane is $p_1x + p_2y + p_3z = 0$. The term order is the block term order $\{x, y, z\} \gg \{p_1, p_2, p_3\}$ with the total degree reverse lexicographic term order on each block.

All results of limiting tangent spaces in this paper have been computed on a PC with [OS: Windows 7 (64bit), CPU: Intel(R) Core i-7-2600 CPU @ 3.40 GHz 3.40 GHz, RAM: 4 GB]. The time is given in second. In Table 6, $> 2h$ means it takes more than 2 hours.

We use the following typical polynomials that define an isolated singularity.

$Q_{11} : x^3 + y^2z + xz^3 + z^5$,
$Q_{12} : x^3 + y^5 + yz^2 + xy^4$,
$S_{11} : x^4 + y^2z + xz^2 + x^3z$,
$S_{12} : x^2y + y^2z + xz^3 + z^5$,
$U_{12} : x^3 + y^3 + z^4 + xyz^2$,
$S_{16} : x^2z + yz^2 + xy^4 + y^6$,
$S_{17} : x^2z + yz^2 + y^6 + y^4z$,
$Q_{16} : x^3 + yz^2 + y^7 + xy^5$,
$Q_{17} : x^3 + yz^2 + xy^5 + y^8$,
$Q_{18} : x^3 + yz^2 + y^8 + xy^6$,
$U_{16} : x^3 + xz^2 + y^5 + x^2y^2$,
$U_{1,3} : x^3 + xz^2 + xy^3 + y^3z^2 + y^4z^2 \ (\mu = 17)$,
$Q_{2,3} : x^3 + yz^2 + x^2y^2 + y^9 + y^{10} \ (\mu = 17)$,
$S_{1,3} : x^2z + yz^2 + x^2y^2 + y^8 + y^9 \ (\mu = 17)$.
$V^\sharp_{1,10} : x^2y + z^3 + y^2z^2 + y^3z^2 + y^4 + z^9 + z^{10} \ (\mu = 25)$.

Singularity	O'Shea	Method 1	Limiting tangent space
Q_{11}	0.6084	0.0156	$p_3(4p_1^3 - 27p_2^2p_3) = 0$
Q_{12}	2.184	0.0312	$p_2(4p_1^3 - 27p_2p_3^2) = 0$
S_{11}	1.466	0.0624	$p_1(4p_1p_3 + p_2^2) = 0$
S_{12}	1.279	0.0312	$p_3(p_1^2 + 4p_2p_3) = 0$
U_{12}	0.4056	0.0312	$p_3 = 0$
S_{16}	3.869	0.1092	$p_2(p_1^2 + 4p_2p_3) = 0$
S_{17}	0.5772	0.0468	$p_2(p_1^2 + 4p_2p_3) = 0$
Q_{17}	345.8	0.0468	$p_2(4p_1^3 - 27p_2p_3^2) = 0$
Q_{18}	5.756	0.156	$p_2(4p_1^3 - 27p_2p_3^2) = 0$
U_{16}	1.966	0.0312	$p_3 = 0$
$U_{1,3}$	70.68	0.1248	$p_2 = 0$
$Q_{2,3}$	>2h	0.078	$p_2(4p_1^3 - 27p_2p_3^2) = 0$
$S_{1,3}$	> 2h	0.0624	$p_2(p_1^2 + 4p_2p_3) = 0$
$V^\sharp_{1,10}$	>2h	5.881	$p_2(27p_2p_1^2 - 4p_3^3) = 0$

Table 6. comparisons of Method 1 with O'Shea's method

As is evident from Table 6, Method 1 results in better performance in contrast to O'Shea's method. O'Shea's method requires Gröbner bases computation whose computational complexity is quite big (double exponential). In contrast, the new method use PLCSs computation that mainly consists of linear algebra computation. This is the big advantage. That's why the new method results in better performance.

Acknowledgments

This work has been partly supported by JSPS Grant-in-Aid for Young Scientists (B) (No.15K17513) and Grant-in-Aid for Scientific Research (C) (No. 15KT0102, 15K04891).

References

[1] M. Artin, 1969, Algebraic approximation of structures over complete local rings, Publ. Math. IHES **36**, pp. 23–58.

[2] A. G. Flores, 2013, Specialization to the tangent cone and Whitney equisingularity. Bull. Soc. math. France **141** (2), pp. 299–p. 342.

[3] T. Gaffney, 1997, Aureoles and integral closure of modules, in Stratifications. singularities and differential equations II, Travaux en Cours, **55**, Hermann, p. 55–p. 62.

[4] A. Grothendieck, 1957, Théorèmes de dualité pour les faisceaux algébriques cohérents. Séminaire Bourbaki **149**.

[5] A. Grothendieck, 1967, Local Cohomology, notes by R. Hartshorne, Lecture Notes in Math., 41, Springer.

[6] A. Hénaut, 1981, Cycles exceptionnel de l'éclatement de Nash d'une hypersurface analytiques complexe à singularité isolée. Bull. Soc. math. France. **109**, pp. 475–481.

[7] J-P. G. Henry and Lê Dũng Tráng, 1977, Limites d'espaces tangent. in Fonctions de plusieurs variables complexes II, Lecture Notes in Mathematics, **482**, Springer, pp. 251–265.

[8] Lê Dũng Tráng, 1981. Limies d'espaces tangents sur les surfaces. Generic section of singularities. Nova Acta Leopodina (N.F.), **52**, no. 240, pp. 119–137.

[9] Lê Dũng Tráng, 1981. Limies d'espaces tangents et obstruction d'Euler des surfaces. in The Euler-Poincaré characteristic, Astérisque, **82-83**, pp. 45–69.

[10] Lê Dũng Tráng, 2007. Generic section of singularities. Singularity Theory, World Sci. Pub., Hackensack, pp. 677–682.

[11] Lê Dũng Tráng and B. Teissier, 1979. Sur la geometrie des surface complexes I -Tangentes exceptionnelles-. American Journal of Mathematics, **101**-2, 1979, pp. 420–452.

[12] Lê Dũng Tráng and B. Teissier, 1988. Limites d'espaces tangents en geometrie analytique. Comment. Math. Helvetici, **63**, 1988, pp. 540–578.

[13] K. Nabeshima and S. Tajima, 2014. On efficient algorithm for computing parametric local cohomology classes associated with semi- quasihomogeneous singularities and standard bases. Proc. International Symposium on Symbolic and Algebraic Computation (ISSAC2014), ACM, pp. 351–358.

[14] K. Nabeshima and S. Tajima, 2017, Algebraic local cohomology with parameters and parametric standard bases for zero-dimensional ideals, Journal of Symbolic Computation, **82**, pp. 91–122.

[15] A. Nobile, 1975, Some properties of the Nash blowing-up, Pasific J. Math. **60**, pp. 297–305.

[16] M. Noro and T. Takeshima, 1992. Risa/Asir - A computer algebra system. Proc. International Symposium on Symbolic and Algebraic Computation(ISSAC1992), ACM, pp. 387–396.
`http://www.math.kobe-u.ac.jp/Asir/asir.html`

[17] D. O'Shea, 1995. Computing limits of tangent spaces: singularities, computation and pedagogy. Singularity theory (Trieste, 1991), World Sci. Publ., River Edge, NJ, pp. 549–573.

[18] S. Tajima, Y. Nakamura and K. Nabeshima, 2009. Standard bases and algebraic local cohomology for zero dimensional ideals. Advanced Studies in Pure Mathematics **56**, pp. 341–361.

[19] B. Teissier, 1973, Cycles evanescents, sections planes et conditions de Whitney, Astérisques **7-8**, Soc. Math. France, pp. 285–362.

[20] B. Teissier, 1977, Variétés polaires I, Invent. Math. **40**, pp. 267–292.

[21] H. Whitney, 1965, Tangents to an analytic variety, Annals of Mathematics, **81**, pp. 496–549.

[22] H. Whitney, 1965, Local properties of analytic varieties, in Differential and Combinatorial Topology, Princeton, pp. 205–244.

Katsusuke Nabeshima
Graduate School of and Science and Technology,
Tokushima University,
2-1 Minamijosanjima, Tokushima, 770-8506, JAPAN
E-mail address: `nabeshima@tokushima-u.ac.jp`

Shinichi Tajima
Graduate School of Pure and Applied Sciences,
University of Tsukuba,
1-1-1 Tennoudai, Tsukuba, 305-8571, JAPAN
E-mail address: `tajima@math.tsukuba.ac.jp`

Advanced Studies in Pure Mathematics 78, 2018
Singularities in Generic Geometry
pp. 345–363

Constant mean curvature surfaces with D_4-singularities

Yuta Ogata and Keisuke Teramoto

Abstract.

We study D_4-singularities of constant mean curvature (CMC) surfaces in Riemannian and semi-Riemannian spaceforms. We will give criteria for D_4^--singularities, which are related to the Hopf differential. We also show the non-existence of (spacelike) CMC surfaces with D_4^+-singularities.

§1. Introduction

Singularities of wave fronts can appear on surfaces via parallel transformations. It is known that generic singularities of wave fronts in 3-spaces are cuspidal edges and swallowtails. Moreover, the singularities of the bifurcations in generic one-parameter families of wave fronts in 3-spaces are cuspidal lips, cuspidal beaks, cuspidal butterflies and D_4-singularities, in addition to the above two (see [1, 15]). There are criteria for these singularities in [16, 17, 20, 26]. On the other hand, in [11], Fukui and Hasegawa studied singularities of the parallel surfaces of regular surfaces in Euclidean 3-space $\mathbb{R}^3$. They gave criteria for cuspidal edges, swallowtails, cuspidal lips, cuspidal beaks, cuspidal butterflies and D_4-singularities by using geometric invariants of the original surfaces, for example principal curvatures.

For constant mean curvature (CMC) surfaces in Riemannian and semi-Riemannian spaceforms, there are several studies. In general, CMC surfaces in semi-Riemannian spaceforms have singularities (see [4, 10, 14, 23, 24, 29, 30], for example). In [30], criteria for cuspidal edges and swallowtails of maximal surfaces were obtained by using Weierstrass data.

Received January 18, 2016.
Revised March 28, 2017.
2010 *Mathematics Subject Classification.* Primary 53A10; Secondary 53B30, 57R45.
Key words and phrases. constant mean curvature surface, parallel transformation, singularity, wave front.

Similarly, for maximal surfaces and CMC 1 surfaces, criteria for corank one singularities were given in [10]. Umeda [29] gave criteria for cuspidal edges, swallowtails and cuspidal cross caps of extended CMC surfaces in Minkowski 3-space $\mathbb{R}^{2,1}$, and in [23, 24], the first author of this paper also studied the analogues of Umeda's criteria for these singularities of extended CMC surfaces in other semi-Riemannian spaceforms.

However, the above previous studies did not consider corank two singularities of such surfaces. It is well-known that the corank two singularities appear even on CMC surfaces in Riemannian spaceforms. Thus in this paper, we consider the criteria for the corank two singularities, especially D_4-singularities.

For CMC $H \neq 0$ surfaces in $\mathbb{R}^3$ and $\mathbb{R}^{2,1}$, the following fact is known. (For the definition of $\hat{f}^t$, see (2.4).)

Fact 1.1 ([11, 13]). *Let f be a conformal (spacelike) CMC surface in $\mathbb{R}^3$ (or $\mathbb{R}^{2,1}$) with mean curvature $H > 0$, unit normal vector ν and the Hopf differential factor Q, and let p be an umbilic point of f. Then for $t = 1/H$, the parallel transform $\hat{f}^t$ of f becomes a conformal (spacelike) CMC surface in $\mathbb{R}^3$ (or $\mathbb{R}^{2,1}$) with mean curvature $-H$ and the Hopf differential factor $-Q$. Moreover, $\hat{f}^t$ has a corank two singularity at p.*

For CMC surfaces in the spherical 3-space $\mathbb{S}^3$, hyperbolic 3-space $\mathbb{H}^3$, de Sitter 3-space $\mathbb{S}^{2,1}$ and anti-de Sitter 3-space $\mathbb{H}^{2,1}$, the following is known. (For definitions of $\hat{f}^t$ and $\check{f}^t$, see (2.5) and (2.6).)

Fact 1.2 ([6, 8]). *Let $f : U \to M^3$ be a conformal (spacelike) CMC H surface and ν its unit normal vector.*

(1) *If $M^3 = \mathbb{H}^3$ or $\mathbb{S}^{2,1}$ with $H > 1$, then for $t = \operatorname{arccoth} H$, $\hat{f}^t$ becomes a conformal (spacelike) CMC surface in M^3 with mean curvature $-H$.*

(2) *If $M^3 = \mathbb{H}^3$ (resp. $\mathbb{S}^{2,1}$) with $0 < H < 1$, then for $t = \operatorname{arctanh} H$, $\check{f}^t = \hat{\nu}^t$ becomes a conformal spacelike CMC surface in $\mathbb{S}^{2,1}$ (resp. $\mathbb{H}^3$) with mean curvature $-H$.*

(3) *If $M^3 = \mathbb{S}^3$ or $\mathbb{H}^{2,1}$ with mean curvature $H > 0$, then for $t = \operatorname{arccot} H$, $\hat{f}^t$ is a conformal (spacelike) CMC surface in M^3 with mean curvature $-H$. Moreover, if f is a conformally immersed minimal (resp. maximal) surface, then ν is a conformal minimal (resp. maximal) surface in M^3 and f is a unit normal vector to ν.*

By these facts, considering parallel transforms of CMC surfaces naturally emphasizes their umbilic points, and parallel transforms play an important role in understanding relations between umbilic points and corank two singularities. Moreover, by Facts 1.1 and 1.2, we can start to consider a regular CMC surface f with an umbilic point instead of a CMC surface $\hat{f}$ (or $\check{f}$) with a D_4-singularity via parallel transform. Thus, in this paper, we study (spacelike) CMC surfaces with D_4-singularities in Riemannian and semi-Riemannian spaceforms, and we give criteria for D_4-singularities in terms of the Hopf differential factors (Theorem 3.2). For minimal surfaces, we give conditions for which they have D_4-singularities by using Weierstrass data (Theorems 3.6 and A.4).

For surfaces in $\mathbb{S}^3$, $\mathbb{H}^3$, $\mathbb{S}^{2,1}$ or $\mathbb{H}^{2,1}$, their unit normal vectors form surfaces. Thus we can compare curvatures of CMC surfaces with curvatures of their unit normal vectors (Proposition 4.1). Moreover, we show a kind of duality between parallel transforms of CMC surfaces and their unit normal vectors (Proposition 4.3).

§2. Preliminaries

2.1. Surfaces in spaceforms

We recall some properties of surfaces in several spaceforms. For more details, see [15, 16, 17].

Let $\mathbb{R}^n = \{(x_1, \ldots, x_n) \mid x_i \in \mathbb{R},\ 1 \le i \le n\}$ be an n-dimensional vector space. For any $\boldsymbol{x} = (x_1, \ldots, x_n)$ and $\boldsymbol{y} = (y_1, \ldots, y_n) \in \mathbb{R}^n$, we define the *pseudo inner product with the signature* $(n-k, k)$ $(0 \le k < n)$ by

$$\langle \boldsymbol{x}, \boldsymbol{y} \rangle = \sum_{i=1}^{n-k} x_i y_i - \sum_{j=n-k+1}^{n} x_j y_j.$$

We denote $\mathbb{R}^{n-k,k} = (\mathbb{R}^n, \langle\,,\rangle)$. We say that a vector $\boldsymbol{x} \in \mathbb{R}^n \setminus \{\boldsymbol{0}\}$ is *spacelike*, *timelike* or *lightlike* if $\langle \boldsymbol{x}, \boldsymbol{x} \rangle > 0$, < 0 or $= 0$ respectively. We note that $\mathbb{R}^{n,0} = \mathbb{R}^n$ is the *Euclidean n-space.* If $k = 1$, we call the space $\mathbb{R}^{n-1,1}$ the *Minkowski n-space.*

Let $\boldsymbol{e}_1, \ldots, \boldsymbol{e}_n$ be the pseudo orthonormal basis of $\mathbb{R}^{n-k,k}$ and $\boldsymbol{x}^i = (x^i_1, \ldots, x^i_n) \in \mathbb{R}^{n-k,k}$ $(1 \le i \le n-1)$. Then we define the *wedge product* $\boldsymbol{x}^1 \wedge \cdots \wedge \boldsymbol{x}^{n-1}$ *with respect to the signature* $(n-k, k)$ by

$$(2.1)\quad \boldsymbol{x}^1 \wedge \cdots \wedge \boldsymbol{x}^{n-1} = \begin{vmatrix} \boldsymbol{e}_1 & \cdots & \boldsymbol{e}_{n-k} & -\boldsymbol{e}_{n-k+1} & \cdots & -\boldsymbol{e}_n \\ x^1_1 & \cdots & x^1_{n-k} & x^1_{n-k+1} & \cdots & x^1_n \\ \vdots & \ddots & \vdots & \vdots & \ddots & \vdots \\ x^{n-1}_1 & \cdots & x^{n-1}_{n-k} & x^{n-1}_{n-k+1} & \cdots & x^{n-1}_n \end{vmatrix}.$$

One can check that $\langle \boldsymbol{x}^i, \boldsymbol{x}^1 \wedge \cdots \wedge \boldsymbol{x}^{n-1} \rangle = 0$ holds for $1 \le i \le n-1$.

Let $n = 4$. Then we define the following spaceforms:

$$\mathbb{S}^3 = \{\boldsymbol{x} \in \mathbb{R}^4 \mid \langle \boldsymbol{x}, \boldsymbol{x} \rangle = 1\}, \qquad \mathbb{H}^3 = \{\boldsymbol{x} \in \mathbb{R}^{3,1} \mid \langle \boldsymbol{x}, \boldsymbol{x} \rangle = -1\},$$
$$\mathbb{S}^{2,1} = \{\boldsymbol{x} \in \mathbb{R}^{3,1} \mid \langle \boldsymbol{x}, \boldsymbol{x} \rangle = 1\}, \qquad \mathbb{H}^{2,1} = \{\boldsymbol{x} \in \mathbb{R}^{2,2} \mid \langle \boldsymbol{x}, \boldsymbol{x} \rangle = -1\}.$$

We call $\mathbb{S}^3$, $\mathbb{H}^3$, $\mathbb{S}^{2,1}$ and $\mathbb{H}^{2,1}$ the *spherical* 3-*space*, the *hyperbolic* 3-*space*, the *de Sitter* 3-*space* and the *anti-de Sitter* 3-*space*, respectively. It is known that $\mathbb{S}^3$ and $\mathbb{S}^{2,1}$ (resp. $\mathbb{H}^3$ and $\mathbb{H}^{2,1}$) have constant sectional curvature 1 (resp. -1).

Let M^3 be a 3-dimensional spaceform one of $\mathbb{R}^3$, $\mathbb{R}^{2,1}$, $\mathbb{S}^3$, $\mathbb{H}^3$, $\mathbb{S}^{2,1}$ or $\mathbb{H}^{2,1}$. Let $f : U \to M^3$ be a surface, where $U \subset (\mathbb{R}^2; u, v)$ is an open set. The surface f is said to be *spacelike* if the induced metric via f is positive definite on U. Then we consider the *unit normal vector* ν to f. If $M^3 = \mathbb{R}^3$ or $\mathbb{R}^{2,1}$, ν is defined as

$$\nu = \frac{f_u \wedge f_v}{|f_u \wedge f_v|} \quad (f_u = \partial f/\partial u,\ f_v = \partial f/\partial v),$$

where $|\boldsymbol{x}| = \sqrt{|\langle \boldsymbol{x}, \boldsymbol{x} \rangle|}$. If M^3 is one of the other spaceforms, ν can be taken as

$$\nu = \frac{f \wedge f_u \wedge f_v}{|f \wedge f_u \wedge f_v|}.$$

In these cases, we use $\langle \cdot, \cdot \rangle$ as the induced metric from the ambient space $\mathbb{R}^{4-k,k}$ $(k = 0, 1, 2)$.

2.2. CMC surface theory

In this section, we explain some basical notations, as in [2] and [5]. Let M^3 be one of $\mathbb{R}^3$, $\mathbb{R}^{2,1}$, $\mathbb{S}^3$, $\mathbb{H}^3$, $\mathbb{S}^{2,1}$ or $\mathbb{H}^{2,1}$. Let $f : U \to M^3$ be a spacelike surface, where U is a simply-connected domain in $\mathbb{C}$ with usual complex coordinate $z = u + iv$ $(i = \sqrt{-1})$. We say that f is a *conformal surface* if there exists a *conformal coordinate system* on U, namely, $\langle f_z, f_z \rangle = \langle f_{\bar z}, f_{\bar z} \rangle = 0$ and $\langle f_z, f_{\bar z} \rangle = 2g^2$ hold for some function $g : U \to \mathbb{R}$, where $\partial_z = (\partial_u - i\partial_v)/2$ and $\partial_{\bar z} = (\partial_u + i\partial_v)/2$. For a conformal surface f, the first fundamental form of f is given as

$$ds^2 = 4g^2(du^2 + dv^2).$$

Take the unit normal vector field ν. Then the mean curvature H and the Hopf differential factor Q are given by

$$H = \frac{1}{2g^2}\langle f_{z\bar z}, \nu \rangle, \quad Q = \langle f_{zz}, \nu \rangle. \tag{2.2}$$

By (2.2), one can check that H and Q change to $-H$ and $-Q$, respectively, when we change ν to $-\nu$. We assume that H is constant. It is known that the Codazzi equation implies that Q is holomorphic. Moreover, the extrinsic Gaussian curvature K is written as

$$K = -\frac{1}{4g^4} Q\bar{Q} + H^2. \tag{2.3}$$

We now define the *parallel transforms $\hat{f}^t$ and $\check{f}^t$* of f. If $M^3 = \mathbb{R}^3$ or $\mathbb{R}^{2,1}$,

$$\hat{f}^t = f + t\nu \tag{2.4}$$

for some constant $t \in \mathbb{R}$. In this case, ν is also a unit normal vector to $\hat{f}^t$. If $M^3 = \mathbb{S}^3$ or $\mathbb{H}^{2,1}$, $\hat{f}^t$ and $\hat{\nu}^t$ are

$$\hat{f}^t = \cos tf + \sin t\nu, \quad \hat{\nu}^t = -\sin tf + \cos t\nu \tag{2.5}$$

for some constant $t \in \mathbb{R}$. If $M^3 = \mathbb{H}^3$ or $\mathbb{S}^{2,1}$, we define

$$\hat{f}^t = \cosh tf + \sinh t\nu, \quad \check{f}^t = \hat{\nu}^t = \sinh tf + \cosh t\nu \tag{2.6}$$

for some constant $t \in \mathbb{R}$ (cf. [6] and [8]).

Remark 2.1. It is a well-known fact that if the spaceforms or the value of H change, then the integrable equation (i.e. Gauss equation) also changes. This change creates a difference in the construction method of the CMC surface f (see [8] for example). However, we will show the existence and non-existence of $D_4^{\pm}$-singularities of the CMC surface without depening on the choice of the spaceform or the value of H (see Theorem 3.2).

2.3. Wave fronts

We recall some notions of wave fronts. For details, see [1, 10, 15, 17, 27].

Let $f : U \to M^3$ be a C^∞ map, where U is a simply-connected domain in $\mathbb{R}^2$ and M^3 is one of $\mathbb{R}^3$, $\mathbb{R}^{2,1}$, $\mathbb{S}^3$, $\mathbb{H}^3$, $\mathbb{S}^{2,1}$ or $\mathbb{H}^{2,1}$. We call f a *wave front* or *front* if for each point $p \in U$ there exists a unit normal vector field ν along f and the map $L = (f, \nu) : U \to T_1M^3$ gives an surface, where T_1M^3 is the unit tangent bundle over M^3. A point $p \in U$ is called a *singular point* if f is not an surface at p. Let $S(f)$ denote the set of singular points of f. We set a function λ on U as

$$f_u \wedge f_v = \lambda\nu \tag{2.7}$$

when $M^3 = \mathbb{R}^3$ or $\mathbb{R}^{2,1}$, and

$$f \wedge f_u \wedge f_v = \lambda\nu \tag{2.8}$$

when $M^3 = \mathbb{S}^3$, $\mathbb{H}^3$, $\mathbb{S}^{2,1}$ or $\mathbb{H}^{2,1}$, where $\wedge$ denotes the wedge product as in (2.1). We call this function λ the *signed area density function.* By definition, $\lambda^{-1}(0) = S(f)$ holds. A singular point p is called *non-degenerate* if the exterior derivative $d\lambda$ does not vanish at p. On a neighborhood of a non-degenerate singular point, there exists a smooth regular curve $\gamma(t)$ satisfying $\gamma(0) = p$ such that $\gamma(t)$ parametrizes the set of singular points. We call this curve γ a *singular curve* and the direction of $\gamma' = d\gamma/dt$ a *singular direction.* The dimension of the kernel $\operatorname{Ker} df_{\gamma(t)}$ of the differential map $df_{\gamma(t)}$ is one and there exists a never-vanishing vector field $\eta(t)$ such that $\langle \eta(t) \rangle_{\mathbb{R}} = \operatorname{Ker} df_{\gamma(t)}$. We call $\eta(t)$ a *null vector field* and the direction of η a *null direction.*

Definition 2.2. Let $f : (U, p) \to (\mathbb{R}^3, f(p))$ be a map-germ around p. Then f has a *cuspidal edge* at p if the map-germ f at p is $\mathcal{A}$-equivalent to the map-germ $(u, v) \mapsto (u, v^2, v^3)$ at $\mathbf{0}$, and f has a *swallowtail* at p if the map-germ f at p is $\mathcal{A}$-equivalent to the map-germ $(u, v) \mapsto (u, 3v^4 + uv^2, 4v^3 + 2uv)$ at $\mathbf{0}$, and f has a $D_4^{\pm}$*-singularity* at p if the map-germ f at p is $\mathcal{A}$-equivalent to the map-germ $(u, v) \mapsto (2uv, \pm u^2 + 3v^2, \pm 2u^2 v + 2v^3)$ at $\mathbf{0}$, where the two map-germs $f, g : (\mathbb{R}^2, \mathbf{0}) \to (\mathbb{R}^3, \mathbf{0})$ are said to be *$\mathcal{A}$-equivalent* if there exist diffeomorphism-germs $\theta : (\mathbb{R}^2, \mathbf{0}) \to (\mathbb{R}^2, \mathbf{0})$ on the source and $\Theta : (\mathbb{R}^3, \mathbf{0}) \to (\mathbb{R}^3, \mathbf{0})$ on the target such that $\Theta \circ f = g \circ \theta$ holds.

We note that cuspidal edges and swallowtails are non-degenerate singular points of fronts. On the other hand, D_4-singularities are degenerate singular points with corank two. There are well-known criteria for cuspidal edges and swallowtails (see [20, Proposition 1.3]). There is a criterion for $D_4^{\pm}$-singularities as well.

Fact 2.3 ([26, Theorem 1.1]). *Let f be a front and λ the signed area density function. A singular point p is a D_4^{+}-singularity (resp. D_4^{-}-singularity) if and only if the following conditions hold:*

(1) $\operatorname{rank} df_p = 0$.

(2) $\det \operatorname{Hess} \lambda < 0$ (*respectively,* $\det \operatorname{Hess} \lambda > 0$) *at p.*

§3. Constant mean curvature surfaces with D_4-singularities

3.1. Surfaces with non-zero constant mean curvature

In this section, we consider the cases such that

$$\begin{cases} H \neq 0 \text{ if } M^3 = \mathbb{R}^3,\ \mathbb{R}^{2,1},\ \mathbb{S}^3 \text{ or } \mathbb{H}^{2,1}, \\ H \neq 0, 1 \text{ if } M^3 = \mathbb{H}^3 \text{ or } \mathbb{S}^{2,1}. \end{cases} \tag{3.1}$$

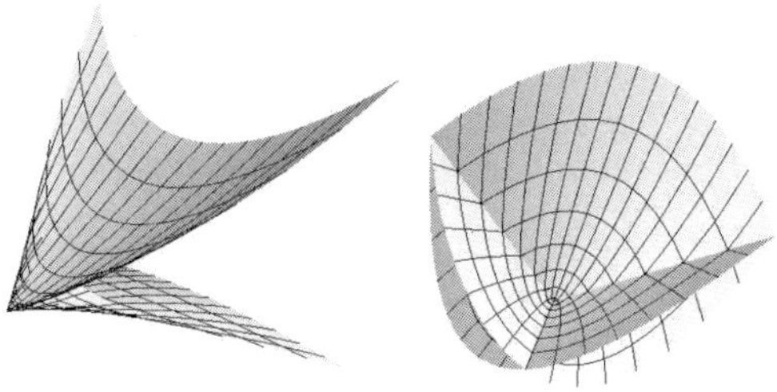

Fig. 1. The left hand side is a D_4^+-singularity and the right hand side is a D_4^-- singularity of a wave front.

Lemma 3.1. *Let $f : (U, z) \to M^3$ be a conformal CMC H surface, ν a unit normal vector to f and p an umbilic point.*

(1) *Suppose that $M^3 = \mathbb{S}^3$ or $\mathbb{H}^{2,1}$ and $H > 0$. Then p is a corank two singular point of $\hat{f}^t$ if and only if $t = \operatorname{arccot} H$.*

(2) *Suppose that $M^3 = \mathbb{H}^3$ or $\mathbb{S}^{2,1}$ and $H > 1$. Then p is a corank two singular point of $\hat{f}^t$ if and only if $t = \operatorname{arccoth} H$.*

(3) *Suppose that $M^3 = \mathbb{H}^3$ or $\mathbb{S}^{2,1}$ and $0 < H < 1$. Then p is a corank two singular point of $\check{f}^t$ if and only if $t = \operatorname{arctanh} H$.*

Proof. We show (2) and (3) in the case of $M^3 = \mathbb{H}^3$. For the case of $M^3 = \mathbb{S}^{2,1}$ and (1), one can show the result in a similar way.

Let $f : U \to M^3 = \mathbb{H}^3$ be a conformal CMC H surface. Suppose that $H > 1$. Then we consider $\hat{f}^t$ as in (2.6). Since $\nu_z = (-2Hf_z - Qg^{-2}f_{\bar{z}})/2$ and $\nu_{\bar{z}} = (-\bar{Q}g^{-2}f_z - 2Hf_{\bar{z}})/2$ by (2.2), we have

$$\hat{f}^t_z = (\cosh t - H \sinh t)f_z - \frac{Q}{2g^2} \sinh t f_{\bar{z}},$$

$$\hat{f}^t_{\bar{z}} = -\frac{\bar{Q}}{2g^2} \sinh t f_z + (\cosh t - H \sinh t)f_{\bar{z}}.$$

Since p is an umbilic point, $Q(p) = \bar{Q}(p) = 0$. Thus $\hat{f}^t_z = \hat{f}^t_{\bar{z}} = \mathbf{0}$ at p if and only if $t = \operatorname{arccoth} H$. Therefore we have the assertion (2).

Next we show (3). Assume that $0 < H < 1$. By direct computations, we see that

$$\check{f}^t_z = (\sinh t - H \cosh t)f_z - \frac{Q}{2g^2} \cosh t f_{\bar{z}},$$

$$\check{f}^t_{\bar{z}} = -\frac{\bar{Q}}{2g^2} \cosh t f_z + (\sinh t - H \cosh t)f_{\bar{z}}.$$

Hence $\check{f}^t_z = \check{f}^t_{\bar{z}} = \mathbf{0}$ at p if and only if $t = \operatorname{arctanh} H$. Q.E.D.

Theorem 3.2. *Let f be a CMC surface in M^3 with mean curvature H satisfying the condition* (3.1), *$Q(z)$ the Hopf differential factor and p a corank two singular point. Then f has a D_4^--singularity at p if and only if $Q_z(p) \neq 0$. Moreover, f does not have a D_4^+-singularity at p.*

Proof. By Facts 1.1 and 1.2, if $f : U \to M^3$ is a CMC H surface, then $\hat{f}^t$ and $\check{f}^t$ are CMC H surfaces on the set of regular points for suitable distance t. Therefore we consider the CMC surface f with an umbilic point p and singularities of its parallel transform $\hat{f}^t$ at p. Since one can prove this similarly by using Lemma 3.1 in other cases, we consider just the case of $M^3 = \mathbb{R}^3$.

Let $f : U \to \mathbb{R}^3$ be a CMC surface and p an umbilic point, where $U \subset \mathbb{C}$ is a simply-connected domain with conformal coordinate $z = u + iv$. By the previous section, $\langle f_z, f_z \rangle = \langle f_{\bar{z}}, f_{\bar{z}} \rangle = 0$ and $\langle f_z, f_{\bar{z}} \rangle = 2g^2$ for some function g on U. We now consider the parallel transformation of f given by $\hat{f}^t = f + t\nu$, where $t \in \mathbb{R}$ is constant. In this case, we can take a unit normal vector $\hat{\nu}^t$ of $\hat{f}^t$ as ν. By Fact 1.1, $\operatorname{rank} d\hat{f}^t(p) = 0$ if and only if $t = 1/H$.

We fix $t = 1/H$. The signed area density function of $\hat{f}^t$ is given by $\hat{\lambda}^t = \langle \hat{f}^t_u \wedge \hat{f}^t_v, \nu \rangle = -2i \langle \hat{f}^t_z \wedge \hat{f}^t_{\bar{z}}, \nu \rangle$. Using $\nu_z = (-2Hf_z - Qg^{-2} f_{\bar{z}})/2$, $\nu_{\bar{z}} = (-2Hf_{\bar{z}} - \bar{Q}g^{-2} f_z)/2$ and (2.3), the signed area density function $\hat{\lambda}^t$ is rewritten as

$$\hat{\lambda}^t = -2i(1 - 2tH + t^2K)\langle f_z \wedge f_{\bar{z}}, \nu \rangle.$$

Since $-2i\langle f_z \wedge f_{\bar{z}}, \nu \rangle \neq 0$, we may regard

$$\tilde{\lambda}^t = 1 - 2tH + t^2K \tag{3.2}$$

as the signed area density function of $\hat{f}^t$. By direct computations, $\tilde{\lambda}^t_z$ and $\tilde{\lambda}^t_{\bar{z}}$ are

$$\tilde{\lambda}^t_z = -t^2 \frac{(Q_z\bar{Q} + Q\bar{Q}_z)g - 4Q\bar{Q}g_z}{4g^5},$$

$$\tilde{\lambda}^t_{\bar{z}} = -t^2 \frac{(Q_{\bar{z}}\bar{Q} + Q\bar{Q}_{\bar{z}})g - 4Q\bar{Q}g_{\bar{z}}}{4g^5}.$$

Since $Q(p) = \bar{Q}(p) = 0$, $\tilde{\lambda}^t_z(p) = \tilde{\lambda}^t_{\bar{z}}(p) = 0$, that is, $d\tilde{\lambda}^t(p) = 0$ holds. We consider the Hessian of $\tilde{\lambda}^t$. The second derivative $\tilde{\lambda}^t_{zz}$ becomes

$$\begin{aligned}(3.3)\quad \tilde{\lambda}^t_{zz} = -\frac{t^2}{4g^5}\{&(Q_{zz}\bar{Q} + 2Q_z\bar{Q}_z + Q\bar{Q}_{zz})g + (Q_z\bar{Q} + Q\bar{Q}_z)g \\ &- 4(Q_z\bar{Q}g_z + Q\bar{Q}_z g_z + Q\bar{Q}g_{zz})\} \\ &- t^2\frac{5\{(Q_z\bar{Q} + Q\bar{Q}_z)g - 4Q\bar{Q}g_z\}g_z}{4g}.\end{aligned}$$

Thus $\tilde{\lambda}^t_{zz}(p) = 0$ holds. Similarly, we see that $\tilde{\lambda}^t_{\bar{z}\bar{z}}(p) = 0$ holds. By direct calculation, we have

(3.4)
$$\begin{aligned}\tilde{\lambda}^t_{z\bar{z}} = -\frac{t^2}{4g^5}\{&(Q_{z\bar{z}}\bar{Q} + Q_z\bar{Q}_{\bar{z}} + Q_{\bar{z}}\bar{Q}_z + Q\bar{Q}_{z\bar{z}})g + (Q_z\bar{Q} + Q\bar{Q}_z)g_{\bar{z}} \\ &- 4(Q_{\bar{z}}\bar{Q}g_z + Q\bar{Q}_{\bar{z}}g_z + Q\bar{Q}g_{z\bar{z}})\} \\ &- t^2\frac{5\{(Q_z\bar{Q} + Q\bar{Q}_z)g - 4Q\bar{Q}g_z\}g_{\bar{z}}}{4g}.\end{aligned}$$

By this equation, $\tilde{\lambda}^t_{z\bar{z}} = -t^2 Q_z\bar{Q}_{\bar{z}}/4g^4$ holds at p. Identifying $\mathbb{C}$ with $\mathbb{R}^2$ by $\mathbb{C} \ni z = u + iv \mapsto (u, v) \in \mathbb{R}^2$, we have

$$\tilde{\lambda}^t_{uu} = \tilde{\lambda}^t_{zz} + 2\tilde{\lambda}^t_{z\bar{z}} + \tilde{\lambda}^t_{\bar{z}\bar{z}},\ \ \tilde{\lambda}^t_{uv} = i(\tilde{\lambda}^t_{zz} - \tilde{\lambda}^t_{\bar{z}\bar{z}}),\ \ \tilde{\lambda}^t_{vv} = -(\tilde{\lambda}^t_{zz} - 2\tilde{\lambda}^t_{z\bar{z}} + \tilde{\lambda}^t_{\bar{z}\bar{z}}).$$

By the above computations, it follows that

$$\tilde{\lambda}^t_{uu} = \tilde{\lambda}^t_{vv} = 2\tilde{\lambda}^t_{z\bar{z}} = -t^2\frac{Q_z\bar{Q}_{\bar{z}}}{2g^4},\quad \tilde{\lambda}^t_{uv} = 0$$

hold at p. Thus we have

$$\det \operatorname{Hess}_{(u,v)}(\tilde{\lambda}^t)_p = \tilde{\lambda}^t_{uu}(p)\tilde{\lambda}^t_{vv}(p) - \tilde{\lambda}^t_{uv}(p)^2 = t^4\frac{(Q_z\bar{Q}_{\bar{z}})^2}{4g^8} \geq 0.$$

This completes the proof of the case $M^3 = \mathbb{R}^3$, by Fact 2.3.

If $M^3 = \mathbb{R}^{2,1}$, we can take $\tilde{\lambda}^t$ as the same as in the case of $\mathbb{R}^3$. If $M^3 = \mathbb{S}^3$ or $\mathbb{H}^{2,1}$, we have the assertion by using the signed area density for $\hat{f}^t$ as in (2.5)

$$\tilde{\lambda}^t = \cos^2 t - 2H\cos t\sin t + K\sin^2 t.$$

If $M^3 = \mathbb{H}^3$ or $\mathbb{S}^{2,1}$, we show this by using

$$\tilde{\lambda}^t = \cosh^2 t - 2H\cosh t\sinh t + K\sinh^2 t$$

for $\hat{f}^t$ and

$$\tilde{\lambda}^t = \sinh^2 t - 2H \cosh t \sinh t + K \cosh^2 t$$

for $\check{f}^t$. Q.E.D.

Examples: Here we construct CMC surfaces with D_4^--singularities in $\mathbb{H}^3$. By Theorem 3.2, we need to choose the Hopf differential factor so that $Q_z(p) \neq 0$ at a point p. Now we fix $Q = -z$ for CMC $H > 1$ or $0 \leq H < 1$ surfaces and we have the 3-legged Smyth-type surfaces as in [3], [23], [24] and [28]. Applying Theorem 3.2, we get the following figures with a D_4^--singularity at the origin $z = 0$:

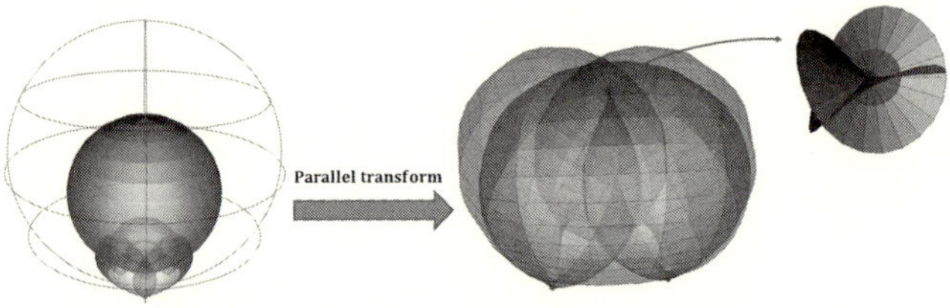

Fig. 2. 3-legged Smyth surface with $H > 1$ in $\mathbb{H}^3$ and its parallel transform with D_4^--singularity.

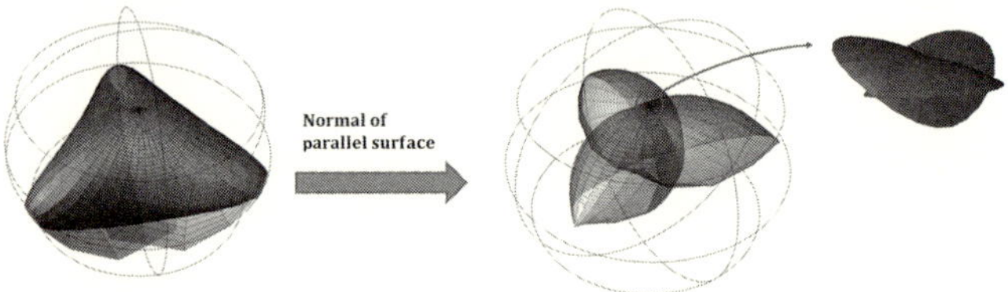

Fig. 3. 3-legged Smyth surface with $0 < H < 1$ in $\mathbb{H}^3$ and the normal vector of its parallel transform into $\mathbb{S}^{2,1}$ with D_4^--singularity.

3.2. Minimal surfaces with D_4-singularities

We now consider the condition that minimal surfaces (resp. maximal surfaces) in $\mathbb{R}^3$ (resp. $\mathbb{R}^{2,1}$) have D_4-singularities. For minimal surfaces, the following representation formula is known.

Fact 3.3 ([25]). *Any simply-connected minimal surface $f : U(\subset \mathbb{C}) \to \mathbb{R}^3$ can be parametrized as*

$$f = \operatorname{Re} \int (1 - g^2, i(1 + g^2), 2g)\omega, \tag{3.5}$$

where $g : U \to \mathbb{C}$ is a meromorphic function and $\omega = \hat{\omega}dz$ and $g^2\hat{\omega}$ are holomorphic.

We call the pair (g, ω) the *Weierstrass data*. On the other hand, the representation formula for maximal surfaces is also known.

Fact 3.4 ([18]). *Any simply-connected maximal surface $f : U(\subset \mathbb{C}) \to \mathbb{R}^{2,1}$ can be parametrized as*

$$f = \operatorname{Re} \int (1 + g^2, i(1 - g^2), -2g)\omega, \tag{3.6}$$

where $g : U \to \mathbb{C}$ is a meromorphic function and $\omega = \hat{\omega}dz$ and $g^2\hat{\omega}$ are holomorphic.

We also call the pair (g, ω) the *Weierstrass data*. We should remark that there are several studies on maximal surfaces (see [7, 10, 22, 30], for example).

Here we consider a relationship between D_4-singularities and minimal (or maximal) surfaces, using the following ansatz: the function g of Weierstrass data (g, ω) is "holomorphic" at a singular point p of f. As you can see in Facts 3.3 and 3.4, the function g is meromorphic in general, and has poles at some $z = q$. However, if a pole q of g coincides with a singular point p of f, then criteria for D_4-singularities become more complicated. (See Appendix A for datails.)

Lemma 3.5. *Let $f : U \to \mathbb{R}^3$ (resp. $\mathbb{R}^{2,1}$) be a minimal surface (resp. a maximal surface) constructed by* (3.5) *(resp.* (3.6)*) with holomorphic functions $g, \hat{\omega}$. Then $p \in U$ is a corank two singular point of f if and only if $\hat{\omega}(p) = 0$. Moreover, f is a front at p if and only if $g_z(p) \neq 0$.*

Proof. Let $f : U \to \mathbb{R}^3$ be a minimal surface with the Weierstrass data $(g, \omega = \hat{\omega}dz^2)$. The differentials of f are

$$f_z = \frac{1}{2}(1 - g^2, i(1 + g^2), 2g)\hat{\omega}, \quad f_{\bar{z}} = \frac{1}{2}(1 - \bar{g}^2, -i(1 + \bar{g}^2), 2\bar{g})\bar{\hat{\omega}}.$$

Thus we have the first assertion.

Next, we show the condition of f to be a front at p. Let p be a corank two singular point of f. Then f is a front at p if and only if its unit normal vector $\nu : U \to \mathbb{S}^2$ gives an surface at p. Under the above settings, the unit normal vector ν to f is given by

$$\nu = \left(\frac{g + \bar{g}}{|g|^2 + 1}, i\frac{\bar{g} - g}{|g|^2 + 1}, \frac{|g|^2 - 1}{|g|^2 + 1} \right).$$

Since g is holomorphic at p, $g_{\bar{z}}(p) = 0$ holds. Differentiating ν, we have

$$\nu_z = g_z \left(\frac{1-\bar{g}^2}{(1+|g|^2)^2}, i\frac{-1-\bar{g}^2}{(1+|g|^2)^2}, \frac{2\bar{g}}{(1+|g|^2)^2} \right),$$
$$\nu_{\bar{z}} = \overline{g_z} \left(\frac{1-g^2}{(1+|g|^2)^2}, i\frac{1+g^2}{(1+|g|^2)^2}, \frac{2g}{(1+|g|^2)^2} \right)$$

at p. Thus we have the conclusion.

For maximal surfaces, one can show this similarly by identifying $\mathbb{R}^{2,1}$ with $\mathbb{R}^3$ and using the Euclidean unit normal vector $\boldsymbol{n}_E$ given as

$$\boldsymbol{n}_E = \frac{1}{\sqrt{(1+|g|^2)^2 + 4|g|^2}} (g+\bar{g}, i(\bar{g}-g), 1+|g|^2)$$

(see [30, (3.3)]). Q.E.D.

Theorem 3.6. *Let $f : U \subset (\mathbb{C}, z) \to \mathbb{R}^3$ (resp. $f : U \subset (\mathbb{C}, z) \to \mathbb{R}^{2,1}$) be a minimal surface (resp. a maximal surface) given by holomorphic functions $g, \hat{\omega}$. Then a point $p \in U$ is a D_4^--singularity of f if and only if $\hat{\omega}(p) = 0$ and $Q_z(p) \neq 0$ (resp. $\hat{\omega}(p) = 0$, $Q_z(p) \neq 0$ and $|g(p)| \neq 1$). Here $Q = g_z\hat{\omega}$ is the Hopf differential factor. Moreover, f does not have a D_4^+-singularity at p.*

Proof. First, we show the case of a minimal surface. The signed area density function λ of f can be given as

$$\lambda = -2i\langle f_z \wedge f_{\bar{z}}, \nu \rangle = (1+|g|^2)^2 |\hat{\omega}|^2.$$

Since $1+|g|^2 \neq 0$, we may treat $\hat{\lambda} = |\hat{\omega}|^2 = \hat{\omega}\bar{\hat{\omega}}$ as the signed area density function. Moreover, since $f_z(p) = f_{\bar{z}}(p) = \mathbf{0}$, p is a corank two singular point. The differentials of $\hat{\lambda}$ in $z, \bar{z} \in U$ are

$$\hat{\lambda}_z = 0,\ \hat{\lambda}_{\bar{z}} = 0,\ \hat{\lambda}_{zz} = 0,\ \hat{\lambda}_{z\bar{z}} = \hat{\omega}_z \bar{\hat{\omega}}_{\bar{z}},\ \hat{\lambda}_{\bar{z}\bar{z}} = 0$$

at p, since $\hat{\omega}(p) = \bar{\hat{\omega}}(p) = \hat{\omega}_{\bar{z}}(p) = \bar{\hat{\omega}}_z(p) = 0$. Identifying $z = u + iv \in \mathbb{C}$ and $(u, v) \in \mathbb{R}^2$, we see that $\hat{\lambda}_{uu}(p) = \hat{\lambda}_{vv}(p) = 2\hat{\omega}_z(p)\bar{\hat{\omega}}_{\bar{z}}(p)$ and $\hat{\lambda}_{uv}(p) = 0$. Hence the Hessian of $\hat{\lambda}$ is

$$\det \mathrm{Hess}_{(u,v)}(\hat{\lambda})_p = 4|\hat{\omega}_z(p)|^4 \geq 0.$$

By Fact 2.3 and Lemma 3.5, f has a D_4^--singularity at p if and only if $g_z(p) \neq 0$ and $\hat{\omega}_z(p) \neq 0$. On the other hand, the derivative of the Hopf differential factor Q is

$$Q_z(p) = g_{zz}(p)\hat{\omega}(p) + g_z(p)\hat{\omega}_z(p) = g_z(p)\hat{\omega}_z(p),$$

since g_{zz} is finite at p. Thus we have the assertion.

Next, we consider a maximal surface that is not a maxface. We identify $\mathbb{R}^3$ with $\mathbb{R}^{2,1}$. The signed area density function is

$$\lambda = -2i\langle f_z \times f_{\bar{z}}, \boldsymbol{n}_{\mathrm{E}}\rangle_{\mathrm{Euc}} = (|g|^2 - 1)|\hat{\omega}|^2\sqrt{(1+|g|^2)^2 + 4|g|^2},$$

where $\times$ and $\langle \cdot, \cdot \rangle_{\mathrm{Euc}}$ mean the Euclidean vector product and the Euclidean inner product of $\mathbb{R}^3$. From Lemma 3.5, we may take $\hat{\lambda} = \hat{\omega}\bar{\hat{\omega}}$. By using similar arguments, we obtain the assertion. Q.E.D.

By the Lawson correspondence, the first fundamental forms of (spacelike) CMC 1 surfaces in $\mathbb{H}^3$ (resp. $\mathbb{S}^{2,1}$) are equal to the first fundamental forms of corresponding minimal surfaces in $\mathbb{R}^3$ (resp. maximal surfaces in $\mathbb{R}^{2,1}$). This means that they have the same signed area density functions. Thus we obtain the condition that (spacelike) CMC 1 surfaces have D_4^--singularities similarly.

On the other hand, when $f : U \to M^3 = \mathbb{S}^3$, $\mathbb{H}^3$, $\mathbb{S}^{2,1}$ or $\mathbb{H}^{2,1}$ is a minimal surface, there is no Weierstrass type representation formula. However, if $p \in U$ is an umbilic point for f, then its unit normal vector ν has a corank two singularity at p. By using similar calculations as in the proof of Theorem 3.2, we see that ν has a D_4^--singularity at p if and only if $Q_z(p) \neq 0$, where Q is the Hopf differential factor of f.

Example: Here we construct CMC 1 surfaces with D_4^--singularities in $\mathbb{H}^3$. Using Theorem 3.6, we fix the Weierstrass data $(g, \omega) = (\cot(z-1), (e^z - 1)dz)$ for a CMC 1 surface in $\mathbb{H}^3$. Applying Theorem 3.6, we get the following figure with a D_4^--singularity at the origin $z = 0$:

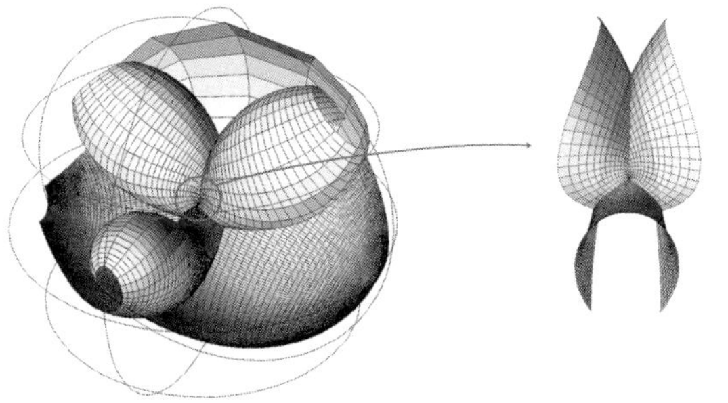

Fig. 4. CMC 1 surface with D_4^--singularity in $\mathbb{H}^3$.

§4. Curvatures of unit normal vector fields to constant mean curvature surfaces

In this section, M^3 denotes one of $\mathbb{S}^3$, $\mathbb{H}^3$, $\mathbb{S}^{2,1}$ or $\mathbb{H}^{2,1}$. We consider relations between a CMC surface $f : U \to M^3$ and a unit normal vector ν to f.

Proposition 4.1. *Let $f : U \to M^3$ be a (spacelike) CMC H surface, and let ν be a unit normal vector to f. Let K denote the extrinsic Gaussian curvature of f. Then the extrinsic Gaussian curvature K_ν and the mean curvature H_ν of ν are*

$$K_\nu = \frac{1}{K}, \quad H_\nu = \frac{H}{K},$$

Moreover, the unit normal vector ν has a constant harmonic mean curvature $1/2H$.

Here the *harmonic mean curvature* HMC is given by

$$HMC = \frac{K}{2H}.$$

Proof. We consider the case $f : U \to \mathbb{H}^3$. One can show other cases similarly. Let (u, v) be a conformal coordinate system on U. Then the determinant of the first fundamental matrix I_ν is given by

$$\det I_\nu = E_\nu G_\nu - F_\nu^2 = K^2 E^2,$$

where

$$I_\nu = \begin{pmatrix} \langle \nu_u, \nu_u \rangle & \langle \nu_u, \nu_v \rangle \\ \langle \nu_v, \nu_u \rangle & \langle \nu_v, \nu_v \rangle \end{pmatrix} = \begin{pmatrix} E_\nu & F_\nu \\ F_\nu & G_\nu \end{pmatrix},$$

and $E = \langle f_u, f_u \rangle = \langle f_v, f_v \rangle$. It follows that the coefficients of the second fundamental form of ν are the same as that of f by definition. Thus we have the assertions by straightforward calculations. Q.E.D.

For $M^3 = \mathbb{S}^3$, $\mathbb{H}^{2,1}$ and $H > 0$, or $M^3 = \mathbb{H}^3$, $\mathbb{S}^{2,1}$ and $H > 1$, the parallel transformations $\hat{f}^t$ and $\hat{\nu}^t$ are defined as in (2.5) and (2.6). If $M^3 = \mathbb{H}^3$, $\mathbb{S}^{2,1}$ and $0 < H < 1$, the parallel transforms of f and ν are $\check{f}^t = \hat{\nu}^t$, $\check{\nu}^t = \hat{f}^t$, respectively. Thus it is sufficient to consider $\hat{f}^t$ and $\hat{\nu}^t$.

Lemma 4.2. *Under the above settings, the extrinsic Gaussian curvatures K^t and K^t_ν, and the mean curvatures H^t and H^t_ν for $\hat{f}^t$ and $\hat{\nu}^t$ are given by the following:*

$$K^t = \begin{cases} \dfrac{\sin^2 t + 2H \cos t \sin t + K \cos^2 t}{\cos^2 t - 2H \cos t \sin t + K \sin^2 t}, \\ \dfrac{\sinh^2 t + 2H \cosh t \sinh t + K \cosh^2 t}{\cosh^2 t - 2H \cosh t \sinh t + K \sinh^2 t}, \end{cases}$$

$$K_\nu^t = \begin{cases} \dfrac{\cos^2 t - 2H\cos t\sin t + K\sin^2 t}{\sin^2 t + 2H\cos t\sin t + K\cos^2 t}, \\ \dfrac{\cosh^2 t - 2H\cosh t\sinh t + K\sinh^2 t}{\sinh^2 t + 2H\cosh t\sinh t + K\cosh^2 t}, \end{cases}$$

$$H^t = \begin{cases} \dfrac{(1-K)\cos t\sin t + H(\cos^2 t - \sin^2 t)}{\cos^2 t - 2H\cos t\sin t + K\sin^2 t}, \\ -\dfrac{(1+K)\cosh t\sinh t - H(\cosh^2 t + \sinh^2 t)}{\cosh^2 t - 2H\cosh t\sinh t + K\sinh^2 t}, \end{cases}$$

$$H_\nu^t = \begin{cases} \dfrac{(1-K)\cos t\sin t + H(\cos^2 t - \sin^2 t)}{\cos^2 t + 2H\cos t\sin t + K\sin^2 t}, \\ -\dfrac{(1+K)\cosh t\sinh t - H(\cosh^2 t + \sinh^2 t)}{\cosh^2 t + 2H\cosh t\sinh t + K\sinh^2 t}. \end{cases}$$

Proof. One can show the above formulas directly by applying the same computations as in [19]. Q.E.D.

By Proposition 4.1 and Lemma 4.2, we immediately have the following:

Proposition 4.3. *Let $f : U \to M^3 = \mathbb{S}^3$, $\mathbb{H}^3$, $\mathbb{S}^{2,1}$, $\mathbb{H}^{2,1}$ be a CMC $H \neq 0$ surface and ν its unit normal vector. A point p is an umbilic point for f if and only if p is a non-flat umbilic point for ν. Moreover, if $\hat{f}^t$ (resp. $\check{f}^t$) has a D_4^--singularity at p, p is a flat umbilic point of $\hat{\nu}^t$ (resp. $\check{\nu}^t$) (see Figures 5, 6).*

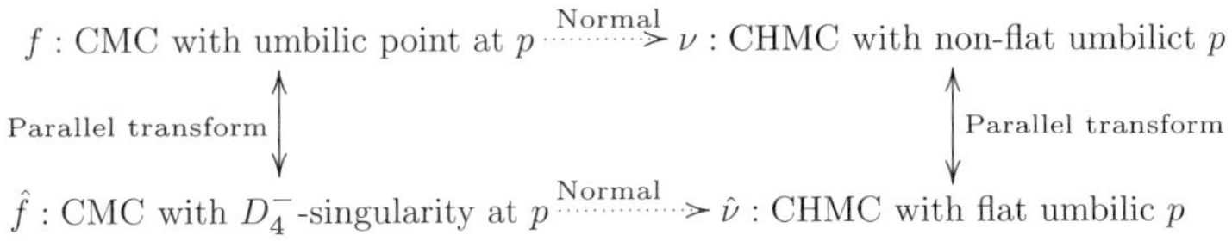

Fig. 5. The case that f is a (spacelike) CMC surface in $\mathbb{S}^3$, $\mathbb{H}^{2,1}$ (resp. $\mathbb{H}^3$, $\mathbb{S}^{2,1}$) with $H > 0$ (resp. $H > 1$)

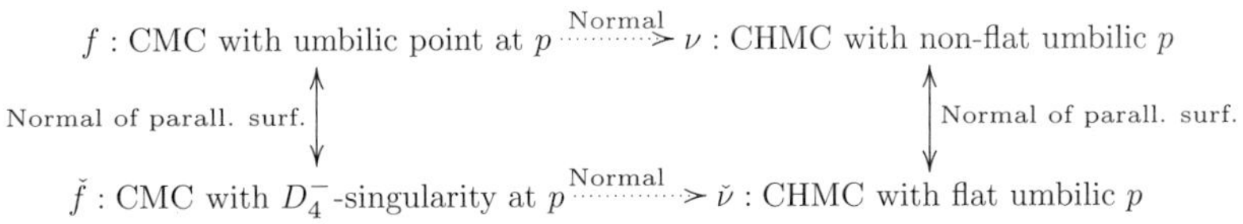

Fig. 6. The case that f is a (spacelike) CMC surface in $\mathbb{H}^3$, $\mathbb{S}^{2,1}$ with $0 < H < 1$

§Appendix A. The D_4-singularity and the poles of meromorphic functions g for minimal surfaces

In Section 3.2, we considered a relationship between D_4-singularities and minimal surfaces, and assumed that the function g of the Weierstrass data (g, ω) is "holomorphic" at a singular point p. Thus, we omitted the case that a pole of the meromorphic function g coincides with a singular point. Here we will consider this remaining case and give the criteria for D_4-singularities of such minimal surfaces. However, for the pole of g, we no longer have the relationship between the criteria for D_4-singularities and the Hopf differential factor Q (see Theorem A.4).

Lemma A.1. *Let f be a minimal surface with the Weierstrass data $(g, \omega = \hat{\omega}dz)$. Suppose that p is a pole of g. Then p is a corank two singular point of f if and only if $g^2\hat{\omega} = 0$ at p.*

Proof. By

$$f_z = \frac{1}{2}(1 - g^2, i(1 + g^2), 2g)\hat{\omega}, \quad f_{\bar{z}} = \frac{1}{2}(1 - \bar{g}^2, -i(1 + \bar{g}^2), 2\bar{g})\bar{\hat{\omega}},$$

we see that $f_z = f_{\bar{z}} = \mathbf{0}$ at p if and only if $\hat{\omega} = g\hat{\omega} = g^2\hat{\omega} = 0$ at p. However $\hat{\omega} = g\hat{\omega} = g^2\hat{\omega} = 0$ is equivalent to $g^2\hat{\omega} = 0$ because $g(p) = \pm\infty$. Q.E.D.

Lemma A.2. *Let f be a minimal surface with the Weierstrass data $(g, \omega = \hat{\omega}dz)$. Suppose that p is a pole of g and corank two singularity of f. Then f is a front at p if and only if $\dfrac{|g_{\bar{z}}|^2 - |g_z|^2}{(|g|^2 + 1)^2} \neq 0$ at p.*

Proof. By $\nu = \left(\dfrac{g + \bar{g}}{|g|^2 + 1}, i\dfrac{\bar{g} - g}{|g|^2 + 1}, \dfrac{|g|^2 - 1}{|g|^2 + 1}\right)$, we have

$$\nu_z \wedge \nu_{\bar{z}} = \frac{2i(|g_{\bar{z}}|^2 - |g_z|^2)}{(|g|^2 + 1)^2}\nu.$$

Thus, we notice that f is a front at p if and only if $\dfrac{|g_{\bar{z}}|^2 - |g_z|^2}{(|g|^2 + 1)^2} \neq 0$ at p. Q.E.D.

Remark A.3. If we write $g = \dfrac{h_2}{h_1}$ by using holomorphic functions h_1 and h_2 such that $h_1(p) = 0$ and $h_2(p)$ is a non-zero value, then

$$\frac{|g_{\bar{z}}|^2 - |g_z|^2}{(|g|^2 + 1)^2} \neq 0 \text{ at } p \Longleftrightarrow (h_1)_z(p) \neq 0.$$

Here we get criteria for D_4-singularities of minimal surfaces when a point p is both a singular point of f and a pole of g.

Theorem A.4. *Let f be a minimal surface with the Weierstrass data $(g, \omega = \hat{\omega} dz)$. Suppose that p is a pole of g. Then the point p is a D_4^--singularity of f if and only if*

$$\begin{cases} g^2\hat{\omega} = 0, \\ \dfrac{|g_{\bar{z}}|^2 - |g_z|^2}{(|g|^2+1)^2} \neq 0 \text{ and} \\ \hat{\omega}_z \neq 0 \text{ or } (g\hat{\omega})_z \neq 0 \text{ or } (g^2\hat{\omega})_z \neq 0 \text{ at } p. \end{cases}$$

Moreover, f does not have a D_4^+-singularity at p.

Proof. The signed area density function λ can be given as

$$\lambda = (1+|g|^2)^2|\hat{\omega}|^2 = |\hat{\omega}|^2 + 2|g\hat{\omega}|^2 + |g^2\hat{\omega}|^2.$$

Then, we have

$$\lambda_z = \hat{\omega}_z\bar{\hat{\omega}} + \hat{\omega}\bar{\hat{\omega}}_z + 2\left((g\hat{\omega})_z\overline{g\hat{\omega}} + g\hat{\omega}(\overline{g\hat{\omega}})_z\right) + (g^2\hat{\omega})_z\overline{g^2\hat{\omega}} + g^2\hat{\omega}(\overline{g^2\hat{\omega}})_z.$$

All terms appearing in the above equation are zero at p. Thus, $\lambda_z(p) = 0$. We can continue computing to get the following

$$\lambda_{\bar{z}}(p) = \lambda_{zz}(p) = \lambda_{\bar{z}\bar{z}}(p) = 0 \text{ and } \lambda_{z\bar{z}}(p) = |\hat{\omega}_z|^2 + 2|(g\hat{\omega})_z|^2 + |(g^2\hat{\omega})_z|.$$

Hence the Hessian of λ is

$$\det \mathrm{Hess}_{(u,v)}(\lambda)_p = 4\left(|\hat{\omega}_z|^2 + 2|(g\hat{\omega})_z|^2 + |(g^2\hat{\omega})_z|\right)^2 \geq 0.$$

By Fact 2.3 and Lemma 3.5, we have the assertion. Q.E.D.

Remark A.5. For the case of maximal surfaces in $\mathbb{R}^{2,1}$, when a point p is both a singular point and a pole of g, we can apply similar computation as in Section 3.2. Thus, we omit it.

Acknowledgements. The authors would like to thank Shintaro Akamine, Wayne Rossman, Kentaro Saji and Masashi Yasumoto for valuable comments. The authors also thank the referee for many helpful suggestions.

References

[1] V. I. Arnol'd, S. M. Gusein-Zade, and A. N. Varchenko, Singularities of differentiable maps, Vol.1, Monographs in Mathematics **82**, Birkhäuser, Boston, 1985.

[2] A. I. Bobenko, Constant mean curvature surfaces and integrable equations, Russian Math. Survays, 46:4, (1991), 1–45.

[3] A. I. Bobenko and A. Its, The Painleve III equation and the Iwasawa decomposition, Manuscripta Math., **87**, (1995), 369–377.

[4] D. Brander, Singularities of spacelike constant mean curvature surfaces in Lorentz-Minkowski space, Math. Proc. Cambridge Philos. Soc. **150** (2011), no. 3, 527–556.

[5] D. Brander, W. Rossman and N. Schmitt, Holomorphic representation of constant mean curvature surfaces in Minkowski space: consequences of non-compactness in loop group methods, Adv. Math., **223**, (2010), no. 3, 949-986.

[6] Q. Chen and Y. Cheng, Spectral transformation of constant mean curvature surfaces in $\mathbf{H}^3$ and Weierstrass representation, Sci. China Ser. A, **45**, (2002), no. 8, 1066–1075.

[7] F. J. M. Estudillo and A. Romero, Generalized maximal surfaces in Lorentz-Minkowski space $\mathbb{L}^3$, Math. Proc. Camb. Philos. Soc. **111** (1992), 515–524.

[8] J. Dorfmeister, J. Inoguchi and S-P. Kobayashi, Constant mean curvature surfaces in hyperbolic 3-space via loop groups, J. Reine Angew. Math., **686**, (2014), 1–36. (arXiv:1108. 1641.)

[9] S. Fujimori, Spacelike CMC 1 surfaces with elliptic ends in de Sitter 3-space, Hokkaido Math. J. **35**, (2006), 289–320.

[10] S. Fujimori, K. Saji, M. Umehara and K. Yamada, Singularities of maximal surfaces, Math. Z. **259**, (2008),827–848.

[11] T. Fukui and M. Hasegawa, Singularities of parallel surfaces, Tohoku Math. J. **64** (2012), 387–408.

[12] A. Honda, Duality of singularities for spacelike CMC surfaces, to appear in Kobe J. Math.

[13] A. Honda, Isometric surfaces with singularities between space forms of the same positive curvature, arXiv:1508.07223.

[14] A. Honda, M. Koiso and K. Saji, Fold singularities on spacelike CMC surfaces in Lorentz-Minkowski space, to appear in Hokkaido Math. J., arXiv:1509.03050.

[15] S. Izumiya, M. C. Romero Fuster, M. A. S. Ruas and F. Tari, Differential Geometry from a Singularity Theory Viewpoint, World Scientific, 2015.

[16] S. Izumiya and K. Saji, The mandala of Legendrian dualities for pseudo-spheres in Lorentz-Minkowski space and "flat" spacelike surfaces. J. Singul. **2**, (2010), 92–127.

[17] S. Izumiya, K. Saji and M. Takahashi, Horospherical flat surfaces in hyperbolic 3-space, J. Math. Soc. Japan **62** (2010), 789–849.

[18] O. Kobayashi, Maximal surfaces in the 3-dimensional Minkowski space L^3, Tokyo J. Math. **6** (1983), 297–309.
[19] M. Kokubu, Surfaces and fronts with harmonic-mean curvature one in hyperbolic three-space, Tokyo J. Math. **32** (2009), no. 1, 177–200.
[20] M. Kokubu, W. Rossman, K. Saji, M. Umehara, and K. Yamada, Singularities of flat fronts in hyperbolic space, Pacific J. Math. **221** (2005), 303–351.
[21] M. Kokubu, W. Rossman, M. Umehara and K. Yamada, Flat fronts in hyperbolic 3-space and their caustics, J. Math, Soc. Japan **59** (2007), 265–299.
[22] M. Kokubu and M. Umehara, Orientability of linear Weingarten surfaces, spacelike CMC-1 surfaces and maximal surfaces, Math. Nachr. **284** (2011), 1903–1918.
[23] Y. Ogata, The DPW method for constant mean curvature surfaces in 3-dimensional Lorentzian spaceforms, with applications to Smyth type surfaces, to appear in Hokkaido J. Math.
[24] Y. Ogata, Spacelike constant mean curvature and maximal surfaces in 3-dimensional de Sitter space via Iwasawa splitting, Tsukuba J. Math. **39**, no. 2, (2015).
[25] R. Osserman, Global properties of minimal surfaces in E^3 and E^n, Ann. of Math. **80** (1964), 340–364.
[26] K. Saji, Criteria for D_4 singularities of wave fronts, Tohoku Math. J. **63** (2011),137–147.
[27] K. Saji, M. Umehara, and K. Yamada, The geometry of fronts, Ann. of Math. **169** (2009), 491–529.
[28] B. Smyth, A generalization of a theorem of Delaunay on constant mean curvature surfaces, Statistical thermodynamics and differential geometry of microstructured materials, IMA Vol. Math. Appl. **51**, Springer, New York, (1993), 123-130.
[29] Y. Umeda, Constant-mean-curvature surfaces with singularities in Minkowski 3-space, Experiment. Math. **18** (2009), no. 3, 311-323.
[30] M. Umehara and K. Yamada, Maximal surfaces with singularities in Minkowski space, Hokkaido Math. J. **35** (2006), 13–40.

Department of Science and Technology, National Institute of Technology, Okinawa College, 905, Henoko, Nago, Okinawa 905-2171, Japan

Department of Mathematics, Graduate School of Science, Kobe University, 1-1, Rokko-dai, Nada-ku, Kobe 657-8501, Japan

E-mail address: `y.ogata@okinawa-ct.ac.jp`
E-mail address: `teramoto@math.kobe-u.ac.jp`

Advanced Studies in Pure Mathematics 78, 2018
Singularities in Generic Geometry
pp. 365–381

Singularities of secant maps on closed plane curves

María Carmen Romero Fuster and Luís Sanhermelando Rodríguez

Abstract.

We study the singularities of secant maps associated to pairs of plane curves providing their geometrical interpretation up to codimension 2. We show that for most pairs of closed plane curves the secant map is a stable map from the torus to the plane. We determine the isotopy type of the singular set of the secant map associated to pairs of convex closed curves in terms of their Whitney indices.

§1. Introduction

The local properties of secant maps associated to curves in 3-space were studied by J.W. Bruce [2], who proved that for generic pairs of curves these map is locally stable and thus may only have isolated cross-cap points. We consider here the secant map associated to closed plane curves $\alpha, \beta : S^1 \to \mathbb{R}^2$ and analyze its singularities from the local and multi-local viewpoints, providing their geometrical characterization up to codimension 2 (Theorem 3.1). As a consequence of Thom's fundamental transversality lemma, we show that the secant map of a generic pair of closed plane curves is a stable map from the torus to the plane (Theorem 4.5). In the particular case $\alpha = \beta$, we see that for most rigid motions ϕ on the plane, the pair $(\alpha, \phi \cdot \alpha)$ is a generic couple of curves.

From a global viewpoint, we prove that the number of singular curves of the secant map of a generic pair of closed convex curves with respective Whitney indexes n and m, is exactly twice the maximum common divisor $\mu_{n,m}$ of n and m (Theorem 5.10). Moreover, all the singular curves are of type $(\frac{n}{\mu_{n,m}}, \frac{m}{\mu_{n,m}})$. As a consequence, we get that

Received January 26, 2016.
Revised January 28, 2017.
Key words and phrases. Singularities, stability, plane curves, Whitney index, isotopy invariants.
Work partially supported by DGCYT and FEDER grant no. MTM2015-64013-P.

given a convex curve α with Whitney index n, the secant map of pair $(\alpha, \phi \cdot \alpha)$ (where ϕ is any rigid motion such that $(\alpha, \phi \cdot \alpha)$ is a good pair of curves) has exactly n singular curves, all of them being toric curves of type $(1,1)$.

The general case, including non necessarily convex curves, will be treated in a forthcoming paper. We would like to thank J. Martínez Alfaro and R. Oset for helpful comments.

§2. Isotopy invariants of closed curves

It is a well known fact that a generic immersion of S^1 into the plane has normal crossings, in other words, a finite number of transverse double points. In fact, standard transversality arguments in jet spaces lead, as a consequence of Thom's Transversality Theorem ([7]), to the fact that the set of regular closed plane curves with normal crossings is open and dense in $C^\infty(S^1, \mathbb{R}^2)$ with the Whitney C^∞-topology. Analogous arguments show that for a curve α lying in an open and dense subset of $C^\infty(S^1, \mathbb{R}^2)$, the inflection points are isolated and do not coincide with the vertices of α. This means that the curvature of α and its derivative do not vanish at the same time. We shall denote by S the open and dense subset of $C^\infty(S^1, \mathbb{R}^2)$ made of closed plane curves with normal crossings and isolated inflection points that satisfy this condition.

An *isotopy* between two regular plane curves α and β is a smooth map

$$\begin{array}{rcl} F : S^1 \times [0,1] & \longrightarrow & \mathbb{R}^2 \\ (s,u) & \longmapsto & F_u(s) \end{array}$$

such that F_u is a regular plane curve for all $u \in [0,1]$ and $F_0 = \alpha$ and $F_1 = \beta$

We say that F is a *stable isotopy* provided F_u is a stable immersion, for all $u \in [0,1]$, that is, F_u is a closed regular plane curve with normal crossings.

The *Gauss map* of a regular curve $\alpha : S^1 \to \mathbb{R}^2$ is given by

$$\begin{array}{rcl} G_\alpha : S^1 & \longrightarrow & S^1 \\ s & \longmapsto & \dfrac{\alpha'(s)}{\|\alpha'(s)\|}. \end{array}$$

Denote by κ_α the curvature function of α. An inflection point s_0 of α is characterized by the condition $\kappa_\alpha(s_0) = 0$, or equivalently $\alpha''(s_0) = 0$.

So we have that s_0 is a singular point of G_α if and only if it is an inflection point of α.

The *Whitney index* of a closed curve α is defined as the degree of the map $G_\alpha : S^1 \to S^1$. We denote it by i_α. The following theorem tells us that this index is a complete isotopy invariant for the space of closed regular plane curves.

Theorem 1 (H. Whitney [10]). *Two closed regular plane curves are isotopic if and only if they have the same Whitney index.*

We recall that the first homology group of the torus $S^1 \times S^1$ is generated by two loops that we shall denote as γ_1 (parallel) and γ_2 (meridian) and we have that any loop γ in the torus is homotopic to $a\gamma_1 + b\gamma_2$, for convenient $a, b \in \mathbb{Z}$, we then say that γ is of type (a, b). Closed curves of type $(0, 0)$ are homotopically trivial. Two curves on the torus are isotopic if and only if they have the same type (a, b).

§3. Singularities of secant maps on plane curves

Given two plane curves $\alpha, \beta : S^1 \to \mathbb{R}^2$, the *secant map* between them is defined as:

$$\begin{array}{rcl} S_{\alpha,\beta} : S^1 \times S^1 & \longrightarrow & \mathbb{R}^2 \\ (s,t) & \longmapsto & (\alpha(s) - \beta(t)). \end{array}$$

Our purpose in this section is to analyze the singularities of secant maps associated to couples of curves (α, β) providing their geometrical interpretation. All the curves we consider along this paper will be immersed curves.

We shall say that a smooth map f from a surface X to the plane is $\mathcal{A}$-*stable* if any element lying in a small enough open neighbourhood of f in $C^\infty(X, \mathbb{R}^2)$ with the Whitney C^∞-topology is $\mathcal{A}$-equivalent to f. A well known theorem of Whitney states that the critical set of stable maps from a surface to the plane is composed of smooth curves made of fold points and isolated cusp points ([7], [11]). From the global viewpoint, we have that the image of the critical set (apparent contour) of a stable map from a closed surface to the plane is a collection of closed plane curves with transverse intersections and isolated singular points (simple cusps) which correspond to the cusps of the map.

Given a smooth map $f : \mathbb{R}^2 \to \mathbb{R}^2$, we denote its singular set by Σf and its apparent contour $f(\Sigma f)$ by $C(f)$. The codimension 1 germs from the plane to the plane were first studied by Gaffney and Ruas (unpublished work) and subsequently by other authors [1, 3, 4, 6, 12]. A complete classification of the simple germs from the plane to the plane, including all the germs of codimensions one and two, was obtained by J. Rieger [9]. The following list contains all these germs up to codimension 2. Observe that all of them have corank at most one.

singularity	**normal form**	**$\mathcal{A}$-codimension**
regular	(x, y)	0
fold	(x, y^2)	0
cusp	$(x, y^3 + xy)$	0
lips and beaks $(4_2^{\pm})$	$(x, y^3 \pm x^2y)$	1
goose (4_3)	$(x, y^3 + x^3y)$	2
swallowtail (5)	$(x, y^4 + xy)$	1
butterfly $(6^{\pm})$	$(x, xy + y^5 \pm y^7)$	2
gulls (11_5)	$(x, xy^2 + y^4 + y^5)$	2

It is well known that the set of stable maps from a closed surface M to the plane is a residual subset of $C^\infty(M, S^1)$.

Given a germ $f : (\mathbb{R}^2, 0) \to (\mathbb{R}^2, 0)$ of corank one, by appropriate changes of variables at the source and the target, we can write it as:

$$\begin{array}{rcl} f : (\mathbb{R}^2, 0) & \longrightarrow & (\mathbb{R}^2, 0) \\ (x, y) & \longmapsto & (x, f_2(x, y)) \end{array}$$

where $j^\infty f_2(x, y) = \sum_{i,j=1}^{\infty} c_{i,j} x^i y^j$.

The following tables provide the characterization of each singularity in terms of the coefficients $c_{i,j}$.

singularity	**normal form**	**conditions**	**frequency**
regular	(x, y)	$c_{0,1} \neq 0$	open dense subset
fold	(x, y^2)	$c_{0,1} = 0$ $c_{0,2} \neq 0$	smooth curves
cusp	$(x, y^3 + xy)$	$c_{0,1} = 0$ $c_{0,2} = 0$ $c_{1,1} \neq 0$ $c_{0,3} \neq 0$	isolated points
swallowtail	$(x, y^4 + xy)$	$c_{0,1} = 0$ $c_{0,2} = 0$ $c_{0,3} = 0$ $c_{1,1} \neq 0$ $c_{0,4} \neq 0$	isolated points in 1-parameter families

singularity	**normal form**	**conditions**	**frequency**
lips beaks lips beaks	$(x, y^3 + x^2y)$ $(x, y^3 - x^2y)$	$c_{0,1} = 0$ $c_{0,2} = 0$ $c_{1,1} = 0$ $c_{0,3} \neq 0$ $3c_{2,1}c_{0,3} > c_{1,2}^2$ $3c_{2,1}c_{0,3} < c_{1,2}^2$	isolated points in 1-parameter families
butterfly (6±) 6^+ 6^-	 $(x, xy + y^5 + y^7)$ $(x, xy + y^5 - y^7)$	$c_{0,1} = c_{0,2} = 0$ $c_{0,3} = c_{0,4} = 0$ $c_{1,1}, c_{0,5} \neq 0$ $c_{0,7} > 0$ $c_{0,7} < 0$	isolated points in 2-parameter families
goose (4_3)	$(x, y^3 + x^3y)$	$c_{0,1} = c_{0,2} = 0$ $c_{1,1} = c_{2,1} = 0$ $c_{0,3}, c_{3,1} \neq 0$	isolated points in 2-parameter families
gulls (11_5)	$(x, xy^2 + y^4 + y^5)$	$c_{0,1} = c_{0,2} = 0$ $c_{1,1} = c_{0,3} = 0$ $c_{1,2}, c_{0,4}, c_{0,5} \neq 0$	isolated points in 2-parameter families

The list of multi-germs from the plane to the plane up to codimension one was first determined by Chíncaro ([5]). A complete list of multi-germs up to codimension 2, obtained by F. Aicardi and T. Ohmoto [8], can be seen in the next table.

singularity	**normal form**	**conditions**	**frequency**
transversal cross of 2 folds	$(x, y^2; \bar{x}, \bar{y}^2)$	$x = \bar{x}$ $y^2 = \bar{y}^2$ $c_{0,1} = 0$ $c_{0,2} \neq 0$ $\bar{c}_{0,1} = 0$ $\bar{c}_{0,2} \neq 0$	isolated points
cross of fold and cusp	$(x, y^2; \bar{x}, \bar{y}^3 + \bar{x}\bar{y})$	$x = \bar{x}$ $y^2 = \bar{y}^3 + \bar{x}\bar{y}$ $c_{0,1} = 0$ $c_{0,2} \neq 0$ $\bar{c}_{0,1} = 0$ $\bar{c}_{0,2} = 0$ $\bar{c}_{1,1} \neq 0$ $\bar{c}_{0,3} \neq 0$	isolated points in 1-parameter families

singularity	**normal form**	**conditions**	**frequency**
cross of 3 folds mutually transversal	$(x, y^2; \bar{x}, \bar{y}^2; \bar{\bar{x}}, \bar{\bar{y}}^2)$	$x = \bar{x} = \bar{\bar{x}}$ $y^2 = \bar{y}^2 = \bar{\bar{y}}^2$ $c_{0,1} = 0$ $c_{0,2} \neq 0$ $\bar{c}_{0,1} = 0$ $\bar{c}_{0,2} \neq 0$ $\bar{\bar{c}}_{0,1} = 0$ $\bar{\bar{c}}_{0,2} \neq 0$	isolated points in 1-parameter families

We analyze now the geometrical meaning of the singularities of the above tables in the particular case of secant maps. Given a couple of plane curves (α, β), consider the Jacobian map of $S_{\alpha,\beta}$:

$$JS_{\alpha,\beta}(s,t) = \begin{pmatrix} \alpha_1'(s) & -\beta_1'(t) \\ \alpha_2'(s) & -\beta_2'(t) \end{pmatrix},$$

where $\alpha(s) = (\alpha_1(s), \alpha_2(s))$ and $\beta(s) = (\beta_1(s), \beta_2(s))$. Since both curves are regular, it follows that rank $JS_{\alpha,\beta}(s,t) \geq 1$, for all $(s,t) \in S^1 \times S^1$. Moreover, we have that $(s,t) \in S^1 \times S^1$ is a singular point of $S_{\alpha,\beta}$ if and only if the tangent vectors $\alpha'(s)$ and $\beta'(t)$ are parallel (we denote this as $\alpha'(s)||\beta'(t)$).

We can write $S_{\alpha,\beta}$ in the standard normal form

$$\begin{array}{rcl} f : (\mathbb{R}^2, 0) & \longrightarrow & (\mathbb{R}^2, 0) \\ (x, y) & \longmapsto & (x, f_2(x, y)) \end{array}$$

where $f_2(x,y)$, x and y are functions of s and t. We shall work locally at a singular point (s,t) that by simplicity of notation we shall consider to be $(0,0)$. Since it is a singular point of $S_{\alpha,\beta}$, we have that $\alpha'(s)||\beta'(t)$, therefore it is possible to make a change of variables in $\mathbb{R}^2$ such that $\alpha(s) = (s, a(s))$ and $\beta(t) = (t, b(t))$ for convenient functions a and b defined in a neighbourhood of $(0,0)$. Then we have,

$$\begin{array}{rcl} S_{\alpha,\beta} : (\mathbb{R}^2, 0) & \longrightarrow & (\mathbb{R}^2, 0) \\ (s, t) & \longmapsto & (s - t, a(s) - b(t)). \end{array}$$

By applying the change of variables $x = s - t, y = s + t$ we get

$$\bar{S}_{\alpha,\beta}(x,y) = (x, f_2(x,y))$$

with $f_2(x,y) = a(\frac{x+y}{2}) - b(\frac{y-x}{2})$. If we write the Taylor series of the functions a and b at the origin as $j^\infty a(0) = \sum_{i=1}^{\infty} a_i s^i$ and $j^\infty b(0) = \sum_{j=1}^{\infty} b_j t^j$, we obtain

$$j^\infty f_2(x,y) = \left(a_1\left(\tfrac{x+y}{2}\right) + a_2\left(\tfrac{x+y}{2}\right)^2 + \cdots\right) - \left(b_1\left(\tfrac{y-x}{2}\right) + b_2\left(\tfrac{y-x}{2}\right)^2 + \cdots\right)$$

$$= \tfrac{a_1+b_1}{2}x + \tfrac{a_1-b_1}{2}y + \tfrac{a_2-b_2}{4}x^2 + 2\tfrac{a_2+b_2}{4}xy + \tfrac{a_2-b_2}{4}y^2 + \cdots = \sum_{i,j=1}^{\infty} c_{i,j}x^i y^j,$$

where $c_{i,j} = \begin{pmatrix} i+j \\ i \end{pmatrix} \frac{a_{i+j} + (-1)^{i+1} b_{i+j}}{2^{i+j}}$.

We can now write the values of the coefficients $c_{i,j}$ in the above tables in terms of the coefficients a_i and b_j. Here we observe that $c_{i,j} = 0$ if and only if $a_{i+j} = (-1)^i b_{i+j}$.

Lemma 2. *The goose* (4_3) *cannot occur as a singularity of secant maps.*

Proof. This follows immediately from the above table by observing that in the particular case of a secant map we have that $c_{2,1} = 0$ if and only if $a_3 = b_3$ if and only if $c_{0,3} = 0$. Q.E.D.

Now, by analyzing the coefficients $c_{i,j}$ in terms of the a_i and b_j, we obtain the following geometrical interpretation for the singularities of a secant map up to codimension 2:

singularity	**normal form**	a_i **and** b_j	**geometrical meaning**
regular	(x, y)	$a_1 \neq b_1$	$\alpha'(s) \nparallel \beta'(t)$
fold	(x, y^2)	$a_1 = b_1$ $a_2 \neq b_2$	$\alpha'(s) \parallel \beta'(t)$ $\kappa_\alpha(s) \neq \kappa_\beta(t)$
cusp	$(x, y^3 + xy)$	$a_1 = b_1$ $a_2 = b_2 \neq 0$ $a_3 \neq b_3$	$\alpha'(s) \parallel \beta'(t)$ $\kappa_\alpha(s) = \kappa_\beta(t) \neq 0$ $\kappa'_\alpha(s) \neq \kappa'_\beta(t)$
swallowtail	$(x, y^4 + xy)$	$a_1 = b_1$ $a_2 = b_2 \neq 0$ $a_3 = b_3 \neq 0$ $a_4 \neq b_4$	$\alpha'(s) \parallel \beta'(t)$ $\kappa_\alpha(s) = \kappa_\beta(t) \neq 0$ $\kappa'_\alpha(s) = \kappa'_\beta(t) \neq 0$ $\kappa''_\alpha(s) \neq \kappa''_\beta(t)$
		$a_1 = b_1$ $a_2 = b_2 = 0$ $a_3 \neq b_3$ $a_3 \neq 0, b_3 \neq 0$	$\alpha'(s) \parallel \beta'(t)$ $\kappa_\alpha(s) = 0 = \kappa_\beta(t)$ $\kappa'_\alpha(s) \neq \kappa'_\beta(t)$ $\kappa'_\alpha(s) \neq 0, \kappa'_\beta(t) \neq 0$
lips	$(x, y^3 + x^2y)$	$a_3 > b_3$	$\kappa'_\alpha(s) > \kappa'_\beta(t)$
beaks	$(x, y^3 - x^2y)$	$a_3 < b_3$	$\kappa'_\alpha(s) < \kappa'_\beta(t)$

singularity	normal form	a_i and b_j	geometrical meaning
butterfly	$(x, xy + y^5 \pm y^7)$	$a_1 = b_1$	$\alpha'(s) \parallel \beta'(t)$
		$a_2 = b_2 \neq 0$	$\kappa_\alpha(s) = \kappa_\beta(t) \neq 0$
		$a_3 = b_3$	$\kappa'_\alpha(s) = \kappa'_\beta(t)$
		$a_4 = b_4$	$\kappa''_\alpha(s) = \kappa''_\beta(t)$
		$a_5 \neq b_5$	$\kappa'''_\alpha(s) \neq \kappa'''_\beta(t)$
6^+		$a_7 > b_7$	$\kappa'''''_\alpha(s) > \kappa'''''_\beta(t)$
6^-		$a_7 < b_7$	$\kappa'''''_\alpha(s) < \kappa'''''_\beta(t)$
gulls	$(x, xy^2 + y^4 + y^5)$	$a_1 = b_1$	$\alpha'(s) \parallel \beta'(t)$
		$a_2 = b_2 = 0$	$\kappa_\alpha(s) = \kappa_\beta(t) = 0$
		$a_3 = b_3 \neq 0$	$\kappa'_\alpha(s) = \kappa'_\beta(t) \neq 0$
		$a_4 \neq b_4$	$\kappa''_\alpha(s) \neq \kappa''_\beta(t)$

We observe that (s,t) is a *cusp point* of the secant map $S_{\alpha,\beta}$ if and only if the translation T_{12} of vector $\beta(t) - \alpha(s)$ takes the tangent line and osculating circle of α at s to the tangent line and osculating circle of β at t. In other words, the curves $T_{12} \cdot \alpha$ and β have a contact of order 2 at the point $p = T_{12} \cdot \alpha(s) = \beta(t)$.

For the codimension 1 phenomena we have:

a) (s,t) is a *swallowtail point* of $S_{\alpha,\beta}$ if and only if the curves $T_{12} \cdot \alpha$ and β have a contact of order 3 at the point $p = T_{12} \cdot \alpha(s) = \beta(t)$ and this is not an inflection point.

b) (s,t) is a *lips* (*or beaks*) *point* of $S_{\alpha,\beta}$ if and only if the curves $T_{12} \cdot \alpha$ and β have a contact of order 2 at the point $p = T_{12} \circ \alpha(s) = \beta(t)$, which is an inflection point of both curves.

By applying these arguments to all the germs of codimension lesser or equal to 2 and we can state:

Theorem 3. *The following table provides the geometrical interpretation of all the possible singularities, up to codimension* 2, *of the secant maps associated to a couple of plane curves* α *and* β, *where* T_{12} *denotes the translation of the plane given by the vector* $\alpha(s) - \beta(t)$:

singularity type	*$\mathcal{A}$-codimensión*	*geometrical interpretation*
regular	*0*	$\alpha'(s) \nparallel \beta'(t)$
fold	*0*	$\alpha'(s) \parallel \beta'(t)$
cusp	*0*	$\alpha'(s) \parallel \beta'(t)$ $\kappa_\alpha(s) = \kappa_\beta(t)$
lips or beaks	*1*	$\alpha'(s) \parallel \beta'(t)$ $\kappa_\alpha(s) = \kappa_\beta(t) = 0$
swallowtail	*1*	$T_{12} \circ \alpha$ *and* β *have contact of order 3*
butterfly	*2*	$T_{12} \circ \alpha$ *and* β *have contact of order 4*
gulls	*2*	$T_{12} \circ \alpha$ *and* β *have contact of order 3 and vanishing curvature*

§4. Stability of secant maps

Given two plane curves $\alpha, \beta : S^1 \longrightarrow \mathbb{R}^2$, consider their secant map

$$\begin{array}{rcl} S_{\alpha,\beta} : S^1 \times S^1 & \longrightarrow & \mathbb{R}^2 \\ (s,t) & \longmapsto & (\alpha(s) - \beta(t)). \end{array}$$

As above, we can take a convenient reparametrization and a change of variables to take this map to the form $\bar{S}_{\alpha,\beta}(x,y) = (x, f_2(x,y))$, where $f_2(x,y) = a(\frac{x+y}{2}) - b(\frac{y-x}{2})$.

We recall the following transversality result due to R. Thom (see [7])

Lemma 4. (***Fundamental transversality lemma***) *Given differentiable manifolds X, B, Y, let W be a submanifold of Y and $j : B \longrightarrow C^\infty(X,Y)$ a (non necessarily continuous) map. Suppose that $\Phi : X \times B \longrightarrow Y$ is a smooth map, such that $\Phi(x,b) = j(b)(x)$ and $\Phi \pitchfork W$. Then the subset $\{b \in B | j(b) \pitchfork W\}$ is dense in B.*

We use this lemma in order to prove the following:

Proposition 5. *Given two plane curves $\alpha, \beta : S^1 \to \mathbb{R}^2$, for almost all (in the sense of an open and dense subset of) Euclidean motions $\phi : \mathbb{R}^2 \to \mathbb{R}^2$, the secant map of the pair $(\alpha, \phi \circ \beta)$ is a locally stable map (that is, it may only have fold curves and isolated cusp singularities) from the torus to the plane.*

Proof. Consider the subset $J^3(S^1,\mathbb{R}^2)\times J^3(S^1,\mathbb{R}^2)$ of pairs of 3-jets of closed plane curves in which we denote the natural local coordinates as $(s,\, t,\, r^1_{0,0},\, r^2_{0,0},\, r^1_{1,0},\, r^2_{1,0},\, r^1_{0,1},\, r^2_{0,1},\, r^1_{2,0},\, r^2_{2,0},\, r^1_{1,1},\, r^2_{1,1},\, r^1_{0,2},\, r^2_{0,2},\, r^1_{3,0},\, r^2_{3,0},\, r^1_{2,1},\, r^2_{2,1},\, r^1_{1,2},\, r^2_{1,2},\, r^1_{0,3},\, r^2_{0,3})$. We take in $J^3(S^1,\mathbb{R}^2)\times J^3(S^1,\mathbb{R}^2)$ the variety

$$T=\{j^3f\in J^3(S^1\times S^1,\mathbb{R}^2)\mid r^i_{k,j}=0 \text{ if } k\neq 0\wedge j\neq 0\}$$

made of 3-jets of maps with vanishing crossed derivatives. Observe that any 3-jet of $J^3(S^1\times S^1,\mathbb{R}^2)$ originated by a linear combination of two closed curves belongs to T. Denote by $O(2)$ the group of all plane rotations and define the map,

$$\begin{array}{rcl}\lambda: O(2) & \longrightarrow & C^\infty(J^3(S^1,\mathbb{R}^2)\times J^3(S^1,\mathbb{R}^2),T)\\ \varphi & \longmapsto & \lambda(\varphi)=S_\varphi\end{array}$$

where

$$\begin{array}{rcl}S_\varphi: J^3(S^1,\mathbb{R}^2)\times J^3(S^1,\mathbb{R}^2) & \longrightarrow & T\\ (j^3\alpha(0),j^3\beta(0)) & \longmapsto & j^3((\varphi\circ\alpha)-\beta)(0).\end{array}$$

It is easy to see that the map

$$\begin{array}{rcl}\Lambda: J^3(S^1,\mathbb{R}^2)\times J^3(S^1,\mathbb{R}^2)\times O(2) & \longrightarrow & T\\ (j^3\alpha(0),j^3\beta(0),\varphi) & \longmapsto & \lambda(\varphi)(j^3\alpha(0),j^3\beta(0))\end{array}$$

is a submersion and hence transversal to any submanifold of T.

We can now take $X=J^3(S^1,\mathbb{R}^2)\times J^3(S^1,\mathbb{R}^2)$, $B=O(2)$, $Y=T$, $j=\lambda$ and $\Phi=\Lambda$ and apply the above Thom's fundamental transversality to the two following submanifolds:

$$W_1=\left\{f\in T\mid \begin{array}{c} r^1_{1,0}r^2_{0,1}=r^2_{1,0}r^1_{0,1}\\ r^1_{2,0}r^2_{0,2}=r^2_{2,0}r^1_{0,2}\\ r^1_{3,0}r^2_{0,3}=r^2_{3,0}r^1_{0,3}\end{array}\right\},$$

$$W_2=\left\{f\in T\mid \begin{array}{c} r^1_{1,0}r^2_{0,1}=r^2_{1,0}r^1_{0,1}\\ r^1_{1,0}r^2_{2,0}=r^1_{2,0}r^2_{1,0}\\ r^1_{0,1}r^2_{0,2}=r^1_{0,2}r^2_{0,1}\end{array}\right\}.$$

These can be respectively considered as the subsets of 3-jets of couples of curves whose secant map has a corank one singularity of swallowtail type or worse and those having a corank two singularity of lips/beaks type or worse. As a consequence of Thom's fundamental transversality lemma, we obtain two dense subsets of rotations ϕ for which the pair $(\phi\circ\alpha,\beta)$ respectively avoids such singularities. Moreover, we can also take each one of these as an open subset, for we observe that if any one of the above requirements holds for a couple (α, β), due to the compactness of the domain, standard topological arguments warranty that they will also hold for any small enough rotation of β. Then the intersection of these two open dense subsets provides the required open and dense subset. Q.E.D.

Lemma 6. *The image of the derivative of $S_{\alpha,\beta}$ at a corank one point (s,t) is a vector parallel to $\alpha'(s)$ (and hence to $\beta'(t)$).*

Proof. Consider parametrisations α and β given by $\alpha(s) = (s, \alpha_2(s))$ and $\beta(t) = (t, \beta_2(t))$, with $j^\infty\alpha_2(0) = \sum_{i=1}^{\infty} a_i s^i$ and $j^\infty\beta_2(0) = \sum_{j=1}^{\infty} b_j t^j$. Then we can write $S_{\alpha,\beta}(s,t) = (s-t, \alpha_2(s) - \beta_2(t))$ and the differential of $S_{\alpha,\beta}$ is given by

$$DS_{\alpha,\beta} = \begin{pmatrix} 1 & -1 \\ \alpha_2'(s) & -\beta_2'(t) \end{pmatrix}.$$

The singular set $\Sigma(S_{\alpha,\beta})$ is a 1-manifold made of the points satisfying the condition $\beta_2'(t) - \alpha_2'(s) = 0$. Consider the orthogonal $(-\beta_2''(t), -\alpha_2''(s))$ to the tangent vector to $\Sigma(S_{\alpha,\beta})$ at (s,t) and take the differential of $S_{\alpha,\beta}$

$$S_{\alpha,\beta*}(-\beta_2''(t), -\alpha_2''(s)) = \begin{pmatrix} -\beta_2''(t) - \alpha_2''(s) \\ -\alpha_2'(s)\beta_2''(t) - \beta_2'(t)\alpha_2''(s) \end{pmatrix}.$$

If (s,t) is a singular point, we have $\beta_2'(t) = \alpha_2'(s)$. So $S_{\alpha,\beta*}(-\beta_2''(t), -\alpha_2''(s)) = (-\beta_2''(t) - \alpha_2''(s))(1, \alpha_2'(s))$. On the other hand, $-\beta_2''(t) - \alpha_2''(s) \neq 0$ because the point is a corank one singularity, therefore we obtain the required result. Q.E.D.

Proposition 7. *Given two plane curves $\alpha, \beta : S^1 \to \mathbb{R}^2$, for almost all Euclidean motion $\phi : \mathbb{R}^2 \to \mathbb{R}^2$ (in the sense of an open and dense subset of motions), the secant map of the pair $(\alpha, \phi \circ \beta)$ is a globally stable map from the torus to the plane.*

Proof. We just need to add appropriate multilocal conditions to the open and dense subset of rotations obtained in the above proposition. Essentially, we need to ensure that the secant map will avoid the following phenomena:

a) Tangencial crossing of 2 folds.
b) Crossing of a fold and a cusp.
c) Crossing of 3 folds.

The arguments run in a similar way as those in Proposition 5. For the condition a) we apply lemma 4 in the following context:

$X = {}_2J^3(S^1,\mathbb{R}^2) \times {}_2J^3(S^1,\mathbb{R}^2)$ (${}_2J^3(S^1,\mathbb{R}^2)$ being the space of couples of 3-jets with different sources), $B = O(2)$, $j = \lambda$, $\Phi = \Lambda$, where

$$Y = T_2 = \{{}_2j^3f \in {}_2J^3(S^1 \times S^1, \mathbb{R}^2) \mid \begin{array}{l} r^i_{a,b} = 0 \text{ if } a \neq 0 \wedge b \neq 0 \\ \bar{r}^i_{a,b} = 0 \text{ if } a \neq 0 \wedge b \neq 0\} \end{array}$$

and

$$W = W_3 = \left\{ f \in T_2 \mid \begin{array}{l} r^1_{0,0} = \bar{r}^1_{0,0} \wedge r^2_{0,0} = \bar{r}^2_{0,0} \\ r^1_{1,0}r^2_{0,1} = r^2_{1,0}r^1_{0,1} \\ \bar{r}^1_{1,0}\bar{r}^2_{0,1} = \bar{r}^2_{1,0}\bar{r}^1_{0,1} \\ r^1_{1,0}\bar{r}^2_{0,1} = \bar{r}^2_{1,0}r^1_{0,1} \end{array} \right\}$$

is the subset of multijets of curves having a tangencial crossing of two folds. Now, as a consequence of the Fundamental Transversality Lemma we can ensure the existence of a residual subset of rotations of the curve β, such that the secant map of the new pair does not meet W_3.

The condition b) is treated through an analogous argument where X, B and Y are as above and

$$W = W_4 = \left\{ f \in T_2 \mid \begin{array}{l} r^1_{0,0} = \bar{r}^1_{0,0} \wedge r^2_{0,0} = \bar{r}^2_{0,0} \\ r^1_{1,0}r^2_{0,1} = r^2_{1,0}r^1_{0,1} \\ \bar{r}^1_{1,0}\bar{r}^2_{0,1} = \bar{r}^2_{1,0}\bar{r}^1_{0,1} \\ \bar{r}^1_{2,0}\bar{r}^2_{0,2} = \bar{r}^2_{2,0}\bar{r}^1_{0,2} \end{array} \right\}.$$

In case c), we also take X, B are as above and put

$$Y = T_3 = \left\{ {}_3j^3f \in {}_3J^3(S^1 \times S^1, \mathbb{R}^2) \mid \begin{array}{ll} r^i_{a,b} = 0 & \text{if } a \neq 0 \wedge b \neq 0 \\ \bar{r}^i_{a,b} = 0 & \text{if } a \neq 0 \wedge b \neq 0 \\ \bar{\bar{r}}^i_{a,b} = 0 & \text{if } a \neq 0 \wedge b \neq 0 \end{array} \right\}$$

and

$$W = W_5 = \left\{ f \in T_3 \mid \begin{array}{l} r^1_{0,0} = \bar{r}^1_{0,0} = \bar{\bar{r}}^1_{0,0} \wedge r^2_{0,0} = \bar{r}^2_{0,0} = \bar{\bar{r}}^2_{0,0} \\ r^1_{1,0}r^2_{0,1} = r^2_{1,0}r^1_{0,1} \\ \bar{r}^1_{1,0}\bar{r}^2_{0,1} = \bar{r}^2_{1,0}\bar{r}^1_{0,1} \\ \bar{\bar{r}}^1_{1,0}\bar{\bar{r}}^2_{0,1} = \bar{\bar{r}}^2_{1,0}\bar{\bar{r}}^1_{0,1} \end{array} \right\}.$$

Finally, arguing as in the previous proposition, we show the existence of open and dense subsets of rotations whose corresponding multi-jets do not cut respectively W_3, W_4 and W_5. By taking the intersections of these subsets we obtain the required open dense subset. Q.E.D.

Theorem 8. *There is an open and dense subset of couples of closed plane curves in $\mathbb{R}^2$ with the Whitney C^∞-topology, whose corresponding secant map is a (globally) stable map from the torus to the plane.*

Proof. As a consequence of the Proposition 7 we can approach any pair of closed plane curves (α, β) by a sequence of pairs $\{(\alpha_n, \beta_n)\}_{n=1}^{\infty}$ whose corresponding secant map is globally stable. From this we can conclude the density. On the other hand, to see the openness of this set, we observe that given a pair of curves (α, β), for which $S_{\alpha,\beta}$ is a globally stable map, any other couple $(\bar{\alpha}, \bar{\beta})$ which is near enough to (α, β) in the C^{∞}-Whitney topology (and thus in the C^3-topology) also satisfies the requirements to be globally stable (for their 3-jets will avoid the subsets W_1, W_2, W_3, W_4 and W_5 constructed in the above propositions). Q.E.D.

Remark 9. *Observe that analogous arguments can be applied when $\alpha = \beta$ in order to show that for almost all Euclidean motions $\phi : \mathbb{R}^2 \to \mathbb{R}^2$, the secant map of the pair $(\alpha, \phi \circ \alpha)$ is a (globally) stable map from the torus to the plane.*

§5. Global viewpoint

Definition 5.1. *Given two closed plane curves α, β we say that (α, β) is a stable pair of curves if the secant map $S_{\alpha,\beta}$ is A-stable.*

The singular set of the secant map of a stable pair of closed plane curves is a finite set of disjoint closed regular curves in the torus. Our aim in this section is to obtain global relations between the isotopy type of the closed plane curves α and β and those of the singular curves of $S_{\alpha,\beta}$ as closed curves in the torus. We denote by $\{\Sigma_i\}_{i=1}^{m}$ the set of singular curves of $S_{\alpha,\beta}$. Since each Σ_i is a closed regular curve in $S^1 \times S^1$, we can choose a continuous regular parametrization $\sigma_i : S^1 \to S^1 \times S^1$, such that $\sigma_i(S^1) = \Sigma_i$. Consider now the two natural projections $\Pi_s, \Pi_t : S^1 \times S^1 \longrightarrow S^1$, respectively given by $\Pi_s(s,t) = s$ and $\Pi_t(s,t) = t$ and denote by G_α and G_β the respective Gauss maps of the closed plane curves α and β. Then it is easy to see that $G_\alpha \circ \Pi_s \circ \sigma_i = G_\beta \circ \Pi_t \circ \sigma_i$. We will denote this map as G_{σ_i} and refer to it as the *Gauss map of the toric curve σ_i*.

Recall that the singularities of the Gauss map G_α (resp. G_β) are the inflection points of the plane curve α (resp. β). Then we have the following characterization for the singularities of the maps G_{σ_i}.

Lemma 10. *The map G_{σ_i} has a singularity at a point $(\alpha(s_0), \beta(t_0)) \in \Sigma_i$ if and only if either s_0 is a singular point of G_α, or t_0 is a singular point of G_β.*

Proof. Clearly, since $G_{\sigma_i} = G_\alpha \circ \Pi_s \circ \sigma_i = G_\beta \circ \Pi_t \circ \sigma_i$, the singular points of G_α and G_β lead to singular points of G_{σ_i}. On the other hand, since the map σ_i has no singular points, we have that a singularity of G_{σ_i} must be either a singular point of G_α or a singular point of Π_s. In case it is not a singular point of G_α, then we have a singular point for Π_s, but this implies that it is not a singular point of Π_t, from which we get that it must be a singular point of G_β. Q.E.D.

Definition 5.2. *We define the tangency function of a pair* (α, β) *as*

$$\begin{array}{rcl} B_{\alpha,\beta} : S^1 \times S^1 & \longrightarrow & \mathbb{R} \\ (s,t) & \longmapsto & det(\alpha'(s), \beta'(t)). \end{array}$$

Lemma 11. *The singular points of the secant map* $S_{\alpha,\beta}$ *are the zeros of* $B_{\alpha,\beta}$.

Proof. We have that $(s,t) \in B_{\alpha,\beta}^{-1}(0) \Leftrightarrow det(\alpha'(s), \beta'(t)) = 0 \Leftrightarrow \alpha'(s) \parallel \beta'(t) \Leftrightarrow (s,t) \in \sum(S_{\alpha,\beta})$. Q.E.D.

Remark 12. *Observe that the above lemma implies that* $B_{\alpha,\beta}^{-1}(0)$ *is a non necessarily connected regular closed curve in the torus.*

Proposition 13. *There is an open and dense subset of stable pairs* (α, β), *for which* 0 *is a regular value of the tangency function* $B_{\alpha,\beta}$.

Proof. We have that 0 fails to be a regular value if there exist (s,t) such that $B_{\alpha,\beta}(s,t) = 0$ and $\frac{\partial B_{\alpha,\beta}}{\partial s}(s,t) = \frac{\partial B_{\alpha,\beta}}{\partial t}(s,t) = 0$. But this implies that $det(\alpha'(s), \beta'(t)) = 0 = det(\alpha''(s), \beta'(t)) = det(\alpha'(s), \beta''(t))$, which implies that rank $\{\alpha'(s), \beta'(t), \alpha''(s), \beta''(t)\} = 1$. This amounts to say that the curves α and β have an inflection point respectively s and t and their tangent vectors are parallel. Clearly, a small rotation of one of the curves will avoid this situation. Q.E.D.

Definition 5.3. *We say that a stable pair closed plane curves* (α, β) *is a good pair if* 0 *is a regular value of its tangency function* $B_{\alpha,\beta}$.

Lemma 14. *Given a good pair of closed plane curves* (α, β),

a) (s,t) *is a singular point of* $\Pi_t|_{B_{\alpha,\beta}^{-1}(0)}$ *if and only if* $\alpha(s)$ *is an inflection point of* α (= *singular point of* G_α).

b) (s,t) *is a singular point of* $\Pi_s|_{B_{\alpha,\beta}^{-1}(0)}$ *if and only if* $\beta(t)$ *is an inflection point of* β (= *singular point of* G_β).

Proof. The singularities of the projections $\Pi_s, \Pi_t : S^1 \times S^1 \to S^1$ on $B_{\alpha,\beta}^{-1}(0)$ are given by the zeros of the partial derivatives of $B_{\alpha,\beta}$. Then we have,

$$s \in \Sigma\Pi_t|_{B_{\alpha,\beta}^{-1}(0)} \Leftrightarrow \frac{\partial B_{\alpha,\beta}}{\partial s}(s,t) = 0.$$

But this means that $\frac{\partial}{\partial s}(det(\alpha'(s), \beta'(t))) = det(\alpha''(s), \beta'(t)) = 0$. Now, $(s,t) \in B^{-1}_{\alpha,\beta}(0)$, so we have that rank $\{\alpha'(s), \beta'(t), \alpha''(s)\} = 1$. Therefore $\alpha(s)$ is an inflection point of α (i.e. a singular point of G_α). Analogously, (s,t) is a singular point of $\Pi_s|_{B^{-1}_{\alpha,\beta}(0)} \Leftrightarrow \beta(t)$ is an inflection point of β (i.e. a singular point of G_β). Q.E.D.

Proposition 15. *The existence of homotopically trivial connected components in the singular set of the secant map $S_{\alpha,\beta}$ implies that both curves are non convex.*

Proof. The restriction of each one of the projections Π_s and Π_t to any homotopically trivial curve in the torus has at least two singular points. Then, as a consequence of Lemma 14 we get that each one of the curves α and β must have at least two inflection points. Q.E.D.

Given two natural numbers n and m, we shall denote by $\mu_{n,m}$ the maximum common divisor of n and m.

Theorem 16. *Given a good pair of closed convex plane curves (α, β) with respective Whitney indices $n, m > 0$, we have that $S_{\alpha,\beta}$ has exactly $2\mu_{n,m}$ singular curves, all of them being toric curves of type $(\frac{n}{\mu_{n,m}}, \frac{m}{\mu_{n,m}})$.*

Proof. We can choose the same orientation in both curves. From the convexity of the curves, we get that each tangent direction occurs exactly $2n$ times in α and $2m$ times in β (for G_α and G_β are both regular maps). Given $v \in S^1$, denote $I_v = \{s \in S^1 : \alpha'(s)||v\}$ and $J_v = \{s \in S^1 : \beta'(s)||v\}$. Once we fix a point in the curve $\alpha(S^1)$ we can label the points of I_v and J_v according to the natural order given by the orientation of α and β respectively. So we can write $I_v = \{s_i\}_{i=1}^{2n}$ and $J_v = \{t_j\}_{j=1}^{2m}$. We observe that there is a unique singular curve of $S_{\alpha,\beta}$ passing through each pair $(s_i, t_j) \in I_v \times J_v$. Moreover, the projections Π_s and Π_t are strictly monotone functions in s and t (for the maxima and minima of Π_s and Π_t correspond respectively to the inflection points of α and β and these curves are assumed to be convex). Therefore, we can assume that the Gauss map G_{σ_i} is strictly increasing on each singular curve σ_i. These curves connect the point (s_i, t_j) to the point (s_{i+1}, t_{j+1}), for $i = 2, \cdots, n-1$ and $j = 2, \cdots, n-1$; moreover, for each j, the point (s_{2n}, t_j) is connected to (s_1, t_{j+1}).

From a global viewpoint, we can decompose the torus into $4nm$ rectangles with vertices at the points $\{(s_i, t_j)\}_{i,j=1}^{2n,2m}$. Since the curves are convex, it follows from Proposition 15 that there are no homotopically trivial singular curves. Then the singular set is made by a union of arcs (without self-intersections) joining the opposite vertices in each one of the rectangles (i.e., arcs joining the point (s_i, t_j) with (s_{i+1}, t_{j+1})

for $i = 1, \cdots, 2n, j = 1, \cdots, 2m$ and (s_{2n}, t_j) with (s_1, t_{j+1}) for $j = 1, \cdots, 2m$). Such curves can be seen, up to isotopy, as the union of all the principal diagonals of the small rectangles. We now prove that these are closed curves of type $(\frac{n}{\mu_{n,m}}, \frac{m}{\mu_{n,m}})$. From the topological viewpoint, we can substitute this grid by another one at which all the rectangles have the same size $= \frac{2\pi}{2n} \times \frac{2\pi}{2m}$. In this last case, the slope of each diagonal curve is given by $\sigma = \frac{m}{n} = \frac{m'}{n'}$, where $n' = \frac{n}{\mu_{m,n}}$ and $m' = \frac{m}{\mu_{m,n}}$. This implies that, up to identification of the torus with the plane rectangle $[0, 2\pi] \times [0, 2\pi]$, each one of these lines is given by the equation $y = \frac{m'}{n'}(x - x_0)$, where x_0 is the initial point of the considered line on the x-axis. Now, we observe that the increment of x along each vertical turn must be given by $\Delta x = 2\pi\frac{m'}{n'}$, so this curve needs to make m' turns in the vertical sense order to reach its initial point again. Analogously, it follows that it needs to make n' turns in the horizontal sense. Suppose now that there is some $m'_1 < m'$ such that $x_0 + m'_1 2\pi\frac{n'}{m'} = x_0 + 2\pi k$. Then we would have that $\frac{m'_1 - n'}{m'} = k$. But this implies that $\frac{m'}{n'} = \frac{k}{m'_1}$ which contradicts the assumption that $\mu_{n,m}$ is the maximum common divisor of m and n. Therefore, we can conclude that the number of vertical turns of each one of these curves is $m' = \frac{m}{\mu_{m,n}}$ and analogously, we get that the number of horizontal turns of each curve is $n' = \frac{n}{\mu_{m,n}}$. It now follows that the total number of curves must be $\mu_{2n,2m} = 2\mu_{n,m}$ as required. Q.E.D.

Finally, we call the attention on the following particular cases:

Corollary 16.1. *Suppose that (α, β) is a good pair of closed convex plane curves, where α is a standard circle and β is a convex plane curve with Whitney index n, then the secant map $S_{\alpha,\beta}$ has exactly 2 singular curves, both of them of type $(1, n)$.*

Corollary 16.2. *Given a convex closed plane curve α with Whitney index $n > 0$, for any rigid motion $\phi : \mathbb{R}^2 \to \mathbb{R}^2$, such that the pair $(\alpha, \phi \cdot \alpha)$ is a good pair of curves, the secant map $S_{\alpha,\phi\cdot\alpha}$ has exactly $2n$ singular curves, all of them of type $(1, 1)$.*

Remark 17. *As a consequence of this last result we have that the topological type of the secant map of the pair $S_{\alpha,\phi\cdot\alpha}$ does not depend on the choice of the rigid motion ϕ, as far as $(\alpha, \phi \cdot \alpha)$ be a good pair of curves.*

References

[1] V.I. Arnol'd, Singularities of systems of rays. Russian Math. Surveys **38**, 1983, pp. 87–176.

[2] J.W. Bruce, Generic Space Curves and Secants. Proc. Royal Soc. Edimburgh A **98**, 1984, pp. 281–289.

[3] J.W. Bruce, Projections and reflections of generic surfaces in $\mathbb{R}^3$. Math. Scand. **54**, 1984, pp. 262–278.

[4] J.W. Bruce and P.J. Giblin, Outlines and their duals, Proc. London Math. Soc. (3) **50**, 1985, pp. 552–570.

[5] E. Chíncaro, Bifurcations of Whitney Maps. Doctoral Thesis, Univ. Fed. Minas Gerais (Brazil), 1978.

[6] T. Gaffney, The structure of $T\mathcal{A}(f)$, classification and an application to differential geometry. Singularities, Part 1 (Arcata, Calif., 1981), pp. 409–427, Proc. Sympos. Pure Math., 40, Amer. Math. Soc., Providence, RI, 1983.

[7] M. Golubitsky and V. Guillemin, Stable Mappings and Their Singularities. GTM **14**, Springer-Verlag, New York, 1973.

[8] T. Ohmoto and F. Aicardi, First order local invariants of apparent contours. Toplogy **45**, 2006, pp. 27–45.

[9] J.H. Rieger, Families of maps from the plane to the plane. J. London Math. Soc. (2) **36**, 1987, pp. 351–369.

[10] H. Whitney, On regular closed curves in the plane. Compositio Math. **4**, 1937, pp. 276–284.

[11] H. Whitney, On singularities of mappings of Euclidean spaces I. Mappings of the plane into the plane. Ann. of Math. **62**, 1955, pp. 374–410.

[12] T. Yoshida, Y. Kabata and T. Ohmoto, Bifurcation of plane-to-plane map-germs of corank 2. Q. J. Math. **66** (2015), no. 1, pp. 369–391.

Departament de Matemàtiques, Facultat de Matemàtiques, Universitat de València, Spain

E-mail address: `carmen.romero@uv.es`

E-mail address: `luis.sanhermelando@uv.es`

Advanced Studies in Pure Mathematics 78, 2018
Singularities in Generic Geometry
pp. 383–410

Discrete linear Weingarten surfaces with singularities in Riemannian and Lorentzian spaceforms

Wayne Rossman and Masashi Yasumoto

Abstract.

In this paper we define and analyze singularities of discrete linear Weingarten surfaces with Weierstrass-type representations in 3-dimensional Riemannian and Lorentzian spaceforms. In particular, we discuss singularities of discrete surfaces with non-zero constant Gaussian curvature, and parallel surfaces of discrete minimal and maximal surfaces, and discrete constant mean curvature 1 surfaces in de Sitter 3-space, including comparisons with different previously known definitions of such singularities.

§1. Introduction

In this paper we examine discrete surfaces with Weierstrass-type representations in spaceforms, taking advantage of the more general setting of Lie sphere geometry and discrete Legendre immersions (see Definition 2.1 here), and with helpful motivations coming from the developing field of Ω surfaces. There are numerous Weierstrass-type representations in 3-dimensional spaceforms in addition to the classical representation for minimal surfaces in $\mathbb{R}^3$, for example, for

---

Received October 13, 2015.
Revised September 29, 2016.
2010 *Mathematics Subject Classification.* Primary 53A10, Secondary 52C99.
Key words and phrases. discrete differential geometry, Weierstrass-type representation, singularity.
The first author was partly supported by the Grant-in-Aid for Scientific Research (C) 15K04845 and (S) 24224001, Japan Society for the Promotion of Science, and the second author was supported by the Grant-in-Aid for JSPS Fellows Number 26-3154. Both authors were supported by the JSPS/FWF bilateral joint project "Transformations and Singularities" between Austria and Japan.

(1) maximal surfaces (spacelike immersion with mean curvature identically 0) in Minkowski 3-space $\mathbb{R}^{2,1}$ by Kobayashi [18], see also works of Fujimori, Saji, Umehara and Yamada [10], [26],
(2) constant mean curvature (CMC, for short) 1 surfaces in hyperbolic 3-space $\mathbb{H}^3$ by Bryant [6] (see also [25]),
(3) flat surfaces in $\mathbb{H}^3$ by Gálvez, Martínez, Milán [11], with separate different representations by Kokubu, Umehara and Yamada [20], [21],
(4) CMC 1 surfaces in de Sitter 3-space $\mathbb{S}^{2,1}$ by Aiyama, Akutagawa [1],
(5) linear Weingarten surfaces of Bryant type (BrLW surfaces, for short) in $\mathbb{H}^3$ by Gálvez, Martínez, Milán [12], and Kokubu and Umehara [19],
(6) linear Weingarten surfaces of Bianchi type (BiLW surfaces, for short) in $\mathbb{S}^{2,1}$ by Aledo, Espinar [2].

Regarding the last two examples above, Izumiya and Saji [17] showed that a necessary and sufficient condition for an immersion in $\mathbb{H}^3$ to be BrLW is that its unit normal vector field is BiLW (see §4).

Recently, there has been work on discretization of the above representations. Bobenko, Pinkall [3] described discrete isothermic surfaces in the Euclidean 3-space $\mathbb{R}^3$, and as an application, they derived the Weierstrass representation for discrete isothermic minimal surfaces in $\mathbb{R}^3$, using integrable systems techniques. In the same vein, Hertrich-Jeromin [13] gave the Weierstrass-type representation for discrete isothermic CMC 1 surfaces in $\mathbb{H}^3$.

Burstall, Hertrich-Jeromin and the first author [8] described discrete linear Weingarten surfaces in any 3-dimensional spaceform using Lie sphere geometry, which we briefly introduce in §5. Using that method, we can treat discrete linear Weingarten surfaces in any 3-dimensional spaceform. They did not consider singularities of discrete surfaces, however, as we will do here.

Returning to smooth surfaces, unlike the minimal and (non-zero) CMC surfaces in $\mathbb{R}^3$, general linear Weingarten surfaces will have singularities. In fact, singularities of maximal surfaces in $\mathbb{R}^{2,1}$, flat and BrLW surfaces in $\mathbb{H}^3$, and BiLW surfaces in $\mathbb{S}^{2,1}$ are investigated in [10, 19, 26]. Thus, it is natural to still consider singularities when discretizing these surfaces. However, difficulties occur with this (Definition 3.1), and overcoming those difficulties is our primary task here.

Hoffmann, Sasaki, Yoshida and the first author [16] described discrete BrLW surfaces in $\mathbb{H}^3$, and furthermore treated singularities of discrete flat surfaces in $\mathbb{H}^3$. For that, they considered the behavior of caustics of smooth flat surfaces at a singular point, via the Weierstrass-type representation. Such a caustic contacts the surface at a singular point, which lead to a natural definition of singularities in the discrete case, i.e. that a singularity of a discrete flat surface is a vertex that contacts the (discrete) caustic. We will define singular vertices in a more direct way that applies to a wider variety of discrete surfaces, and show equivalence of the definitions in the case considered just above (Theorem 6.1).

The second author [28] described discrete maximal surfaces in the Minkowski 3-space $\mathbb{R}^{2,1}$ and analyzed their singular faces, that is, non-spacelike faces (Definition 8.1). This is also a natural way to define singular behavior, because the tangent plane of a smooth maximal surface is non-spacelike precisely at singular points.

Thus, singularities of discrete surfaces could be either vertices or faces, and two of our primary results here are about relating those two viewpoints, in particular, in the cases of discrete maximal surfaces in $\mathbb{R}^{2,1}$ and discrete CMC 1 surfaces in $\mathbb{S}^{2,1}$.

Smooth 2-dimensional Legendre immersions in Lie sphere geometry project to surfaces in spaceforms that can have singularities. However, those surfaces considered together with their unit normal maps become immersions (by definition), and they are called fronts. The most typical singularities on fronts are cuspidal edges of 3/2 type, and next perhaps are swallowtails. At such singularities, exactly one of the principal curvatures will diverge (see [24]), and equivalently, one of the principal curvature spheres will become a point sphere. Using that the notion of principal curvature spheres in Lie sphere geometry is independent of the choice of projection to a 3-dimensional spaceform, we define singular vertices on projections of discrete Legendre immersions.

While typical singularities on smooth surfaces can be found by locating the points where one principal curvature blows out to infinity, on discrete surfaces the principal curvatures are discrete functions from the set of edges to the real numbers, and thus we can only identify the vertices at which a principal curvature changes sign. As a result, it is not so immediate to distinguish the points that are singular from the points that are parabolic (at which exactly one of the two bounded principal curvature becomes zero) or flat (at which both principal curvatures become zero). This is why we will use a particular terminology "FPS vertices" in Definition 3.1. This is the first of our three goals:

(1) We will find and examine cases where the distinction between singular and parabolic or flat points is possible. Such cases include surfaces of constant Gaussian curvature (CGC) $K \neq 0$ (see §6), and some particular discrete linear Weingarten surfaces for which Weierstrass type representations exist (§7, §8, §9).
(2) We will confirm that the discrete Weierstrass type representations are compatible with other ways of defining discrete surfaces with specific curvature properties. In particular, they are compatible with the definitions given by Burstall, Hertrich-Jeromin and the first author in [8] (Proposition 5.1).
(3) We will find relationships between singular vertices and singular faces in particular cases (Theorem 8.1, Theorem 9.1, Corollary 9.1).

§2. Discrete Legendre immersions

First we recall smooth Legendre immersions in the context of Lie sphere geometry, that is, maps Λ of 2-manifolds M^2 into the collection of null planes in $\mathbb{R}^{4,2}$, with metric signature $(-,+,+,+,+,-)$, i.e.

$$(X,Y)_{\mathbb{R}^{4,2}} = (X,Y) := -x_1y_1 + x_2y_2 + x_3y_3 + x_4y_4 + x_5y_5 - x_6y_6$$

for $X = (x_1, x_2, x_3, x_4, x_5, x_6)^t$, $Y = (y_1, y_2, y_3, y_4, y_5, y_6)^t \in \mathbb{R}^{4,2}$. Then

$$\mathbb{L}^5 := \{X \in \mathbb{R}^{4,2} | (X,X) = 0\}$$

denotes the light cone of $\mathbb{R}^{4,2}$.

Let $\Lambda \subset \mathbb{L}^5$ be a 2-dimensional null subspace, which projectivizes to a line in the projectivized light cone $\mathbb{PL}^5$ called a *contact element.* This line will represent a family of spheres (a pencil) that are all tangent (with same orientation) at one point.

If Λ is a (smooth) map from $M = M^2$ to the collection of null planes in $\mathbb{R}^{4,2}$, where M is a 2-dimensional manifold, then Λ is a *Legendre immersion* if,

(1) for any pair of sections X_1, X_2 of Λ,

$$dX_1 \perp X_2 \quad \text{(contact condition), and}$$

(2) for any $m \in M$ and any choice of $Y \in T_mM$, $dX(Y) \in \Lambda(m)$ for all sections X of $\Lambda(m)$ implies $Y = 0$ (immersion condition).

The immersion condition can be restated in terms of a basis of sections for the null planes Λ as follows: If

$$\Lambda = \text{span}\{X_1, X_2\} ,$$

with basis $X_1, X_2 : M^2 \to \mathbb{L}^5$, then the immersion condition is equivalent to

$$dX_1(Y), dX_2(Y) \in \Lambda(m) \quad \text{implies} \quad Y = 0$$

for all $Y \in T_m M$, and one can then check that this condition is independent of the choice of basis X_1, X_2.

By choosing two nonzero perpendicular vectors p, q in $\mathbb{R}^{4,2}$ (p not null), we can project Λ to a surface $f : M^2 \to M^3$ in the 3-dimensional spaceform

$$M^3 = M^3_{p,q} := \{X \in \mathbb{R}^{4,2} \mid (X, X) = (X, p) = 0, (X, q) = -1\}$$

with sectional curvature $-(q, q)$, by taking $f \in \mathrm{Sec}(\Lambda)$ such that

$$(f, p) = 0 \quad \text{and} \quad (f, q) = -1 \ , \tag{1}$$

where $\mathrm{Sec}(\Lambda)$ denotes the set of all sections of Λ. Note that, when we choose a constant timelike (resp. spacelike) vector $p \in \mathbb{R}^{4,2}$ and a constant vector $q \in \mathbb{R}^{4,2}$, M^3 becomes a 3-dimensional Riemannian (resp. Lorentzian) spaceform. For details, see [23], and for a particular choice of p and q, see Section 4.

Let n denote the unit normal to f in M^3, i.e. $n \in \mathrm{Sec}(\Lambda)$ and

$$(n, q) = 0 \quad \text{and} \quad (n, p) = -1.$$

The sections of $\Lambda = \mathrm{span}\{f, n\}$ represent the sphere congruences of f, and then f, resp. n, is the point sphere, resp. tangent geodesic plane, congruence. Let s_α for $\alpha = 1, 2$ be sections of Λ that represent the principal curvature sphere congruences, which can be defined by

$$s_\alpha = \kappa_\alpha f + n$$

using the principle curvatures κ_α of f, or equivalently by the directional derivative conditions that $D_{\vec{v}_\alpha} s_\alpha \in \mathrm{Sec}(\Lambda)$ for some tangent vector fields $\vec{v}_\alpha$ on M^2.

For Λ above to be a Legendre immersion, both immersion and contact conditions must be satisfied. For a discrete Legendre map Λ as in Definition 2.1 below, discretized versions of the immersion and contact conditions are needed. We also assume the existence of "discrete curvature line coordinates", that is, we require that the four vertices of each quadrilateral be concircular, which is called a *principal net*. In this way, the properties of smooth Legendre immersions motivate the following definition of discrete Legendre immersions:

Definition 2.1 ([8]). *A map*

$$\Lambda : \mathbb{Z}^2 \text{ (or some subdomain of } \mathbb{Z}^2) \to \{\text{null planes in } \mathbb{R}^{4,2}\}$$

is a discrete Legendre immersion *if, for any quadrilateral, with vertices i, j, k, ℓ ordered counterclockwise about the quadrilateral and with i in the lower left corner in $\mathbb{Z}^2$, and with corresponding surface vertices f_i, f_j, f_k, f_l defined like in* (1),

(1) (*principal net condition*) $\dim(span\{f_i, f_j, f_k, f_\ell\}) = 3$,
(2) (*first immersion condition*) *There exist p, q such that the difference of any two of f_i, f_j, f_k, f_ℓ is non-null,*
(3) (*second immersion condition*) *For some p, q as in item* (2) *above, $f_k - f_i$ and $f_\ell - f_j$ are not parallel,*
(4) (*contact condition*) $\Lambda_i \cap \Lambda_j \neq \{\vec{0}\}$, $\Lambda_i \cap \Lambda_\ell \neq \{\vec{0}\}$.

Remark. Item (1) in Definition 2.1 and $(f_*, q) = -1$ imply

$$f_i, f_j, f_k, f_\ell$$

all lie in some 2-dimensional plane. Item (3) implies any two or three vertices amongst f_i, f_j, f_k, f_ℓ span a 2 or 3 dimensional subspace of $\mathbb{R}^{4,2}$, respectively, with nondegenerate induced metric $(+, -)$ or $(+, +, -)$.

§3. FPS vertices of projections of discrete Legendre immersions

Generically, a smooth surface (section) $f \in \mathrm{Sec}(\Lambda)$ will have a singularity when one of the principal curvature spheres s_α becomes a point sphere [24], i.e. when $s_\alpha \perp p$ for $\alpha = 1$ or 2. Also, where f does not have a singularity, it will have a parabolic or flat point if one of the s_α becomes a tangent geodesic plane, i.e. $s_\alpha \perp q$.

In the case of discrete Legendre immersions, the domain becomes $\mathbb{Z}^2$, or some subdomain of $\mathbb{Z}^2$, rather than M^2. We define the curvature spheres as those spheres represented by nonzero vectors ([8])

$$s_1 \in \Lambda_i \cap \Lambda_j \quad \text{and} \quad s_2 \in \Lambda_i \cap \Lambda_\ell\ .$$

Thus we have spheres in M^3, associated to edges, that lie in both of the sphere pencils defined at the two endpoints of the edges. In particular the normal geodesics (i.e. the geodesics through the vertices and perpendicular to the spheres in the sphere pencils) emanating from the adjacent vertices, when they do intersect, will intersect at equal distances from the two vertices.

Thus, $s_1 = s_{(m,n),(m+1,n)}$ will be defined on horizontal edges from vertex $i = (m,n) \in \mathbb{Z}^2$ to vertex $j = (m+1,n) \in \mathbb{Z}^2$ as the representative (for a sphere) that is common to both the null planes Λ_i and Λ_j, and $s_2 = s_{(m,n),(m,n+1)}$ is defined analogously on vertical edges from i to $\ell = (m, n+1)$. We then define the principal curvatures by ([4], [8])

$$\kappa_{ij} = \frac{(s_1, q)}{(s_1, p)}, \quad \kappa_{i\ell} = \frac{(s_2, q)}{(s_2, p)}. \tag{2}$$

As the principal curvature spheres s_α and principal curvatures $\kappa_{\alpha\beta}$ are defined on edges, not vertices, we lose the ability to look for points in the domain where s_α is exactly perpendicular to p or q. Thus we reformulate the conditions for singularities and parabolic or flat points by finding vertices in the domain at which the $\kappa_{\alpha\beta}$ change sign in at least one direction:

Definition 3.1. *For a Λ as in Definition 2.1, together with a choice of spaceform determined from a choice of p and q, we say that (m,n) is a* flat-or-parabolic-or-singular (FPS) vertex *if*

$$\kappa_{(m-1,n),(m,n)} \cdot \kappa_{(m,n),(m+1,n)} \leq 0 \ \text{or}\ \kappa_{(m,n-1),(m,n)} \cdot \kappa_{(m,n),(m,n+1)} \leq 0.$$

When both p and q are non-null, switching p and q will result in the projected surface f changing to its Gauss map n. In the smooth case, generically, a parabolic or flat point on one of the two surfaces corresponds to a singular point on the other, thus it is not surprising that these notions appear together in Definition 3.1.

In certain special cases, we can distinguish the singular points from the parabolic or flat points, which we will see here.

As another approach for considering singularities on discrete surfaces, motivated by the second author's work [28], we can define singular faces. We come back to this in Definition 8.1, and examine criteria for singular faces, and also their relationships with singular vertices in some special cases.

§4. Smooth linear Weingarten surfaces of Bryant and Bianchi type in $\mathbb{H}^3$ and $\mathbb{S}^{2,1}$

We include this section to motivate the discretizations in §5. In $\mathbb{R}^{3,1}$ with signature $(-,+,+,+)$, with points $(x_0, x_1, x_2, x_3) \in \mathbb{R}^{3,1}$ described in matrix form as

$$X = \begin{pmatrix} x_0 + x_3 & x_1 - ix_2 \\ x_1 + ix_2 & x_0 - x_3 \end{pmatrix},$$

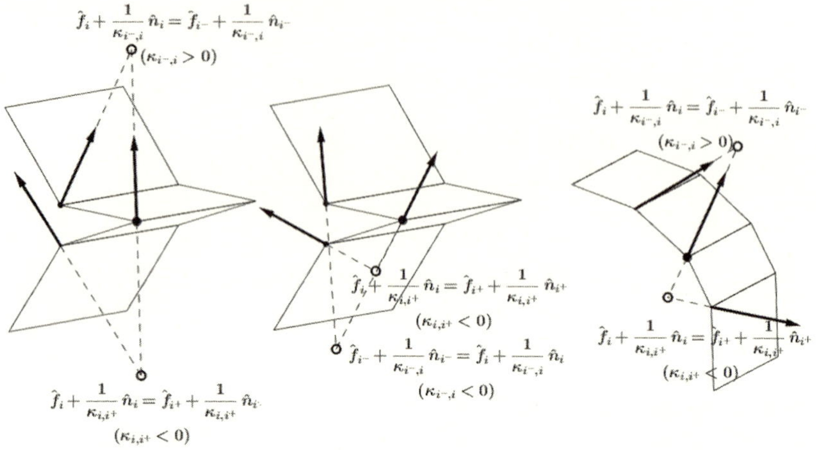

Fig. 1. Examples of FPS vertices in $\mathbb{R}^3$ on the left and right, and a non-example in the middle. The figure on the left shows a situation we should regard as a singularity, and the figure on the right shows what should be regarded as a flat or parabolic point. The figure in the middle is neither. Here, if $i = (m, n)$, then we have either $i^+ = (m+1, n)$, $i^- = (m-1, n)$ or $i^+ = (m, n+1)$, $i^- = (m, n-1)$.

the metric is $\langle X, Y\rangle = \frac{-1}{2}\mathrm{tr}\left(X\begin{pmatrix}0 & -i\\ i & 0\end{pmatrix}Y^t\begin{pmatrix}0 & -i\\ i & 0\end{pmatrix}\right)$. We define

$$\mathbb{H}^3 := \{X \in \mathbb{R}^{3,1} | \langle X, X\rangle = -1\} = \{\pm F\overline{F}^t | F \in \mathrm{SL}_2\mathbb{C}\},$$
$$\mathbb{S}^{2,1} := \{X \in \mathbb{R}^{3,1} | \langle X, X\rangle = 1\} = \{F\begin{pmatrix}1 & 0\\ 0 & -1\end{pmatrix}\overline{F}^t | F \in \mathrm{SL}_2\mathbb{C}\}.$$

We call a surface $\hat{f}$ in $\mathbb{H}^3$ a *linear Weingarten surface of Bryant type* (*BrLW surface*, for short) if $\hat{f}$ satisfies

$$2t(H_{\hat{f}} - 1) + (1 - t)(K_{ext,\hat{f}} - 1) = 0, \tag{3}$$

where $K_{ext,\hat{f}}$ and $H_{\hat{f}}$ are the extrinsic Gaussian and mean curvatures of $\hat{f}$ with respect to $\mathbb{H}^3$, and we call a surface $\hat{n}$ in $\mathbb{S}^{2,1}$ a *linear Weingarten surface of Bianchi type* (*BiLW surface*, for short) if $\hat{n}$ satisfies

$$2t(H_{\hat{n}} - 1) - (1 + t)(K_{ext,\hat{n}} - 1) = 0. \tag{4}$$

Solving

$$dE = E \begin{pmatrix} 0 & g' \\ (g')^{-1} & 0 \end{pmatrix} dz$$

for $E \in \mathrm{SL}_2(\mathbb{C})$, where g is a holomorphic function with nonzero derivative $g' = \partial_z g$ on a Riemann surface M^2 with local coordinate z, we take, for any constant $t \in \mathbb{R}$,

$$(5)\qquad \begin{cases} L = \begin{pmatrix} 0 & \sqrt{\mathcal{T}} \\ \frac{-1}{\sqrt{\mathcal{T}}} & \frac{-t\bar{g}}{\sqrt{\mathcal{T}}} \end{pmatrix} \quad (\mathcal{T} := 1 + tg\bar{g}), \\ \hat{f} = \mathrm{sgn}(\mathcal{T}) E L \overline{EL}^t, \quad \hat{n} = \mathrm{sgn}(\mathcal{T}) E L \begin{pmatrix} 1 & 0 \\ 0 & -1 \end{pmatrix} \overline{EL}^t, \end{cases}$$

making the genericity assumption $\mathcal{T} \neq 0$.

Then $\hat{f}$ is a BrLW surface in $\mathbb{H}^3$ with unit normal vector field $\hat{n}$, since $\langle \hat{f}, \hat{n} \rangle = \langle d\hat{f}, \hat{n} \rangle = 0$. Moreover, $\hat{n}$ is a BiLW surface in $\mathbb{S}^{2,1}$. Here we outline a proof of this.

The three fundamental forms of $\hat{f}$ become, with $h := |g'|^{-2}\mathcal{T}^{-2}$,

$$\begin{aligned} I &= h\left\{((1-t)|g'|^2 + \mathcal{T}^2)^2 dx^2 + ((1-t)|g'|^2 - \mathcal{T}^2)^2 dy^2\right\}, \\ II &= -h\left\{(|g'|^4 - (t|g'|^2 - \mathcal{T}^2)^2) dx^2 + (|g'|^4 - (t|g'|^2 + \mathcal{T}^2)^2) dy^2\right\}, \\ III &= h\left\{((1+t)|g'|^2 - \mathcal{T}^2)^2 dx^2 + ((1+t)|g'|^2 + \mathcal{T}^2)^2 dy^2\right\}. \end{aligned}$$

The principal curvatures of $\hat{f}$ and $\hat{n}$ are then

$$\begin{aligned} &k_{1,\hat{f}} = -\frac{(1+t)|g'|^2 - \mathcal{T}^2}{(1-t)|g'|^2 + \mathcal{T}^2}, \quad k_{2,\hat{f}} = -\frac{(1+t)|g'|^2 + \mathcal{T}^2}{(1-t)|g'|^2 - \mathcal{T}^2}, \\ &k_{1,\hat{n}} = \frac{1}{k_{1,\hat{f}}}, \quad k_{2,\hat{n}} = \frac{1}{k_{2,\hat{f}}}, \quad H_{\hat{f}} = \frac{H_{\hat{n}}}{K_{ext,\hat{n}}}, \quad K_{ext,\hat{f}} = \frac{1}{K_{ext,\hat{n}}}, \end{aligned}$$

and so $\hat{f}$ satisfies Equation (3) and $\hat{n}$ satisfies Equation (4). In fact, all BrLW and BiLW surfaces without umbilics (g' would be zero at umbilics) can be constructed this way, using holomorphic functions g.

Thus sufficient conditions for $\hat{f}$ and $\hat{n}$, respectively, to have singularities are

$$\mathcal{T}^4 = (1-t)^2|g'|^4, \quad \mathcal{T}^4 = (1+t)^2|g'|^4,$$

respectively. For certain special values of t these conditions simplify as follows:

$$\begin{aligned} &\hat{f} \text{ with } t = 0: && |g'| = 1, \\ &\hat{n} \text{ with } t = 0: && |g'| = 1, \\ &\hat{f} \text{ with } t = 1: && \text{null condition}, \\ &\hat{n} \text{ with } t = -1: && |g| = 1. \end{aligned}$$

Because $\hat{f}$ and $\hat{n}$ are smooth well-defined maps that can have singularities, it is natural to lift to Lie sphere geometry in $\mathbb{R}^{4,2}$, with

$$f = (\hat{f}, 1, 0)^t, \quad n = (\hat{n}, 0, 1)^t \tag{6}$$

determined by

$$p = (0,0,0,0,0,1)^t, \quad q = (0,0,0,0,-1,0)^t.$$

For a BrLW surface $\hat{f} \in \mathbb{H}^3 = M^3_{p,q}$ with BiLW normal $\hat{n} \in \mathbb{S}^{2,1} = M^3_{q,p}$, we can define the Legendre lift $\Lambda = \text{span}\{s_+, s_-\}$ for

$$s_\pm = b_\pm f + n \quad \text{with} \quad b_+ = 1 \quad \text{and} \quad b_- = \frac{t+1}{t-1},$$

and then $s_\pm$ have constant conserved quantities

$$q_+ = (0,0,0,0,1,1)^t, \quad q_- = (0,0,0,0,t-1,t+1)^t.$$

in the sense that $(s_\pm, q_\pm) = 0$, equivalently the equations $\Gamma^\pm q_\pm = 0$ for the associated families of flat connections hold (see [7], [23]). Furthermore, because $b_\pm$ are constant and because the elements g_{ij} of the first fundamental form of $\hat{f}$ satisfy (using Equation (3))

$$\pm\frac{\sqrt{g_{11}}}{\sqrt{g_{22}}} = \frac{1-\kappa_2}{1-\kappa_1} = \frac{-t-1+(t-1)\kappa_2}{t+1-(t-1)\kappa_1},$$

all of Equations (4.5) and (4.10) and (4.11) in [23] hold, and so $s_\pm$ are isothermic sphere congruences. Thus Λ is an Ω surface with a pair of constant conserved quantities.

Conversely, if we start with an Ω surface with constant conserved quantities $q_\pm$ for isothermic sphere congruences $s_\pm = b_\pm f + n$ respectively, we can reverse the above arguments to see that we obtain a BrLW surface $\hat{f}$ with BiLW normal $\hat{n}$ in the spaceforms $M^3_{p,q}$ and $M^3_{q,p}$, with p and q as above.

This proves the next lemma, which was already understood in [7]:

Lemma 4.1 ([7]). *All smooth BrLW and BiLW surfaces in* $\mathbb{H}^3$ *and* $\mathbb{S}^{2,1}$ *are projections of* Ω *surfaces with constant conserved quantities, at least one of which is lightlike. Conversely, for any smooth* Ω *surface with constant conserved quantities*[1] $q_\pm$, *at least one of which is lightlike, its projections* $\hat{f}$ *and* $\hat{n}$ *given by choosing* $p, q \in span\{q_\pm\}$ *are BrLW and BiLW surfaces, respectively.*

The same result holds for general linear Weingarten $\hat{f}$ and $\hat{n}$, even without the condition that at least one of the $q_\pm$ is lightlike, again see [7]. However, here we consider only the cases given in Lemma 4.1.

§5. Discrete surfaces with Weierstrass-type representations

First we give Weierstrass-type representations for discrete surfaces using the more symmetric form of the base equation as in §6 of [16].

Let $g : \mathbb{Z}^2 \to \mathbb{C}$ be a function satisfying

$$cr(g_i, g_j, g_k, g_\ell) := \frac{(g_i - g_j)(g_k - g_\ell)}{(g_j - g_k)(g_\ell - g_i)} = \frac{\alpha_{ij}}{\alpha_{i\ell}} < 0 \ ,$$

where α_{ij} (resp. $\alpha_{i\ell}$) is a scalar function defined on the horizontal edges (resp. vertical edges) and unchanging with respect to vertical (resp. horizontal) shifts. A complex-valued function g satisfying the above condition is called a *discrete holomorphic function* and α_{ij}, $\alpha_{i\ell}$ are called *cross ratio factorizing functions*. Now we assume the discrete analog of $g' \neq 0$, i.e. $dg_{ij} := g_j - g_i \neq 0$ and $dg_{i\ell} \neq 0$ for all quadrilaterals. We again make the genericity assumption

$$\mathcal{T}_i := 1 + t g_i \overline{g_i} \neq 0$$

for all vertices i, for the chosen constant $t \in \mathbb{R}$. Take $\lambda \in \mathbb{R}$ to be any non-zero constant so that $1 - \lambda\alpha_{ij} \neq 0$ on all edges. Solving

$$E_i^{-1} E_j = \frac{1}{\sqrt{1 - \lambda\alpha_{ij}}} \begin{pmatrix} 1 & dg_{ij} \\ \frac{\lambda\alpha_{ij}}{dg_{ij}} & 1 \end{pmatrix}$$

and the analogous equation with j replaced by ℓ, for $E_i \in \mathrm{SL}_2\mathbb{C}$ for all i, and defining

$$L_i = \begin{pmatrix} 0 & \sqrt{\mathcal{T}_i} \\ \frac{-1}{\sqrt{\mathcal{T}_i}} & \frac{-t\bar{g}_i}{\sqrt{\mathcal{T}_i}} \end{pmatrix},$$

[1]We assume q_+, q_- are not parallel, and that $\mathrm{span}\{q_\pm\}$ is not a null plane.

the surface $\hat{f}$ and its normal $\hat{n}$

$$\hat{f}_i = \mathrm{sgn}(\mathcal{T}_i) E_i L_i \overline{E_i L_i}^t, \quad \hat{n}_i = \mathrm{sgn}(\mathcal{T}_\mathrm{i}) E_i L_i \begin{pmatrix} 1 & 0 \\ 0 & -1 \end{pmatrix} \overline{E_i L_i}^t,$$

we will see that these are discrete BrLW surfaces and BiLW surfaces in $\mathbb{H}^3$ and $\mathbb{S}^{2,1}$, respectively. Direct computations confirm the following lemma:

Lemma 5.1. *For any choice of* t*, we have the following*:

- $d\hat{f}_{ij} || d\hat{n}_{ij}$, $d\hat{f}_{i\ell} || d\hat{n}_{i\ell}$ *in* $\mathbb{R}^{3,1}$ *for all edges* ij, $i\ell$, *and the principal curvatures* κ_{i*} *satisfy*

$$dn_{i*} = -\kappa_{i*} df_{i*} \ ,$$

and furthermore

$$\kappa_{i*} = \frac{-|dg_{i*}|^2(1+t) + (1+t|g_i|^2)(1+t|g_*|^2)\lambda\alpha_{i*}}{-|dg_{i*}|^2(-1+t) + (1+t|g_i|^2)(1+t|g_*|^2)\lambda\alpha_{i*}} \ , \tag{7}$$

for $* = j, \ell$.
- $1 + t|g_i|^2 > 0$, *resp.* $1 + t|g_i|^2 < 0$, *if and only if* $\hat{f}_i$ *lies in the upper, resp. lower, sheet of* $\mathbb{H}^3$.
- $\hat{f}_i$, $\hat{f}_j$, $\hat{f}_k$, $\hat{f}_\ell$ *lie in a plane in* $\mathbb{R}^{3,1}$, *and thus are concircular in* $\mathbb{H}^3$.

Corollary 5.1. *For any choice of* t*, the parallel surfaces*

$$\cosh\theta \cdot \hat{f} + \sinh\theta \cdot \hat{n} \ , \quad \cosh\theta \cdot \hat{n} + \sinh\theta \cdot \hat{f}$$

are concircular for all $\theta \in \mathbb{R}$.

Proof. $d\hat{f}_{i*} || d\hat{n}_{i*}$ and the fact that corresponding quadrilaterals of f and n lie in parallel planes imply that corresponding quadrilaterals of $\cosh\theta \cdot \hat{f} + \sinh\theta \cdot \hat{n}$ also lie in parallel planes, proving the corollary. Q.E.D.

Like in Equation (6), we can lift $\hat{f}$ and $\hat{n}$ to $f, n \in \mathbb{R}^{4,2}$, producing a discrete Legendre immersion $\Lambda = \mathrm{span}\{f, n\}$. We define

$$\mathcal{A}(f,f)_{ijk\ell} := \frac{1}{2} df_{ik} \wedge df_{j\ell} \ ,$$

and we can define real-valued functions $H = H_{\hat{f}}$ and $K = K_{\hat{f}}$ on faces by

$$\mathcal{A}(f,n)_{ijk\ell} := \frac{1}{4}\{df_{ik} \wedge dn_{j\ell} + dn_{ik} \wedge df_{j\ell}\} = -H \cdot \mathcal{A}(f,f)_{ijk\ell} \ , \tag{8}$$

$$\mathcal{A}(n,n)_{ijk\ell} = \frac{1}{2} dn_{ik} \wedge dn_{j\ell} = K \cdot \mathcal{A}(f,f)_{ijk\ell} \,. \tag{9}$$

We have the following definition:

Definition 5.1 ([8]). *We call* $K_{\hat{f}}$ *and* $H_{\hat{f}}$ *the (extrinsic)* Gaussian *and* mean curvature *of the projection* $\hat{f}$ *of* Λ *to the spaceform given by* p, q.

Proven similarly to the corresponding result for $\mathbb{R}^3$ in [4], using item 1 of Lemma 5.1, we have:

Lemma 5.2. *For all choices of spaceforms, we have*

$$H_{\hat{f}} = \frac{\kappa_{ij}\kappa_{k\ell} - \kappa_{i\ell}\kappa_{jk}}{\kappa_{ij} - \kappa_{i\ell} - \kappa_{jk} + \kappa_{k\ell}} \,,$$

$$K_{\hat{f}} = \frac{\kappa_{ij}\kappa_{jk}\kappa_{k\ell}\kappa_{i\ell}}{\kappa_{ij} - \kappa_{i\ell} - \kappa_{jk} + \kappa_{k\ell}} \left(-\frac{1}{\kappa_{ij}} + \frac{1}{\kappa_{jk}} + \frac{1}{\kappa_{i\ell}} - \frac{1}{\kappa_{k\ell}} \right).$$

Proof. The compatibility condition $\hat{n}_{ij} + \hat{n}_{jk} = \hat{n}_{i\ell} + \hat{n}_{\ell k}$ for $\hat{n}$ implies

$$\kappa_{ij} d\hat{f}_{ij} + \kappa_{jk} d\hat{f}_{jk} = \kappa_{i\ell} d\hat{f}_{i\ell} + \kappa_{k\ell} d\hat{f}_{\ell k} \,.$$

Thus

$$\kappa_{ij} d\hat{f}_{ij} + \kappa_{jk}(d\hat{f}_{ik} - d\hat{f}_{ij}) = \kappa_{i\ell} d\hat{f}_{i\ell} + \kappa_{k\ell}(d\hat{f}_{ik} - d\hat{f}_{i\ell})$$
$$\Rightarrow d\hat{f}_{ik} = c_1 d\hat{f}_{ij} + c_2 d\hat{f}_{i\ell} \,,$$

where

$$c_1 = \frac{-\kappa_{ij} + \kappa_{jk}}{\kappa_{jk} - \kappa_{k\ell}}, \quad c_2 = \frac{-\kappa_{lk} + \kappa_{i\ell}}{\kappa_{jk} - \kappa_{k\ell}} \,. \tag{10}$$

Similarly, by the compatibility condition for $\hat{f}$ and the condition $dn_{**} = -\kappa_{**} df_{**}$, we have $d\hat{n}_{ik} = c_3 df_{ij} + c_4 df_{il}$, where

$$c_3 = \frac{\kappa_{k\ell}(\kappa_{jk} - \kappa_{ij})}{\kappa_{k\ell} - \kappa_{jk}} \,, \quad c_4 = \frac{\kappa_{jk}(\kappa_{i\ell} - \kappa_{k\ell})}{\kappa_{k\ell} - \kappa_{jk}} \,. \tag{11}$$

Note that $d\hat{f}_{j\ell} = d\hat{f}_{i\ell} - d\hat{f}_{ij}$, $d\hat{n}_{j\ell} = d\hat{n}_{i\ell} - d\hat{n}_{ij}$, and we have

$$H_{\hat{f}} = \frac{\kappa_{i\ell} c_1 + \kappa_{ij} c_2 - c_3 - c_4}{2(c_1 + c_2)} \,, \quad K_{\hat{f}} = -\frac{\kappa_{i\ell} c_3 + \kappa_{ij} c_4}{c_1 + c_2} \,. \tag{12}$$

Substituting Equations (10) and (11) into Equation (12), we have $H_{\hat{f}}$ and $K_{\hat{f}}$ as in Lemma 5.2. Q.E.D.

We can similarly define the Gaussian and mean curvatures $K_{\hat{n}}$, $H_{\hat{n}}$ of $\hat{n}$, and we see that

$$K_{\hat{n}} = \frac{1}{K_{\hat{f}}}, \quad H_{\hat{n}} = \frac{H_{\hat{f}}}{K_{\hat{f}}}, \quad \kappa_{ij,\hat{n}} = \frac{1}{\kappa_{ij,\hat{f}}}, \quad \kappa_{il,\hat{n}} = \frac{1}{\kappa_{il,\hat{f}}}. \tag{13}$$

One can confirm the next lemma via Lemma 5.2 and Equations (7), (13):

Lemma 5.3. *The mean and Gaussian curvatures $H_{\hat{f}}$ and $K_{\hat{f}}$ of a discrete surface $\hat{f}$ with Weierstrass-type representation in $\mathbb{H}^3$ satisfy*

$$2t(H_{\hat{f}} - 1) + (1-t)(K_{\hat{f}} - 1) = 0 \ , \tag{14}$$

and the mean and Gaussian curvatures $H_{\hat{n}}$ and $K_{\hat{n}}$ of a discrete surface $\hat{n}$ with Weierstrass-type representation in $\mathbb{S}^{2,1}$ satisfy

$$2t(H_{\hat{n}} - 1) - (1+t)(K_{\hat{n}} - 1) = 0 \ . \tag{15}$$

Thus we know that the discrete surfaces with Weierstrass-type representations defined here are included amongst the discrete BrLW and BiLW surfaces defined in [8], by the following Proposition 5.1 from [8]. This proposition also includes discrete minimal surfaces in $\mathbb{R}^3$ and their parallel surfaces in $\mathbb{R}^3$, as well as parallel surfaces of discrete maximal surfaces in $\mathbb{R}^{2,1}$.

Proposition 5.1 ([8]). *All discrete BrLW and BiLW surfaces in $\mathbb{H}^3$ and $\mathbb{S}^{2,1}$, and all parallel surfaces of discrete minimal surfaces in $\mathbb{R}^3$ and discrete maximal surfaces in $\mathbb{R}^{2,1}$, are projections of discrete Ω surfaces with constant conserved quantities, at least one of which is lightlike. Conversely, for any discrete Ω surface with constant conserved quantities[2] $q_\pm$, at least one of which is lightlike, its projections $\hat{f}$ and $\hat{n}$ given by choosing $p, q \in \mathrm{span}\{q_\pm\}$ are discrete BrLW and BiLW surfaces, respectively, or $\hat{f}$ is either a parallel surface of a discrete minimal surface in $\mathbb{R}^3$ or maximal surface in $\mathbb{R}^{2,1}$.*

In the smooth case, as mentioned in [19], parallel surfaces of BrLW surfaces in $\mathbb{H}^3$ are also BrLW surfaces, and BrLW surfaces are classified into the following three types:

(1) flat surfaces (BrLW surfaces with $t = 0$),
(2) linear Weingarten surfaces of hyperbolic type (BrLW surfaces with $t > 0$),

[2] Again we assume q_+, q_- are not parallel, and that $\mathrm{span}\{q_\pm\}$ is not a null plane.

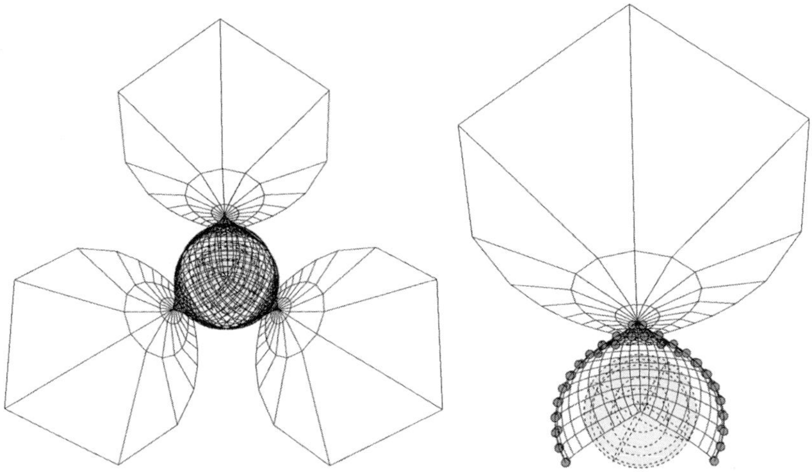

Fig. 2. One example of a discrete BrLW surface in $\mathbb{H}^3$ with $t = -2$. Solid gray vertices are the FPS vertices of the surface. In order to draw the surface, we use stereographic projection from the south pole $(0, 0, 0, -1)$. When $1 + t|g_i|^2 > 0$, the surface is projected to the inside of the unit ball $\mathbb{B}^3$, whose boundary is drawn in gray above. When $1 + t|g_i|^2 < 0$, it is projected to the outside of $\mathbb{B}^3$. One-third of the surface is shown on the right.

(3) linear Weingarten surfaces of de Sitter type (BrLW surfaces with $t < 0$).

Parallel surfaces of each type belong to the same type. Thus parallel surfaces of a flat front are also flat. Likewise, parallel surfaces of the other two types again belong to the same types.

Here we see that the same result as in the smooth case holds also in the discrete case. Let $\hat{f}$ be a discrete BrLW surface in $\mathbb{H}^3$. From [22], we have that the Gaussian and mean curvatures $K^\theta_{\hat{f}}$, $H^\theta_{\hat{f}}$ of the parallel surface $\hat{f}^\theta$ at oriented distance θ are

$$K^\theta_{\hat{f}} = \frac{K_{\hat{f}} \cosh^2\theta - H_{\hat{f}} \sinh 2\theta + \sinh^2\theta}{\cosh^2\theta - H_{\hat{f}} \sinh 2\theta + K_{\hat{f}} \sinh^2\theta},$$

$$H^\theta_{\hat{f}} = -\frac{(K_{\hat{f}} + 1)\sinh(2\theta) - 2H_{\hat{f}} \cosh(2\theta)}{2\{\cosh^2\theta - H_{\hat{f}} \sinh(2\theta) + K_{\hat{f}} \sinh^2\theta\}}.$$

Observing that $\hat{f} = (\hat{f}^\theta)^{-\theta}$, we have

$$(16)\qquad \begin{cases} K_{\hat{f}} = \dfrac{K^\theta_{\hat{f}} \cosh^2\theta + H^\theta_{\hat{f}} \sinh 2\theta + \sinh^2\theta}{\cosh^2\theta + H^\theta_{\hat{f}} \sinh 2\theta + K^\theta_{\hat{f}} \sinh^2\theta}, \\ H_{\hat{f}} = \dfrac{(K^\theta_{\hat{f}} + 1)\sinh(2\theta) + 2H^\theta_{\hat{f}} \cosh(2\theta)}{2\{\cosh^2\theta + H^\theta_{\hat{f}} \sinh(2\theta) + K^\theta_{\hat{f}} \sinh^2\theta\}}. \end{cases}$$

Substituting Equation (16) into Equation (14), we have

$$2T(H^\theta_{\hat{f}} - 1) + (1 - T)(K^\theta_{\hat{f}} - 1) = 0,$$

where $T = \mathrm{e}^{-2\theta}t$. Thus discrete BrLW surfaces in $\mathbb{H}^3$ are classified into the three types (1) – (3) mentioned above. Similarly, discrete BiLW surfaces in $\mathbb{S}^{2,1}$ are classified into three types.

§6. Singular vertices on discrete nonzero CGC surfaces in M^3

When a smooth surface has CGC $K = \kappa_1\kappa_2 \neq 0$, then when one of the κ_α passes through zero, the other passes through infinity, and we can always call this a singular point. This is precisely what allowed for the description of singularities of discrete flat (i.e. $K \equiv 1$) surfaces in $\mathbb{H}^3$ as given in [16]. Here we develop that into a definition without reliance on a Weierstrass type representation, extending it to all discrete surfaces in any M^3 with nonzero constant Gaussian curvature.

Definition 6.1. *Consider Λ as in Definition 2.1, together with a choice of spaceform determined by choosing p and q, that has projection $\hat{f}$ with nonzero constant discrete Gaussian curvature $K_{\hat{f}}$. We say that (m, n) is a* singular vertex *of $\hat{f}$ if*

$$\kappa_{(m-1,n),(m,n)} \cdot \kappa_{(m,n),(m+1,n)} \leq 0 \;\; \textit{or} \;\; \kappa_{(m,n-1),(m,n)} \cdot \kappa_{(m,n),(m,n+1)} \leq 0.$$

For a $K \equiv 1$ surface with Weierstrass-type representation in $\mathbb{H}^3$, it was shown in [16] that, without loss of generality, $|\kappa_{(m,n),(m,n+1)}| > 1$ and $|\kappa_{(m,n),(m+1,n)}| < 1$ for all m and n, which we note in the following theorem:

Theorem 6.1. *In the case of a $K \equiv 1$ surface in $\mathbb{H}^3$ with Weierstrass-type representation so that the horizontal edges have principal curvatures with absolute value greater than 1, the first inequality in Definition 6.1 is equivalent to the definition of singular vertices for discrete flat surfaces in $\mathbb{H}^3$ as given in [16].*

Proof. By Lemma 5.1, for $t = 0$ we have

$$\kappa_{i*} = \frac{-|dg_{i*}|^2 + \lambda\alpha_{i*}}{|dg_{i*}|^2 + \lambda\alpha_{i*}} .$$

Let p_-, p and p_+ be three consecutive vertices in one direction in the lattice domain. We can define singularities on discrete flat (i.e. $K \equiv 1$) surfaces in $\mathbb{H}^3$, now without referring to caustics as in [16], by simply using the condition

$$\frac{-|dg_{p_-p}|^2 + \lambda\alpha_{p_-p}}{|dg_{p_-p}|^2 + \lambda\alpha_{p_-p}} \cdot \frac{-|dg_{pp_+}|^2 + \lambda\alpha_{pp_+}}{|dg_{pp_+}|^2 + \lambda\alpha_{pp_+}} < 0,$$

as understood in [16]. Q.E.D.

However, our definition allows the second inequality in Definition 6.1, which allows us to include more singular vertices (see Figure 3).

§7. Discrete minimal surfaces and their parallel surfaces

7.1. Smooth minimal surfaces in $\mathbb{R}^3$

We can always take a smooth constant mean curvature (CMC) surface in a 3-dimensional Riemannian spaceform to have local isothermic coordinates $z = u + iv$ on M^2, $u, v \in \mathbb{R}$ (away from umbilic points), and then the Hopf differential becomes rdz^2 for some real constant r. Rescaling the coordinate z by a constant real factor, we may assume $r = 1$. So we now assume we have an isothermic minimal surface in $\mathbb{R}^3$ with Hopf differential $Q = dz^2$. Then

$$\frac{Q}{dg} = \frac{dz}{g'}, \tag{17}$$

where g is the stereographic projection of the Gauss map to the complex plane, and $g' = dg/dz$. The map g taking z in the domain of the immersion (of the surface) to $\mathbb{C}$ is holomorphic. We avoid umbilics, so we have $g' \neq 0$. We are concerned with only local behavior, so we can ignore the possibility that g has poles. Then the Weierstrass representation is

$$\hat{f} = \mathrm{Re} \int_{z_0}^{z} (1 - g^2, i + ig^2, 2g)^t \frac{dz}{g'} , \tag{18}$$

with the last factor coming from (17). The metric of $\hat{f}$ is

$$\frac{(1 + |g|^2)^2}{|g'|^2} dz d\bar{z} . \tag{19}$$

By direct computation, we have:

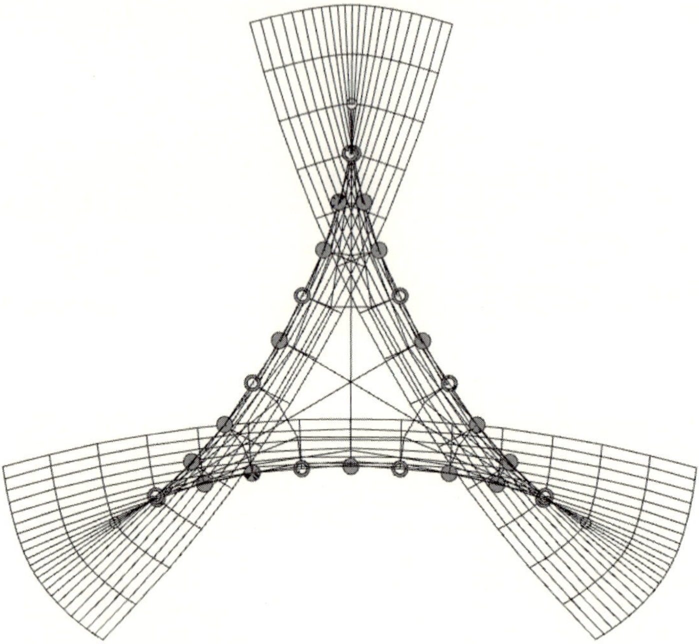

Fig. 3. A flat surface in $\mathbb{H}^3$ with its singular vertices in the sense of [16] shown with solid gray dots, and the extra singular vertices that would be included by Definition 6.1 shown with hollowed-out dots.

Lemma 7.1. *For a smooth minimal surface $\hat{f}$ as given in Equation* (18), *the partial derivatives of $\hat{f}$ are*

$$\hat{f}_u = \mathrm{Re}\left(\frac{1-g^2}{g_u}, \frac{i(1+g^2)}{g_u}, \frac{2g}{g_u}\right)^t,$$
$$\hat{f}_v = -\mathrm{Re}\left(\frac{1-g^2}{g_v}, \frac{i(1+g^2)}{g_v}, \frac{2g}{g_v}\right)^t.$$

Furthermore, the principal curvatures of the surface are

$$\pm\kappa_1 = \mp\kappa_2 = \frac{2|g'|^2}{(1+|g|^2)^2}. \tag{20}$$

The next lemma will be used as motivation for the discussion about discrete minimal surfaces that follows it:

Lemma 7.2. *Any parallel surface of a minimal surface in* $\mathbb{R}^3$ *without umbilics will have constant harmonic mean curvature, and will have neither parabolic nor flat points.*

Proof. For surfaces in $\mathbb{R}^3$ with Gaussian and mean curvatures K and H, the parallel surfaces at distance ρ have Gaussian and mean curvatures

$$\hat{K} = \frac{K}{1 - 2\rho H + \rho^2 K}, \quad \hat{H} = \frac{H - \rho K}{1 - 2\rho H + \rho^2 K},$$

so when we have a minimal surface ($H = 0$) in $\mathbb{R}^3$,

$$\hat{K} = \frac{K}{1 + \rho^2 K}, \quad \hat{H} = \frac{-\rho K}{1 + \rho^2 K}.$$

If the minimal surface has no umbilics, then $K \neq 0$, which implies no parallel surface can have any parabolic or flat points.

The parallel surfaces all have constant harmonic mean curvature since $\frac{\hat{H}}{\hat{K}} = -\rho$. Q.E.D.

7.2. Discrete minimal surfaces in $\mathbb{R}^3$

Analogously to the smooth case, a suitable representation (or definition, see [3], [13]) for discrete minimal surfaces (equivalently, defined as discrete surfaces with $H_{\hat{f}} \equiv 0$) is, with $* = j, \ell$,

$$\text{(21)} \qquad \hat{f}_* - \hat{f}_i = \frac{\alpha_{i*}}{2}\mathrm{Re}\left(\frac{1 - g_* g_i}{g_* - g_i}, \frac{\sqrt{-1}(1 + g_* g_i)}{g_* - g_i}, \frac{g_* + g_i}{g_* - g_i}\right)^t,$$

where the map g from a domain in $\mathbb{Z}^2$ to $\mathbb{C}$ is a discrete holomorphic function with cross ratio factorizing function α_{i*}. As in the smooth case where we avoided umbilics, likewise here we assume

$$g_* - g_i \neq 0 .$$

Example 7.1. The discrete holomorphic function $g = c(m+n\sqrt{-1})$ for c a complex constant will produce a discrete minimal Enneper surface. The discrete holomorphic function $g = e^{c_1 m + c_2 n\sqrt{-1}}$ for choices of real constants c_1 and c_2 so that the cross ratio is identically -1 will produce a discrete minimal catenoid. (See [3] for graphics.)

Furthermore, the principal curvatures κ_{i*} defined on edges (similarly to (20)) are

$$\kappa_{i*} = -\frac{4|dg_{i*}|^2}{\alpha_{i*}(1 + |g_i|^2)(1 + |g_*|^2)} .$$

Based on Lemma 7.2, we can justify the following definition:

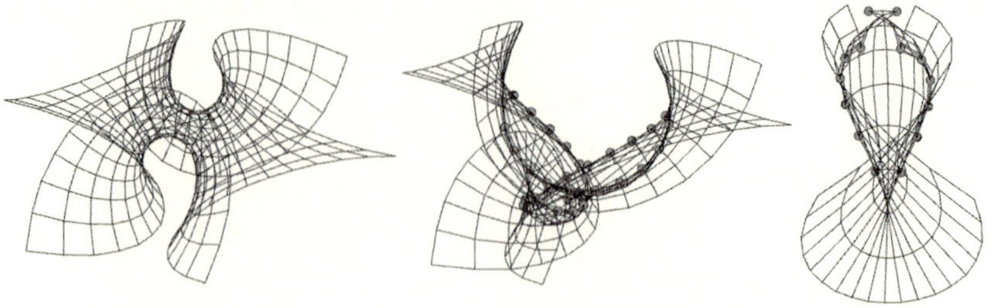

Fig. 4. A discrete higher-order Enneper minimal surface in $\mathbb{R}^3$, its parallel surface at distance 20, and a one-third piece of the parallel surface, with singular vertices marked.

Definition 7.1. *For any discrete minimal surface, we say that (m, n) is a* singular vertex *of any given parallel surface if the principal curvatures κ_{**} of that parallel surface satisfy*

$$\kappa_{(m-1,n),(m,n)} \cdot \kappa_{(m,n),(m+1,n)} \leq 0 \quad or$$
$$\kappa_{(m,n-1),(m,n)} \cdot \kappa_{(m,n),(m,n+1)} \leq 0 \ .$$

§8. Discrete maximal surfaces and their parallel surfaces

Here we give the analogous situation as in §7, but now in Lorentz 3-space.

8.1. Smooth maximal surfaces in $\mathbb{R}^{2,1}$

First we briefly review smooth maximal surfaces. Let

$$\mathbb{R}^{2,1} := (\{(x_1, x_2, x_0)^t | x_j \in \mathbb{R}\}, \langle \cdot, \cdot \rangle)$$

be 3-dimensional Minkowski space with the Lorentz metric signature $(+, +, -)$.

Note that, for fixed $d \in \mathbb{R}$ and vector $n \in \mathbb{R}^{2,1} \setminus \{0\}$, a plane $\mathcal{P} = \{x \in \mathbb{R}^{2,1} \mid \langle x, n \rangle = d\}$ is *spacelike* or *timelike* or *lightlike* when n is timelike or spacelike or lightlike, respectively. Furthermore, a smooth surface in $\mathbb{R}^{2,1}$ is called spacelike if its tangent planes are spacelike. Let

$$\hat{f} : M^2 \to \mathbb{R}^{2,1}$$

be a conformal immersion, where M^2 is a simply-connected domain in $\mathbb{C}$ with complex coordinate $z = u + iv$ $(u, v \in \mathbb{R})$. $\hat{f}$ is a maximal surface if it is spacelike (which follows automatically from the conformality condition) with mean curvature identically 0.

Defining

$$\mathbb{H}^2_+ := \{x = (x_1, x_2, x_0)^t \in \mathbb{R}^{2,1} | \langle x, x \rangle = -1,\ x_0 > 0\},$$
$$\mathbb{H}^2_- := \{x = (x_1, x_2, x_0)^t \in \mathbb{R}^{2,1} | \langle x, x \rangle = -1,\ x_0 < 0\},$$

we have the following statement, analogous to the case of smooth minimal surfaces in $\mathbb{R}^3$, as in (18) (and having a similar proof): Away from umbilic points, smooth maximal surfaces lie in the class of isothermic surfaces, and each such surface can be represented with isothermic coordinates (u, v), $z = u + iv$, as

$$\hat{f} = \mathrm{Re} \int (1 + g^2, i(1 - g^2), -2g)^t \frac{dz}{g'} \tag{22}$$

for some choice of smooth holomorphic function $g : M^2 \to \mathbb{C}$. The Gauss map of $\hat{f}$ lies in $\mathbb{H}^2_+ \cup \mathbb{H}^2_-$, and its stereographic projection to $\mathbb{C}$ is g.

Differentiating Equation (22) gives the following equations (analogous to Lemma 7.1):

$$\hat{f}_u = \mathrm{Re} \left(\frac{1 + g^2}{g_u}, \frac{i(1 - g^2)}{g_u}, -\frac{2g}{g_u} \right)^t,$$
$$\hat{f}_v = -\mathrm{Re} \left(\frac{1 + g^2}{g_v}, \frac{i(1 - g^2)}{g_v}, -\frac{2g}{g_v} \right)^t.$$

Remark. Unlike the case of the Weierstrass representation for minimal surfaces in $\mathbb{R}^3$, smooth maximal surfaces in $\mathbb{R}^{2,1}$ have singularities when $|g| = 1$, because the metrics

$$\frac{(1 - |g|^2)^2}{|g'|^2} dz d\bar{z} \tag{23}$$

of the smooth maximal surfaces can degenerate, due to the minus sign in the numerator in Equation (23), unlike the plus sign we have for the metrics of minimal surfaces in $\mathbb{R}^3$, as in (19).

The principal curvatures of $\hat{f}$ are (analogous to (20))

$$\pm\kappa_1 = \mp\kappa_2 = \frac{2|g'|^2}{(1 - |g|^2)^2}.$$

By exactly the same proof as for Lemma 7.2, we have:

Lemma 8.1. *Any parallel surface of a maximal surface in* $\mathbb{R}^{2,1}$ *without umbilics will have constant harmonic mean curvature, and will have neither parabolic nor flat points.*

8.2. Discrete maximal surfaces in $\mathbb{R}^{2,1}$

The following theorem was proven in [28] (analogous to (21)):

Proposition 8.1. *Discrete maximal surfaces* $\hat{f}$ *(defined as discrete surfaces with* $H_{\hat{f}} \equiv 0$ *in* $\mathbb{R}^{2,1}$*), maps from* $\mathbb{Z}^2$ *(or some subdomain) to* $\mathbb{R}^{2,1}$*, can be constructed using discrete holomorphic functions* g *from the same domain to the complex plane* $\mathbb{C}$ *by solving*

$$\hat{f}_* - \hat{f}_i = \frac{\alpha_{i*}}{2}\mathrm{Re}\left(\frac{1+g_*g_i}{g_*-g_i}, \frac{\sqrt{-1}(1-g_*g_i)}{g_*-g_i}, -\frac{g_*+g_i}{g_*-g_i}\right)^t, \tag{24}$$

with α_{i*} *the cross ratio factorizing functions for* g*. Conversely, any discrete maximal surface satisfies* (24) *for some discrete holomorphic function* g*.*

Lemma 8.2. *The principal curvatures* κ_{i*} *of* $\hat{f}$ *defined on edges are*

$$\kappa_{i*} = -\frac{4|dg_{i*}|^2}{\alpha_{i*}(1-|g_i|^2)(1-|g_*|^2)}\,.$$

We recall the following definition of singular faces as in [28]:

Definition 8.1. *A face of* $\hat{f}$ *with vertices* $\hat{f}_i, \hat{f}_j, \hat{f}_k, \hat{f}_\ell$ *is* singular *if those four vertices lie in a non-spacelike plane.*

It was proven in [28] that a quadrilateral of $\hat{f}$ is singular if and only if the corresponding circumcircle of g intersects the unit circle $\mathbb{S}^1 \subset \mathbb{C}$. From this we can conclude the following theorem:

Theorem 8.1. *Let* p_-*,* p *and* p_+ *be three consecutive vertices in one direction in the lattice domain of a maximal surface* $\hat{f}$ *in* $\mathbb{R}^{2,1}$*, with corresponding values* g_-*,* g *and* g_+ *for the discrete holomorphic function in the Weierstrass type representation* (24)*. Suppose* p *is an FPS vertex. Then the pair of faces adjacent to the edge* p_-p *are singular, or the pair of faces adjacent to the edge* pp_+ *are singular, including the possibility that all four faces are singular.*

Proof. Because

$$\kappa_{p_-p}\kappa_{pp_+} = (\text{nonnegative term})(1-|g_{p_-}|^2)(1-|g_{p_+}|^2)\ ,$$

$\kappa_{p_-p}\kappa_{pp_+} < 0$ implies at least one of $(1-|g_p|^2)(1-|g_{p_-}|^2)$ or $(1-|g_p|^2)(1-|g_{p_+}|^2)$ is negative, and so at least one of the edges p_-p or pp_+ has two adjacent singular faces. Q.E.D.

Theorem 8.1 indicates one reason why we should regard, in the case of discrete maximal surfaces, all FPS vertices as singular. In fact, like in the case of discrete minimal surfaces, Lemma 8.1 indicates we can say the same of parallel surfaces of discrete maximal surfaces as well:

Definition 8.2. *For any discrete maximal surface, we say that* (m,n) *is a* singular vertex *of any given parallel surface (allowing also for the initial maximal surface itself) if the principal curvatures of the parallel surface satisfy*

$$\kappa_{(m-1,n),(m,n)} \cdot \kappa_{(m,n),(m+1,n)} \leq 0 \quad or$$
$$\kappa_{(m,n-1),(m,n)} \cdot \kappa_{(m,n),(m,n+1)} \leq 0 \ .$$

§9. Singular faces on discrete CMC 1 surfaces with Weierstrass-type representations in $\mathbb{S}^{2,1}$

As in Definition 8.1, a quadrilateral of a discrete CMC 1 surface $\hat{n}$ with Weierstrass-type representation in $\mathbb{S}^{2,1}$ is singular if it does not lie in a spacelike plane. We give a geometric condition (Theorem 9.1) for when a quadrilateral of $\hat{n}$ is singular, analogous to a condition in the case of discrete maximal surfaces (see [28]). We then prove a relation (Corollary 9.1) between FPS vertices and singular faces on discrete CMC 1 faces in $\mathbb{S}^{2,1}$ (similar to Theorem 8.1), a relation that helps indicate which of the FPS vertices are actually singular.

The condition for a singular face to occur is

$$(d\hat{f}_{ij}, d\hat{f}_{ij})(d\hat{f}_{i\ell}, d\hat{f}_{i\ell}) - (d\hat{f}_{ij}, d\hat{f}_{i\ell})^2 \leq 0 \ . \tag{25}$$

In the smooth CMC 1 case, with g as in §4, the singularities occur exactly where $|g| = 1$. The condition is still $|g| = 1$ even under the coordinate transformation $z \to \sqrt{\lambda\alpha}z$. The next proposition and theorem are the corresponding condition in the discrete case to $|g| = 1$, and can be proven by computationally spelling out Equation (25). We define

$$h_1 = (1 - |g_j|^2)|dg_{i\ell}|^2(1 - \lambda\alpha_{ij}) \ ,$$
$$h_2 = (1 - |g_\ell|^2)|dg_{ij}|^2(1 - \lambda\alpha_{i\ell}) \ ,$$
$$h_3 = (1 - |g_i|^2)|dg_{j\ell}|^2 \ .$$

Proposition 9.1. *A face of a discrete CMC* 1 *surface* $\hat{n}$ *with Weierstrass-type representation in* $\mathbb{S}^{2,1}$ *is singular if and only if*

$$\mathcal{H} := h_1^2 + h_2^2 + h_3^2 - (h_2 - h_1)^2 - (h_3 - h_1)^2 - (h_3 - h_2)^2 \leq 0 \ .$$

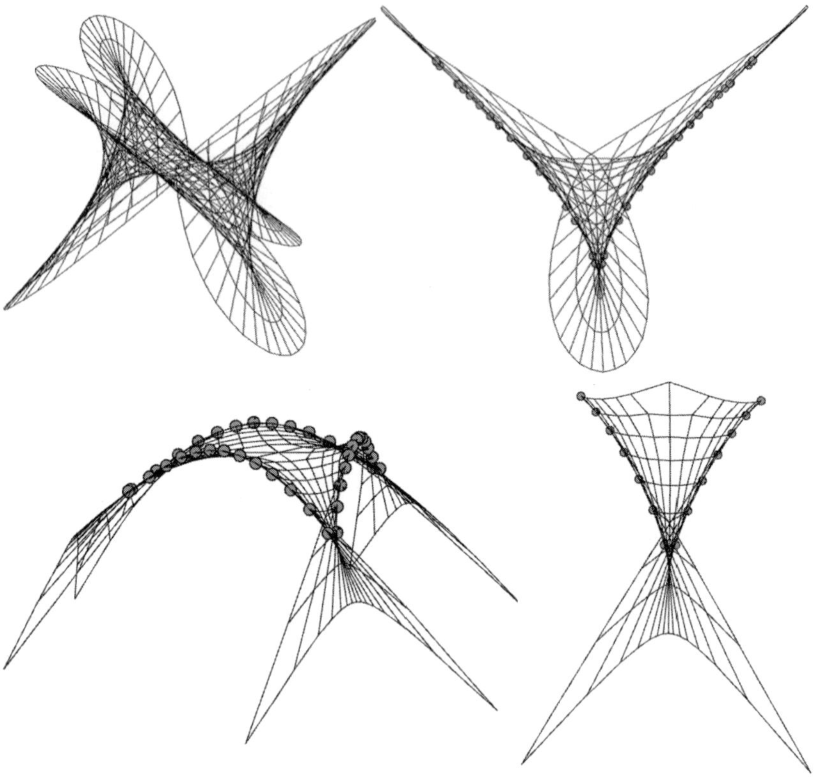

Fig. 5. A discrete higher-order Enneper-type maximal surface in $\mathbb{R}^{2,1}$, its parallel surface at distance 20, and a one-third piece of the parallel surface, with singular vertices marked.

Theorem 9.1. *A quadrilateral of* $\hat{n}$ *as in Proposition 9.1 is singular for all* λ *sufficiently close to zero if the corresponding circumcircle of* g *intersects* $\mathbb{S}^1$ *transversally. The converse holds as well under the generic assumption that* $\partial_\lambda \mathcal{H} \neq 0$.

Proof. If the four points g_i, g_j, g_k, g_l lie on a circle with radius $r \in \mathbb{R}$ and center $p \in \mathbb{C}$, the condition for $\mathcal{H} < 0$ at $\lambda = 0$ is

$$(|p|^2 - (r-1)^2)(|p|^2 - (r+1)^2) < 0 .$$

The result follows. Q.E.D.

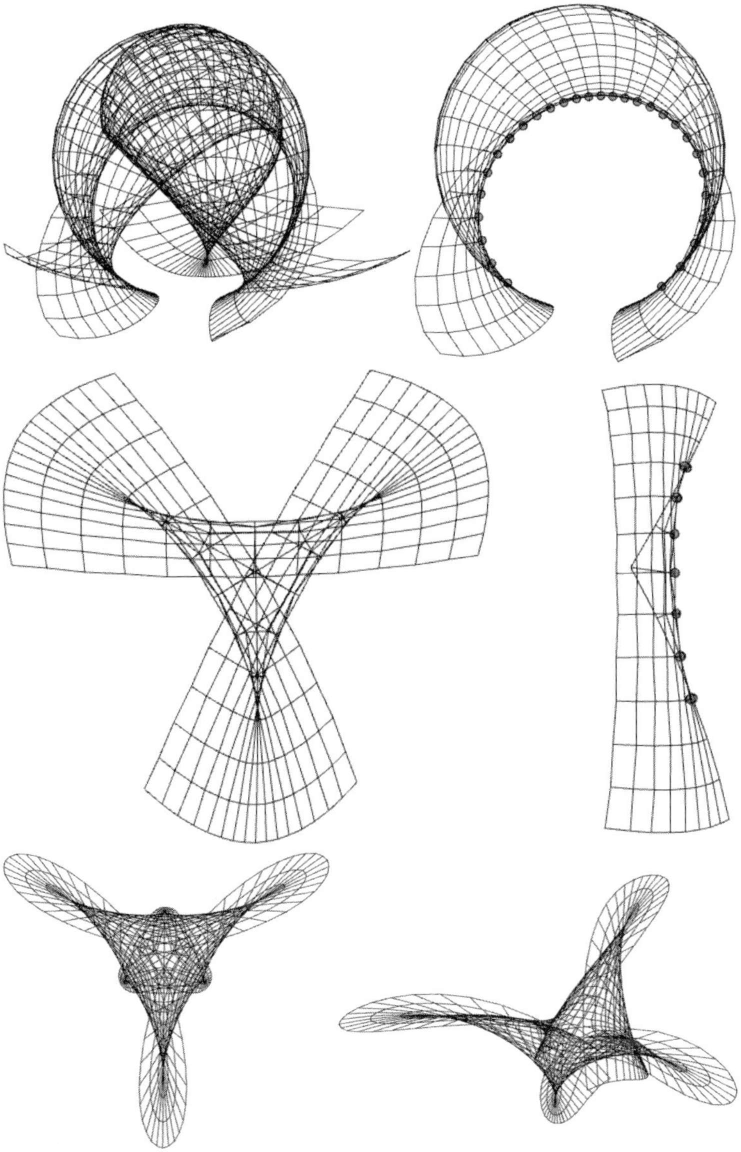

Fig. 6. From top to bottom: A discrete harmonic mean curvature 1 surface in $\mathbb{S}^{2,1}$, a discrete flat surface in $\mathbb{S}^{2,1}$ and a discrete CMC 1 surface in $\mathbb{S}^{2,1}$, each shown twice. In order to draw the surfaces, we project to the hollow ball model for $\mathbb{S}^{2,1}$. (For an explanation of the hollow ball model, see [9] for example.)

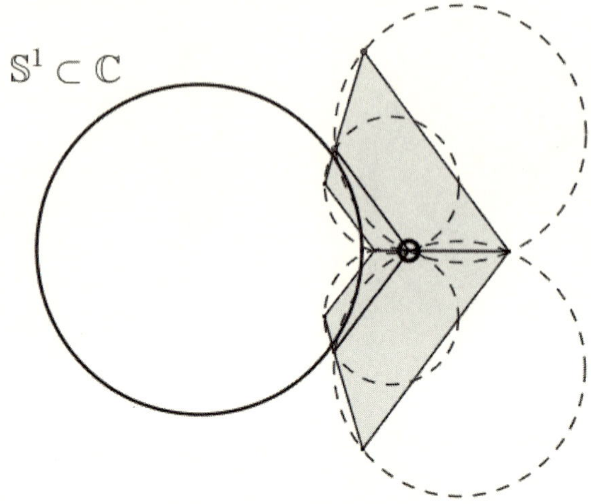

Fig. 7. A counterexample to the converse in Corollary 9.1. Numerical data for a discrete holomorphic function is shown. The four faces of a discrete CMC 1 surface determined from the four gray faces above are singular faces. On the other hand, for sufficiently small λ, the marked vertex is not singular.

Theorem 9.2. *Let p_-, p and p_+ be three consecutive vertices in one direction in the lattice domain of a CMC 1 surface $\hat{n}$ with Weierstrass-type representation in $\mathbb{S}^{2,1}$, with corresponding values g_-, g and g_+ for the discrete holomorphic function in the Weierstrass type representation. Under the genericity assumption $|g| \neq 1$, then $\kappa_{p_-p} \cdot \kappa_{pp_+} < 0$ for all λ sufficiently close to zero if and only if exactly one of $|g_-|^2$ and $|g_+|^2$ has value less than 1 and the other has value greater than 1.*

Proof. Because the surface is CMC 1 in $\mathbb{S}^{2,1}$, we have $t = -1$. Then Equations (7) and (13) imply the result. Q.E.D.

This theorem tells us that we will find FPS vertices roughly where g (discretely) crosses $\mathbb{S}^1$. Because of Theorem 9.2, we can now regard these points as singular vertices and not parabolic nor flat points. Combining Theorems 9.1 and 9.2, the following rigorous statement is immediate:

Corollary 9.1. *Under the conditions of Theorem 9.2, for all λ sufficiently close to zero, the pair of faces adjacent to the edge p_-p are singular, or the pair of faces adjacent to the edge pp_+ are singular, including the possibility that all four faces are singular.*

The converse of this corollary does not hold, that is, it is possible to have four singular faces (for all λ sufficiently close to 0) adjacent to a given vertex that is non-singular for all λ sufficiently close to 0 (see Figure 7). Furthermore, taking $\lambda \to 0$, the example in Figure 7 demonstrates that the converse to Theorem 8.1 also does not hold.

References

[1] R. Aiyama and K. Akutagawa, *Kenmotsu-Bryant type representation formulas for constant mean curvature surfaces in* $\mathbb{H}^3(-c^2)$ *and* $\mathbb{S}^3_1(c^2)$, Ann. Global Anal. Geom. (1) **17** (1998), 49–75.

[2] J. A. Aledo and J. M. Espinar, *A conformal representation for linear Weingarten surfaces in the de Sitter space*, J. Geom. and Phys. **57** (2007), 1669–1677.

[3] A. Bobenko and U. Pinkall, *Discrete isothermic surfaces*, J. Reine Angew. Math., **475** (1996), 187–208.

[4] A. Bobenko, H. Pottmann and J. Wallner, *A curvature theory for discrete surfaces based on mesh parallelity*, Math. Ann. 348 (2010), 1–24.

[5] A. Bobenko and Y. Suris, *Discrete differential geometry, integrable structure*, Graduate Textbooks in Mathematics **98**, A.M.S., 2008.

[6] R. Bryant, *Surfaces of mean curvature one in hyperbolic* 3*-space*, Asterisque **154-155** (1987), 321–347.

[7] F.E. Burstall. U. Hertrich-Jeromin and W. Rossman, *Lie geometry of linear Weingarten surfaces*, C. R. Acad. Sci. Paris, Ser. I **350** (2012), 413–416.

[8] F.E. Burstall. U. Hertrich-Jeromin and W. Rossman, *Discrete linear Weingarten surfaces*, to appear in Nagoya Math J.

[9] S. Fujimori, *Spacelike CMC* 1 *surfaces with elliptic ends in de Sitter 3-Space*, Hokkaido Math. J. **35** (2006), 289–320.

[10] S. Fujimori, K. Saji, M. Umehara and K. Yamada, *Singularities of maximal surfaces*, Math. Zeit. **259** (2008), 827–848.

[11] J. A. Gálvez, A. Martínez and F. Milán, *Flat surfaces in hyperbolic* 3*-space*, Math. Ann. **316** (2000), 419–435.

[12] J. A. Gálvez, A. Martínez and F. Milán, *Complete linear Weingarten surfaces of Bryant type, a Plateau problem at infinity*, Trans. A.M.S. **356** (2004), 3405–3428.

[13] U. Hertrich-Jeromin, *Transformations of discrete isothermic nets and discrete cmc-*1 *surfaces in hyperbolic space*, Manusc. Math. **102** (2000), 465–486.

[14] U. Hertrich-Jeromin, *Introduction to Möbius differential geometry*, London Mathematical Society Lecture Note Series **300**, 2003.

[15] U. Hertrich-Jeromin, T. Hoffmann and U. Pinkall, *A discrete version of the Darboux transform for isothermic surfaces*, Oxf. Lect. Ser. Math. Appl. **16** (1999), 59–81.

[16] T. Hoffmann, W. Rossman, T. Sasaki, M. Yoshida, *Discrete flat surfaces and linear Weingarten surfaces in hyperbolic 3-space*, Trans. A.M.S. **364** (2012), 5605–5644.

[17] S. Izumiya and K. Saji, *The mandala of Legendrian dualities for pseudo-spheres in Lorentz-Minkowski space and "flat" spacelike surfaces*, J. of Singularities **2** (2010), 92–127.

[18] O. Kobayashi, *Maximal surfaces in the 3-dimensional Minkowski space* $\mathbb{L}^3$, Tokyo J. Math. **6** (1983), 297–309.

[19] M. Kokubu and M. Umehara, *Orientability of linear Weingarten surfaces, spacelike CMC-1 surfaces and maximal surfaces*, Math. Nachr. **284**, 14–15 (2011), 1903–1918.

[20] M. Kokubu, M. Umehara and K. Yamada, *An elementary proof of Small's formula for null curves in* $\mathrm{PSL}(2, C)$ *and an analogue for Legendrian curves in* $\mathrm{PSL}(2, C)$, Osaka J. Math. **40** (2003), no. 3, 697–715.

[21] M. Kokubu, M. Umehara and K. Yamada, *Flat fronts in hyperbolic 3-space.* Pacific J. Math. **216** (2004), no.1, 149–175.

[22] Y. Ogata and M. Yasumoto, *Construction of discrete constant mean curvature surfaces in Riemannian spaceforms and applications*, Diff. Geom. Appl. **54** (2017), Part A, 264–281.

[23] W. Rossman, *Isothermic surfaces in Möbius and Lie sphere geometries*, Rokko Lec. Series Math. **22** (2014), 1–138.

[24] K. Teramoto, *Parallel and dual surfaces of cuspidal edges*, Diff. Geom. Appl. **44** (2016), 52–62.

[25] M. Umehara, K. Yamada, *Complete surfaces of constant mean curvature* 1 *in the hyperbolic 3-space*, Annals of Math. **137** (1993), 611–638.

[26] M. Umehara, K. Yamada, *Maximal surfaces with singularities in Minkowski space*, Hokkaido Math. J. **35** (2006), 13–40.

[27] W. Wunderlich, *Zur Differenzengeometrie der Flachen konstanter negativer Krummung*, Osterreich. Akad. Wiss. Math.-Nat. Kl. S.-B. IIa., **160** (1951), 39–77.

[28] M. Yasumoto, *Discrete maximal surfaces with singularities in Minkowski space*, Diff. Geom. Appl. **43** (2015), 130–154.

(W. Rossman, M. Yasumoto) *Department of mathematics, Faculty of science, Kobe University, Rokkodai-cho 1-1, Nada-ku, Kobe, 657-8501, Japan*

Current address, M. Yasumoto:
Osaka City University Advanced Mathematical Institute, 3-3-138, Sugimoto, Sumiyoshi-ku, Osaka, 558-8585, Japan

E-mail address, W. Rossman: `wayne@math.kobe-u.ac.jp`
E-mail address, M. Yasumoto: `myasu@math.kobe-u.ac.jp`

Advanced Studies in Pure Mathematics 78, 2018
Singularities in Generic Geometry
pp. 411–429

On pairs of geometric foliations on a cuspidal edge

Kentaro Saji

Abstract.

We study the topological configurations of the lines of principal curvature, the asymptotic and characteristic curves on a cuspidal edge, in the domain of a parametrization of this surface as well as on the surface itself. Such configurations are determined by the 3-jets of a parametrization of the surface.

§1. Introduction and preliminaries about cuspidal edges

A singular point x of a map $f : (\boldsymbol{R}^2, x) \to (\boldsymbol{R}^3, 0)$ is called a *cuspidal edge* if the map-germ f at x is $\mathcal{A}$-equivalent to $(u, v) \mapsto (u, v^2, v^3)$ at 0. (Two map-germs $f_1, f_2 : (\boldsymbol{R}^n, 0) \to (\boldsymbol{R}^m, 0)$ are *$\mathcal{A}$-equivalent* if there exist diffeomorphisms $S : (\boldsymbol{R}^n, 0) \to (\boldsymbol{R}^n, 0)$ and $T : (\boldsymbol{R}^m, 0) \to (\boldsymbol{R}^m, 0)$ such that $f_2 \circ S = T \circ f_1$.) If the singular point x of f is a cuspidal edge, then f at x is a front in the sense of [1] (see also [21]), and furthermore, they are one of two types of generic singularities of fronts (the other one is a *swallowtail* which is a singular point u of f satisfying that f at u is $\mathcal{A}$-equivalent to $(u, v) \mapsto (u, u^2v + 3u^4, 2uv + 4u^3)$ at 0).

It is shown in [22] that a cuspidal edge can locally be parametrized after smooth changes of coordinates in the source and isometries in the

Received May 31, 2016.
Revised January 23, 2017.
2010 *Mathematics Subject Classification.* Primary 53A05, 58F14; Secondary 58K05, 34A09, 53C03, 58F10, 58F18.
Key words and phrases. Cuspidal edge, principal configuration, lines of curvature.
Partly supported by the Japan Society for the Promotion of Science (JSPS) and the Coordenadoria de Aperfeiçoamento de Pessoal de Nível Superior under the Japan-Brazil research cooperative program and the Grant-in-Aid for Scientific Research (Young Scientists (B)) No. 26400087, from JSPS.

target by

$$f(u,v) = \left(u, a_1(u) + \frac{v^2}{2}, b_2(u) + v^2 b_3(u) + v^3 b_4(u,v)\right), \tag{1.1}$$

where $a_1(u), b_2(u), b_3(u), b_4(u,v)$ are C^∞-functions satisfying $a_1(0) = a_1'(0) = b_2(0) = b_2'(0) = b_3(0) = 0$, $b_2''(0) > 0$, $b_4(0,0) \neq 0$. Writing $a_1(u) = a_{20}u^2/2 + a_{30}u^3/6 + u^4 h_1(u)$, $b_2(u) = b_{20}u^2/2 + b_{30}u^3/6 + u^4 h_2(u)$, $b_3(u) = b_{12}u/2 + u^2 h_3(u)$, $b_4(u,v) = b_{03}/6 + uh_4(u) + vh_5(u,v)$, we have

$$\begin{aligned} f(u,v) = \Big(u, \frac{a_{20}}{2}u^2 + \frac{a_{30}}{6}u^3 + \frac{v^2}{2},& \\ \frac{b_{20}}{2}u^2 + \frac{b_{30}}{6}u^3 + \frac{b_{12}}{2}uv^2 + \frac{b_{03}}{6}v^3\Big) &+ h(u,v), \end{aligned} \tag{1.2}$$

where $b_{03} \neq 0$, $b_{20} \geq 0$ and

$$h(u,v) = \big(0,\ u^4 h_1(u), u^4 h_2(u) + u^2 v^2 h_3(u) + uv^3 h_4(u) + v^4 h_5(u,v)\big),$$

with $h_1(u), h_2(u), h_3(u), h_4(u), h_5(u,v)$ smooth functions. Several differential geometric invariants of cuspidal edges are investigated ([20, 22, 23, 25, 27, 28]), and coefficients of (1.2) are such invariants. According to [22], it is known that a_{20} coincides with the singular curvature κ_s, b_{20} coincides with the limiting normal curvature κ_ν, b_{03} coincides with the cuspidal curvature κ_c and b_{12} coincides with the cusp-directional torsion κ_t at the origin. The singular curvature is the geodesic curvature of the singular set with sign, and the limiting normal curvature is the normal curvature of the singular set, and they relates to the shape of cuspidal edge (see [28]). The cuspidal curvature measures the wideness of the cusp, and the cusp-directional torsion measures the rotating ratio of the cusp along the singular set (see [22, 23]).

On the other hand, let $U \subset \boldsymbol{R}^2$ be an open subset and (u,v) a coordinate system on U. Let

$$\omega = a(u,v)dv^2 + 2b(u,v)dudv + c(u,v)du^2 \tag{1.3}$$

be a 2-tensor on U, where a, b, c are smooth functions, called the *coefficients* of ω. We call $\omega = 0$ a *binary differential equation* (*BDE*) *corresponding to* ω. If $b^2 - ac > 0$ at $x \in U$, then $\omega(x) = 0$ defines two directions in T_xU, and integral curves of these directions for two smooth and transverse foliations, called *foliations with respect to* ω. If $b^2 - ac = 0$ at $x \in U$, then generically $\omega(x) = 0$ defines a single direction, and the integral curves form in general a family of cusps. Thus we are mainly interested in behavior of integral curves of a BDE near a point where $b^2 - ac$ vanishes. We call *discriminant* of a BDE the set where $b^2 - ac = 0$. If the single direction is transverse to the discriminant, then the BDE is equivalent to $dv^2 + udu^2 = 0$ ([7, 8]). The normal form for the stable cases when the single direction is tangent to the discriminant is obtained in [9, 10]. Topological classifications of generic families of BDEs are obtained in [3, 5, 6, 26, 29, 32]. On the other hand, BDEs as geometric foliations on surfaces in three space is studied in [2, 12, 15, 16, 17, 29]. See [11, 18, 19] for other approaches for geometric foliations.

In this paper, following [12, 13, 15, 26, 29], we stick to special BDEs from differential geometry of surface in $\boldsymbol{R}^3$. There are three fundamental BDEs on a regular surface in $\boldsymbol{R}^3$. Let $f : (\boldsymbol{R}^2, 0) \to (\boldsymbol{R}^3, 0)$ be a regular surface with a unit normal vector field ν. Let ω_{lc}, ω_{as} and ω_{ch} be 2-tensors defined by

$$\begin{aligned}
\omega_{lc} &= (EM - FL)dv^2 + (EN - GL)dudv + (FN - GM)du^2,\\
\omega_{as} &= N\,dv^2 + 2M\,dudv + L\,du^2,\\
\omega_{ch} &= \big(2M(GM - FN) - N(GL - EN)\big)\,dv^2\\
&\qquad + 2\big(M(GL + EN) - 2FLN\big)\,dudv\\
&\qquad\qquad + \big(L(GL - EN) - 2M(FL - EM)\big)\,du^2,
\end{aligned}$$

where (u, v) is a coordinate system on $(\boldsymbol{R}^2, 0)$, and

$$\begin{pmatrix} E & F \\ F & G \end{pmatrix} = \begin{pmatrix} \langle f_u, f_u\rangle & \langle f_u, f_v\rangle \\ \langle f_v, f_u\rangle & \langle f_v, f_v\rangle \end{pmatrix}, \quad \begin{pmatrix} L & M \\ M & N \end{pmatrix} = \begin{pmatrix} \langle f_{uu}, \nu\rangle & \langle f_{uv}, \nu\rangle \\ \langle f_{uv}, \nu\rangle & \langle f_{vv}, \nu\rangle \end{pmatrix},$$

where $\langle\ ,\ \rangle$ stands for the standard inner product of $\boldsymbol{R}^3$. Each integral curve of the foliations with respect to ω_{lc} is called a *line of curvature*, each integral curve with respect to ω_{as} is called an *asymptotic curve* and each integral curve with respect to ω_{ch} is called a *characteristic curve* or *harmonic mean curvature curve*. Asymptotic curves appear only on a domain where the Gaussian curvature K of f is non-negative, and characteristic curves appear only on a domain where the Gaussian curvature K of f is non-positive. Since $\omega_{ch} = 0$ can be deformed as

$$\begin{aligned}
&(NH - GK)\,dv^2 + 2(MH - FK)\,dudv + (LH - EK)\,du^2 = 0\\
\Leftrightarrow\quad & \frac{N\,dv^2 + 2M\,dudv + L\,du^2}{G\,dv^2 + 2F\,dudv + E\,du^2} = \frac{K}{H}\left(= \frac{2}{\kappa_1^{-1} + \kappa_2^{-1}}\right),
\end{aligned}$$

where K is the Gaussian curvature, H is the mean curvature, and κ_1, κ_2 are the principal curvatures of f, we see that along the characteristic curve, the normal curvature of it is equal to the harmonic mean of the principal curvatures (see [13], for example).

Acknowledgements. This paper is prepared while the author was visiting Luciana Martins at IBILCE - UNESP. He would like to thank Luciana Martins for fruitful discussions. He would also like to thank Farid Tari for valuable comments. He would also like to thank the referee for careful reading and helpful suggestions.

§2. Preliminaries on BDEs

In this section, following [5, 6], we introduce a method to study the configurations of the solution curves of a BDE. Let $\omega(u,v)$ be the 2-tensor on $(U;(u,v))\subset \boldsymbol{R}^2$ as in (1.3). If $(a,b,c)\neq(0,0,0)$ at $x\in U$, then ω is called of *Type* 1 at x, and if $(a,b,c)=(0,0,0)$ at $x\in U$, then ω is called of *Type* 2 at x. If ω is of Type 2 at x, then $\delta=b^2-ac$ has a critical point at x. Since we are interested in local behavior of ω, we set $x=(0,0)$. If ω is of Type 1, then ω defines a single direction at points on Δ, and if it is of Type 2, then all directions in the plane are solutions of $\omega=0$ at that point. Moreover, if ω is of Type 2 at x, then Δ is not a smooth curve. We are interested in the configurations of the foliations of $\omega=0$. We define the following equivalence.

Definition 2.1. Two binary differential equations $\omega_1=0$ and $\omega_2=0$ are *equivalent* if there exist a diffeomorphism germ $\Phi:(\boldsymbol{R}^2,0)\to(\boldsymbol{R}^2,0)$ and a non-zero function $\rho:(\boldsymbol{R}^2,0)\to\boldsymbol{R}$ such that $\rho\big(\Phi^*\omega_1\big)=\omega_2$ holds. If Φ is a homeomorphism such that Φ takes the integral curves of ω_1 to those of ω_2, they are called *topologically equivalent.*

If two binary differential equations are equivalent then the configurations of their foliations can be regarded the same. To obtain the topological configurations, we use the following method developed in [5, 6, 12, 15, 29, 30]. We separate our consideration into the following three cases:

- Case 1: $(a(0),b(0),c(0))\neq(0,0,0)$ and $\delta(0)\neq 0$ (Type 1).
- Case 2: $(a(0),b(0),c(0))\neq(0,0,0)$ and $\delta(0)=0$ (Type 1).
- Case 3: $(a(0),b(0),c(0))=(0,0,0)$ (Type 2).

Consider the associated surface to ω

$$\mathcal{M}=\{(u,v,[\alpha,\beta])\in(\boldsymbol{R}^2,0)\times\boldsymbol{RP}^1\mid a\beta^2+2b\alpha\beta+c\alpha^2=0\}.$$

Then $\mathcal{M}$ is a smooth manifold if $b^2-ac\neq 0$, or if $a_u+2b_up+c_up^2=0$ and $a_v+2b_vp+c_vp^2=0$ do not have any common root. The second condition is equivalent to

$$\det \begin{pmatrix} a_u & 2b_u & c_u & 0 \\ 0 & a_u & 2b_u & c_u \\ a_v & 2b_v & c_v & 0 \\ 0 & a_v & 2b_v & c_v \end{pmatrix}(0) \neq 0. \tag{2.1}$$

Consider the projection $\pi : \mathcal{M} \to \boldsymbol{R}^2$, $\pi(u, v, [\alpha : \beta]) = (u, v)$. Then $\pi^{-1}(u, v)$ consists of two points if $b^2 - ac > 0$, and is empty if $b^2 - ac < 0$. Let us set $p = \beta/\alpha$ (we need to consider the case $q = \alpha/\beta$ for some cases) and $\mathcal{F}(u, v, p) = ap^2 + 2bp + c$. If $\mathcal{F}_p(0) \neq 0$, then π is a local diffeomorphism, and if $\mathcal{F}_p(0) = 0$ and $\mathcal{F}_{pp}(0) \neq 0$ hold, then π is a fold (a map-germ $h : (\boldsymbol{R}^2, 0) \to (\boldsymbol{R}^2, 0)$ is a *fold* if h is $\mathcal{A}$-equivalent to $(u, v) \mapsto (u, v^2)$). Let us consider the vector field

$$\xi(u, v, p) = p\mathcal{F}_p(u, v, p)\partial_u + \mathcal{F}_p(u, v, p)\partial_v - \big(p\mathcal{F}_u(u, v, p) + \mathcal{F}_v(u, v, p)\big)\partial p.$$

Then ξ is tangent to $\mathcal{M}$, and the projections $d\pi(\xi_1), d\pi(\xi_2)$ satisfy

$$\omega(d\pi(\xi_i), d\pi(\xi_i)) = 0 \quad (i = 1, 2),$$

where $\xi_i = \xi(u, v, p_i)$ and $\pi^{-1}(u, v) = \{(u, v, p_1), (u, v, p_2)\}$, or $\pi^{-1}(u, v) = \{(u, v, p_1)\}$. To study geometric the foliations of ω, we use ξ on $\mathcal{M}$.

2.1. Case 1

We assume that $b^2 - ac > 0$ at 0. Then one can easily see that the BDE (1.3) is equivalent to a BDE $a\,dv^2 + 2b\,dudv + c\,du^2 = 0$ which satisfies $(a(0, 0), b(0, 0), c(0, 0)) = (1, 0, -1)$. Furthermore, it holds that for any $k > 0$, a BDE $a\,dv^2 + 2b\,dudv + c\,du^2 = 0$ which satisfies $(a(0, 0), b(0, 0), c(0, 0)) = (1, 0, -1)$ is equivalent to a BDE $a\,dv^2 + 2b\,dudv + c\,du^2 = 0$ whose k-jet of (a, b, c) at $(0, 0)$ is $(1, 0, -1)$, moreover, it is equivalent to a BDE

$$\omega_{reg} = dv^2 - du^2 = 0 \tag{2.2}$$

([6, Proposition 4.4]). Therefore the configuration is a pair of transverse smooth foliations.

2.2. Case 2

We consider the case 2, namely $(a(0), b(0), c(0)) \neq (0, 0, 0)$ and $\delta(0) = 0$.

Lemma 2.2. *We assume that* $(a(0), b(0), c(0)) \neq (0, 0, 0)$ *and* $\delta(0) = 0$. *If* ω *as in* (1.3) *satisfies*

$$j^1(a, b, c)(0, 0) = (0, 0, \alpha_0) + (\alpha_1 v, \alpha_2 v, \alpha_3 u + \alpha_4 v),$$

then it is equivalent to a BDE $a\,dv^2 + 2b\,dudv + c\,du^2 = 0$ *which satisfies* $(j^1a(0,0), j^1b(0,0), j^1c(0,0)) = (1,0,u)$ *when* $\alpha_0\alpha_1 \neq 0$.

Proof. We assume that $\alpha_0\alpha_1 \neq 0$. Consider $u = V + \beta_1U^2 + \beta_2V^2$, $v = U$ and $R = 1 + \beta_3U$, where $\beta_1, \beta_2, \beta_3 \in \boldsymbol{R}$. Then $R\omega(U,V)$ is given by

$$\big(\alpha_1U + O(2)\big)\,dV^2 + \big(2(\alpha_2 + 2\alpha_0\beta_1)U + O(2)\big)\,dUdV \\ + \big(\alpha_0 + (\alpha_4 + \alpha_0\beta_3)U + (\alpha_3 + 4\alpha_0\beta_2)V + O(2)\big)\,dU^2,$$

where $O(2)$ stands for remainders of order 2. Setting $\beta_1 = -\alpha_2/(2\alpha_0)$, $\beta_2 = -\alpha_3/(4\alpha_0)$ and $\beta_3 = -\alpha_4/\alpha_0$, and re-scaling, we get the desired result. Q.E.D.

Any BDE of the form (1.3) with $(j^1a(0,0), j^1b(0,0), j^1c(0,0)) = (1,0,u)$ is smoothly equivalent to

$$\omega_{cusp} = dv^2 + u\,du^2 = 0 \tag{2.3}$$

([8], see also [6, Section 4.2], [30, Proposition 3.3-2]). The solutions form a family of cusps.

2.3. Case 3

We assume $(a(0), b(0), c(0)) = (0,0,0)$. In this case, δ has a critical point at 0. We assume $\mathcal{M}$ is a smooth manifold. Since $\mathcal{F}_p(0,0,p) = 0$ holds, ξ has a zero at $(0,0,p)$ if and only if $p\mathcal{F}_u(0,0,p) + \mathcal{F}_v(0,0,p) = 0$. This is a cubic equation for p. Set

$$\varphi_\omega(p) = \varphi(p) = p\mathcal{F}_u(0,0,p) + \mathcal{F}_v(0,0,p). \tag{2.4}$$

Let $D_\omega = D$ denotes the discriminant of this equation. If $D > 0$ then $\varphi(p) = 0$ has three distinct real roots, and if $D < 0$ then $\varphi(p) = 0$ has one real and two distinct imaginary roots.

When $D > 0$, let p_1, p_2, p_3 be the solutions of $\varphi(p) = 0$ and $p_1 < p_2 < p_3$. When $D < 0$, let p_1 be the solution of $\varphi(p) = 0$. If $\varphi(0) = \mathcal{F}_v(0,0,0) = a_v(0) \neq 0$ then $p_i \neq 0$ ($i = 1,2,3$ or $i = 1$) holds. We need to understand the singularity of ξ near p_i. If $\mathcal{F}_u(0,0,p_i) \neq 0$, then $\mathcal{M}$ is parameterized by v as $(u(v,p), v, p)$ near $(0,0,p_i)$. We denote the linear part of ξ by

$$j^1\xi(0,0,p_i) = \big(\xi_{11}v + \xi_{12}(p-p_i)\big)\partial v + \big(\xi_{21}v + \xi_{22}(p-p_i)\big)\partial p.$$

Also we remark that since $\mathcal{F}(u(v,p), v, p) \equiv 0$, it holds that $\mathcal{F}_uu_v + \mathcal{F}_v \equiv 0$, where $\equiv$ means that the equality holds identically. On the other hand, p_i is a solution of $\varphi(p) = 0$, and $\varphi(p) = p\mathcal{F}_u + \mathcal{F}_v$, we have

$$u_v(0,p_i) = p_i.$$

Furthermore, $\mathcal{F}_p = 0$ at $(u,v) = (0,0)$, it holds that $u_p = 0$ at $(v,p) = (0,p_i)$. We have

$$\begin{aligned} \xi_{11} &= \mathcal{F}_{up}u_v + \mathcal{F}_{pv} \\ \xi_{12} &= \mathcal{F}_{up}u_p + \mathcal{F}_{pp} \\ &= 0 \quad (\text{at } (0,0,p_i)) \\ \xi_{22} &= -(\mathcal{F}_u + p\mathcal{F}_{uu}u_p + p\mathcal{F}_{up} + \mathcal{F}_{uv}u_p + \mathcal{F}_{vp}) \end{aligned}$$

Thus the eigenvalues of the linear part of ξ are
(2.5)
$\alpha(p_i) = \mathcal{F}_{up}u_v + \mathcal{F}_{pv}$ and $-\varphi'(p_i) = -(\mathcal{F}_u + p\mathcal{F}_{uu}u_p + p\mathcal{F}_{up} + \mathcal{F}_{uv}u_p + \mathcal{F}_{vp})$.

The configuration of the integral curves of ω is determined by these information. The following theorem is known. Let $\det \operatorname{Hess} \delta(0,0)$ be the determinant of the Hesse matrix of $\delta(u,v)$ at $(0,0)$.

Theorem 2.3. [5, Theorem 4.1] *Let ω be a 2-tensor as in (1.3) and satisfies $(a,b,c)(0) = (0,0,0)$, $\det \operatorname{Hess} \delta(0,0) < 0$, $D \neq 0$ and $\varphi(p)$ and $\alpha(p)$ do not have any common roots. Then the BDE $\omega = 0$ is topologically equivalent to one of the following BDEs:*

- *The case $D > 0$: (Then $\varphi(p) = 0$ has 3 roots p_1, p_2, p_3.)*
  - *$\omega_{3s} = vdv^2 + 2ududv + vdu^2 = 0$ (3 saddles) when $-\varphi'(p_i)\alpha(p_i)$ are negative for all $i = 1,2,3$.*
  - *$\omega_{2s1n} = vdv^2 + 2(v-u)dudv + vdu^2 = 0$ (2 saddles + 1 node) when $-\varphi'(p_i)\alpha(p_i)$ are two negative and one positive for $i = 1,2,3$.*
  - *$\omega_{1s2n} = 3vdv^2 - 4ududv + 3vdu^2 = 0$ (1 saddle + 2 nodes) when $-\varphi'(p_i)\alpha(p_i)$ are one negative and two positive for $i = 1,2,3$.*
- *The case $D < 0$: (Then $\varphi(p) = 0$ has 1 root p_1.)*
  - *$\omega_{1s} = vdv^2 + 2ududv + vdu^2 = 0$ (1 saddle) when $-\varphi'(p_1)\alpha(p_1)$ is positive.*
  - *$\omega_{1n} = 2vdv^2 - ududv + 2vdu^2 = 0$ (1 node) when $-\varphi'(p_1)\alpha(p_1)$ is negative.*

Note that in the case of $D > 0$, all $-\varphi'(p_i)\alpha(p_i)$ $i = 1,2,3$ cannot be positive, see [5]. The integral curves of the above BDEs are in Figures 1, 2, 3 which are taken from [5].

§3. Geometric binary differential equations on a cuspidal edge

Let $f : (\boldsymbol{R}^2, 0) \to (\boldsymbol{R}^3, 0)$ be a parametrization of a cuspidal edge. We take f as in (1.2). Then the coefficients of the first fundamental

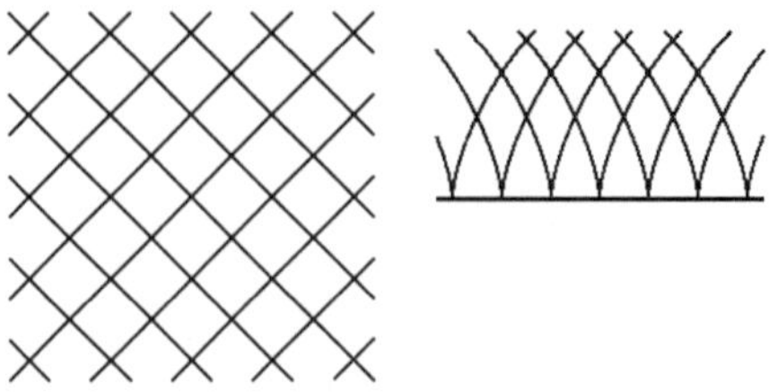

Fig. 1. Integral curves of ω_{reg} and ω_{cusp}.

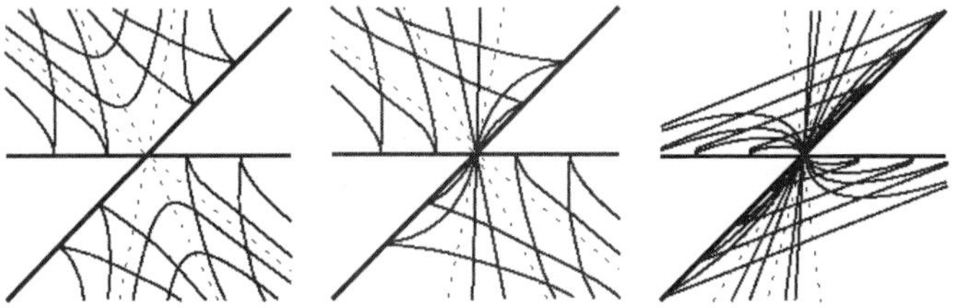

Fig. 2. Integral curves of ω_{3s}, ω_{2s1n} and ω_{1s2n}. Here and in the rest of the paper, the dashed curves are separatrices, i.e., they are integral curves that separate distinct topological sectors.

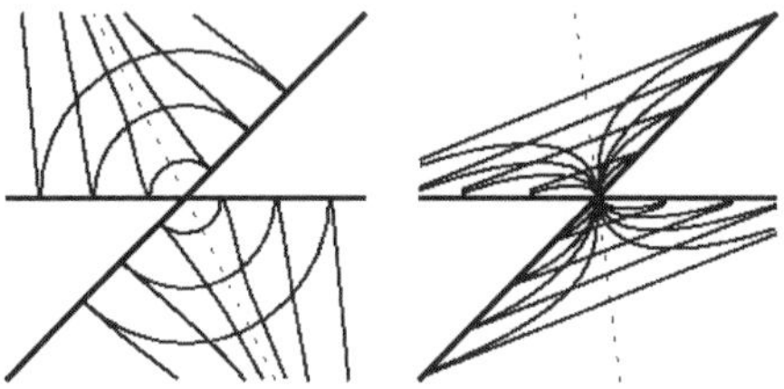

Fig. 3. Integral curves of ω_{1s} and ω_{1n}.

form of the cuspidal edge with respect to f are

$$
\begin{aligned}
E \;=\;& 1+(a_{20}^2+b_{20}^2)u^2+(a_{20}a_{30}+b_{20}b_{30})u^3+b_{12}b_{20}uv^2\\
&+\frac{1}{4}\Big(a_{30}^2+b_{30}^2+2a_{20}p(0)+2b_{20}q_1(0)\Big)u^4\\
&+\frac{1}{2}\Big(b_{12}b_{30}+8b_{20}q_2(0)\Big)u^2v^2+2b_{20}q_3(0)uv^3\\
&+\frac{1}{4}b_{12}^2v^4+O(5),
\end{aligned}
\tag{3.1}
$$

$$
(3.2)\quad
\begin{aligned}
F \;=\;& a_{20}uv + \left(\frac{1}{2}a_{30} + b_{12}b_{20}\right)u^2v + \frac{1}{2}b_{03}b_{20}uv^2\\
&+ \frac{1}{4}b_{03}b_{12}v^4 + \left(\frac{1}{2}b_{12}^2 + 4b_{20}q_4(0,0)\right)uv^3\\
&+ \frac{1}{4}\Big(b_{03}b_{30} + 12b_{20}q_3(0)\Big)u^2v^2\\
&+ \left(\frac{1}{2}b_{12}b_{30} + 4p(0) + 2b_{20}q_2(0)\right)u^3v + O(5),
\end{aligned}
$$

$$
(3.3)\quad G \;=\; v^2 + \frac{1}{4}b_{03}^2v^4 + b_{03}b_{12}uv^3 + b_{12}^2u^2v^2 + O(5),
$$

where $O(n)$ stands for remainders of order n $(n = 1, 2, \ldots)$. Since f is as in (1.2), we see $f_u \times (f_v/v) \neq 0$. We set $\nu_2 = f_u \times (f_v/v)$, and $L_2 = \langle f_{uu}, \nu_2\rangle$, $M_2 = \langle f_{uv}, \nu_2\rangle$, $N_2 = \langle f_{vv}, \nu_2\rangle$. Then we have:

$$
(3.4)\quad
\begin{aligned}
L_2 \;=\;& b_{20} + (b_{30} - a_{20}b_{12})u - \frac{1}{2}a_{20}b_{03}v\\
&- \big(a_{30}b_{12} - 12q_1(0) + 2a_{20}q_2(0)\big)u^2\\
&- \big(a_{30}b_{03} + 6a_{20}q_3(0)\big)uv\\
&+ \big(2q_2(0) - 4a_{20}q_4(0,0)\big)v^2 + O(3)\\
M_2 \;=\;& b_{12}v + 4q_2(0)uv + 3q_3(0)v^2 + O(3)\\
N_2 \;=\;& \frac{1}{2}b_{03}v + 3q_3(0)uv + 8q_4(0,0)v^2 + O(3).
\end{aligned}
$$

We have $L_2 = L|\nu_2|$, $M_2 = M|\nu_2|$, $N_2 = N|\nu_2|$. It should be remarked that there exist C^∞-functions $\widetilde{F}, \widetilde{G}, \widetilde{N}, \widetilde{M}$ such that $G = v^2\widetilde{G}$, $F = v\widetilde{F}$, $N_2 = v\widetilde{N}$ and $M_2 = v\widetilde{M}$ holds. We set

$$
(3.5)\quad \widetilde{E} = E,\ \widetilde{F} = \frac{F}{v},\ \widetilde{G} = \frac{G}{v^2},\ \widetilde{L} = L_2,\ \widetilde{M} = \frac{M_2}{v},\ \widetilde{N} = \frac{N_2}{v}.
$$

3.1. Lines of principal curvature

In this subsection we consider the BDE $\omega_{lc} = 0$. Using (3.5), $\omega_{lc} = 0$ is equivalent to

$$
v^2\left(\widetilde{F}\widetilde{N} - v\widetilde{G}\widetilde{M}\right)dv^2 + v\left(\widetilde{E}\widetilde{N} - v^2\widetilde{G}\widetilde{L}\right)dudv + v\left(\widetilde{E}\widetilde{M} - \widetilde{F}\widetilde{L}\right)du^2 = 0.
$$

To determine the topological configuration of $\omega_{lc} = 0$, we factor out v and consider $\widetilde{\omega}_{lc} = 0$, where

$$
\widetilde{\omega}_{lc} = v\left(\widetilde{F}\widetilde{N} - v\widetilde{G}\widetilde{M}\right)dv^2 + \left(\widetilde{E}\widetilde{N} - v^2\widetilde{G}\widetilde{L}\right)dudv + \left(\widetilde{E}\widetilde{M} - \widetilde{F}\widetilde{L}\right)du^2.
$$

We have the following proposition.

Proposition 3.1. *The BDE* $\widetilde{\omega}_{lc} = 0$ *is equivalent to the BDE* $\omega_{reg} = 0$.

This proposition implies that the lines of principal curvature of a cuspidal edge form a pair of smooth and transverse foliations in the domain of a parametrization.

Proof of Proposition 3.1. Set

$$(a, b, c) = \left(v\widetilde{F}\widetilde{N} - v^2\widetilde{G}\widetilde{M},\ \frac{1}{2}\left(\widetilde{E}\widetilde{N} - v\widetilde{G}\widetilde{L}\right),\ \widetilde{E}\widetilde{M} - \widetilde{F}\widetilde{L} \right).$$

Then since $b(0) = \widetilde{E}(0)\widetilde{N}(0)$, $\widetilde{E}(0) = \langle f_u, f_u \rangle (0) \neq 0$ and $\widetilde{N}(0) = \langle f_{vvv}, \nu_2 \rangle (0) = b_{03} \neq 0$ hold, $\widetilde{\omega}_{lc}$ is as in Case 1. Moreover, since $a(0) = 0$ and $b(0) \neq 0$, we see that $b^2 - ac > 0$ at 0. Hence $\widetilde{\omega}_{lc}$ is equivalent to $\omega_{reg} = 0$ (See Section 2.1). Q.E.D.

The fact of the existence of the curvature line coordinate system at cuspidal edge is also shown in [24].

An example of picture of this configuration on the cuspidal edges is in Figures 4. Since one family of the integral curves are tangent to the null direction on singular curve, one family of the integral curves near singular curve form the (2, 3)-cusps. A map-germ $(\boldsymbol{R}, 0) \to (\boldsymbol{R}^3, 0)$ is an $(2, 3)$-*cusp* if it is $\mathcal{A}$-equivalent to $t \mapsto (t^2, t^3)$.

Fig. 4. Integral curves of $\widetilde{\omega}_{lc}$ on images of cuspidal edges.

3.2. Asymptotic curves

In this subsection we consider $\omega_{as} = 0$, and it is equivalent to $\widetilde{\omega}_{as} = N_2\, dv^2 + 2M_2\, dudv + N_2\, du^2 = 0$. Since $M_2(u, 0) = 0$, and $N_2(u, 0) = 0$, the singular set i.e., the cuspidal edge curve is part of discriminant of $\widetilde{\omega}_{as}=0$, and, ∂_v is a solution to $\widetilde{\omega}_{as}=0$ on the singular set. By (3.4),

$$\begin{aligned}\delta(u,v) &= -\frac{1}{2}b_{20}b_{03}v+\left(\frac{1}{4}a_{20}b_{03}^2+b_{12}^2-8b_{20}q_4(0,0)\right)v^2\\ &\quad+\frac{1}{2}b_{03}\Big(a_{20}b_{12}-b_{30}-2b_{20}q_3(0)\Big)uv\\ &\quad+\Big(-b_{03}q_2(0)+6b_{12}q_3(0)\\ &\qquad\qquad+6a_{20}b_{03}q_4(0,0)-7b_{20}(q_4)_v(0,0)\Big)v^3\\ &\quad+\left(\frac{1}{4}a_{30}b_{03}^2+3a_{20}b_{03}q_3(0)-8b_{30}q_4(0,0)\right.\\ &\qquad\qquad\left.+8b_{12}q_2(0)+8a_{20}b_{12}q_4(0,0)\right)uv^2\\ &\quad+\left(\frac{1}{2}a_{30}b_{03}b_{12}-6b_{03}q_1(0)+a_{20}b_{03}q_2(0)\right.\\ &\qquad\qquad\left.+3a_{20}b_{12}q_3(0)-3b_{30}q_3(0)\right)u^2v+O(4).\end{aligned}\tag{3.6}$$

Thus if $b_{20}\neq 0$, then the BDE is in Case 2. In this case, by (3.4), $\omega_{as}=0$ is equivalent to $\omega_{cusp}=0$ (see Subsection 2.2). By [28, Corollary 3.6], $b_{20}\neq 0$ implies that the Gaussian curvature is unbounded and changes sign between the two sides of the cuspidal edge. This means that in this case, the singular set of cuspidal edge plays the same role as the parabolic curve on regular surfaces. Since $b_{20}\neq 0$, then the BDE is $\omega_{cusp}=0$, this implies that the folded saddle, the folded node, the folded focus in Davydov's classification [9] (see also [30, Section 3.2]) does not appear not only cuspidal edges, but also all singularities written in the form (1.1) (for instance, cuspidal cross cap).

We assume now $b_{20}=0$. Then the BDE is in Case 3. We have the following. If $a_{20}b_{12}-b_{30}\neq 0$, then δ is a Morse function near 0. We study the geometric foliation near 0 as in Subsection 2.3. We consider

$$\mathcal{F}(u,v,p)=N_2+2M_2p+L_2p^2.$$

Then we have

$$\begin{aligned}(\mathcal{F}_u,\mathcal{F}_v,\mathcal{F}_p)(u,v,p) &= \big((N_2)_u+2(M_2)_up+(L_2)_up^2,\\ &\qquad (N_2)_v+2(M_2)_vp+(L_2)_vp^2,2M_2+2L_2p\big),\\ (\mathcal{F}_u,\mathcal{F}_v,\mathcal{F}_p)(0,0,p) &= \big((L_2)_up^2,(N_2)_v+2(M_2)_vp+(L_2)_vp^2,2Lp\big)\\ &=\left((b_{30}-a_{20}b_{12})p^2,\frac{1}{2}(b_{03}+4b_{12}p-a_{20}b_{03}p^2),0\right).\end{aligned}$$

In this case, the left hand side of (2.1) is $(b_{30}-a_{20}b_{12})^2b_{03}^2/4$. Thus if $b_{30}-a_{20}b_{12}\neq 0$ at 0, then $\mathcal{M}=\{\mathcal{F}=0\}$ is a smooth manifold. We have $\varphi_{as}=\varphi_{\omega_{as}}$ and $D_{as}=D_{\omega_{as}}$ defined in Subsection 2.3 as follows

$$\begin{aligned}\varphi_{as}(p) &= (b_{30}-a_{20}b_{12})p^3-\frac{1}{2}a_{20}b_{03}p^2+2b_{12}p+\frac{1}{2}b_{03},\\ 4D_{as} &= a_{20}^3b_{03}^4+13a_{20}^2b_{03}^2b_{12}^2-b_{30}(128b_{12}^3+27b_{03}^2b_{30})\\ &\quad +2a_{20}(64b_{12}^4+9b_{03}^2b_{12}b_{30}).\end{aligned}$$

Furthermore, $\alpha(p)$ is given by

$$\alpha(p)=2(b_{30}-a_{20}b_{12})p^2-a_{20}b_{03}p+2b_{12}.$$

If p_i is a solution of $\varphi_{as}(p)=0$ and $p\alpha(p)-2\varphi_{as}(p)=-b_{03}-2b_{12}p$ holds, then $\alpha(p_i)\neq 0$ if and only if $-b_{03}-2b_{12}p_i\neq 0$. Assume that $b_{12}\neq 0$. Substituting $p=-b_{03}/(2b_{12})$ into $\varphi_{as}(p)$, we get

$$\varphi_{as}\left(\frac{-b_{03}}{2b_{12}}\right)=-\frac{1}{8b_{12}^3}b_{03}(4+b_{03}^2b_{30}).$$

If $b_{12}=0$, then $p\alpha(p)-2\varphi_{as}(p)=-b_{03}\neq 0$. Since we assume that $b_{30}-a_{20}b_{12}\neq 0$, if $b_{12}=0$ then $b_{30}\neq 0$. Thus we get $\alpha(p_i)\neq 0$ if and only if

$$4b_{12}^3+b_{03}^2b_{30}\neq 0.$$

We can now use Theorem 2.3 to obtain the following result.

Proposition 3.2. *If $b_{20}\neq 0$, then ω_{as} is equivalent to $\omega_{cusp}=0$. If $b_{20}=0$, $b_{30}-a_{20}b_{12}\neq 0$, $D_{as}\neq 0$ and $4b_{12}^3+b_{03}^2b_{30}\neq 0$, then ω_{as} is topologically equivalent to one of the following:*

- *The case $D_{as}>0$: (Then $\varphi_{as}(p)=0$ has 3 roots p_1,p_2,p_3)*
  - ω_{3s} *($-\varphi'_{as}(p_i)\alpha(p_i)$ are negative for all $i=1,2,3$).*
  - ω_{2s1n} *($-\varphi'_{as}(p_i)\alpha(p_i)$ are two negative and one positive for $i=1,2,3$).*
  - ω_{1s2n} *($-\varphi'_{as}(p_i)\alpha(p_i)$ are one negative and two positive for $i=1,2,3$).*
- *The case $D_{as}<0$: (Then $\varphi_{as}(p)=0$ has 1 root p_1)*
  - ω_{1s} *($-\varphi'_{as}(p_1)\alpha(p_1)$ is negative).*
  - ω_{1n} *($-\varphi'_{as}(p_1)\alpha(p_1)$ is positive).*

Remark 3.3. We observe that by the Proposition 3.2, b_{20}, $b_{30}-a_{20}b_{12}$ and $4b_{12}^3+b_{03}^2b_{30}$ have geometric meanings. In fact, b_{20} is the limiting normal curvature and $b_{30}-a_{20}b_{12}$ coincides with the derivation of limiting normal curvature (see [23]). The invariant $4b_{12}^3+b_{03}^2b_{30}$ is related to the singularities of parallel surfaces of the cuspidal edge (see [31, Lemma 3.1]).

3.3. Characteristic curves

We consider the BDE $\omega_{ch} = 0$. Using (3.5), we show that $\omega_{ch} = 0$ is equivalent to $v(a\,dv^2 + 2b\,dudv + c\,du^2) = 0$, where

$$\begin{aligned} a &= v\Big(\widetilde{E}\widetilde{N}^2 + \big(- \widetilde{G}\widetilde{L}\widetilde{N} - 2\widetilde{F}\widetilde{M}\widetilde{N}\big)v + 2\widetilde{G}\widetilde{M}^2v^2\Big) \\ b &= v\big(- 2\widetilde{F}\widetilde{L}\widetilde{N} + \widetilde{E}\widetilde{M}\widetilde{N} + \widetilde{G}\widetilde{L}\widetilde{M}v\big) \\ c &= -\widetilde{E}\widetilde{L}\widetilde{N} + \big(\widetilde{G}\widetilde{L}^2 - 2\widetilde{F}\widetilde{L}\widetilde{M} + 2\widetilde{E}\widetilde{M}^2\big)v. \end{aligned}$$

We factor out v, so $\omega_{ch} = 0$ is topologically equivalent to $\tilde{\omega}_{ch} = a\,dv^2 + 2b\,dudv + c\,du^2 = 0$. Since $a(u,0) = 0$, and $b(u,0) = 0$, the singular set is a part of discriminant of $\tilde{\omega}_{ch} = 0$, and ∂_v is a solution to $\tilde{\omega}_{ch} = 0$ on the singular set. The function $\delta = b^2 - ac$ is given by

$$\begin{aligned} v\Big[&\widetilde{E}^2\widetilde{L}\widetilde{N}^3 + \big(4\widetilde{F}^2\widetilde{L}^2 - 2\widetilde{E}\widetilde{G}\widetilde{L}^2 - 4\widetilde{E}\widetilde{F}\widetilde{L}\widetilde{M} - 2\widetilde{E}^2\widetilde{M}^2\big)\widetilde{N}^2v \\ &+\big(- \widetilde{G}^2\widetilde{L}^3 - 4\widetilde{F}\widetilde{G}\widetilde{L}^2\widetilde{M} - 4\widetilde{F}^2\widetilde{L}\widetilde{M}^2 + 6\widetilde{E}\widetilde{G}\widetilde{L}\widetilde{M}^2 + 4\widetilde{E}\widetilde{F}\widetilde{M}^3\big)\widetilde{N}v^2 \\ &+\big(\widetilde{G}\widetilde{L}^2 - 4\widetilde{F}\widetilde{L}\widetilde{M} + 4\widetilde{E}\widetilde{M}^2\big)\widetilde{G}\widetilde{M}^2v^3\Big]. \end{aligned}$$

When f is taken as in (1.2), we have

$$\begin{aligned} a &= \frac{1}{4}b_{03}^2v + O(2), \\ b &= \frac{1}{2}b_{12}b_{03}v + O(2), \\ c &= -\frac{1}{2}b_{20}b_{03} + \frac{1}{2}\big(a_{20}b_{12}b_{03} - b_{30}b_{03} - 6b_{20}q_3(0)\big)u \\ &\qquad + \frac{1}{4}\Big[a_{20}b_{03}^2 + 8b_{12}^2 + 4b_{20}\big(b_{20} - 8q_4(0,0)\big)\Big]v + O(2), \\ \delta &= \frac{1}{8}b_{03}^3b_{20}v + \frac{1}{8}b_{03}^2\big(- a_{20}b_{03}b_{12} + b_{03}b_{30} + 18b_{20}q_3(0)\big)uv \\ &\quad - \frac{1}{16}b_{03}^2\Big[a_{20}b_{03}^2 + 4\Big[b_{12}^2 + 2b_{20}\big(b_{20} - 12q_4(0,0)\big)\Big]\Big]v^2 + O(3). \end{aligned}$$

Since $b_{03} \neq 0$, if $b_{20} \neq 0$, then $\widetilde{\omega}_{ch}$ is as in Case 2, and if $b_{20} = 0$, then it is as in Case 3. In the following, we divide our consideration into these two cases.

The case $b_{20} \neq 0$: By the argument in Subsection 2.2, $\widetilde{\omega}_{ch} = 0$ is equivalent to $\omega_{cusp} = 0$. In this case, by (3.4), $\omega_{as} = 0$ is equivalent to $\omega_{cusp} = 0$ (see Subsection 2.2). Like as the case of $\omega_{as} = 0$, the singular set of cuspidal edge plays the same role as the parabolic curve on regular surfaces. Moreover, the folded saddle, the folded node, the folded focus do not appear.

The case $b_{20}=0$: The left hand side of (2.1) is $b_{03}^6(a_{20}b_{12}-b_{30})^2/64$. Thus $\mathcal{M}$ is a smooth manifold if $a_{20}b_{12}-b_{30}\neq 0$ at 0. Furthermore, δ is of Morse type if and only if $a_{20}b_{12}-b_{30}\neq 0$. This is exactly the same conditions as the case of asymptotic curves. We assume that $a_{20}b_{12}-b_{30}\neq 0$. We need to consider $\mathcal{F}(u,v,p)=a+2bp+cp^2$ and $\varphi_{ch}(p)=(p\mathcal{F}_u+\mathcal{F}_v)(0,0,p)$. We have

$$
\begin{aligned}
4\varphi_{ch}(p)=\ &b_{03}^2+4b_{03}b_{12}p\\
&+(a_{20}b_{03}^2+8b_{12}^2)p^2+(2a_{20}b_{03}b_{12}-2b_{03}b_{30})p^3,
\end{aligned}
\tag{3.7}
$$

and the discriminant $D_{ch}=D_{\omega_{ch}}$ of the cubic φ_{ch} is given by

$$
\begin{aligned}
D_{ch}=-\frac{b_{03}^2}{64}\Big(&a_{20}^3b_{03}^6+11a_{20}^2b_{03}^4b_{12}^2\\
&-2a_{20}(16b_{03}^2b_{12}^4+9b_{03}^4b_{12}b_{30})\\
&\qquad+256b_{12}^6+160b_{03}^2b_{12}^3b_{30}+27b_{03}^4b_{30}^2\Big).
\end{aligned}
\tag{3.8}
$$

Furthermore, 4α is given by

$$
4\alpha(p)=2b_{03}(a_{20}b_{12}-b_{30})p^2+(a_{20}b_{03}^2+8b_{12}^2)p+2b_{12}b_{03}.
$$

Then we have

$$
\begin{aligned}
4\varphi'_{ch}(p)\alpha(p)=\ &4b_{03}^2b_{12}^2+4b_{03}b_{12}(a_{20}b_{03}^2+8b_{12}^2)p\\
&+(a_{20}^2b_{03}^4+26a_{20}b_{03}^2b_{12}^2+64b_{12}^4-10b_{03}^2b_{12}b_{30})p^2\\
&+5b_{03}(a_{20}b_{03}^2+8b_{12}^2)(a_{20}b_{12}-b_{30})p^3\\
&+6b_{03}^2(-a_{20}b_{12}+b_{30})^2p^4,
\end{aligned}
\tag{3.9}
$$

and the condition that $\varphi_{ch}(p)$ and $\alpha(p)$ do not have any common roots is given by $4b_{12}^3+b_{03}^2b_{30}\neq 0$. We summerize the above discussion in the following proposition.

Proposition 3.4. *If* $b_{20}\neq 0$, *then* ω_{ch} *is equivalent to* $\omega_{cusp}=0$.
If $b_{20}=0$, $b_{30}-a_{20}b_{12}\neq 0$, $D_{ch}\neq 0$ *and* $4b_{12}^3+b_{03}^2b_{30}\neq 0$, *then* ω_{ch} *is topologically equivalent to one of the following*:

- *The case* $D_{ch}>0$: *(Then* $\varphi_{ch}(p)=0$ *has 3 roots* p_1,p_2,p_3*)*
  - ω_{3s} *(*$-\varphi'_{ch}(p_i)\alpha(p_i)$ *are negative for all* $i=1,2,3$*).*
  - ω_{2s1n} *(*$-\varphi'_{ch}(p_i)\alpha(p_i)$ *are two negative and one positive for* $i=1,2,3$*).*
  - ω_{1s2n} *(*$-\varphi'_{ch}(p_i)\alpha(p_i)$ *are one negative and two positive for* $i=1,2,3$*).*
- *The case* $D_{ch}<0$: *(Then* $\varphi_{ch}(p)=0$ *has 1 root* p_1*)*
  - ω_{1s} *(*$-\varphi'_{ch}(p_1)\alpha(p_1)$ *is negative).*
  - ω_{1n} *(*$-\varphi'_{ch}(p_1)\alpha(p_1)$ *is positive).*

Remark 3.5. We remark that if $b_{12} = 0$, then $D_{as} = D_{ch}$. Namely, the configrations of foliations with respect to ω_{as} and ω_{ch} are of the same type.

Examples of pictures of these configrations on the cuspidal edges are in Figures 5, 6 and 7. Since the integral curves emanate from singular curve along the null direction, integral curves near the singular curve do not form the $(2,3)$-cusp but form the $(3,4)$-cusps (see Appendix A).

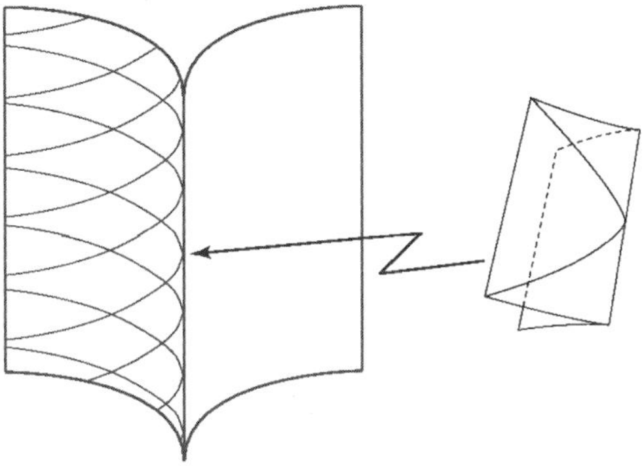

Fig. 5. Integral curves of ω_{cusp} on images of cuspidal edges.

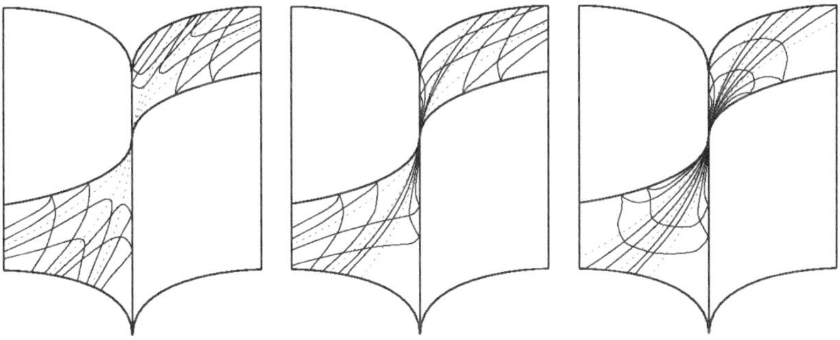

Fig. 6. Integral curves of ω_{3s}, ω_{2s1n} and ω_{1s2n} on images of cuspidal edges.

§4. Generic foliations

In Propositions 3.2 and 3.4, all the conditions are written in terms of the 3-jet of (1.2). We can state a genericity result for cuspidal edge.

Fig. 7. Integral curves of ω_{1s} and ω_{1n} on images of cuspidal edges.

By (1.2), we identify the set of jets of parametrization of cuspidal edges with $(0,0)$

$$\mathcal{C}_k = \Big\{ \big(j^k a_1(0), j^k b_2(0), j^k b_3(0), j^k b_4(0,0)\big) \in J^k(1,1)^3 \times J^k(2,1) \,\Big|$$

$$a_1(0) = a_1'(0) = b_2(0) = b_2'(0) = b_3(0) = 0,\ b_2'(0) > 0,\ b_4(0,0) \neq 0 \Big\},$$

for $k \geq 3$. With notation as before, consider

$$(a_{20}, a_{30}, \ldots, b_{20}, b_{30}, \ldots, b_{12}, \ldots, b_{03}, \ldots)$$

as a coordinate system of $\mathcal{C}_k$ (cf. (1.2)). Define a subset of $\mathcal{C}_k$ by

$$\begin{aligned}\mathcal{N} = \{b_{20} = 0,\ b_{30} - a_{20}b_{12} = 0\} \cup \{b_{20} = 0,\ D_{asy} = 0\} \\ \cup \{b_{20} = 0,\ 4b_{12}^3 - b_{03}^2 b_{30} = 0\} \cup \{b_{20} = 0,\ D_{hmc} = 0\}.\end{aligned}$$

Then $\mathcal{N}$ is an algebaric subset of $\mathcal{C}_k$ of codimension 2. Since the singular set of a cuspidal edge is a curve, generically it will avoid the set $\mathcal{N}$. This implies that for generic cuspidal edges the configuration of $\omega_{lc}, \omega_{as}, \omega_{ch}$ are those in Propositions 3.1, 3.2, 3.4.

§Appendix A. Criteria for (3, 4) and (3, 4, 5)-cusp

In this section, we state criteria for $(3,4)$ and $(3,4,5)$-cusp. Set

$$c_1(t) = (t^3, t^4, 0, \ldots, 0), \quad c_2(t) = (t^3, t^4, t^5, 0, \ldots, 0),$$

and a map-germ $\gamma : (\boldsymbol{R}, 0) \to (\boldsymbol{R}^n, 0)$, where $n \geq 2$ (respectively, $n \geq 3$) is called $(3,4)$-*cusp* (respectively, $(3,4,5)$-*cusp*) if γ is $\mathcal{A}$-equivalent to the map-germ c_1 (respectively, c_2) at 0.

Proposition A.1. *A map-germ* $\gamma : (\boldsymbol{R}, 0) \to (\boldsymbol{R}^n, 0)$, *where* $n \geq 3$ *(respectively,* $n \geq 2$*) is* $(3,4,5)$*-cusp (respectively,* $(3,4)$*-cusp) if and only if*

(i) $\gamma'(0) = \gamma''(0) = 0$,

(ii) $\gamma^{(3)}(0), \gamma^{(4)}(0)$ *and* $\gamma^{(5)}(0)$ *are linearly independent (respectively, linearly dependent, and* $\gamma^{(3)}(0)$ *and* $\gamma^{(4)}(0)$ *are linearly independent) where* $(\ )^{(i)} = d^i/dt^i$.

Although this proposition is known [4, 14], we give a sketch of proof for the readers who are not familiar with it.

Proof. Since

$$\Phi(\gamma(t))^{(i)} = d\Phi_0(\gamma^{(i)}(0)) \quad (i = 3, 4, 5)$$

holds for a map $\Phi : (\boldsymbol{R}^n, 0) \to (\boldsymbol{R}^n, 0)$ under the assumption $\gamma'(0) = \gamma''(0) = 0$, it is obvious that the conditions do not depend on the parameter and the coordinate system on $\boldsymbol{R}^n$.

To show the proposition, it is enough to show that $(t^3 + a_1 t^5, t^4 + a_2 t^5, a_3 t^5, 0, \dots, 0) + O(6)$ is $\mathcal{A}$-equivalent to $(t^3, t^4, a_3 t^5, 0, \dots, 0) + O(6)$, where $a_1, a_2, a_3 \in \boldsymbol{R}$. Considering the parameter change $t \mapsto t - a_2 t^2/4 - (16a_1 + 3a_2^2)t^3/48$, it can be proved. Q.E.D.

Using Proposition A.1, we have the following:

Proposition A.2. *Let* $f : (\boldsymbol{R}^2, 0) \to (\boldsymbol{R}^3, 0)$ *be a cuspidal edge and* $\gamma : (\boldsymbol{R}, 0) \to (\boldsymbol{R}^2, 0)$ *an ordinary cusp such that* $df_0(\gamma''(0)) = 0$. *Then* $\hat{\gamma} = f \circ \gamma$ *is a* $(3,4)$*-cusp at* 0.

Proof. Without loss of generality, one can assume that f is given by the form (1.2), and $\gamma(t) = (t^3 + a_4 t^4 + a_5 t^5 + O(6), t^2)$ $(a_4, a_5 \in \boldsymbol{R})$, because $df_0(\partial v) = 0$. Then $\hat{\gamma}(t) = (t^3 + a_4 t^4 + a_5 t^5, t^4/2, 0) + O(6)$. By Proposition A.1, we have the conclusion. Q.E.D.

References

[1] V. I. Arnol'd, S. M. Gusein-Zade and A. N. Varchenko, *Singularities of differentiable maps, Vol.* 1, Monogr. Math. **82**, Birkhäuser Boston, Inc., Boston, MA, 1985.

[2] J. W. Bruce and , D. L. Fidal, *On binary differential equations and umbilics*, Proc. Roy. Soc. Edinburgh Sect. A **111** (1989), no. 1-2, 147–168.

[3] J. W. Bruce, G. J. Fletcher and F. Tari, *Bifurcations of implicit differential equations*, Proc. Roy. Soc. Edinburgh Sect. A **130** (2000), no. 3, 485–506.

[4] J. W. Bruce and T. J. Gaffney, *Simple singularities of mappings* $\boldsymbol{C}, 0 \to \boldsymbol{C}^2, 0$, J. London Math. Soc. (2) **26** (1982), no. 3, 465–474.

[5] J. W. Bruce and F. Tari, *On binary differential equations*, Nonlinearity **8** (1995), 255–271.

[6] J. W. Bruce and F. Tari, *Implicit differential equations from the singularity theory viewpoint*, Banach Center Publ. **33** (1996), 23–38.

[7] M. Cibrario, *Sulla reduzione a forma delle equationi lineari alle derviate parziale di secondo ordine di tipo misto*, Accademia di Scienze e Lettere, Instituto Lombardo Redicconti **65** (1932), 889–906.

[8] L. Dara, *Singularités génériques des équations differentielles multiformes*, Bol. Soc. Brasil Math. **6** (1975), 95–128.

[9] A. A. Davydov, *Normal forms of differential equations unresolved with respect to derivatives in a neighbourhood of its singular point*, Functional Anal. Appl. **19** (1985), 1–10.

[10] A. A. Davydov, *Qualitative Theory of Control Systems*, Translations of Mathematical Monographs **142**, American Mathematical Society, Providence, R.I. Moscow 1994.

[11] A. A. Davydov, G. Ishikawa, S. Izumiya and W. Sun, *Generic singularities of implicit systems of first order differential equations on the plane*, Jpn. J. Math. **3** (2008), no. 1, 93–119.

[12] R. Garcia, C. Gutierrez, and J. Sotomayor, *Lines of principal curvature around umbilics and Whitney umbrellas*, Tohoku Math. J., **52** (2000), 163–172.

[13] R. Garcia and J Sotomayor, *Harmonic mean curvature lines on surfaces immersed in* $\boldsymbol{R}^3$, Bull. Braz. Math. Soc. **34** (2003), 303–331.

[14] C. G. Gibson and C. A. Hobbs, *Simple singularities of space curves*, Math. Proc. Cambridge Philos. Soc. **113** (1993), no. 2, 297–310.

[15] V. Guíñez, *Positive quadratic differential forms and foliations with singularities on surfaces*, Trans. Amer. Math. Soc. **309** (1988), 477–502.

[16] V. Guíñez, *Locally stable singularities for positive quadratic differential forms*, J. Differential Equations **110** (1994), no. 1, 1–37.

[17] V. Guíñez, C. Gutierrez, *Rank* 1 *codimension one singularities of positive quadratic differential forms*, J. Differential Equations **206** (2004), no. 1, 127–155.

[18] A. Hayakawa, G. Ishikawa, S. Izumiya and K. Yamaguchi, *Classification of generic integral diagrams and first order ordinary differential equations*, Internat. J. Math. **5** (1994), no. 4, 447–489.

[19] S. Izumiya and F. Tari, *Apparent contours in Minkowski* 3*-space and first order ordinary differential equations*, Nonlinearity **26** (2013), no. 4, 911–932.

[20] S. Izumiya, K. Saji and N. Takeuchi, *Flat surfaces along cuspidal edges*, J. Sing. **16** (2017), 73–100.

[21] M. Kokubu, W. Rossman, K. Saji, M. Umehara and K. Yamada, *Singularities of flat fronts in hyperbolic* 3*-space*, Pacific J. Math. **221** (2005), no. 2, 303–351.

[22] L. F. Martins and K. Saji, *Geometric invariants of cuspidal edges*, Canad. J. Math. **68** (2016), 445–462.

[23] L. F. Martins, K. Saji, M. Umehara and K. Yamada, *Behavior of Gaussian curvature and mean curvature near non-degenerate singular points on wave fronts*, Geometry and Topology of Manifolds **154** Springer Proc. Math. and Statistics, 247–281.

[24] S. Murata and M. Umehara, *Flat surfaces with singularities in Euclidean 3-space*, J. Differential Geom. **82** (2009), no. 2, 279–316.

[25] K. Naokawa, M. Umehara and K. Yamada, *Isometric deformations of cuspidal edges*, Tohoku Math. J. (2) **68** (2016), 73–90.

[26] J. M. Oliver, *On pairs of foliations of a parabolic cross-cap*, Qual. Theory Dyn. Syst. **10** (2011), 139–166.

[27] R. Oset Sinha and F. Tari, *On the flat geometry of the cuspidal edge*, to appear in Osaka J. Math., arXiv:1610.08702.

[28] K. Saji, M. Umehara, and K. Yamada, *The geometry of fronts*, Ann. of Math. **169** (2009), 491–529.

[29] F. Tari, *On pairs of geometric foliations on a cross-cap*, Tohoku Math. J. **59** (2007), no. 2, 233–258.

[30] F. Tari, *Pairs of foliations on surfaces*, Real and complex singularities, London Math. Soc. Lecture Note Ser. **380**, 305–337.

[31] K. Teramoto, *Parallel and dual surfaces of cuspidal edges*, Differential Geom. Appl. **44** (2016), 52–62.

[32] R. Thom, *Sur les équations différentielles multiformes et leurs intégrales singulières*, Bol. Soc. Brasil. Mat. **3** (1972), no. 1, 1–11.

Department of Mathematics, Graduate School of Science, Kobe University, Rokko, Nada, Kobe 657-8501, Japan
E-mail address: saji@math.kobe-u.ac.jp

Advanced Studies in Pure Mathematics 78, 2018
Singularities in Generic Geometry
pp. 431–448

Implicit Hamiltonian systems and singular curves of distributions

Asahi Tsuchida

§1. Introduction

The subject of the sub-Riemannian geometry is to study a triple $(M, \mathcal{D}, g)$ of a manifold M, a distribution $\mathcal{D}$ on M and a bi-linear positive definite form g on $\mathcal{D}$, which is called a *sub-Riemannian manifold*. Here, a distribution is a sub-bundle of the tangent bundle TM of M. The object appears naturally as a collapsing Riemannian manifold and is a generalization of a Riemannian manifold. However properties of sub-Riemannian manifolds are much different from those of Riemannian manifolds. In fact, not much is known about the properties of exponential maps and even about smoothness of minimizers. For example there is a problem which is open for decades: "Are all locally minimizers smooth on any sub-Riemannian manifold?"

In this article we give a new clue to study of minimizers on sub-Riemannian manifolds.

For a distribution $\mathcal{D}$, there is an important class of curves called horizontal curves. A *horizontal curve* is an absolutely continuous curve $\gamma\colon I \to M$ such that $\dot{\gamma}(t)$ is a measurable and bounded map which satisfies $\dot{\gamma}(t) \in \mathcal{D}_{\gamma(t)}$ for almost every $t \in I$.

According to Chow–Rashevsky's theorem, if a distribution $\mathcal{D}$ on a connected manifold M satisfies Hörmander's condition, every two points are connected by a horizontal curve. For a sub-Riemannian manifold

---

Received July 9, 2016.
Revised January 28, 2017.
2010 *Mathematics Subject Classification.* Primary 53C17; Secondary 70F25, 70H45.

which satisfies Hörmander's condition, we may define a distance

$$d_{CC}(p,q) := \inf_{\gamma}\Big\{L(\gamma) := \int_{[a,b]} \sqrt{g(\dot\gamma(t),\dot\gamma(t))}dt \mid \\ \gamma\colon [a,b] \to M : horizontal, \gamma(a) = p, \gamma(b) = q\Big\}$$

which is called a *Carnot–Carathéodory* (or *sub-Riemannian*) *distance*. It is known that the topology from Carnot–Carathéodory distance agrees with the original one (ball-box theorem[10]).

A horizontal curve γ connecting p and q is called a *minimizer* if $d_{CC}(p,q) = L(\gamma)$. A horizontal curve $\gamma\colon I \to M$ is a *local minimizer* if for any $t_0 \in I$, there exists $\varepsilon > 0$ such that for all closed sub-interval J of $[t_0 - \varepsilon, t_0 + \varepsilon]$, $\gamma\mid_{J\cap I}$ is a minimizer between the end points. Note that any minimizer is necessarily a local minimizer.

To consider local minimizers, we classify horizontal curves on a sub-Riemannian manifold $(M,\mathcal{D},g)$ by using the end-point mapping.

For a bounded measurable curve $c\colon [0,T] \to \mathcal{D}$, if a curve $\gamma := \pi_{\mathcal{D}} \circ c\colon [0,T] \to M$ satisfies $\dot\gamma(t) = c(t)$ for almost everywhere on $[0,T]$, then γ is a horizontal curve and c is called an *admissible velocity*. Here $\pi_{\mathcal{D}}\colon \mathcal{D} \to M$ is the canonical projection.

Expressing $c(t)$ as $c(t) = u_1(t)X_1(\gamma(t)) + \cdots + u_k(t)X_k(\gamma(t))$ with respect to a local framing $\{X_1,\dots,X_k\}$ of the distribution $\mathcal{D}$ on an open subset $O \subset M$, the condition is written as

$$\dot\gamma(t) = u_1(t)X_1(\gamma(t)) + \cdots + u_k(t)X_k(\gamma(t)),$$

with the fibre coordinates $(u_1,\dots,u_k)$.

For any point q_0 on M, the set of admissible velocities

$$\mathcal{V}_{q_0} := \{c \mid c\colon [0,T] \to \mathcal{D} : admissible\ velocity,\ \gamma(0) = q_0\}$$

will be a Banach manifold. The map

$$\mathrm{End}(q_0)\colon \mathcal{V}_{q_0} \to M, c \mapsto \gamma(T)$$

is called an *end-point mapping* and is differentiable by means of Fréchet derivative. A singular (resp. regular) point of the end-point mapping is called a *singular* (*resp. regular*) *velocity*. The trajectory corresponds to singular velocity is called a *singular* (*resp. regular*) *curve*, i.e. the differential map

$$d\,\mathrm{End}(q_0)_c\colon T_c\mathcal{V}_{q_0} \to T_{\gamma(T)}M$$

is not surjective (resp. surjective) at singular (resp. regular) velocity c. Since every curve is either regular or singular, every minimizer is also either a regular curve or a singular curve.

For a sub-Riemannian manifold, there is a geodesic equation given as Hamiltonian formulation, not as Lagrangian. We give a function

$$H_E(x,p) = -\frac{1}{2}\sum_{i,j} g^{ij}(x)\langle p, X_i(x)\rangle\langle p, X_j(x)\rangle$$

on T^*M, where $g_{ij} = g(X_i, X_j)$ and $(g^{ij})_{i,j}$ is the inverse matrix of $(g_{ij})_{i,j}$. We may consider a Hamiltonian vector field associated to H_E with canonical symplectic form on the cotangent bundle T^*M. An ordinary differential equation related to the Hamiltonian vector field is given by

$$\dot{x}(t) = \frac{\partial H_E}{\partial p}(x(t), p(t)), \quad \dot{p}(t) = -\frac{\partial H_E}{\partial x}(x(t), p(t))$$

with Darboux coordinates (x, p) of T^*M. A solution of this equation (which is called the geodesic equation) is called a *normal bi-extremal* and its projection to M is called a *normal extremal* (or a *normal geodesic*).

It is known that regular local minimizers are normal geodesics, and so they are smooth since they are solutions of the geodesic equation. A singular minimizer is sometimes also a normal geodesic depending on a metric g. A singular (local) minimizer which is not a normal geodesic is called a *strictly singular (local) minimizer*. Examples of singular minimizers which are not normal on Martinet distribution are given by R. Montgomery in 1994 [8].

Although the examples given by R. Montgomery guarantee the existence of strictly singular minimizers, we know few other examples. So, we are interested in finding more examples of strictly singular minimizers. As a strategy, we divide the problem into two problems; to detect singular horizontal curves which are not normal geodesics and to find local minimizers among them. In this paper we concentrate on the former one and give new examples of singular horizontal curves which are not normal geodesics.

Now, we introduce a result of foothold in the study of singular curves. Take a function $H\colon T^*M \times_M \mathcal{D} \to \mathbb{R}, H(x,p,u) := \langle p, u\rangle$ for $x \in M, p \in T_x^*M$ and $u \in \mathcal{D}_x$, here $\times_M$ is a fibre product. Let $\{X_1, \ldots, X_k\}$ be a local framing of $\mathcal{D}$ on an open neighborhood U_{x_0} of x_0 in M and $(u_1, \ldots, u_k)$ the fibre coordinates related to the local framing. Then we have locally

$$H(x,p,u) = \sum_{i=1}^{k} u_i\langle p, X_i(x)\rangle.$$

Then singular curves are characterized by a constrained Hamiltonian system;

Proposition 1.1 ([6], p.567). *A horizontal curve $x(t)$ on M with rank k distribution $\mathcal{D}$ is a singular curve if and only if there exist a positive number $\varepsilon > 0$, a curve $p(t)$ on $T^*_{x(t)}M\setminus\{0\}$ and $u(t) \in \mathcal{D}_{x(t)}$ such that the curve $(x(t), p(t), u(t))$ satisfies the following equation (which is called the constrained Hamiltonian system) for almost all $t \in [0, \varepsilon)$;*

$$\begin{cases} \dot{x}(t) = \dfrac{\partial H}{\partial p}(x(t), p(t), u(t)), \\ \dot{p}(t) = -\dfrac{\partial H}{\partial x}(x(t), p(t), u(t)), \\ \dfrac{\partial H}{\partial u_i}(x(t), p(t), u(t)) = 0 \ (1 \leq i \leq k). \end{cases}$$

A solution $(x(t), p(t))$ on T^*M of the constrained Hamiltonian system in Proposition 1.1 is called an *abnormal bi-extremal* and the projection of an abnormal bi-extremal to M is called an *abnormal extremal* (or a *singular curve*). It is known that a local minimizer is either a normal extremal or an abnormal extremal; and the two possibilities are not mutually exclusive. Abnormal extremals are not local minimizers in general.

We may consider a constrained Hamiltonian system for any distribution. However it is not known whether there exists a solution or does not, in general. To study the existence of solutions, we intend to apply the theory of solvability of implicit differential systems. The theory, which is introduced in §2, is given by T. Fukuda and S. Janeczko. We give a refinement of Fukuda–Janeczko's theory in the case of two parameters in §3. In §4, an application of the results in §3 to rank two distribution is given. Then we have the following main result in this paper.

Theorem 1. *Let $(M, \mathcal{D}, g)$ be a sub-Riemannian smooth manifold with a distribution $\mathcal{D}$ of rank two. Suppose that $\mathcal{D}_1 := \mathcal{D} + [\mathcal{D}, \mathcal{D}]$ is a sub-bundle of rank three and $\mathcal{D}_2 := \mathcal{D}_1 + [\mathcal{D}, \mathcal{D}_1]$ is a sub-bundle of rank four. Then for any point q in M, there exist an open neighborhood U_q of q in M and a C^∞ immersive singular curve $x(t)$ which is not a normal geodesic in U_q defined on a small interval.*

Theorem 1 is proved in §4 as an application of the results in §3. It is not known whether the singular curve in Theorem 1 is a minimizer or not.

§2. Preliminary

2.1. Implicit differential systems and their solvability

An implicit differential system is a generalization of a differential equation defined by a vector field. We define some basic notions. Let $\pi\colon TM \to M$ be a canonical projection and N a submanifold of M.

Definition 2.1. An *implicit differential system* on M is a subset S of tangent bundle TM.

A C^1 curve $\gamma\colon (a,b) \to N$ is called a *solution of S over N* if $(\gamma(t), \dot\gamma(t)) \in S \cap \pi^{-1}(N)$ for all $t \in (a,b)$.

A point $(x_0, \dot x_0) \in S$ is a *solvable point of S over N* if there exist a positive number $\varepsilon > 0$ and a solution $\gamma\colon (-\varepsilon, \varepsilon) \to N$ such that $(\gamma(0), \dot\gamma(0)) = (x_0, \dot x_0)$.

From now on we consider the case where S is a smooth submanifold of TM. A point $(x_0, \dot x_0) \in S$ is a *smoothly solvable point of S over N* if there exist an open neighborhood W in $S \times \mathbb{R}$ of $(x_0, \dot x_0, 0)$ and C^∞ map $\bar\gamma\colon W \to N$ such that $\gamma_{(x,\dot x)}(t) := \bar\gamma(x, \dot x, t)$ is a solution of S over N with $(\gamma(0), \dot\gamma(0)) = (x, \dot x)$ for all $(x, \dot x) \in \pi_1(W)$, where $\pi_1\colon S \times \mathbb{R} \to S$ is a natural projection.

An implicit differential system S over N is called a *smoothly solvable submanifold over N* if S consists only of smoothly solvable points of S over N.

When a submanifold N is M itself, definitions above are given in Fukuda and Janeczko's papers [4][5]. A solution of S over M is just called a *solution of S*. We say, simply, an implicit differential system S *is (smoothly) solvable* if S is (smoothly) solvable over M.

2.2. Implicit Hamiltonian systems

Let (M, ω) be a symplectic manifold. Then there is the induced symplectic structure $\dot\omega$ on the tangent bundle TM: We have an bundle isomorphism induced from interior product $\flat\colon TM \to T^*M, \flat_x(v_q) = \iota_{v_q}\omega_q$ for each point $q \in M$. The symplectic structure $\dot\omega$ is given by the pullback of the Liouville form θ on T^*M, i.e., $\dot\omega := \flat^* d\theta$. The induced symplectic structure $\dot\omega$ is locally written by

$$\dot\omega = \sum_{i=1}^{n} d\dot p_i \wedge dx_i - d\dot x_i \wedge dp_i$$

with the canonical coordinates $(x, p, \dot x, \dot p)$ of tangent bundle TM related to Darboux coordinates $(x, p) = (x_1, \dots, x_n, p_1, \dots, p_n)$ for the standard symplectic form $\omega = \sum_{i=1}^n dp_i \wedge dx_i$ of M.

We will define the notion of implicit Hamiltonian systems as Lagrangian submanifolds of $(TM, \dot{\omega})$. Then they are regarded as a generalization of Hamiltonian vector fields, i.e., of Hamiltonian dynamical systems. In what follows we set $M = \mathbb{R}^{2n}$ with the standard symplectic form ω as above.

Definition 2.2 ([4, 5]). A Lagrangian submanifold L of $(T\mathbb{R}^{2n}, \dot{\omega})$ (i.e., $\dim L = 2n$ and $\omega|_L = 0$) is called an *implicit Hamiltonian system.*

There is a well-known result that a Lagrangian submanifold is locally generated by a Morse family;

Theorem 2.3 ([1]). *Let L be a Lagrangian submanifold of $T\mathbb{R}^{2n}$ and $(q_0, \dot{q}_0) = (x_0, p_0, \dot{x}_0, \dot{p}_0) \in L$. Suppose*

$$\operatorname{corank} d(\pi\,|_L)(q_0, \dot{q}_0) = k > 0.$$

Then there exist an open neighborhood O of $(q_0, \dot{q}_0)$ in $T\mathbb{R}^{2n}$, an open neighborhood W of $(q_0, 0) \in \mathbb{R}^{2n} \times \mathbb{R}^k$ and a smooth function $F\colon W \to \mathbb{R}$ such that

$$L \cap O = \Big\{(x_0, p_0, \dot{x}_0, \dot{p}_0) \in O \mid \exists u \in \mathbb{R}^k s.t. (x, p, u) \in W, \frac{\partial F}{\partial u_l}(x, p, u) = 0,$$
$$\dot{x}_i = \frac{\partial F}{\partial p_i}(x, p, u), \dot{p}_i = -\frac{\partial F}{\partial x_i}(x, p, u), 1 \le i \le n, 1 \le l \le k\Big\},$$

and that

$$\operatorname{rank}\left(\frac{\partial^2 F}{\partial x_i \partial u_l}(q_0, 0), \frac{\partial^2 F}{\partial p_i \partial u_l}(q_0, 0)\right)_{1 \le i \le n, 1 \le l \le k} = k, \quad \frac{\partial^2 F}{\partial u_r \partial u_s}(q_0, 0) = 0$$

for $1 \le r, s \le k$.

Recall that the family of functions $F\colon \mathbb{R}^{2n} \times \mathbb{R}^k \to \mathbb{R}$ with $2n$ parameters $(x_1, \dots, x_n, p_1, \dots, p_n)$ on $(\mathbb{R}^k; u_1, \dots, u_k)$ is called a *Morse family* if $0 \in \mathbb{R}^k$ is a critical point of the map $F(q_0, u)$ and the map

$$\left(\frac{\partial F}{\partial u_1}, \dots, \frac{\partial F}{\partial u_k}\right) : \mathbb{R}^{2n} \times \mathbb{R}^k \to \mathbb{R}^k$$

is submersive at $(q_0, 0)$. We denote L_F a Lagrangian submanifold generated by Morse family $F\colon \mathbb{R}^{2n} \times \mathbb{R}^k \to \mathbb{R}$. That is, for the catastrophe set

$$C(F) = \left\{(x, p, u) \in \mathbb{R}^{2n} \times \mathbb{R}^k \mid \frac{\partial F}{\partial u_i}(x, p, u) = 0, i = 1, \dots, k\right\}$$

of F and C^∞ map $\phi_F\colon \mathbb{R}^{2n} \times \mathbb{R}^k \to T\mathbb{R}^{2n}$ defined by

$$\phi(x,p,u) = \left(x, p, \frac{\partial F}{\partial p_i}(x,p,u), -\frac{\partial F}{\partial x_i}(x,p,u) \right),$$

we set $L_F = \phi_F(C(F))$.

The following propositions are given in [4] which are a necessary condition and a sufficient condition for L_F to be (smoothly) solvable in the sense of Definition 2.1.

Proposition 2.4 ([4]). *Let $(x,p,\dot{x},\dot{p})$ be a solvable point of L_F. We set $\phi_F(x,p,u) = (x,p,\dot{x},\dot{p})$. Then there exists a real vector $\mu = (\mu_1,\dots,\mu_k)$ in $\mathbb{R}^k$ such that*

$$\begin{pmatrix} \frac{\partial^2 F}{\partial u_1 \partial u_1}(x,p,u) & \cdots & \frac{\partial^2 F}{\partial u_1 \partial u_k}(x,p,u) \\ \vdots & \ddots & \vdots \\ \frac{\partial^2 F}{\partial u_k \partial u_1}(x,p,u) & \cdots & \frac{\partial^2 F}{\partial u_k \partial u_k}(x,p,u) \end{pmatrix} \begin{pmatrix} \mu_1 \\ \vdots \\ \mu_k \end{pmatrix} = \begin{pmatrix} \{\frac{\partial F}{\partial u_1}, F\}(x,p,u) \\ \vdots \\ \{\frac{\partial F}{\partial u_k}, F\}(x,p,u) \end{pmatrix}.$$

Here the bracket $\{,\}$ is the Poisson bracket associated to the symplectic form ω.

Proposition 2.5 ([4]). *A point $(x,p,\dot{x},\dot{p})$ in L_F is (smoothly) solvable if a linear equation*

$$\begin{pmatrix} \frac{\partial^2 F}{\partial u_1 \partial u_1}(x,p,u) & \cdots & \frac{\partial^2 F}{\partial u_1 \partial u_k}(x,p,u) \\ \vdots & \ddots & \vdots \\ \frac{\partial^2 F}{\partial u_k \partial u_1}(x,p,u) & \cdots & \frac{\partial^2 F}{\partial u_k \partial u_k}(x,p,u) \end{pmatrix} \begin{pmatrix} \mu_1(x,p,u) \\ \vdots \\ \mu_k(x,p,u) \end{pmatrix} = \begin{pmatrix} \{\frac{\partial F}{\partial u_1}, F\}(x,p,u) \\ \vdots \\ \{\frac{\partial F}{\partial u_k}, F\}(x,p,u) \end{pmatrix}$$

has a (smooth) solution on a neighborhood of $(x,p,u) = \phi_F(x,p,\dot{x},\dot{p})$ in $C(F)$.

The differences between the necessary condition and the sufficient condition appear as those of the domain and smoothness of the solution μ.

Now we consider a Morse family of particular type:

$$F\colon \mathbb{R}^{2n} \times \mathbb{R}^k \to \mathbb{R}, \quad F(x,p,u) = \sum_{j=1}^{k} a_j(x,p)u_j + b(x,p).$$

Note that functions $a_1,\dots,a_k$ are independent, i.e., differential one forms $da_1(x,p),\dots,da_k(x,p)$ are linearly independent in $T^*_{(x,p)}\mathbb{R}^{2n}$ at each point $(x,p) \in \mathbb{R}^{2n}$ because F is a Morse family. The catastrophe set of F is given by $C(F) = K \times \mathbb{R}^k$ with

$$K := \{(x,p) \in \mathbb{R}^{2n} \mid a_i(x,p) = 0, i = 1,\dots,k\}.$$

Applying Proposition 2.4 and Proposition 2.5, we know that L_F is smoothly solvable if and only if $\{a_i, a_j\}(x,p) = 0$, $\{b, a_i\}(x,p) = 0$ on K, $(1 \leq i, j \leq k)$([5]). Considering conditions in the propositions, we set

$$\widetilde{S_F} := \Big\{(x,p,u) \in C(F) \mid \sum_{j=1}^{k} \{a_i, a_j\}(x,p)u_j = \{b, a_i\}(x,p), 1 \leq i \leq k\Big\},$$

$$S_F := \phi_F(\widetilde{S_F}).$$

Then we see that every smoothly solvable submanifold of L_F is contained in S_F ([5]). Moreover S_F itself may be smoothly solvable:

Theorem 2.6 ([5]). *S_F is a smoothly solvable submanifold of L_F if*

$$\operatorname{rank}\left(\{a_i, a_j\}(x,p)\right)_{1 \leq i,j \leq k} = r\,(\textit{constant}) \quad \text{and}$$

$$\begin{pmatrix} \{b, a_1\}(x,p) \\ \vdots \\ \{b, a_k\}(x,p) \end{pmatrix} \in \operatorname{Im}\left(\{a_i, a_j\}(x,p)\right)_{1 \leq i,j \leq k},$$

holds for every $(x,p) \in K = \{(x,p) \in \mathbb{R}^{2n} \mid \frac{\partial F}{\partial u_i}(x,p,u) = 0, 1 \leq i \leq k\}$.

§3. Main results

Now we pose a question; for which submanifold S of L_F does there exist a submanifold A of K such that S is smoothly solvable over A? In this section, we will give an answer to this question in the case $k = 2$.

Let $F \colon \mathbb{R}^{2n} \times \mathbb{R}^2 \to \mathbb{R}$ be a Morse family which is defined by

$$F(x,p,u) = a_1(x,p)u_1 + a_2(x,p)u_2.$$

We consider solvability of the Lagrangian submanifold L_F which is generated by F. Moreover we consider solvability of submanifolds of L_F. In detail, we consider a map $\phi_F \colon \mathbb{R}^{2n} \times \mathbb{R}^2 \to T(\mathbb{R}^{2n})$ which is defined by

$$\phi_F(x,p,u) = \left(x, p, \frac{\partial F}{\partial p}(x,p,u), -\frac{\partial F}{\partial x}(x,p,u)\right)$$

and the catastrophe set

$$C(F) = \{(x,p,u) \mid a_1(x,p) = a_2(x,p) = 0\} = K \times \mathbb{R}^2$$

of F for $K = \{(x,p) \mid a_1(x,p) = a_2(x,p) = 0\} \subset \mathbb{R}^{2n}$, then we define L_F by the image $\phi_F(C(F))$. According to Fukuda–Janeczko's Theorem 2.6, L_F is smoothly solvable if and only if $\{a_1, a_2\} = 0$ locally on K.

Now we consider the cases where the assumptions of Theorem 2.6 are not fulfilled. We consider the family of vector fields $X_u \colon K \to T(\mathbb{R}^{2n})$ along K with parameter u

$$X_u(x,p) = \left(x, p, \frac{\partial F}{\partial p}(x,p,u), -\frac{\partial F}{\partial x}(x,p,u)\right),$$

and detect submanifolds of K on which X_u are tangent to K. We are going to give smoothly solvable submanifolds over submanifolds of K in the following series of Propositions 3.1 – 3.6.

Let $A_0 = K$. The vector field X_u is tangent to A_0 if and only if

$$X_u(a_1) = X_u(a_2) = 0 \quad i.e.\ u_1\{a_1, a_2\} = u_2\{a_1, a_2\} = 0.$$

Hence $\phi_F(A_0 \times \mathbb{R}^2)$ is smoothly solvable if and only if $\{a_1, a_2\} = 0$ on $A_0 = K$. This fact is also obtained as a corollary of Theorem 2.6.

Then we consider a submanifold A_1 of A_0 consisting of points at which X_u is tangent to A_0;

$$A_1 := \{(x,p) \mid a_1(x,p) = a_2(x,p) = \{a_1, a_2\}(x,p) = 0\}.$$

We assume that the functions $a_1, a_2, \{a_1, a_2\}$ are independent. The vector field X_u is tangent to A_1 if and only if

$$X_u(\{a_1, a_2\}) = 0 \quad i.e.\ u_1\{a_1, \{a_1, a_2\}\}(x,p) + u_2\{a_2, \{a_1, a_2\}\}(x,p) = 0$$

on A_1. Note that we have $X_u(a_1) = X_u(a_2) = 0$ from the definition of A_1. Let $\mathcal{E}_{\mathbb{R}^{2n}, q_0}$ be the $\mathbb{R}$ -algebra of C^∞ function germs at q_0 on $\mathbb{R}^{2n}$. We denote by $\langle a_1, a_2, \{a_1, a_2\}\rangle_{\mathcal{E}_{\mathbb{R}^{2n}, q_0}}$ the $\mathcal{E}_{\mathbb{R}^{2n}, q_0}$-module generated by a_1, a_2 and $\{a_1, a_2\}$. We set

$$\xi_1 := \{a_1, \{a_1, a_2\}\}, \ \xi_2 := \{a_2, \{a_1, a_2\}\}.$$

Then the vector field X_u is tangent to A_1 if the functions ξ_1 and ξ_2 belong to the $\mathcal{E}_{\mathbb{R}^{2n}, q_0}$-module $\langle a_1, a_2, \{a_1, a_2\}\rangle_{\mathcal{E}_{\mathbb{R}^{2n}, q_0}}$ for any point $q_0 \in A_1$.

Proposition 3.1. *Assume that a_1, a_2, and $\{a_1, a_2\}$ are independent. Then $\phi_F(A_1 \times \mathbb{R}^2)$ is a smoothly solvable submanifold of L_F over A_1 if and only if $\xi_1, \xi_2 \in \langle a_1, a_2, \{a_1, a_2\}\rangle_{\mathcal{E}_{\mathbb{R}^{2n}, q_0}}$ for any point q_0 in A_1.*

To find smoothly solvable submanifolds of L_F over submanifolds of A_1, we construct fiber bundles as follows. Let

$$C_{(x,p)} := \{(u_1, u_2) \mid u_1\{a_1, \{a_1, a_2\}\}(x,p) + u_2\{a_2, \{a_1, a_2\}\}(x,p) = 0\}$$

for $(x,p) \in K$ and define line bundles

$$\begin{aligned}
\overline{A_2^1}^1 &:= \{(x,p,u) \mid u \in C^1_{(x,p)}, (x,p) \in A_2^1\},\\
\overline{A_2^1}^2 &:= \{(x,p,u) \mid u \in C^2_{(x,p)}, (x,p) \in A_2^1\},\\
\overline{A_2^2}^1 &:= \{(x,p,u) \mid u \in C^1_{(x,p)}, (x,p) \in A_2^2\},\\
\overline{A_2^2}^2 &:= \{(x,p,u) \mid u \in C^2_{(x,p)}, (x,p) \in A_2^2\},\\
\overline{A_{1,1}}^2 &:= \{(x,p,u) \mid u \in C^2_{(x,p)}, (x,p) \in A_{1,1}\},\\
\overline{A_{1,2}}^1 &:= \{(x,p,u) \mid u \in C^1_{(x,p)}, (x,p) \in A_{1,2}\},\\
\overline{A_{1,(1,2)}}^{1,2} &:= \{(x,p,u) \mid u \in C^{1,2}_{(x,p)}, (x,p) \in A_{1,(1,2)}\},
\end{aligned}$$

with

$$\begin{aligned}
A_2^1 &:= A_1 \cap \{(x,p) \mid \xi_1 = 0\}, & C^1_{(x,p)} &= \{(u_1,0) \in C_{(x,p)}\},\\
A_2^2 &:= A_1 \cap \{(x,p) \mid \xi_2 = 0\}, & C^2_{(x,p)} &= \{(0,u_2) \in C_{(x,p)}\},\\
A_{1,1} &:= A_1 \cap \{(x,p) \mid \xi_1 \neq 0\}, & C^{1,2}_{(x,p)} &= C_{(x,p)} \setminus \{0\},\\
A_{1,2} &:= A_1 \cap \{(x,p) \mid \xi_2 \neq 0\}, & &\\
A_{1,(1,2)} &:= A_1 \cap \{(x,p) \mid \xi_1 \neq 0, \xi_2 \neq 0\}. & &
\end{aligned}$$

Let us consider the case that one of ξ_1 and ξ_2 belongs to the $\mathcal{E}_{\mathbb{R}^{2n},q_0}$-module $\langle a_1, a_2, \{a_1,a_2\}\rangle_{\mathcal{E}_{\mathbb{R}^{2n},q_0}}$ and the other does not.

Proposition 3.2. *Assume that a_1, a_2 and $\{a_1,a_2\}$ are independent. Assume also that*

$$\xi_2 \in \langle a_1, a_2, \{a_1,a_2\}\rangle_{\mathcal{E}_{\mathbb{R}^{2n},q_0}} \text{ and } \xi_1 \notin \langle a_1, a_2, \{a_1,a_2\}\rangle_{\mathcal{E}_{\mathbb{R}^{2n},q_0}}$$

at every point q_0 of A_1. Then the followings hold.

(1) *$\phi_F(\overline{A_{1,1}}^2)$ is a smoothly solvable submanifold of L_F over $A_{1,1}$.*
(2) *Assume, furthermore, that $\xi_1, a_1, a_2, \{a_1,a_2\}$ are independent.*
  (a) *$\phi_F(\overline{A_2^1}^2)$ is a smoothly solvable submanifold of L_F over A_2^1.*
  (b) *$\phi_F(A_2^1 \times \mathbb{R}^2)$ is a smoothly solvable submanifold of L_F over A_2^1 if $\{a_1, \xi_1\} \in \langle a_1, a_2, \{a_1,a_2\}, \xi_1\rangle_{\mathcal{E}_{\mathbb{R}^{2n},q_0}}$ for any point q_0 in A_2^1.*

Proof. (1): $A_{1,1}$ is an open submanifold of A_1 from the definition. Since there exist β_1, β_2 and $\beta_3 \in \mathcal{E}_{\mathbb{R}^{2n},q_0}$ such that $\xi_2 = \beta_1 a_1 + \beta_2 a_2 + \beta_3\{a_1,a_2\}$, the vector $X_u(x,p)$ with $u \in C^2_{(x,p)}$ is tangent to $A_{1,1}$ at each point $(x,p) \in A_{1,1}$ because

$$\begin{aligned} &X_u(a_1) = u_2\{a_2, a_1\} = 0, \\ &X_u(a_2) = 0 \cdot \{a_1, a_2\} = 0, \\ &X_u(\{a_1, a_2\}) = 0 \cdot \xi_1 + u_2\xi_2 = u_2\xi_2 = \beta_1 a_1 + \beta_2 a_2 + \beta_3\{a_1, a_2\} = 0 \end{aligned}$$

on $A_{1,1}$.

(2)-(a): We check the condition that $X_u(x,p)$ is tangent to A_2^1 with $u \in C^2_{(x,p)}$ at each point $(x,p) \in A_2^1$. Note that $\{a_1, \xi_2\} = \{a_2, \xi_1\}$ from Jacobian identity:

$$\begin{aligned} \{a_1, \{a_2, \{a_1, a_2\}\}\} &= \{a_2, \{a_1, \{a_1, a_2\}\}\} + \{\{a_1, a_2\}, \{a_1, a_2\}\} \\ &= \{a_2, \{a_1, \{a_1, a_2\}\}\}. \end{aligned}$$

Then

$$X_u(\{a_1, a_2\}) = 0 \cdot \xi_1 + u_2\xi_2 = 0,$$

$$\begin{aligned} X_u(\xi_1) &= u_1\{a_1, \xi_1\} + u_2\{a_2, \xi_1\} = u_1\{a_1, \xi_1\} + u_2\{a_1, \xi_2\} \qquad \cdots (\star) \\ &= u_2(\{a_1, \beta_1 a_1\} + \{a_1, \beta_2 a_2\} + \{a_1, \beta_3\{a_1, a_2\}\}) \\ &= u_2(a_1\{a_1, \beta_1\} + \beta_2\{a_1, a_2\} + a_2\{a_1, \beta_2\} \\ &\qquad\qquad + \beta_3\xi_1 + \{a_1, a_2\}\{a_1, \beta_3\}) \\ &= 0 \end{aligned}$$

on A_2^1.

(2)-(b): From an equality $(\star)$ we have

$$X_u(\xi_1) = u_1\{a_1, \xi_1\} + u_2\{a_2, \xi_1\} = u_1\{a_1, \xi_1\} + u_2\{a_1, \xi_2\}.$$

Since $\xi_2 \in \langle a_1, a_2, \{a_1, a_2\}\rangle_{\mathcal{E}_{\mathbb{R}^{2n}, q_0}}$, it holds that

$$\{a_1, \xi_2\} \in \langle a_1, a_2, \{a_1, a_2\}, \xi_2\rangle_{\mathcal{E}_{\mathbb{R}^{2n}, q_0}}.$$

Consequently, by using $\{a_1, \xi_1\} \in \langle a_1, a_2, \{a_1, a_2\}, \xi_2\rangle_{\mathcal{E}_{\mathbb{R}^{2n}, q_0}}$, we obtain $X_u(\xi_1) = 0$ on A_2^1. Q.E.D.

In the same way we have the counterpart of Proposition 3.2.

Proposition 3.3. *Assume that a_1, a_2 and $\{a_1, a_2\}$ are independent. Assume also that*

$$\xi_1 \in \langle a_1, a_2, \{a_1, a_2\}\rangle_{\mathcal{E}_{\mathbb{R}^{2n}, q_0}} \text{ and } \xi_2 \notin \langle a_1, a_2, \{a_1, a_2\}\rangle_{\mathcal{E}_{\mathbb{R}^{2n}, q_0}}$$

at every point q_0 of A_1. Then the followings hold.

(1) $\phi_F(\overline{A_{1,2}}^1)$ *is a smoothly solvable submanifold of* L_F *over* $A_{1,2}$.
(2) *Assume, furthermore, that* $\xi_2, a_1, a_2, \{a_1, a_2\}$ *are independent.*
   (a) $\phi_F(\overline{A_2^2}^1)$ *is a smoothly solvable submanifold of* L_F *over* A_2^2.
   (b) $\phi_F(A_2^2 \times \mathbb{R}^2)$ *is a smoothly solvable submanifold of* L_F *over* A_2^2 *if* $\{a_2, \xi_2\} \in \langle a_1, a_2, \{a_1, a_2\}, \xi_2\rangle_{\mathcal{E}_{\mathbb{R}^{2n}, q_0}}$ *for any point* q_0 *in* A_2^2.

In the case $\xi_1, \xi_2 \notin \langle a_1, a_2, \{a_1, a_2\}\rangle_{\mathcal{E}_{\mathbb{R}^{2n}, q_0}}$ we have

Proposition 3.4. *Assume that* a_1, a_2 *and* $\{a_1, a_2\}$ *are independent. Then* $\phi_F(\overline{A_{1,(1,2)}}^{1,2})$ *is a smoothly solvable submanifold of* L_F *over* $A_{1,(1,2)}$ *if* $\xi_1, \xi_2 \notin \langle a_1, a_2, \{a_1, a_2\}\rangle_{\mathcal{E}_{\mathbb{R}^{2n}, q_0}}$ *for every point* q_0 *in* $A_{1,(1,2)}$.

Proof. $A_{1,(1,2)}$ is an open submanifold of A_1 from the definition. The vector $X_u(x,p)$ with $u \in C^{1,2}_{(x,p)}$ is tangent to $A_{1,(1,2)}$ at each point $(x,p) \in A_{1,(1,2)}$ since

$$\begin{aligned} X_u(a_1) &= u_2\{a_2, a_1\} = 0, \\ X_u(a_2) &= u_1\{a_1, a_2\} = 0, \\ X_u(\{a_1, a_2\}) &= u_1\xi_1 + u_2\xi_2 = 0 \end{aligned}$$

on $A_{1,(1,2)}$. Q.E.D.

In Proposition 3.1-3.4, we gave sufficient conditions for existence of smoothly solvable submanifolds of L_F over $A_{1,1}$, A_2^1, $A_{1,2}$, A_2^2 and $A_{1,(1,2)}$ and examples of smoothly solvable submanifolds of L_F over them. The following two propositions give different sufficient conditions for existence of smoothly solvable submanifolds of L_F over A_2^1 and A_2^2 and examples of smoothly solvable submanifolds over them respectively.

Proposition 3.5. *Assume that* a_1, a_2, $\{a_1, a_2\}$ *and* ξ_1 *are independent. Then* $\phi_F(\overline{A_2^1}^1)$ *is a smoothly solvable submanifold of* L_F *over* A_2^1 *if* $\{a_1, \xi_1\} \in \langle a_1, a_2, \{a_1, a_2\}, \xi_1\rangle_{\mathcal{E}_{\mathbb{R}^{2n}, q_0}}$ *for any point* q_0 *in* A_2^1.

Proof. Since $a_1, a_2, \{a_1, a_2\}$ and ξ_1 are independent, A_2^1 is a submanifold of K. For the vector field X_u along A_2^1 with $(u_1, 0)$,

$$\begin{aligned} X_u(a_1) &= 0 \cdot \{a_1, a_2\} = 0, \\ X_u(a_2) &= u_1\{a_2, a_1\} = 0, \\ X_u(\{a_1, a_2\}) &= u_1\xi_1 = 0 \end{aligned}$$

hold on A^1_2. Since $\{a_1, \xi_1\} \in \langle a_1, a_2, \{a_1, a_2\}, \xi_1 \rangle_{\mathcal{E}_{\mathbb{R}^{2n},q_0}}$, there exist $\beta_1, \beta_2, \beta_3$ and $\beta_4 \in \mathcal{E}_{\mathbb{R}^{2n},q_0}$ such that $\{a_1, \xi_1\} = \beta_1 a_1 + \beta_2 a_2 + \beta_3\{a_1, a_2\} + \beta_4 \xi_1$. Hence we have

$$X_u(\xi_1) = u_1\{a_1, \xi_1\} = u_1(\beta_1 a_1 + \beta_2 a_2 + \beta_3\{a_1, a_2\} + \beta_4 \xi_1) = 0$$

on A^1_2. Thus the vector field X_u with $(u_1, 0)$ is tangent to A^1_2. Q.E.D.

In the same way we have

Proposition 3.6. *Assume that* a_1, a_2, $\{a_1, a_2\}$ *and* ξ_2 *are independent. Then* $\phi_F(\overline{A^2_2}^2)$ *is a smoothly solvable submanifold of* L_F *over* A^2_2 *if* $\{a_2, \xi_2\} \in \langle a_1, a_2, \{a_1, a_2\}, \xi_2 \rangle_{\mathcal{E}_{\mathbb{R}^{2n},q_0}}$ *for any point* q_0 *in* A^2_2.

§4. Application

We apply the results obtained in §3, to study distributions and its singular curves and we prove Theorem 1. Now we recall some basic notations for the study of distributions. The *Lie flag* of a distribution $\mathcal{D}$ is the sequence $\mathcal{D}_0 \subset \mathcal{D}_1 \subset \cdots$ defined inductively by

$$\mathcal{D}_0 := \mathcal{D}, \quad \mathcal{D}_{i+1} := \mathcal{D}_i + [\mathcal{D}_0, \mathcal{D}_i], \quad i \geq 0.$$

The *small growth vector* of a distribution $\mathcal{D}$ at $q \in M$ is the sequence of the dimension of Lie flags;

$$(\dim \mathcal{D}_0(q), \dim \mathcal{D}_1(q), \dim \mathcal{D}_2(q), \ldots).$$

For example a contact distribution $\mathcal{D} \subset TM$ on a manifold M of dimension $2n+1$, has small growth vector $(2n, 2n+1)$. An Engel distribution $\mathcal{D}$ on 4 dimensional manifold M has small growth vector $(2, 3, 4)$.

In this section we consider distributions with small growth vector $(2, 3, 4, \ldots)$. The following lemma plays an essential role throughout the section.

Lemma 4.1. *Let* $\mathcal{D}$ *be a rank two distribution with small growth vector* $(2, 3, 4, \ldots)$ *at any point in an open neighborhood of* $q \in M$ *and* g *a bi-linear positive definite form on* $\mathcal{D}$. *Then there exist an open neighborhood* U_q *and local orthonormal frame* X_1, X_2 *of* $\mathcal{D}$ *on* U_q *such that*

$$X_1, X_2, [X_1, X_2], [X_1, [X_1, X_2]]$$

are linearly independent at q *and* $[X_2, [X_1, X_2]]$ *is a functional linear combination of* X_1, X_2 *and* $[X_1, X_2]$ *on* U_q.

Proof. $\mathcal{E}_{M,q}$ denotes the $\mathbb{R}$ -algebra of C^∞ function germs at q on M. Let X_1, X_2 be any local frame of $\mathcal{D}$ around q. We may suppose that $X_1, X_2, [X_1, X_2], [X_1, [X_1, X_2]]$ are linearly independent at q. From the assumption, $[X_2, [X_1, X_2]] \in \langle X_1, X_2, [X_1, X_2], [X_1, [X_1, X_2]]\rangle_{\mathcal{E}_{M,q}}$. Then there exists a $\lambda \in \mathcal{E}_{M,q}$ such that

$$[X_2, [X_1, X_2]] \equiv \lambda[X_1, [X_1, X_2]] \mod \langle X_1, X_2, [X_1, X_2]\rangle_{\mathcal{E}_{M,q}}$$

Set $\tilde{X_2} = X_2 - \lambda X_1$. Then $(X_1, \tilde{X_2})$ is a local frame of $\mathcal{D}$ around q. Then

$$\begin{aligned}
[\tilde{X_2}, [X_1, \tilde{X_2}]] &= [\tilde{X_2}, [X_1, X_2 - \lambda X_1]] \\
&= [\tilde{X_2}, [X_1, X_2]] - [\tilde{X_2}, X_1(\lambda)X_1] \\
&= [\tilde{X_2}, [X_1, X_2]] - X_1(\lambda)[\tilde{X_2}, X_1] - \tilde{X_2}(X_1(\lambda))X_1 \\
&\equiv [\tilde{X_2}, [X_1, X_2]] = [X_2 - \lambda X_1, [X_1, X_2]] \\
&\equiv 0 \mod \langle X_1, \tilde{X_2}, [X_1, \tilde{X_2}]\rangle_{\mathcal{E}_{M,q}}.
\end{aligned}$$

For functions

$$g_{11} := g(X_1, X_1),\ g_{12} := g(X_1, \tilde{X_2}) \text{ and } g_{22} := g(\tilde{X_2}, \tilde{X_2}),$$

we set

$$X_1' = \frac{\sqrt{g_{22}}}{\sqrt{g_{11}g_{22} - g_{12}^2}}\left(X_1 - \frac{g_{12}}{g_{22}}\tilde{X_2}\right),\ X_2' = \frac{1}{\sqrt{g_{22}}}\tilde{X_2}.$$

Then (X_1', X_2') is a local orthonormal basis of $\mathcal{D}$ around q i.e.,

$$g(X_1', X_1') = 1,\ g(X_1', X_2') = 0,\ g(X_2', X_2') = 1,$$

and

$$X_1', X_2', [X_1', X_2'], [X_1', [X_1', X_2']]$$

are linearly independent since X_1' and X_2' is a functional linear combination of X_1 and $\tilde{X_2}$. Moreover (X_1', X_2') satisfies

$$[X_2', [X_1', X_2']] \equiv 0 \mod \langle X_1', X_2', [X_1', X_2']\rangle_{\mathcal{E}_{M,q}}$$

because of the following for functions

$$\alpha_1 = \frac{\sqrt{g_{22}}}{\sqrt{g_{11}g_{22} - g_{12}^2}},\ \alpha_2 = -\frac{\sqrt{g_{22}}}{\sqrt{g_{11}g_{22} - g_{12}^2}}\frac{g_{12}}{g_{22}} \text{ and } \alpha_3 = \frac{1}{\sqrt{g_{22}}};$$

$$\begin{aligned}
[X_2', [X_1', X_2']] &= [\alpha_3 \tilde{X}_2, [\alpha_1 X_1, \alpha_3 \tilde{X}_2]] - [\alpha_3 \tilde{X}_2, [\alpha_2 \tilde{X}_2, \alpha_3 \tilde{X}_2]] \\
&= [\alpha_3 \tilde{X}_2, \alpha_1 \alpha_3 [X_1, \tilde{X}_2]] + [\alpha_3 \tilde{X}_2, \alpha_1 X_1(\alpha_3) \tilde{X}_2] \\
&\qquad\qquad - [\alpha_3 \tilde{X}_2, \alpha_3 \tilde{X}_2(\alpha_1) X_1] \\
&\equiv [\alpha_3 \tilde{X}_2, \alpha_1 \alpha_3 [X_1, \tilde{X}_2]] \mod \langle X_1', X_2', [X_1', X_2'] \rangle_{\mathcal{E}_{M,q}} \\
&\equiv \alpha_1 \alpha_3^2 [\tilde{X}_2, [X_1, \tilde{X}_2]] \mod \langle X_1', X_2', [X_1', X_2'] \rangle_{\mathcal{E}_{M,q}} \\
&\equiv 0 \mod \langle X_1', X_2', [X_1', X_2'] \rangle_{\mathcal{E}_{M,q}}
\end{aligned}$$

Q.E.D.

Let $\{X_1, X_2\}$ be a local frame of $\mathcal{D}$ on U_q for any $q \in M$ with the property of Lemma 4.1 and define a function $H \colon T^*U_q \times_{U_q} \mathcal{D} \to \mathbb{R}$ for the distribution $\mathcal{D}$ locally by

$$H(x, p, u) = u_1 \langle p, X_1(x) \rangle + u_2 \langle p, X_2(x) \rangle.$$

For functions $a_1(x,p) := \langle p, X_1(x) \rangle$ and $a_2(x,p) := \langle p, X_2(x) \rangle$, Proposition 3.2-(2)-(a) can be applied and we obtain the following

Proposition 4.2. *For a rank 2 distribution $\mathcal{D}$ with small growth vector $(2, 3, 4, \ldots)$ at each point q in M, there exist an open neighborhood U_q of q, a frame $\{X_1, X_2\}$ of $\mathcal{D}$ on U_q and an abnormal bi-extremal $(x(t), p(t))$ in $T^*U_q \setminus \{o\}$ such that*

$$\dot{x}(t) = X_2(x(t)), \quad \dot{p}(t) = -\frac{\partial \langle p, X_2(x) \rangle}{\partial x}(x(t), p(t))$$

and

$$\begin{aligned}
\langle p(t), X_1(x(t)) \rangle = 0, &\qquad \langle p(t), X_2(x(t)) \rangle = 0, \\
\langle p(t), [X_1, X_2](x(t)) \rangle = 0, &\quad \langle p(t), [X_1, [X_1, X_2]](x(t)) \rangle = 0.
\end{aligned}$$

Proof. From the property in Lemma 4.1 we take a local frame $\{X_1, X_2\}$ of $\mathcal{D}$ on an open neighborhood U_q of q in M such that

$$X_1, X_2, [X_1, X_2], [X_1, [X_1, X_2]]$$

are linearly independent and $[X_2, [X_1, X_2]]$ is a functional linear combination of X_1, X_2 and $[X_1, X_2]$. We take the function $H \colon T^*U_q \times_{U_q} \mathcal{D} \to \mathbb{R}$ as

$$H(x, p, u) = \langle p, X_1(x) \rangle u_1 + \langle p, X_2(x) \rangle u_2$$

and set $a_1(x,p) := \langle p, X_1(x) \rangle$ and $a_2(x,p) := \langle p, X_2(x) \rangle$. Then H is a Morse family because X_1 and X_2 are linearly independent at each point.

In conformity with the property of vector fields X_1 and X_2, functions $a_1, a_2, \{a_1, a_2\}$ and $\{a_1, \{a_1, a_2\}\}$ are independent and we have

$$a_1, a_2, \{a_1, a_2\}, \{a_2, \{a_1, a_2\}\} \in \langle a_1, a_2, \{a_1, \{a_1, a_2\}\}\rangle_{\mathcal{E}_{T^*U_q,Q}}$$

for any $Q \in T^*U_q$.

According to the proof of Proposition 3.2, X_u is a tangent vector field to a submanifold

$$\begin{aligned} A_2^1 = \{(x,p) \in T^*U_q \mid a_1(x,p) = a_2(x,p) &= \{a_1, a_2\}(x,p) \\ &= \{a_1, \{a_1, a_2\}\}(x,p) = 0\} \end{aligned}$$

of T^*U_q for $u = (0,1)$. Thus there exists an integral curve $(x(t), p(t))$ of X_u starting from a point in A_2^1, that is, $(x(t), p(t))$ satisfies ordinary differential equations

$$\begin{aligned} \dot{x}(t) &= \frac{\partial H}{\partial p}(x(t), p(t), (0,1)) = \frac{\partial \langle p, X_2(x)\rangle}{\partial p}(x(t), p(t)) = X_2(x(t)) \\ \dot{p}(t) &= -\frac{\partial H}{\partial x}(x(t), p(t), (0,1)) = -\frac{\partial \langle p, X_2(x)\rangle}{\partial x}(x(t), p(t)) \end{aligned}$$

and following conditions;

$$\begin{aligned} a_1(x(t), p(t)) = \langle p(t), X_1(x(t))\rangle = 0, \quad & \langle p(t), X_2(x(t))\rangle = 0, \\ \langle p(t), [X_1, X_2](x(t))\rangle = 0, \quad & \langle p(t), [X_1, [X_1, X_2]](x(t))\rangle = 0. \end{aligned}$$

Q.E.D.

Theorem 4.3. *Let M be a smooth manifold and $\mathcal{D}$ be a rank two distribution. Suppose that the distribution $\mathcal{D}$ has small growth vector $(2, 3, 4, \ldots)$ everywhere in an open neighborhood of any point q in M. Then for any point q in M, there exist an open neighborhood U_q of q in M and a C^∞ immersive singular curve $x(t)$ in U_q which is defined on a small interval.*

Proof. From Proposition 4.2, for any point q in M, there exist an open neighborhood U_q of q and an abnormal bi-extremal $(x(t), p(t))$ on $A_2^1 \subset T^*U_q$. Therefore the projection of $(x(t), p(t))$ by canonical projection $\pi_M\colon T^*M \to M$ is a singular curve $x(t)$ with admissible velocity directed to X_2 in $\pi_M(A_2^1)$. Q.E.D.

We now prove Theorem 1.

Proof of Theorem 1. Let q be a point of M and let $x(t)$ be the C^∞ immersive singular horizontal curve in a neighborhood U_q of q obtained in Theorem 4.3. Let $(x(t), p(t))$ be the abnormal bi-extremal considered in the proof of Theorem 4.3 which is obtained in Proposition 4.2. We are going to prove that this curve $x(t)$ is not a normal extremal. From Lemma 4.1 we may take a local orthonormal frame $\{X_1, X_2\}$ of $\mathcal{D}$ on U_q. We consider the Hamiltonian function in terms of the orthonormal frame $\{X_1, X_2\}$;

$$H_E(x,p) = -\frac{1}{2}\sum_{i=1}^{2}\langle p, X_i(x)\rangle^2.$$

Suppose that $x(t)$ is a normal extremal. Then there must exist a normal bi-extremal of the form $(x(t), \tilde{p}(t))$ which satisfies the following differential equation;

$$\dot{x}(t) = \frac{\partial H_E}{\partial p}(x(t), \tilde{p}(t)) = -\sum_{i=1}^{2} X_i(x(t)),$$

$$\dot{\tilde{p}}(t) = -\frac{\partial H_E}{\partial x}(x(t), \tilde{p}(t)) = \sum_{i=1}^{2}\frac{\partial\langle p, X_i(x)\rangle}{\partial x}(x(t), \tilde{p}(t)).$$

Since the abnormal extremal $x(t)$ satisfies $\dot{x}(t) = X_2(t)$ by Proposition 4.2,

$$X_2(x(t)) = \dot{x}(t) = -X_1(x(t)) - X_2(x(t))$$

holds. Thus $X_1(x(t)) + 2X_2(x(t)) = 0$ holds. This is a contradiction to $\{X_1, X_2\}$ being a local frame of $\mathcal{D}$. Q.E.D.

Remark 4.4 ([2], Theorem 2.8). *It is known that there are no singular minimizer for a generic sub-Riemannian manifold (the genericity is used for the distribution as map-germs) with rank greater than 2.*

Remark 4.5 ([7], Proposition 11). *There is a result given by Liu–Sussmann which is similar type to Theorem 1, however the method for the proof is different from ours.*

Acknowledgements. It is pleasure to acknowledge fruitful discussions with Takuo Fukuda and Goo Ishikawa. The author is grateful to the referee for careful reading and helpful comments.

References

[1] Arnold. V. I, Varchenko. A, Gusein-Zade. S.M, *Singularities of Differentiable Maps, Vol. I*, Springer, (1985).

[2] Chitour. Y, Jean. F, Trélat. E, Genericity results for singular curves, *J. differential geometry*, **73** (2006), 45–73.

[3] Bonnard. B, Chyba. M, *Singular trajectories and their role in control theory*, Springer, (2003).

[4] Fukuda. T, Janeczko. S, Singularities of implicit differential systems and their integrability, *Banach center publications*, **65** (2004), 23–47.

[5] Fukuda. T, Janeczko. S, *A résumé on Workshop on singularities, geometry, topology and related topics (2014, September 1st – 3rd)*, personal communication.

[6] Hsu. L, Calculus of variations via the Griffiths formalism, *J. Diff. Geom*, **36** (1992), 551–589.

[7] Liu. W, Sussman.H. J, *Shortest Paths for sub-Riemannian metrics on rank-two distributions*, American Mathematical Society, Memoirs of the AMS, vol. **118**, no. 564 (1995).

[8] Montgomery. R, Abnormal minimizers, SIAM, *J. Control Optim.* **32** 6, (1994), 1605–1620.

[9] Montgomery. R, A tour of subriemannian geometries, their geodesics and applications, *American Mathematical Society* Mathematical surveys and Monographs vol. **91** (2002).

[10] Nagel. A, Stein. E. M, Wainger. S, Balls and metrics defined by vector fields I: Basic properties, *Acta Mathematica*, **155**, Issue 1 (1985), 103–147.

[11] Tsuchida. A, Smooth solvability of implicit Hamiltonian systems and existence of singular control for affine control systems (in Japanese), *RIMS Kôkyûroku* **1948** (2015), 153–159.

Department of Mathematics, Hokkaido University,
kita 10, Nishi 8, kita-ku,
Sapporo, Hokkaido, 060-0810, Japan.
E-mail address: asahi-t@math.sci.hokudai.ac.jp

Advanced Studies in Pure Mathematics 78, 2018
Singularities in Generic Geometry
pp. 449–469

Weierstrass-type representations for timelike surfaces

Masashi Yasumoto

Abstract.

In this paper we give the Weierstrass-type representation for Lorentz conformal minimal surfaces in Minkowski 3-space that was derived by Konderak, and a new one for Lorentz conformal constant mean curvature 1 surfaces in anti de Sitter 3-space, using integrable systems techniques. As an application, we analyze their singularities. Finally, we describe first steps toward discretization of these timelike surfaces.

§1. Introduction

The study of zero mean curvature (ZMC, for short) surfaces in Minkowski 3-space $\mathbb{R}^{2,1}$ is one recent topic of research on the differential geometry of surfaces. Kobayashi [11] derived a Weierstrass-type representation for conformal immersions with mean curvature identically 0 in $\mathbb{R}^{2,1}$, called conformal maximal surfaces. Magid [16] derived a Weierstrass-type representation for timelike immersions with mean curvature identically 0 and null coordinate systems, which are called *timelike minimal surfaces* in $\mathbb{R}^{2,1}$, and Inoguchi, Toda [9] derived its normalized version using a loop groups formulation, which is closely related to integrable systems techniques. Unlike the case of minimal surfaces in Euclidean 3-space $\mathbb{R}^3$, ZMC surfaces in $\mathbb{R}^{2,1}$ have certain singularities (see [6], [7], [18], [20] for example). Very recently, Akamine [1]

Received April 1, 2016.
Revised April 28, 2017.
2010 *Mathematics Subject Classification.* Primary 53A10, Secondary 52C99.
The author was supported by the Grant-in-Aid for JSPS Fellows Number 26-3154, JSPS/FWF bilateral joint project "Transformations and Singularities" between Austria and Japan, and the JSPS Program for Advancing Strategic International Networks to Accelerate the Circulation of Talented Researchers "Mathematical Science of Symmetry, Topology and Moduli, Evolution of International Research Network based on OCAMI".

analyzed behaviors of two null curves generating a timelike minimal surface and its Gaussian curvature near singular points.

Stepping away from ZMC surfaces in $\mathbb{R}^{2,1}$, here we also see timelike constant mean curvature (CMC, for short) surfaces in other Lorentzian spaceforms. As mentioned in [14], there exists a correspondence between timelike CMC surfaces in different Lorentzian spaceforms, which is called the Lawson-type correspondence. In particular, there exists a Lawson-type correspondence between timelike minimal surfaces in $\mathbb{R}^{2,1}$ and timelike CMC 1 surfaces in anti de Sitter 3-space $\mathbb{H}^{2,1}$. So, like in the case of conformal CMC 1 surfaces in $\mathbb{H}^3$ (see [3], [19] for example), it is natural to expect that there exists a Weierstrass-type representation for timelike CMC 1 surfaces in $\mathbb{H}^{2,1}$. In fact, Lee [14] derived the Weierstrass-type representation for timelike CMC 1 surfaces in $\mathbb{H}^{2,1}$ parametrized by null coordinate systems.

In the "smooth" (or, continuous) case, we can reparametrize surfaces. In particular, we can reparametrize timelike surfaces with null coordinate systems to timelike surfaces with Lorentz conformal coordinate systems (see [9] for example). However, as typified by the work in [2], it is generally difficult to reparametrize discrete surfaces (cf. [8]). For this reason, when discretizing surfaces, we would like to find suitable coordinate systems that are compatible with discretization. In [21], we describe discrete surfaces with Lorentz conformal curvature line coordinate systems in $\mathbb{R}^{2,1}$ called *discrete timelike isothermic surfaces*. In the case of isothermic surfaces, away from umbilic points, CMC (possibly, zero mean curvature) surfaces can be reparametrized to be isothermic. In contrast to that, timelike minimal and CMC surfaces are not necessarily timelike isothermic. As mentioned in [15], there are three kinds of Lorentz isothermic surfaces, so discretizing the other cases is a remaining problem.

In this paper, as preparation for [21], we derive a Weierstrass-type representation for timelike minimal surfaces in $\mathbb{R}^{2,1}$ parametrized by Lorentz conformal coordinate systems (Lorentz conformal minimal surfaces in $\mathbb{R}^{2,1}$, for short) via integrable systems techniques. Using paracomplex analysis introduced in [10], we derive the Lax pair for Lorentz conformal CMC surfaces in $\mathbb{R}^{2,1}$ and $\mathbb{H}^{2,1}$ in Propositions 3.1, 5.1. As an application, we obtain the Weierstrass-type representations for Lorentz conformal minimal surfaces in $\mathbb{R}^{2,1}$ (see Theorem 1), which was derived by Konderak [13], and for Lorentz conformal CMC 1 surfaces in $\mathbb{H}^{2,1}$ (see Theorem 2). Furthermore, we analyze the singularities of these surfaces. In the case of Lorentz conformal minimal surfaces in $\mathbb{R}^{2,1}$, Takahashi [18] gave explicit criteria for cuspidal edge singularities, swallowtail singularities and cuspidal cross cap singularities to appear (see Theorem 3 here).

Here we introduce an alternative proof of this. In addition, we give explicit criteria for the same types of singularities to appear on Lorentz conformal CMC 1 surfaces in $\mathbb{H}^{2,1}$, in Theorem 4. In the last appendix, we briefly introduce discrete timelike isothermic surfaces in $\mathbb{R}^{2,1}$. In particular, we introduce the Weierstrass-type representation for discrete timelike isothermic minimal surfaces in $\mathbb{R}^{2,1}$.

§2. Lorentz conformal CMC surfaces in $\mathbb{R}^{2,1}$

First we consider Lorentz conformal CMC surfaces in $\mathbb{R}^{2,1}$ using Lax pairs. In this section we introduce 3×3 Lax pairs for Lorentz conformal CMC surfaces in $\mathbb{R}^{2,1}$ and derive the compatibility condition for Lorentz conformal CMC surfaces in $\mathbb{R}^{2,1}$. Let

$$\mathbb{R}^{n,1} := \left(\{x = (x_1, \cdots, x_n, x_0)^t; x_i \in \mathbb{R}\}, \langle \cdot, \cdot \rangle\right)$$

be the $(n+1)$-dimensional Minkowski space with Lorentz metric $\langle x, y \rangle = x_1 y_1 + \cdots + x_n y_n - x_0 y_0$, and let $\mathbb{C}'$ be the set of para-complex numbers, that is,

$$\mathbb{C}' := \{a + j'b; a, b \in \mathbb{R},\ j' \text{: non-real s.t. } (j')^2 = +1\},$$

where j' is called the *para-complex imaginary unit*. Then $\mathbb{C}'$ can be identified with the Minkowski plane $\mathbb{R}^{1,1}$ by the identification

$$\begin{array}{ccc} \mathbb{R}^{1,1} & \longrightarrow & \mathbb{C}' \\ \Psi & & \Psi \\ (x, y) & \longmapsto & x + j'y \end{array}.$$

Let $g : \Sigma(\subset \mathbb{C}') \to \mathbb{C}'$ be a map. Take $z = x + j'y \in \Sigma$ and set $\partial_z := \frac{1}{2}(\partial_x + j'\partial_y)$, $\partial_{\bar{z}} := \frac{1}{2}(\partial_x - j'\partial_y)$, then g is a *p-holomorphic function* if $g_{\bar{z}} = \partial_{\bar{z}} g = 0$ holds, that is, the Cauchy-Riemann type equation holds (see Theorem V.1 in [10]). In this paper we abbreviate $|z|_*^2 := z\bar{z}$ for $z \in \mathbb{C}'$.

Let $f : \Sigma \subset \mathbb{C}' \to \mathbb{R}^{2,1}$ be a Lorentz conformal immersion satisfying

$$\langle f_z, f_z \rangle = \langle f_{\bar{z}}, f_{\bar{z}} \rangle = 0,\ \langle f_z, f_{\bar{z}} \rangle = 2\mathrm{e}^{2u}$$

and $N : \Sigma \to \mathbb{S}^{1,1}$ its spacelike unit normal vector field, where

$$\mathbb{S}^{1,1} := \{X \in \mathbb{R}^{2,1}; \langle X, X \rangle = 1\},$$

and $u : \Sigma \to \mathbb{R}$ is a real-valued function. In the above equation, $\langle \cdot, \cdot \rangle$ denotes the split-complex bilinear extension of the usual $\mathbb{R}^{2,1}$ Lorentz inner product (which is no longer an actual Lorentz inner product). The mean curvature H of f is $H = \frac{1}{2}e^{-2u}\langle f_{z\bar{z}}, N\rangle$. Setting $Q := \langle f_{zz}, N\rangle$, we have

$$\begin{aligned}
&\langle f_z, f_{zz}\rangle = \langle f_z, f_{z\bar{z}}\rangle = \langle f_{\bar{z}}, f_{z\bar{z}}\rangle = \langle f_{\bar{z}}, f_{\bar{z}\bar{z}}\rangle = \langle N, N_z\rangle = \langle N, N_{\bar{z}}\rangle = 0,\\
&\langle f_{\bar{z}}, f_{zz}\rangle = 4u_z e^{2u},\ \langle f_z, f_{\bar{z}\bar{z}}\rangle = 4u_{\bar{z}} e^{2u},\ \langle N_z, f_z\rangle = -Q,\\
&\langle N_{\bar{z}}, f_{\bar{z}}\rangle = -\bar{Q},\ \langle N, f_{z\bar{z}}\rangle = -\langle N_z, f_{\bar{z}}\rangle = -\langle N_{\bar{z}}, f_z\rangle = 2He^{2u},
\end{aligned}$$

where Q is called the (coefficient of the) Hopf differential of f. The Hopf differential will play an important role in distinguishing the three types of Lorentz conformal immersions in Appendix A. Then we have the following Gauss-Weingarten type equations:

$$\begin{aligned}
&f_{zz} = 2u_z f_z + QN,\ f_{z\bar{z}} = 2He^{2u}N,\\
&f_{\bar{z}\bar{z}} = 2u_{\bar{z}} f_{\bar{z}} + \bar{Q}N,\ N_z = -\frac{1}{2}(2Hf_z + Qe^{-2u}f_{\bar{z}}),\\
&N_{\bar{z}} = -\frac{1}{2}(2Hf_{\bar{z}} + \bar{Q}e^{-2u}f_z).
\end{aligned}$$

Here we define $e_1 := (f_z + f_{\bar{z}})/(2e^u)$, $e_2 := j'(f_z - f_{\bar{z}})/(2e^u)$ and then $\mathcal{F} := (N, e_1, e_2)$ is an orthogonal frame of the surface satisfying

$$\begin{aligned}
&\mathcal{F}_z = \mathcal{F}(\Theta + \Upsilon_z),\quad \mathcal{F}_{\bar{z}} = \mathcal{F}(\bar{\Theta} + \Upsilon_{\bar{z}})\quad \text{with}\\
&\Theta = \begin{pmatrix} 0 & \alpha & \beta \\ -\alpha & 0 & 0 \\ \beta & 0 & 0 \end{pmatrix},\ \Upsilon = \begin{pmatrix} 0 & 0 & 0 \\ 0 & 0 & j'u \\ 0 & j'u & 0 \end{pmatrix},\ \text{where}\\
&\alpha = \frac{1}{2}(Qe^{-u} + 2He^u),\ \beta = \frac{j'}{2}(Qe^{-u} - 2He^u).
\end{aligned}$$

The compatibility condition $\mathcal{F}_{z\bar{z}} = \mathcal{F}_{\bar{z}z}$ gives

$$\begin{aligned}
4u_{z\bar{z}} + 4H^2e^{2u} - e^{-2u}Q\bar{Q} = 0 \quad &\text{(Gauss)},\\
Q_{\bar{z}} = 2H_z e^{2u} \quad &\text{(Codazzi)}.
\end{aligned}$$

The Codazzi equation implies that Q is p-holomorphic when H is constant. Henceforth, we assume that H is constant.

§3. The 2 × 2 Lax pair for timelike CMC surfaces

Here we identify each element in $\mathbb{R}^{2,1}$ as follows:

$$\begin{array}{ccc}
\mathbb{R}^{2,1} & \longrightarrow & \mathfrak{su}'_2 \\
\Psi & & \Psi \\
(x_1, x_2, x_0) & \longmapsto & \begin{pmatrix} j'x_1 & -j'x_2 - x_0 \\ -j'x_2 + x_0 & -j'x_1 \end{pmatrix}
\end{array}, \tag{1}$$

where $\mathfrak{su}_2'$ is the Lie algebra of the Lie group

$$\mathrm{SU}_2' := \left\{ \begin{pmatrix} a & b \\ -\bar{b} & \bar{a} \end{pmatrix}; a, b \in \mathbb{C}',\ |a|_*^2 + |b|_*^2 = 1 \right\}.$$

Then the metric becomes $\langle X, Y \rangle = \dfrac{1}{2}\mathrm{trace}(XY)$, with X and Y identified with the matrix forms above. So the metric does not change under the adjoint action $X \mapsto FXF^{-1}$, $Y \mapsto FYF^{-1}$. Now, the adjoint action $F\begin{pmatrix} j'x_1 & -j'x_2 - x_0 \\ -j'x_2 + x_0 & -j'x_1 \end{pmatrix}F^{-1}\left(F = \begin{pmatrix} a + j'b & c + j'd \\ -c + j'd & a - j'b \end{pmatrix} \in \mathrm{SU}_2'\right)$ induces the associated linear map

$$(x_1, x_2, x_0)^t \mapsto A(x_1, x_2, x_0)^t, \text{ where}$$
$$A = \begin{pmatrix} a^2 - b^2 - c^2 + d^2 & -2ac + 2bd & -2bc + 2ad \\ 2ac + 2bd & a^2 + b^2 - c^2 - d^2 & 2ab + 2cd \\ 2bc + 2ad & 2ab - 2cd & a^2 + b^2 + c^2 + d^2 \end{pmatrix}$$
$$\in \mathrm{SO}_{2,1}^+ := \left\{ M = (a_{ij}) \in \mathrm{SL}_3\mathbb{R}; MI'M^t = I', a_{33} > 0 \right\},$$

with $I' = \begin{pmatrix} 1 & 0 & 0 \\ 0 & 1 & 0 \\ 0 & 0 & -1 \end{pmatrix}$. Let $F \in \mathrm{SU}_2'$ be the matrix satisfying

$$e_1 = F\begin{pmatrix} 0 & -j' \\ -j' & 0 \end{pmatrix}F^{-1},\ e_2 = F\begin{pmatrix} 0 & -1 \\ 1 & 0 \end{pmatrix}F^{-1},\ N = F\begin{pmatrix} j' & 0 \\ 0 & -j' \end{pmatrix}F^{-1}.$$

Then,

$$f_z = -2j'\mathrm{e}^u F\begin{pmatrix} 0 & 1 \\ 0 & 0 \end{pmatrix}F^{-1},\ f_{\bar{z}} = -2j'\mathrm{e}^u F\begin{pmatrix} 0 & 0 \\ 1 & 0 \end{pmatrix}F^{-1}.$$

Set $U := F^{-1}F_z = \begin{pmatrix} U_{11} & U_{12} \\ U_{21} & U_{22} \end{pmatrix}$, $V := F^{-1}F_{\bar{z}} = \begin{pmatrix} V_{11} & V_{12} \\ V_{21} & V_{22} \end{pmatrix}$. Our task here is to compute U and V explicitly. The compatibility condition $f_{z\bar{z}} = f_{\bar{z}z}$ gives $U_{12} = -V_{21}$, $u_z + U_{22} - U_{11} = 0$, $u_{\bar{z}} + V_{11} - V_{22} = 0$. And the condition $f_{z\bar{z}} = 2H\mathrm{e}^{2u}N$ implies $U_{12} = -V_{21} = -H\mathrm{e}^u$. By conditions $f_{zz} = 2u_z f_z + QN$, $f_{\bar{z}\bar{z}} = 2u_{\bar{z}} f_{\bar{z}} + \bar{Q}N$, we have $U_{21} = \frac{1}{2}Q\mathrm{e}^{-u}$, $V_{12} = -\frac{1}{2}\bar{Q}\mathrm{e}^{-u}$. Finally, by the trace-free conditions of U and V, we conclude the following proposition, which is called the 2×2 Lax pair for timelike CMC surfaces:

Proposition 3.1. *For some $F_{z_0 \in \Sigma} \in \mathrm{SU}_2'$, we have the solution $F \in \mathrm{SU}_2'$ by solving $F_z = FU$, $F_{\bar{z}} = FV$, where $F_z = FU$, $F_{\bar{z}} = FV$, and*

$$U = \frac{1}{2}\begin{pmatrix} u_z & -2H\mathrm{e}^u \\ Q\mathrm{e}^{-u} & -u_z \end{pmatrix},\quad V = \frac{1}{2}\begin{pmatrix} -u_{\bar{z}} & -\bar{Q}\mathrm{e}^{-u} \\ 2H\mathrm{e}^u & u_{\bar{z}} \end{pmatrix}.$$

§4. Weierstrass-type representation for Lorentz conformal minimal surfaces in $\mathbb{R}^{2,1}$

Applying the Lax pair for timelike CMC surfaces, we will derive a Weierstrass-type representation for Lorentz conformal minimal surfaces in $\mathbb{R}^{2,1}$. Defining functions $a,\ b : \Sigma \to \mathbb{C}'$ such that

$$F = \mathrm{e}^{-\frac{u}{2}} \begin{pmatrix} a & \bar{b} \\ -b & \bar{a} \end{pmatrix} \quad (a\bar{a} + b\bar{b} = \mathrm{e}^{u}),$$

one can compute that

$$F_{\bar{z}} = \frac{\mathrm{e}^{-u/2}}{2} \begin{pmatrix} -u_{\bar{z}}a + 2a_{\bar{z}} & -u_{\bar{z}}\bar{b} + 2\bar{b}_{\bar{z}} \\ u_{\bar{z}}b - 2b_{\bar{z}} & -u_{\bar{z}}\bar{a} + 2\bar{a}_{\bar{z}} \end{pmatrix}.$$

On the other hand, by Proposition 3.1, we have

$$FV = \frac{\mathrm{e}^{-u/2}}{2} \begin{pmatrix} -u_{\bar{z}} & j'\bar{Q}\mathrm{e}^{-u}a + u_{\bar{z}}\bar{b} \\ u_{\bar{z}}b & -j'\bar{Q}\mathrm{e}^{-u}b + u_{\bar{z}}\bar{a} \end{pmatrix}.$$

So, $a_{\bar{z}} = b_{\bar{z}} = 0$. Thus, a and b are p-holomorphic functions. Now,

$$f_z = -2j'\mathrm{e}^{u} F \begin{pmatrix} 0 & 1 \\ 0 & 0 \end{pmatrix} F^{-1} = -2j' \begin{pmatrix} ab & a^2 \\ -b^2 & -ab \end{pmatrix}.$$

Then we have $f_z = (-2ab, a^2 - b^2, j'(a^2 + b^2))^t$ in the standard $\mathbb{R}^{2,1}$ coordinate via the identification (1). Here we introduce the following lemma, which immediately follows from the Cauchy-Riemann type equation (compare with Remark 3.4.2 in [5]).

Lemma 4.1. *For any real-valued $\phi : \Sigma \to \mathbb{R}$ and p-holomorphic function $\Psi : \Sigma \to \mathbb{C}'$,*

$$\psi_z = \Psi_z \Leftrightarrow \psi = \mathrm{Re}\Psi + c \quad (c:\ \textit{constant}).$$

Using Lemma 4.1, we have $\mathrm{Re} \displaystyle\int f_z dz = \frac{1}{2} f + \vec{c}$ for some constant $\vec{c} \in \mathbb{R}^{2,1}$. Then,

$$\begin{aligned} f &= 2\mathrm{Re} \int (-2ab, a^2 - b^2, j'(a^2 + b^2))^t dz \\ &= \mathrm{Re} \int (2g, 1 - g^2, -j'(1 + g^2))^t \omega, \end{aligned}$$

where $g = a/b$ and $\omega = -2b^2 dz = \hat{\omega} dz$ ($\hat{\omega} := -2b^2$). Note that, by holomorphicity of a and b, g is a p-meromorphic function and ω is a p-holomorphic 1-form. Thus we have the following theorem, which was already described by Konderak [13]:

Theorem 1. *Let (g, ω) be a pair consisting of a p-meromorphic function and a p-holomorphic 1-form. Then a Lorentz conformal minimal surface f in $\mathbb{R}^{2,1}$ can be locally constructed by*

$$f = \mathrm{Re} \int (2g, 1 - g^2, -j'(1 + g^2))^t \omega \tag{2}$$

with metric $(1 + g\bar{g})\omega\bar{\omega}$. Conversely, any Lorentz conformal minimal surface f in $\mathbb{R}^{2,1}$ is locally described in this manner.

Remark. We have three remarks here:

- The coefficient $\hat{\omega}$ of ω is a map from Σ to $\{X \in \mathbb{C}'; |X|_*^2 > 0,\ \mathrm{Re}X < 0\}$. On the other hand, even when $\hat{\omega}$ takes values in $\{X \in \mathbb{C}'; |X|_*^2 > 0,\ \mathrm{Re}X > 0\}$, f is still Lorentz conformal minimal.
- When $\hat{\omega}$ takes values in $\{X \in \mathbb{C}'; |X|_*^2 < 0\}$, f is still Lorentz conformal minimal, but causalities of f_x and f_y switch. This can be interpreted as follows: When $|\hat{\omega}|_*^2 < 0$, we set $\tilde{g} := j'g$ and $\tilde{\omega} := j'\omega$. Then Equation (2) can be rewritten as

$$f = \mathrm{Re} \int \left(2\tilde{g}, j'(1 - \tilde{g}^2), -1 - \tilde{g}^2\right)^t \tilde{\omega}. \tag{3}$$

  The metric of f is $-(1 - |\tilde{g}|_*^2)^2 |\tilde{\omega}|_*^2$. That form is the same as the Weierstrass-type representation for Lorentz conformal minimal surfaces in $\mathbb{R}^{2,1}$ derived in [18]. Also in our approach, if we assume that $\langle f_z, f_{\bar{z}} \rangle = -2\mathrm{e}^{2u}$ in Section 2, we have the same form as in Equation (3).
- Let f be a Lorentz conformal minimal surface given as in Theorem 1. Then a surface $f^\sharp := \mathrm{Im} \int (2g, j'(1 - g^2), -1 - g^2)^t \omega$ is also a Lorentz timelike minimal surface with metric $-(1 + g\bar{g})\omega\bar{\omega}$. The $f^\sharp$ given in this way is called the *conjugate* Lorentz conformal minimal surface of f.

Here we see the Gauss map N of a Lorentz conformal minimal surface f described by Equation (2). N is described as follows:

$$N = F \begin{pmatrix} j' & 0 \\ 0 & -j' \end{pmatrix} F^{-1} = j'\mathrm{e}^{-u} \begin{pmatrix} |a|_*^2 - |b|_*^2 & -2a\bar{b} \\ -2\bar{a}b & -|a|_*^2 + |b|_*^2 \end{pmatrix}.$$

Thus we have $N = \dfrac{1}{1 + g\bar{g}}(-1 + g\bar{g}, 2\mathrm{Re}(g), 2\mathrm{Im}(g))^t$. This tells us that the Gauss maps of Lorentz conformal timelike minimal surfaces can be expressed via the inverse of stereographic projection of g.

§5. Lorentz conformal CMC surfaces in $\mathbb{H}^{2,1}$

Here we consider Lorentz conformal CMC surfaces in $\mathbb{H}^{2,1}$. Let $\mathbb{R}^{2,2} := (\{x = (x_1, x_2, x_3, x_4)^t; x_i \in \mathbb{R}\}, \langle\cdot,\cdot\rangle_*)$ be the 4-dimensional Lorentz space with metric $\langle x, y\rangle_* = x_1y_1 + x_2y_2 - x_3y_3 - x_4y_4$. Then

$$\mathbb{H}^{2,1} := \{x \in \mathbb{R}^{2,2}; \langle x, x\rangle_* = -1\}$$

denotes the 3-dimensional anti de Sitter space. Let $f : \Sigma \to \mathbb{H}^{2,1}$ be a Lorentz conformal immersion satisfying $\langle f_z, f_z\rangle_* = \langle f_{\bar z}, f_{\bar z}\rangle_* = 0$, $\langle f_z, f_{\bar z}\rangle_* = 2\mathrm{e}^{2u}$. Here $N : \Sigma \to \mathbb{S}^{1,2}$ denotes the unit normal vector field satisfying $\langle f, N\rangle_* = \langle f_z, N\rangle_* = \langle f_{\bar z}, N\rangle_* = 0$, where $\mathbb{S}^{1,2} := \{x \in \mathbb{R}^{2,2}; \langle x, x\rangle_* = 1\}$. By a similar computation as in Section 2, we have

$$\begin{aligned}
&f_{zz} = 2u_z f_z + QN,\ f_{z\bar z} = -2\mathrm{e}^{2u} f - 2H\mathrm{e}^{2u} N,\\
&f_{\bar z\bar z} := 2u_{\bar z} f_{\bar z} + \bar{Q}N,\ N_z = -\frac{1}{2}(2Hf_z + Q\mathrm{e}^{-2u} f_{\bar z}),\\
&N_{\bar z} = -\frac{1}{2}(2Hf_{\bar z} + \bar{Q}\mathrm{e}^{-2u} f_z).
\end{aligned}$$

Setting $\mathcal{F} := (f, f_z, f_{\bar z}, N)$, we have $\mathcal{F}_z = \mathcal{F}\mathcal{U}$, $\mathcal{F}_{\bar z} = \mathcal{F}\mathcal{V}$, where

$$\mathcal{U} = \begin{pmatrix} 0 & 0 & 2\mathrm{e}^{2u} & 0 \\ 1 & 2u_z & 0 & -H \\ 0 & 0 & 0 & -\frac{1}{2}Q\mathrm{e}^{-2u} \\ 0 & Q & 2H\mathrm{e}^{2u} & 0 \end{pmatrix},$$
$$\mathcal{V} = \begin{pmatrix} 0 & 2\mathrm{e}^{2u} & 0 & 0 \\ 0 & 0 & 0 & -\frac{1}{2}\bar{Q}\mathrm{e}^{-2u} \\ 1 & 0 & 2u_{\bar z} & -H \\ 0 & 2H\mathrm{e}^{2u} & \bar{Q} & 0 \end{pmatrix}.$$

The compatibility condition $\mathcal{F}_{z\bar z} = \mathcal{F}_{\bar z z}$ gives

$$\begin{aligned}
4u_{z\bar z} + 4\mathrm{e}^{2u}(H^2 - 1) - Q\bar{Q}\mathrm{e}^{-2u} = 0 \quad &\text{(Gauss equation)},\\
Q_{\bar z} = 2H_z\mathrm{e}^{2u} \quad &\text{(Codazzi equation)}.
\end{aligned}$$

From here, we consider 2×2 Lax pairs for Lorentz conformal CMC surfaces in $\mathbb{H}^{2,1}$. First we identify

$$\begin{array}{ccc}
\mathbb{R}^{2,2} & \longrightarrow & \left\{\begin{pmatrix} x_1 + x_4 & x_3 - j'x_2 \\ x_3 + j'x_2 & x_1 - x_4 \end{pmatrix}; x_i \in \mathbb{R}\right\} \\
\cup\!\!\!\text{I} & & \cup\!\!\!\text{I} \\
x = (x_1, x_2, x_3, x_4)^t & \longmapsto & X = \begin{pmatrix} x_1 + x_4 & x_3 - j'x_2 \\ x_3 + j'x_2 & x_1 - x_4 \end{pmatrix},
\end{array}$$

and the Lorentz metric becomes

$$\langle X, Y\rangle_{\mathbb{R}^{2,2}} = -\frac{1}{2}\mathrm{trace}\left(X\begin{pmatrix}0 & -1\\ 1 & 0\end{pmatrix}Y\begin{pmatrix}0 & -1\\ 1 & 0\end{pmatrix}\right)$$

for X, Y considered in matrix form.

Here we give a description of rigid motions of $\mathbb{R}^{2,2}$. For a given point $X \in \mathbb{R}^{2,2}$, consider the adjoint action $X \mapsto A \cdot X \cdot \bar{A}^t$, where $A = \begin{pmatrix}a & b\\ c & d\end{pmatrix} \in \mathrm{SL}_2\mathbb{C}'$ for $a = a_1 + j'a_2$, $b = b_1 + j'b_2$, $c = c_1 + j'c_2$, $d = d_1 + j'd_2$ with $a_i, b_i, c_i, d_i \in \mathbb{R}$. By a straightforward computation, one can show that the adjoint action induces the following matrix R:

Lemma 5.1. *Set* $\mathrm{SO}_{2,2} := \{G \in \mathrm{SL}_4\mathbb{R}; GI''G^t = I''\}$, *where*

$$I'' = \begin{pmatrix}1 & 0 & 0 & 0\\ 0 & 1 & 0 & 0\\ 0 & 0 & -1 & 0\\ 0 & 0 & 0 & -1\end{pmatrix}.$$

Then, by an adjoint action $X \mapsto A \cdot X \cdot \bar{A}^t$, x, *the vector form of* X, *maps to* $R \cdot x$, *where*

$$R = \begin{pmatrix}\alpha_1 & \mathrm{Im}(\bar{a}b + \bar{c}d) & \mathrm{Re}(\bar{a}b + \bar{c}d) & \alpha_2\\ \mathrm{Im}(\bar{a}c + \bar{b}d) & \mathrm{Re}(\bar{a}d - \bar{b}c) & \mathrm{Im}(\bar{a}d + \bar{b}c) & \mathrm{Im}(\bar{a}c - \bar{b}d)\\ \mathrm{Re}(\bar{a}c + \bar{b}d) & \mathrm{Im}(\bar{a}d - \bar{b}c) & \mathrm{Re}(\bar{a}d + \bar{b}c) & \mathrm{Re}(\bar{a}c - \bar{b}d)\\ \alpha_3 & \mathrm{Im}(\bar{a}b - \bar{c}d) & \mathrm{Re}(\bar{a}b - \bar{c}d) & \alpha_4\end{pmatrix} \in \mathrm{SO}_{2,2}$$

with

$$\alpha_1 = \frac{1}{2}(a\bar{a} + b\bar{b} + c\bar{c} + d\bar{d}),\ \alpha_2 = \frac{1}{2}(a\bar{a} - b\bar{b} + c\bar{c} - d\bar{d}),$$

$$\alpha_3 = \frac{1}{2}(a\bar{a} + b\bar{b} - c\bar{c} - d\bar{d}),\ \alpha_4 = \frac{1}{2}(a\bar{a} - b\bar{b} - c\bar{c} + d\bar{d}).$$

Using the matrix form, $\mathbb{H}^{2,1}$ is expressed as

$$\mathbb{H}^{2,1} = \{X | X = \bar{X}^t,\ \langle X, X\rangle_{\mathbb{R}^{2,2}} = -1\}.$$

Moreover, we have another expression of $\mathbb{H}^{2,1}$ as in Lemma 5.2. The proof of Lemma 5.2 is almost the same as the proof of Lemma 5.2.1 in [5], so here we mention only the result.

Lemma 5.2. $\mathbb{H}^{2,1}$ *can be written as* $\left\{A\begin{pmatrix}1 & 0\\ 0 & -1\end{pmatrix}\bar{A}^t; A \in \mathrm{SL}_2\mathbb{C}'\right\}$.

In this setting, there exists an $F \in \mathrm{SL}_2\mathbb{C}'$ such that

$$f = F \begin{pmatrix} 1 & 0 \\ 0 & -1 \end{pmatrix} \bar{F}^t, \ e_1 = F \begin{pmatrix} 0 & -j' \\ j' & 0 \end{pmatrix} \bar{F}^t,$$
$$e_2 = F \begin{pmatrix} 0 & 1 \\ 1 & 0 \end{pmatrix} \bar{F}^t, \ N = F\bar{F}^t,$$

where $e_1 := \dfrac{f_u}{2\mathrm{e}^u}$, $e_2 := \dfrac{f_v}{2\mathrm{e}^u}$. Our task is to determine such $F = F(z, \bar{z}) \in \mathrm{SL}_2\mathbb{C}'$. Defining $U := F^{-1}F_z = \begin{pmatrix} U_{11} & U_{12} \\ U_{21} & U_{22} \end{pmatrix}$, $V := F^{-1}F_{\bar{z}} = \begin{pmatrix} V_{11} & V_{12} \\ V_{21} & V_{22} \end{pmatrix}$, we can write

$$f_u = 2j'\mathrm{e}^u F \begin{pmatrix} 0 & -1 \\ 1 & 0 \end{pmatrix} \bar{F}^t, \ f_v = 2\mathrm{e}^u F \begin{pmatrix} 0 & 1 \\ 1 & 0 \end{pmatrix} \bar{F}^t,$$

and we have

$$f_z = 2j'\mathrm{e}^u F \begin{pmatrix} 0 & 0 \\ 1 & 0 \end{pmatrix} \bar{F}^t, \ f_{\bar{z}} = -2j'\mathrm{e}^u F \begin{pmatrix} 0 & 1 \\ 0 & 0 \end{pmatrix} \bar{F}^t.$$

The compatibility condition $f_{z\bar{z}} = f_{\bar{z}z}$ of f implies that $V_{12} = -\overline{V_{12}}$, $u_{\bar{z}} + V_{22} + \overline{U_{11}} = 0$, $U_{21} = -\overline{U_{21}}$. By the condition $f_{z\bar{z}} = 2\mathrm{e}^{2u}f + 2H\mathrm{e}^{2u}N$, we have $V_{12} = j'\mathrm{e}^u(1+H)$, $U_{21} = j'\mathrm{e}^u(1-H)$, and the condition $f_{zz} = 2u_z f_z + QN$ implies $U_{12} = \frac{j'}{2}Q\mathrm{e}^{-u}$, $u_z - U_{22} - \overline{V_{11}} = 0$, $V_{21} = -\frac{j'}{2}\bar{Q}\mathrm{e}^{-u}$, and also the condition $N_z = -Hf_z - \frac{1}{2}Q\mathrm{e}^{-2u}f_{\bar{z}}$ gives $U_{11} = -\overline{V_{11}}$, $U_{22} = -\overline{V_{22}}$. In conclusion, we have the following proposition.

Proposition 5.1. *For some $F_{z_0 \in \Sigma} \in \mathrm{SL}_2\mathbb{C}'$, we obtain the solution $F \in \mathrm{SL}_2\mathbb{C}'$ by solving $F_z = FU$, $F_{\bar{z}} = FV$, where*

$$U = \frac{1}{2}\begin{pmatrix} -u_z & j'Q\mathrm{e}^{-u} \\ 2j'\mathrm{e}^u(1-H) & u_z \end{pmatrix},$$
$$V = \frac{1}{2}\begin{pmatrix} u_{\bar{z}} & 2j'\mathrm{e}^u(1+H) \\ -j'\bar{Q}\mathrm{e}^{-u} & -u_{\bar{z}} \end{pmatrix}.$$

Proposition 5.1 implies that we obtain any CMC H surface in $\mathbb{H}^{2,1}$ by solving the equations in Proposition 5.1 and inserting the solution into $f = F\begin{pmatrix} 1 & 0 \\ 0 & -1 \end{pmatrix}\bar{F}^t$. We should remark that, for any $B \in \mathrm{SU}'_{1,1} := \left\{ \begin{pmatrix} p & q \\ \bar{q} & \bar{p} \end{pmatrix} \middle| \, p, q \in \mathbb{C}', \ |p|_*^2 - |q|_*^2 = 1 \right\}$, replacing F in $f = F\begin{pmatrix} 1 & 0 \\ 0 & -1 \end{pmatrix}\bar{F}^t$

with FB does not change the resulting surface. Like in the case of conformal CMC 1 surfaces in $\mathbb{H}^3$, we would like to determine the $B \in \mathrm{SU}'_{1,1}$ so that FB is anti p-holomorphic i.e. $(FB)_z = 0$. The condition $(FB)_z = 0$ implies $B_z = -UB$. Setting $W := B_{\bar{z}}B^{-1}$, we have the following expression

$$B_x = (W - U)B,\ B_y = -j'(W + U)B.$$

Then we must choose two matrices $W - U,\ -j'(W + U) \in \mathrm{su}'_{1,1}$ so that $B \in \mathrm{SU}'_{1,1}$, where $\mathrm{su}'_{1,1}$ is the Lie algebra of $\mathrm{SU}'_{1,1}$. By a simple computation, we can show that

$$W - U,\ -j'(W + U) \in \mathrm{su}'_{1,1} \Leftrightarrow W = \begin{pmatrix} 1 & 0 \\ 0 & -1 \end{pmatrix} \bar{U}^t \begin{pmatrix} 1 & 0 \\ 0 & -1 \end{pmatrix}.$$

Here we assume that $H \equiv 1$. Then the compatibility condition $U_{\bar{z}} + W_z + [U, W] = 0$ for B does hold. Setting $B = \begin{pmatrix} p & q \\ \bar{q} & \bar{p} \end{pmatrix} \in \mathrm{SU}'_{1,1}$, we have

$$\begin{aligned} (FB)^{-1}(FB)_{\bar{z}} &= B^{-1}\left\{V + \begin{pmatrix} 1 & 0 \\ 0 & -1 \end{pmatrix} \bar{U}^t \begin{pmatrix} 1 & 0 \\ 0 & -1 \end{pmatrix}\right\} B \\ &= 2j'\mathrm{e}^u \begin{pmatrix} \bar{p}\bar{q} & \bar{p}^2 \\ -\bar{q}^2 & -\bar{p}\bar{q} \end{pmatrix}. \end{aligned}$$

So $(\bar{F}\bar{B})^{-1}(\bar{F}\bar{B})_z = -2j'\mathrm{e}^u \begin{pmatrix} pq & p^2 \\ -q^2 & -pq \end{pmatrix} = 2\mathrm{e}^u p^2 \begin{pmatrix} -\frac{j'q}{p} & -j' \\ j'\left(\frac{j'q}{p}\right)^2 & \frac{j'q}{p} \end{pmatrix}$.

Setting $g = \dfrac{j'q}{p}$, $\hat{\omega} = 2\mathrm{e}^u p^2$, $\hat{F} = \bar{F}\bar{B}$, and we have

$$\hat{F}_z = \hat{F} \begin{pmatrix} -g & -j' \\ j'g^2 & g \end{pmatrix} \hat{\omega}.$$

Setting $\tilde{F} := (\hat{F}^{-1})^t$, we have $\tilde{F}_z = \tilde{F} \begin{pmatrix} g & -j'g^2 \\ j' & -g \end{pmatrix} \hat{\omega}$. Note that, writing $\hat{f} = \hat{F} \begin{pmatrix} 1 & 0 \\ 0 & -1 \end{pmatrix} \left(\overline{\hat{F}}\right)^t = (x_1, x_2, x_3, x_4)^t$ in the vector form, we have $\tilde{f} = \tilde{F} \begin{pmatrix} 1 & 0 \\ 0 & -1 \end{pmatrix} \left(\overline{\tilde{F}}\right)^t = (-x_1, -x_2, x_3, x_4)^t$. So $\hat{f}$ and $\tilde{f}$ coincide, up to a rigid motion of $\mathbb{R}^{2,2}$. In conclusion, replacing $\tilde{F}$ with F, we have the following theorem, which we call the Weierstrass-type representation for Lorentz conformal CMC 1 surfaces in $\mathbb{H}^{2,1}$.

Theorem 2. *Any Lorentz conformal CMC 1 surface in* $\mathbb{H}^{2,1}$ *can be locally constructed in the following way*:

(1) *Solve*

$$F_z = F\begin{pmatrix} g & -j'g^2 \\ j' & -g \end{pmatrix}\hat{\omega}$$

with some initial condition $F_{z_0 \in \Sigma} \in \mathrm{SL}_2\mathbb{C}'$.

(2) *Substitute* F *in* (1) *into* $f = F\begin{pmatrix} 1 & 0 \\ 0 & -1 \end{pmatrix}\bar{F}^t$.

Furthermore, the metric of f *becomes* $df^2 = (1 + |g|_*^2)^2|\omega|_*^2 dz d\bar{z}$.

Like in the case of Lorentz conformal minimal surfaces in $\mathbb{R}^{2,1}$, replacing g and $\hat{\omega}$ with $j'g$ and $j'\hat{\omega}$, we have another expression for the Weierstrass-type representation for Lorentz conformal CMC 1 surfaces in $\mathbb{H}^{2,1}$, as follows:

Proposition 5.2. *Any Lorentz conformal CMC 1 surface in* $\mathbb{H}^{2,1}$ *can be locally constructed in the following way*:

(1) *Solve*

$$F_z = F\begin{pmatrix} g & -g^2 \\ 1 & -g \end{pmatrix}\hat{\omega}$$

with some initial condition $F_{z_0 \in \Sigma} \in \mathrm{SL}_2\mathbb{C}'$.

(2) *Substitute* F *in* (1) *into* $f = F\begin{pmatrix} 1 & 0 \\ 0 & -1 \end{pmatrix}\bar{F}^t$.

Furthermore, the metric of f *becomes* $df^2 = -(1 - |g|_*^2)^2|\hat{\omega}|_*^2 dz d\bar{z}$.

Like in the case of conformal CMC 1 surfaces in 3-dimensional de Sitter space, surfaces described by Theorem 2 (or Proposition 5.2) generally have singularities. Their singularities are analyzed in Section 7.

§6. Singularities of Lorentz conformal minimal surfaces in $\mathbb{R}^{2,1}$

Here we analyze singularities of Lorentz conformal minimal surfaces in $\mathbb{R}^{2,1}$. Note that surfaces described by Equation (2) are locally Lorentz conformal minimal immersions in $\mathbb{R}^{2,1}$, but this is not the case globally. In fact, the following proposition holds.

Lemma 6.1. *Let* f *be a surface in* $\mathbb{R}^{2,1}$ *given by Equation* (2). *Then, away from* $|\omega|_*^2 = 0$, f *has singularities if and only if* $|g|_*^2 = -1$.

Proof. By a direct computation, we have

$$f_x = \left(g\omega + \bar{g}\bar{\omega}, \frac{1}{2}(\omega + \bar{\omega} - g^2\omega - \bar{g}^2\bar{\omega}), -\frac{j'}{2}(\omega - \bar{\omega} + g^2\omega - \bar{g}^2\bar{\omega})\right)^t,$$

$$f_y = j'\left(g\omega - \bar{g}\bar{\omega}, \frac{1}{2}(\omega - \bar{\omega} - g^2\omega + \bar{g}^2\bar{\omega}), -\frac{j'}{2}(\omega + \bar{\omega} + g^2\omega + \bar{g}^2\bar{\omega})\right)^t.$$

Now we regard f_x and f_y as vectors in $\mathbb{R}^3$ and we take the vector product $f_x \times_{\mathbb{R}^3} f_y$ of f_x and f_y in $\mathbb{R}^3$. Then we have

$$f_x \times_{\mathbb{R}^3} f_y = (1 + |g|_*^2)|\omega|_*^2(-1 + |g|_*^2, 2\mathrm{Re}(g), -2\mathrm{Im}(g))^t.$$

Since we assume that $|\omega|_*^2 \neq 0$, f has singularities when $|g|_*^2 = -1$. Q.E.D.

Remark. By the proof of Lemma 6.1, we can take the unit normal vector field $\nu : D \to \mathbb{S}^2$ of f, as a surface in $\mathbb{R}^3$, as

$$\nu = \frac{(-1 + |g|_*^2, 2\mathrm{Re}(g), -2\mathrm{Im}(g))^t}{\sqrt{(|g|_*^2 - 1)^2 + 4(\mathrm{Re}(g)^2 + \mathrm{Im}(g)^2)}}. \tag{4}$$

Here we introduce useful criteria for an image of a singular point to be $\mathcal{A}$-equivalent to a cuspidal edge, swallowtail, or cuspidal cross cap, which was shown in [7], [12] (see also [18], [20]).

Proposition 6.1 ([7], [12]). *Let $p = \gamma(0) \in D \subset \mathbb{R}^2$ be a non-degenerate singular point of a front $f : D \to \mathbb{R}^3$, let $\gamma(t)$ be a singular curve around p, and let $\eta(t)$ be a vector field of null directions along $\gamma(t)$. Then we have the following*:

(1) *The image $f(p)$ is $\mathcal{A}$-equivalent to a cuspidal edge if and only if $\eta(0)$ is not proportional to $\dot{\gamma}$, where $\dot{\gamma} = d\gamma/dt$.*

(2) *The image $f(p)$ is $\mathcal{A}$-equivalent to a swallowtail if and only if $\eta(0)$ is proportional to $\dot{\gamma}$, and*

$$\frac{d}{dt}\det(\dot{\gamma}, \eta(t))\Big|_{t=0} \neq 0.$$

(3) *Let $f : D \to \mathbb{R}^3$ be a frontal with normal vector field ν, and let $\gamma(t)$ be a singular curve on D passing through a non-degenerate singular point $p = \gamma(0)$. Then the image $f(p)$ is $\mathcal{A}$-equivalent to a cuspidal cross cap if and only if*
   - $\det(\dot{\gamma}(0), \eta(0)) \neq 0$,
   - $\det(df(\dot{\gamma}(0)), \nu(0), d\nu(\eta(0))) = 0$,
   - $\displaystyle\frac{d}{dt}\det(df(\dot{\gamma}(0)), \nu(0), d\nu(\eta(0)))\Big|_{t=0} \neq 0.$

Applying these useful criteria for such singularities, we give explicit conditions for cuspidal edge, swallowtail and cuspidal cross cap singularities of Lorentz conformal minimal surfaces to appear, which was obtained by Takahashi [18]:

Theorem 3. *Let $f : \Sigma \to \mathbb{R}^{2,1}$ be a surface given by Equation (2), and let $p \in \Sigma$ be a singular point of f. Then*

(1) *$f(p)$ is $\mathcal{A}$-equivalent to a cuspidal edge if and only if*

$$\mathrm{Re}\left(\frac{g'}{g^2\hat{\omega}}\right) \neq 0, \quad \mathrm{Im}\left(\frac{g'}{g^2\hat{\omega}}\right) \neq 0,$$

(2) *$f(p)$ is $\mathcal{A}$-equivalent to a swallowtail if and only if*

$$\frac{g'}{g^2\hat{\omega}} \in \mathbb{R} \setminus \{0\}, \quad \mathrm{Re}\left\{\frac{g}{g'}\left(\frac{g'}{g^2\hat{\omega}}\right)'\right\} \neq 0,$$

(3) *$f(p)$ is $\mathcal{A}$-equivalent to a cuspidal cross cap if and only if*

$$\frac{g'}{g^2\hat{\omega}} \in j'\mathbb{R} \setminus \{0\}, \quad \mathrm{Im}\left\{\frac{g}{g'}\left(\frac{g'}{g^2\hat{\omega}}\right)'\right\} \neq 0,$$

where $' = d/dz$.

In [18], Takahashi substituted the Weierstrass-type representation (3) for another expression as in Theorem 4.3 in [16] (in [18], this expression is called the real representation for timelike minimal surfaces in $\mathbb{R}^{2,1}$). After that, he derived explicit criteria for singularities of timelike minimal surfaces [16] in $\mathbb{R}^{2,1}$. As a corollary, he showed Theorem 3 here. In this paper we give Theorem 3 more directly. In order to show Theorem 3, we give the following lemma.

Lemma 6.2. *Let f be a surface described by Equation* (2) *and let p be a point in Σ. Then f is a front on a neighborhood of p, and p is a non-degenerate singular point if and only if* $\mathrm{Re}\left(\dfrac{g'}{g^2\hat{\omega}}\right) \neq 0$.

Proof. Here we assume that $|g|_*^2 = -1$ and $|\hat{\omega}|_*^2 \neq 0$. Then

$$\begin{aligned} df &= \frac{1}{2}\left(2, \frac{1}{g} - g, -j\left(\frac{1}{g} + g\right)\right)^t g\omega + \frac{1}{2}\left(2, \frac{1}{\bar{g}} - \bar{g}, j\left(\frac{1}{\bar{g}} + \bar{g}\right)\right)^t \bar{g}\bar{\omega} \\ &= \frac{1}{2}\left(2, \frac{1}{g} - g, -j\left(\frac{1}{g} + g\right)\right)^t (g\omega + \bar{g}\bar{\omega}) \\ &= (1, -\mathrm{Re}(g), -\mathrm{Im}(g))^t (g\omega + \bar{g}\bar{\omega}). \end{aligned}$$

In particular, $\eta = \dfrac{j}{g\hat{\omega}}$ gives the null direction at p with the following identification:

$$(a, b) \in \mathbb{R}^2 \leftrightarrow z := a + j'b \in \mathbb{C}' \leftrightarrow a\partial_x + b\partial_y \leftrightarrow z\partial_z + \bar{z}\partial_{\bar{z}}.$$

Take ν as in Equation (4), and we have

$$d\nu = \mathrm{sgn}(\mathrm{Im}(g))\frac{j}{2\sqrt{2}}\left(\frac{dg}{g} - \frac{d\bar{g}}{\bar{g}}\right)\frac{1}{(\mathrm{Im}(g))^2}(\mathrm{Re}(g), 1, 0)^t.$$

If $dg(p) = 0$, the pair $(f, \nu) : \Sigma \to \mathbb{R}^3 \times \mathbb{S}^2$ is not immersed. Assume that $dg \neq 0$, and we have the null direction of $d\nu$ at p which is proportional to $\mu := \overline{\left(\frac{g'}{g}\right)}$. On the other hand, f is a front on a neighborhood of p if and only if η and μ are linearly independent, implying

$$\det(\mu, \eta) = \mathrm{Im}(\bar{\mu}\eta) = \mathrm{Im}\left(\frac{g'}{g} \cdot \frac{j'}{g\hat{\omega}}\right) = \mathrm{Re}\left(\frac{g'}{g^2\hat{\omega}}\right) \neq 0.$$

Define the signed area density as follows:

$$\lambda := (f_x \times f_y) \cdot \nu = (1 + |g|_*^2)|\omega|_*^2\sqrt{(|g|_*^2 - 1)^2 + 4(\mathrm{Re}(g)^2 + \mathrm{Im}(g)^2)},$$

where $\cdot$ in the above equation denotes the ordinary Euclidean inner product. When p is a singular point, since $|g(p)|_*^2 = -1$, we have

$$d\lambda = -2\sqrt{2}|\mathrm{Im}(g)||\hat{\omega}|_*^2\left(\frac{dg}{g} + \frac{d\bar{g}}{\bar{g}}\right).$$

Thus we have that $d\lambda(p) \neq 0$ if and only if $dg(p) \neq 0$. Therefore, if $\mathrm{Re}\left(\frac{g'}{g^2\hat{\omega}}\right)$ holds at p, p is non-degenerate, since $dg(p) \neq 0$. Q.E.D.

Here we go back to the proof of Theorem 3. First we assume that $\mathrm{Re}\left(g'/(g^2\hat{\omega})\right) \neq 0$ at a singular point p. This condition implies that f is a front and p is a non-degenerate singular point. Since the set of singular points must satisfy $|g|_*^2 = -1$, the singular curve $\gamma(t)$ with $\gamma(0) = p$ satisfies $g(\gamma(t))\overline{g(\gamma(t))} = -1$. Differentiating this equation with respect to t implies $\mathrm{Re}\left(\frac{g'}{g}\dot{\gamma}\right) = 0$, where $\dot{\gamma} := \frac{d\gamma}{dt}$. This implies

$$\dot{\gamma} \perp \overline{\left(\frac{g'}{g}\right)} \Rightarrow \dot{\gamma} \parallel j'\overline{\left(\frac{g'}{g}\right)}.$$

So we can parametrize γ as $\dot{\gamma}(t) = j'\overline{\left(\frac{g'}{g}\right)}(\gamma(t))$. Applying item (1) in Proposition 6.1, we have

$$\det(\dot{\gamma}, \eta) = \mathrm{Im}(\bar{\dot{\gamma}}\eta) = -\mathrm{Im}\left(\frac{g'}{g^2\hat{\omega}}\right) \neq 0.$$

Thus we have proven item (1) in Theorem 3. Next we assume that $\operatorname{Im}\left(\frac{g'}{g^2\hat{\omega}}\right)\Big|_{t=0} = 0$. Then

$$\begin{aligned}
\frac{d}{dt}\det(\dot{\gamma},\eta)\Big|_{t=0} &= -\operatorname{Im}\left\{\left(\frac{g'}{g^2\hat{\omega}}\right)'\frac{d\gamma}{dt}\right\} \\
&= -\operatorname{Im}\left\{j'\left(\frac{g'}{g^2\hat{\omega}}\right)'\overline{\left(\frac{g'}{g}\right)}\right\} = -\operatorname{Re}\left\{\left(\frac{g'}{g^2\hat{\omega}}\right)'\overline{\left(\frac{g'}{g}\right)}\right\} \\
&= -\left|\frac{g'}{g}\right|_*^2 \operatorname{Re}\left\{\frac{g}{g'}\left(\frac{g'}{g^2\hat{\omega}}\right)'\right\} \neq 0.
\end{aligned}$$

Applying item (2) in Proposition 6.1, we have the condition of item (2) in Theorem 3. Finally we show item (3) in Theorem 3. First we have

$$\det(\dot{\gamma}(0),\eta(0)) = \operatorname{Im}(\bar{\dot{\gamma}},\eta) = -\operatorname{Im}\left(\frac{g'}{g^2\hat{\omega}}\right) \neq 0,$$
$$\det(df(\dot{\gamma}),\nu,d\nu(\eta)) = \operatorname{Re}\left(\frac{g'}{g^2\hat{\omega}}\right)\cdot\psi_0 = 0,$$

where ψ_0 is a smooth function on a neighborhood of p satisfying $\psi_0(p) \neq 0$. Thus we have the first condition of item (3) in Theorem 3. By the last condition of item (3) in Proposition 6.1, we have

$$\begin{aligned}
\frac{d}{dt}\det(df(\dot{\gamma}),\nu,d\nu(\eta))\Big|_{t=0} &= \frac{d}{dt}\operatorname{Re}\left(\frac{g'}{g^2\hat{\omega}}\right)\Big|_{t=0} \\
&= \left|\frac{g'}{g}\right|_*^2 \operatorname{Im}\left\{\frac{g}{g'}\left(\frac{g'}{g^2\hat{\omega}}\right)'\right\} \neq 0.
\end{aligned}$$

This completes the proof of Theorem 3.

§7. Singularities of Lorentz conformal CMC 1 surfaces in $\mathbb{H}^{2,1}$

Here we introduce our second result about criteria for Lorentz conformal CMC 1 surfaces in $\mathbb{H}^{2,1}$. Again we assume that $|\hat{\omega}|_*^2 \neq 0$. We can use the Weierstrass-type representation in Proposition 5.2 to prove Theorem 4, but we will omit the complete proof, as the following result is analogous to that of Theorem 3.

Theorem 4. *Let $f : \Sigma \to \mathbb{H}^{2,1}$ be a surface in $\mathbb{H}^{2,1}$ described by Proposition 5.2 and let p be a non-degenerate singular point of f. Then f has singularities if and only if $|g(p)|_*^2 = 1$, and*

(1) *$f(p)$ is $\mathcal{A}$-equivalent to a cuspidal edge if and only if*

$$\mathrm{Re}\left(\frac{g'}{g^2\hat{\omega}}\right) \neq 0, \quad \mathrm{Im}\left(\frac{g'}{g^2\hat{\omega}}\right) \neq 0,$$

(2) *$f(p)$ is $\mathcal{A}$-equivalent to a swallowtail if and only if*

$$\frac{g'}{g^2\hat{\omega}} \in \mathbb{R} \setminus \{0\}, \quad \mathrm{Re}\left\{\frac{g}{g'}\left(\frac{g'}{g^2\hat{\omega}}\right)'\right\} \neq 0,$$

(3) *$f(p)$ is $\mathcal{A}$-equivalent to a cuspidal cross cap if and only if*

$$\frac{g'}{g^2\hat{\omega}} \in j'\mathbb{R} \setminus \{0\}, \quad \mathrm{Im}\left\{\frac{g}{g'}\left(\frac{g'}{g^2\hat{\omega}}\right)'\right\} \neq 0.$$

Due to the replacement of $\hat{F}$ with $\tilde{F}$ in Section 5 (see the above argument for Theorem 2), the proof of Theorem 4 is almost the same as of Theorem 3.4 in [7], so here we remark on only the differences.

Lemma 7.1. *Let f be a surface in $\mathbb{H}^{2,1}$ described by Proposition 5.2. Away from $|\hat{\omega}|_*^2 \neq 0$, f has singular points if and only if $|g(p)|_*^2 = 1$.*

Proof. Define $\xi := F\bar{F}^t$ and a 3-form Ω on $\mathbb{H}^{2,1}$ by

$$\Omega(X_1, X_2, X_3) := \det(f, X_1, X_2, X_3)$$

for arbitrary vector fields X_1, X_2, X_3 of $\mathbb{H}^{2,1}$, where f denotes the position vector in $\mathbb{H}^{2,1}$. Then Ω gives a volume element on $\mathbb{H}^{2,1}$, since

$$\begin{aligned}
\Omega(f_x, f_y, \xi) &= \det(f, f_x, f_y, \xi) \\
&= \det\begin{pmatrix} 0 & 2\mathrm{Re}(g\hat{\omega}) & 2\mathrm{Im}(g\hat{\omega}) & 1 \\ 0 & \mathrm{Im}((1-g^2)\hat{\omega}) & \mathrm{Re}((1-g^2)\hat{\omega}) & 0 \\ 0 & \mathrm{Re}((1+g^2)\hat{\omega}) & \mathrm{Im}((1+g^2)\hat{\omega}) & 0 \\ 1 & 0 & 0 & 0 \end{pmatrix} \\
&= (1-|g|_*^2)(1+|g|_*^2)|\hat{\omega}|_*^2.
\end{aligned}$$

Note that, unlike the case of CMC 1 faces in $\mathbb{S}^{2,1}$, f might have singularities if $|g|_*^2 = -1$. On the other hand, we can confirm that f does not have singularities in that case, proving the lemma. Q.E.D.

Except for that point, if we admit the same η and μ as in the case of Lorentz conformal minimal surfaces in $\mathbb{R}^{2,1}$ (see the proof of Lemma 6.2), the proof of Theorem 4 is almost the same as the one of Theorem 3.4 in [7]. Thus we omit the proof here.

Finally, we see duality of singularities for Lorentz conformal minimal surfaces in $\mathbb{R}^{2,1}$. As a direct consequence of Theorem 3, we immediately have the following property.

Corollary 7.1 ([18]). *Let f be a surface in $\mathbb{R}^{2,1}$ described by Equation (2), let $f^\sharp$ be its conjugate surface and let p be a singular point of f and $f^\sharp$. Then $f(p)$ is $\mathcal{A}$-equivalent to a swallowtail if and only if $f^\sharp(p)$ is $\mathcal{A}$-equivalent to a cuspidal cross cap.*

Remark. Since the criteria for the three types of singularities in Theorems 3, 4 are exactly the same, the same duality holds in the case of Lorentz conformal CMC 1 faces in $\mathbb{H}^{2,1}$.

§Appendix A. Introduction to discrete timelike isothermic surfaces in $\mathbb{R}^{2,1}$

In this appendix we briefly introduce discrete timelike isothermic surfaces in $\mathbb{R}^{2,1}$. First we define two kinds of smooth timelike surfaces in $\mathbb{R}^{2,1}$ and $\mathbb{H}^{2,1}$.

Definition A.1. *Let f be a timelike immersion into $\mathbb{R}^{2,1}$. Then f is timelike isothermic (resp. anti isothermic) if f admits Lorentz conformal curvature line coordinates (resp. Lorentz conformal asymptotic coordinates).*

Magid [15] considered Lorentz isothermic surfaces in $\mathbb{R}^{n-j,j}$. As mentioned in [15], there are three kinds of Lorentz isothermic surfaces in $\mathbb{R}^{n-j,j}$. Here we only consider the case $n = 2$, $j = 1$. In this case we can characterize such Lorentz isothermic surfaces in $\mathbb{R}^{2,1}$ using the notion of Hopf differential Q (several terminologies can be found in [15]).

Proposition A.1. *Let $f : D(\subset \mathbb{C}') \to \mathbb{R}^{2,1}$ be a Lorentz conformal immersion parametrized by para-complex coordinates $z = x + j'y$ with spacelike unit normal vector field ν. Then f has an umbilic point at $p \in D$ if and only if $Q = 0$ at p. Moreover, f is timelike isothermic (resp. anti isothermic) if and only if Q defined in Section 2 is a non-zero real function (resp. pure para-imaginary unit times non-zero real function).*

Remark. As mentioned in [15], there is another kind of timelike isothermic surface, which has not yet been named. Here we refer to these types of surfaces as timelike surfaces of the third kind. A Lorentz conformal surface is a timelike surface of the third kind if and only if Q is $(1 \pm j')$ times a non-zero real function.

When f is an umbilic-free timelike isothermic surface, we can reparametrize f so that $Q \equiv 1$. In particular, when we consider timelike isothermic minimal surfaces in $\mathbb{R}^{2,1}$, we can reparametrize Lorentz conformal minimal surfaces f described by Equation (2) so that $Q = g'\hat{\omega} \equiv 1$. So the Weierstrass-type representation for timelike isothermic minimal surfaces in $\mathbb{R}^{2,1}$ can be written as

$$f = \mathrm{Re} \int \left(\frac{2g}{g'}, \frac{1-g^2}{g'}, \frac{-j'(1+g^2)}{g'} \right)^t dz.$$

Our first main result in [21] is a discrete analogue of the Weierstrass-type representation for discrete timelike isothermic minimal surfaces (see Theorem 5 here).

We now briefly introduce discrete timelike isothermic surfaces in $\mathbb{R}^{2,1}$. In particular, we introduce discrete timelike (isothermic) minimal surfaces in $\mathbb{R}^{2,1}$. As in Section 3, each element in $\mathbb{R}^{2,1}$ is identified with the matrix as in Equation (1), and we denote $p = (m, n)$, $q = (m+1, n)$, $r = (m+1, n+1)$, $s = (m, n+1)$. Then we define discrete timelike isothermic surfaces in $\mathbb{R}^{2,1}$ as follows (the reason why we define them in this way can be found in [21]):

Definition A.2. *Let $F : \mathbb{Z}^2 \to \mathbb{R}^{2,1}$ be a discrete surface. Then*

- *F is called a* discrete timelike isothermic surface *if*
  - *each quadrilateral (F_p, F_q, F_r, F_s) with vertices F_p, F_q, F_r, F_s lies in a timelike plane,*
  - *all quadrilaterals (F_p, F_q, F_r, F_s) are convex,*
  - *all quadrilaterals satisfy $cr(F_p, F_q, F_r, F_s) = 1$.*
- *A discrete timelike isothermic surface $g : \mathbb{Z}^2 \to \mathbb{R}^{1,1} \cong \mathbb{C}'$ is called a* discrete p-holomorphic function.

Roughly speaking, in Definition A.2, the first two conditions are the discrete counterpart of an immersion condition for a smooth timelike surface.

We have a Weierstrass-type representation for discrete timelike isothermic minimal surfaces in $\mathbb{R}^{2,1}$. Details can be found in [21].

Theorem 5. *A discrete timelike minimal surface $F : \mathbb{Z}^2 \to \mathbb{R}^{2,1}$ can be locally constructed using a discrete p-holomorphic function $g : \mathbb{Z}^2 \to \mathbb{C}'$ by solving*

$$F_q - F_p = \frac{1}{2}\mathrm{Re} \left(\frac{g_p + g_q}{g_q - g_p}, \frac{1 - g_p g_q}{g_q - g_p}, -\frac{j'(1 + g_p g_q)}{g_q - g_p} \right)^t,$$

$$F_s - F_p = \frac{1}{2}\mathrm{Re} \left(\frac{g_p + g_s}{g_s - g_p}, \frac{1 - g_p g_s}{g_s - g_p}, -\frac{j'(1 + g_p g_s)}{g_s - g_p} \right)^t.$$

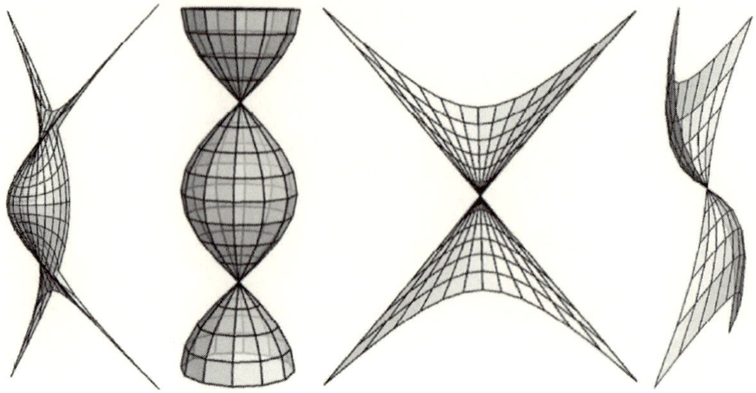

Fig. 1. Examples of discrete timelike isothermic minimal surfaces

Conversely, any discrete timelike minimal surface locally satisfies the above equations for some discrete p-holomorphic function g.

In Figure 1, we show several examples of discrete timelike isothermic minimal surfaces in $\mathbb{R}^{2,1}$. These pictures obviously have certain configurations of singularities. In [21], their singularities are analyzed.

Finally, we introduce several open problems related to the topics in this paper.

- How can we describe discrete timelike isothermic non-zero CMC surfaces in $\mathbb{R}^{2,1}$? If we can describe such discrete surfaces, is there any construction of discrete timelike isothermic CMC surfaces like in [17]?
- Is there a Weierstrass-type representation for discrete timelike isothermic CMC 1 surfaces in $\mathbb{H}^{2,1}$? If yes, do discrete timelike isothermic CMC 1 surfaces in $\mathbb{H}^{2,1}$ have singularities?
- How can we describe discrete anti-isothermic surfaces?

References

[1] S. Akamine, *Behavior of the Gaussian curvature of timelike minimal surfaces with singularities*, to appear in Hokkaido Math. J.

[2] A.I. Bobenko and U. Pinkall, *Discrete isothermic surfaces*, J. Reine Angew. Math. **475** (1996), 187–208.

[3] R. Bryant, *Surfaces of mean curvature one in hyperbolic space*, Astérisque No. 154–155 (1987), **12**, 321–347, 353 (1988).

[4] J. Dorfmeister, J. Inoguchi and M. Toda, *Weierstraß-type representation of timelike surfaces with constant mean curvature*, Differential geometry and integrable systems (Tokyo, 2000), 77–99, Contemp. Math., **308**, Amer. Math. Soc., Providence, RI, 2002.

[5] S. Fujimori, S. Kobayashi and W. Rossman, *Loop group methods for constant mean curvature surfaces*, Rokko Lecture Series **17** (2005).

[6] S. Fujimori, W. Rossman, M. Umehara, K. Yamada and S.-D. Yang, *Embedded triply periodic zero mean curvature surfaces of mixed type in Lorentz-Minkowski 3-space*, Michigan Math. J. **63** (2014), 189–207.

[7] S. Fujimori, K. Saji, M. Umehara and K. Yamada, *Singularities of maximal surfaces*, Math. Z. **259** (2008), no. 4, 827–848.

[8] T. Hoffmann, A. O. Sageman-Furnas and M. Wardetzky, *A discrete parametrized surface theory in* $\mathbb{R}^3$, to appear in IMRN.

[9] J. Inoguchi and M. Toda, *Timelike minimal surfaces via loop groups*, Acta Appl. Math. **83** (2004), no. 3, 313–355.

[10] A. Khrennikov and G. Segre, *An introduction to hyperbolic analysis*, preprint, arXiv:math-ph/0507053.

[11] O. Kobayashi, *Maximal surfaces in the 3-dimensional Minkowski space* L^3. Tokyo J. Math. **6** (1983), no. 2, 297–309.

[12] M. Kokubu, W. Rossman, M. Umehara, K. Saji and K. Yamada, *Singularities of flat fronts in hyperbolic space*, Pacific J. Math. **221** (2005), no. 2, 303–351.

[13] J.J. Konderak, *A Weierstrass representation theorem for Lorentz surfaces*, Complex Var. Theory Appl. **50** (2005), no. 5, 319–332.

[14] S. Lee, *Timelike surfaces of constant mean curvature* ± 1 *in anti-de Sitter 3-space* $\mathbb{H}^3_1(-1)$, Ann. Global Anal. Geom. **29** (2006), no. 4, 361–407.

[15] M.A. Magid, *Lorentzian isothermic surfaces in* $\mathbf{R}^n_j$, Rocky Mountain J. Math. **35** (2005), no. 2, 627–640.

[16] M.A. Magid, *Timelike surfaces in Lorentz 3-space with prescribed mean curvature and Gauss map*, Hokkaido Math. J. **20** (1991), no. 3, 447–464.

[17] Y. Ogata and M. Yasumoto, *Construction of discrete constant mean curvature surfaces in Riemannian spaceforms*, Diff. Geom. Appl. **54** (2017), Part A, 264–281.

[18] H. Takahashi, *Timelike minimal surfaces with singularities (in Japanese)*, master degree thesis, 2012.

[19] M. Umehara and K. Yamada, *Complete surfaces of constant mean curvature 1 in the hyperbolic 3-space*, Ann. of Math. (2) **137** (1993), no. 3, 611–638.

[20] M. Umehara and K. Yamada, *Maximal surfaces with singularities in Minkowski space*, Hokkaido Math. J. 35 (2006), no. 1, 13–40.

[21] M. Yasumoto, *Discrete timelike minimal surfaces and discrete wave equations*, in preparation.

Osaka City University Advanced Mathematical Institute,
3-3-138 Sugimoto, Sumiyoshi-ku Osaka 558
E-mail address, M. Yasumoto: `yasumoto@sci.osaka-cu.ac.jp`

Advanced Studies in Pure Mathematics

A SERIES OF UP-TO-DATE GUIDES OF LASTING INTEREST
TO ADVANCED MATHEMATICS

Volume 16 Conformal Field Theory and Solvable Lattice Models.
Edited by M. Jimbo, T. Miwa and A. Tsuchiya.
June, 1988

Volume 17 Algebraic Number Theory—in honor of K. Iwasawa.
Edited by J. Coates, R. Greenberg, B. Mazur and I. Satake. August, 1989

Volume 18 I—Recent Topics in Differential and Analytic Geometry.
Edited by T. Ochiai. November, 1990
II—Kähler Metric and Moduli Spaces.
Edited by Edited by T. Ochiai. November, 1990

Volume 19 Integrable Systems in Quantum Field Theory and Statistical Mechanics.
Edited by M. Jimbo, T. Miwa and A. Tsuchiya.
November, 1989

Volume 20 Aspects of Low Dimensional Manifolds.
Edited by Y. Matsumoto and S. Morita.
December, 1992

Volume 21 Zeta Functions in Geometry.
Edited by N. Kurokawa and T. Sunada.
December, 1992

Volume 22 Progress in Differential Geometry.
Edited by K. Shiohama. September, 1993

Volume 23 Spectral and Scattering Theory and Applications.
Edited by K. Yajima. February, 1994

Volume 24 Progress in Algebraic Combinatorics.
Edited by E. Bannai and A. Munemasa. May, 1996

Volume 25 CR-Geometry and Overdetermined Systems.
Edited by T. Akahori, G. Komatsu, K. Miyajima, M. Namba and K. Yamaguchi. July, 1997

Volume 26 Analysis on Homogeneous Spaces and Representation Theory of Lie Groups, Okayama-Kyoto.
Edited by M. Kashiwara, T. Kobayashi, T. Matsuki, K. Nishiyama and T. Oshima. April, 2000

Volume 27 Arrangements – Tokyo 1998.
Edited by M. Falk and H. Terao. August, 2000

Volume 28 Combinatorial Methods in Representation Theory.
Edited by M. Kashiwara, K. Koike, S. Okada, I. Terada, and H.-F. Yamada. December, 2000

Volume 29 Singularities – Sapporo 1998.
Edited by J.-P. Brasselet and T. Suwa.
December, 2000

Volume 30 Class Field Theory – Its Centenary and Prospect.
Edited by K. Miyake. April, 2001

Volume 31 Taniguchi Conference on Mathematics Nara '98.
Edited by M. Maruyama and T. Sunada. May, 2001

Volume 32 Groups and Combinatorics — in memory of Michio Suzuki.
Edited by E. Bannai, H. Suzuki, H. Yamaki and T. Yoshida. October, 2001

Volume 33 Computational Commutative Algebra and Combinatorics.
Edited by T. Hibi. February, 2002

Volume 34 Minimal Surfaces, Geometric Analysis and Symplectic Geometry.
Edited by K. Fukaya, S. Nishikawa and J. Spruck.
June, 2002

Volume 35 Higher Dimensional Birational Geometry.
Edited by S. Mori and Y. Miyaoka. July, 2002

Volume 36 Algebraic Geometry 2000, Azumino.
Edited by S. Usui, M. Green, L. Illusie, K. Kato, E. Looijenga, S. Mukai and S. Saito. November, 2002

Volume 37 Lie Groups, Geometric Structures and Differential Equations — One Hundred Years after Sophus Lie —.
Edited by T. Morimoto, H. Sato and K. Yamaguchi.
December, 2002

Volume 38 Operator Algebras and Applications.
Edited by H. Kosaki. January, 2004

Volume 39 Stochastic Analysis on Large Scale Interacting Systems.
Edited by T. Funaki and H. Osada. March, 2004

Volume 40 Representation Theory of Algebraic Groups and Quantum Groups.
Edited by T. Shoji, M. Kashiwara, N. Kawanaka, G. Lusztig and K. Shinoda. March, 2004

Volume 41 Stochastic Analysis and Related Topics in Kyoto
In honour of Kiyosi Itô.
Edited by H. Kunita, S. Watanabe and Y. Takahashi.
March, 2004

Volume 52	Groups of Diffeomorphisms in honor of Shigeyuki Morita on the occasion of his 60th birthday. Edited by R. Penner, D. Kotschick, T. Tsuboi, N. Kawazumi, T. Kitano and Y. Mitsumatsu. November, 2008
Volume 53	Advances in Discrete Dynamical Systems. Edited by S. Elaydi, K. Nishimura, M. Shishikura and N. Tose. June, 2009
Volume 54	Algebraic Analysis and Around in honor of Professor Masaki Kashiwara's 60th birthday. Edited by T. Miwa, A. Matsuo, T. Nakashima and Y. Saito. February, 2009
Volume 55	Noncommutativity and Singularities Proceedings of French–Japanese symposia held at IHÉS in 2006. Edited by J.-P. Bourguignon, M. Kotani, Y. Maeda and N. Tose. July, 2009
Volume 56	Singularities — Niigata–Toyama 2007. Edited by J.-P. Brasselet, S. Ishii, T. Suwa and M. Vaquie. November, 2009
Volume 57	Probabilistic Approach to Geometry. Edited by M. Kotani, M. Hino and T. Kumagai. March, 2010
Volume 58	Algebraic and Arithmetic Structures of Moduli Spaces (Sapporo 2007). Edited by I. Nakamura and L. Weng. June, 2010
Volume 59	New Developments in Algebraic Geometry, Integrable Systems and Mirror Symmetry (RIMS, Kyoto, 2008). Edited by M.-H. Saito, S. Hosono and K. Yoshioka. August, 2010
Volume 60	Algebraic Geometry in East Asia — Seoul 2008. Edited by J. Keum, S. Kondō, K. Konno and K. Oguiso. September, 2010
Volume 61	Exploring New Structures and Natural Constructions in Mathematical Physics. Edited by K. Hasegawa, T. Hayashi, S. Hosono and Y. Yamada. October, 2011
Volume 62	Arrangements of Hyperplanes—Sapporo 2009. Edited by H. Terao and S. Yuzvinsky. March, 2012

Volume 63	Galois–Teichmüller Theory and Arithmetic Geometry. Edited by H. Nakamura, F. Pop, L. Schneps and A. Tamagawa. October, 2012
Volume 64	Nonlinear Dynamics in Partial Differential Equations. Edited by S.-I. Ei, S. Kawashima, M. Kimura and T. Mizumachi. April, 2015
Volume 65	Algebraic Geometry in East Asia — Taipei 2011. Edited by J. A. Chen, M. Chen, Y. Kawamata and J. Keum. April, 2015
Volume 66	Singularities in Geometry and Topology 2011. Edited by V. Blanlœil and O. Saeki. May, 2015
Volume 67	Variational Methods for Evolving Objects. Edited by L. Ambrosio, Y. Giga, P. Rybka and Y. Tonegawa. June, 2015
Volume 68	School on Real and Complex Singularities in São Carlos, 2012. Edited by O. Saeki, R. N. Araújo dos Santos, T. Nishimura and V. H. Jorge Pérez. July, 2016
Volume 69	Development of Moduli Theory—Kyoto 2013. Edited by O. Fujino, S. Kondō, A. Moriwaki, M.-H. Saito and K. Yoshioka. July, 2016
Volume 70	Minimal Models and Extremal Rays (Kyoto, 2011). Edited by J. Kollár, O. Fujino, S. Mukai and N. Nakayama. September, 2016
Volume 71	Schubert Calculus — Osaka 2012. Edited by H. Naruse, T. Ikeda, M. Masuda and T. Tanisaki. December, 2016
Volume 72	Geometry, Dynamics, and Foliations 2013 In honor of Steven Hurder and Takashi Tsuboi on the occasion of their 60th birthdays. Edited by T. Asuke, S. Matsumoto and Y. Mitsumatsu. August, 2017
Volume 73	Hyperbolic Geometry and Geometric Group Theory. Edited by K. Fujiwara, S. Kojima and K. Ohshika. September, 2017
Volume 74	Higher Dimensional Algebraic Geometry in honour of Professor Yujiro Kawamata's sixtieth birthday. Edited by K. Oguiso, C. Birkar, S. Ishii and S. Takayama. November, 2017

Volume 75 Algebraic Varieties and Automorphism Groups.
Edited by K. Masuda, T. Kishimoto, H. Kojima, M. Miyanishi and M. Zaidenberg. December, 2017

Volume 76 Representation Theory, Special Functions and Painlevé Equations — RIMS 2015.
Edited by H. Konno, H. Sakai, J. Shiraishi, T. Suzuki and Y. Yamada. June, 2018

Volume 77 The 50th Anniversary of Gröbner Bases.
Edited by T. Hibi. August, 2018

Volume 78 Singularities in Generic Geometry.
Edited by S. Izumiya, G. Ishikawa, M. Yamamoto, K. Saji, T. Yamamoto and M. Takahashi. September, 2018

TO BE CONTINUED